A FIELD GUID S
OF BERM
AND THE

T0094361

"The exquisite artwork of Val Kells represents the gold-standard in scientific fish illustration. Combined with the depth of information on identifying features and breadth of ecological insights provided by Luiz Rocha and Carole Baldwin, this book represents yet another masterpiece in the *A Field Guide to Coastal Fishes* series that is sure to excite and inspire."

—*Martin Arostegui, Postdoctoral Scholar, Biology Department, Woods Hole Oceanographic Institution*

"The attention to detail in every one of Val's meticulous illustrations elevates *A Field Guide to Coastal Fishes* from a standard field guide into a work of art. All proportions, fin ray counts, and nuances in coloration are accurately captured for each species. I have found her work to be very helpful, especially in differentiating between similar species. In my estimation, Val Kells' guide is to fish what Roger Tory Peterson's guide is to birds. This is an indispensable reference that will bring you years of enjoyment whether you are a beginner in fish identification or a professional ichthyologist."

—*Carlos Estapé, Contributor, Smithsonian Tropical Research Institute*

"Fishes and fishing will always enthrall and fascinate humankind. With rich and beautiful illustrations, and practical, easy to comprehend information for species identification, *A Field Guide to Coastal Fishes of Bermuda, Bahamas, and the Caribbean Sea* does so much to further that interest."

—*Rob Robins, Collection Manager, Florida Museum of Natural History*

"As with her other recent field guides, Val Kells and her team of accomplished ichthyologists (in this case the brilliant Drs. Luiz Rocha and Carole Baldwin) have outdone themselves with their attention to detail on the spectacular fishes of the Caribbean and adjacent waters. Of course, the sharks and gorgeous reef fishes are perfectly represented here in every color of the rainbow; but even tiny mosquitofishes get the same loving treatment. This is a must-have field guide for everyone who loves fishes, and it is perhaps the crown jewel of the series."

—*Prosanta Chakrabarty, Ph.D., Professor/Curator of Fishes, LSU Museum of Natural Science*

"Citizen scientists, scuba divers, and snorkelers alike will delight in Val Kells' beautiful illustrations and attention to detail as they seek field markings to help identify the mystery fishes they encounter on their underwater adventures. Val and her co-authors provide plenty of information about each species and where they can be found, plus the detailed illustrations help fish geeks differentiate similar-looking pecies. They've knocked this one out of the (marine) park!"

—Janna Nichols, Citizen Science Program Manager, Reef Environmental Education Foundation (REEF)

"The field guide is exquisite in both imagery and description. This is a must for any avid fisher, diver, or naturalist visiting or living in Bermuda, the Bahamas, and the Caribbean Sea."

—Neil Hammerschlag, Research Associate Professor, Director of the Shark Research & Conservation Program, University of Miami

"This field guide is a fabulous addition to any fishwatcher's library. Val Kells' talent as a scientific illustrator and amazing attention to detail shines through, representing almost 1,300 species in life-like detail. Together with Rocha and Baldwin, she has created a beautiful and informative resource."

—Christy Pattengill-Semmens, Ph.D. Co-executive Director, Reef Environmental Education Foundation (REEF)

"I've spent my lifetime fishing for whatever species I could land, obscure or glamorous, big or small. This quest has taken me to 94 countries and to an assortment of venues, from offshore reefs to hotel fountains, and Val's books have been essential companions to identify exactly what I caught. This new book will undoubtedly spark new adventures as I realize how many fishes I aspire to catch and identify in the Caribbean and beyond."

—Steve Wozniak, 2,049 fish species and counting, IGFA Representative and World Record Holder

"The Greater Caribbean Region is home to arguably the most well-studied fish faunas in the world. Rocha and Baldwin are two of the leading experts in the region, and this is readily apparent throughout this brilliant summary of an enormous body of data. And once again, Val Kells has spoiled us all by bringing an entire fauna of fishes to life through her spectacular watercolor illustrations. As those of us in the world of fishes have now come to expect from Kells, the world-class illustrations vividly capture every detail, from the smallest patch of pigment on a tiny, camouflaged blenny to the subtle metallic tones on the flanks of the tunas and mackerels."

—Luke Tornabene, Assistant Professor, School of Aquatic and Fishery Sciences, Curator of Fishes, Burke Museum of Natural History and Culture, University of Washington

A FIELD GUIDE TO COASTAL FISHES OF BERMUDA, BAHAMAS, AND THE CARIBBEAN SEA

A Field Guide to Coastal Fishes

OF BERMUDA, BAHAMAS, AND THE CARIBBEAN SEA

Val Kells
Luiz A. Rocha · Carole C. Baldwin

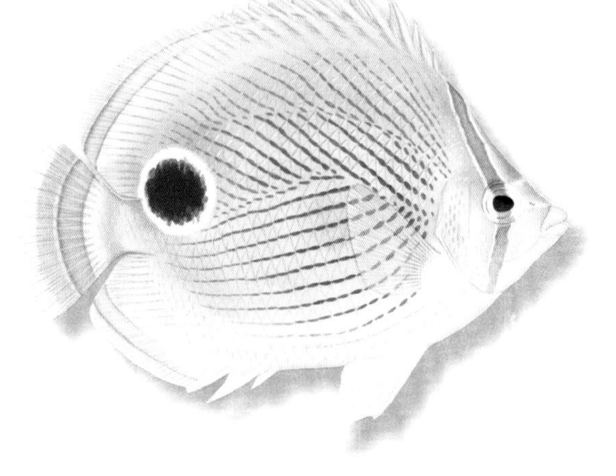

Johns Hopkins University Press
Baltimore

© 2022 Val Kells, Luiz A. Rocha, and Carole C. Baldwin
Illustrations © 2022 Val Kells
All rights reserved.
Published 2022
Printed in Canada on acid-free paper
2 4 6 8 9 7 5 3 1

Johns Hopkins University Press
2715 North Charles Street
Baltimore, Maryland 21218-4363
www.press.jhu.edu

Library of Congress Control Number: 2022934125

ISBN-13: 978-1-4214-4468-0 (pbk. : alk. paper)
ISBN-10: 1-4214-4468-2 (pbk. : alk. paper)

A catalog record for this book is available from the British Library.

*Special discounts are available for bulk purchases of this book.
For more information, please contact Special Sales at
specialsales@jh.edu.*

To my father, David Girvin Kells, Jr.,
and my grandfather, David Girvin Kells
—VK

To my wife, Claudia,
and my children, Gabriel and Sophia
—LR

To my dad and mom,
the late David E. and Alexandra S. Baldwin
—CB

Contents

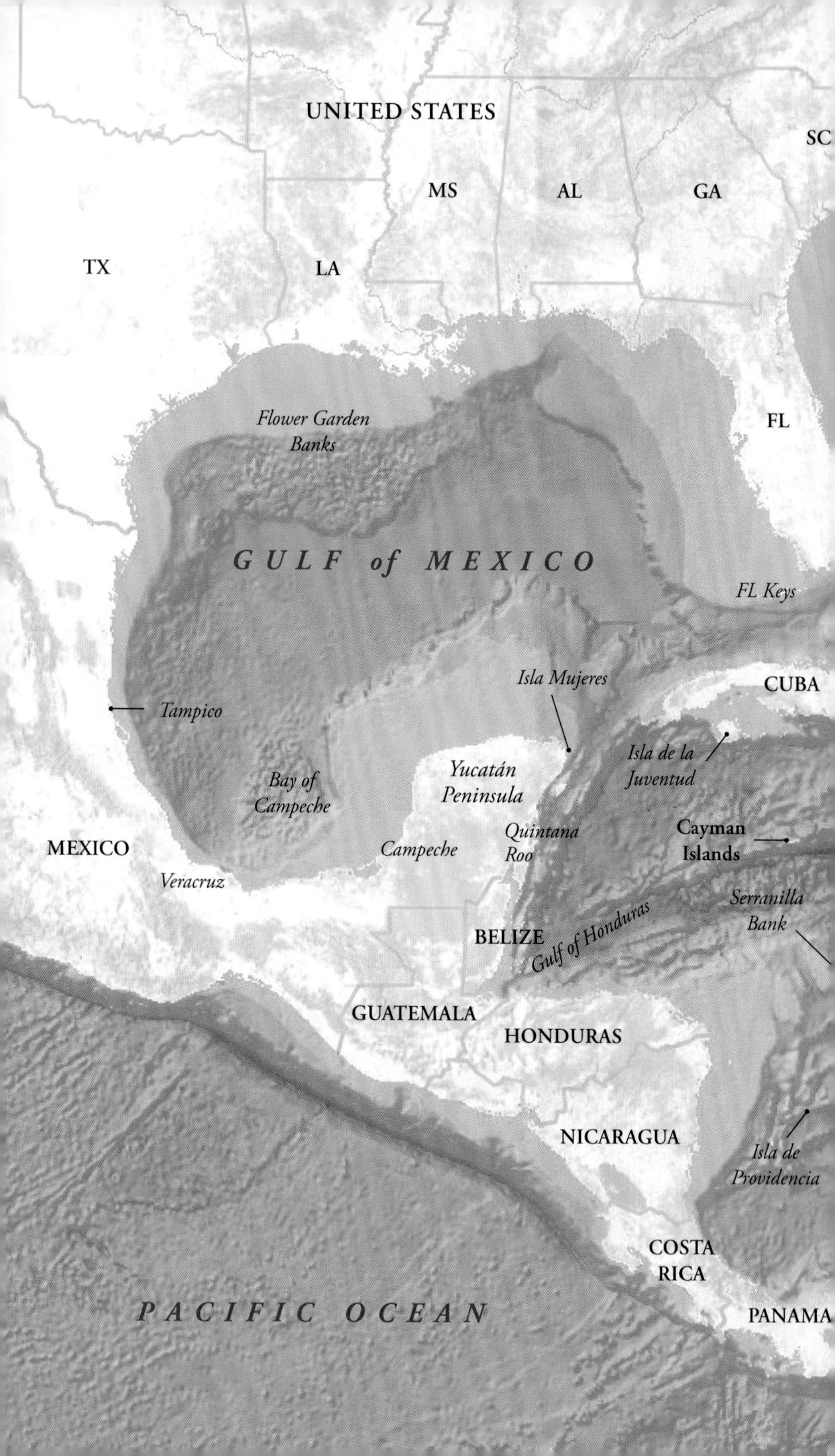

UNITED STATES
SC
MS
AL
GA
TX
LA
FL
Flower Garden Banks
GULF of MEXICO
FL Keys
Isla Mujeres
CUBA
Tampico
Isla de la Juventud
Yucatán Peninsula
Bay of Campeche
Cayman Islands
Quintana Roo
MEXICO
Campeche
Veracruz
Serranilla Bank
BELIZE
Gulf of Honduras
GUATEMALA
HONDURAS
NICARAGUA
Isla de Providencia
COSTA RICA
PACIFIC OCEAN
PANAMA

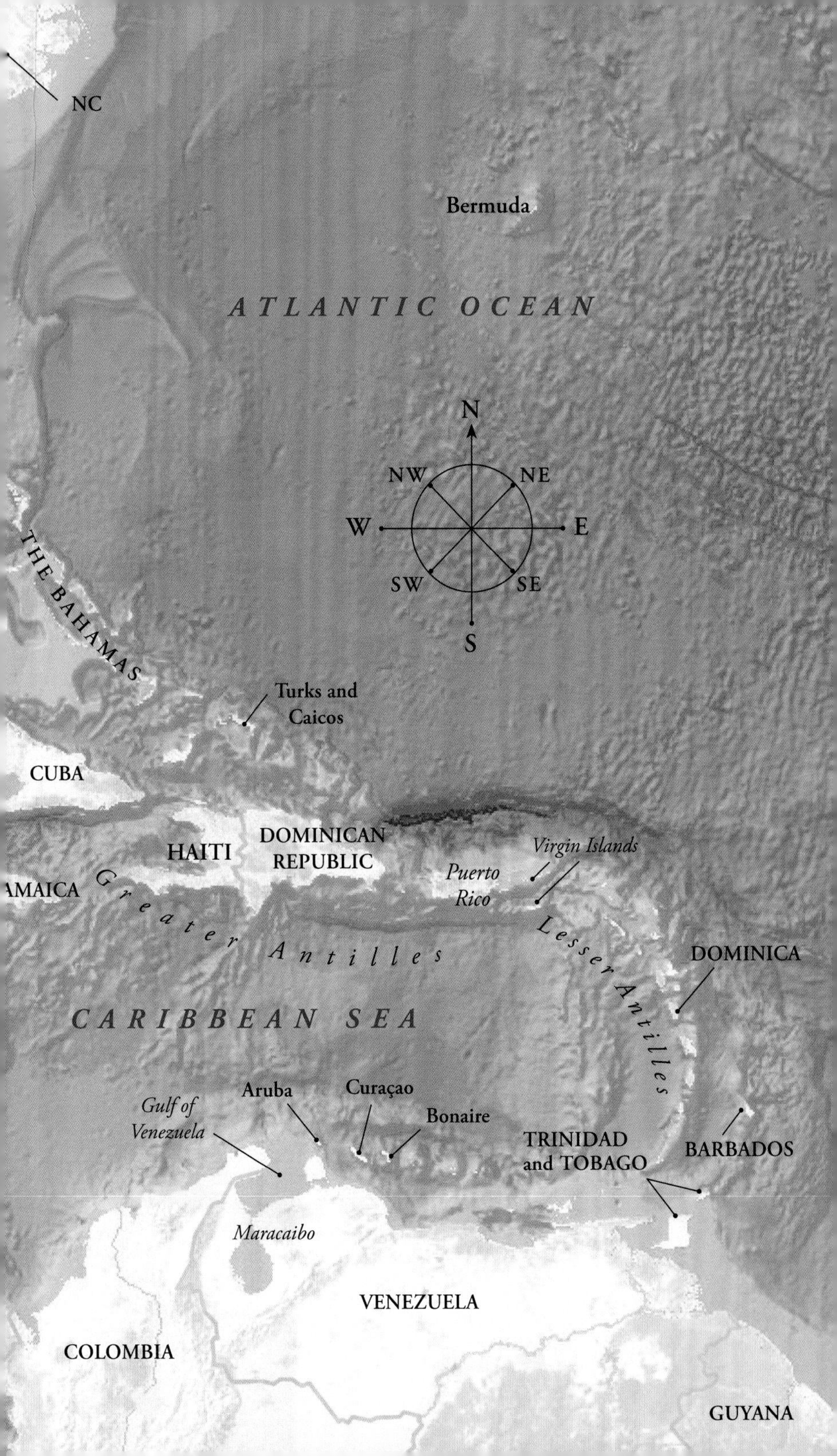

NC

Bermuda

ATLANTIC OCEAN

N
NW
NE
W
E
SW
SE
S

THE BAHAMAS

Turks and Caicos

CUBA

HAITI

DOMINICAN REPUBLIC

JAMAICA

Virgin Islands

Puerto Rico

Greater Antilles

Lesser Antilles

DOMINICA

CARIBBEAN SEA

Aruba

Curaçao

Bonaire

Gulf of Venezuela

TRINIDAD and TOBAGO

BARBADOS

Maracaibo

VENEZUELA

COLOMBIA

GUYANA

Acknowledgments

After five and a half years in development, *A Field Guide to Coastal Fishes: From Maine to Texas* was published in the spring of 2011 and immediately embraced by divers, fishermen, students, researchers, teachers, naturalists, and scientists. Soon after publication, Luiz Rocha, Curator of Fishes at the California Academy of Sciences, reached out to Val and suggested they collaborate on a West Coast edition. Val immediately signed on, and she enlisted co-author Larry Allen, then a Chair and Professor of Biology at California State University, Northridge. Five years later, *A Field Guide to Coastal Fishes: From Alaska to California* was published to great acclaim. Both books have had multiple printings and have set the standard for field guides to fishes.

It seemed a natural progression of lives, careers, and ambitions to fill a need by adding a third guide to the series. In your hands is the result of the combined efforts of three experts in their respective fields. All information is up to date as of publication, with current range information, numerous and new juvenile illustrations, accounts and illustrations of newly described species, taxonomy based on the latest research, and comprehensive sections detailing many divers' favorites: fascinating Blennies and multitudinous Gobies.

∞

No book is solely the result of the authors' work. A successful and comprehensive book results from the collaboration and effort of many people. We are blessed with a wide network of friends, associates, and colleagues—many of whom graciously and freely gave their time, expertise, and support.

We would like to thank the following persons for contributions of information, knowledge, referrals, and assistance: Aaron Adams, Florida Atlantic University; Martini Arostegui, Woods Hole Oceanographic Institution; Ricardo Betancur, University of Oklahoma; Dave Catania, California Academy of Sciences; Carl J. Ferraris, California Academy of Sciences; Jon Fong, California Academy of Sciences; Ben Frable, Scripps Institution of Oceanography; Phil Hastings, Scripps Institution of Oceanography; Mysi Hoang, California Academy of Sciences; John D. McEachran, Texas A&M University; Edward Pfeiler, Unidad Guaymas; Diane Pitassy, Smithsonian National Museum of Natural History; D. Ross Robertson, Smithsonian Tropical Research Institute; David G. Smith, Smithsonian National Museum of Natural History; Emily Stump, University of British Columbia; Luke Tornabene, University of Washington; Albert M. Van der Heiden, Centro de Investigacíon en Alimentacíon y Desarrollo; James Van Tassell, American Museum of Natural History; Benjamin C. Victor, Ocean Science Foundation; Liz Wallace, Florida Fish and Wildlife Conservation Commission; Jeff Williams, Smithsonian National Museum of Natural History.

We also gratefully acknowledge the California Academy of Sciences and California State University, Northridge, for generously underwriting Val's watercolor paper and the production of the range maps.

Acknowledgments

For providing valuable photographic reference necessary to illustrate obscure or poorly documented species and newly described species, or for directing us toward sources of photo reference, we extend deep gratitude to Carol D. Cox and Carlos and Allison Estapé.

We thank our many friends and family who gave their unwavering encouragement and friendship throughout the long production of this book. Many people touched this project in ways small to very large. They include Evette Barton; Lane Becken; Tory Blackford; Gardy Bloemers; Wendy Browning-Lynch; Vince Burke; Carlin Stargell Camp; Candice Cobb and Martha McMillan; Sarah DuBose; Joasha Dundas; John Graves; Kyra Hagge; Terri Kirby Hathaway; Lisa Huizenga Hogan; Dave Kells; John Kells; Gail Leonard; the Migliaccio family; Norman Miller; Mary Michaela Murray; Elizabeth Neff; Robin Noonan; Rosemary and Charles Robinson; Bob Yates; and Anne Wolf. Cheers!

A large number of online friends were supportive and enthusiastic about this project. It was a great source of motivation to read encouraging notes from people we've known for years and others we have yet to meet in person. These notes particularly helped propel Val through the fun and frustrations of Blennies and Gobies. For taking the time, we send our thanks . . . you know who you are!

A special shout-out goes to Luke Tornabene, Jim Van Tassell, and Ben Victor for reviewing and editing the Blenny and Goby master lists and providing their invaluable expertise on these two large and fascinating families of fishes. Special thanks also go to Ross Robertson for his efforts in compiling the "Shorefishes of the Greater Caribbean" website, a valuable resource for our efforts here.

We owe an enormous amount of gratitude to the directors, editors, and staff of Johns Hopkins University Press. In particular, we thank our editor, Tiffany Gasbarrini, and our copyeditor, Carrie Love. Carrie spent many hours meticulously refining the manuscript. Her keen eyes and hard work added a final, priceless polish. Tiffany had the trust, faith, and vision to champion our project. With unflinching support, she helped to bring it from concept to reality. Thank you, Tiffany—this book is another life work come true.

Finally, we would like to express our deepest thanks to our families. This book was another enormous undertaking made much easier by their steadfast patience, encouragement, and understanding. Thank you, Drew and Dave. Thank you, Claudia, Gabriel, and Sophia. Thank you, Dad and Mom—the late David E. and Alexandra (Stevie) Baldwin.

—Val, Luiz, and Carole

Preface

About This Book

We developed this book to fulfill the need for a comprehensive, current, and compact field guide to fishes of Bermuda, Bahamas, and the Caribbean Sea and as a companion to *A Field Guide to Coastal Fishes: From Maine to Texas* and *A Field Guide to Coastal Fishes: From Alaska to California*. We hope this book continues the good work.

Area and Species Covered

The species included are brackish and marine fishes that are encountered from the southern Texas border to Yucatán, Bermuda, Bahamas, and the Caribbean Sea. This area generally extends from the intertidal zone to depths of about 660 feet. We provide identification and natural history information for most fishes we know to have stable populations within this range. This includes all native and non-native fishes that spend all or part of their adult lives in marine waters.

We describe species that are predominantly freshwater inhabitants but can also be found in low-salinity waters. For comprehensiveness, some rare and deep-water species are included. Very rare species and those generally occurring below 660 feet have been mostly omitted. Other poorly documented species were excluded for lack of information such as specimen photographs or video clips for live color reference.

Many of the fishes that occur from Bermuda to the Caribbean Sea also occur at other, often distant, locations. Strays and waifs may be found in areas outside of the species' typical range. Wherever possible, those locations are noted in the text. Most of the fishes found in the Caribbean Sea also occur in southern coastal US waters; thus, the range map includes this broader area.

Names and Sequence of Species

The Latin and common names of the families in this book follow those presented in *Fishes of the World*, fourth edition, by Joseph S. Nelson. The common names of species are taken from *Common and Scientific Names of Fishes from the United States, Canada, and Mexico*, seventh edition, published by the American Fisheries Society (AFS), Special Publication 34. With a few exceptions, Latin names, authority, and date follow Eschmeyer's most recent online *Catalog of Fishes*. The sequence of families of fishes in this guide are organized following the fourth edition of Nelson's authoritative work, *Fishes of the World*. We have chosen to follow this traditional ordering for consistency with most currently published field guides; however, taxonomy and phylogenetics are in a transition phase. Recent studies support changes to traditional phylogenetic sequences, and we encourage readers to seek out these studies for evolving information. The inclusion of Scaridae in Labridae and splitting of Serranidae and Epinephelidae follow the most recent research. Sequences of species within each family follow the alphabetical order of genus and species names, rather than the alphabetical order of common names, with Pipefishes being the only exception.

The first letter of each word in single- and multiple-word names is capitalized, except after a hyphen, unless that word requires capitalization as a proper noun. This is in accordance with recent changes adopted by the American Fisheries Society and as published in *Common and Scientific Names of Fishes from the United States, Canada, and Mexico*, seventh edition, Special Publication 34, AFS. Although we elected to use the most recently accepted AFS common names for the individual species, other commonly used local names or those accepted by the Smithsonian Tropical Research Institute are also mentioned wherever possible. Whenever we encountered errors or conflicting information in regard to Latin or common names, we made appropriate corrections and inclusions based on the most recently published documentation.

Organization and Presentation

We have arranged this book into three primary sections: Introduction, Families, and Species. These sections are supported by supplemental materials, which include a glossary of terms, a list of additional resources, and an index.

The Introduction provides an overview of the evolution, diversity, and features of fishes. It also includes information that will help the user identify fishes. Each family section describes, in concise terms, each of the 161 families of fishes that are found along the coasts of Bermuda, Bahamas, and the Caribbean Sea.

The species section is the largest section in this book and includes descriptions of 1,263 individual species, based on the most recently published scientific information available. An additional 77 rare species are described in the online appendix. A condensed summary of range and habitat is provided for each species. To save room, names of states are abbreviated. For example, North Carolina reads: NC. North, South, East, and West are also sometimes abbreviated and read, respectively: N, S, E, W. The biologic description provides a very brief summary of the species' behavior, diet, and/or ecology, where room allowed. The depths provided are approximate maximum recorded depths, although many fishes may be more common at shallower depths. The lengths given for each species are the approximate maximum recorded adult total length. The International Union for Conservation of Nature (IUCN) species conservation status categories at time of printing included Near Threatened, Vulnerable, Endangered, and Critically Endangered. Those entries with no status listed are Least Concern or Data Deficient, or they have not been evaluated.

Each account is accompanied by a large, full-color illustration of the adult species, unless otherwise noted. Each fish is shown in living color, as it would appear in hand, or at the surface in clear water. While no two fish of the same species are exactly alike, the illustration intends to closely represent the species as a whole. Great care was taken to accurately portray the correct placement and proportion of anatomical features. The illustrations are presented in a "size relative" fashion, meaning those in the same genus and on the same page are shown at a size relative to the longest fish in the genus. Juveniles can be much smaller than adults; they are shown larger than relative size for clarity. All are shown facing left—except one right-facing family—or from above with fins displayed.

Diversity and Classification

Fishes are the most diverse group of vertebrates on Earth. There are currently over 32,000 known living species of fishes and many thousands of others that have become extinct in the 500 million years since the early fish ancestors first swam in the seas. Over 1,690 marine fish species live in the Greater Caribbean. They inhabit bays, inlets, estuaries, rocky and coral reefs, sandy bottoms, and mangrove areas, as well as the open ocean. These species range in size from the Whale Shark, which may grow to 50 feet, to the tiny Leopard Goby, which as an adult grows to 0.6 inches long. Caribbean coastal fishes include species that represent the most early diverging evolutionary lineages of ichthyofauna—like the Nurse Shark—to much more recently diverging forms, such as the Balloonfish.

The scientific classification of fishes is, and will likely always be, an ongoing process subject to debate and change as new information unfolds. However, many scientists divide fishes into five recognized classes: Myxini, the Hagfishes; Petromyzontida, the Lampreys; Chondrichthyes, the Cartilaginous Fishes; Actinopterygii, the Ray-finned Fishes; and Sarcopterygii, the Lobe-finned Fishes. Of these five classes, only the Lobe-finned Fishes are not known to occur along the coasts of the Americas and Caribbean islands.

The table below shows how three representative fishes are classified in three families of fishes that occur along the Caribbean coasts:

	Whale Shark	**Green Moray**	**Queen Angelfish**
Family:	Rhincodontidae	Muraenidae	Pomacanthidae
Genus:	*Rhincodon*	*Gymnothorax*	*Holacanthus*
Species:	*typus*	*funebris*	*ciliaris*

The "jawless" fishes—Hagfishes and Lampreys—are often referred to as primitive. They lack true jaws, do not have paired fins, and exhibit a simple, cartilaginous skeletal structure. They have a single nostril located on the top of the head. Their form of locomotion is simple and eel-like. The simplicity of the jaws in these classes of fishes limits them to rasping prey. However, this characteristic has not prohibited them from succeeding. Through scavenging and parasitizing, jawless fishes have survived, evolved, and prospered for hundreds of millions of years.

The Cartilaginous Fishes of the Class Chondrichthyes—the familiar sharks, skates, and rays, as well as the less familiar chimaeras—are more structurally advanced than jawless fishes, even while lacking true bones. They have true jaws, and their nostrils are located on both sides of the head, generally under the snout. The skull and jaws are constructed of large single units, rather than multiple pieces as seen in the Ray-finned Fishes. Cartilaginous Fishes lack a swim bladder and most rely on large oily livers for buoyancy. Unlike jawless fishes, they possess paired fins: the pectoral and pelvic fins.

Another differentiating feature is that all cartilaginous species practice internal fertilization; the females produce either egg cases or live young.

The largest and most diverse class of fishes is, by far, the Ray-finned Fishes of the Class Actinopterygii. This group possesses a bony, rather than cartilaginous, skeleton. Like the Cartilaginous Fishes, bony fishes also possess true jaws. However, their jaws are composed of many small bones rather than large cartilaginous units. The skull is also a complex structure of small bones. The wide array of jaw and tooth types in this class has spawned a large variety of feeding systems, including biting, crushing, filter-feeding, sucking, picking, and scraping. Nostrils are found on the upper part of both sides of the head. Bony fishes usually have swim bladders, many of which are complex in structure. They also possess paired fins, but in some the pectoral and pelvic fins may be absent. Their methods of reproduction are wide and varied.

Adaptations to Life in Water

Living completely submerged in water presents a host of challenges. Fishes need to regulate the amount of salt and water in their bodies and extract dissolved oxygen from the water. Their senses are adapted to aquatic life. Beyond the basics of sight and smell, most fishes have a lateral line, a sensory organ that is highly developed to detect the minutest motions in the water. Fishes have also developed many ways to communicate with each other under water. Some grind their pharyngeal teeth, while others grunt by manipulating their air bladder. Some fishes have light-producing organs that they may use to locate each other or attract prey in the darkness.

Water can be over 900 times more dense than air. Many fishes cruise through this dense solution by undulating their body, caudal peduncle, and tail to create forward thrust. Others move by flapping, fanning, or sculling their fins. Eels move much like snakes, winding their way through the water and over the bottom. Skates undulate their pectoral-fin lobes, whereas rays flap them. Many fishes have developed ingenious forms of locomotion. Searobins "walk" across the bottom by using their free pectoral-fin rays. Remoras, while able to swim freely, have a modified dorsal fin that forms a suction disk. This disk allows the Remora to "hitch a ride" on its host. Some fishes hardly swim at all. While capable of slow and awkward swimming, the Longlure Frogfish spends most of its life perched among sponges and corals awaiting prey.

Identifying Fishes

All fishes change shape and most change color as they develop. Juveniles can be drastically different from their adult counterparts. Adults of the same sex and species often have subtle differences. Numerous fishes are sexually dimorphic, meaning males and females are different in color and form. Many fishes change color and pattern depending on the time of day, time of migration, diet, depth, mood, or breeding phase. Some change color when they are hunting; others change color to appear intimidating. In addition, almost every fish changes color when it is caught, when it is in distress, or after it has died.

Introduction

Depending on the subject at hand, identification can be an easy or daunting task. Some species of fishes are so distinctive that they do not resemble any other and are thus easy to identify. Others are so similar in appearance that only subtle nuances distinguish one from another. Even though the variety and changes in the color of fishes is great, observed colors and patterns are the most common tools used for identification.

Below are several examples of commonly observed color patterns in Bahamian, Bermudan, and Caribbean coast fishes.

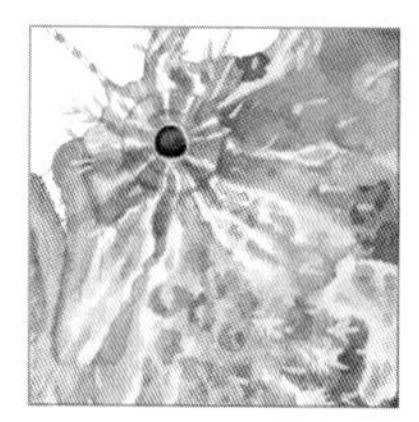

bands radiating from eye

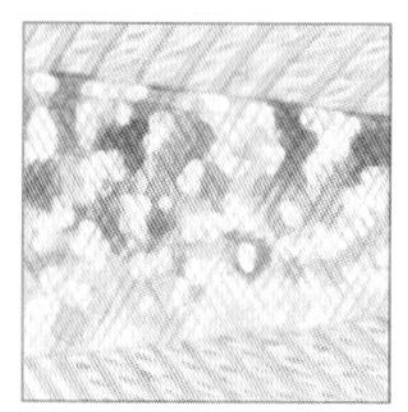

mottled pattern

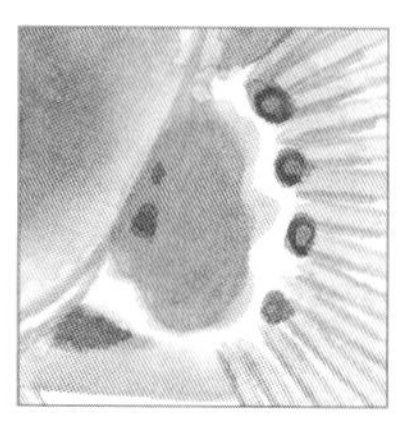

white semi-circle

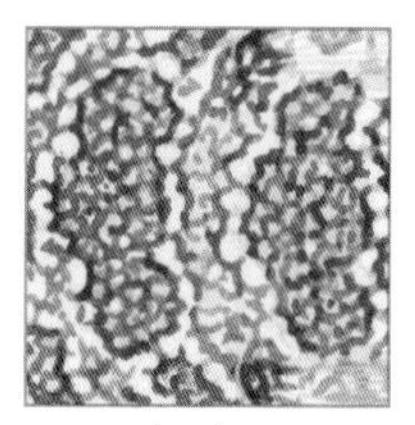

vermiculated pattern

alternating red and yellow stripes

oblique lines

cross-hatch pattern

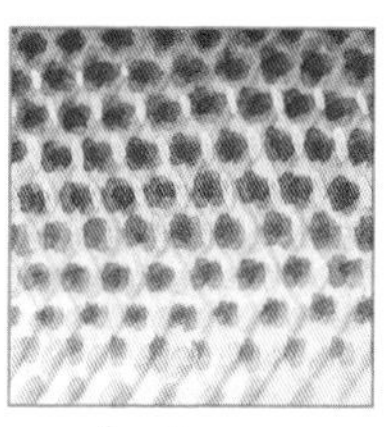

spots forming rows along scales

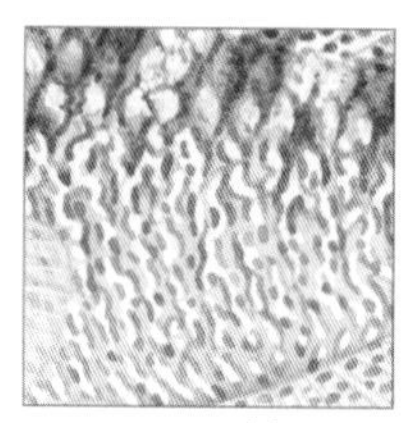

wavy spots and lines

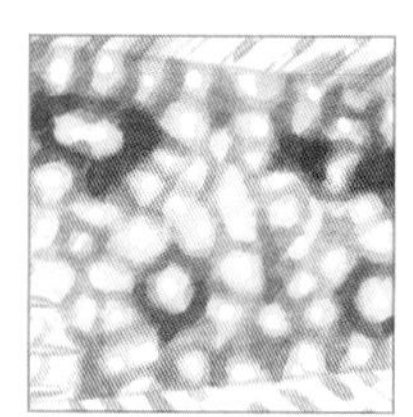

honeycomb pattern

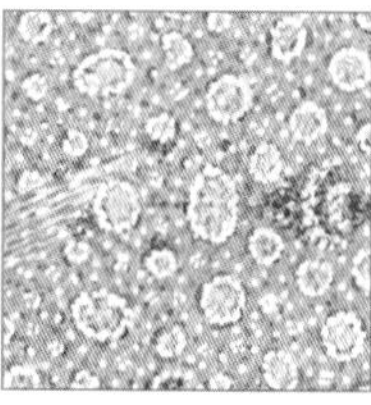

rosettes

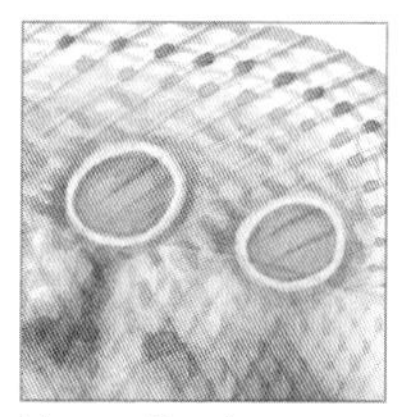

blue ocellated spots

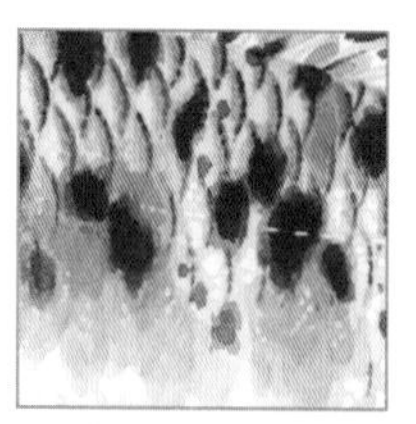

speckles, spots, and blotches

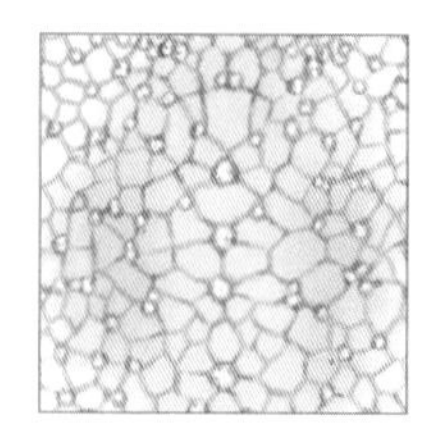

reticulated pattern

brown saddle extending onto fin

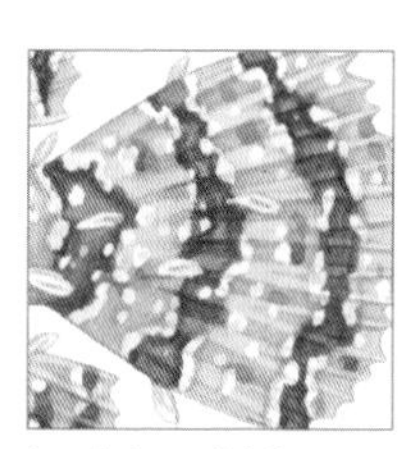

banded caudal fin

When a fish is not identifiable by either color or pattern, the fish's anatomy can help to secure an identification. Shape, size, and placement of anatomical features vary from one species to the next and thus distinguish one from another. The following illustrations show the primary external features of several cartilaginous and bony fishes that are commonly used as tools for identification.

Cartilaginous Fishes

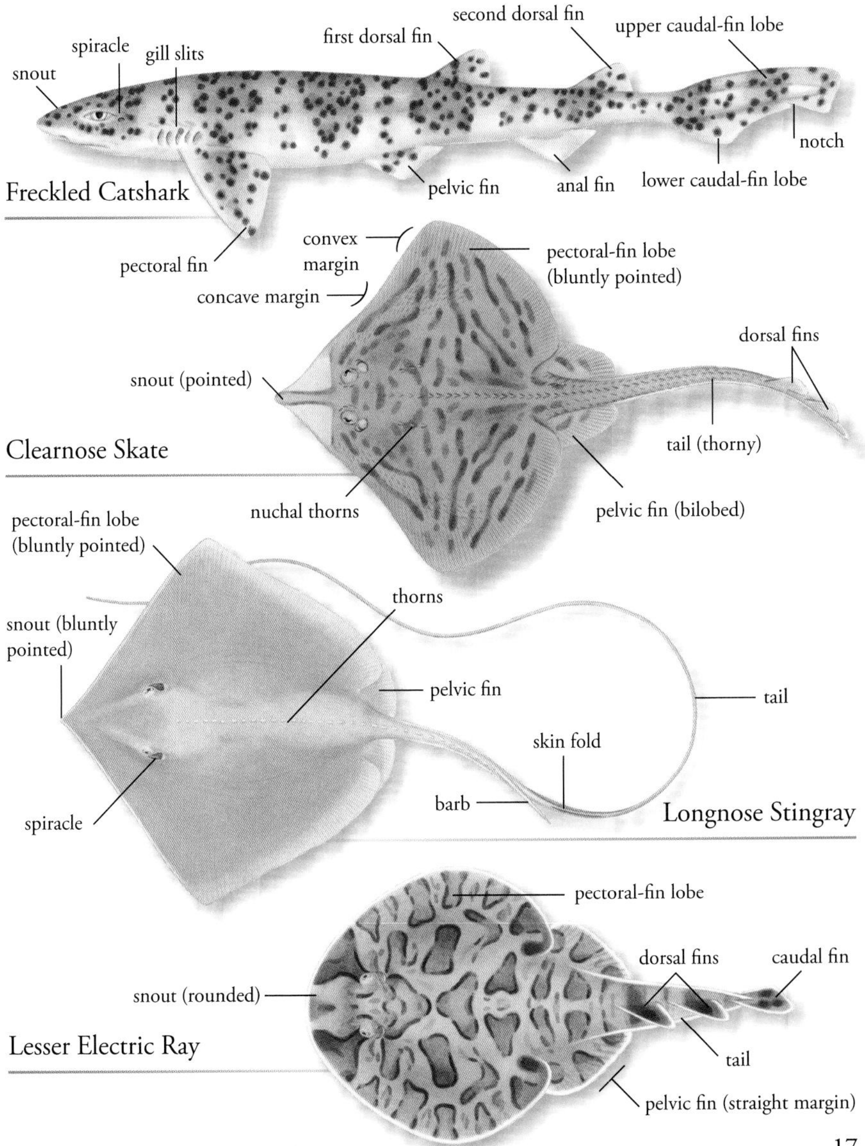

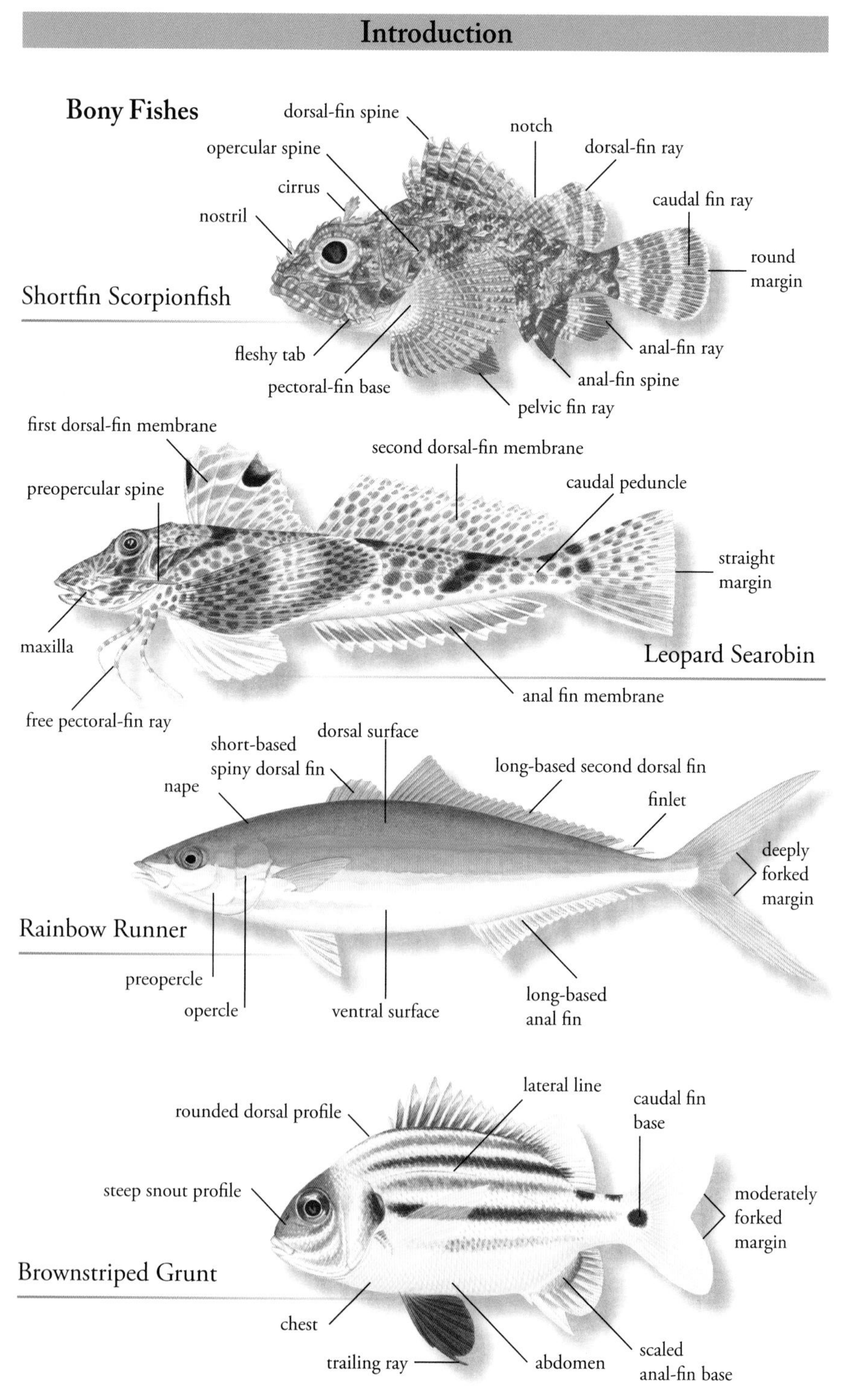
Bony Fishes
dorsal-fin spine
notch
opercular spine
dorsal-fin ray
cirrus
caudal fin ray
nostril
round margin
Shortfin Scorpionfish
anal-fin ray
fleshy tab
anal-fin spine
pectoral-fin base
pelvic fin ray
first dorsal-fin membrane
second dorsal-fin membrane
preopercular spine
caudal peduncle
straight margin
maxilla
Leopard Searobin
anal fin membrane
free pectoral-fin ray
dorsal surface
short-based spiny dorsal fin
long-based second dorsal fin
nape
finlet
deeply forked margin
Rainbow Runner
preopercle
opercle
ventral surface
long-based anal fin
lateral line
caudal fin base
rounded dorsal profile
steep snout profile
moderately forked margin
Brownstriped Grunt
chest
trailing ray
abdomen
scaled anal-fin base

Lengths and proportions will also help in differentiating one fish from another. Total lengths are used in this book. To determine total length, measure from the tip of the snout or the tip of the lower jaw to the tip of the caudal fin. If the caudal fin is forked, press the upper and lower tips toward each other. The lengths provided in the accounts give the reader a general idea of how big a species may become. The illustration below shows positions of specific identifying features and indicates lengths and depth.

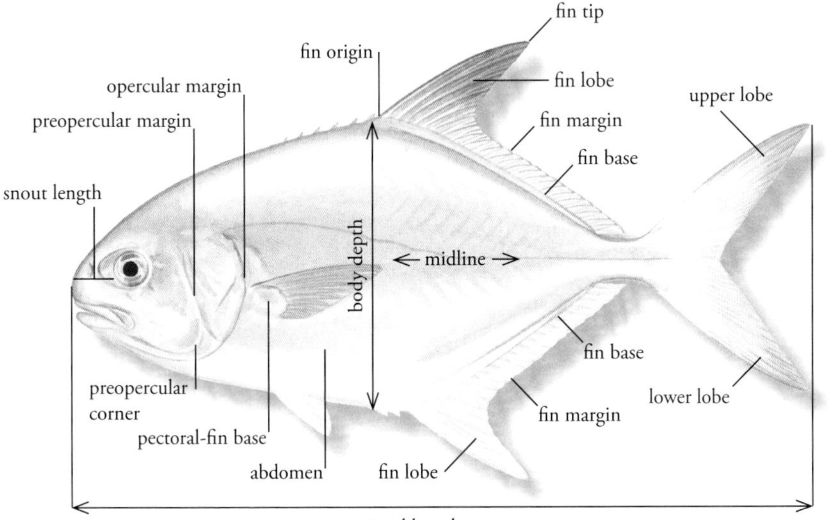

The overall shape of a fish's body can also help in identification. Shapes of the fishes presented in this book are in profile, as a fish would appear from the side. Proportions of a fish's depth relative to its length are important identifiers. A fish is said to be "deep-bodied" when the measurement of depth is great relative to length, such as in the Banded Butterflyfish. A fish is said to be elongate when its depth is small relative to length, such as in the Bonefish. Some examples of profile are shown below.

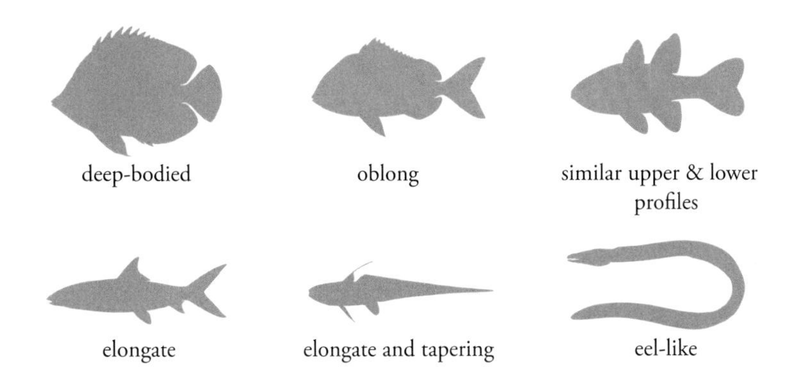

The cross-sectional shape of a fish is the shape of the body as it appears head on. A fish that is flattened from side to side is said to be laterally compressed. A fish that is flattened from top to bottom is described as flattened or depressed. Below are some simplified cross-sectional views.

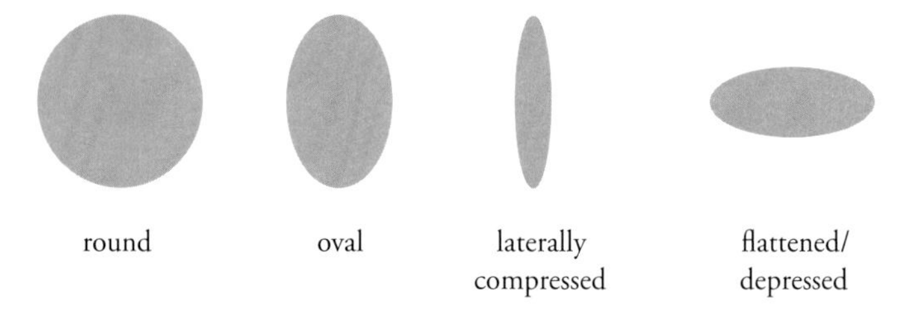

Sometimes it is necessary to go further and count spines, rays, scales, or other countable features to determine the identity of a fish. When counting spines and rays, the norm is to count anteriorly (front) to posteriorly (toward tail). Even if a spine is very small or directed forward, it is still counted. Additionally, if the last ray on the dorsal or anal fin is split to a unified base, it is still counted as a single ray.

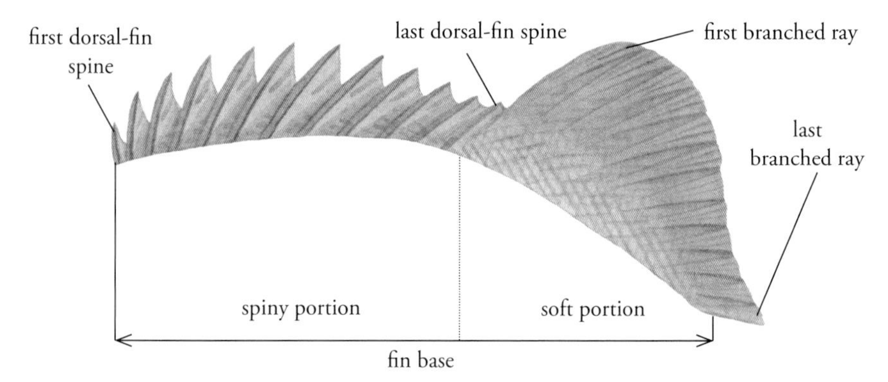

There are times when lateral-line scales need to be counted. These are counted from the first pored, lateral-line scale behind the opercle to the last pored scale that corresponds to the crease in the caudal peduncle when the caudal fin is moved from side to side. At other times, scales above the lateral line are counted. These are counted from the highest arch of the lateral line diagonally backward to the base of the dorsal fin.

The habitat, geographic area, and depth range may also help in identification. For example, the Large-eye Toadfish and the Cotuero Toadfish have very similar appearances and features. However, their ranges barely overlap. When using depth as an identifier, it should be noted that many recorded depths are those taken by trawl or line. These records reflect the deepest point of the trawl or line and not necessarily the deepest level at which a fish may swim. Depth records may also reflect seasonal migrations or the preferred depth of juveniles or adults.

Conservation

Marine environments face many threats, including those from run-off, pollution, destruction of habitat, and overfishing. It cannot be overstated how important it is to reduce, and possibly reverse, these harms. If each person takes one small step toward conservation and preservation, the overall effect could be tremendous.

Many fishes included in this book are currently Threatened or Endangered. The IUCN (International Union for Conservation of Nature) Red List of Threatened Species contains an evaluation of the extinction risk of global plants and animals. This information plays a significant role in guiding conservation activities and serves to monitor changes in the conservation status of species. Of the many fish species that are monitored, sharks are one of the most vulnerable. They have been heavily exploited, are largely misunderstood and misrepresented, and are overfished by commercial and recreational fisheries. Sharks grow slowly, have a long gestational period, and do not produce a large quantity of offspring. Therefore, it is very difficult for them to recover from the depletion they have suffered.

On the positive side, there are several examples of population recovery due to conservation. One is the Nassau Grouper, *Epinephelus striatus*. For decades, this fish was severely overfished, and populations declined by as much as 80% or more. However, after targeted conservation efforts, Nassau Grouper has shown signs of recovery in the Cayman Islands. In spite of this promising news, and because overall populations are still a fraction of their once healthy numbers, it is still listed as Critically Endangered on the IUCN Red List.

We encourage our readers to educate themselves about the fishes they encounter. Most fishes have defense mechanisms meant to protect them from other fishes; these mechanisms can also harm people. Many fishes will defend themselves if threatened, and some are venomous. We also encourage our readers to

- Handle fishes gently.
- Respect seasonal and catch limits.
- When diving, do not touch any portion of a reef.
- Obtain proper permits and licenses, as many countries use sales revenues to fund conservation and enforcement.
- Do not release non-native fishes into any open body of water.
- Whenever possible, use circle hooks, practice catch-and-release, and respirate the fish before releasing it back into the water.
- Release native live baitfishes back into the water.
- Collect and dispose of waste properly.
- Anchor in designated areas only.
- Participate in fisheries management, reef surveys, and non-native fish roundups.

In short, if we care for fishes and their environment, they will be here in the future for us and future generations to admire.

FAMILIES

Families

Note: Please refer to the Introduction for information regarding the organization of families referenced throughout.

Family Ginglymostomatidae - Nurse Sharks

Nurse Sharks are almost cylindrical in shape. The snout is short and rounded. Nostrils bear obvious barbels. Gill slits are small and spiracles are smaller than eyes. Dorsal fins are similar in size and shape, and located near the tail. The caudal fin is long and low. Nurse sharks are bottom-dwelling and occur worldwide in tropical to subtropical seas. They are primarily nocturnal and social, often resting on the bottom in small groups. One species in the area. Page 58.

Family Rhincodontidae - Whale Shark

The Whale Shark is the largest fish on Earth. The mouth is wide and, when open, exposes five rows of very long gill plates. Teeth are very small. The eyes are small and positioned just behind the mouth. Spiracles are similar in size and are set just behind the eyes. Three dorsal ridges are present, with the lowest ridge becoming a strong keel at the caudal peduncle. The first dorsal fin is larger than the second, and the caudal fin is very tall. The Whale Shark occurs circumglobally in warm seas. They filter-feed mostly on plankton. One species in this family. Page 58.

Family Odontaspididae - Sand Tiger Sharks

Sand Tiger Sharks are stout with a conical, depressed snout. The large mouth extends beyond the eyes and contains protruding teeth. Gill slits are low on the body and anterior to the pectoral fins. Dorsal, pelvic, and anal fins are similar in size. Pectoral fins are relatively small. They occur worldwide in warm marine waters. Sand Tiger Shark embryos feed on lesser developed embryos, fertilized eggs, and unfertilized eggs before birth. Two species in the area. Page 58.

Family Pseudocarchariidae - Crocodile Shark

The Crocodile Shark has very large eyes, long gill slits, and highly protrusible jaws. The dorsal fins are low, the pectoral fins are small, the caudal fin is asymmetrical, and there are low keels on the caudal peduncle. It occurs circumglobally in tropical to subtropical seas, usually well offshore from near surface to about 2,000 ft. In the western Atlantic, it is found from Lesser Antilles and Venezuela to Brazil. Embryos feed on unfertilized eggs and possibly other embryos. One species in this family. Page 60.

Family Alopiidae - Thresher Sharks

Thresher Sharks have a round, streamlined body, and a very long, asymmetrical caudal fin that can be as long or longer than the body. The head is short, and the snout is pointed. The fourth and fifth gill slits are above the pectoral-fin base. Pectoral fins are as long or longer than the head. Second dorsal fin and anal fin are very small. Thresher sharks occur worldwide in tropical to cold seas. The elongated tail is used to disrupt schooling fishes and to stun prey. Two species in the area. Page 60.

Family Lamnidae - Mackerel Sharks

Mackerel Sharks have a round body and a conical snout. The large mouth has sharp, triangular teeth. Gill slits are anterior to pectoral fins. Gills lack gill rakers. The first dorsal fin is angular and erect. Second dorsal and anal fins are small. Caudal peduncle is heavily keeled. Caudal-fin lobes are similarly sized. Mackerel Sharks occur worldwide in tropical to temperate seas. Their specialized circulatory systems keep them warmer than the ambient water temperature. Ancestors of this family grew to 65 ft. Four species in the area. Page 60.

Family Scyliorhinidae - Catsharks

Catsharks are typically small and cigar-shaped with a caudal fin that is low and asymmetrical. The snout is short and depressed. The mouth extends behind the posterior margin of the small, oval to slit-like eyes. The first dorsal fin is small and originates above or posterior to the origin of the pelvic fins. Catsharks are bottom-dwelling in circumglobal temperate to tropical marine waters from intertidal areas to about 6,500 ft. There are 65 known species in the Catshark family. Two shallow-water species in the area. Page 62.

Family Triakidae - Hound Sharks

Hound Sharks are slender with a low, asymmetrical caudal fin, a depressed snout, and slit-like eyes. Gill slits are small and spiracles are close to the eyes. Dorsal and ventral fins are similarly shaped. All have small to long labial furrows. They are primarily demersal and occur worldwide in warm temperate and tropical seas, rarely in fresh water. They occur from near shore to the outer continental shelf and feed on invertebrates and bony fishes. Five species in the area. Page 62.

Family Carcharhinidae - Requiem Sharks

The Requiem Shark family is diverse and consists of about 50 species with several common traits. The head is neither flattened nor laterally expanded. The eyes are circular or oval with well-developed nictitating membranes. Spiracles are absent or very small. The mouth is usually large and extends well beyond the eyes. The caudal fin is long and strongly asymmetrical with a rippled or undulating dorsal margin. Requiem Sharks are circumglobal in a wide variety of marine to freshwater habitats. They are strong and active. Twenty-two species in the area. Page 64.

Family Sphyrnidae - Hammerhead Sharks

Hammerhead Sharks are moderately slender with a greatly flattened and laterally expanded head. The eyes and nostrils are on the outer margins of the modified head. Spiracles are absent. First dorsal fin is tall. Hammerhead Sharks occur worldwide in warm marine to brackish waters. The flattened head creates more surface area for increased electrochemical perception. The position of the eyes enhances vision. Most Hammerhead Sharks are shy and difficult to approach. Six species in the area. Page 72.

Family Hexanchidae - Cow Sharks

Cow Sharks are slender to stout in shape and are distinguished by the presence of six or seven long gill slits. Most other shark families have five gill slits. The mouth is long with comb-shaped teeth. Cow Sharks have only one dorsal fin, which is located close to the caudal fin. The caudal fin is long and low. Cow Sharks are circumglobal in shallow to deep marine waters from bays to submarine canyons. Females bear live young. Three species in the area. Page 76.

Family Squalidae - Dogfish Sharks

Dogfish Sharks are moderately slender with a conical or depressed head. The spiracles are relatively large and close to the eyes. Gill slits are small and low on the body. Dorsal fins are relatively small, with concave posterior margins, and have a small or prominent grooveless spine. The anal fin is absent. The caudal fin is low and asymmetrical. Dogfish Sharks occur worldwide in arctic to tropical seas. They have strong jaws and prey mainly on fishes and invertebrates. Some form large schools. Four species in the area. Page 76.

Family Centrophoridae - Gulper Sharks

Gulper Sharks are small to moderate in size and cylindrical in shape. The eyes are large. Spiracles are relatively large and just behind eyes. Gill slits are low on the body. Dorsal fins have a strong, grooved spine, which may be short to long. Anal fin is absent. Caudal fin is asymmetrical and notched. Gulper Sharks are demersal in worldwide tropical to warm temperate seas. They prey on a wide variety of bottom-dwelling organisms. One shallow-water species in the area. Page 78.

Family Etmopteridae - Lantern Sharks

Lantern Sharks are small and cylindrical in shape. The snout is short to moderately long. Eyes are large and well developed. Spiracles are large and set just behind eyes. The snout is short. Gill slits are small and low on the body or near body midline. Dorsal fins possess a strong, grooved spine. First dorsal fin is usually smaller than second. Anal fin is absent. Caudal fin is low and notched. Lantern Sharks are darkly colored and usually have luminescent organs on the abdomen, over the pelvic fin, and on the caudal peduncle and base. They occur worldwide in deep water over continental shelves and slopes. Some form large schools. Family account only.

Family Dalatiidae - Kitefin Sharks

Kitefin Sharks are very small to moderately large. The head is short and the eyes are comparatively large. Some have modified jaws to take crater-like bites out of prey. The gill slits are small and anterior to the pectoral fins. Dorsal fins lack spines, or the first dorsal fin may have a small, grooved spine. Luminous organs are present on the ventral surface. Kitefin Sharks are typically demersal in deep water over circumglobal continental slopes. Two relatively shallow-water species in area. Page 78.

Family Squatinidae - Angel Sharks

Angel Sharks have dorsally flattened heads, bodies, and fins. Dorsal profile of head, pectoral fins, and trunk creates a diamond to square shape. The eyes and spiracles are on top of the head. The nostrils and mouth are at the front of a rounded snout. Pectoral fins are separate and triangular in shape. Dorsal fins are near the end of the tail. Angel Sharks occur worldwide at or near bottom in tropical to temperate seas, from shore to about 4,200 ft. Three species in the area. Page 78.

Family Torpedinidae - Torpedo Electric Rays

Torpedo Electric Rays are dorsally flattened with head, body, and pectoral fins forming an almost circular disk when viewed from above. The eyes and spiracles are on top of the head. Pelvic fins have rounded margins. First dorsal fin is larger than second. Caudal fin is triangular. Large, kidney-shaped electric organs are visible from above. They are circumglobal in tropical to temperate seas. Can discharge up to 45 volts to stun prey and defend against predators. Two species in the area. Page 80.

Family Narcinidae - Numbfishes

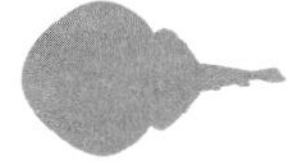

Numbfishes are dorsally flattened with head, body, and pectoral fins forming an oblong disk when viewed from above. The eyes and spiracles are on top of the head. Pelvic fins have almost straight margins. Dorsal fins are large and almost equal in size. Caudal fin is triangular. A pair of large electric organs is visible from above. Numbfishes are demersal and circumglobal in tropical to warm temperate seas. Electric charge is used as defense and to stun prey. Four species in area. Page 80.

Family Pristidae - Sawfishes

Sawfishes are long and dorsally flattened. The eyes and spiracles are on top of the head. The skin is rough and covered in denticles. The pectoral, pelvic, and dorsal fins are triangular in shape. The rostrum is modified into a long, narrow, flattened "saw." The saw is armed on each side with embedded and modified teeth-like denticles and is used to flush benthic prey or to slash and disable schooling fishes. Sawfishes are demersal in worldwide tropical to subtropical coastal waters. Two species in area. Page 82.

Family Rhinobatidae - Guitarfishes

Guitarfishes are moderately flattened with a wedge-shaped snout. The pectoral fins are moderately broad and rounded. Eyes and spiracles are on top of the head. Dorsal fins are located on top of an elongated caudal peduncle. The skin is rough and covered in denticles. Their small, rounded teeth are used to crush bottom-dwelling crustaceans and mollusks. Guitarfishes are demersal over shallow sandy and muddy bottoms of coastal areas in tropical to warm temperate seas. One species in the area. Page 82.

Family Rajidae - Skates

Skates are dorsally flattened with broadly expanded pectoral fins. The body and fins form a square to diamond shape when viewed from above. The snout may be elongate or blunt. The pelvic fins are typically bilobed. The tail is moderately slender and with or without dorsal and caudal fins. Eyes and spiracles are on top of the head. Most have denticles and thorns dorsally. Skates are bottom-dwelling in worldwide polar to tropical seas. Female Skates deposit large fertilized eggs in rectangular, leathery cases. They swim by undulating the pectoral fins. Over 240 species in the family. Eleven shallow-water species in the area. Page 84.

Family Urotrygonidae - American Round Stingrays

American Round Stingrays are small to moderate in size. The dorsally flattened body and pectoral fins form an oval to almost round disk when viewed from above. Eyes and spiracles are on top of the head. The tail is moderately long and has one or more venomous spines and a distinct caudal fin. Dorsal fins are absent. American Round Stingrays occur in warm, shallow waters of the Atlantic and eastern Pacific. Conceal themselves in sand or mud. Three species in the area. Page 88.

Family Dasyatidae - Whiptail Stingrays

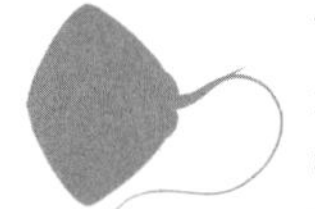

Whiptail Stingrays are moderate to very large in size. The body is dorsally flattened, expanded, and rhomboid or oval in shape when viewed from above. The pectoral fins are very broad and extend to the tip of the snout. The snout may be pointed or blunt. Eyes and spiracles are on top of head. Pelvic fins are single-lobed. Caudal fin is absent. The tail is long and whip-like and possesses one or more serrated, venomous spines. The spine is used in self-defense. Whiptail Stingrays occur worldwide in tropical to warm temperate seas. Most are demersal. They prey on fishes and invertebrates. Eight species in the area. Page 90.

Family Gymnuridae - Butterfly Rays

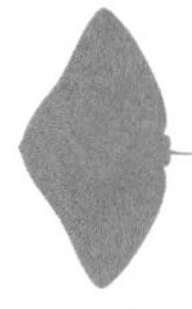

Butterfly rays are very dorsally flattened and laterally expanded. The pectoral fins and body form a diamond-shaped disk. Eyes and spiracles are on top of the head. The tail is short and pointed and may have a serrated spine. Butterfly Rays occur worldwide in tropical to temperate seas. Demersal over shallow coastal sandy and muddy bottoms, also in estuaries and river mouths. Feed on benthic invertebrates and fishes. Two species in the area. Page 92.

Family Myliobatidae - Eagle Rays

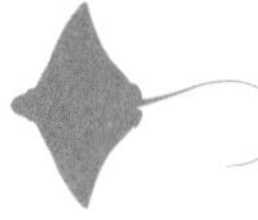

Eagle rays are moderate to large in size, dorsally flattened, and diamond-shaped. Head is elevated with eyes and spiracles on the sides. The pectoral fins are broad and pointed with front edges forming a single, projecting lobe under the snout. Tail may be long and whip-like, often with a serrated spine. Eagle Rays occur worldwide in tropical to temperate seas. They swim in a flapping motion and may leap from water in pursuit of prey. Previously grouped with Rhinopteridae and Mobulidae. Three species in the area. Page 94.

Family Rhinopteridae - Cownose Rays

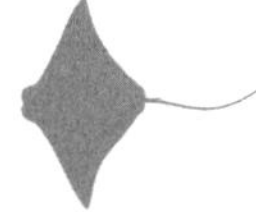

Cownose Rays are moderate to large in size, dorsally flattened, and rhomboid to diamond-shaped when viewed from above. The head is elevated with eyes and spiracles on the sides. Jaws have a series of pavement-like teeth. The pectoral fins are broad and pointed with a separate, bilobed projection under the snout. Tail may be long and whip-like, often with a serrated spine. Cownose Rays occur worldwide in tropical to temperate seas. They swim in a flapping motion. Most are bottom feeders. Previously grouped with Myliobatidae and Mobulidae. Two species in the area. Page 94.

Family Mobulidae - Mantas and Devil Rays

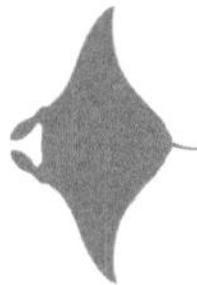

Mantas and Devil Rays are moderate to very large, dorsally flattened, and rhomboid to diamond-shaped when viewed from above. The head is elevated with eyes and spiracles on the sides. Teeth are minute. The pectoral fins are broad and pointed. A separate pair of cephalic fins are located on either side of the head and extend in front of the mouth. Tail is moderate to whip-like, some with a serrated spine. Mantas and Devil Rays occur worldwide in tropical to temperate seas. They swim in a flapping motion. All are filter feeders. Previously grouped with Myliobatidae and Mobulidae. Three described and one undescribed species in the area. Page 96.

Family Lepisosteidae - Gars

Gars are elongate and cylindrical in cross-section. The body is covered in strong, smooth diagonal and rhomboid-shaped scales. The jaws are long, round or flattened in cross-section, and forceps-like. A single dorsal fin is close to the caudal fin. The caudal fin is attached to the tail at an angle. Gars occur in fresh and brackish waters. They have an air bladder that allows them to gulp air. Eggs are reported to be toxic. Family represented by one species in the area around northern Cuba. Family account only.

Family Elopidae - Tenpounders

Tenpounders are elongate and cylindrical in cross-section. The mouth is large and the upper jaw extends past the eyes. The single dorsal fin has a concave rear margin. The caudal fin is deeply forked. Tenpounders occur primarily in coastal waters of tropical and subtropical oceans. Some enter brackish or fresh water. Tenpounder larvae are transparent and ribbon-like. Two species in the area. Page 98.

Family Megalopidae - Tarpons

Tarpons are elongate and moderately compressed with a single dorsal fin and a deeply forked caudal fin. The mouth is large and upturned, with the upper jaw extending past the eyes. Anal fin is long-based. Scales are large. The tarpon family consists of two species: one in the Indo-Pacific, the second in the Atlantic Ocean. They are primarily marine but may enter fresh water. Juveniles are often in estuaries and around mangroves. Tarpons are able to gulp air from the water surface. One species in the area. Page 98.

Family Albulidae - Bonefishes

Bonefishes are elongate and cylindrical in shape with a single dorsal fin and a deeply forked caudal fin. The snout is conical and the mouth is subterminal. The body is translucent. Bonefishes occur near shore in worldwide tropical to warm temperate seas. They are active fishes that forage over sandy and muddy bottoms for invertebrates and small fishes. Bonefish larvae are transparent and ribbon-like. A thick coating of slime covers the skin. Three described and at least one undescribed species in the area. Page 98.

Family Anguillidae - Freshwater Eels

Freshwater Eels are long and round in cross-section. The dorsal and anal fins are confluent with the caudal fin. The snout is short and acute to rounded. Lower jaw protrudes. The lips are thick and the teeth are small. Pectoral fins are well developed. The body is covered in embedded scales in adults. Freshwater Eels occur in tropical to temperate eastern Atlantic, Indian, and western Pacific oceans. They are usually catadromous, living in fresh water and spawning at sea. Larvae remain at sea for up to two years, move back to shore, mature, and migrate into fresh water. One species in the area. Page 100.

Family Heterenchelyidae - Mud Eels

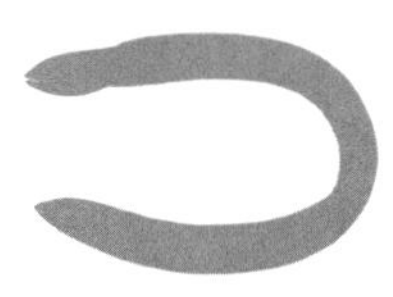

Mud Eels are moderately elongate and round in cross-section. The head is blunt and the jaws are large. Eyes are tiny and covered with skin. The dorsal-fin origin is over the gill opening, and dorsal and anal fins are low. Scales, lateral line, pores on head, and pectoral fins are absent. They burrow head first in bottom sediment. Mud Eels occur in the tropical Atlantic and eastern Pacific oceans, and the Mediterranean Sea. One species in the area. Page 100.

Family Moringuidae - Spaghetti Eels

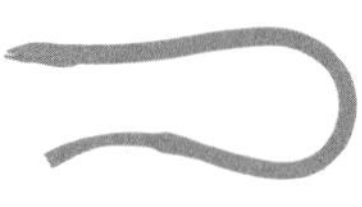

Spaghetti Eels are very long to extremely long and slender. The jaws are moderately large and the eyes are small and covered with skin. Pectoral fins are small to feeble, and dorsal and anal fins are reduced to low folds. The lateral line is complete, and head pores are present on lower jaw only. Scales are absent. They burrow head first in bottom sediment. Spaghetti Eels occur the tropical western Atlantic and Indo-Pacific oceans. Rarely in fresh water. Two species in the area. Page 100.

Family Chlopsidae - False Moray Eels

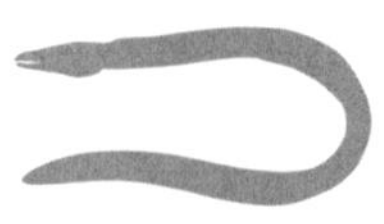

False Moray Eels are elongate with moderate to large jaws and large eyes. Rear nostrils are either just above top lip or opening downward through the lip. Lateral-line pores are present on head but not on body. Pectoral fins are absent in some; when present, pectoral fins are above the gill opening. Dorsal and anal fins are confluent with tail. False Moray Eels are found in the subtropical and tropical Atlantic, Indian, and Pacific oceans. Most are cryptic and burrow in bottom sediment or live in reef crevices. Seven species in the area. Page 100.

Family Muraenidae - Moray Eels

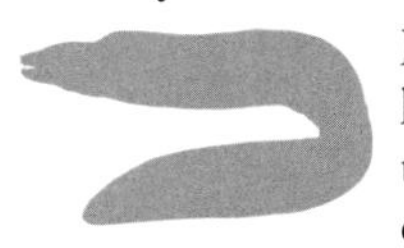

Moray Eels are elongate, somewhat laterally compressed, and long and slender to robust. The dorsal and anal fins merge with the caudal fin. The snout is short to relatively long. Teeth are well developed. Pectoral fins are absent. Moray Eels occur worldwide in primarily tropical to subtropical seas. Most are marine, inhabiting rocky or coralline holes and crevices. Some live around mangroves and tidal creeks. They are predators and scavengers. The flesh of some is toxic. Twenty-three species in the area. Page 104.

Family Synaphobranchidae - Cutthroat Eels

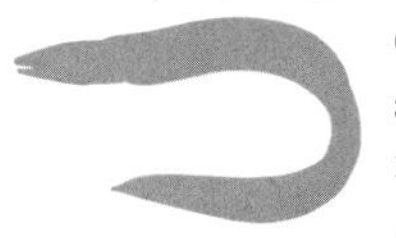

Cutthroat Eels are stout to elongate with typically large jaws. Eyes are small to well developed. Gill opening low on the body. Pectoral fins are present in most. The anus is located well before midbody. Dorsal and anal fins are well developed and confluent with tail. Scales are present or absent. Lateral line is complete. Cutthroat Eels are bottom-dwelling in the deep waters of the Atlantic, Pacific, and Indian oceans. One species in the area. Page 112.

Family Ophichthidae - Snake Eels and Worm Eels

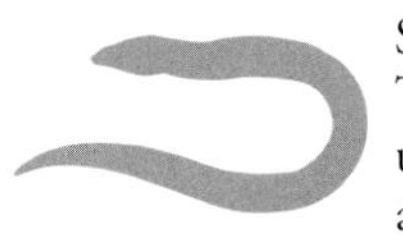

Snake and Worm Eels are elongate and serpentine or worm-like. The snout is short to moderately elongate. Anterior nostrils are usually tube-like and may be very long and fleshy. Pectoral fins are present or absent. The caudal fin is usually absent with the tail ending as a hardened tip. When present, dorsal, anal, and caudal fins are confluent. Scales are absent. Snake and Worm Eels burrow head or tail first. They occur in a variety of worldwide tropical to warm temperate marine and brackish habitats. Thirty species in the area. Page 112.

Family Muraenesocidae - Pike Congers

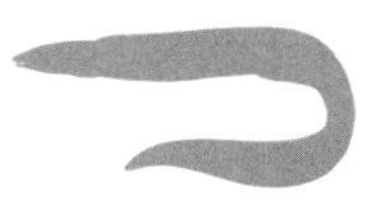

Pike Congers are large and heavy bodied with large teeth and strong jaws. The front nostrils are tubular; rear nostril is a hole in front of mid-eye. The eyes are large and skin-covered. Gill openings are large and nearly meet below. Pectoral fins are well developed. Dorsal and anal fins are well developed and confluent with tail. Lateral line is conspicuous with a system of multiple pores. Pike Congers are found in circumtropical waters of the Atlantic, Pacific, and Indian oceans. One species in the area. Page 122.

Family Congridae - Conger Eels

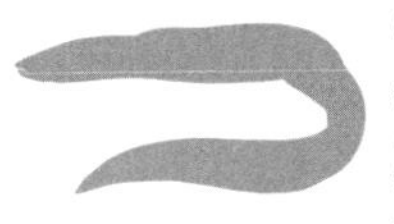

Conger Eels are elongate, serpentine, and medium to very large in size. The snout is short to long and usually longer than the lower jaw. Well-developed flanges occur on one or both lips. Eyes are large and well developed. Pectoral fins are usually present. Dorsal, anal, and caudal fins are confluent. They occur at or near bottom in worldwide tropical to warm temperate seas. Some species form large or small colonies. Many burrow or seek shelter during the day and forage at night. Sixteen species in the area. Page 122.

Family Nettastomatidae - Duckbill Eels

Duckbill Eels are elongate, serpentine, and medium to large in size. The snout is long and narrow and projects over lower jaw. The head is slender. Jaws extend to rear margin of eyes. Some teeth are exposed when mouth is closed. Dorsal and anal fins are confluent with long, slender tail. Duckbill Eels occur near or on bottom in moderately deep water of the Atlantic, Pacific, and Indian oceans. One species in the area. Page 126.

Family Pristigasteridae - Longfin Herrings

Longfin Herrings are relatively small and slender to deep-bodied. Lower jaw slightly projects. Mouth is usually upturned and may have a notch at the upper jaw tip. Anterior jaw teeth are enlarged and canine-like. The single dorsal fin lacks spines, and the abdomen is lined with scutes that form a distinct keel. Pelvic fins are small or absent. As the common name implies, the anal fin is very long-based. Longfin Herrings occur circumglobally in tropical to subtropical marine and brackish waters, some in fresh waters. Four species in the area. Page 128.

Family Engraulidae - Anchovies

Anchovies are relatively small, with a blunt and rounded snout, and a single dorsal fin. The snout extends beyond the jaws. Jaws are long and slender and extend almost to end of gill cover. Eyes are large. Lateral line is absent. Most are translucent with a thin to wide silvery stripe on each side. Scales are delicate and easily shed. Anchovies occur worldwide in tropical to temperate marine and brackish waters. Most are schooling and filter-feed on plankton. Eighteen species in the area. Page 130.

Family Clupeidae - Herrings

Herrings are cylindrical in shape or laterally compressed. The body is typically silvery. The mouth is usually upturned and may have a notch at the upper jaw tip. Most have adipose lids on eyes. A row of scutes is usually present along the chest and abdomen. The dorsal fin is single; the caudal fin is deeply forked. Herrings occur worldwide in tropical to polar seas. They are usually marine, coastal, and schooling. Some tolerate low salinities; others are anadromous. Most are plankton filter feeders with numerous gill rakers. Thirteen species in the area. Page 134.

Family Aspredinidae - Banjo Catfishes

Banjo Catfishes are wide and depressed anteriorly, long and slender posteriorly. Eyes are very small; gill opening is a small slit. Long barbel present at corners of mouth, and short barbels present under head and on abdomen. Pectoral fins with a strong, barbed spine. Anal fin long-based; adipose fin absent. Banjo Catfishes occur in salt, brackish, and fresh water of coastal areas of northern and northeastern South America. Two species in the area. Page 140.

Family Auchenipteridae - Driftwood Catfishes

Driftwood Catfishes are moderately large and slightly depressed anteriorly. The head is rounded with a very large bony shield. Usually three pairs of barbels around mouth. Dorsal and pectoral fins possess strong spines; dorsal-fin spine swollen at base in adults. Lateral line is branched and wavy. Driftwood Catfishes occur from Panama to Argentina. One brackish water species in the area. Page 140.

Family Ariidae - Sea Catfishes

Sea Catfishes are moderately elongate with long barbels around a broad mouth. The head is depressed and has a bony shield. Dorsal and pectoral fins possess serrated spines. An adipose fin is always present. The skin is scaleless. Sea Catfishes occur worldwide in tropical to warm temperate marine, brackish, and fresh water. Some form schools. In most species, the male mouth-broods eggs. Eight species in the area. Page 142.

Family Argentinidae - Argentines or Herring Smelts

Argentines are elongate and somewhat laterally compressed. The mouth is small and the eyes are large. The single dorsal fin is located at the midbody line. An adipose fin is always present. Scales are easily shed. They are pelagic or demersal over outer shelves and upper slopes of worldwide tropical to warm temperate seas. Prey on planktonic invertebrates and small fishes. Two shallow-water species in the area. Page 144.

Family Aulopidae - Flagfins

Flagfins are elongate and oval in cross-section. The mouth is large, wide, and toothy. The single dorsal fin is expanded and originates on the anterior one-third of the body. Pelvic fins are usually laterally expanded. An adipose fin is present. Flagfins are demersal over continental shelves and slopes of worldwide tropical to warm temperate seas. One species in the area. Page 146.

Family Synodontidae - Lizardfishes

Lizardfishes are elongate and cylindrical in shape. The mouth is wide and toothy. The single dorsal fin is located over the midbody line. Pelvic fins are usually laterally expanded. An adipose fin is present. Lizardfishes are demersal over a variety of bottoms in the Atlantic, Pacific, and Indian oceans. Some occur in brackish waters. They are swift, voracious predators. Twelve species in the area. Page 146.

Family Chlorophthalmidae - Greeneyes

Greeneyes are small and slender. The mouth is large. Eyes are large; pupil is teardrop-shaped. Dorsal fin inserts on anterior third of body. Adipose fin is present. They are demersal in deep tropical to temperate Atlantic, Pacific, and Indian oceans. One species in the area. Page 150.

Family Lampridae - Opahs

Opahs are deep-bodied, laterally compressed, and oval in profile. The mouth is small and protrusible. The pectoral fins are vertically oriented. Dorsal and anal fins are long-based. Caudal fin is forked. Scales are minute and reflective. Opahs are pelagic and found worldwide in tropical to temperate seas. They may wander north during summer months. They swim by flapping their pectoral fins. One species in the area. Page 150.

Family Regalecidae - Oarfishes

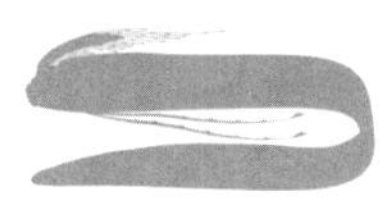

Oarfishes are very long and laterally compressed. The head is angular. The mouth is large, oblique, and highly protrusible. The dorsal fin runs from the top of the head to the tail with anterior rays that are long and trailing. The caudal fin is usually absent in large specimens. Anal fin is absent. Pelvic fins are long, trailing, and single rayed with paddle-like projections. Oarfishes occur circumglobally in warm temperate to tropical seas from surface to about 3,300 ft. They swim vertically in water column. One species in the area. Page 152.

Family Polymixiidae - Beardfishes

Beardfishes are relatively deep-bodied and compressed. The snout is short and rounded, and the eyes are large. A pair of long barbels are present under the lower jaw. The single dorsal fin is long-based. The caudal fin is forked. Beardfishes live near bottom over continental shelves and upper slopes of worldwide tropical to temperate seas. They use barbels to locate benthic prey. Two species in the area. Page 152.

Family Bregmacerotidae - Codlets

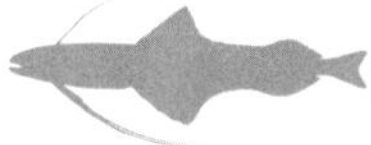

Codlets are small and moderately elongate. The snout is rounded; the eyes are large. The first dorsal fin consists of a single flexible ray located on top of the head. The second dorsal and anal fins are long-based with tall anterior rays and short middle rays. Pelvic fins are located under the head and consist of long filamentous rays. Codlets occur worldwide in tropical to subtropical seas. Most are pelagic from near surface to mid-depths of about 6,500 ft. Some occur in coastal waters and estuaries. One species in the area. Page 152.

Family Moridae - Deepsea Cods

Deepsea Cods are relatively elongate. The snout is rounded; the eyes are relatively large. There may be one, two, or rarely, three dorsal fins. They have one or two anal fins. The pelvic fins are small or filamentous. The caudal peduncle is narrow and the caudal fin is small. Chin barbel is present or absent. Some possess a light organ. Deepsea Cods are found near bottom of worldwide, deep continental shelves and slopes to about 8,200 ft. One relatively shallow-water species in the area. Page 152.

Family Merlucciidae - Merlucciid Hakes

Merlucciid Hakes are relatively elongate and laterally compressed posteriorly. The snout is long and depressed. The jaws are large and have strong, pointed teeth. There is a V-shaped ridge on the head. The first dorsal fin is short-based; the second is long-based and notched at midlength. Chin barbel is absent. Merluccid Hakes occur in the northern and southern Atlantic, eastern Pacific, and western South Pacific oceans. Most are found near bottom over continental shelves and upper slopes. Some occur inshore. Some are of commercial importance. One species in the area. Page 154.

Family Steindachneriidae - Luminous Hake

The Luminous Hake is elongate, deep-bodied anteriorly, narrow and tapered posteriorly. The mouth and eyes are large. First dorsal fin is short-based; second dorsal fin is long-based and tapered. Anal fin with a tall anterior lobe, followed by very long and low portion. Caudal fin is absent. Luminescent tissue is present along ventral length of body. The Luminous Hake occurs over deep continental shelves and upper slopes from Gulf of Mexico to French Guiana. One species in this family. Note: Some authors treat this family as a subfamily of Merluciidae. Page 154.

Family Phycidae - Phycid Hakes

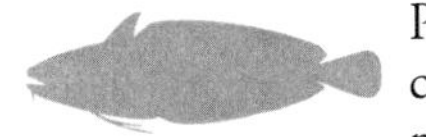

Phycid Hakes are relatively elongate, soft-bodied, and rounded in cross-section anteriorly. The mouth is large. The snout is rounded to moderately long. Chin barbel is present. Two dorsal fins are usually present. Rarely, the first dorsal fin is a single ray followed by a series of short rays and a long-based third fin. Anal fin is long-based. Pelvic fins are typically long and slender. Caudal fin is well developed. Phycid Hakes occur in the Atlantic, western Pacific, and western Indian oceans. They are demersal primarily over soft bottoms from near shore to upper continental slopes. Three species in the area. Page 154.

Family Carapidae - Pearlfishes

Pearlfishes are elongate and scaleless. The body tapers to a pointed tail. The anal fin usually originates under pectoral fins with taller rays than the dorsal-fin rays. Chin barbel, opercular spines, and pelvic fins absent. Pearlfishes occur worldwide in tropical to warm temperate seas. They are found from near shore to about 5,200 ft. Some are free-swimming. Others live commensally in sea cucumbers, starfishes, clams, or tunicates. Two shallow-water species in the area. Page 156.

Family Ophidiidae - Cusk-eels

Cusk-eels are elongate and scaled. The anal fin originates posteriorly to the pectoral-fin tips. Dorsal and anal fins are continuous with caudal fin and have rays of similar height. Chin barbel and opercular spine may be present or absent. Pelvic fins are usually present and are located on or near the throat. Cusk-eels are bottom-dwelling in shallow to deep waters of worldwide tropical to temperate seas. Eleven well-recorded species in the area. Page 156.

Family Bythitidae - Viviparous Brotulas

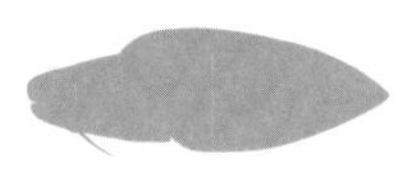

Viviparous Brotulas are moderately elongate. Scales are present in most. Dorsal and anal fins are continuous with caudal fin or, rarely, free. Dorsal rays are taller than anal rays. Chin barbel is absent. Opercle has a well-developed spine. Pelvic fins, when present, have one or two rays. Male sex organ lacks bony pseudoclaspers. Viviparous Brotulas are bottom-dwelling in circumtropical seas. Some occur in freshwater caves. Females give birth to live young. Six species in the area. Note: Previously grouped with Dinematichthyidae. Page 160.

Family Dinematichthyidae - Brotulas

Brotulas are moderately elongate. Dorsal and anal fins separate from the caudal fin. Scales cover head and body. Chin barbel is absent. Eyes vestigial in some. Opercle has a well-developed spine. Pelvic fins have one thread-like ray and are inserted under rear margin of operculum. Male sex organ with one or two (rarely three) pairs of bony pseudoclaspers. Bottom-dwelling in circumtropical seas. Some occur in freshwater caves. Females give birth to live young. Nineteen species in the area. Note: Previously grouped with Bythitidae. Page 162.

Family Batrachoididae - Toadfishes

Toadfishes are small to medium in size. The head is broad and dorsally flattened, with eyes on top. Barbels or fleshy tabs are often present around the mouth and head. Many possess one or more sharp spines on the rear portion of the opercle. The first dorsal fin has two or three spines. Pelvic fins are thoracic. Toadfishes are demersal and occur worldwide in tropical to temperate waters. Most are coastal; some live in fresh water or on continental shelves. They are sluggish but voracious predators. Twenty-two species in the area. Page 164.

Family Lophiidae - Goosefishes

Goosefishes are moderately to greatly dorsally flattened with a broad head and a wide body that tapers to the tail. The mouth is very broad, toothy, and usually bordered by numerous fleshy tabs. Gill openings are located behind the pectoral fins. The first dorsal-fin spine may be isolated on the snout and act as a lure. Goosefishes are bottom-dwelling in worldwide tropical to temperate seas. They are ambush predators capable of capturing very large prey. One shallow-water species in the area. Page 172.

Family Antennariidae - Frogfishes

Frogfishes are very small to medium-sized with deep and rounded bodies. The mouth is large, oblique, and toothy. Gill openings are located behind the pectoral fin. The first dorsal-fin spine is separate and modified and bears a small to large lure. Pectoral fins are elongate and leg-like. Frogfishes occur circumglobally in tropical to subtropical seas. They use their pectoral fins to cling to or "walk" across substrate. They are usually well camouflaged, often blending with surroundings. Eight species in the area. Page 172.

Family Ogcocephalidae - Batfishes

Batfishes are dorsally flattened and laterally expanded. The head and body form a triangular to circular disk. The rostrum is short to long. A cavity under the snout contains a lure. Gill openings are located behind the limb-like pectoral fins. Pelvic fins are located under the body. Tubercles and/or bucklers cover the body. Batfishes are demersal in circumglobal tropical to subtropical seas to about 3,000 ft. Use pelvic and pectoral fins to "walk" across the bottom. Ten species in the area. Page 176.

Family Mugilidae - Mullets

Mullets are medium to large in size. The head is typically broad and dorsally flattened. The eyes are usually partly covered by adipose lids. The snout is short. The mouth is small or moderate in size and usually oriented upward. First dorsal fin has four spines. Pectoral fins are high on the body. Mullets occur worldwide in tropical to temperate salt, brackish, and fresh water. Most tolerate varying salinities. They feed by filtering detritus or plankton. Eleven species in the area. Page 180.

Family Atherinopsidae - New World Silversides

New World Silversides are typically small and translucent with a silvery stripe on each side. The upper jaw is protractile. They have two well-separated dorsal fins, the first with two to nine spines. The pectoral fins are located high on the body. New World Silversides occur in marine to fresh water of North, Central, and South America. They form small to large schools near the surface. Usually omnivorous planktivores. Six species in the area. Page 184.

Family Atherinidae - Old World Silversides

Old World Silversides are small and translucent with a silvery stripe on each side. The upper jaw is not protractile. They have two well-separated dorsal fins, the first with two to five spines. The pectoral fins are located high on the body. Old World Silversides mainly occur in warm marine waters of the Atlantic and Indo-West Pacific oceans. Some occur in estuaries or fresh water. They form large schools and feed on a variety of plankton. Two species in the area. Page 186.

Family Exocoetidae - Flyingfishes

Flyingfishes are small to medium-sized with very long and expanded pectoral fins that are set high on the body and almost always extend past the dorsal-fin origin. Pelvic fins are usually expanded. The caudal fin is deeply forked with the lower lobe longer than the upper lobe. Flyingfishes live at or near the surface from inshore to well offshore in all tropical to subtropical oceans. Known for leaping and gliding for long distances. Feed on zooplankton. Fourteen species in the area. Page 186.

Family Hemiramphidae - Halfbeaks

Halfbeaks are elongate and slender. The upper jaw is short; the lower jaw is usually very long with a fleshy tip. Single dorsal and anal fins are near the tail. The pectoral fins are short to long and set high on the body. Scales are easily shed. Lateral line runs near lower margin of body. Halfbeaks occur near surface of all tropical to warm temperate seas. They are typically omnivorous and feed on a variety of grasses, invertebrates, and small fishes. Some are actively sought as baitfish for game fishing. Eight species in the area. Page 192.

Family Belonidae - Needlefishes

Needlefishes are long and slender with long, pointed jaws. The mouth has many sharp teeth. The single dorsal and anal fins are located near the tail. The caudal fin may be emarginate or asymmetrical with a long lower lobe. Body is either oval or round in cross-section. Needlefishes occur worldwide in tropical to warm temperate seas. Some enter brackish and fresh water. They are usually found near the surface and feed primarily on small, schooling fishes. Seven species in the area. Page 194.

Family Rivulidae - New World Rivulines

New World Rivulines are small, elongate, and cylindrical in shape. The head is somewhat flattened; the mouth is small. The single dorsal fin is located close to the caudal fin, its base shorter than the anal-fin base. Pectoral fins are broad and low on the body. Males have larger anal fins than females. New World Rivulines occur in quiet tropical to subtropical fresh and brackish waters of the western Atlantic Ocean. Hermaphrodism is common. Popular in the aquarium trade. This family has been separated from Aplocheilidae. Two species in the area. Page 196.

Family Fundulidae - Topminnows

Topminnows are small with elongate to moderately deep bodies. The head is usually flattened. The snout is short and the mouth is protrusible. The lower jaw protrudes beyond the upper. The single dorsal fin is located posterior to midlength, its origin near horizontal to anal-fin origin. Males have larger anal fins than females. Females are larger than males. They occur in tropical to temperate fresh to coastal marine waters of the western Atlantic. Many adapt to a wide range of salinities and temperatures. One species in the area. Page 198.

Family Cyprinodontidae - Pupfishes

Pupfishes are small and robust to short and deep-bodied. The single dorsal fin is located posteriorly, its origin well before anal-fin origin. Pectoral fins are set low on the body. Caudal peduncle and caudal fin are broad. Females are larger than males. Pupfishes occur in tropical to temperate fresh, brackish, and marine waters of the Americas and West Indies. Most occur in shallow water. Pupfishes adapt to varying conditions. Two species in the area. Page 198.

Family Anablepidae - Four-eyed Fishes

Four-eyed Fishes are small to medium and elongate with a depressed head. The mouth is small. The eyes are large, protruding from the top of the head, and divided horizontally by an opaque band. The pupil is similarly divided, allowing for vision above and below the water line. The anal fin in males is modified into a scaled, tubular reproductive organ. Four-eyed Fishes occur in fresh and brackish waters of northern South America and western Central America. Two species in the area. Page 198.

Family Poeciliidae - Livebearers

Livebearers are small with elongate to moderately deep bodies. The head is flattened. The single dorsal fin is located posteriorly and may be fan-like, its origin usually behind anal-fin origin. Caudal peduncle is elongate; the caudal fin is broad. Livebearers occur in quiet fresh, brackish, and marine coastal waters of the Americas. Most live near the surface. Most males have a modified anal fin used for internal fertilization of females. Females give birth to live young. Some hybridize and produce only females. Popular in aquarium trade. Thirteen species in the area. Page 200.

Family Anomalopidae - Flashlight Fishes

Flashlight Fishes are small, blackish, and deep-bodied. A large light organ below eyes is visible when a membranous lid is retracted, and hidden when lid is closed. Scutes present along abdomen. Enlarged white scales along lateral line and at soft dorsal- and anal-fin bases. Flashlight Fishes are scattered in warm waters of the eastern Pacific, Indo-Pacific, and Caribbean Sea. Spend days in caverns of deep reefs; ascend to shallow waters at night. Use light organ to search for prey at night. One species in the area. Page 206.

Family Trachichthyidae - Roughies

Roughies are deep-bodied with large mucous cavities and sensory canals on the head. The cavities may be visible through the skin. Head is large, snout is short, and the mouth is large and oblique. Preopercle has a triangular spine at corner. Dorsal fin is single or deeply notched. Large, keeled scutes present along abdomen. Roughies occur worldwide in tropical to temperate seas, near bottom in deep water over continental shelves and around seamounts. Two species in the area. Page 206.

Family Berycidae - Alfonsinos

Alfonsinos are moderate-sized with moderately to deeply compressed bodies. The mouth is large and oblique. Eyes are very large. Preopercle is spineless. Dorsal fin is single, short-based, and located over midbody. Alfonsinos occur in the Atlantic, Indian, and Pacific oceans on or near the bottom of continental shelves and slopes. They are also found over seamounts. Sought commercially in many locations. One species in the area. Page 206.

Family Holocentridae - Squirrelfishes

Squirrelfishes are small to medium-sized with oval to moderately elongate bodies. The eyes are large. The head has ridges and mucous channels. The gill covers are serrated or spiny. Some are reported to have venomous spines on the preopercle. Many are nocturnal and hide in crevices during the day and forage at night. Squirrelfishes are found in the tropical Atlantic, Pacific, and Indian oceans. Most occur around hard bottoms and reefs. They are typically nocturnal. Eleven species in the area. Page 206.

Family Parazenidae - Smooth Dories

Smooth Dories are deep-bodied to moderately slender and laterally compressed. Eyes are large. The mouth is large and highly protrusible. Dorsal and anal fins may have a bony ridge along their bases. Pelvic fins may be greatly expanded. Body scales are minute and silvery. Scutes may be present along abdomen. Occur in deep water in scattered areas of the western Atlantic and Indo-Pacific oceans. Two species in the area. Page 212.

Family Grammicolepididae - Tinselfishes

Tinselfishes are deep-bodied and laterally compressed. The head is comparatively small. The mouth is small. First dorsal- and anal-fin spines may be elongate and shorten with age. Scales cover head and body and are silvery, narrow, and vertically elongate. Found scattered in warm, deep Atlantic and Pacific marine waters. Little is known of their habits. Two species in the area. Page 212.

Family Zeidae - Dories

Dories are deep-bodied, laterally compressed, and oval in profile. The mouth is large, protrusible, and oblique. Body scales are minute or absent. May have large buckler scales along bases of dorsal and anal fins. Scutes may be present along abdomen. Pelvic fins may be greatly expanded. Dories occur in relatively deep marine waters over continental shelves and upper slopes of the Atlantic, Indian, and Pacific oceans. Commercially important. One species in the area. Page 212.

Family Syngnathidae - Pipefishes and Seahorses

Pipefishes and Seahorses are included in the same family despite the outward difference in appearance. All are elongate with a body covered in a series of bony rings. The mouth is small; the snout is tube-like. Dorsal fin is single. Anal fin, when present, is very small. Pelvic fins are absent in all. Pipefishes and Seahorses are found worldwide in warm temperate to tropical marine and brackish waters. Some are found in fresh water. Females deposit eggs into males' pouch or attach eggs to males' abdomen. Males then care for eggs until eggs hatch. Twenty-three species in the area. Page 214.

Families

Family Aulostomidae - Trumpetfishes

Trumpetfishes are elongate, somewhat slender, and slightly laterally compressed in shape. The mouth is small and snout is tube-like. The chin bears a small barbel. First dorsal fin is series of isolated spines. Caudal fin is rounded. Occur in tropical marine waters of the Atlantic and Indo-Pacific. Usually occur around coral reefs. May hover facing downward. Often hunt alongside other fishes. One species in the area. Page 220.

Family Fistulariidae - Cornetfishes

Cornetfishes are slender, elongate, and slightly laterally compressed in shape. The mouth is small and the snout is tube-like. First dorsal fin is absent. Second dorsal and anal fins are similar in shape. Caudal fin is forked, with a long filament trailing from the middle rays. Found worldwide in tropical to subtropical seas. Occur along coastal areas, over seagrass beds, rubble, and soft bottoms, and around reefs. They may change color to match surroundings. Feed on fishes and shrimps. Two species in the area. Page 220.

Family Macroramphosidae - Snipefishes

Snipefishes are small, laterally compressed, and moderately deep-bodied with a long tube-like snout. Teeth are absent. First dorsal fin has five to eight spines. The second spine is enlarged and serrated along the rear margin. Body has bony plates above pectoral fins and/or along ventral midline. Occur worldwide in tropical to subtropical seas. Juveniles are pelagic; adults live near bottom. Two species in the area. Page 220.

Family Dactylopteridae - Flying Gurnards

Flying Gurnards are elongate and squarish in cross-section with a bony head, a blunt snout, and expanded pectoral fins. The top and sides of the head are covered in bony plates. Keeled spines extend from nape. The preopercle bears a long spine. Flying Gurnards are demersal over soft bottoms in temperate to tropical waters of the Atlantic and Indo-West Pacific oceans. They use pelvic-fin rays to "walk" across the bottom and to locate prey. Pectoral fins are spread when the fish is alarmed. Able to change color to blend with surroundings. One species in the area. Page 222.

Family Scorpaenidae - Scorpionfishes

Scorpionfishes are small to moderate in size with a relatively large head bearing numerous spines and ridges. The pectoral fins are rounded to fan-like. The dorsal fin is continuous and notched. Venom glands are usually present at base of dorsal-, pelvic-, and anal-fin spines. Venom may be slightly to highly toxic. Many have fleshy tabs over eyes, on cheeks, and sides of the body. They are demersal or pelagic and occur worldwide in tropical to temperate seas. Most are well camouflaged and are adept at ambushing prey. Twenty-four species in the area. Page 222.

Family Triglidae - Searobins

Searobins are moderately elongate with a large, bony head that is armored with plates, ridges, and spines. Venom glands and chin barbels are absent. Pectoral fins are small or very broad and expanded. First three pectoral-fin rays are free and fleshy. They are demersal and occur worldwide from near shore to continental shelves of tropical to warm temperate brackish and marine waters. Free pectoral-fin rays are used for locomotion and to detect demersal prey. Use swim bladder for sound production, particularly during spawning. Sixteen species in the area. Page 230.

Family Peristediidae - Armored Searobins

Armored Searobins are moderately elongate. The head is large and armored with plates, ridges, and spines. Rows of spiny scutes cover the body. Flattened projections extend from snout and sides of head. Lip and chin barbels are usually present. First two pectoral-fin rays are free and fleshy. They are demersal and occur worldwide on continental and insular slopes in deep tropical to temperate seas. Free pectoral-fin rays are used for locomotion and barbels are used to locate prey. Seven shallow-water species in the area. Page 236.

Family Centropomidae - Snooks

Snooks are moderately elongate and laterally compressed. The mouth is large with a protruding lower jaw. Preopercle has a serrated lower margin. Head profile is almost straight to concave. Lateral line is well developed and extends onto caudal fin. Dorsal fins are separate. Anal fin has three strong spines, the second being the stoutest. Occur worldwide in shallow, coastal tropical to subtropical marine, brackish, and sometimes fresh water. They are voracious predators sought for sport and food. Six species in the area. Page 238.

Family Acropomatidae - Lanternbellies

Lanternbellies are oblong and moderately compressed. The snout is short and rounded. Upper jaw is protrusible. Eyes are large. Opercle has one to two spines on rear margin. Dorsal fins are nearly to completely separate. Three species have light organs. They occur in the water column over continental shelves and slopes of the Atlantic, Indian, and Pacific oceans. Previously called Temperate Ocean-basses. One species in the area. Page 240.

Family Symphysanodontidae - Slopefishes

Slopefishes are small and slender to moderately deep-bodied. Upper margin of jaw covered by suborbital bone when the mouth is closed. Most of the head and maxillae are covered with scales. Opercle with two spines. Dorsal fin continuous or with a slight notch, and usually with 9 spines and 10 soft rays. Anterior pelvic-fin rays and upper and lower caudal-fin rays can be very long in some males. Found near bottom over continental shelves, on deep reefs, and on upper slopes in circumglobal seas. Two species in the area. Page 240.

Family Polyprionidae - Wreckfishes

Wreckfishes are large, oblong, and moderately compressed. The head has a bony knob between the eyes and one at the nape. The opercle has two spines, with the lower spine at the end of a long, horizontal ridge. Found in the Atlantic, Indian, and Pacific oceans and in the Mediterranean. Adults are found over continental shelves and seamounts. Juveniles are pelagic. One species in the area. Page 242.

Family Epinephelidae - Groupers

Groupers are usually robust and comparatively deep-bodied with large mouths. The opercle has three spines, with the middle spine usually most conspicuous. The dorsal fin is usually continuous, sometimes deeply notched. Anal fin has three spines. Lateral line runs comparatively low and extends only to end of caudal peduncle. Groupers occur circumglobally from near shore to moderately deep tropical to temperate seas. A few live in fresh water. The majority are hermaphroditic and mature first as females and then change into males. Thirty-nine species in the area. Page 242.

Family Serranidae - Sea Basses

Sea Basses are usually robust with large mouths. The opercle has three spines with the middle spine most conspicuous. The dorsal fin is usually continuous, sometimes deeply notched, and some with trailing anterior rays. Anal fin has three spines. Lateral line extends only to the end of caudal peduncle. Sea Basses occur circumglobally from near shore to moderately deep tropical to temperate seas. A few live in fresh water. The majority are hermaphrodites that mature first as females and then change into males. Some function simultaneously as both sexes. Forty-eight species in the area. Page 258.

Family Grammatidae - Basslets

Basslets are small and moderately elongate. The snout is short and the eyes are large. Opercle has up to two small spines. Dorsal fin is continuous. Pelvic fins are moderately long to very long and filamentous. Lateral line is interrupted or absent. Basslets occur primarily over coral reefs, rocky ledges, and dropoffs of the tropical western Atlantic. They feed on plankton and are often oriented upside down on substrate. Sixteen species in the area. Page 276.

Family Opistognathidae - Jawfishes

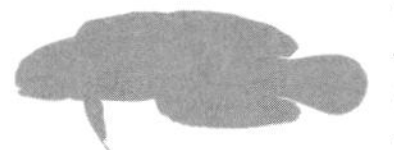

Jawfishes are small and oblong to moderately elongate. The head is rounded. The eyes are large and the jaws are very large. The dorsal fin is continuous, and the pelvic fins are located anterior to the pectoral fins. Jawfishes occur circumglobally in tropical to subtropical seas on muddy, sandy, and rubble bottoms. Most are found near coral reefs. All live in vertical burrows that each constructs and maintains by moving sediment with its mouth. Males mouth-brood eggs until hatching. Fifteen species in the area. Page 282.

Family Priacanthidae - Bigeyes

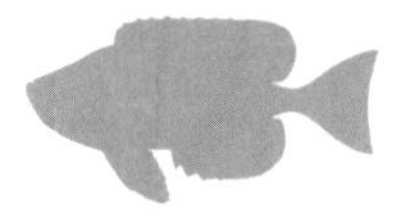

Bigeyes are moderately deep-bodied and laterally compressed with oblique jaws and very large, reflective eyes. The preopercle has a single spine at the lower corner. The dorsal fin is continuous. Pelvic fins are small to greatly expanded with inner rays attached to the abdomen by a membrane. Head, body, and irises are shades of red; may also display pale blotches. Bigeyes occur circumglobally in tropical to warm temperate seas. They are generally found near bottom around reefs and rock formations, occasionally in open water. Most Bigeyes are nocturnal. Four species in the area. Page 286.

Family Apogonidae - Cardinalfishes

Cardinalfishes are small, oblong, and moderately compressed. The snout is short; the mouth and eyes are large. The caudal peduncle is long. Coloration is highly variable and may become pale at night. Cardinalfishes occur circumglobally in shallow tropical to warm temperate seas, usually around coral reefs and seaweed beds. Some are in brackish or fresh water. They are generally secretive and nocturnal. Some are commensal. Males mouth-brood eggs. Twenty-two species in the area. Page 288.

Family Malacanthidae - Tilefishes

Tilefishes are oblong to elongate with a single, low dorsal fin. The head profile is gently to steeply sloping. The mouth is moderately large and fleshy. A prominent to reduced predorsal ridge is present in most. The opercle has a single blunt or notched spine. Tilefishes are demersal and occur circumglobally in tropical to warm temperate seas from shore to about 1,600 ft. They inhabit caves and crevices or may construct mounds or burrows. Some authors divide Tilefishes into two families. Nine species in the area. Page 298.

Family Pomatomidae - Bluefish

The Bluefish is moderately elongate and compressed with separate dorsal fins and a forked caudal fin. The jaws have prominent, sharp teeth. The lower jaw protrudes slightly, and the opercle has a single broad, flat spine. The Bluefish occurs worldwide in eight populations in subtropical to warm temperate seas from shore to over continental shelves. It is schooling, swift, and ravenous and follow schools of squids and small fishes. Females carry up to 1 million eggs. Important sportfish and foodfish. There is one species in this family. Page 300.

Family Coryphaenidae - Dolphinfishes

Dolphinfishes are elongate and laterally compressed. The head profile is rounded in females, steeply sloping in adult males. The spineless dorsal fin originates on the head and reaches to the caudal peduncle. The caudal fin is deeply forked. Dolphinfishes occur circumglobally in tropical to warm temperate seas. They live in open ocean waters over continental shelves and slopes from surface to about 660 ft. They are often associated with *Sargassum* seaweed, flotsam, and oil rigs. Sought as foodfish. Two species in this family. Page 300.

Family Rachycentridae - Cobia

The Cobia is elongate and round in cross-section with a broad mouth and a flattened head. Six to nine short spines precede the second dorsal fin. The pectoral fins are long and pointed; the caudal fin is forked. The Cobia occurs in tropical to warm temperate waters of the Atlantic and Indo-Pacific over a variety of habitats, including shallow coral reefs, rocky shores, and continental shelves to about 165 ft. The Cobia is a fast-growing carnivore of crabs, benthic invertebrates, and fishes. There is one species in this family. Page 300.

Family Echeneidae - Remoras

Remoras are elongate and round in cross-section. The head is broad and flattened. The first dorsal fin is modified into an oval-shaped cephalic disk that the Remora uses to attach itself to a host. Remoras occur worldwide in tropical to warm temperate seas. They are associated with a variety of hosts, including sharks, rays, tarpons, seabasses, cobias, jacks, parrotfishes, billfishes, molas, dolphins, and whales. Some are host-specific, some attach to a range of hosts, and others are free-swimming. Eight species in the area. Page 302.

Family Carangidae - Jacks and Pompanos

Jacks and Pompanos are small to large, and deep and compressed to elongate and fusiform. The eyes have an adipose lid that is either poorly or strongly developed. First dorsal fin may be well developed or a series of spines. Most have a forked caudal fin. First one or two anal-fin spines are separate and may be embedded. Many have bony scutes along lateral line. Body shape and coloring change dramatically with age. Jacks and Pompanos occur circumglobally in tropical to warm temperate brackish to oceanic habitats. Most are schooling and pursue pelagic prey. Thirty-three species in the area. Page 304.

Family Bramidae - Pomfrets

Pomfrets are medium to large, laterally compressed, and round to teardrop-shaped. The eyes are large; the lower jaw protrudes. Dorsal and anal fins are long-based and may be either low, lobed, or fan-like. Scales covering head and body are large and often keeled. Pomfrets occur worldwide in temperate to warm temperate seas. Most are pelagic and schooling. Six species in the area. Page 318.

Family Emmelichthyidae - Rovers

Rovers are moderately elongate with similarly shaped dorsal and ventral profiles. The mouth is highly protrusible, with the lower jaw projecting beyond the upper. Teeth, when present, are conical and located anteriorly. Most of the head and maxilla are covered in scales. The dorsal fin is either deeply notched or divided. Color is usually reddish dorsally, silvery below. Rovers occur worldwide in tropical to warm temperate seas near bottom over continental shelves and upper slopes. Two species in the area. Page 320.

Family Lutjanidae - Snappers

Snappers are elongate and moderately deep-bodied. The upper jaw is moderately protrusible. Well-developed canine-like teeth are usually present. The cheeks are scaled; the snout and suborbital area are scaleless. Preopercular margin is serrated. Dorsal fin is unnotched to slightly notched. Color is often reddish but may be yellowish or purplish to grayish with stripes or bars. Snappers occur circumglobally in tropical to warm temperate seas from shore to about 1,800 ft. Most are demersal; some live in fresh water. Most are nocturnal. Females spawn several times per season. Eighteen species in the area. Page 322.

Family Lobotidae - Tripletails

Tripletails are oval to oblong, deep-bodied, and laterally compressed. The head profile is steeply sloping. The snout is short and blunt. Second dorsal, anal, and caudal fins are broad and rounded. Tripletails occur circumglobally in tropical to warm temperate seas from coastal waters to well offshore. Some live in brackish or fresh water. They are sluggish fishes, often floating on their sides with flotsam or *Sargassum* seaweed. Juveniles are believed to mimic mangrove leaves. One species in the area. Page 328.

Family Gerreidae - Mojarras

Mojarras are moderately slender to deep-bodied and laterally compressed. The upper jaw is highly protrusible. The dorsal fin is tall anteriorly with a slight to deep notch between the spiny and soft portions. The bases of the dorsal and anal fins are scaled. Caudal fin is deeply forked. Mojarras occur circumglobally in tropical to warm temperate seas. Most are found along the coast, many enter brackish water, and some live in fresh water. Thirteen species in the area. Page 330.

Family Haemulidae - Grunts

Grunts are oblong, moderately deep-bodied, and compressed. The dorsal head profile is almost straight to convex. The snout is moderately short to long and scaleless. Mouth is small to moderate with thick lips. Canine teeth absent. Preopercular margin serrated. Dorsal fin is continuous and either notched or unnotched. Grunts occur circumglobally in tropical to warm temperate seas in shallow coastal, brackish, and occasionally, fresh water. All are capable of sound production. Twenty-five species in the area. Page 334.

Family Sparidae - Porgies

Porgies are small to medium, oblong to oval in profile, and moderately to deeply compressed. The dorsal head profile is usually steep. Snout and suborbital area are scaleless. The mouth is small with a slightly protrusible upper jaw. Front teeth are conical or incisor-like; teeth in sides of jaws molar-like. Preopercular margin smooth. Dorsal fin is continuous and weakly to slightly notched. Porgies are found worldwide in tropical to warm temperate seas. Most occur over continental shelves; some enter estuaries or live in fresh water. Some are hermaphroditic. Fifteen species in the area. Page 346.

Family Polynemidae - Threadfins

Threadfins are moderately elongate with widely separated dorsal fins and a deeply forked caudal fin. The snout is short and projects beyond the large, inferior mouth. The pectoral fin is low on the body with lower rays that are separate, long, and filamentous. Threadfins are circumglobal in shallow tropical to warm temperate seas. Some enter estuaries and rivers. Some are hermaphroditic. Two species in the area. Page 352.

Family Sciaenidae - Drums

Drums are small to large with a gently to steeply sloping dorsal head profile. The body may be short and deep to moderately elongate. Snout and chin have conspicuous pores; some have single or multiple barbels on chin. The lateral line extends onto the caudal fin. The spiny dorsal fin is short-based and usually continuous with the long-based soft dorsal fin. A gas bladder is always present and is used to produce drumming sounds. Drums are found in a wide range of habitats in circumglobal tropical to warm temperate seas. Some live in estuaries or fresh water. Many are migratory and use estuaries as nursery grounds. Many sought as sportfish and foodfish. Twenty-three species in the area. Page 352.

Family Mullidae - Goatfishes

Goatfishes are moderately elongate with a convex dorsal profile and nearly straight ventral profile. The ventral portion of the head and chest nearly flat. The dorsal fins are separate; the caudal fin is forked. Two well-developed barbels extend from chin. The upper jaw is slightly protrusible. Goatfishes are found near bottom over continental shelves and slopes of circumglobal tropical to warm temperate seas. Some enter estuaries. Use barbels to locate prey. Four species in the area. Page 364.

Family Pempheridae - Sweepers

Sweepers are small, compressed, and deep-bodied. The mouth and eyes are large. The dorsal fin is single and short-based; the anal fin is long-based. Posterior profile of back and caudal peduncle is nearly horizontal. A few possess luminous organs. Sweepers occur over inner continental shelves in tropical to subtropical marine and brackish waters of the western Atlantic, Indian, and Pacific oceans. They are common over coral reefs and in caves and crevices. Nocturnal and schooling. Two species in the area. Page 366.

Family Kyphosidae - Sea Chubs

Sea Chubs are oval to oblong in profile and moderately compressed. The head is short; the snout is blunt. Some have a distinct hump in front of eyes. The mouth is small and horizontal. The dorsal fin is continuous, with the spiny portion depressible into a scaled groove. Sea Chubs are found circumglobally in tropical to subtropical seas. They are associated with coral reefs, rocky bottoms, hard structures, and *Sargassum* seaweed. Four species in the area. Page 366.

Family Chaetodontidae - Butterflyfishes

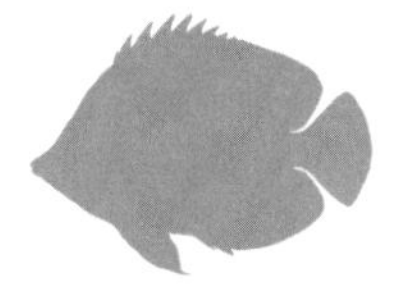

Butterflyfishes are oval, round, or rhomboid in profile, and deep-bodied and compressed. The mouth is small and the snout is blunt to long and pointed. Preopercular margin may be serrated, but a prolonged spine is absent at the lower angle. Dorsal fin is continuous to slightly notched. Spiny portion may be tall and deeply incised. Butterflyfishes occur circumglobally in shallow tropical to subtropical seas. Most are reef associated. Some are found in brackish or fresh water. They graze on small invertebrates and plankton. Butterflyfishes usually occur singly or in pairs. Popular in the aquarium trade. Seven species in the area. Page 368.

Family Pomacanthidae - Angelfishes

Angelfishes are oval to round in profile, deep-bodied, and compressed. The mouth is small and the snout is blunt. Preopercular margin is often serrated, and a prolonged spine is always present at the lower angle. Dorsal fin is continuous. Second dorsal fin may have long, trailing middle rays. Angelfishes occur circumglobally in shallow tropical to subtropical seas. They are generally associated with coral reefs, where they graze on sponges, invertebrates, and algae. Often seen singly or in pairs. Popular in the aquarium trade. Seven species in the area. Page 372.

Family Cirrhitidae - Hawkfishes

Hawkfishes are small and oval to oblong in shape. The snout is blunt to pointed. Dorsal fin is continuous with cirri projecting from tips of spiny dorsal-fin spines. The lower pectoral fin is incised with long, unbranched rays. Hawkfishes occur circumglobally in shallow tropical seas. Most are found in the Indo-Pacific. They use the thickened lower pectoral-fin rays to cling to and move over substrate. One species in the area. Page 374.

Family Cichlidae - Cichlids

Cichlids are small to large, moderately deep and compressed, and highly variable in profile. Upper jaw is protrusible. Dorsal fin is continuous, often with incised spiny membranes and elongate middle soft rays. Lateral line is in two sections. Cichlids are the third-most populous fish family on Earth and occur in the Americas and Africa to India. Most live in fresh water; some in brackish water. Many have been introduced outside of their native ranges. Four shallow, brackish water species in the area. Page 376.

Family Pomacentridae - Damselfishes

Damselfishes are small, oval to oblong, and laterally compressed. The mouth is small and oblique. The lateral line is either incomplete or interrupted. The dorsal fin is continuous, with the spiny base longer than the soft base. Caudal fin is shallowly to deeply forked. Damselfishes are found circumglobally in tropical to warm temperate seas over reefs, sandy, rubble, and seagrass bottoms. Most occur in the Indo-Pacific. Most are territorial and will defend eggs and feeding areas. Nineteen species in the area. Page 378.

Family Labridae - Wrasses and Parrotfishes

Wrasses and Parrotfishes are small to very large and highly variable in profile. Snout can be pointed, rounded, or steeply sloping. Lips are often fleshy. Teeth are separate or partially fused. Dorsal fin is continuous. Second dorsal fin may have elongated middle rays. Scales are relatively large. Wrasses and Parrotfishes are circumglobal in shallow tropical to warm temperate seas. Many are reef associated; others occur over seagrass beds and sandy to rocky bottoms. Most are hermaphroditic. Color and pattern often change with sex. Some bury themselves in sand; others secrete a mucous cocoon in which they rest. This is the second-largest family of marine fishes. Thirty-six species in the area. Note: The family Scaridae (Parrotfishes) was recently grouped into Labridae. Page 386.

Family Percophidae - Duckbills

Duckbills are elongate with a flattened head and snout. The mouth is very large. The eyes are large and closely set on top of head. Spiny dorsal fin (when present) is short-based; soft dorsal fin is long-based. Duckbills are demersal over outer continental shelves and upper slopes of the Atlantic, Indo-West Pacific, and southeast Pacific oceans. They feed on shrimps and small fishes. Three shallow-water species in the area. Page 402.

Family Uranoscopidae - Stargazers

Stargazers have a large head and a robust body. The mouth is large and oblique to nearly vertical. Lips are often fringed. Eyes are either on or near top of head. First dorsal fin is present or absent. Pectoral fins are fan-like. A blunt or sharp spine is present over the pectoral fins. These spines are venomous in most species. The body is naked or covered in small, smooth scales. Stargazers are demersal in circumglobal tropical to temperate seas from shore to about 3,000 ft. Some species have a lure-like filament in the mouth. One genus has an electric organ used to stun prey. Three species in the area. Page 404.

Family Tripterygiidae - Triplefin Blennies

Triplefin Blennies are small and cryptic fishes. The snout is short; the eyes are large. Nostrils may bear a tentacle or cirri. Eyes may have single or multiple cirri. Cirri absent on nape. There are three distinct dorsal fins: the first two are spiny, the third soft. Head and abdomen scaleless in most specimens. Triplefin Blennies are demersal over coral reefs and rocky areas from shore to about 1,800 ft. in circumtropical seas. Six species in the area. Page 404.

Family Dactyloscopidae - Sand Stargazers

Sand Stargazers are small and moderately elongate. Eyes are set on top of the head, are very small to somewhat large, and are stalked, unstalked, and/or fringed. Snout is very short. Mouth is oblique to almost vertical and may have fringed lips. Upper edge of opercle is often fringed. The lower portion of the opercle is expanded. The dorsal fin may be continuous, incised anteriorly, or notched. Sand Stargazers are demersal in tropical to warm temperate western Atlantic and eastern Pacific oceans. Thirteen species in the area. Page 408.

Family Blenniidae - Combtooth Blennies

Combtooth Blennies are small and scaleless. The head is usually very blunt. Nostrils sometimes with fleshy flaps. Eyes often with cirri, fleshy flaps, or tentacles. Mouth is usually small, with a closely packed, single row of comb-like teeth that are fixed or movable. Some have canine teeth. Pectoral-fin rays are unbranched. Caudal-fin rays are branched or unbranched. Combtooth Blennies occur circumglobally in tropical to warm temperate seas. All are demersal; most are in shallow water. Fourteen species in the area. Page 412.

Family Labrisomidae - Labrisomid Blennies

Labrisomid Blennies are small and usually scaled. The head is blunt to pointed. Cirri are often present on nostrils and eyes. Nape may have a single cirrus, multiple cirri, or fleshy tabs. Mouth is small to moderate with fleshy lips and teeth that are variable in size and arrangement. All fin rays, including caudal, are unbranched. Labrisomid Blennies are demersal in the tropical to warm temperate western Atlantic and eastern Pacific oceans. They are generally reef associated in shallow water. Males and females often colored and patterned differently. Sixty species in the area. Page 418.

Family Chaenopsidae - Tube Blennies

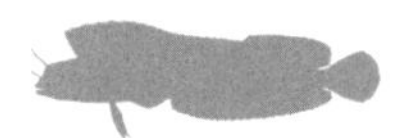

Tube Blennies are small, elongate, and scaleless. Cirri on nostrils and over eyes vary in complexity or may be entirely absent. Lateral line is absent. Caudal fin may be continuous or separate from dorsal and anal fins. Body may be transparent to darkly pigmented. Males are generally darker or more brightly colored than females. Some male species have a much higher anterior first dorsal-fin lobe than females. Tube Blennies are demersal in the tropical to subtropical western Atlantic and eastern Pacific oceans. Many live in abandoned worm tubes and invertebrate shells. Forty-five species in the area. Page 438.

Family Gobiesocidae - Clingfishes

Clingfishes are small and scaleless. The head and anterior portion of the body are generally rounded and flattened. Eyes are on top of the head. Dorsal fin is single. Pelvic fins are modified into an adhesive sucking disk. Clingfishes occur circumglobally in shallow tropical to warm temperate seas and in brackish and fresh water. Most are bottom-dwelling. Use adhesive disk to attach to hard substrates and plants. Twenty-seven species in the area. Page 452.

Family Callionymidae - Dragonets

Dragonets are elongate and variably flattened anteriorly. The head may be moderately to greatly expanded. Gill openings are small pores behind upper head. Preopercle has a strong, variably shaped spine. First dorsal fin is often tall and sometimes filamentous. Pelvic fins are attached by a membrane to the pectoral-fin base. Color may be bright to cryptic. Dragonets are circumglobal in shallow to deep tropical to temperate seas. A few enter fresh water. Three species in the area. Page 458.

Family Eleotridae - Sleepers

Sleepers are small to moderately sized and elongate to somewhat stout. The head is usually flattened on top. Eyes are widely separated. Two dorsal fins present, with second dorsal-fin base shorter than the distance between its rear origin and the caudal-fin origin. Pelvic-fin bases are close together or united, but fins are separate. Sleepers are found in most shallow tropical to subtropical fresh, brackish, and marine waters. Five species in the area. Page 460.

Family Gobiidae - Gobies

Gobies are small to very small and elongate. The head is short and broad. First dorsal fin, when present, is separate from second dorsal fin. Second dorsal-fin base is usually longer than the distance between its rear origin and caudal-fin origin. Pelvic fins are separate, somewhat connected, or completely united to form a disk. Lateral line is absent. Body is scaleless, partially scaled, or entirely scaled. Gobies occur circumglobally in a wide variety of tropical to warm temperate waters. They are most diverse over reefs. Most are marine and demersal. Some are free-swimming. Gobies are the largest family of marine fishes, with more than 1,500 species. One hundred twenty-six species in the area. Page 462.

Family Microdesmidae - Wormfishes

Wormfishes are small, elongate to eel-like, and often laterally compressed. The snout is blunt and the lower jaw protrudes. The dorsal fin is single and long-based. Caudal fin is separate or continuous with dorsal and anal fins. Pelvic fins are small and separate. Lateral line is absent. Wormfishes occur circumglobally in shallow, nearshore, tropical to warm temperate seas. They are found around coral reefs, over muddy bottoms, and in tidepools. Wormfishes often burrow into soft muddy and sandy bottoms. Three species in the area. Page 494.

Family Ptereleotridae - Dartfishes

Dartfishes are small and elongate. The mouth is almost vertical, and the lower jaw protrudes. The first dorsal fin is short-based, with six spines, and separate from the second dorsal fin. Caudal fin is pointed to rounded or oblong. Dartfishes occur circumglobally in tropical to subtropical seas. Some live in fresh water. Many are reef associated. This family was previously grouped with Gobies. Two species in the area. Page 496.

Family Ephippidae - Spadefishes

Spadefishes are deep-bodied and circular to oblong in profile. The head is short and the body is deeply compressed. Jaws have slender, movable, brush-like teeth. First dorsal fin is distinct from second and is usually notched in most. Preopercle is smooth to serrate and lacks a prominent spine at corner. Spadefishes occur circumglobally in tropical to warm temperate seas and rarely in brackish waters. They form schools over coral and artificial reefs and rocky areas. Juveniles are pelagic; some mimic floating leaves or debris. One species in the area. Page 496.

Family Acanthuridae - Surgeonfishes

Surgeonfishes are deep-bodied, laterally compressed, and round to oval in profile. The head profile is steep. The mouth is small and not protrusible. The eyes are set high on the head. Dorsal fin is continuous. The caudal peduncle has one or more collapsible spines or keeled and bony plates laterally. Surgeonfishes occur circumglobally in tropical to subtropical seas. They are usually associated with shallow coral and rocky reefs. Three species in the area. Page 496.

Family Sphyraenidae - Barracudas

Barracudas are elongate and small to moderately large. The head is long, the snout is pointed, and the lower jaw protrudes. The large jaws and roof of the mouth have numerous sharp conical or flattened teeth. Dorsal fins are short-based and widely separated. Barracudas occur circumglobally in tropical to warm temperate seas. They occur in a wide variety of coastal habitats from surface to near bottom. Small species and juveniles of large species form schools. Barracudas are swift, voracious predators. Three species in the area. Page 498.

Family Gempylidae - Snake Mackerels

Snake Mackerels are very elongate to moderately deep-bodied. The lower jaw protrudes. Teeth are strong, some are fang-like, and some project from lower jaw. First dorsal fin is long-based; second dorsal fin is short-based. Finlets follow second dorsal and anal fins in some species. Pelvic fins are small, rudimentary, or absent. Lateral line is single or double. Snake Mackerels occur circumglobally in tropical to subtropical seas. Seven shallow-water species in the area. Page 498.

Family Trichiuridae - Cutlassfishes

Cutlassfishes are elongate, strongly compressed, and ribbon-like. The snout is pointed and the lower jaw protrudes. Large, fang-like teeth are usually present. The dorsal fin is extremely long-based and is notched in some. Pelvic fins are very small or absent. Anal fin is sometimes reduced to a series of small, projecting spines. Caudal fin is small or absent. Cutlassfishes occur circumglobally in tropical to warm temperate seas from surface to about 6,500 ft. One shallow-water species in the area. Page 500.

Family Scombridae - Mackerels and Tunas

Mackerels and Tunas are medium to large and elongate to very robust. The body is somewhat compressed to rounded in cross-section. The snout is conical. The first and second dorsal fins are short-based, well separated, and insert into grooves. A series of finlets follow second dorsal and anal fins. Two oblique keels at caudal-fin base are present in all. Many also with a lateral keel on the very narrow caudal peduncle. Some species have a corselet of enlarged scales around the pectoral-fin area. Mackerels and Tunas occur circumglobally in tropical to temperate seas. Most are marine, pelagic, and highly migratory. Some occur coastally. Important foodfish. Sixteen species in the area. Page 502.

Family Xiphiidae - Swordfish

The Swordfish is large, robust anteriorly, and rounded in cross-section. The upper jaw forms a long, flattened, sword-like bill. Teeth are absent in adults. First dorsal fin is short-based in adults and widely separated from second dorsal fin. First dorsal fin is long-based and elevated in juveniles. The caudal peduncle is deeply notched at the caudal-fin base and has strong lateral keels. Pelvic fins absent. Swordfish is oceanic in worldwide offshore tropical to temperate seas. Highly migratory. Important foodfish. There is one species in this family. Page 508.

Family Istiophoridae - Billfishes

Billfishes are elongate and moderately compressed in cross-section. The upper jaw forms a long, spear-like bill that is round in cross-section. First dorsal fin is long-based in all and sail-like in one species. The first dorsal and first anal fins insert into grooves. Pelvic fins are long and narrow. The caudal peduncle is deeply notched at the caudal-fin base and has strong lateral keels. Billfishes occur circumglobally in tropical to warm temperate seas. They are usually offshore in upper water column. Swift predators. Important sportfishes. Five species in the area. Page 508.

Family Nomeidae - Driftfishes

Driftfishes are moderate to large, slender to deep-bodied, and laterally compressed. The mouth is small, the snout is moderately blunt to very blunt. First dorsal fin and pelvic fins insert into a groove. Second dorsal and anal fins are about the same shape and are scaled at the base. Caudal keels are absent. Driftfishes occur circumglobally in tropical to subtropical oceanic waters. Adults inhabit midwaters or are demersal over continental slopes. Juveniles are associated with jellyfishes. Three species in the area. Page 510.

Family Ariommatidae - Ariommatids

Ariommatids are small, slender to moderately deep-bodied, and laterally compressed to oval to rounded in cross-section. The mouth is small; the snout is short and blunt. First dorsal fin and pelvic fins insert into a groove. Second dorsal and anal fins are about the same shape. Two low, fleshy keels on caudal-fin base. Ariommatids occur circumglobally in tropical to subtropical oceanic waters. Absent from eastern Pacific. Adults are offshore in deep water. Juveniles occur near surface. Two species in the area. Page 512.

Family Stromateidae - Butterfishes

Butterfishes are deep-bodied and laterally compressed. Snout is short; mouth is small. Eyes are covered by adipose eyelids. Dorsal and anal fins are single, long-based, scaled, and similar in shape. Anterior fin lobes may be elongate. Pelvic fins are absent. Butterfishes are found circumglobally in tropical to warm temperate seas. Absent from southwestern Pacific and Indian oceans. They are pelagic along the coast, over continental shelves, and sometimes in estuaries. Two species in the area. Page 512.

Family Paralichthyidae - Sand Flounders

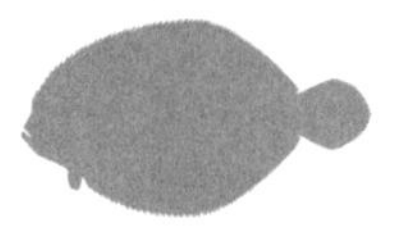

Sand Flounders are deep-bodied and laterally compressed. The mouth is large and protrusible. Eyes are large to relatively small and usually left-facing. Anterior dorsal-fin rays may be long and mostly free of membrane. Pelvic fins are symmetrical in shape and short-based. Sand Flounders occur worldwide in tropical to temperate seas. Some in fresh water. All are demersal and usually over soft bottoms. Usually burrow into sediment to await prey. Many are quick color-changers. Twelve species in the area. Page 514.

Family Bothidae - Lefteye Flounders

Lefteye Flounders are somewhat to very deep-bodied and laterally compressed. The mouth is moderate to large and protrusible. Eyes are left-facing (rarely right-facing) and are close-set or widely separated. Pelvic fin on eyed side is larger and with longer base than blind-side fin. Lefteye Flounders occur circumglobally in tropical to warm temperate seas. Demersal, usually over soft bottoms. In many species males have separated eyes and elongate pectoral fin. Five species in the area. Page 518.

Family Achiridae - American Soles

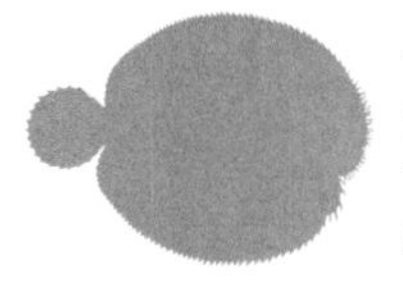

American Soles are deep-bodied and laterally compressed. The mouth is small. The eyes are small to minute and right-facing (rarely left-facing). The preopercular margin is concealed or appears as a groove. Pectoral fins are small to absent. Eyed-side pelvic fin may be free or joined to anal fin. American Soles occur in the western Atlantic along the North, Central, and South American coasts. Most are found near shore over soft bottoms. Seven species in the area. Page 520.

Family Cynoglossidae - Tonguefishes

Tonguefishes are moderately deep-bodied and laterally compressed. The body is lance- or tongue-shaped and tapers to a blunt or pointed tail. The mouth is small. Eyes are small and left-facing. Dorsal and anal fins are confluent with caudal fin. Pectoral fins are absent. Eyed-side pelvic fin, when present, is confluent with anal fin. Lateral line is absent. Tonguefishes occur circumglobally in tropical to temperate seas. They are demersal over a variety of bottoms from shore to about 5,000 ft. Fourteen species in the area. Page 522.

Family Balistidae - Triggerfishes

Triggerfishes are deep-bodied and moderately compressed. The mouth is small with a single row of strong teeth. Gill openings are short and slit-like. A patch of enlarged scales above the pectoral fins is usually present. First dorsal fin with three stout spines; second locks first upright. Pelvic fins reduced and encased. Scales plate-like and geometric. Triggerfishes occur circumglobally in tropical to warm temperate seas from surface to bottom in a variety of habitats. Five species in the area. Page 526.

Family Monacanthidae - Filefishes

Filefishes are moderate to deep-bodied and laterally compressed. The mouth is small with moderately strong teeth. Gill openings are short and slit-like. First dorsal fin with one or two spines; second spine may lock first spine in upright position. Pelvic fins absent or consisting of a movable pelvic bone. Pelvic flap present in some. Caudal peduncle may have spines. Scales minute and prickly. Filefishes occur circumglobally in tropical to temperate seas from shore to about 650 ft. Ten species in the area. Page 528.

Family Ostraciidae - Boxfishes

Boxfishes are triangular and square to pentangular in cross-section. The mouth is small with fleshy lips and moderately strong teeth. Gill openings are slit-like. Spiny dorsal fin and pelvic skeleton are absent. The body is encased in close-set hexagonal to polygonal plates. Plates are either visible or concealed by skin. Plates may form "horns" over eyes in some. Boxfishes occur circumglobally in tropical to warm temperate seas. Demersal over rocky and coral reefs, sandy and seagrass bottoms. Five species in the area. Page 532.

Family Tetraodontidae - Puffers

Puffers are slender to somewhat robust, becoming rounded when inflated. The mouth is small with four fused teeth forming a "beak." Gill openings are slit-like. Spiny dorsal fin and pelvic skeleton are absent. Body is covered in tough, elastic skin. Most with small, spine-like prickles on abdomen, sides, and back. Puffers occur circumglobally in tropical to warm temperate seas. Most in shallow, nearshore waters. Some in brackish or fresh water. Many have highly toxic organs and flesh. Fifteen species in the area. Page 534.

Family Diodontidae - Porcupinefishes

Porcupinefishes are somewhat robust, becoming round when inflated. The mouth is small, with fused teeth forming a strong "beak." Gill openings are slit-like. Spiny dorsal fin and pelvic skeleton are absent. Body is covered in tough, elastic skin and short to long spine-like scales. Some with spines always erect; others with spines erect only when inflated. Porcupinefishes occur circumglobally in tropical to warm temperate seas. Most are demersal and solitary; one is pelagic, offshore, and schooling. Seven species in the area. Page 540.

Family Molidae - Molas

Molas are large, deep-bodied, and laterally compressed. The body profile is rounded or oval to oblong. Mouth and gill openings are small. Dorsal and anal fins are set back on the body and are erect and similar in shape. A caudal peduncle and true caudal fin are absent. Caudal fin is replaced by a leathery structure that is supported by dorsal- and anal-fin rays. Pelvic fins are absent. Molas are pelagic in circumglobal tropical to warm temperate seas. Often seen swimming on their sides at the surface. Juveniles are schooling. Three species in the area. Page 542.

SPECIES

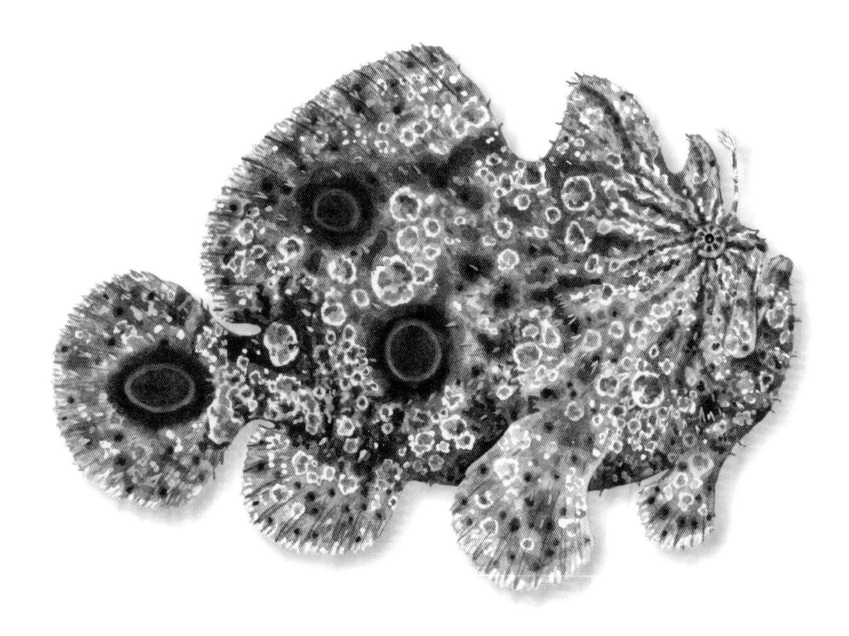

Ginglymostomatidae - Nurse Sharks

Nurse Shark - *Ginglymostoma cirratum* (Bonnaterre, 1788)

FEATURES: Brown, gray, or yellowish dorsally; paler below. Small juveniles with vague dark areas and small, dark spots. Two nasal barbels and grooves connected to mouth. Snout short, rounded. Dorsal fins closely spaced, located posteriorly. Caudal fin lacks distinct lower lobe. HABITAT: RI (rare) to FL, Bermuda, Bahamas, and Caribbean Sea to Brazil. Occur over sand flats, around mangroves, channels, and coral reefs from near shore to about 426 ft. BIOLOGY: Social and nocturnal. Return to favored caves and crevices to rest. Often form large congregations. Feed on variety of invertebrates and fishes. IUCN: Vulnerable.

Rhincodontidae - Whale Shark

Whale Shark - *Rhincodon typus* Smith, 1828

FEATURES: Dark gray, greenish gray, or reddish dorsally. Pale below. Pattern of white or yellow spots and bars covers upper body and fins. Spots form bars on and lines on midbody. Snout short; mouth broad. Back with long ridges; lowest ridges become strong caudal keels. Largest fish in the world. HABITAT: Circumglobal in tropical and warm temperate seas. In the W Atlantic from NY to FL, Gulf of Mexico, Bermuda, Bahamas, and Caribbean Sea to Brazil. Pelagic, at or near surface, near shore or in the open sea. BIOLOGY: Slow-moving and docile filter feeders. Females can carry up to 300 young. IUCN: Endangered.

Odontaspididae - Sand Tiger Sharks

Sand Tiger - *Carcharias taurus* Rafinesque, 1810

FEATURES: Grayish to gray brown dorsally. Pale below. Irregular dark spots and blotches on upper and lower sides and fins. Snout pointed. Teeth protrude. First dorsal-fin origin well behind pectoral fins. HABITAT: Circumglobal. In northwestern Atlantic from Gulf of Maine to Yucatán. In shallow, inshore waters on or near bottom, from surf zone to continental shelf. Often associated with caves, gullies, and reefs. BIOLOGY: Females gestate for about 9–12 months. Litters contain only one or two young born every other year. IUCN: Vulnerable.

Ragged-tooth Shark - *Odontaspis ferox* (Risso, 1810)

FEATURES: Brownish to grayish brown dorsally. Pale below. Often with dark spots and blotches on upper and lower sides and fins. Snout broadly pointed. Teeth protrude. First dorsal-fin origin over rear pectoral-fin margin. HABITAT: Scattered circumglobally in temperate waters. In western Atlantic from NC to Brazil including northern Gulf of Mexico and off eastern Yucatán. Rare to uncommon. Occur near surface to about 1,740 ft. Often on or near bottom over continental shelves and upper slopes. BIOLOGY: Feed on fishes, squids, and shrimps. IUCN: Vulnerable.

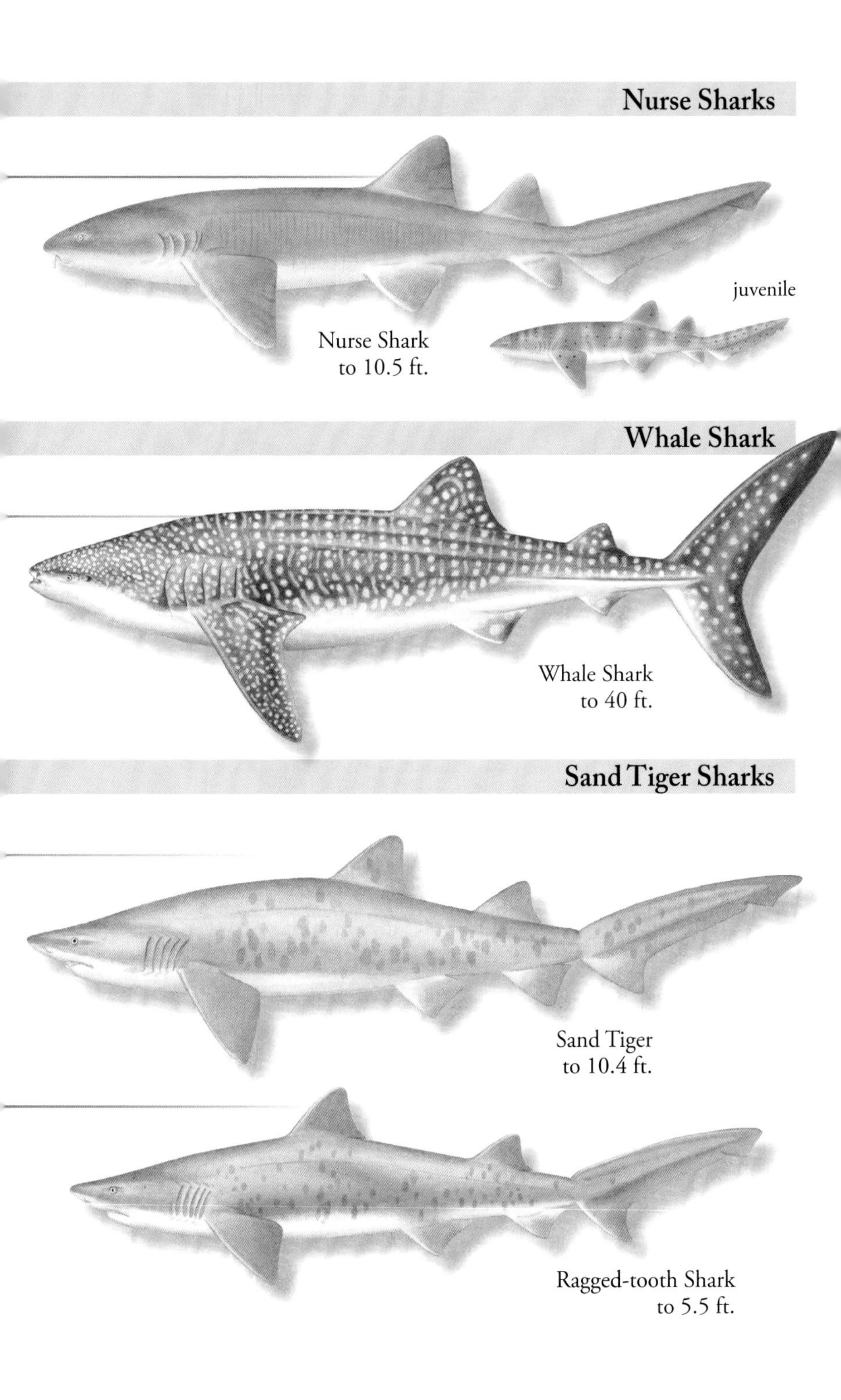

Nurse Shark
to 10.5 ft.

Whale Shark
to 40 ft.

Sand Tiger
to 10.4 ft.

Ragged-tooth Shark
to 5.5 ft.

Pseudocarchariidae - Crocodile Shark

Crocodile Shark - *Pseudocarcharias kamoharai* (Matsubara, 1936)

FEATURES: Shades of gray dorsally; pale below. Some with white flecks. Eyes comparatively large. Jaws highly protrusible. Dorsal, ventral, and pectoral fins short and with transparent to white margins. HABITAT: Circumtropical. In western Atlantic from Venezuela, Lesser Antilles to Brazil. Oceanic and pelagic. Near surface to about 2,000 ft. IUCN: Near Threatened.

Alopiidae - Thresher Sharks

Bigeye Thresher - *Alopias superciliosus* (Lowe, 1841)

FEATURES: Bluish brown dorsally. Pale below. Skin has iridescent highlights. Forehead prominently indented over very large eyes. Tail almost as long as body. HABITAT: Circumglobal in tropical and warm temperate seas. In western Atlantic from NY to FL, Gulf of Mexico, Bahamas, Greater Antilles, and Venezuela. Found coastally over the continental shelf to deep water. BIOLOGY: Feed on fishes and squids. Use tail to stun prey. Litters usually contain only two young. Commercially fished for flesh and oil. IUCN: Vulnerable.

Common Thresher Shark - *Alopias vulpinus* (Bonnaterre, 1788)

FEATURES: Brown, gray, blackish, or blue gray dorsally and under snout. Two distinct white patches ventrally. Tail almost as long as body. HABITAT: Circumglobal in tropical to cold temperate seas. In western Atlantic from Newfoundland to FL and Gulf of Mexico; Venezuela to Argentina. Coastal and pelagic from surface to about 1,200 ft. BIOLOGY: Active and strong; often leap out of the water. Feed on small schooling fishes, squids, octopods, and crustaceans. Herd prey into tight groups. Use tail to stun prey. Sometimes hunt in groups. IUCN: Vulnerable.

Lamnidae - Mackerel Sharks

Shortfin Mako - *Isurus oxyrinchus* Rafinesque, 1810

FEATURES: Shades of gray blue, purplish, or dark blue dorsally. White ventrally. Snout pointed. Pectoral fins shorter than head. Prominent caudal keels present. HABITAT: Circumglobal in tropical to temperate seas. In western Atlantic from Nova Scotia to Brazil. Pelagic. BIOLOGY: Active and strong swimmers. Acrobatic and bold. Feed on schooling fishes and other sharks. IUCN: Endangered.

Longfin Mako - *Isurus paucus* Guitart Manday, 1966

FEATURES: Slate blue or gray black dorsally and on sides. Dark chin strap. Snout pointed. Pectoral fins broad, longer than head. Broad caudal keels present. HABITAT: Worldwide in tropical to warm temperate seas. In western Atlantic from Georges Bank to S Brazil. Offshore. BIOLOGY: Litters contain two to eight young. Taken by longlines, nets, and as bycatch. IUCN: Endangered.

ALSO IN THE AREA: *Carcharodon carcharias*, *Lamna nasus*, see p. 547.

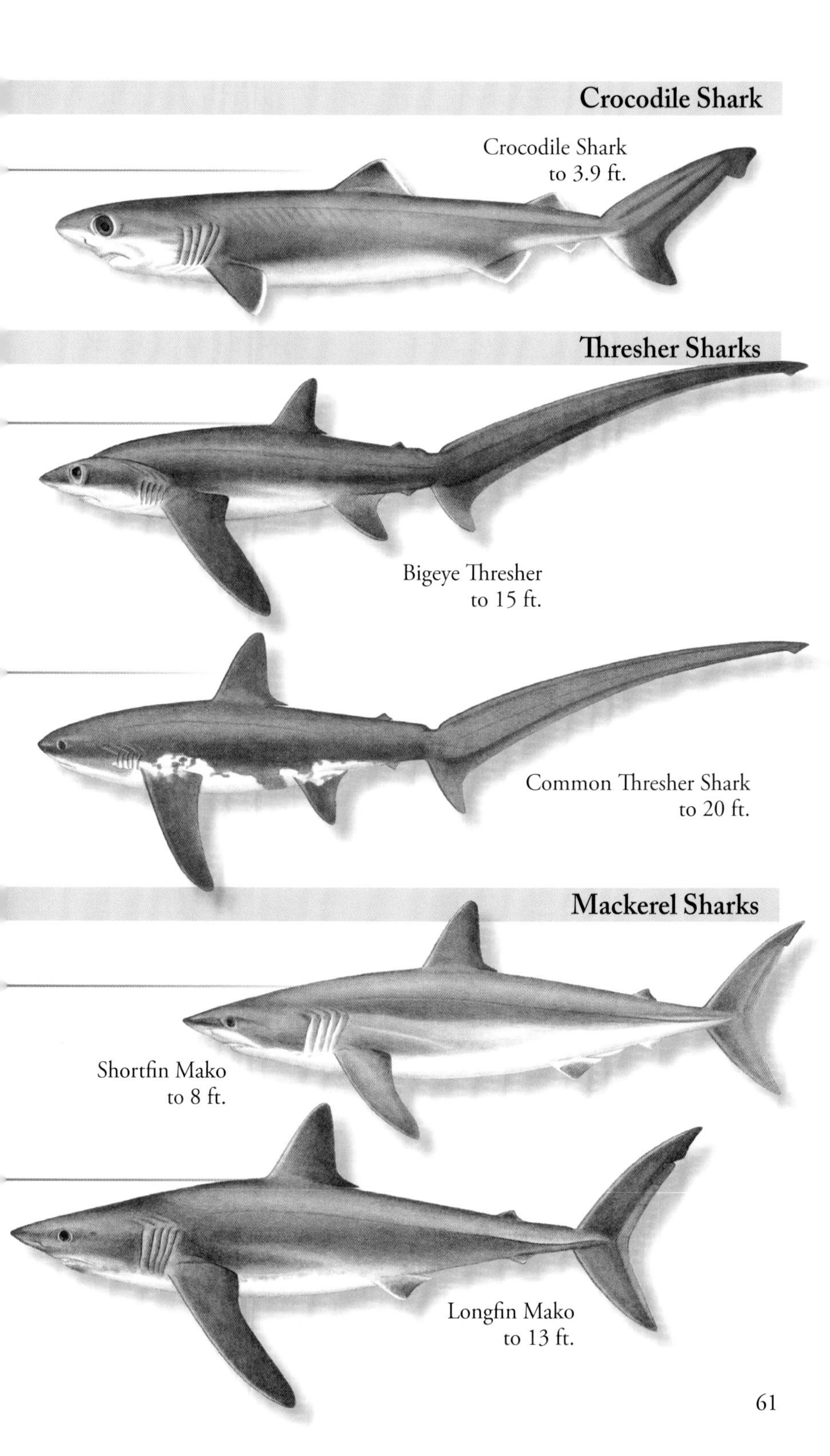
Crocodile Shark
Crocodile Shark
to 3.9 ft.
Thresher Sharks
Bigeye Thresher
to 15 ft.
Common Thresher Shark
to 20 ft.
Mackerel Sharks
Shortfin Mako
to 8 ft.
Longfin Mako
to 13 ft.

Scyliorhinidae - Catsharks

Freckled Catshark - *Scyliorhinus haeckelii* (Miranda Ribeiro, 1907)

FEATURES: Color, pattern variable. Shades of tan with seven to eight dusky saddles, and small, blackish spots on saddles. Some with rows of spots between saddles and white spots posteriorly. Dorsal fins set posteriorly. First dorsal fin triangular to square at tip. HABITAT: Eastern Colombia to Uruguay. On or near bottom, primarily around reefs and over algae beds from about 120 to 1,300 ft.

Chain Dogfish - *Scyliorhinus retifer* (Garman, 1881)

FEATURES: Dusky pale brown with dark "chain" pattern over body. Alternating dark saddles within chain pattern. Dorsal fins set posteriorly. HABITAT: MA to FL, Gulf of Mexico to southern Caribbean Sea. Bottom-dwelling on the outer continental slope, from about 240 to 2,000 ft. BIOLOGY: Sluggish. Feed on bottom fishes and invertebrates. Females lay eggs that attach to bottom materials. Young incubate for about seven months.

Triakidae - Hound Sharks

Smooth Dogfish - *Mustelus canis* (Mitchill, 1815)

FEATURES: Uniformly slate gray or gray dorsally. May pale or darken. White or yellowish ventrally. Fins edged with white. Eyes moderately large. Pectoral fins broad. Dorsal fins with deeply concave rear margin. Caudal fin deeply notched. Teeth lack cusps. HABITAT: Bay of Fundy to FL, Gulf of Mexico, and Caribbean Sea to Argentina. Coastal. On muddy bottoms to about 300 ft. BIOLOGY: Active and predatory. Feed on variety of invertebrates. Often kept in public aquariums. Threatened by heavy fishing. IUCN: Near Threatened.

Smalleye Smoothhound - *Mustelus higmani* (Springer & Lowe, 1963)

FEATURES: Pale gray dorsally with golden to brassy reflections—some more uniformly bronze. Pale below. Eyes comparatively small. Pectoral fins moderately broad, pointed. First dorsal fin slightly larger than second, both with concave rear margins. Teeth bluntly rounded with low, blunt cusps. HABITAT: Northern Gulf of Mexico and southern Caribbean Sea to Brazil. Over soft bottoms from shoreline to about 4,200 ft. Enter estuaries and lagoons. BIOLOGY: Feed mainly on crustaceans. Litters usually with one to seven pups.

Dwarf Smoothhound - *Mustelus minicanis* Heemstra, 1997

FEATURES: Shades of gray dorsally, pale below. Head short, flattened on top. Eyes comparatively large. Pectoral fins moderately broad. First dorsal fin slightly taller than second, both with concave rear margins. Lower caudal-fin lobe poorly developed. Teeth bluntly rounded with a weak cusp. HABITAT: NE Colombia and Venezuela. Demersal over soft bottoms from about 230 to 600 ft. BIOLOGY: Rare. Known from only a few specimens.

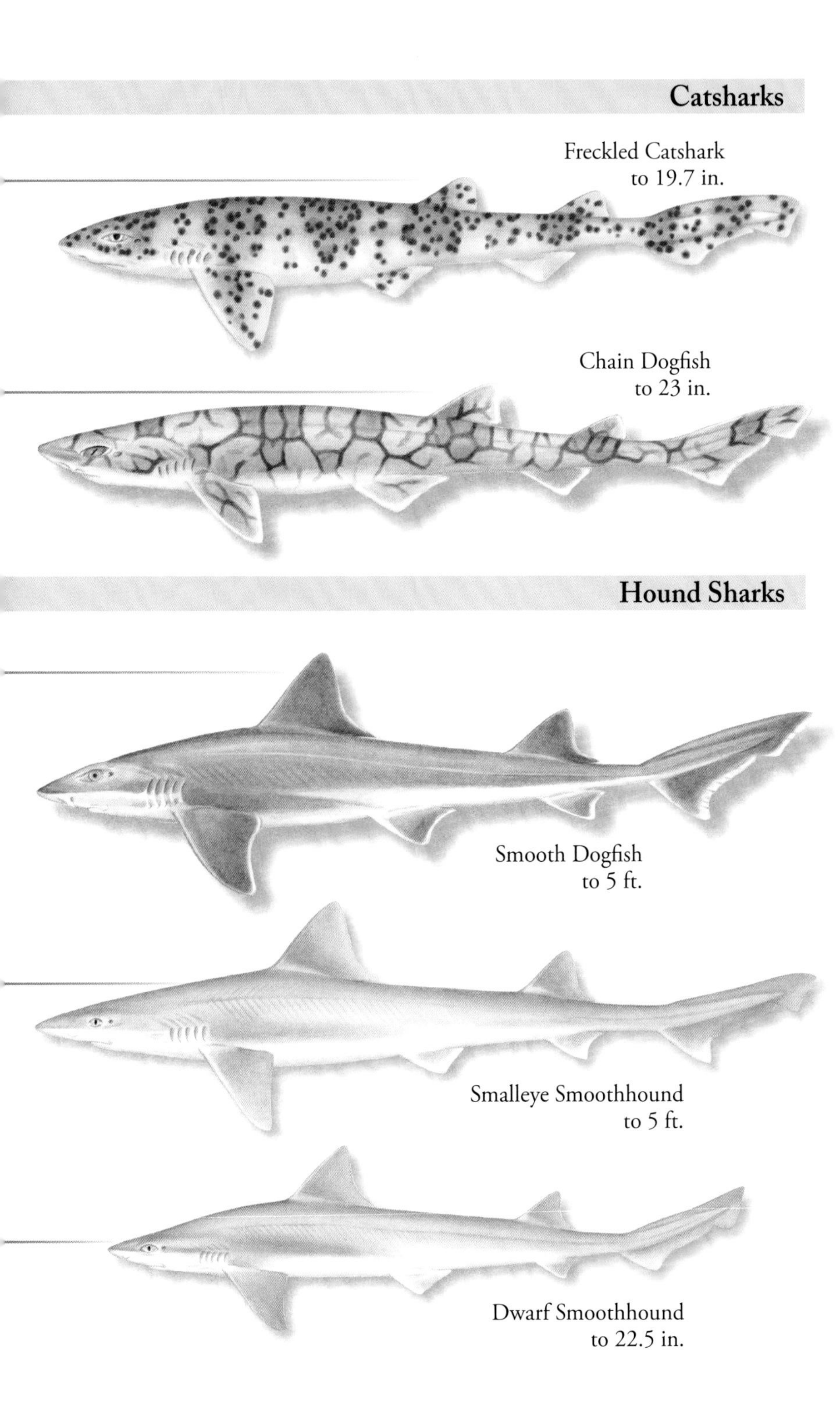
Catsharks
Freckled Catshark
to 19.7 in.
Chain Dogfish
to 23 in.
Hound Sharks
Smooth Dogfish
to 5 ft.
Smalleye Smoothhound
to 5 ft.
Dwarf Smoothhound
to 22.5 in.

Florida Smoothhound - *Mustelus norrisi* Springer, 1939

FEATURES: Uniformly gray or grayish brown dorsally. Pale ventrally. Eyes comparatively small. Pectoral fins comparatively narrow. Dorsal fins with deeply concave rear margins. Caudal fin deeply notched. Teeth small, low, each with a blunt cusp. HABITAT: Eastern Gulf of Mexico to Venezuela. Found over mud and sandy bottoms, near shore to about 250 ft. BIOLOGY: Common and migratory; move inshore in winter. Found offshore otherwise. Prey include shrimps, crabs, and small fishes.

Gulf Smoothhound - *Mustelus sinusmexicanus* Heemstra, 1997

FEATURES: Gray to grayish brown dorsally, pale below. Young may have dusky tips on dorsal and caudal fins. Eyes moderately large. First dorsal fin taller than second, both with concave margins. Teeth small, low, and bluntly rounded with a strong cusp and weak cusplet. HABITAT: Northern Gulf of Mexico and southern Bay of Campeche. Demersal over soft bottoms from about 120 to 750 ft. Usually between 135 and 300 ft.

Carcharhinidae - Requiem Sharks

Blacknose Shark - *Carcharhinus acronotus* (Poey, 1860)

FEATURES: Yellowish, greenish gray, yellowish brown, or bronze dorsally. White or pale ventrally. Tip of snout has dusky blotch. Blotch paler on adults. Caudal-fin tip dusky. HABITAT: NC to FL, Gulf of Mexico, Bahamas, and Caribbean Sea to Brazil. Inshore to continental slopes and on sandy or coralline bottoms. BIOLOGY: Feed on small fishes. Females give birth to three to six young. Fished heavily by longlines and gill nets. IUCN: Near Threatened.

Bignose Shark - *Carcharhinus altimus* (Springer, 1950)

FEATURES: Gray to bronze dorsally. White ventrally. All fins with dusky tips except pelvic fins. Inner corners of pectoral fins blackish. Snout large, bluntly rounded. First dorsal fin origin over posterior end of pectoral-fin base. HABITAT: Spotty distribution in all warm temperate and tropical seas. Recorded from FL, Bahamas, Cuba, N Gulf of Mexico to Venezuela. Bottom-dwelling offshore from about 75 to 1,300 ft. BIOLOGY: Prey on small fishes, rays, sharks, and cephalopods. Litters contain 3–15 young.

Narrowtooth Shark - *Carcharhinus brachyurus* (Günther, 1870)

FEATURES: Bronze, coppery, or grayish brown with greenish highlights dorsally. Pale below. Prominent white streak on sides. Snout broad, bluntly pointed. First dorsal-fin origin over tip of pectoral inner margin. Second dorsal fin low, nearly triangular, and located over anal fin. HABITAT: Scattered circumglobally in warm temperate seas. Recorded from Gulf of Mexico and southern Brazil to Argentina. Occur inshore to over continental shelves. BIOLOGY: Active and seasonally migratory. OTHER NAMES: Bronze Whaler, Copper Shark. IUCN: Near Threatened.

Hound Sharks

Florida Smoothhound
to 3 ft.

Gulf Smoothhound
to 4.6 ft.

Requiem Sharks

Blacknose Shark
to 4.6 ft.

Bignose Shark
to 9.8 ft.

Narrowtooth Shark
to 9.6 ft.

Spinner Shark - *Carcharhinus brevipinna* (Valenciennes, 1839)

FEATURES: Gray dorsally. White ventrally. Dorsal, pectoral, anal, and lower lobe of caudal fins tipped dusky or black. Snout long and pointed. First dorsal-fin origin over or just behind posterior tip of pectoral-fin inner margin. HABITAT: NC to FL, Gulf of Mexico, Bahamas, and Cuba. Also Guyana. Wide ranging in warm coastal waters to about 250 ft. BIOLOGY: Feed on small schooling fishes, sharks, rays, and squids. Often leap and spin out of the water while feeding. Active, schooling, migratory. IUCN: Near Threatened.

Silky Shark - *Carcharhinus falciformis* (Müller & Henle, 1839)

FEATURES: Dark gray, gray brown, or blackish dorsally. White ventrally. Fins with inconspicuous, dusky tips. Snout moderately long and rounded. Long, narrow pectoral fins. First dorsal fin small, originates behind rear margin of pectoral fin. HABITAT: Worldwide in tropical and subtropical seas. MA to FL, Gulf of Mexico, and Caribbean Sea to southern Brazil. Coastal and oceanic. BIOLOGY: Feed on fishes and some invertebrates. They are bold, active, inquisitive, and sometimes aggressive. Sought commercially. IUCN: Vulnerable.

Galapagos Shark - *Carcharhinus galapagensis* (Snodgrass & Heller, 1905)

FEATURES: Brownish to brownish gray dorsally. White ventrally. Most fin tips dusky, especially those under the pectoral fins. Snout moderately long and broadly rounded. Pectoral fins long. First dorsal fin originates over posterior inner pectoral-fin margin. HABITAT: Circumtropical in warm seas. Occur in northwestern Atlantic around Bermuda and in the northern and eastern Caribbean Sea. Usually associated with oceanic islands. BIOLOGY: Often occur in aggregations. May be confused with *Carcharhinus obscurus*. IUCN: Near Threatened.

Finetooth Shark - *Carcharhinus isodon* (Valenciennes, 1839)

FEATURES: Gray to bluish gray dorsally. White ventrally. Snout pointed. Gill slits comparatively long. Pectoral fins small. First dorsal fin comparatively small. Teeth very small. HABITAT: Rarely NY to NC. Usually from SC to FL and Gulf of Mexico; Trinidad and Guyana to southern Brazil. Occur in coastal waters from intertidal zone to about 65 ft. BIOLOGY: Highly active. Schooling and migratory with changing water temperatures. Feed on fishes and shrimp.

Bull Shark - *Carcharhinus leucas* (Valenciennes, 1839)

FEATURES: Gray dorsally. White ventrally. Fins with dusky tips. Eyes well forward on head. Snout very short, wide, and blunt. Body stout. First dorsal fin triangular to falcate, its origin over rear pectoral insertion. HABITAT: Circumglobal in tropical to warm temperate seas. In western Atlantic from MA to FL, Gulf of Mexico, Bahamas, and Caribbean Sea to Argentina. Generally near shore. May enter bays, estuaries, and rivers. BIOLOGY: Powerful and may be aggressive. Have the ability to thrive in fresh water. Diet is wide and varied. IUCN: Near Threatened.

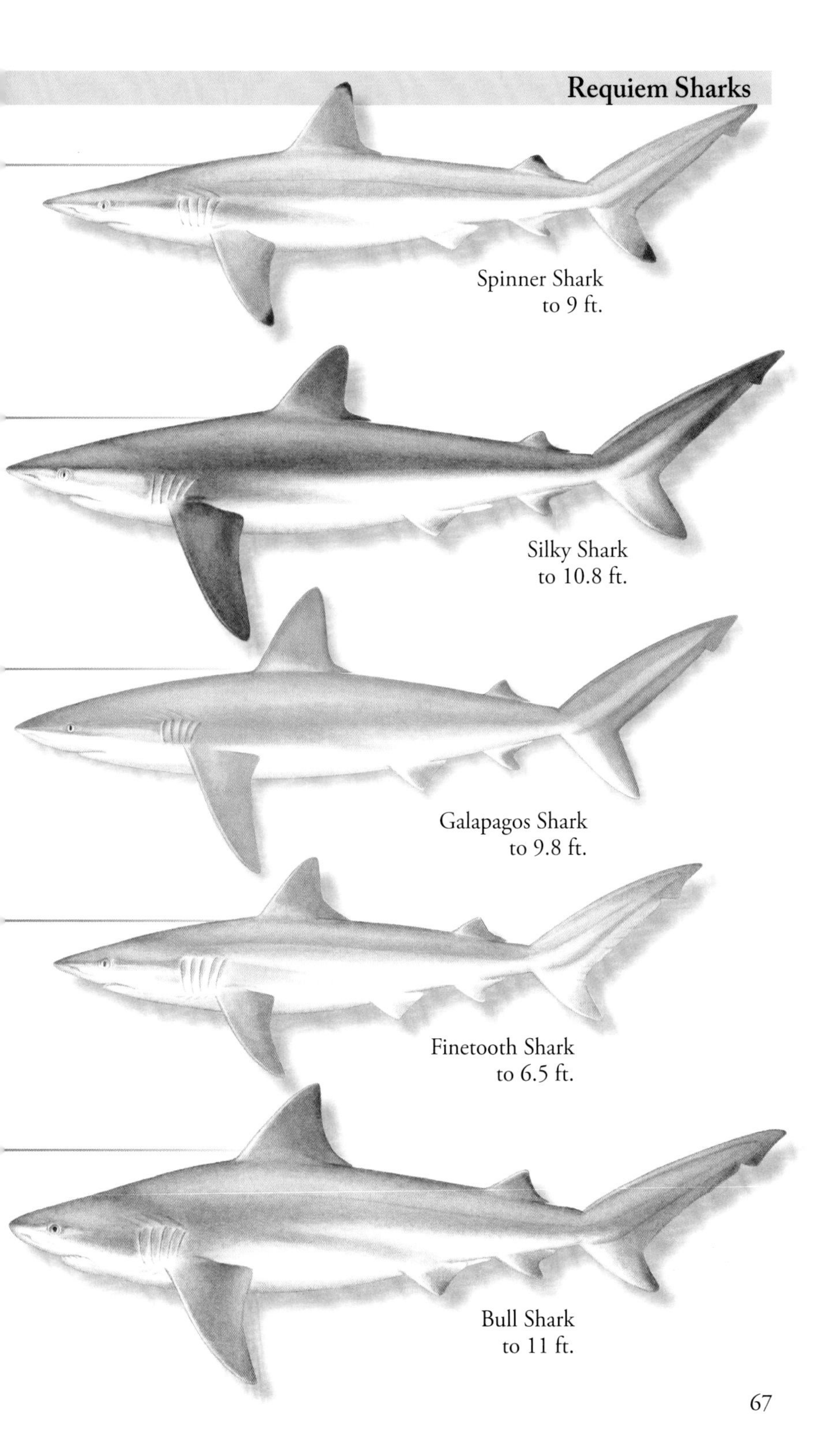
Spinner Shark
to 9 ft.
Silky Shark
to 10.8 ft.
Galapagos Shark
to 9.8 ft.
Finetooth Shark
to 6.5 ft.
Bull Shark
to 11 ft.

Blacktip Shark - *Carcharhinus limbatus* (Valenciennes, 1839)

FEATURES: Gray, brown gray, or blue gray dorsally. Pale ventrally. Dorsal fins, pectoral fins, and ventral lobe of caudal fin tipped black. Tips of pelvic and anal fins sometimes blackish. Snout moderately long and narrowly pointed. Dorsal-fin origin over posterior pectoral-fin insertion. HABITAT: Circumglobal in warm, shallow seas. In western Atlantic from MA to FL, Gulf of Mexico, Bahamas, and Caribbean Sea to southern Brazil. Usually inshore. IUCN: Near Threatened.

Oceanic Whitetip Shark - *Carcharhinus longimanus* (Poey, 1861)

FEATURES: Gray or brown dorsally. White below. Distal portions of first dorsal, pectoral, pelvic and caudal fins mottled white. Black blotches present on fin tips of immature specimens. Dorsal fin very tall and rounded at top. Pectoral fins large and paddle-like. Body large and stout. HABITAT: Circumglobal. In western Atlantic from ME to FL, Gulf of Mexico, and Caribbean Sea to Argentina. Pelagic over edges of continental shelves to mid-ocean. BIOLOGY: Feed primarily on fishes but also invertebrates, sea birds, and carrion. IUCN: Vulnerable.

Dusky Shark - *Carcharhinus obscurus* (Lesueur, 1818)

FEATURES: Shades of gray to bronze dorsally. White ventrally. Most fins dusky-tipped. Pectoral fins long and pointed. First dorsal-fin origin over or slightly before posterior end of inner pectoral-fin margin. HABITAT: Circumglobal in tropical to warm temperate seas. In western Atlantic from MA to FL, Gulf of Mexico, Bahamas, and Caribbean Sea. Occur from shoreline to outer continental shelves. BIOLOGY: Active and migratory. Prey include fishes, other sharks, rays, carrion, and a variety of invertebrates. Protected in the United States. IUCN: Vulnerable.

Reef Shark - *Carcharhinus perezii* (Poey, 1876)

FEATURES: Gray brown, dark gray, or olive gray dorsally. White to yellow ventrally. No distinct markings on body or fins. Snout short, bluntly rounded. First dorsal fin rounded at tip, its origin over or slightly before posterior end of inner pectoral-fin margin. HABITAT: FL, Bermuda, Bahamas, and Caribbean Sea to Brazil. Also Flower Garden Banks. Shallow coastal waters. Usually bottom-dwelling. BIOLOGY: The most common shark on Caribbean reefs. Capable of lying motionless on the bottom. IUCN: Near Threatened.

Sandbar Shark - *Carcharhinus plumbeus* (Nardo, 1827)

FEATURES: Grayish dorsally. White ventrally. Snout broad, rounded, and short. Eyes small. Body medium-sized and stout. First dorsal fin very tall, triangular, with origin over rear pectoral insertion. HABITAT: Circumglobal in tropical to warm temperate seas. In western Atlantic from MA to FL, Gulf of Mexico, Bahamas, and Caribbean Sea to Brazil. Occur over sandy or muddy bottoms, mostly near shore. BIOLOGY: Slow-growing and late-maturing with small litters. They have been heavily fished for meat, skins, and fins. IUCN: Vulnerable.

Blacktip Shark
to 8.4 ft.
Oceanic Whitetip Shark
to 13 ft.
Dusky Shark
to 12 ft.
Reef Shark
to 9.7 ft.
Sandbar Shark
to 7.8 ft.

Smalltail Shark - *Carcharhinus porosus* (Ranzani, 1839)

FEATURES: Bluish gray or gray dorsally. Pale to white ventrally. Pectoral, dorsal, and caudal fins dusky edged or unmarked. Snout long, pointed. Eyes comparatively large. First dorsal fin comparatively small, triangular, its origin over or behind posterior pectoral inner margin. Second dorsal-fin origin behind anal-fin origin. HABITAT: Northern Gulf of Mexico to southern Brazil. Also eastern Pacific. Along coasts, in estuaries, and over muddy bottoms to about 100 ft. BIOLOGY: Feed on small fishes, invertebrates, and other sharks. Litters are small and produce only 2–7 young. IUCN: Vulnerable.

Night Shark - *Carcharhinus signatus* (Poey, 1868)

FEATURES: Brownish dorsally, pale below. Others described as grayish blue dorsally. Rear pectoral-fin margin pale. Eyes green. Snout long and pointed. Gill slits comparatively short. First dorsal fin comparatively small. HABITAT: DE to FL, Gulf of Mexico, Bahamas, and eastern Caribbean Sea to Brazil. Semi-oceanic in deep water, usually over outer continental shelves. Usually below about 900 ft. during the day, migrate closer to surface at night. IUCN: Vulnerable.

Tiger Shark - *Galeocerdo cuvier* (Péron & Lesueur, 1822)

FEATURES: Shades of gray with darker bars and spots dorsally. Markings muted in older sharks. Snout short, blunt. Body large and stout. Trailing fin tips pointed. Small caudal keel present. HABITAT: Circumglobal in tropical to temperate seas. In western Atlantic from MA to FL, Gulf of Mexico, Bermuda, Bahamas, and Caribbean Sea to Uruguay. Coastal and offshore; at surface or bottom. BIOLOGY: Voracious predators with an indiscriminate and varied diet. Prey include invertebrates, fishes, sharks, marine reptiles, birds, and mammals. Litters with 10–82 young. IUCN: Near Threatened.

Daggernose Shark - *Isogomphodon oxyrhynchus* (Müller & Henle, 1839)

FEATURES: Yellowish gray dorsally. White below. Snout very long, flat, and pointed. Eyes very small. Pectoral fins large, paddle-like. Second dorsal fin about half as tall as first. HABITAT: Occur along the South American coast from eastern Venezuela to northern Brazil. Over soft bottoms in turbid waters. In estuaries, around mangroves, along beaches from shore to about 130 ft. BIOLOGY: Feed on small schooling fishes. IUCN: Critically Endangered.

Lemon Shark - *Negaprion brevirostris* (Poey, 1868)

FEATURES: Pale yellow brown. Snout short. Body large and stocky. Dorsal fins almost equal in size. Pectoral fins broad and short. Caudal fin low. HABITAT: NJ to FL, Gulf of Mexico, Bermuda, Bahamas, and Caribbean Sea to southern Brazil. Also in eastern Pacific. Around coral, mangroves, docks, bays, or river mouths. Inshore and coastal to about 300 ft. BIOLOGY: May occasionally enter river mouths in search of prey. Occur singly or in groups of about 20. They are most active at dusk and dawn. IUCN: Near Threatened.

Smalltail Shark
to 4.4 ft.

Night Shark
to 9 ft.

Tiger Shark
to 18 ft.

Daggernose Shark
to 11 ft.

Lemon Shark
to 11 ft.

Carcharhinidae - Requiem Sharks, *cont.*

Blue Shark - *Prionace glauca* (Linnaeus, 1758)

FEATURES: Shades of blue dorsally. White ventrally. Snout long and pointed. Body slender. Pectoral fins very long and slender. First dorsal fin well behind pectoral fins. Caudal fin long and tall. HABITAT: Circumglobal in all tropical to temperate seas. In western Atlantic from Newfoundland to FL, Gulf of Mexico, Bermuda, Bahamas, and Caribbean Sea to Argentina. Oceanic and pelagic. BIOLOGY: Inquisitive and brazen. They are most active in early evening and night. Migrate long distances following prey, sometimes crossing the Atlantic. IUCN: Near Threatened.

Brazilian Sharpnose Shark - *Rhizoprionodon lalandii* (Valenciennes, 1839)

FEATURES: Grayish brown dorsally. White below. Spots on sides absent. Pectoral fins dark with posterior pale margins. Caudal fin with black margin. Labial furrows well developed. Pectoral fins short, broad. First dorsal-fin origin over or just behind posterior pectoral fin inner margin. Second dorsal-fin origin over or slightly before anal-fin rear insertion. HABITAT: Eastern Honduras to southern Brazil. Over sandy and muddy bottoms in shallow coastal waters from near shore to about 230 ft.

Caribbean Sharpnose Shark - *Rhizoprionodon porosus* (Poey, 1861)

FEATURES: Brown to grayish brown dorsally. White below. Spots on sides absent. Pectoral-fin margins white; caudal-fin margin blackish. Labial furrows well developed. First dorsal-fin origin over posterior pectoral-fin inner margin. Second dorsal-fin origin over or slightly behind anal-fin midpoint. HABITAT: Bahamas and Caribbean Sea to Brazil. Occur in estuaries, river mouths, and along the coast from shoreline to about 1,600 ft.

Atlantic Sharpnose Shark - *Rhizoprionodon terraenovae* (Richardson, 1837)

FEATURES: Gray or brownish gray dorsally. White ventrally. Dorsal fins with dusky tips. Adults with scattered white spots dorsally and white pectoral-fin margins. Labial furrows well developed. First dorsal-fin origin over posterior pectoral-fin inner margin. Second dorsal-fin origin over or just behind anal-fin midpoint. HABITAT: New Brunswick to FL, and Gulf of Mexico to Honduras. Common in bays, estuaries, sounds, river mouths, and surf zone of sandy beaches. Typically from intertidal zone to about 35 ft.

Sphyrnidae - Hammerhead Sharks

Scalloped Hammerhead - *Sphyrna lewini* (Griffith & Smith, 1834)

FEATURES: Brownish gray or gray dorsally. White ventrally. Head hammer-shaped in dorsal view; anterior edge curved, with three notches near middle. Fins usually tipped dusky or black. HABITAT: Circumglobal in warm seas over continental shelves. In western Atlantic from NJ to FL, Gulf of Mexico, Bahamas, and Caribbean Sea to Brazil. BIOLOGY: Occur singly, in pairs, or in huge congregations. Migratory. Prey on fishes, crustaceans, and squids. IUCN: Critically Endangered.

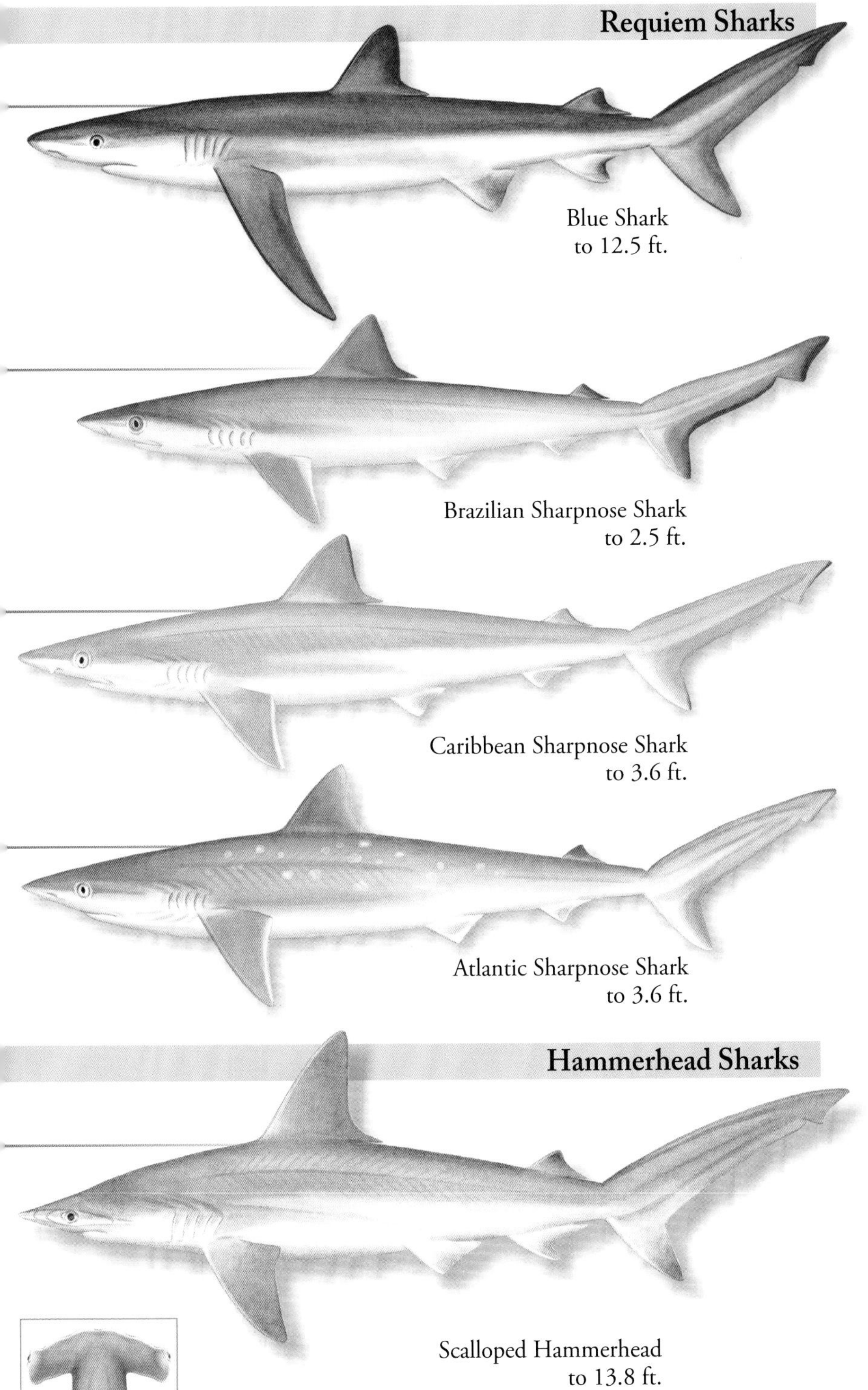

Blue Shark
to 12.5 ft.

Brazilian Sharpnose Shark
to 2.5 ft.

Caribbean Sharpnose Shark
to 3.6 ft.

Atlantic Sharpnose Shark
to 3.6 ft.

Hammerhead Sharks

Scalloped Hammerhead
to 13.8 ft.

Scoophead - *Sphyrna media* Springer, 1940

FEATURES: Grayish brown dorsally. Pale below. Fins unmarked. Head mallet-shaped in dorsal view, its anterior margin broadly curved with very shallow indentations and a weak depression in front of nostrils. First dorsal fin high with concave rear margin. Pectoral fins short and broad. Anal-fin base long. HABITAT: Panama to Brazil. Occur on or near bottom over mud, sand, gravel, seagrasses, and around mangroves. Found from shoreline to over continental shelves. Also in eastern Pacific.

Great Hammerhead - *Sphyrna mokarran* (Rüppell, 1837)

FEATURES: Gray or gray brown dorsally. Pale ventrally. Body, fins unmarked. Head T-shaped in dorsal view; large specimens with a notch at center. Dorsal fin very tall and sickle-shaped. All fins pointed, with concave rear margins. HABITAT: Circumglobal in warm seas. In western Atlantic from NC to FL, Gulf of Mexico, Bahamas, and Caribbean Sea to Brazil. Occur from near shore to over continental shelves. BIOLOGY: Powerful, migratory, and nomadic. Feed on variety of fishes, squids, and crustaceans. Litters contain 13–42 young. Common target or bycatch of fisheries. IUCN: Endangered.

Bonnethead - *Sphyrna tiburo* (Linnaeus, 1758)

FEATURES: Gray or grayish brown dorsally; may have small dark spots. Pale to white ventrally. Head spade-shaped in dorsal view. Second dorsal and anal fins similar in size. HABITAT: RI to FL, Gulf of Mexico, Bahamas, Cuba, and southern Caribbean Sea. Also eastern Pacific. Occur in warm, shallow, coastal waters over sand and mud, in river mouths, and around reefs. BIOLOGY: Social and migratory. Prey primarily on crabs and shrimps. Common bycatch of shrimp fisheries. Popular in public aquariums.

Smalleye Hammerhead - *Sphyrna tudes* (Valenciennes, 1822)

FEATURES: Gray brown to golden above. Pale below. Fins unmarked. Head mallet-shaped in dorsal view, anterior margin with distinct indentation at midline and shallow depressions in front of nostrils. First dorsal fin very tall and erect, originates just behind rear pectoral-fin insertion. Anal fin comparatively long-based. HABITAT: Eastern Colombia to Uruguay. Occur near bottom and in water column over soft bottoms from estuaries to about 130 ft. IUCN: Vulnerable.

Smooth Hammerhead - *Sphyrna zygaena* (Linnaeus, 1758)

FEATURES: Gray or brownish gray dorsally. Pale or white ventrally. Fins unmarked or tipped dusky or black. Head hammer-shaped in dorsal view; anterior margin curved with no notch at center. HABITAT: Circumglobal in tropical to temperate seas. In western Atlantic from Nova Scotia to FL, and Antilles to Argentina. Occur close inshore to offshore. Near surface in deep water. BIOLOGY: Strong and seasonally migratory. Young form very large schools. Prey include fishes, other sharks, rays, crustaceans, and squids. IUCN: Vulnerable.

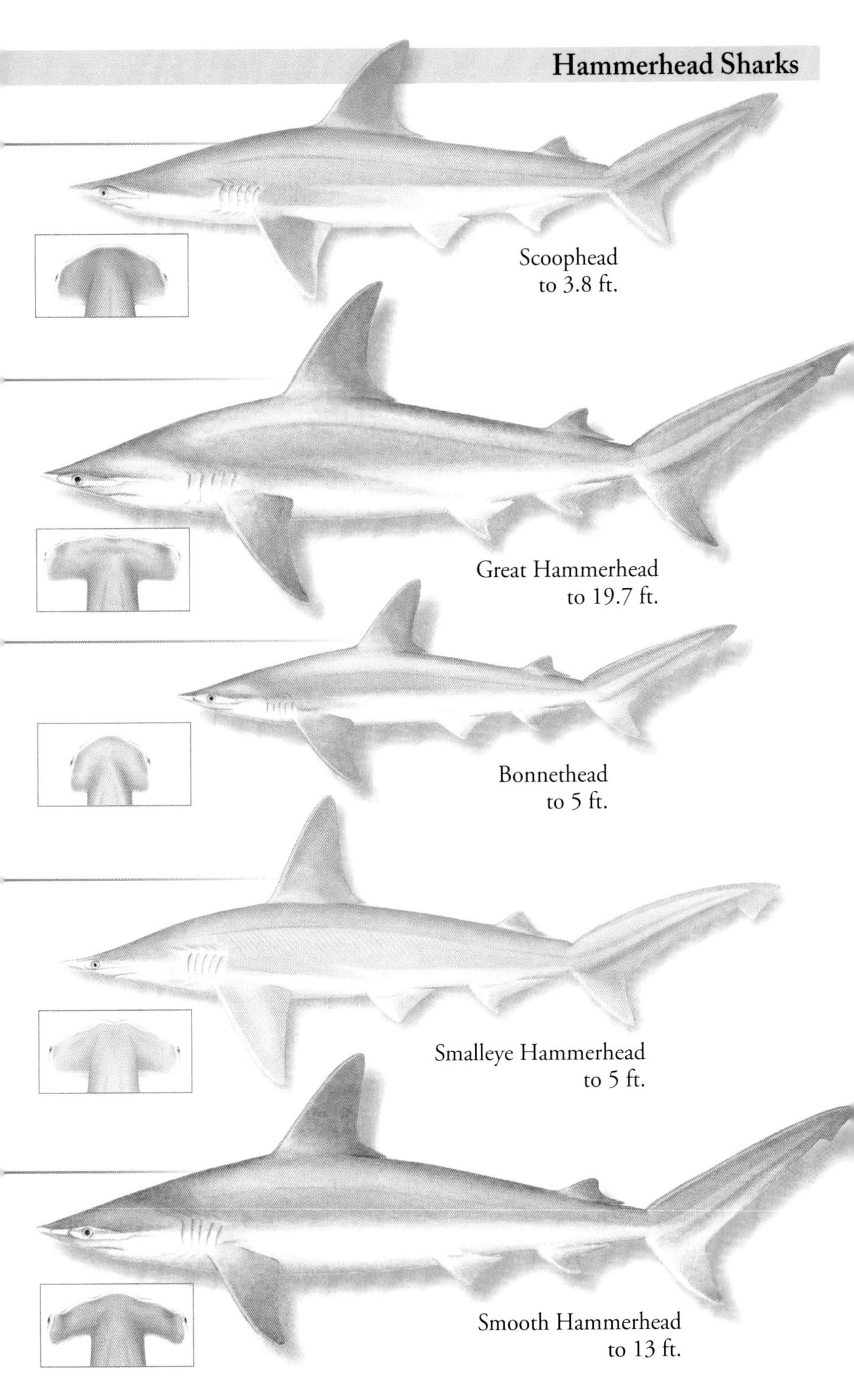

Scoophead
to 3.8 ft.

Great Hammerhead
to 19.7 ft.

Bonnethead
to 5 ft.

Smalleye Hammerhead
to 5 ft.

Smooth Hammerhead
to 13 ft.

Hexanchidae - Cow Sharks

Sharpnose Sevengill Shark - *Heptranchias perlo* (Bonnaterre, 1788)

FEATURES: Brownish gray dorsally, pale below. Tips of dorsal and caudal fins black in young, fading with age. Head comparatively narrow and pointed. Seven long gill slits. Single dorsal fin. HABITAT: Circumglobal in deep tropical to temperate seas. In western Atlantic from NC to northern Gulf of Mexico and Cuba; eastern Caribbean Sea and Venezuela to Argentina. Demersal over outer continental shelves and upper slopes. BIOLOGY: Prey on bony fishes, crustaceans, and squids. Litters contain 9–20 young. IUCN: Near Threatened.

Bluntnose Sixgill Shark - *Hexanchus griseus* (Bonnaterre, 1788)

FEATURES: Brownish to grayish dorsally, paler below. Body usually unmarked, but may have dark spots on sides. Lateral line usually pale. Head comparatively broad, snout blunt. Six long gill slits. Single dorsal fin. HABITAT: Scattered circumglobally. In western Atlantic from NC to Gulf of Mexico, and eastern Caribbean Sea to Argentina. Occur from surface to about 8,200 ft. Usually between 1,600 and 3,600 ft. BIOLOGY: Prey on squids and fishes. IUCN: Near Threatened.

Bigeye Sixgill Shark - *Hexanchus vitulus* Springer & Waller, 1969

FEATURES: Shades of gray dorsally and on sides. Pale ventrally. Rear fin margins usually white, sometimes dark. Snout bluntly pointed. Eyes comparatively large. Six long gill slits. Single dorsal fin. HABITAT: Circumglobal in tropical to temperate seas. In western Atlantic from FL to Gulf of Mexico, Bahamas, and Caribbean Sea to Argentina. Usually near bottom on soft continental shelves and upper slopes from about 50 to 2,000 ft. Also in the water column.

Squalidae - Dogfish Sharks

Spiny Dogfish - *Squalus acanthias* Linnaeus, 1758

FEATURES: Gray or bluish gray dorsally. Pale ventrally. Rows of white spots usually on upper sides. Pectoral fins with pale margins. Snout pointed. Dorsal fins with single, round spines. First dorsal-fin origin behind pectoral fins. Anal fin absent. Caudal peduncle with a low keel. HABITAT: Greenland to Argentina. Also eastern Atlantic and Indo-Pacific. Found from intertidal zone to continental slopes. BIOLOGY: Slow-growing and long-lived. Seasonally migratory or residential; schooling. May form large feeding packs. IUCN: Vulnerable.

Cuban Dogfish - *Squalus cubensis* Howell Rivero, 1936

FEATURES: Gray dorsally. Pale to white ventrally. Dorsal and caudal fins with dark inner margins, distinct white distal margins. Snout blunt. Dorsal fins with single, strong spine. First dorsal-fin origin over posterior pectoral-fin inner margin. Anal fin absent. Caudal peduncle with an obscure keel. HABITAT: NC to FL; Gulf of Mexico to Argentina. Bottom-dwelling from about 195 to 1,200 ft. BIOLOGY: Form large, dense schools.

Cow Sharks

Sharpnose Sevengill Shark
to 4.5 ft.

Bluntnose Sixgill Shark
to 6.5 ft.

Bigeye Sixgill Shark
to 6 ft.

Dogfish Sharks

Spiny Dogfish
to 6.5 ft.

Cuban Dogfish
to 3.6 ft.

Squalidae - Dogfish Sharks, *cont.*

Shortspine Dogfish - *Squalus mitsukurii* Jordan & Snyder 1903

FEATURES: Shades of gray dorsally. Pale below. Pectoral-, anal-, and caudal-fin margins pale. Snout parabolic. Dorsal fins with a single, strong spine. First dorsal-fin origin over inner pectoral-fin margin. Anal fin absent. Caudal peduncle with a low keel. HABITAT: NC to Gulf of Mexico and Caribbean Sea to Brazil. Over continental and insular slopes from about 450 to 2,460 ft.

Centrophoridae - Gulper Sharks

Gulper Shark - *Centrophorus granulosus* (Bloch & Schneider, 1801)

FEATURES: Light brown dorsally. Paler ventrally. Eyes large. Gill slits relatively long. Dorsal fins with a single long and grooved spine. Pectoral fins have long posterior lobes. Skin moderately smooth. HABITAT: Circumglobal in tropical to temperate seas. In western Atlantic from FL, northern Gulf of Mexico, and Caribbean Sea to Brazil. On or near bottom to about 4,000 ft. BIOLOGY: Solitary. Prey on fishes, squids, and crustaceans. Litters are small with only one or two young.

Dalatiidae - Kitefin Sharks

Kitefin Shark - *Dalatias licha* (Bonnaterre, 1788)

FEATURES: Uniform dark brownish or blackish. Posterior margins of fins may be translucent. Snout short, blunt. Eyes and spiracles large. Gill slits small and low. Dorsal fins lack spines. HABITAT: Scattered circumglobally. Georges Bank to N Gulf of Mexico and Hispaniola. On or near bottom in deep water over continental shelves. BIOLOGY: Large, buoyant liver allows hovering in the water column. Litters with 10–16 young. IUCN: Near Threatened.

Cookiecutter Shark - *Isistius brasiliensis* (Quoy & Gaimard, 1824)

FEATURES: Shades of brown. Abdomen paler with luminescent organs. Broad, dark band behind head. Fins with dark inner margins, pale outer margins. Eyes very large. Mouth with sharp, triangular lower teeth. Dorsal fins low. Body cylindrical. HABITAT: Scattered circumglobally. In western Atlantic from Gulf of Mexico, Bahamas, and Caribbean Sea to Brazil. Oceanic. BIOLOGY: Migrate vertically at night. Take deep, round bites from large fishes, cetaceans, and squids.

Squatinidae - Angel Sharks

Atlantic Angel Shark - *Squatina dumeril* Lesueur, 1818

FEATURES: Gray to brownish dorsally; may have darker spots. White ventrally. Thorny denticles on snout, above eyes, and along midline. Pectoral fins broad, squared at tips, together with head form a rhomboid shape. Dorsal fins small, oval, and similar in shape. HABITAT: MA to FL, and Gulf of Mexico to southern Caribbean Sea. Bottom-dwelling, close inshore to about 4,000 ft.

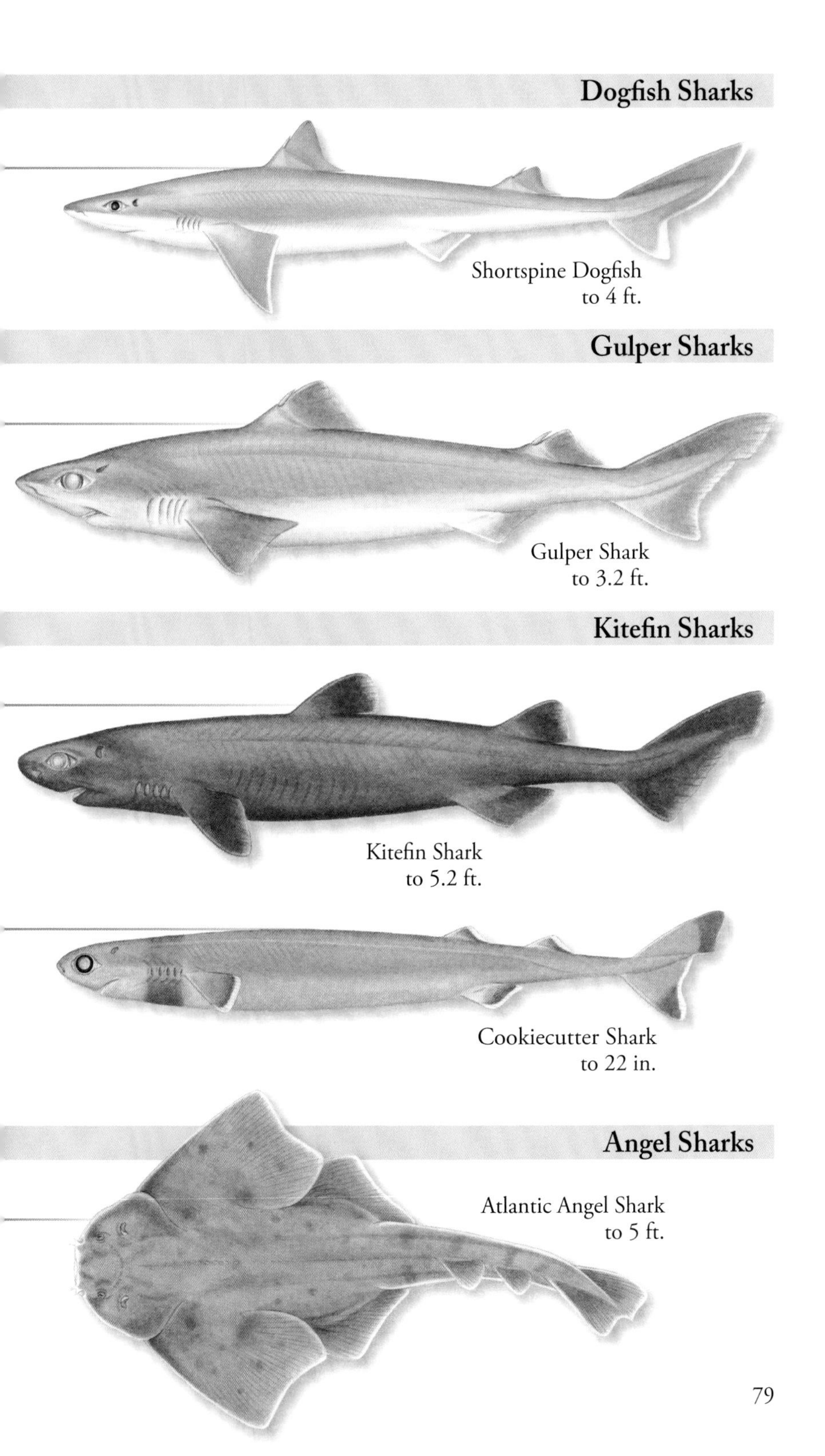
Dogfish Sharks
Shortspine Dogfish
to 4 ft.
Gulper Sharks
Gulper Shark
to 3.2 ft.
Kitefin Sharks
Kitefin Shark
to 5.2 ft.
Cookiecutter Shark
to 22 in.
Angel Sharks
Atlantic Angel Shark
to 5 ft.

Squatinidae - Angel Sharks, *cont.*

Disparate Angel Shark - *Squatina heteroptera* Castro-Aguirre, Espinosa Pérez & Huidobro Campos, 2007

FEATURES: Shades of brown dorsally with small, scattered pale spots and a large black blotch near anterior pectoral-fin margin. Tail with several dark bars. Pectoral fins broad, squared at tips. Dorsal fins small, well separated, with second taller and less erect than first. Both with a base of about 50% of height. HABITAT: Western Gulf of Mexico from S Texas to off Yucatán Peninsula. Demersal over soft bottoms from about 330 to 540 ft. NOTE: May be a synonym of *Squatina dumeril.*

Mexican Angel Shark - *Squatina mexicana* Castro-Aguirre, Espinosa Pérez & Huidobro Campos, 2007

FEATURES: Shades of brown to gray dorsally with small, scattered dark spots and a large black blotch near anterior pectoral-fin margin. Pectoral fins broad, squared at tips. Dorsal fins small, closely set, and similar in size and shape. Both with a base of about 57% of height. HABITAT: In Gulf of Mexico from W FL to off Yucatán Peninsula. Demersal over soft bottoms from about 230 to 590 ft. NOTE: May be a synonym of *Squatina dumeril.*

Torpedinidae - Torpedo Electric Rays

Florida Torpedo - *Torpedo andersoni* Bullis, 1962

FEATURES: Pale tan with numerous irregular, reddish brown to greenish spots and blotches dorsally. Pale cream ventrally. Head, body, and fins form a round disk. Two large, kidney-shaped electric organs on sides of head. First dorsal fin larger than second. Caudal fin triangular. HABITAT: FL, Bahamas, and Caribbean Sea. Demersal over reefs, rocks, sand, and gravel from about 36 to 750 ft.

Atlantic Torpedo - *Tetronarce nobiliana* (Bonaparte, 1835)

FEATURES: Uniform shades of pale to dark gray or brown dorsally; with or without darker spots. Ventral area pale with dark margins. Head, body, and fins form round disk. Two large, kidney-shaped electric organs on sides of head. First dorsal fin larger than second. Pectoral fins rounded. Caudal fin triangular. HABITAT: Nova Scotia to FL and Gulf of Mexico to southern Caribbean Sea. Also eastern Atlantic. Bottom-dwelling from near shore to about 2,000 ft.

Narcinidae - Numbfishes

Brownband Numbfish - *Diplobatis guamachensis* Martín Salazar, 1957

FEATURES: Tan to golden tan dorsally with numerous irregular, dark brown, wavy lines and cross bands. White ventrally. Head, body, and fins form oval shape. Dorsal fins similarly shaped. Caudal fin triangular. HABITAT: Southern Caribbean Sea. Most common in Gulf of Venezuela region. Demersal over muddy and sandy bottoms from about 100 to 600 ft. IUCN: Vulnerable.

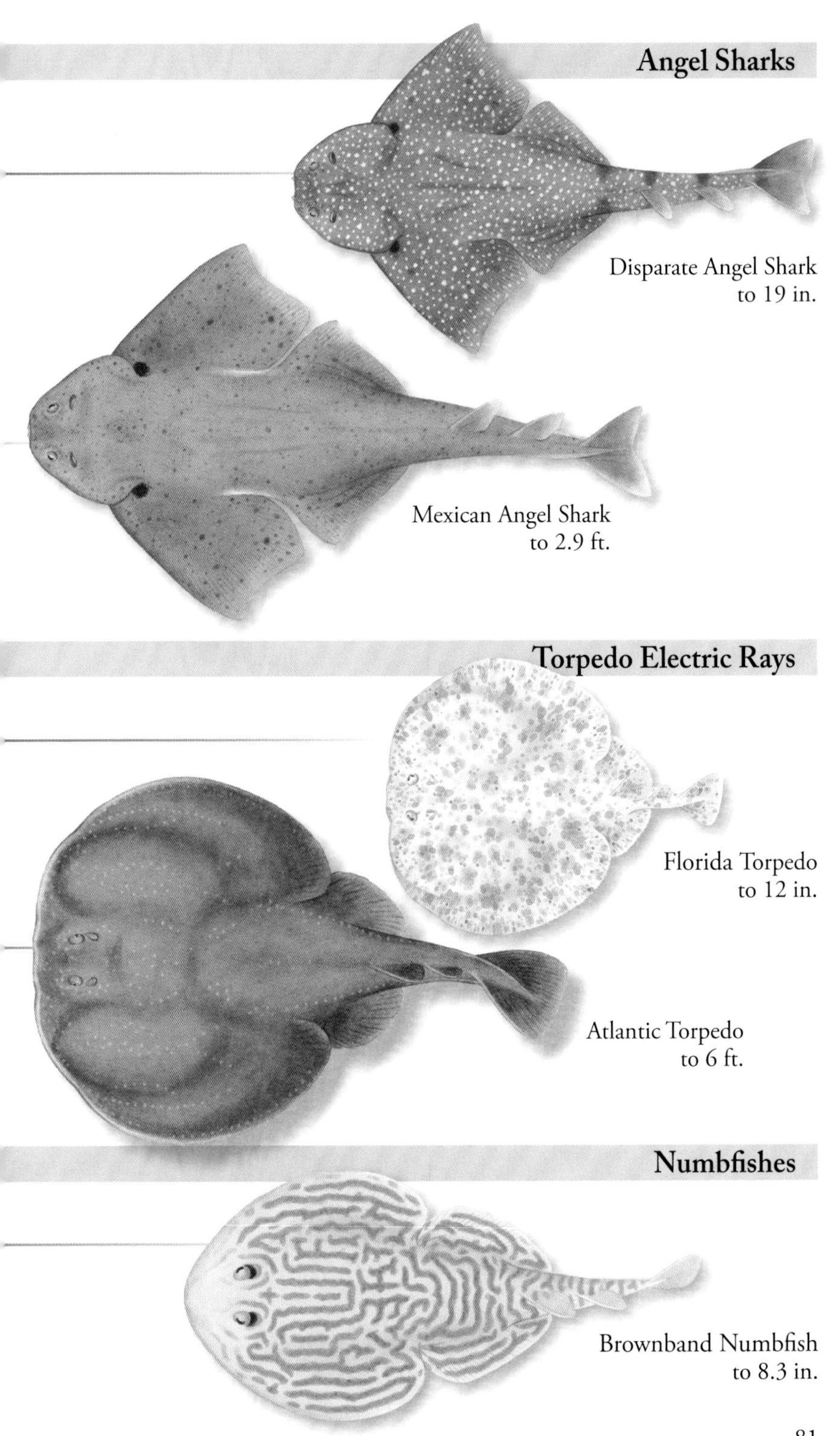
Angel Sharks
Disparate Angel Shark
to 19 in.
Mexican Angel Shark
to 2.9 ft.
Torpedo Electric Rays
Florida Torpedo
to 12 in.
Atlantic Torpedo
to 6 ft.
Numbfishes
Brownband Numbfish
to 8.3 in.

Narcinidae - Numbfishes, *cont.*

Variegated Electric Ray - *Diplobatis picta* Palmer, 1950

FEATURES: Shades of tan dorsally with numerous and highly variable spots, mottling, and ocelli that may be larger at disk edges. Ventral surface white. Dorsal fins similarly sized. Caudal fin triangular to rounded. HABITAT: Southern Caribbean Sea to northern Brazil. Demersal over muddy and sandy bottoms from near shore to about 426 ft. IUCN: Vulnerable.

Lesser Electric Ray - *Narcine bancroftii* (Griffith & Smith, 1834)

FEATURES: Color and pattern vary. Shades of brown dorsally with darker, irregular ocellated blotches. Whitish ventrally. Snout always with dark blotches. Head, body, and fins form oval shape. Pelvic-fin margins almost straight. Caudal fin triangular. HABITAT: NC to FL, Gulf of Mexico, and Caribbean Sea to northern Brazil. Demersal between surf and about 120 ft. BIOLOGY: Prey on invertebrates and fishes. Females carry up to 18 young of various developmental stages. Seasonally migratory. IUCN: Critically Endangered.

ALSO IN THE AREA: *Diplobatis colombiensis*, see p. 547.

Pristidae - Sawfishes

Smalltooth Sawfish - *Pristis pectinata* Latham, 1794

FEATURES: Dark brownish gray to blackish dorsally. Pale to white ventrally. Rostrum elongate with 24–32 pairs of teeth. Body and fins flattened. First dorsal fin almost directly above pelvic fins. HABITAT: NY to FL, Gulf of Mexico, Bermuda, Bahamas, and Caribbean Sea to Argentina. Near shore and in bays, estuaries, and rivers. BIOLOGY: Saw used to dislodge, injure prey. IUCN: Critically Endangered.

Largetooth Sawfish - *Pristis pristis* (Linnaeus, 1758)

FEATURES: Dark gray to yellowish brown dorsally. Grayish ventrally. Rostrum elongate with 16–20 pairs of teeth. Body and fins flattened. First dorsal-fin origin anterior to pelvic fins. HABITAT: Circumglobal in tropical to warm temperate seas. In western Atlantic from S FL and Gulf of Mexico to Brazil. Inshore. Occur in bays, estuaries, and rivers. BIOLOGY: Use saw to dislodge and injure prey. IUCN: Critically Endangered.

Rhinobatidae - Guitarfishes

Brazilian Guitarfish - *Pseudobatos horkelii* (Müller & Henle, 1841)

FEATURES: Uniform olive gray to chocolate brown dorsally. Dark blotch on snout tip. Pale ventrally. Rostral cartilage not expanded or exposed at tip and/or without rounded or flat tubercles. Nostrils large; lengths of openings nearly 1.5 times the distance between them. HABITAT: Scattered records from Lesser Antilles to Brazil. Demersal over sandy and soft bottoms. Occur from estuaries to about 100 ft. IUCN: Critically Endangered.

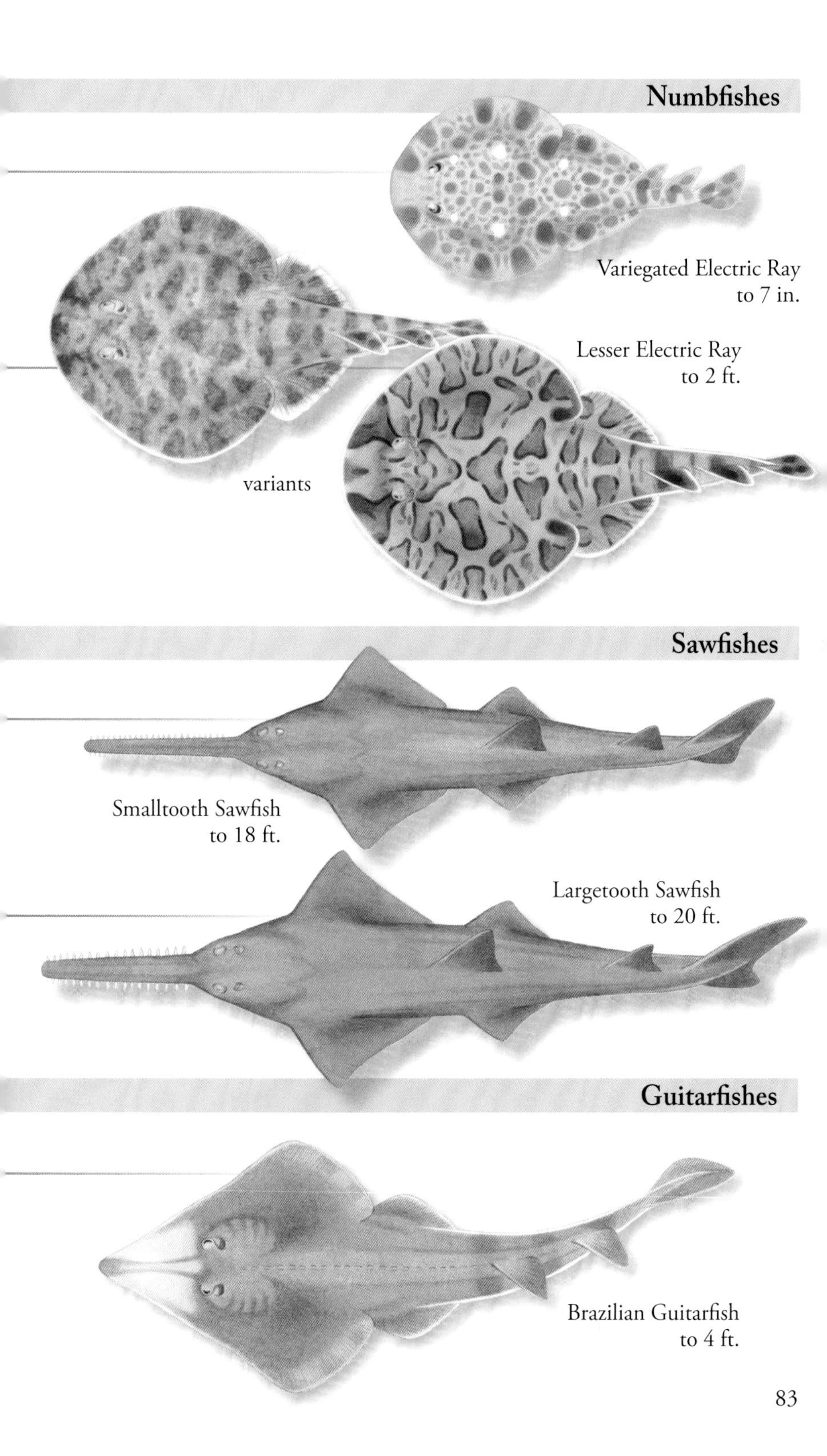
Numbfishes
Variegated Electric Ray
to 7 in.
Lesser Electric Ray
to 2 ft.
variants
Sawfishes
Smalltooth Sawfish
to 18 ft.
Largetooth Sawfish
to 20 ft.
Guitarfishes
Brazilian Guitarfish
to 4 ft.

Rhinobatidae - Guitarfishes, *cont.*

Atlantic Guitarfish - *Pseudobatos lentiginosus* (Garman, 1880)

FEATURES: Color and pattern highly variable. Plain grayish or brownish dorsally; generally with many small, pale spots. Ventral area pale. Rostral cartilage expanded and flattened at tip with a few large, conical tubercles (absent in juveniles). HABITAT: NC to Yucatán. Coastal and bottom-dwelling over sandy, weedy, or muddy bottoms in shallow waters. BIOLOGY: Bury in bottom sediment. IUCN: Near Threatened.

Chola Guitarfish - *Pseudobatos percellens* (Walbaum, 1792)

FEATURES: Olive gray to brown or reddish dorsally, occasionally with darker brown blotches and pale spots about equal to eye diameter. Pale yellowish below. Rostral cartilage not expanded at tip, generally without tubercles. Lengths of nostril openings about equal to or slightly greater than distance between them. HABITAT: Caribbean Sea to Brazil. Demersal over soft bottoms from shore to about 360 ft. IUCN: Near Threatened.

Rajidae - Skates

Clark's Fingerskate - *Dactylobatus clarkii* (Bigelow & Schroeder, 1958)

FEATURES: Pale brown dorsally with dark markings and symmetrically arranged, white ocellated spots. Ventral area white with irregular grayish band along posterior margins. Blunt snout set between concave anterior margins. Disk heart-shaped. Band of thornlets from snout to pectoral tips. Thorns on snout, around eyes, at shoulder, and down tail. HABITAT: E FL; scattered in Gulf of Mexico and Caribbean Sea to Brazil. Demersal on continental slopes to 3,000 ft. BIOLOGY: Lay eggs on or near bottom.

Spreadfin Skate - *Dipturus olseni* (Bigelow & Schroeder, 1951)

FEATURES: Dark brown to olive brown dorsally with many small, dark, obscure spots. Ventral area gray to black. Thorns around eyes and along tail midline. Dorsal fins distinctly separate. Caudal fin bilobed. HABITAT: Gulf of Mexico from FL to Yucatán Peninsula. Demersal in waters from about 180 to 1,250 ft. BIOLOGY: Lay their eggs on or near bottom. Egg cases have horn-like projections.

Prickly Brown Ray - *Dipturus teevani* (Bigelow & Schroeder, 1951)

FEATURES: Pale brown dorsally. Creamy to dusky ventrally. Dorsal and caudal fins black. Snout very long, pointed. Middle anterior margins of pectoral fins concave. Trailing tips of pectoral fins slightly concave. Dorsal fins connected to one another at base. Few thorns around eyes. Single row of thorns on tail. HABITAT: NC to FL Keys, N Gulf of Mexico, Bahamas, and Caribbean Sea. Demersal along continental slope to about 2,400 ft. BIOLOGY: Egg cases have horn-like projections.

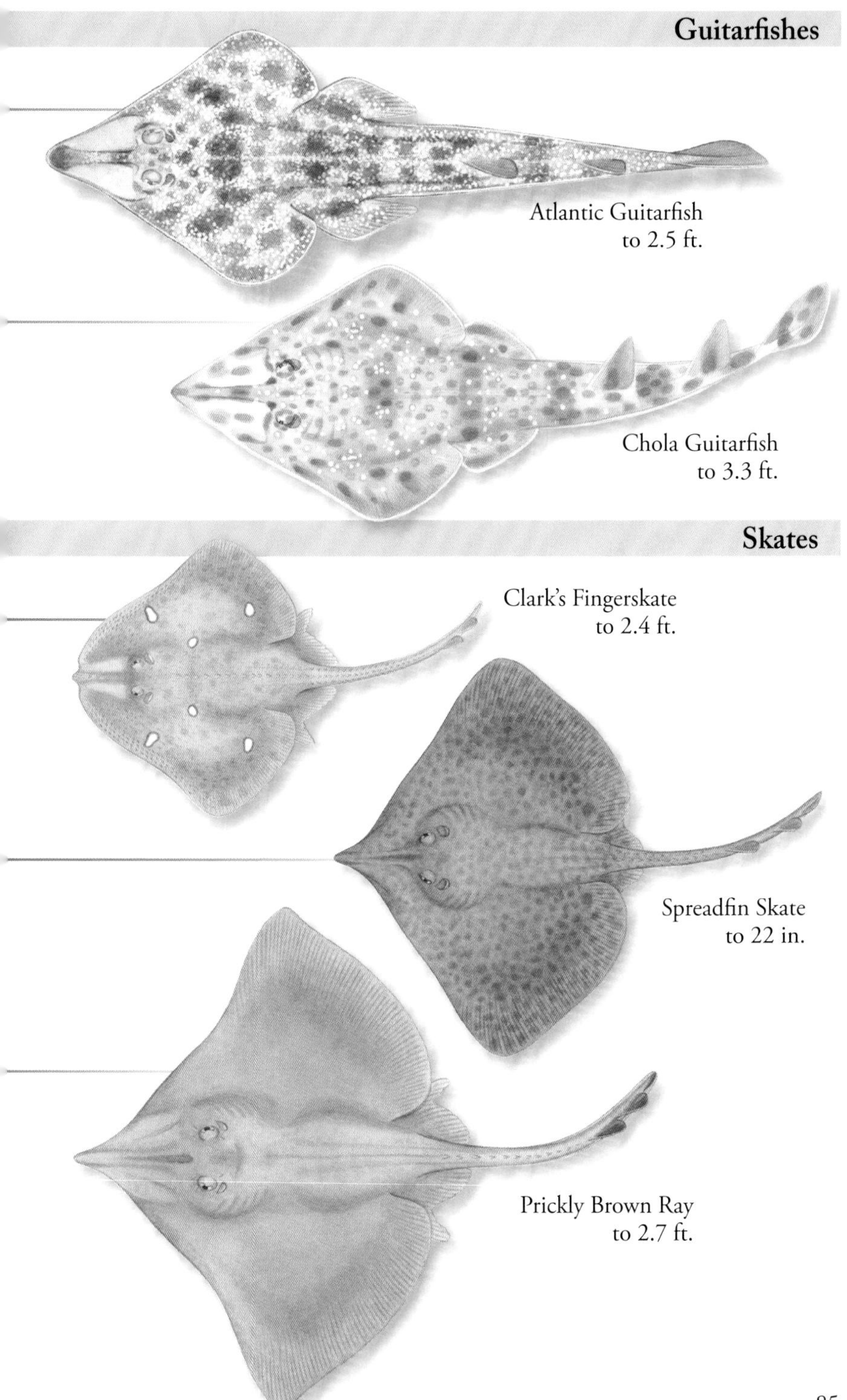

Guitarfishes
Atlantic Guitarfish
to 2.5 ft.
Chola Guitarfish
to 3.3 ft.
Skates
Clark's Fingerskate
to 2.4 ft.
Spreadfin Skate
to 22 in.
Prickly Brown Ray
to 2.7 ft.

Underworld Windowskate - *Fenestraja plutonia* (Garman, 1881)

FEATURES: Yellowish brown, grayish brown, or purplish brown dorsally with dark spots and blotches. Ventral area yellowish white. Long tail has dark bands. Snout short, blunt. Thorns around eyes, between spiracles, at shoulder, and along midline. Tail covered in rows of thorns and denticles. Disk heart-shaped. Tail long. HABITAT: NC to FL Keys, northern Gulf of Mexico. Also off Cuba, Costa Rica, and northern South America. Demersal along continental slopes to about 3,000 ft. BIOLOGY: Egg cases have horn-like projections.

Gulf Windowskate - *Fenestraja sinusmexicanus* (Bigelow & Schroeder, 1950)

FEATURES: Reddish brown to purplish brown dorsally; may have irregular dark blotches. Yellowish white below. Thorns around eyes, along anterior pectoral margins, and on back. Three rows of thorns from back and along tail. Dorsal surface covered with small denticles. Disk heart-shaped. Tail long. HABITAT: Southern FL and Gulf of Mexico to southern Caribbean Sea. Also off Bahamas. Demersal over continental shelves and slopes from about 173 to 3,595 ft.

Rosette Skate - *Leucoraja garmani* (Whitley, 1939)

FEATURES: Tan to brown dorsally with small, dark, and pale spots that form rosettes around a dark center. Spots form bands on tail. Ventral area whitish. Patch of thorns on shoulder. Thorns and denticles along tail. Disk heart-shaped. Tail long. HABITAT: MA to Venezuela, including Gulf of Mexico. Demersal on soft bottoms of outer continental shelves and upper slopes. BIOLOGY: Have two similar-looking subspecies: *Leucoraja garmani virginica*, which occurs from MA to NC, and *Leucoraja garmani garmani*, from NC to FL. Classification subject to change pending ongoing studies.

Freckled Skate - *Leucoraja lentiginosa* (Bigelow & Schroeder, 1951)

FEATURES: Shades of tan to brown dorsally with scattered, dark freckles and indistinct, pale spots. Ventral area whitish. Patch of thorns on shoulder. Thorns and denticles along tail. Disk is heart-shaped. Tail is long. HABITAT: NW FL to Yucatán. Demersal on soft bottoms along outer continental shelves and upper slopes. BIOLOGY: Feed on bottom-dwelling invertebrates and fishes. Formerly considered a subspecies of *Leucoraja garmani.*

Finspot Skate - *Rostroraja cervigoni* (Bigelow & Schroeder, 1964)

FEATURES: Shades of uniform brown dorsally with a pair of ocelli that have either two central spots or a central ring. May also have numerous small, pale blotches. Ventral surface whitish to grayish, may be darker at margins. Thorns around eyes, on nuchal region, and along tail. Tail relatively short. HABITAT: Colombia to Suriname. Demersal over soft bottoms from about 120 to 570 ft. IUCN: Near Threatened.

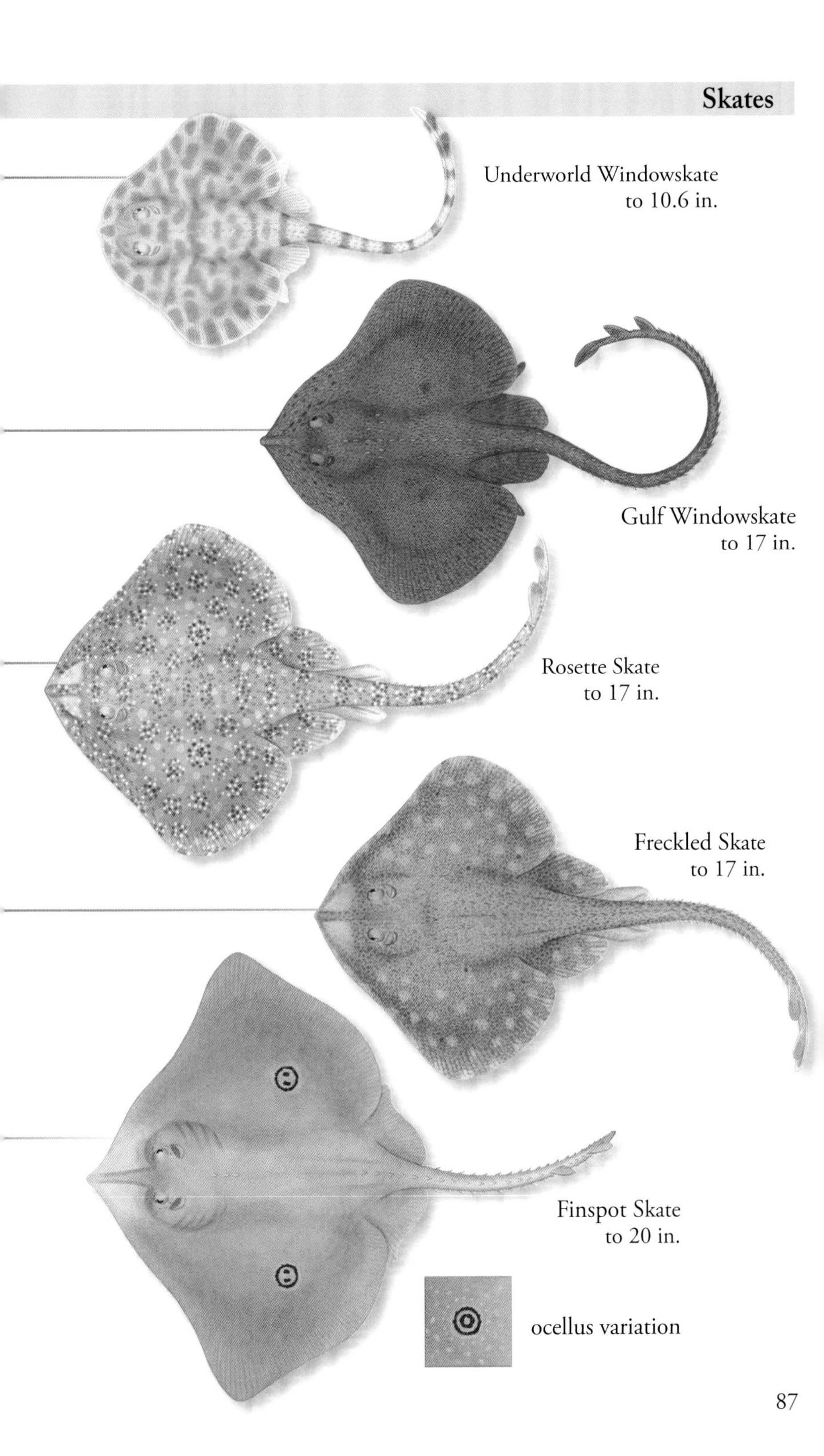
Underworld Windowskate
to 10.6 in.
Gulf Windowskate
to 17 in.
Rosette Skate
to 17 in.
Freckled Skate
to 17 in.
Finspot Skate
to 20 in.
ocellus variation

Rajidae - Skates, *cont.*

Clearnose Skate - *Rostroraja eglanteria* (Bosc, 1800)

FEATURES: Shades of brown to gray dorsally with numerous darker spots and lines that radiate from dorsal midline. May also have scattered pale spots and blotches. Whitish to yellowish ventrally. Small thorns around eyes and spiracles. Rows of thorns down back and tail. HABITAT: MA to FL and Gulf of Mexico to Yucatán. Rare in northern Gulf of Mexico. Occur on soft bottoms from shore to about 1,000 ft. BIOLOGY: Seasonally migratory.

Roundel Skate - *Rostroraja texana* (Chandler, 1921)

FEATURES: Brown dorsally with ocellated eye spots on pectoral fins. Each eyespot a brown to black circle with a bright yellow ring. Ventral area white. Denticles on snout tip, around eyes, down back and tail. HABITAT: SE FL to Yucatán. Demersal over soft bottoms of continental shelves from about 50 to 360 ft. BIOLOGY: Feed primarily on bottom-dwelling crustaceans. Caught in trawls and marketed.

ALSO IN THE AREA: *Rostroraja ackleyi*, see p. 547.

Urotrygonidae - American Round Stingrays

Yellow Stingray - *Urobatis jamaicensis* (Cuvier, 1816)

FEATURES: Color, pattern highly variable. Mottled tan to brown dorsally and covered in small, irregular gold to white or brownish spots. Spots cluster to form larger blotches. Body and fins form oval disk shape. Single spine on tail. Caudal fin rounded. HABITAT: NC to FL, Gulf of Mexico, Bahamas, and Caribbean Sea to northern South America. Demersal over nearshore sandy, muddy, and seagrass bottoms, around coral bottoms, and in bays and estuaries. BIOLOGY: Females gather on eel grass beds to give birth to live young. Young are born tail first, with barbs safely sheathed.

Smalleye Round Ray - *Urotrygon microphthalmum* Delsman, 1941

FEATURES: Shades of gray brown dorsally; white below. Caudal fin dark to blackish. Snout long, pointed, with margins on either side concave. Eyes comparatively small. Body and fins form a round disk. Thorns absent along dorsal midline. Large, venomous spine on tail. HABITAT: Occur in coastal waters from Venezuela to the mouth of the Amazon River. Demersal over sandy and muddy bottoms, in estuaries, and brackish water from about 32 to 177 ft.

Venezuelan Round Ray - *Urotrygon venezuelae* Schultz, 1949

FEATURES: Shades of gray brown dorsally; creamy to white below. Caudal fin dark to blackish. Snout with a small point, margin on either side slightly concave. Body and fins form nearly round disk. Row of about 45 thorns along dorsal midline and onto tail. Large, venomous spine on tail. HABITAT: Occur in coastal waters from Colombia to mouth of the Amazon River. Demersal over soft bottoms from shoreline to about 100 ft. IUCN: Near Threatened.

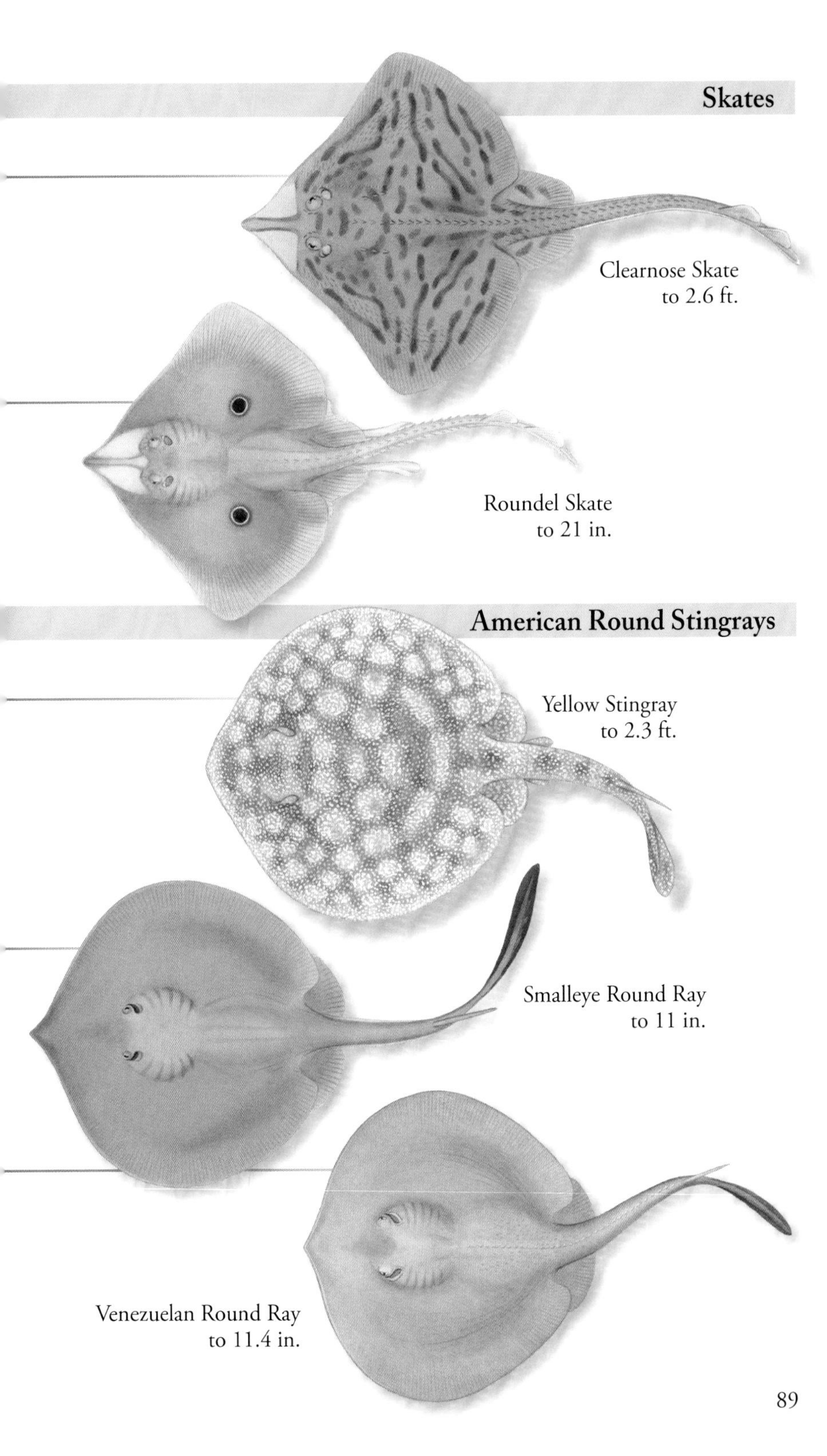
Skates
Clearnose Skate
to 2.6 ft.
Roundel Skate
to 21 in.
American Round Stingrays
Yellow Stingray
to 2.3 ft.
Smalleye Round Ray
to 11 in.
Venezuelan Round Ray
to 11.4 in.

Dasyatidae - Whiptail Stingrays

Roughtail Stingray - *Bathytoshia centroura* (Mitchill, 1815)

FEATURES: Dark brown to olive dorsally. Whitish ventrally. Snout projects slightly from barely convex fin margins. Denticles on snout, behind spiracles. Tubercles scattered on shoulder, at pectoral-fin base, along back, and down tail. Tail with a strong ventral fold. HABITAT: MA to FL, northern Gulf of Mexico, and Bahamas, and Brazil to Argentina. Demersal along continental shelves to about 300 ft. BIOLOGY: Largest stingray in the area. Prey on a variety of invertebrates and bony fishes. Litters contain two to six young. NOTE: Previously known as *Dasyatis centroura*.

Sharpsnout Stingray - *Fontitrygon geijskesi* (Boeseman, 1948)

FEATURES: Shades of brown dorsally. Ventral surface whitish; may have brownish margins. Snout very long, pointed, and angular. Eyes very small. Disk very thin, rhomboid in shape. Pelvic fins pointed and wing-like. Tail very long and whip-like with one or two spines near base. Single row of tubercles along back and onto tail. HABITAT: Occur off Belize and along the South American coast from Venezuela to Brazil. Demersal over sandy bottoms from shore to about 260 ft. NOTE: Previously known as *Dasyatis geijskesi*. IUCN: Near Threatened.

Southern Stingray - *Hypanus americanus* (Hildebrand & Schroeder, 1928)

FEATURES: Color varies with bottom substrate. May be gray, brown, or olive dorsally. Small, pale spot between eyes. Ventral area white with darker margins. Snout continuous with almost straight anterior fin margins. Disk angular. Series of denticles on shoulder. Row of denticles down back to base of tail. Tail with one or two spines and a ventral skin fold. HABITAT: NJ to FL, Gulf of Mexico, and Caribbean Sea to Brazil. Demersal and near shore in shallow water. BIOLOGY: Bury in sandy bottoms. Feed mainly on bivalves and worms. Migrate northward during summer. NOTE: Previously known as *Dasyatis americana*.

Longnose Stingray - *Hypanus guttata* (Bloch & Schneider, 1801)

FEATURES: Shades of gray, brown, or olive dorsally, sometimes with dark spots. White to yellowish below. Snout long, pointed, and angular. Anterior fin margins nearly straight. Disk angular. Row of tubercles along back to tail spine. Tail very long and whip-like with a ventral ridge. HABITAT: Southern Gulf of Mexico and Caribbean Sea to Brazil. Demersal and coastal over soft bottoms from shore to about 120 ft. NOTE: Previously known as *Dasyatis guttata*.

Atlantic Stingray - *Hypanus sabinus* (Lesueur, 1824)

FEATURES: Shades of brown dorsally; occasional dark stripe on midline. Ventral area whitish; may have darker margins. Snout projects from concave anterior fin margins. Pectoral fins rounded. Tubercles from midline to serrated barb on tail. Dorsal and ventral folds on tail. HABITAT: VA to FL; Gulf of Mexico to Yucatán. Demersal along the coastal and in estuaries. BIOLOGY: Feed on bottom-dwelling invertebrates and fishes. NOTE: Previously known as *Dasyatis sabina*.

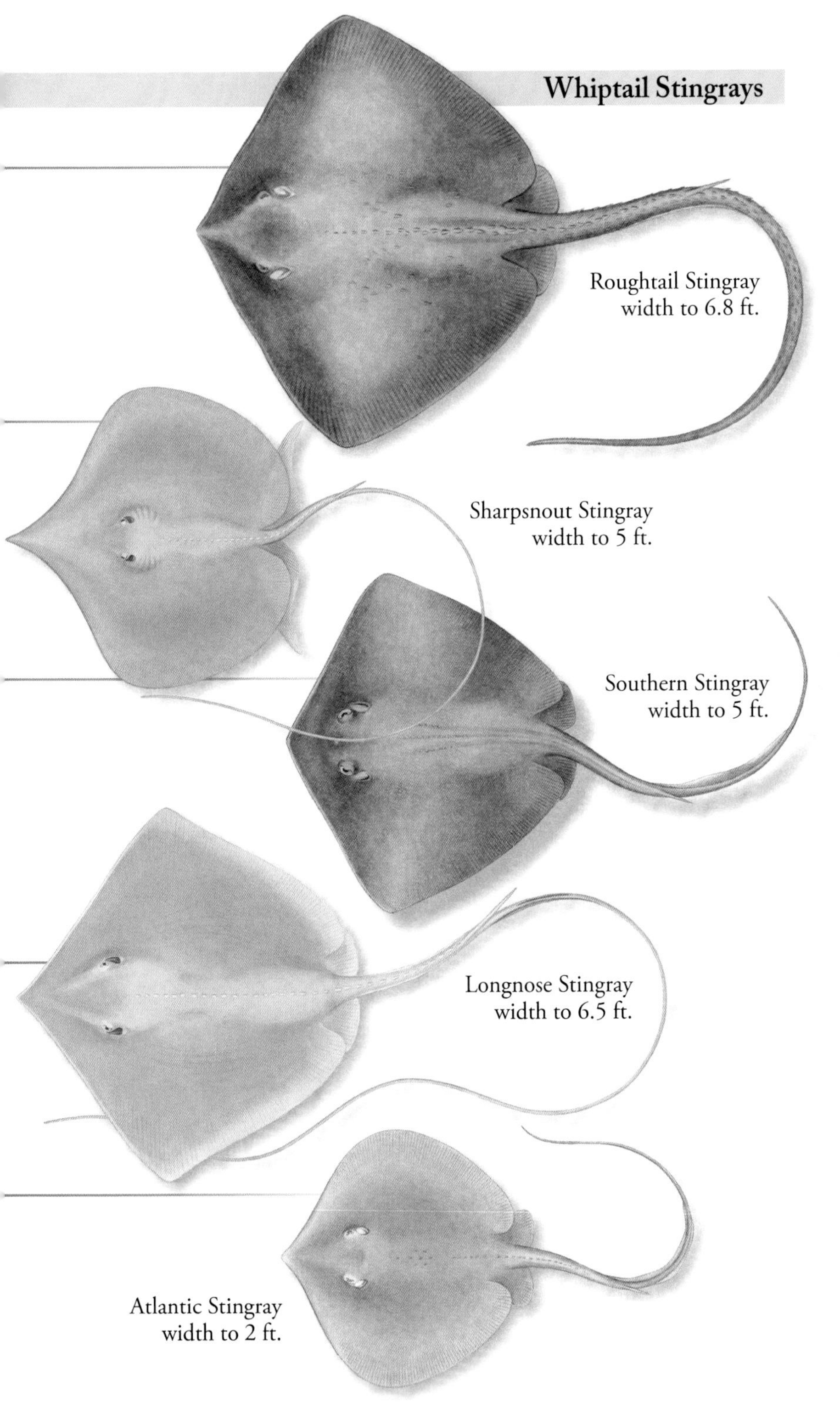
Roughtail Stingray
width to 6.8 ft.
Sharpsnout Stingray
width to 5 ft.
Southern Stingray
width to 5 ft.
Longnose Stingray
width to 6.5 ft.
Atlantic Stingray
width to 2 ft.

Dasyatidae - Whiptail Stingrays, *cont.*

Bluntnose Stingray - *Hypanus say* (Lesueur, 1817)

FEATURES: Grayish brown, olive brown, reddish brown to dusky green dorsally. Ventral area whitish; may have blotches or dark margin. Snout projects slightly from slightly convex anterior fin margins. Pectoral-fin tips slightly rounded. Rows of denticles or thorns at shoulder and down tail. Dorsal and ventral folds on tail. HABITAT: MA and NJ (rare), commonly VA to FL, Gulf of Mexico, and Caribbean Sea to Brazil. Occur from shore to about 65 ft. NOTE: Previously known as *Dasyatis say*.

Pelagic Stingray - *Pteroplatytrygon violacea* (Bonaparte, 1832)

FEATURES: Dark bluish gray to purplish black dorsally. Ventral area is a paler shade of dorsal color; may be mottled. Snout short and barely protrudes from rounded anterior fin margins. Tail thick at base, narrow beyond the long spine. Ventral fold present from base to spine. HABITAT: Probably circumglobal. In western Atlantic from Newfoundland to FL, Gulf of Mexico, and eastern Caribbean Sea to Brazil. Pelagic over continental shelves to oceanic waters.

Caribbean Whiptail Stingray - *Styracura schmardae* (Werner, 1904)

FEATURES: Shades of brown, olive, or sepia to black dorsally. Yellowish to whitish below, with or without spots. Tip of snout with a very small point. Anterior disk margins evenly rounded to nearly straight. Disk ovoid, widest anteriorly. Two pairs of tubercles on shoulder. Tail long, with a low ventral keel, and one or two spines located posteriorly. HABITAT: Southern Gulf of Mexico, Bahamas, and Caribbean Sea to Brazil. Demersal over soft bottoms from shore to about 80 ft. Common around mangrove lagoons.

Gymnuridae - Butterfly Rays

Lessa's Butterfly Ray - *Gymnura lessae* Yokita & de Carvalho, 2017

FEATURES: Shades of brown to gray with small, pale spots and ocelli on dorsal surface. Darker spots form irregular blotches. Ventral area creamy white. Pectoral-fin tips bluntly rounded, somewhat trailing. Tail with skin folds on dorsal and ventral surfaces. Barbs absent from tail. HABITAT: VA to FL, Gulf of Mexico, western Bahamas, and Jamaica. Demersal over sandy bottoms from near shore to about 180 ft. BIOLOGY: Feed on fishes, crustaceans, and mollusks. IUCN: Vulnerable.

Brazilian Butterfly Ray - *Gymnura micrura* (Bloch & Schneider, 1801)

FEATURES: Brown, gray, olive, or purplish dorsally with darker and lighter irregular spots and blotches. Ventral area whitish with darker margins. Tail banded. Middle anterior disk margins weakly concave. Pectoral-fin tips blunt. Tail barb absent. HABITAT: Venezuela to southern Brazil. Demersal over sandy and muddy bottoms, close to shore, and in estuaries. BIOLOGY: Feed on bivalves, crustaceans, and other fishes. IUCN: Near Threatened.

ALSO IN THE AREA: *Gymnura altavela*, see p. 547.

Whiptail Stingrays

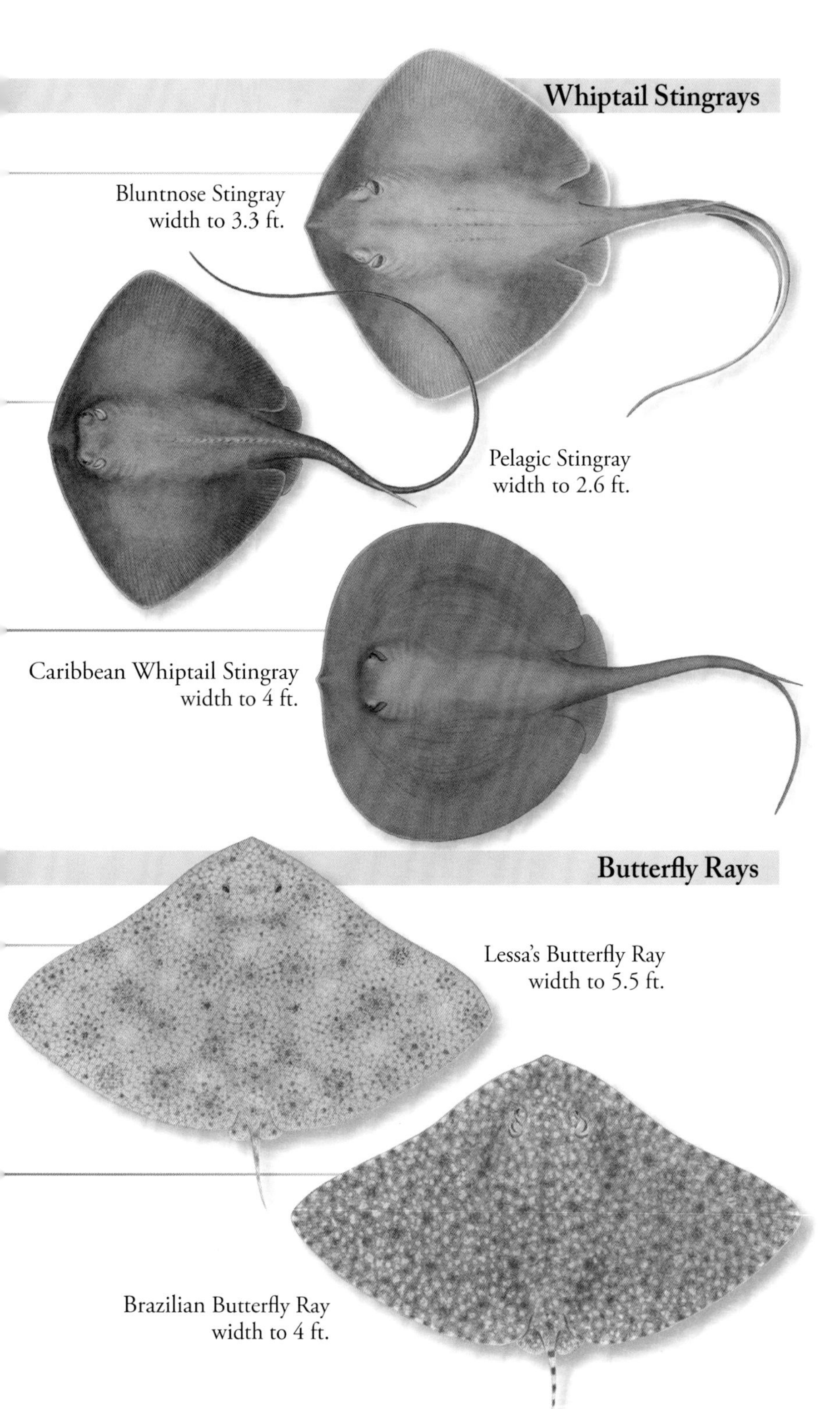

Myliobatidae - Eagle Rays

Spotted Eagle Ray - *Aetobatus narinari* (Euphrasen, 1790)

FEATURES: Dorsal surface blackish, gray, olive gray, or brownish with numerous pale spots, crescents, or circles. Ventral area whitish. Snout long, protruding, and narrow. Jaws with single rows of long, curved, plate-like teeth. Pectoral-fin tips pointed; trailing edges slightly concave. Dorsal fin located between pelvic fins. HABITAT: Circumglobal in temperate to tropical seas. In western Atlantic from NC to FL, Gulf of Mexico, Bahamas, and Caribbean Sea to Brazil. Occur coastally from surface and near shore to about 262 ft. Common in estuaries. BIOLOGY: Active, acrobatic, and migratory. Often leap from the water during spawning or in pursuit of prey. May congregate in large numbers for spawning. Prey include mollusks, squids, shrimps, and fishes. NOTE: Recent studies indicate this species may comprise several different species in different geographic areas. IUCN: Near Threatened.

Bullnose Ray - *Myliobatis freminvillei* Lesueur, 1824

FEATURES: Reddish brown, dark brown, or grayish dorsally with numerous irregular pale spots. Ventral area white. Snout broad, protruding. Jaws with seven flat rows of hexagonal-shaped, plate-like teeth; central row very wide. Pectoral-fin tips pointed; trailing edges slightly concave. Dorsal fin located just behind rear pelvic-fin margin. HABITAT: MA (rare) to FL, northern Gulf of Mexico, and coast of South America. Occur in coastal waters and estuaries to about 33 ft. BIOLOGY: Migratory, may travel long distances, and may leap from surface. Feed on invertebrates.

Southern Eagle Ray - *Myliobatis goodei* Garman, 1885

FEATURES: Uniform dark brown to grayish brown dorsally. Ventral area brownish white with dusky margins. Snout slightly protruding, rounded. Jaws with seven flat rows of hexagonal-shaped, plate-like teeth; central row very wide. Pectoral-fin tips pointed; trailing edges slightly concave. Small dorsal fin located posterior to pelvic fins. HABITAT: NC to S FL, Quintana Roo, and southern Caribbean Sea to Argentina. Occur at bottom to near surface of coastal waters. BIOLOGY: Feed on crustaceans and bivalves. They are capable of traveling long distances. IUCN: Vulnerable.

Rhinopteridae - Cownose Rays

Cownose Ray - *Rhinoptera bonasus* (Mitchill, 1815)

FEATURES: Uniform brown to reddish brown dorsally. Ventral surface white to yellowish. Snout bilobed, indented at front. Jaws with six to eight (usually seven) rows of hexagonal-shaped, plate-like teeth; central row widest. Pectoral fins pointed; trailing edges concave. Dorsal fin located between pelvic fins. HABITAT: MA to FL, Gulf of Mexico, and southern Caribbean Sea to Argentina. Also NW Cuba. Found at bottom or near surface from shore to over continental shelves. May enter estuaries. BIOLOGY: Congregate in salty bays and inshore shelves during summer. Prey on mollusks and crustaceans. Use plate-like teeth to crush hard shells. IUCN: Near Threatened.

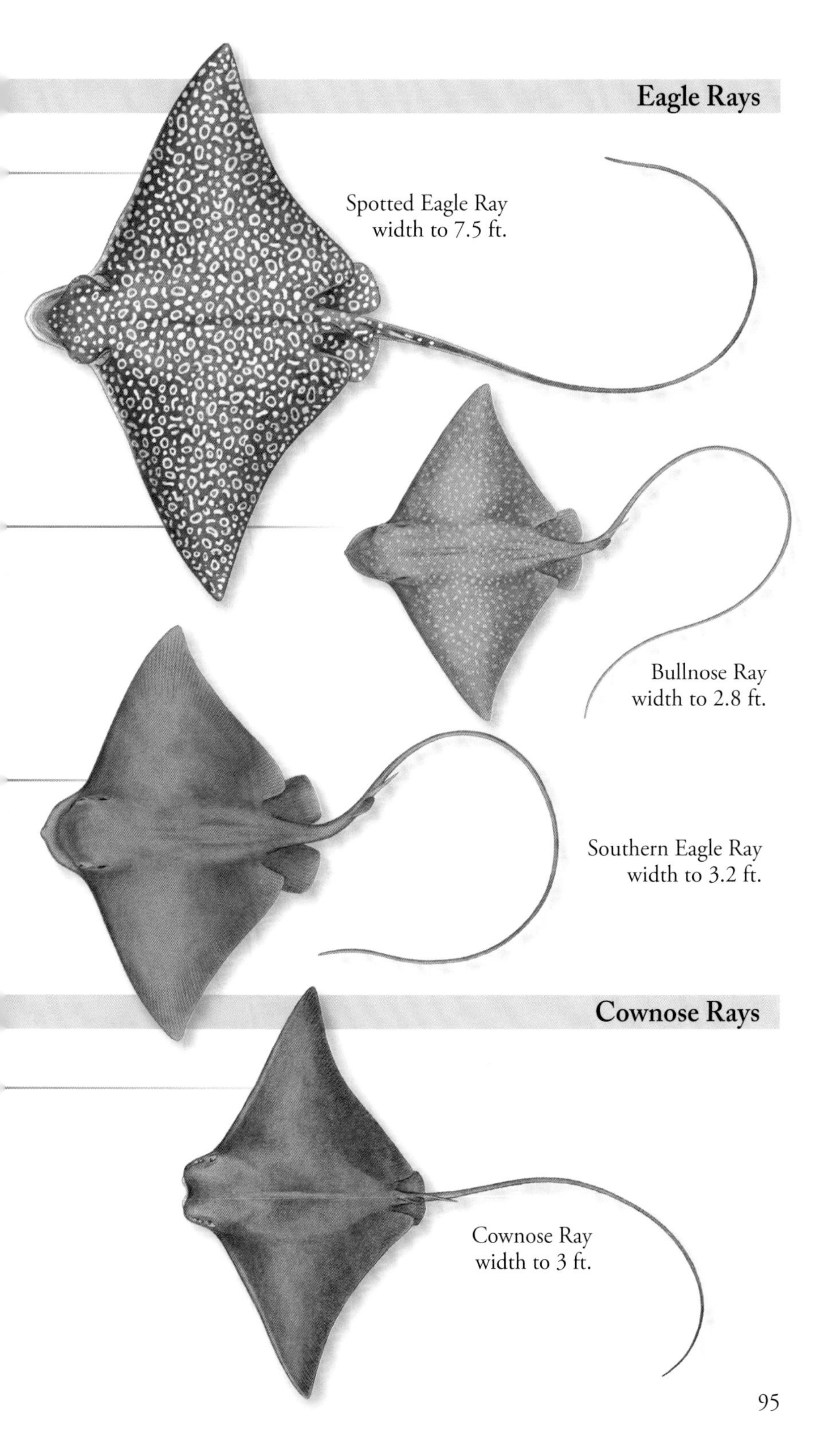
Eagle Rays
Spotted Eagle Ray
width to 7.5 ft.
Bullnose Ray
width to 2.8 ft.
Southern Eagle Ray
width to 3.2 ft.
Cownose Rays
Cownose Ray
width to 3 ft.

Rhinopteridae - Cownose Rays, *cont.*

Ticon Cownose Ray - *Rhinoptera brasiliensis* Müller, 1836

FEATURES: Uniform brown to reddish brown dorsally. Ventral surface white with dark margins. Snout bilobed, indented at front. Jaws with 8–10 (usually 9) rows of comparatively narrow, hexagonal-shaped, plate-like teeth; central three rows very wide. Pectoral fins pointed; trailing edges concave. Dorsal fin located between pelvic fins. HABITAT: Recorded from NC to southwestern Gulf of Mexico and northern South America. Demersal and pelagic from shoreline to about 82 ft. BIOLOGY: Prey on bony fishes, crustaceans, gastropods, and bivalves. Use plate-like teeth to crush hard shells. IUCN: Endangered.

Mobulidae - Mantas and Devil Rays

Giant Manta - *Mobula birostris* (Walbaum, 1792)

FEATURES: Color and pattern vary. Blackish, reddish, olivaceous to brown dorsally. Dorsal surface either uniform or with pale patches, spots, or chevrons. Marks may be faint to very bright. Ventral area either white or with black blotches, which, when present, are common behind last gill opening; ventral area may also appear almost solid black. Head very broad with mouth at front. Pectoral fins wide, arched, and pointed. A separate pair of cephalic fins extends in front of eyes and mouth. HABITAT: Circumglobal in tropical to warm temperate seas over continental shelves. MA (rare), NC to FL, Gulf of Mexico, Bermuda, Bahamas, and Caribbean Sea to Brazil. BIOLOGY: Use cephalic fins to direct flow of water into mouth. Use specialized gill plates to strain food. Migratory. May be seen leaping from the water. NOTE: Previously *Manta birostris*. IUCN: Vulnerable.

Devil Ray - *Mobula hypostoma* (Bancroft, 1831)

FEATURES: Uniform dark bluish black to dark brownish black dorsally. Ventral area whitish or yellowish. Head broad, slightly projecting with a lunate anterior margin. Pectoral fins wide, arched, and pointed. A separate pair of cephalic fins extends in front of eyes and mouth. Cephalic fin width about 50% of length. HABITAT: NC to FL, Gulf of Mexico, and Caribbean Sea to Argentina. Also in eastern Atlantic. Pelagic. Found near surface of warm, tropical waters over continental shelves. BIOLOGY: May travel singly, in pairs, or in schools. Use cephalic fins to direct water over gills. May leap into air.

Sicklefin Devil Ray - *Mobula tarapacana* (Philippi, 1892)

FEATURES: Olivaceous to brownish black dorsally. Ventral surface white anteriorly, abruptly gray posteriorly. Head strongly projecting, with a concave anterior margin. Pectoral fins strongly curved backward. A separate pair of cephalic fins extends in front of eyes and mouth. Cephalic fin width about 60% of length. Tail less than half of disk width. HABITAT: Circumtropical in warm seas. In western Atlantic from northern Gulf of Mexico to Brazil. Pelagic in coastal to offshore waters, from surface to about 5,250 ft. Reef associated. BIOLOGY: Use cephalic fins to direct water over gills. IUCN: Endangered.

ALSO IN THE AREA: Mobula species A, see p. 547.

Cownose Rays

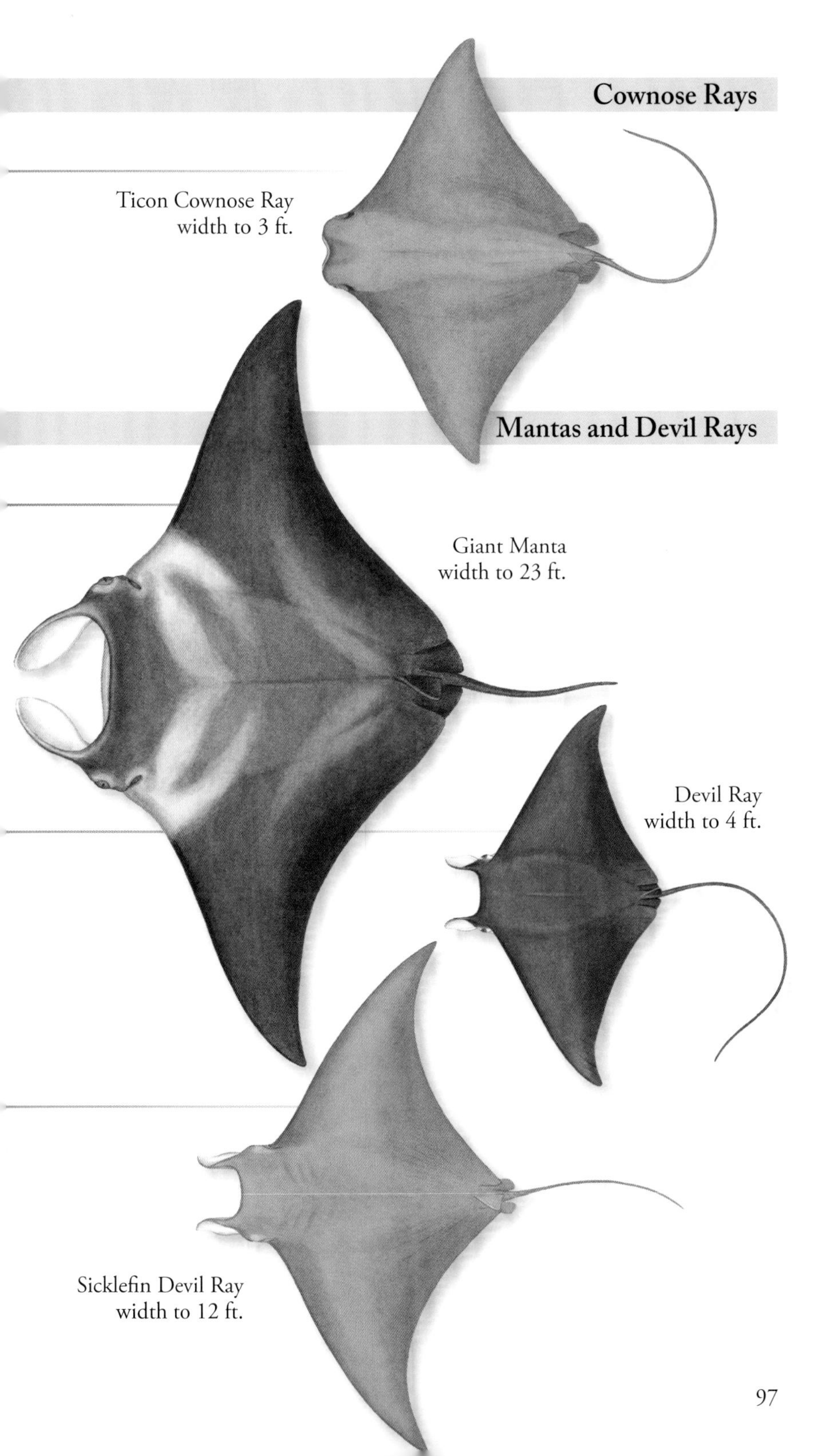

Mantas and Devil Rays

Elopidae - Tenpounders

Southern Ladyfish - *Elops smithi* McBride et al., 2010

FEATURES: Greenish gray to bluish dorsally. Silvery on sides and below. Some gold on cheeks. Snout short, pointed. Mouth and eyes large. Dorsal-fin origin over pectoral-fin origin. Caudal fin deeply forked. Body elongate. Vertebrae number 75–78. HABITAT: Bahamas and Caribbean Sea to Brazil. Inshore in bays, lagoons, estuaries, and around mangroves. NOTE: Previously known as *Elops saurus*.

ALSO IN THE AREA: *Elops saurus*, see p. 547.

Megalopidae - Tarpons

Tarpon - *Megalops atlanticus* Valenciennes, 1847

FEATURES: Bluish to greenish dorsally. Silvery below. Snout short, upturned. Large mouth juts upward. Posterior dorsal-fin ray thin, trailing. Anal fin with long base and concave margin. HABITAT: Nova Scotia (rare) to FL, Gulf of Mexico, Bahamas, Caribbean Sea to Brazil. Also Bermuda. Occur in inshore marine, brackish, and fresh waters. Also offshore. BIOLOGY: Prey on fishes and invertebrates. Spawn during late spring and summer. Sought as gamefish. IUCN: Vulnerable.

Albulidae - Bonefishes

Shafted Bonefish - *Albula nemoptera* (Fowler, 1911)

FEATURES: Head and body silvery. Many with a black anchor mark on snout tip. Yellow wash on pectoral and pelvic fins. Snout long, conical. Mouth terminal; extends to under mid-eye. Last dorsal and anal rays filamentous, more so in dorsal fin. Scales comparatively small. Lateral line with 76–84 scales, usually 80–82. HABITAT: Southern Caribbean Sea from Belize to Brazil. Possibly Dominican Republic and Puerto Rico. Occur primarily over soft bottoms in estuaries and in river outflows along mountainous shores. BIOLOGY: Feed on demersal invertebrates and fishes. NOTE: Previously known as *Albula pacifica*.

Bonefish - *Albula vulpes* (Linnaeus, 1758)

FEATURES: Silvery bluish to greenish dorsally, often with faint to dark saddles. Silvery below. Small, dark blotch at tip of long snout. Black blotch at pectoral-fin base. Mouth subterminal, does not extend below eyes. Lateral line with 65–71 scales. HABITAT: In warm, coastal waters over sandy or muddy bottoms. Usually over sandy flats. Bay of Fundy (rare) to FL, Gulf of Mexico, Bahamas, Caribbean Sea to Brazil. Also Bermuda. BIOLOGY: Forage singly or in groups for invertebrates and fishes. Prized game fish. IUCN: Near Threatened.

NOTE: Taxonomy is undergoing revision. Other known species in the area include *Albula goreensis* and *Albula* sp. cf. *vulpes*. *A. goreensis* is indistinguishable from *A. vulpes* in appearance but primarily inhabit channels. A variant in the area exhibits a short snout, large eyes, and a yellow blotch on the pectoral-fin base.

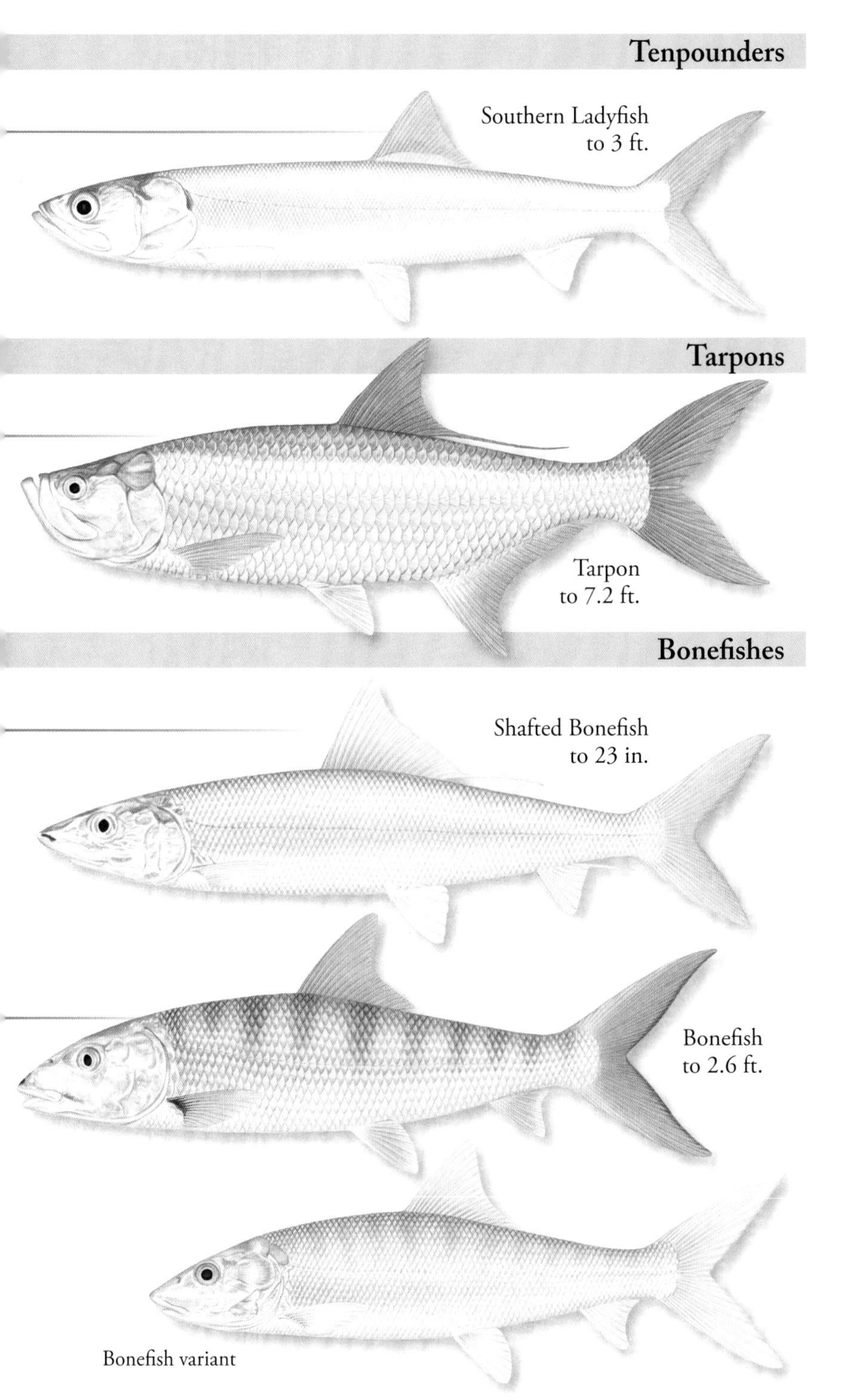

Tenpounders
Southern Ladyfish
to 3 ft.
Tarpons
Tarpon
to 7.2 ft.
Bonefishes
Shafted Bonefish
to 23 in.
Bonefish
to 2.6 ft.
Bonefish variant

Anguillidae - Freshwater Eels

American Eel - *Anguilla rostrata* (Lesueur, 1817)

FEATURES: Brownish, greenish to black or gray dorsally. Ventral area whitish. Lower jaw protrudes beyond upper jaw. Mouth has fleshy "lips." Body long, round, slender. Pectoral fin well developed. Dorsal fin originates between gill and anus. HABITAT: Greenland to FL, Gulf of Mexico, Bahamas, Antilles to Trinidad. Also Bermuda. On or near bottom in fresh, estuarine, and marine environments. Inshore and coastal. BIOLOGY: Nocturnal foragers. Catadromous. IUCN: Endangered.

Heterenchelyidae - Mud Eels

Caribbean Mud Eel - *Pythonichthys sanguineus* Poey, 1868

FEATURES: Skin translucent and pinkish due to blood vessels. Posterior anal and caudal fins blackish. Snout blunt, eyes very small, mouth large, head short, and gill opening crescent-shaped. Dorsal fin low anteriorly, tall at tail, and confluent with anal fin. Body round in cross-section anteriorly, compressed at tail. Pectoral fins and lateral line absent. HABITAT: Cuba, Puerto Rico, and Colombia to Suriname. Demersal in muddy bottoms along coast from about 32 to 177 ft.

Moringuidae - Spaghetti Eels

Spaghetti Eel - *Moringua edwardsi* (Jordan & Bollman, 1889)

FEATURES: Yellowish brown to blackish dorsally. White below. Juveniles reddish anteriorly, yellowish posteriorly. Lower jaw protrudes. Eyes large in adults. Pectoral fins small. Males with lobed dorsal and anal fins and slightly forked tail. Females with low dorsal and anal fins, rounded tail. Body very long, worm-like. HABITAT: Bermuda, S FL, Bahamas, and Caribbean Sea to Brazil. Burrows in shallow, sandy bottoms from brackish waters to around reefs, from shore to about 98 ft.

ALSO IN THE AREA: *Neoconger mucronatus*, see p. 547.

Chlopsidae - False Moray Eels

Seagrass Eel - *Chilorhinus suensonii* Lütken, 1852

FEATURES: Purplish brown dorsally; somewhat paler below. Underside of head pale. Body covered with minute, brownish flecks. Older individuals with scattered white. Mouth small; eyes large. Pectoral fins tiny to absent. Dorsal and anal fins increase in height posteriorly, confluent at tail. Lateral line reduced to pores on head and jaw. Body short, stout. HABITAT: Bermuda, Bahamas, and Caribbean Sea to Brazil. Demersal over algal and seagrass beds. Coastal, usually in clear water from shore to about 1,476 ft.

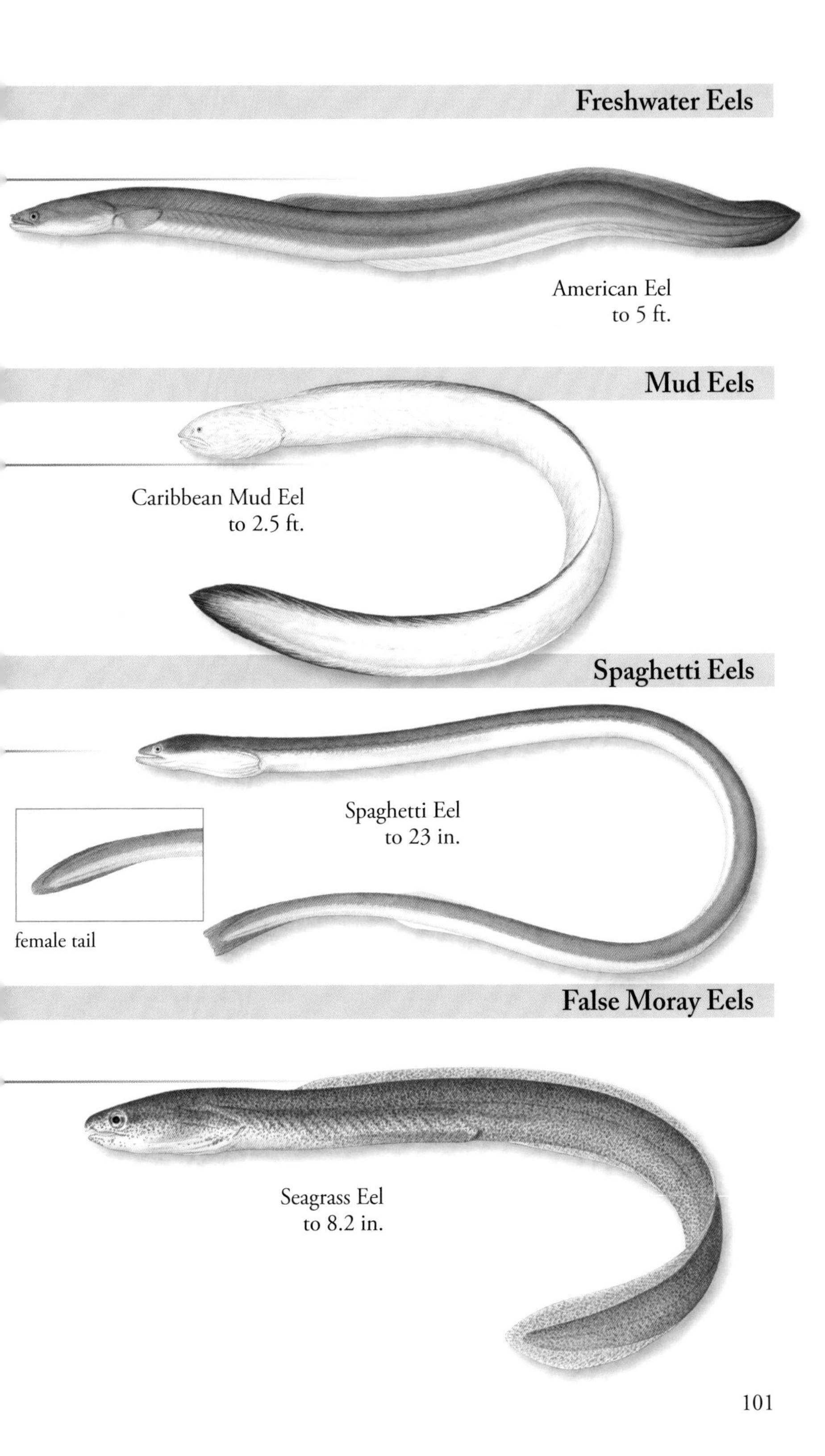
Freshwater Eels
American Eel
to 5 ft.
Mud Eels
Caribbean Mud Eel
to 2.5 ft.
Spaghetti Eels
Spaghetti Eel
to 23 in.
female tail
False Moray Eels
Seagrass Eel
to 8.2 in.

Bicolor Eel- *Chlopsis bicolor* Rafinesque, 1810

FEATURES: Shades of brown dorsally; abruptly white below. White tapers posteriorly to uniformly brown tail. Snout long and pointed. Front nostrils point downward. Eyes large. Gill opening small. Dorsal and anal fins confluent at tail. Pectoral fins absent. Body long, slightly compressed. HABITAT: Scattered records from North Carolina, south Florida, Bermuda, Bahamas, Yucatán, Venezuela, Dominica, and US Virgin Islands to Brazil. Also in eastern Atlantic and Mediterranean Sea. Occur over rocky and rubble bottoms from about 104 to 1,151 ft. IUCN: Least Concern.

Mottled False Moray - *Chlopsis dentatus* (Seale, 1917)

FEATURES: Pinkish to grayish with brownish blotches that cluster and form loose bars and marbling. Fins irregularly banded. Snout bluntly pointed. Eyes large. Dorsal-fin origin slightly behind gill opening. Dorsal and anal fins confluent at tail. Pectoral fins absent. Five to six pores on lower jaw; one on upper gill area. Body moderately long and slightly compressed. HABITAT: Scattered records from Bermuda, Bahamas, NW Cuba, Quintana Roo, Colombia, Guadalupe, Dominica, and Barbados. Also in eastern Atlantic. Demersal over rock and rubble from about 209 to 1,200 ft. BIOLOGY: Feed on other fishes.

False Moray - *Kaupichthys hyoproroides* (Strömman, 1896)

FEATURES: Head and body uniformly shades of brown to gray brown. Lower head may be paler. Pale "collar" or blotches behind eyes absent. Eyes large. Posterior nostril on upper lip concealed by an upper flap. Mouth extends to slightly past rear eye margin. Pectoral fins well developed. Dorsal-fin origin over or slightly behind gill opening. Dorsal and anal fins confluent at tail. Lateral line reduced to two pores above gill area and six to seven pores on and behind lower jaw. HABITAT: Bermuda, Bahamas, and Caribbean Sea. Found coastally from shore to about 524 ft. Also in Central Indo-Pacific.

Collared Eel - *Kaupichthys nuchalis* Böhlke, 1967

FEATURES: Overall brownish with paler fins and a distinct to faint "collar" or blotches behind eyes and on lower jaw. Snout moderately long and blunt. Eyes large. Mouth extends to slightly past rear eye margin. Dorsal-fin origin over or slightly behind gill opening. Dorsal and anal fins confluent at tail. Pectoral fins present and broadly rounded. Six to seven pores on lower jaw; two pores high above gill area. Body moderately long. HABITAT: Bermuda, Bahamas, and Caribbean Sea to Venezuela. Demersal over reefs from shore to about 462 ft. Often associated with sponges.

ALSO IN THE AREA: *Catesbya pseudomuraena*, *Robinsia catherinae*, see p. 547.

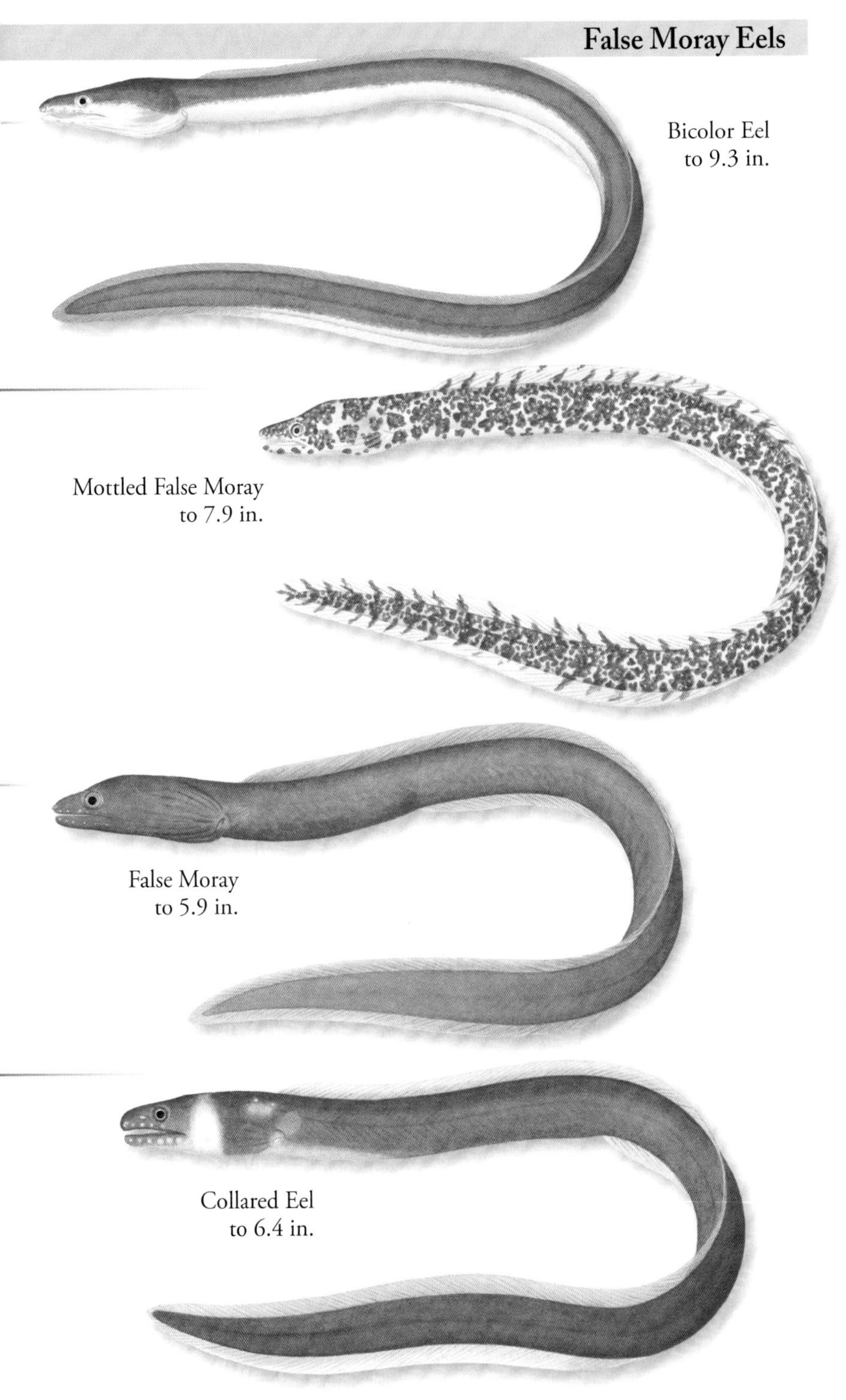
Bicolor Eel
to 9.3 in.
Mottled False Moray
to 7.9 in.
False Moray
to 5.9 in.
Collared Eel
to 6.4 in.

Muraenidae - Moray Eels

Pygmy Moray - *Anarchias similis* (Lea, 1913)

FEATURES: Color varies. Uniformly dark brown to dark brown with an alternating pattern of whitish to yellowish starburst-shaped marks along body. Tip of tail yellow. Lower jaw with pale blotches. Snout short. Rear nostril a hole above center of eye. Tail tip hard. Dorsal and anal fins originate near end of tail and are confluent with tail. HABITAT: Georgia to Florida, northern Gulf of Mexico, Bermuda, Bahamas, and Caribbean Sea to Brazil. Reef associated from about 18 to 590 ft. BIOLOGY: Feed on fishes and crustaceans.

Broadbanded Moray - *Channomuraena vittata* (Richardson, 1845)

FEATURES: Tannish, reddish brown, or brownish gray with 13–16 broad, darker bars on head and body. Jaws large, with lower jaw protruding slightly. Teeth in small, dense rows. Dorsal and anal fins at tip of tail only. Body robust; skin wrinkled. HABITAT: Scattered circumtropically. In the western Atlantic from Bermuda and Bahamas to Venezuela and other localities. Demersal around reefs, rocks, and in crevices from shore to about 330 ft. BIOLOGY: Feed on fishes and crustaceans.

Chain Moray - *Echidna catenata* (Bloch, 1795)

FEATURES: Brownish black irregular bars and spots with yellowish chain-like pattern cover head and body. Pattern is variable and colors may be reversed in larger specimens. Dorsal fin begins on head. Body stout; tail tip rounded. HABITAT: S FL, Bermuda, Bahamas, and Caribbean Sea to Brazil. Prefer shallow water. Found on coral reefs, in rocky areas, and over sandy bottoms. BIOLOGY: Chain Moray are solitary. Observed leaving water for short periods in pursuit of prey. Feed on crustaceans and small fishes.

Fangtooth Moray - *Enchelycore anatina* (Lowe, 1838)

FEATURES: Color and pattern variable. Head shades of yellow fading to a dark brown, blackish, or reddish brown ground color. Body covered in patterns of variable and alternating yellowish rosettes or wavy blotches. Snout long. Jaws hooked with numerous long, sharp teeth. Eyes moderate, located over middle upper jaw. Dorsal fin originates above or just behind gill opening. HABITAT: Scattered in warm waters of eastern and western Atlantic. In western Atlantic from NE FL, Bermuda, St. Paul's Rocks, and off Rio de Janeiro, Brazil.Reef associated between about 32 to 1,213 ft. Usually below 164 ft.

Chestnut Moray - *Enchelycore carychroa* Böhlke & Böhlke, 1976

FEATURES: Head and body uniformly dark brown; darker toward tail. Blackish corners of mouth, brachial grooves, eye margins, and dorsal-fin folds. White spots around head pores. Snout long, narrow, and arched. Jaws close only at tips. Body moderately compressed. HABITAT: FL, Flower Garden Banks, Bermuda, Bahamas, Caribbean Sea to Brazil. From shore to shallow rocky bottoms and on coral reefs.

Pygmy Moray
to 8 in.
Broadbanded Moray
to 5 ft.
Chain Moray
to 2.3 ft.
Fangtooth Moray
to 3.6 ft.
Chestnut Moray
to 12 in.

Muraenidae - Moray Eels, *cont.*

Viper Moray - *Enchelycore nigricans* (Bonnaterre, 1788)

FEATURES: Head and body uniformly or slightly mottled in shades of brown. Young are pale with reticulated pattern. Snout is long, narrow, and arched. Jaws close only at tips. Dorsal fin begins over gill opening. Body moderately compressed; tail tapered. HABITAT: NW Gulf of Mexico, S FL, Bermuda, Bahamas, and Caribbean Sea to Brazil. Also eastern Atlantic. Around coral reefs and rocky shorelines to about 80 ft.

Saddled Moray - *Gymnothorax conspersus* Poey, 1867

FEATURES: Body shades of brown with paler head, becoming blackish brown at tail. White spots cover head and body; spots tiny on head, become larger posteriorly, and very large and single on tail tip. Dorsal fin blackish with white saddles. Anal fin blackish. Snout and mouth moderate. Lower jaw straight. Dorsal-fin origin on head. HABITAT: NC, FL, Bahamas, and Caribbean Sea to Brazil. Occur over soft bottoms from about 164 to 2,624 ft. Usually below 656 ft.

Green Moray - *Gymnothorax funebris* Ranzani, 1839

FEATURES: Head and body uniform shades of green to brown. Head pores, gill openings, and anus may be dark. Eyes may be reddish. Dorsal fin begins between eye and gill opening. Body stout. HABITAT: S FL, Gulf of Mexico, Bahamas, and Caribbean Sea to Brazil. Also western Gulf of Mexico and Bermuda. Found in shallow tide pools, in rocky crevices, and on coral and rocky reefs. Occasionally in brackish tidal creeks and around mangroves. BIOLOGY: The coloring of the Green Moray is a result of yellow mucus overlying gray blue skin. May be defensive.

Lichen Moray - *Gymnothorax hubbsi* Böhlke & Böhlke, 1977

FEATURES: Overall shades of brown to reddish brown. Head with small, dense, irregular yellowish spots and wavy lines. Body with orange brown branching blotches arranged loosely in vertical rows. End of tail yellowish. Snout blunt. Mouth short, closes completely. Dorsal-fin origin on head. HABITAT: NC to northern Gulf of Mexico, western Bahamas, northern Cuba, and along coast of Colombia. Occur over loose bottoms from about 193 to 600 ft.

Blacktail Moray - *Gymnothorax kolpos* Böhlke & Böhlke, 1980

FEATURES: Head and body greenish brown, grading to almost black at tail. Small, pale ocellated spots cover head, body, and dorsal fin. Spots become larger and more separated toward tail. Only three or four large spots on tail tip. Anal fin dark, unmarked. Body stout; tail tapered. HABITAT: NC and GA; Gulf of Mexico to Campeche Bay. Occur on or near muddy or sandy bottoms and banks from about 150 to 750 ft. BIOLOGY: Blacktail Moray are harmless. Caught by trawl, trap, and hook-and-line.

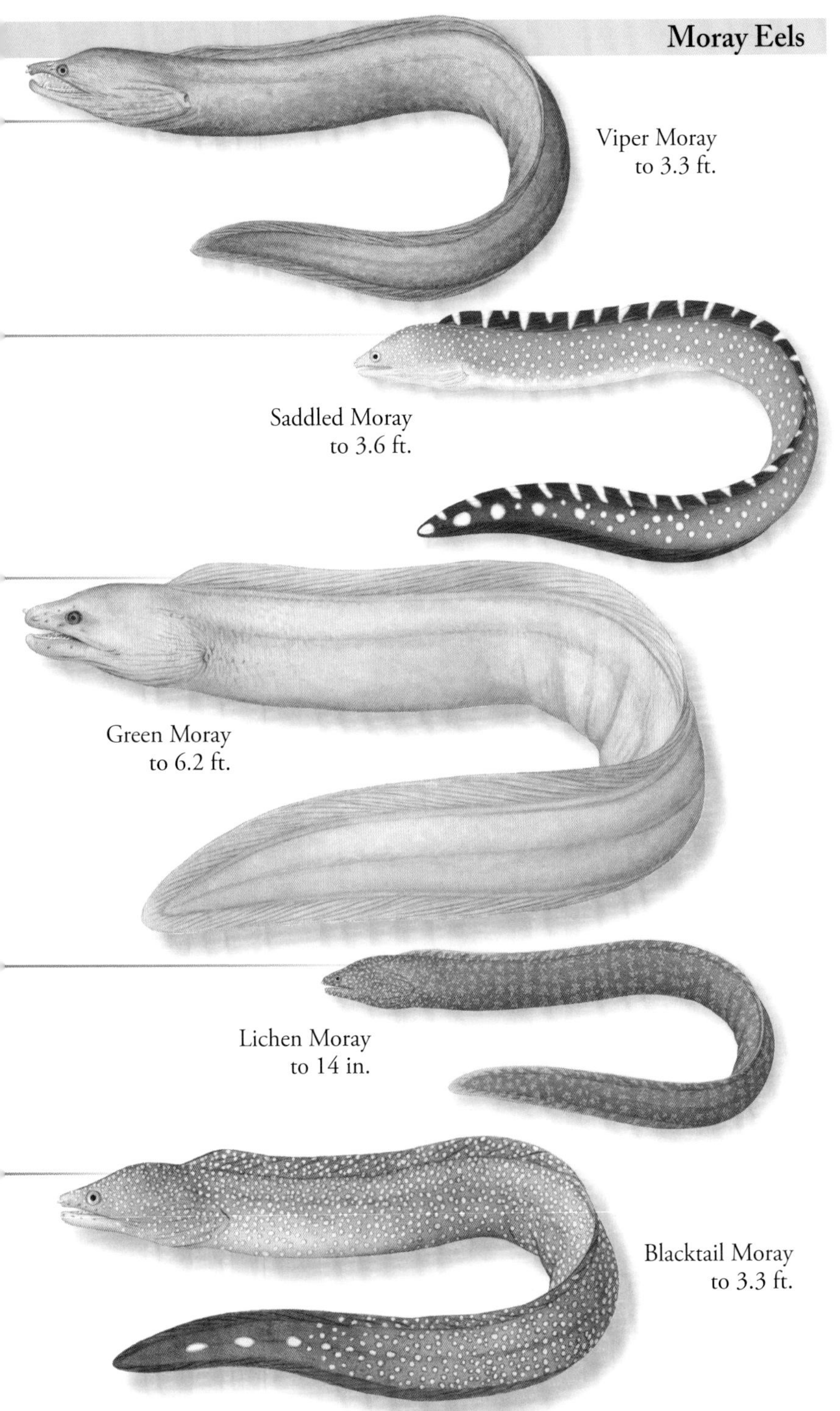
Viper Moray
to 3.3 ft.
Saddled Moray
to 3.6 ft.
Green Moray
to 6.2 ft.
Lichen Moray
to 14 in.
Blacktail Moray
to 3.3 ft.

Sharktooth Moray - *Gymnothorax maderensis* (Johnson, 1862)

FEATURES: Head greenish, fading to brownish ground color. May be covered in lime green slime. Posterior head, body, and fins covered in minute to tiny, whitish to pale yellow spots that may become larger toward tail. Dorsal- and anal-fin margins whitish. Snout moderately long, blunt. Head large. Dorsal-fin origin on head. HABITAT: Widely scattered in warm waters of eastern and western Atlantic. Occur over soft and rocky bottoms from about 278 to 930 ft.

Goldentail Moray - *Gymnothorax miliaris* (Kaup, 1856)

FEATURES: Color and pattern highly variable. May have brownish background with closely scattered, pale yellowish spots or dots. Spots may be minute, small, or large and blending. Others are uniformly pale yellowish with few to many dark markings (*inset*). Pattern continuous over entire body in all. Iris with a golden ring. Tail always pale. HABITAT: S FL, Bermuda, Bahamas, and Caribbean Sea to Brazil. On coral reefs and rocky shorelines to about 164 ft. BIOLOGY: Goldentail Moray are solitary. May hunt with other predators. Prey include fishes and invertebrates.

Spotted Moray - *Gymnothorax moringa* (Cuvier, 1829)

FEATURES: Pattern and intensity of color vary. Head, body, and fins whitish with small, overlapping dark spots. Some very dark; others sparsely spotted. Pores on lower jaw often in white spots. Dorsal fin often with anterior dark margin and pale posterior margin. Snout moderately long; jaws close completely. HABITAT: NC to FL, Bahamas, and Caribbean Sea to Brazil. Also Bermuda. Found in shallow coral and rocky reefs and seagrass bed habitats. BIOLOGY: May be defensive or aggressive when threatened. Feed on fishes and crustaceans.

Blackedge Moray - *Gymnothorax nigromarginatus* (Girard, 1858)

FEATURES: Head and body brown with small, well-separated, whitish to yellow spots. Abdomen pale. Eyes with a dark ring. Dorsal fin often with undulating, dark margin. Anal fin with uniformly dark margin. Snout and jaws short. HABITAT: SC to FL; Gulf of Mexico to Panama. Also Bermuda, Puerto Rico, British and US Virgin Islands. Occur over seagrass beds and banks from about 33 to 62 ft. Also found around inshore jetties. BIOLOGY: May inhabit large, abandoned snail shells. Caught as trawl bycatch of shrimp fisheries.

Ocellated Moray - *Gymnothorax ocellatus* Agassiz, 1831

FEATURES: Head and body shades of tan to brown with irregular pale spots that increase in size posteriorly. Abdomen pale. Dark ring around eyes. Dorsal and anal fins with undulating white marks on a black outer edge. Snout and jaws short. HABITAT: Caribbean Sea to Brazil. Occur in coastal waters over soft bottoms from shore to about 524 ft. BIOLOGY: Feed on fishes and invertebrates.

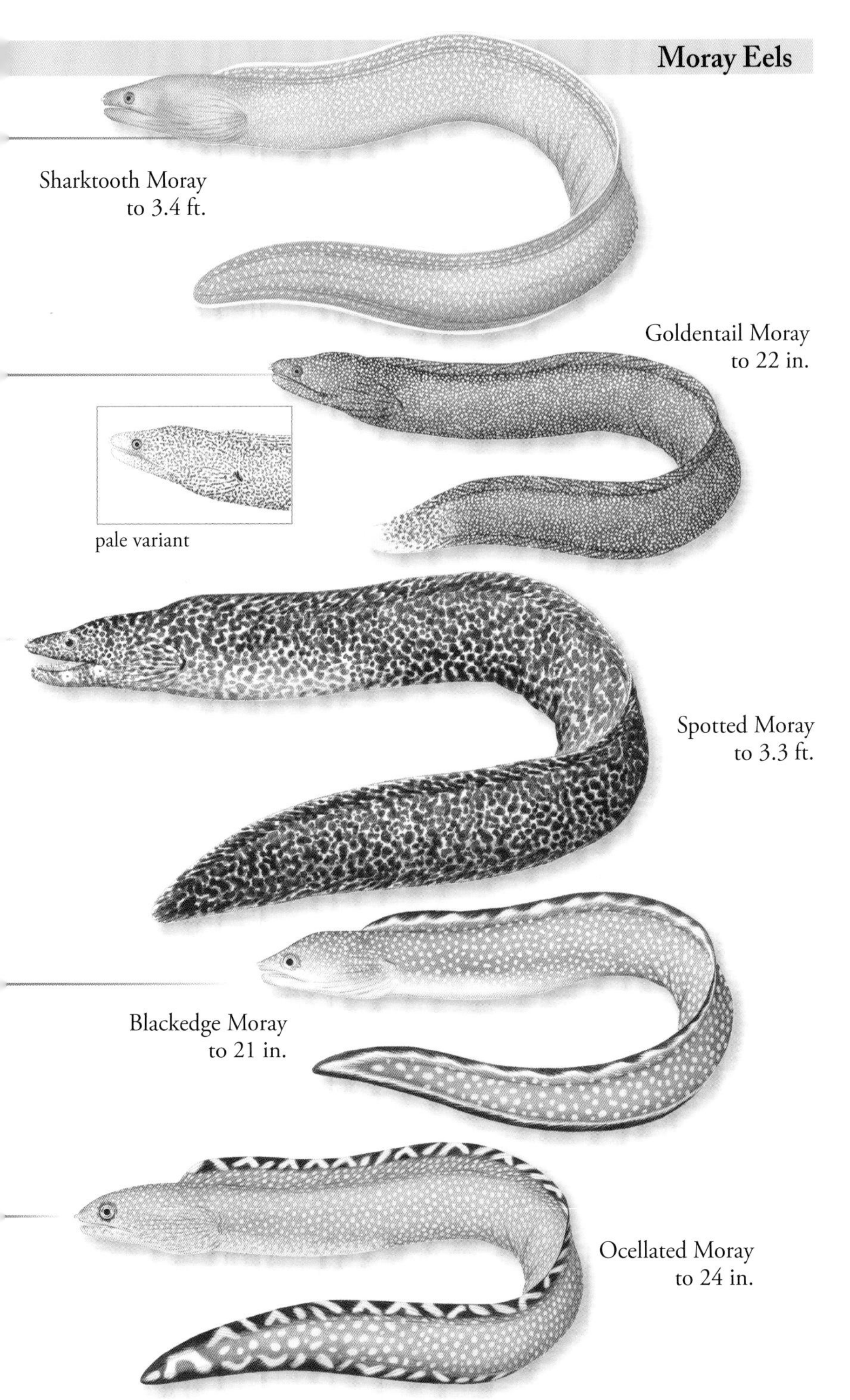

Sharktooth Moray
to 3.4 ft.

Goldentail Moray
to 22 in.

Spotted Moray
to 3.3 ft.

Blackedge Moray
to 21 in.

Ocellated Moray
to 24 in.

Polygon Moray - *Gymnothorax polygonius* Poey, 1875

FEATURES: Shades of brown with clusters of pale vermiculations loosely forming irregular polygon-like shapes on body. Head spotted. Irises golden. Inside mouth and gill opening blackish. Edges of fins at tail blackish. Snout elongate. Dorsal-fin origin on head. Body robust. HABITAT: NC to FL, Bermuda, Bahamas, Cuba, and scattered in Caribbean Sea to Brazil. Demersal over a variety of bottoms from about 295 to 885 ft. BIOLOGY: Feed on bony fishes and cephalopods.

Honeycomb Moray - *Gymnothorax saxicola* Jordan & Davis, 1891

FEATURES: Head and body brown with irregular, pale yellow spots. Spots become larger and less numerous toward tail. Midbody often has row of larger, regularly spaced spots. Dorsal fin has undulating black and white marginal saddles. Anal fin begins with black margin, ends in saddles. Snout short, blunt. Jaws short. Body compressed posteriorly. HABITAT: NC to FL. In eastern Gulf of Mexico from FL to Mobile Bay, AL. Also Bermuda. Inhabit seagrass beds and banks. BIOLOGY: Reported to forage at night. Taken as bycatch.

Purplemouth Moray - *Gymnothorax vicinus* (Castelnau, 1855)

FEATURES: Color varies. May be mottled with irregular, dark spots or almost uniformly purplish brown with darker freckles. Dark mark at each corner of mouth. Inside of mouth grayish purple. Posterior one-third of dorsal fin and entire anal fin with pale edge. HABITAT: NC, northern Gulf of Mexico, S FL, Bahamas, and Caribbean Sea to Brazil. Also Bermuda. Demersal over shallow coral and rocky reefs and seagrass habitats. BIOLOGY: Purplemouth Moray are solitary. More active at night.

Redface Moray - *Monopenchelys acuta* (Parr, 1930)

FEATURES: Head red to orange red. White under lower jaw and on throat. Body brownish red to brownish orange. Irises golden. Tip of tail yellow. Juveniles paler than adults. Eyes moderately large. Dorsal and anal fins originate far back on body, behind anus, and are confluent with tail. HABITAT: S FL, Bahamas, and Caribbean Sea. Also in tropical western Atlantic and Central Indo-Pacific oceans. Reef associated from about 32 to 230 ft.

Whitespotted Moray - *Muraena pavonina* Richardson, 1845

FEATURES: Color and pattern vary. Snout may be purplish. Body and fins shades of brown to purplish black and densely covered in large to small white spots. Gill opening black. Rear nostrils long and fleshy and located just above eyes. Juveniles blackish with fewer and larger white spots. Anal-fin origin anterior to midbody. HABITAT: Eastern Caribbean Sea to Brazil. Also Central Atlantic around Ascension Island. Reef associated from shore to about 196 ft.

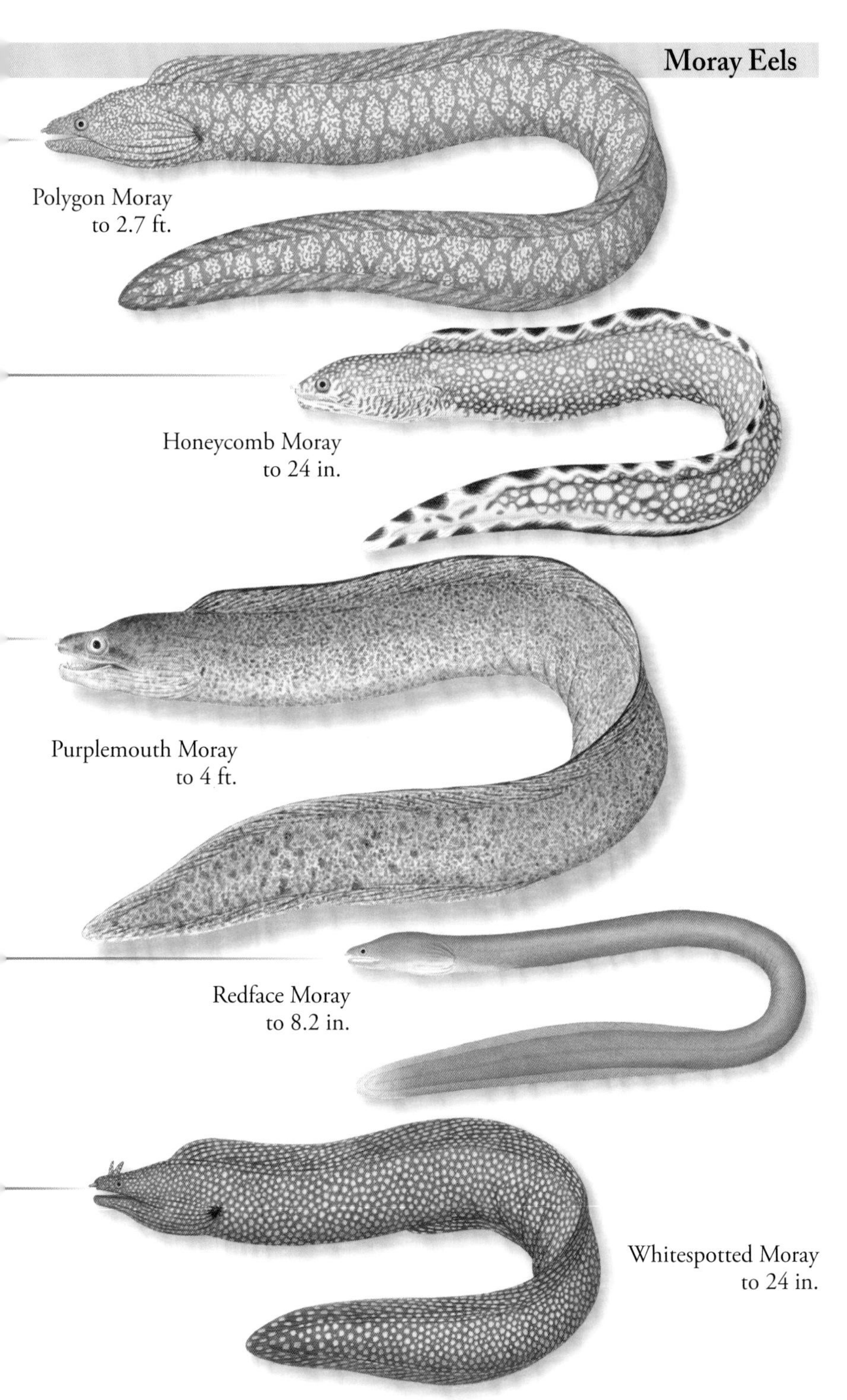
Polygon Moray
to 2.7 ft.
Honeycomb Moray
to 24 in.
Purplemouth Moray
to 4 ft.
Redface Moray
to 8.2 in.
Whitespotted Moray
to 24 in.

Muraenidae - Moray Eels, *cont.*

Reticulate Moray - *Muraena retifera* Goode & Bean, 1882

FEATURES: Body brown to dark brown with pale spots variably arranged in rosette pattern. Rosettes peppered with white dots. Pattern may be inconspicuous. Head brown with pale spots. Dark blotch over gill opening. Body stout, tapers toward tail. HABITAT: NC to west coast of FL. Also Bay of Campeche. Occur over continental coastal waters over muddy or sandy bottoms to about 250 ft. BIOLOGY: Uncommon in the area. Caught in trawls and by hook-and-line. Flesh may be poisonous.

Stout Moray - *Muraena robusta* Osório, 1911

FEATURES: Color and pattern vary. Head generally shades of reddish orange, fading to purplish gray. Body with large orange brown spots and blotches that become darker posteriorly. Inside mouth yellow. Jaws and head large. Body robust, quickly tapers toward tail. Dorsal-fin origin on head. HABITAT: NC to FL, and southern Caribbean Sea. Primarily found in eastern Atlantic. Reef associated from shore to about 223 ft. Usually below 100 ft.

Marbled Moray - *Uropterygius macularius* (Lesueur, 1825)

FEATURES: Reddish brown to dark brown. Head with irregular white spots concentrated on lower part. Body with pale, scattered mottling. Tip of tail yellow. Snout short; body long and round in cross-section, compressed near tail. Dorsal and anal fins originate near and are confluent with tail. HABITAT: FL, Bermuda, Bahamas, and Caribbean Sea. Demersal over coral and rocky reefs from near shore to about 754 ft.

Synaphobranchidae - Cutthroat Eels

Shortbelly Eel - *Dysomma anguillare* Barnard, 1923

FEATURES: Tan dorsally; white below. Dorsal and anal fin whitish; tail area dark. Snout bulbous, fleshy, and overhangs mouth. Pectoral fins well developed. Anus just behind pectoral-fin tips. Anal-fin origin just behind anus. Caudal fin distinct but confluent with dorsal and anal fins. HABITAT: Bahamas, Gulf of Mexico, and from FL Keys to French Guiana. Demersal over muddy bottoms from about 100 to 885 ft. Often found near the mouths of large rivers.

Ophichthidae - Snake Eels and Worm Eels

Key Worm Eel - *Ahlia egmontis* (Jordan, 1884)

FEATURES: Pale tan to yellowish with minute dark specks. Larger specimens darker with red brown band behind head. Snout overhangs mouth. Eyes and pectoral fins well developed. Dorsal-fin origin behind anus. Caudal fin confluent with dorsal and anal fins. Body very long and slender. HABITAT: NC, FL, and Gulf of Mexico to Brazil. Occur over seagrass beds in bays and around mangroves, also over sandy bottoms around reefs.

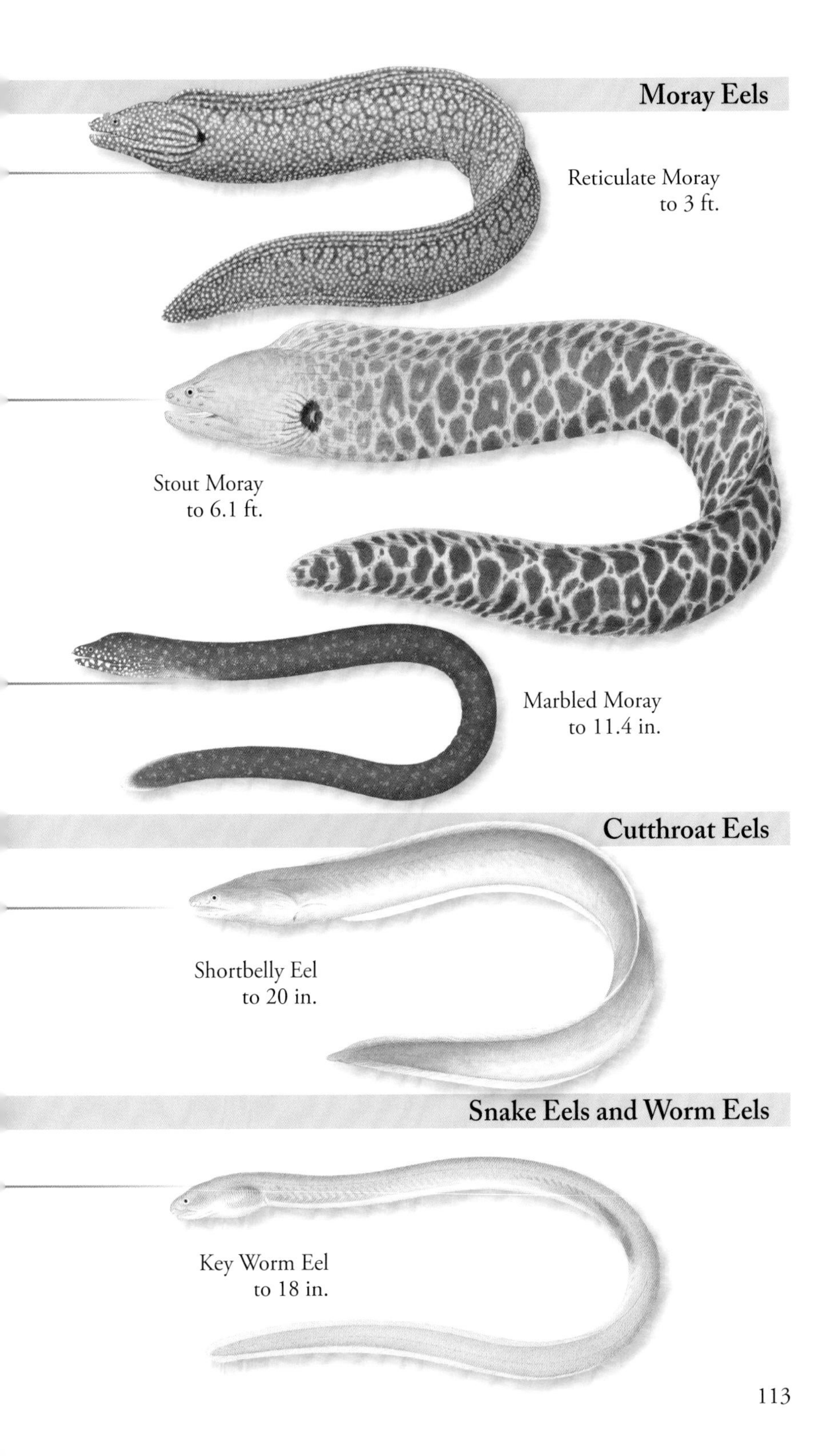
Moray Eels
Reticulate Moray
to 3 ft.
Stout Moray
to 6.1 ft.
Marbled Moray
to 11.4 in.
Cutthroat Eels
Shortbelly Eel
to 20 in.
Snake Eels and Worm Eels
Key Worm Eel
to 18 in.

Tusky Eel - *Aplatophis chauliodus* Böhlke, 1956

FEATURES: Tannish to grayish brown and covered in fine brownish fleck and speckles. Flecks very small on head. Dorsal- and anal-fin margins may be black. Snout short, upturned. Lower jaw long, projects beyond upper. Anterior teeth in upper and lower jaws fang-like, extend outside mouth. Eyes small, low on head. Dorsal fin low, originates behind pectoral fins. Tail blunt, finless. HABITAT: Northern Gulf of Mexico, S FL, Cuba, Puerto Rico, and Panama to French Guiana. Demersal over soft bottoms from estuaries to about 300 ft.

Striped Eel - *Aprognathodon platyventris* Böhlke, 1967

FEATURES: Pale with a blackish stripe along upper sides and a narrow stripe along belly that fades into spots toward anus. Head spotted. Dorsal fin blackish, or with a blackish margin. Snout overhangs mouth; underside lacks a groove. Dorsal and anal fins low. Dorsal-fin origin on head. Pectoral fins absent. Anus far back on body. Tip of tail finless. HABITAT: S FL, Bahamas, Greater and Lesser Antilles, Quintana Roo, Panama, and off Venezuela. Demersal over shallow sandy and seagrass bottoms. Occur from shore to about 55 ft.

Academy Eel - *Apterichtus ansp* (Böhlke, 1968)

FEATURES: Pinkish to shades of tan with very fine sprinkling of specks dorsally. Whitish band through eyes, and whitish areas on lower jaw. Lateral line in pale stripe. Snout sharply pointed, overhangs lower jaw; underside flattened and grooved. Eyes small, over tip of lower jaw. All fins absent. Body long, cylindrical, pointed at tail. HABITAT: NC to FL; Bahamas to Virgin Islands and Brazil. Demersal over sandy bottoms from shore to about 130 ft.

Sooty Eel - *Bascanichthys bascanium* (Jordan, 1884)

FEATURES: Greenish brown to sooty brown dorsally and on anterior head. Whitish below. Dorsal and anal fins pale. Snout short, blunt, grooved underneath. Eyes small. Dorsal-fin origin on head. Dorsal and anal fins low. Pectoral fins small, located near center of gill opening. Tail tip hard, blunt, and finless. Body very long. HABITAT: NC to FL; Gulf of Mexico to Belize. Also Panama, Puerto Rico, and San Blas. Demersal along sandy beaches from shore to about 80 ft.

Twostripe Snake Eel - *Callechelys bilinearis* Kanazawa, 1952

FEATURES: Blackish with three white stripes: along dorsal-fin base, along abdomen onto anal-fin base, and along lateral line. Lateral-line stripe may be solid or a series of connected spots. Sides of head may be pale. Snout pointed, slightly overhangs mouth. Dorsal-fin origin over or just behind eyes. Tip of tail a hard, finless point. Pectoral fins absent. HABITAT: Bermuda and Bahamas to Brazil, and Yucatán to Colombia. Also south-central Atlantic islands. Demersal along open and semi-protected beaches, primarily around islands. Occur from shore to about 120 ft.

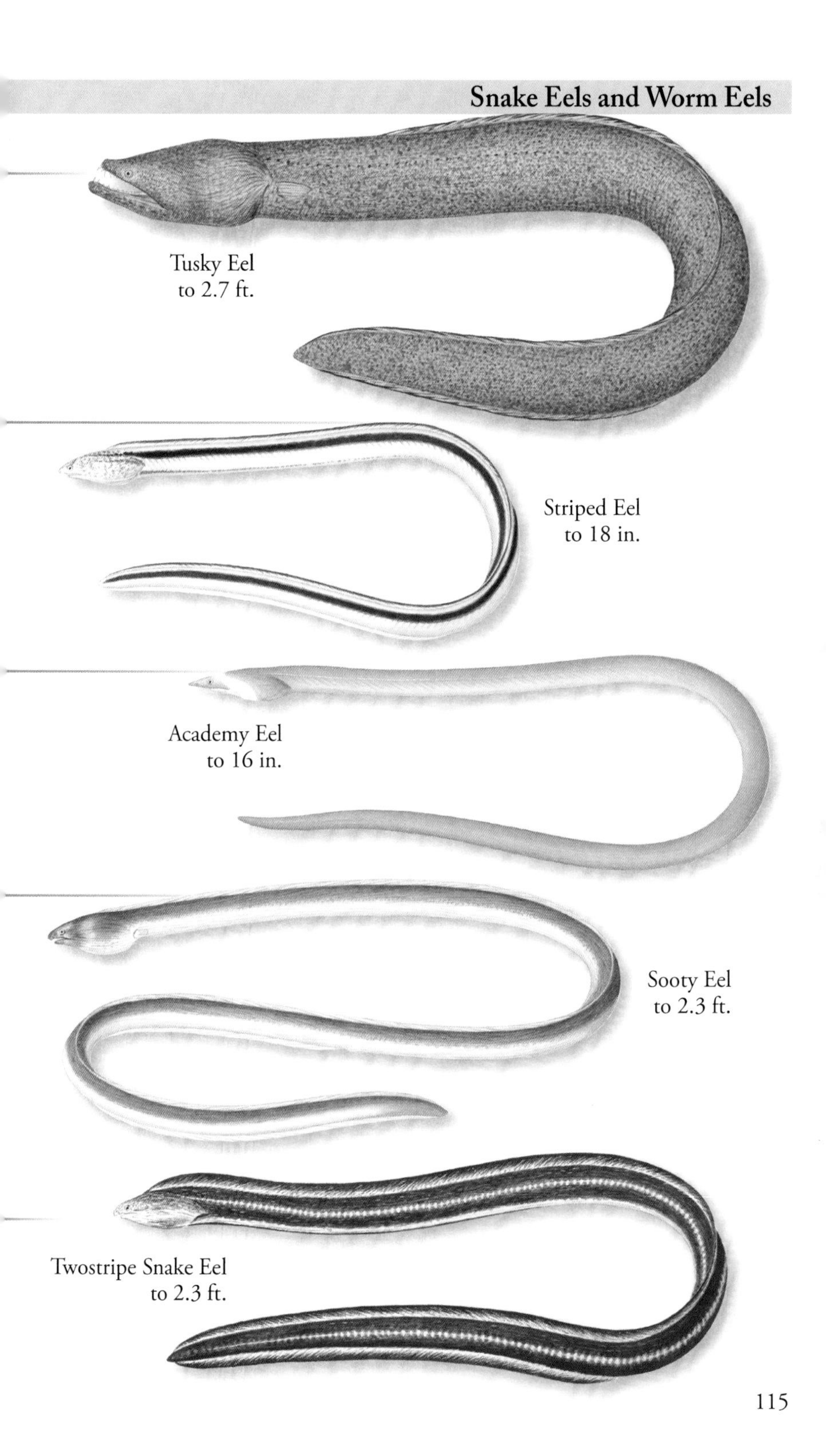
Tusky Eel
to 2.7 ft.
Striped Eel
to 18 in.
Academy Eel
to 16 in.
Sooty Eel
to 2.3 ft.
Twostripe Snake Eel
to 2.3 ft.

Shorttail Snake Eel - *Callechelys guineensis* (Osório, 1893)

FEATURES: Shades of white or pale blue to yellowish with numerous, irregular dark brown to blackish spots, dots, and blotches on head, body, and fins. Spots on head and snout very small. Snout long, pointed, overhangs lower jaw, with groove underneath. Dorsal-fin origin on head. Anal fin originates far back on body. Tip of tail a hard and finless point. Pectoral fins absent. Body very long. HABITAT: SC to north and south Gulf of Mexico. Bahamas and Caribbean Sea to Venezuela. Demersal and buried in bottom sediment of sandy and seagrass bottoms. Occur from shore to about 120 ft.

Slantlip Eel - *Caralophia loxochila* Böhlke, 1955

FEATURES: Brown dorsally; pale tan below. Dorsal fin pale. Snout conical, without groove underneath, overhanging lower jaw. Front nostril star-shaped, may be divided. Dorsal-fin origin on head. Pectoral fins absent. Dorsal and anal fins very low. Tip of tail a hard and finless point. Body very long. HABITAT: Florida and Bahamas to Brazil. Also Panama. Demersal and buried in sandy bottoms around reefs and seagrass beds from shore to about 40 ft.

Spotted Spoon-nose Eel - *Echiophis intertinctus* (Richardson, 1848)

FEATURES: Creamy to pale yellow dorsally. Ventral area pale. Head with small dark spots; upper body with irregular, large to small, dark brown spots. Fins pale with dark margins. Head and mouth large. Tail finless. HABITAT: NC to FL, Bahamas, and Caribbean Sea to Brazil. In the Gulf of Mexico from FL to LA and Yucatán. Over sandy bottoms. Near shore to about 210 ft. BIOLOGY: Spotted Spoon-nose Eels forage at night. They burrow and hide in bottom sediment, then ambush prey.

Snapper Eel - *Echiophis punctifer* (Kaup, 1859)

FEATURES: Creamy, pale yellow, tan, or pinkish dorsally. White to creamy below. Top of head densely stippled. Body with irregular dark brown to black spots about equal to eye diameter. Fins pale with dark margins. Head and mouth large. Tail finless. HABITAT: Gulf of Mexico, off Colombia and Venezuela, and from Caribbean islands to Brazil. Occur over soft bottoms from shore to about 330 ft. Also in eastern Atlantic. BIOLOGY: Burrow and hide in bottom sediment, then ambush prey.

Surf Eel - *Ichthyapus ophioneus* (Evermann & Marsh, 1900)

FEATURES: Overall tannish or orangish to yellowish. May be paler under head. Snout sharply pointed with a toothed groove underneath, overhangs mouth. Front nostril a flat opening. Eyes minute. All fins absent. Tip of tail hard and pointed. Body very long. HABITAT: S Florida, Bahamas, Bermuda, and Caribbean Sea to Brazil. Found over sandy and rocky bottoms from shore to about 164 ft. Often found off unprotected beaches. BIOLOGY: Burrow and hide in bottom sediment, then ambush prey.

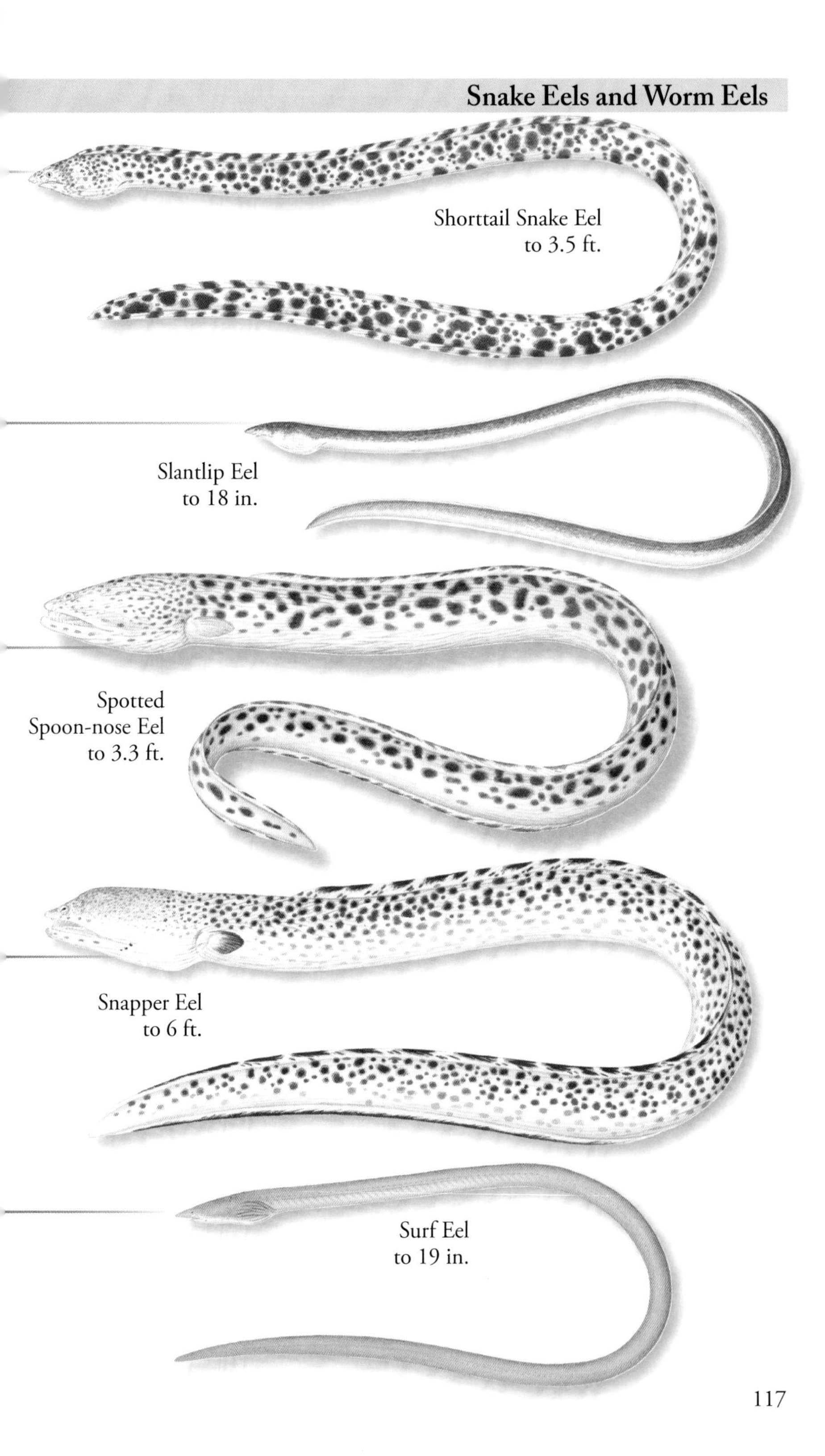
Shorttail Snake Eel
to 3.5 ft.
Slantlip Eel
to 18 in.
Spotted
Spoon-nose Eel
to 3.3 ft.
Snapper Eel
to 6 ft.
Surf Eel
to 19 in.

Sharptail Eel - *Myrichthys breviceps* (Richardson, 1848)

FEATURES: Color varies. Shades of blue gray, greenish gray, tan, or dark brown. All with small white spots with yellow centers on head. Body with rows of large, diffuse white spots. Some spots may have yellow centers. Snout short, conical. Tail a blunt, finless point. Body serpentine. HABITAT: Bermuda, S FL, Bahamas, W Gulf of Mexico, and Caribbean Sea to Brazil. Demersal over seagrass beds and over sandy and coral rubble bottoms from shore to about 65 ft.

Goldspotted Eel - *Myrichthys ocellatus* (Lesueur, 1825)

FEATURES: Yellowish to greenish tan with two rows of regularly spaced bright yellow spots inside black rings that have diffuse margins. Snout short, conical. Dorsal fin begins at head. Tail a blunt, finless point. HABITAT: S FL, Bahamas, south to Brazil. Also Bermuda. Demersal over reefs and seagrass beds from near shore to about 500 ft. Also in lagoons and areas of sandy or rocky rubble. BIOLOGY: Usually forage at night and bury themselves in the bottom during the day.

Broadnose Worm Eel - *Myrophis platyrhynchus* Breder, 1927

FEATURES: Tannish dorsally and on sides. Abdomen pale. Tiny black spots pepper upper head, body, and tail. Snout blunt, broadly depressed. Eyes large with two pores behind each. Dorsal-fin origin about midway between tip of snout and anus. Pectoral fins well developed. Dorsal and anal fins confluent with tail. Lateral-line pores visible on head and trunk only. Body comparatively compressed. HABITAT: Bermuda, Bahamas, and Caribbean Sea to Brazil. Also off FL and Bay of Campeche. Demersal over soft bottoms from shore to about 720 ft. Usually shallower than 33 ft.

Speckled Worm Eel - *Myrophis punctatus* Lütken, 1852

FEATURES: Dark brown to tan dorsally. Numerous tiny, dark spots dorsally and on sides. Abdomen pale. Color darkens with size. Snout blunt, broadly depressed, and overhangs mouth. Eyes large with one pore behind each. Dorsal fin begins above middle of abdomen. Body long, slender, worm-like, and laterally compressed toward tail. HABITAT: NC to FL, Gulf of Mexico, Bahamas, Antilles to Brazil. Also Bermuda. Occur on seagrass beds, around mangroves, in estuaries, and over reefs to about 23 ft. BIOLOGY: Adults spawn offshore.

Shrimp Eel - *Ophichthus gomesii* (Castelnau, 1855)

FEATURES: Color varies. Slate gray, greenish gray, grayish brown, olive brown, or blackish dorsally with very faint and fine specks. Pale ventrally. Head pores inconspicuous and pale or dark-margined. Dorsal-fin margin dark. Snout broad, overhangs mouth. Dorsal-fin origin over rear third of pectoral fin. Pectoral fins long and pointed to rounded. Tip of tail a blunt, finless point. HABITAT: SC to Brazil. Also N Gulf of Mexico, West Indies, and Bermuda. From near shore to about 295 ft. BIOLOGY: A common inshore fish where it occurs. Commonly caught as bycatch of shrimp fisheries.

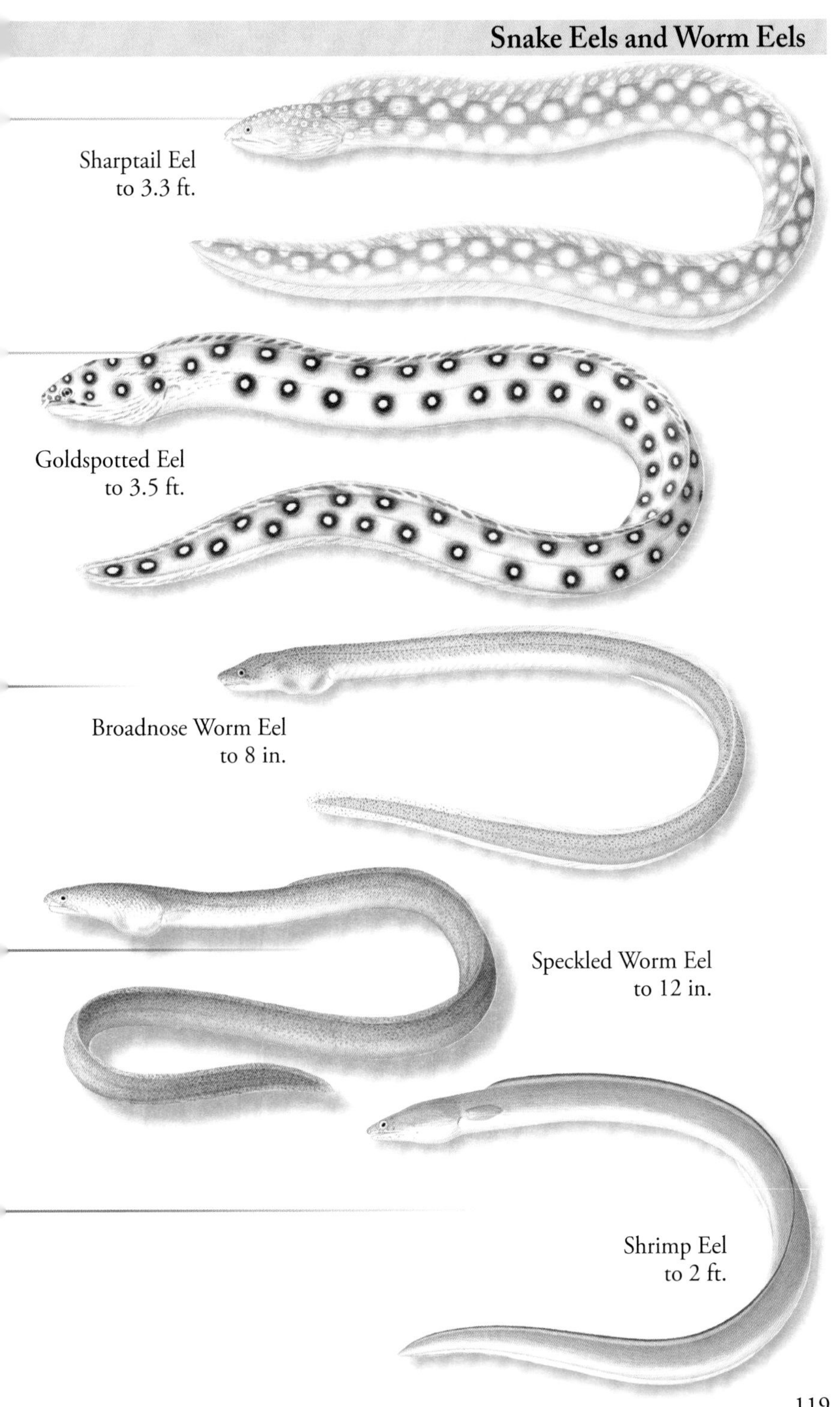
Sharptail Eel
to 3.3 ft.
Goldspotted Eel
to 3.5 ft.
Broadnose Worm Eel
to 8 in.
Speckled Worm Eel
to 12 in.
Shrimp Eel
to 2 ft.

Faintsaddled Snake Eel - *Ophichthus hyposagmatus* McCosker & Böhlke, 1984

FEATURES: Whitish with 16 greenish tan saddles on body. Head greenish tan. Dorsal-fin margin dark. Snout overhangs mouth. Head flattened above eyes. Eyes large, bulging. Dorsal-fin origin over pectoral-fin tips. Pectoral fins well developed. Tip of tail blunt and finless. HABITAT: NE Gulf of Mexico, off Quintana Roo, Haiti, Dominican Republic, and southern Caribbean Sea. Demersal over soft bottoms from about 288 to 960 ft.

Spotted Snake Eel - *Ophichthus ophis* (Linnaeus, 1758)

FEATURES: Tannish, creamy, or white with rows of large, irregular, brownish to blackish spots along midside and under dorsal-fin base. Head with smaller spots and a dark bar across nape. Anterior nostrils with a distinct barbell. Eyes large, protruding. Dorsal-fin origin over rear pectoral fin. Pectoral fins large. Tip of tail blunt and finless. HABITAT: FL, N Gulf of Mexico, Bermuda, Bahamas, and Caribbean Sea to Brazil. Demersal and buried in sandy, gravel, and rubble bottoms from shore to about 164 ft. BIOLOGY: Burrows in sediment, often only exposes head.

Palespotted Eel - *Ophichthus puncticeps* (Kaup, 1859)

FEATURES: Brown, dark gray, or gray dorsally. Thin, dark stripe and series of pale spots along lateral line. Small, pale spots on head. Pale ventrally. Snout slightly overhangs lower jaw. Tail finless. HABITAT: NC to Suriname, including West Indies. In Gulf of Mexico from FL to Texas. Demersal from near shore to about 700 ft. BIOLOGY: Reported to have toxic flesh.

King Snake Eel - *Ophichthus rex* Böhlke & Caruso, 1980

FEATURES: Dusky yellow or slate gray to brownish with 14–15 brownish faint to distinct saddles. White below. Dorsal and anal fins faintly banded. Snout short. Eyes moderate. Dorsal-fin origin behind pectoral-fin tips. Pectoral fins well developed. Tip of tail a blunt, finless point. HABITAT: Gulf of Mexico from W FL to Bay of Campeche. Demersal over mud, sand, and gravel from about 50 to 1,200 ft. Usually shallower than 500 ft.

Blackspotted Snake Eel - *Quassiremus ascensionis* (Studer, 1889)

FEATURES: Pale grayish to pale brownish dorsally. Pale below. Pattern varies. Upper body with two rows of large, black spots with diffuse reddish rings. Smaller spots often in between. Head with small spots and dark areas between and behind eyes. Dorsal and anal fins spotted. Snout conical, overhangs mouth. Pectoral fins minute. Dorsal-fin origin behind gill opening. Tip of tail a blunt and finless point. HABITAT: Bermuda, Bahamas, S FL, Dominica to St. Vincent, Quintana Roo, Honduras, and off Brazil. Also around Ascension Island. Demersal and buried in sand, rubble, and seagrass bottoms from near shore to about 70 ft.

NOTE: There are 12 other poorly recorded Snake and Worm Eels in the area.

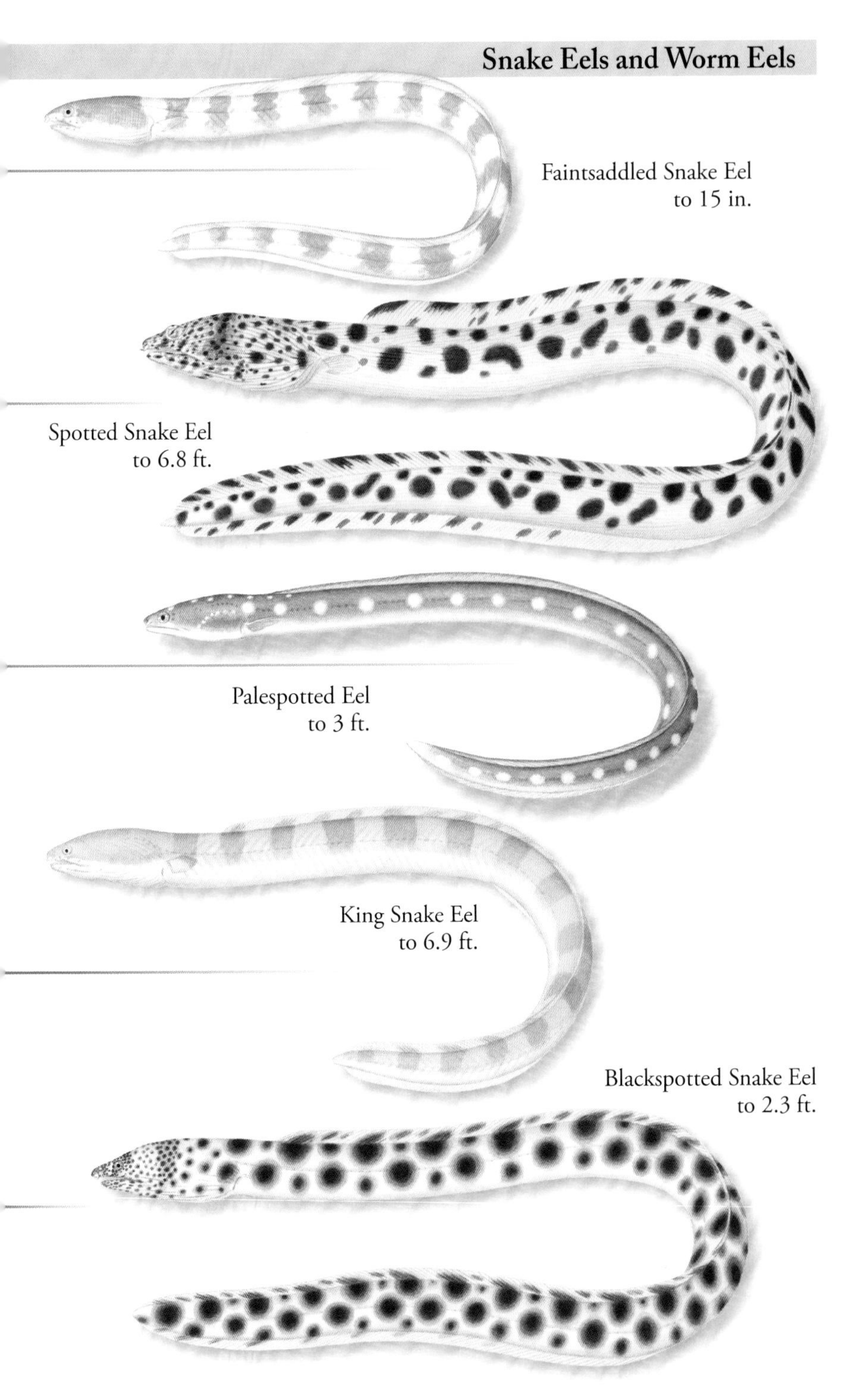
Faintsaddled Snake Eel
to 15 in.
Spotted Snake Eel
to 6.8 ft.
Palespotted Eel
to 3 ft.
King Snake Eel
to 6.9 ft.
Blackspotted Snake Eel
to 2.3 ft.

Muraenesocidae - Pike Congers

Guyana Pike Conger - *Cynoponticus savanna* (Bancroft, 1831)

FEATURES: Described as blackish gray dorsally, lighter below. Observed specimens grayish dorsally, yellowish on sides, silvery below. Pectoral fins black. Other fin margins black. Snout projects beyond lower jaw. Front nostril a tube; rear nostril a large pore in front of eye. Mouth with large, conical teeth. Gill slits nearly meet ventrally. Pectoral fins large. Lateral line with conspicuous branches. HABITAT: Cuba, Haiti, Dominican Republic, Puerto Rico, Virgin Islands, Nicaragua to Brazil. Over soft bottoms from near shore and in bays and estuaries to about 330 ft.

Congridae - Conger Eels

Bandtooth Conger - *Ariosoma balearicum* (Delaroche, 1809)

FEATURES: Shades of brown dorsally. Sides with golden or silvery reflections. Paler below. Dorsal and anal fins with dark margins. Snout long and projecting over jaws. Flanges on both lips. Eyes large. Lateral line with 46 to 52 pores. HABITAT: NC to FL, Gulf of Mexico, Caribbean Sea to Brazil. Also Bermuda, Mediterranean Sea, and Indian Ocean. Found on sandy and muddy bottoms from bays and shallow coastal waters to about 2,400 ft. May also occur at the surface.

Bullish Conger - *Bathycongrus bullisi* (Smith & Kanazawa, 1977)

FEATURES: Grayish or grayish brown to brownish. Abdomen white. Dorsal and anal fins with narrow black margins posteriorly. Stomach and intestines black. Snout fleshy and projecting over jaws. Flange on lower lip only. Dorsal-fin origin slightly posterior to pectoral-fin base. Pectoral fins well developed. HABITAT: Northern Gulf of Mexico to Amazon River. Absent from Greater and Lesser Antilles. Occur over soft bottoms from about 180 to 1,800 ft.

Gray Conger - *Conger esculentus* Poey, 1861

FEATURES: Shades of gray to brown dorsally. Abdomen pale. Upper half of dorsal- and anal-fin margins black. Snout slightly projecting. Mouth reaches under mid-eye, with flanges on upper and lower lips. Eyes large. Pores behind eyes absent. Dorsal-fin origin usually over pectoral-fin tips. Anal-fin origin anterior to midbody. Lateral line with series of 32–36 pores before anus. HABITAT: Bermuda, Bahamas, Greater and Lesser Antilles to Venezuela. Demersal over hard bottoms from near shore to about 2,000 ft.

Conger Eel - *Conger oceanicus* (Mitchill, 1818)

FEATURES: Shades of gray to brown dorsally. Abdomen slightly paler. Dorsal and anal fins with dark margins. Snout long, fleshy, and projecting over jaws. Mouth reaches to rear or rear margin of eye. Flanges on both lips; upper flange wide. Pores behind eyes absent. Dorsal-fin origin usually over pectoral-fin tips. Lateral line with series of 37–42 pores before anus. HABITAT: MA to E Gulf of Mexico. Found on or near bottom from shoreline to edge of continental shelves and about 2,000 ft. BIOLOGY: Seasonally migratory and spawn offshore. Feed primarily on fishes.

Pike Congers
Guyana Pike Conger
to 6.5 ft.
Conger Eels
Bandtooth Conger
to 13 in.
Bullish Conger
to 24 in.
Gray Conger
to 5.2 ft.
Conger Eel
to 6.5 ft.

Manytooth Conger - *Conger triporiceps* Kanazawa, 1958

FEATURES: Bluish gray to purplish gray on head and body. May also be brownish. All with black margins on dorsal and anal fins. Pectoral fins may appear dark. Snout slightly projecting. Flanges on upper and lower lips. One or two pores behind eyes. Dorsal-fin origin over pectoral-fin tips. Lateral line with 37–44 pores before anus. HABITAT: S FL and Gulf of Mexico, Bermuda, Bahamas, and Caribbean Sea to Venezuela. Demersal over sandy bottoms, rocky rubble, and around reefs from near shore to about 540 ft.

Giraffe Spotted Garden Eel - *Heteroconger camelopardalis* (Lubbock, 1980)

FEATURES: Lips black. Top of head brown with pale wavy lines and two vertical rows of pale spots behind eyes. A blackish, triangular-shaped blotch on opercle followed posteriorly by a long, narrow, blackish, V-shaped mark. Rest of body whitish with irregular brownish spots that are darker dorsally. Snout short. Mouth oblique. Eyes large. Body very long and whip-like. HABITAT: Scattered in E Caribbean Sea to Brazil. At least one large colony off Tobago. Also around S Atlantic islands. Burrow vertically in sand and rubble plains from about 50 to 170 ft.

Brown Garden Eel - *Heteroconger longissimus* Günter, 1870

FEATURES: Eyes dark above, white-rimmed below. Shades of reddish brown to brown or grayish brown with small pale flecks fading to yellow at tail. Anterior portion of abdomen may be pale or similar to rest of body. Dorsal fin white or with white margin. Snout short. Mouth oblique. Eyes large. Body very long and whip-like. HABITAT: Bermuda, FL Keys, Bahamas, and Caribbean Sea to Venezuela. Burrow vertically in sandy bottoms from about 30 to 196 ft.

Margintail Conger - *Paraconger caudilimbatus* (Poey, 1867)

FEATURES: Shades of brown to gray dorsally; paler below. Snout and lower jaw dark. Dark blotch on gill chamber. Large white blotch around pectoral fins. Upper and lower jaws with a flange. Eyes very large; irises dark. Dorsal-fin origin over pectoral fins. Pectoral fins well developed. Upper edge of gill opening above pectoral-fin base. Tip of tail stiff. HABITAT: Florida, Gulf of Mexico, Bermuda, Bahamas, and Caribbean Sea. Occur over soft bottoms and around reefs from shore to about 230 ft.

Splendid Conger - *Pseudophichthys splendens* (Lea, 1913)

FEATURES: Shades of gray dorsally; pale below. Inside mouth, gill area, and stomach black. Intestines pale. Snout swollen and projecting. Mouth small, extends to under anterior portion of eye. Eyes large. Flange on lower lip only. Dorsal-fin origin just behind pectoral-fin tips. Dorsal and anal fins confluent with tail. HABITAT: SC to FL, Gulf of Mexico, and Caribbean Sea to the Guianas. Also in western Atlantic. Occur over soft bottoms from about 120 to 5,400 ft.

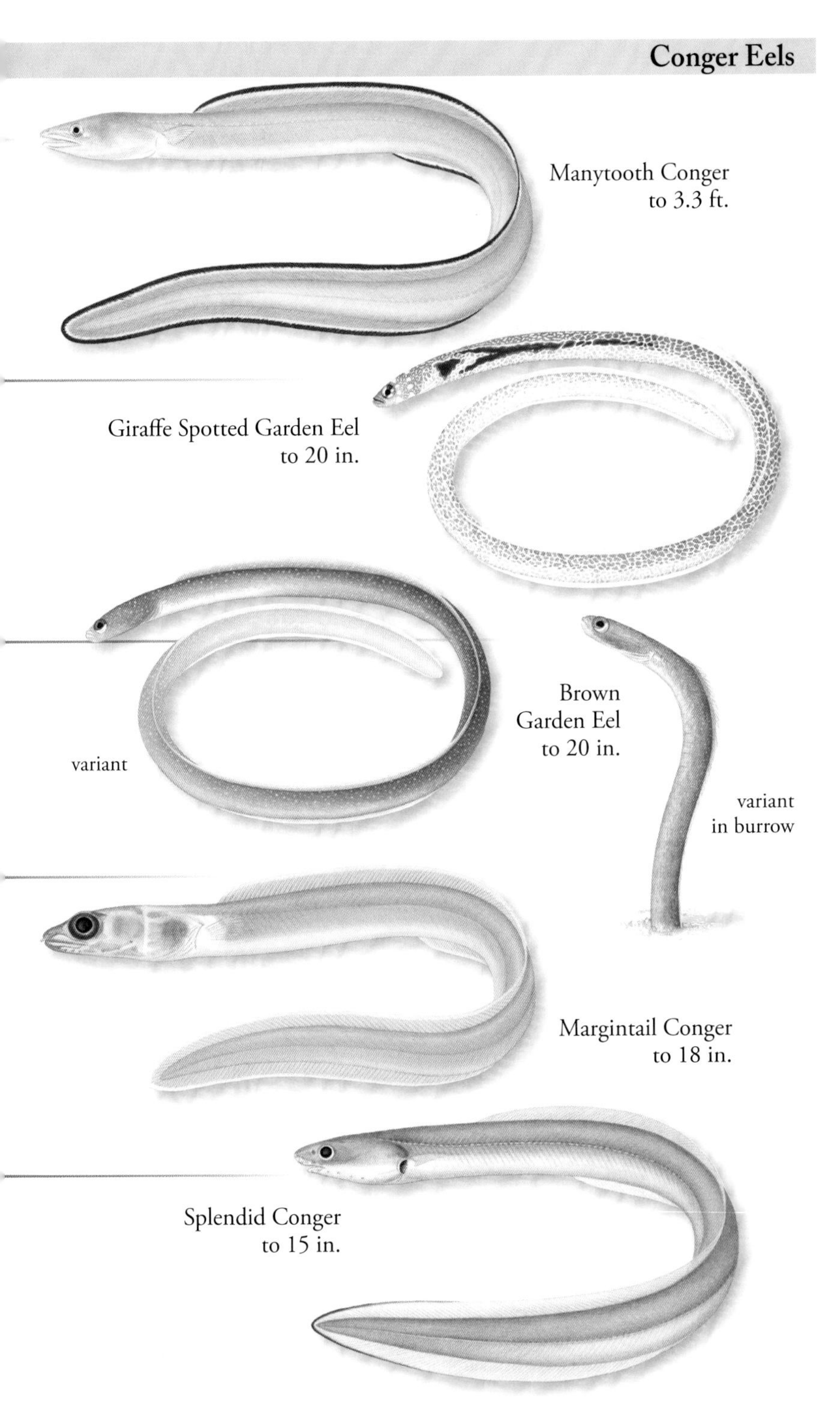
Manytooth Conger
to 3.3 ft.
Giraffe Spotted Garden Eel
to 20 in.
variant
Brown
Garden Eel
to 20 in.
variant
in burrow
Margintail Conger
to 18 in.
Splendid Conger
to 15 in.

Congridae - Conger Eels, *cont.*

Yellow Conger - *Rhynchoconger flavus* (Goode & Bean, 1896)

FEATURES: Shades of yellowish, yellowish brown, greenish, or brownish. Abdomen pale. Stomach and intestine pale. Dorsal- and anal-fin margins black posteriorly. Snout long, fleshy, and projecting. Upper and lower lips with flanges. Underside of snout with a round tooth patch that is exposed when mouth is closed. Eyes large. Dorsal-fin origin over or just anterior to pectoral-fin base. HABITAT: Gulf of Mexico and Caribbean Sea to mouth of Amazon River. Occur on soft bottoms of coastal waters from about 85 to 600 ft.

Threadtail Conger - *Uroconger syringinus* Ginsburg, 1954

FEATURES: Gray to brown dorsally. Silvery white along abdomen. Dorsal- and anal-fin margins black posteriorly. Snout long, projecting, with front teeth exposed. Flange on lower jaw only. Eyes large. Pectoral fins well developed. Dorsal-fin origin over pectoral-fin base. Top of gill opening below top of pectoral-fin base. Tail very long, tapered, and whip-like. HABITAT: FL, Bahamas, Gulf of Mexico, and Caribbean Sea to Suriname. Also in eastern Atlantic. Demersal over soft bottoms from about 140 to 1,260 ft.

ALSO IN THE AREA: *Ariosoma anale*, see p. 547.

Nettastomatidae - Duckbill Eels

Freckled Pike Conger - *Hoplunnis macrura* Ginsburg, 1951

FEATURES: Dusky dorsally; silvery on sides and below. Small brown spots pepper top of head and back. Dorsal- and anal-fin margins black posteriorly. Inner surface of gill chamber black. Stomach pale. Snout long; jaws large. Dorsal-fin origin anterior to gill opening. Dorsal and anal fins confluent with tail. Pectoral fins present. Anus located anterior to midbody. Body very long and slender. HABITAT: FL and Gulf of Mexico to French Guiana. Absent from western Caribbean Sea. Demersal over silty and muddy bottoms from about 65 to 1,000 ft.

Blackfin Sorcerer - *Nettastoma melanura* Rafinesque, 1810

FEATURES: Silvery to pearly white in life. Brown after capture. Dorsal and anal fins transparent; black posteriorly. Snout long, cylindrical anteriorly, broad and flattened at rear. Jaws large with bands of sharp teeth. Eyes large. Dorsal-fin origin over or behind gill opening. Dorsal and anal fins confluent with tail. Pectoral fins absent. Body very long, serpent-like, and very slender at tail. HABITAT: FL, Gulf of Mexico, and coasts of Central and South America to Brazil. Also Virgin Islands and eastern Bahamas. Demersal over soft silty and muddy bottoms from about 120 to 5,400 ft. Usually deeper than 1,000 ft.

Conger Eels

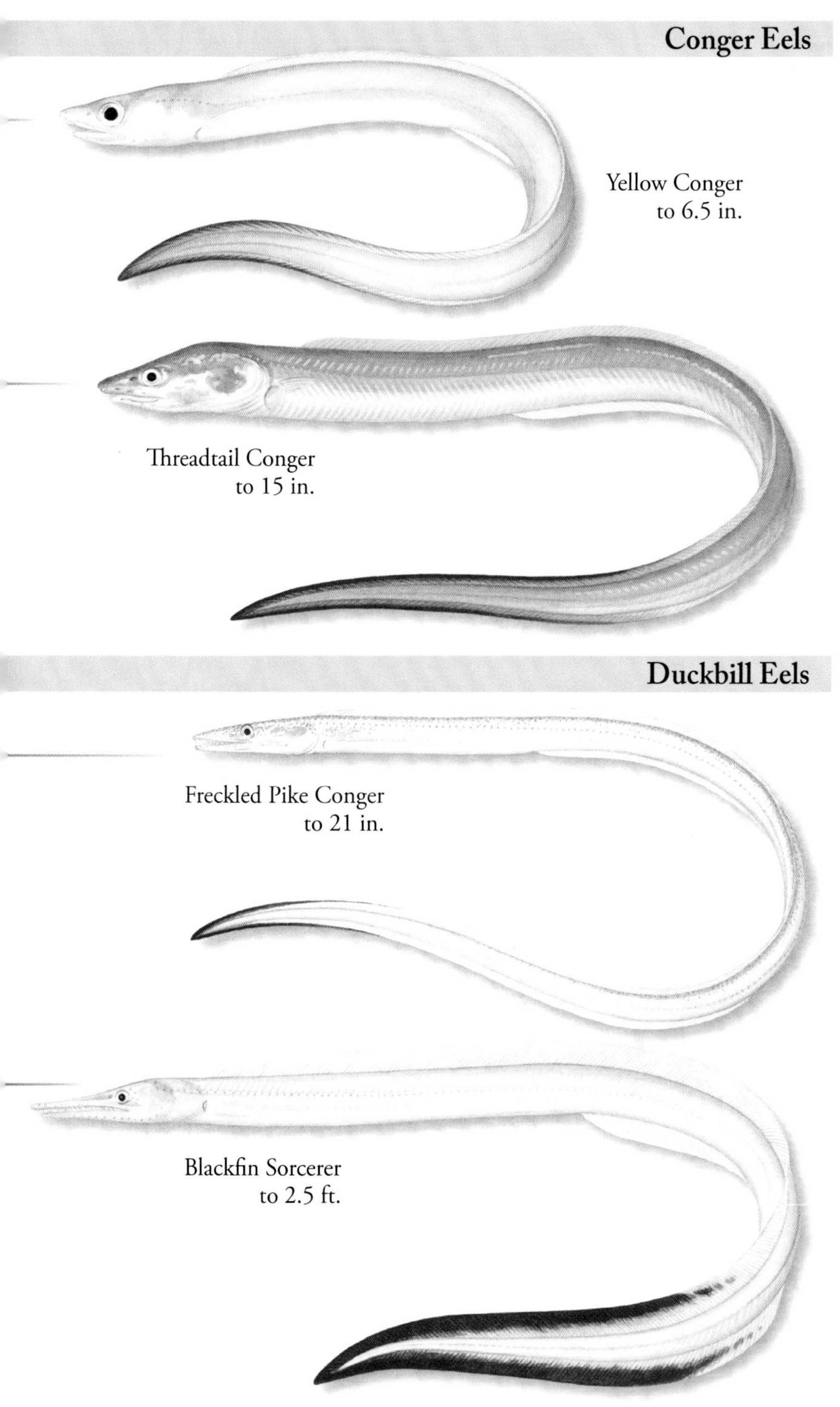

Yellow Conger
to 6.5 in.
Threadtail Conger
to 15 in.
Duckbill Eels
Freckled Pike Conger
to 21 in.
Blackfin Sorcerer
to 2.5 ft.

Pristigasteridae - Longfin Herrings

Dogtooth Herring - *Chirocentrodon bleekerianus* (Poey, 1867)

FEATURES: Translucent with a silvery band along midside that is about three-quarters of eye diameter at its deepest. Head and abdominal area silvery. A row of fine, dark spots present above anal fin and below dorsal fin. Fins mostly colorless. Jaws very large with enlarged canine-like teeth at tip and a row of large and small sharp teeth along lower edge of maxilla. Pectoral fins small. Pelvic fins small. Dorsal-fin origin behind anal-fin origin. Anal fin long-based with 35–41 rays. Abdomen with a row of scutes. HABITAT: Greater Antilles and eastern Honduras to Brazil. In estuaries, river mouths, lagoons, and along the coast from surface to about 130 ft.

Caribbean Longfin Herring - *Odontognathus compressus* Meek & Hildebrand, 1923

FEATURES: Translucent with a narrow, silvery band along midside. Head and abdominal area silvery. A row of fine, dark spots above anal fin and a dark line along midline of back present. Caudal-fin base dark, otherwise fins mostly colorless. Jaws moderately large, oblique. Maxilla pointed at rear tip. Top of head concave. Pectoral fins large. Pelvic fins absent. Dorsal fin over middle of anal fin. Anal fin long-based with 58–62 rays. Abdomen with a row of scutes. HABITAT: Costa Rica to Suriname. Found over muddy bottoms in estuaries and along inshore waters from surface to about 100 ft.

Guiana Longfin Herring - *Odontognathus mucronatus* Lacepède, 1800

FEATURES: Translucent with narrow whitish band along midside. Head and abdominal area silvery. Midline of head and back dark. Anal fin dark; other fins colorless. Jaws moderately large and oblique. Maxilla sharply pointed at rear tip. Top of head concave. Pectoral fins large. Pelvic fins absent. Dorsal fin small, located above middle of anal fin. Anal fin very long-based with 70–85 rays. Caudal peduncle comparatively narrow. Abdomen with a row of scutes. HABITAT: Gulf of Paria and Trinidad to southern Brazil. Found over sandy and muddy bottoms in estuaries and shallow coastal waters from surface to about 100 ft. Also ascends rivers.

American Coastal Pellona - *Pellona harroweri* (Fowler, 1917)

FEATURES: Bluish gray dorsally; silvery on sides and below. Dorsal, anal, and caudal fins yellowish. Dorsal-fin tip and caudal-fin margin dusky. Mouth oblique; lower jaw protrudes. Teeth minute. Eyes large. Pectoral fins low on the body. Pelvic fins small. Dorsal fin over midbody. Anal fin low and long-based with 36–38 rays. Upper head profile nearly straight. Abdomen with a row of scutes. Body deep and compressed. Scales large and easily shed. HABITAT: Honduras to southern Brazil. Occur over muddy bottoms in estuaries and inshore waters from surface to about 115 ft.

ALSO IN THE AREA: *Neoopisthopterus cubanus*, see p. 547.

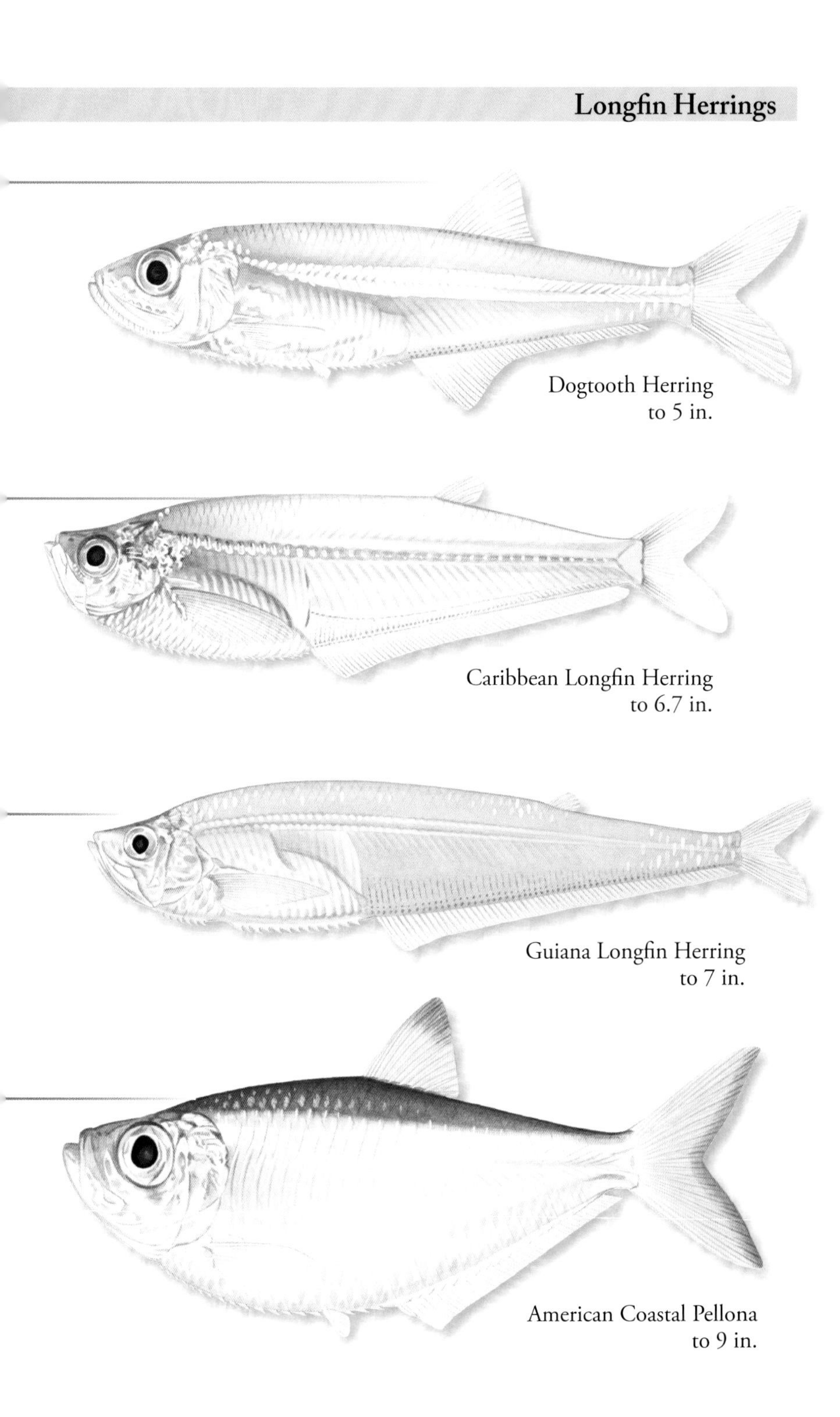

Dogtooth Herring
to 5 in.

Caribbean Longfin Herring
to 6.7 in.

Guiana Longfin Herring
to 7 in.

American Coastal Pellona
to 9 in.

Engraulidae - Anchovies

Key Anchovy - *Anchoa cayorum* (Fowler, 1906)

FEATURES: Translucent with silver on head and jaws. Width of silvery midbody stripe slightly less than eye diameter. Snout conical, its length shorter than eye diameter. Maxilla long, pointed at rear, with tip reaching almost to gill opening. Anal fin long-based with 21–25 rays, its origin anterior to middle of dorsal fin. HABITAT: FL Keys, Bahamas, and Caribbean Sea. Pelagic over continental shelves and in clear, oceanic waters. BIOLOGY: Often in large schools. Feed on zooplankton.

Narrowstriped Anchovy - *Anchoa colonensis* Hildebrand, 1943

FEATURES: Translucent grayish green to bluish green dorsally; translucent below. Silvery on head and jaws. Width of silvery midbody stripe about equal to pupil diameter. Scale margins on back stippled with black. Snout pointed, about as long as three-quarters of eye diameter. Maxilla extends to lower margin of preopercle. Anal fin short-based with 17–21 rays, its origin under middle of dorsal fin. Anus closer to anal-fin origin than pelvic-fin tips. HABITAT: Jamaica, Haiti, Dominican Republic, Puerto Rico, and Bay of Campeche to Trinidad. Occur coastally over sandy and gravel bottoms in shallow waters. BIOLOGY: Feed on zooplankton.

Striped Anchovy - *Anchoa hepsetus* (Linnaeus, 1758)

FEATURES: Dusky, translucent dorsally. Abdomen and ventral area pale. Silvery on head and jaws. Width of silvery midbody stripe about equal to eye diameter. Caudal fin may have dark or dusky margin. Maxilla extends past preopercular margin. Anal fin short-based with 16–23 rays, its origin under rear of dorsal fin. HABITAT: Nova Scotia to FL, northern Gulf of Mexico, Cuba, Venezuela to S Brazil. In bays, in estuaries, and along coast from shore to about 230 ft. BIOLOGY: Schooling and tolerant of varying salinities. Feed on zooplankton.

Bigeye Anchovy - *Anchoa lamprotaenia* Hildebrand, 1943

FEATURES: Translucent gray dorsally. Silvery on head and jaws. Silvery midbody stripe with dark upper margin, its width about three-quarters of eye diameter. Scales along back with black flecks. Snout pointed, about as long as three-quarters of eye diameter. Maxilla extends to posterior preopercular margin. Anal-fin base moderately long with 18–23 rays, its origin under middle of dorsal fin. Anus closer to pelvic-fin tips than anal-fin origin. HABITAT: S FL, Bahamas, and Caribbean Sea to Brazil. Occur coastally over continental shelves in clear waters. Not found in low-salinity waters.

Dusky Anchovy - *Anchoa lyolepis* (Evermann & Marsh, 1900)

FEATURES: Translucent gray dorsally; silvery on head and jaws. Silvery midbody stripe narrow anteriorly, depth about equal to eye diameter at midbody, its upper margin often dark. Snout long, prominent. Head comparatively long. Maxilla extends almost to gill opening. Anal-fin base moderately long with 19–24 rays, its origin under rear portion of dorsal fin. HABITAT: NY to FL, northern Gulf of Mexico to Venezuela, and Antilles. Occur coastally in salt water from shore to about 75 ft. BIOLOGY: Often form dense schools. Feed on plankton. Used as bait.

Key Anchovy
to 3.5 in.

Narrowstriped Anchovy
to 3.5 in.

Striped Anchovy
to 6 in.

Bigeye Anchovy
to 4.7 in.

Dusky Anchovy
to 4.7 in.

Bay Anchovy - *Anchoa mitchilli* (Valenciennes, 1848)

FEATURES: Dusky, translucent with greenish tinge dorsally. Abdomen translucent. Sides whitish or pale. Silvery on head and jaws. Width of silvery midbody stripe about as wide as pupil diameter. Fins speckled with tiny spots. Snout blunt and short. Head comparatively short. Maxilla extends to preopercular opening. Anal fin long-based with 23–30 rays, its origin under anterior dorsal-fin rays. HABITAT: ME to FL and Gulf of Mexico to Yucatán. Near shore, coastal, and in bays and estuaries to about 118 ft. BIOLOGY: Bay Anchovy are schooling and feed on zooplankton. An important prey of larger fishes and birds.

Little Anchovy - *Anchoa parva* (Meek & Hildebrand, 1923)

FEATURES: Translucent gray dorsally; silvery on head and jaws. Width of silvery midbody stripe about equal to pupil diameter. Snout pointed, about as long as half of eye diameter. Maxilla reaches just past lower preopercular margin. Anal-fin base moderately long, with 21–25 rays, its origin under middle of dorsal fin. Anus closer to pelvic-fin tips than anal-fin origin. HABITAT: Jamaica, Haiti, Dominican Republic, Puerto Rico, and Bay of Campeche to Trinidad. Occur in brackish lagoons and shallow coastal waters. Also in fresh water.

Spicule Anchovy - *Anchoa spinfer* (Valenciennes, 1848)

FEATURES: Bluish to greenish dorsally. Silvery head, sides, and below. Small specimens with a distinct silvery stripe on sides that fades with age. Caudal fin yellowish. Tip of dorsal and margin of caudal fin dark. Snout bluntly pointed. Maxilla extends almost to gill opening. Sharp "spicule" at lower edge of opercle. Anal fin long-based with 36–40 rays, its origin under middle of dorsal fin. HABITAT: Honduras to Brazil. Occur in estuaries, river mouths, lagoons, and along coast from shore to about 180 ft. Occasionally in fresh water.

Broadband Anchovy - *Anchoviella lepidentostole* (Fowler, 1911)

FEATURES: Translucent gray dorsally; silvery on head and jaws. Silvery midbody stripe narrow anteriorly, equal to eye diameter at its widest. Caudal fin yellowish with dark margin. Tip of maxilla not reaching past preopercular margin. Axillary scale above pectoral fins nearly reaches tips. Anal fin long-based with 22–25 rays, its origin under middle of dorsal fin. HABITAT: Colombia to Brazil. Occur in coastal waters, estuaries, and river mouths from shore to about 164 ft.

Flat Anchovy - *Anchoviella perfasciata* (Poey, 1860)

FEATURES: Translucent gray dorsally; silvery on head and jaws. Silvery midbody stripe three-quarters of eye diameter or equal to eye diameter. Snout pointed, its length about three-quarters of eye diameter. Maxilla not reaching lower preopercular margin. Anal fin short-based with 16–19 rays, its origin under rear of dorsal fin. Body elongate with area of back above midbody stripe more narrow than stripe. HABITAT: NC to FL, Gulf of Mexico, Greater and Lesser Antilles, and northern coast of South America. Occur in shallow coastal waters.

Bay Anchovy
to 4 in.

Little Anchovy
to 3 in.

Spicule Anchovy
to 9.5 in.

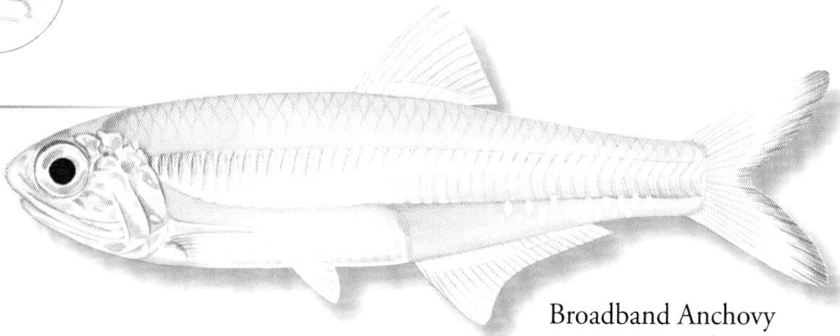

Broadband Anchovy
to 4.3 in.

Flat Anchovy
to 4.3 in.

Engraulidae - Anchovies, *cont.*

Atlantic Anchoveta - *Cetengraulis edentulus* (Cuvier, 1829)

FEATURES: Dark greenish blue dorsally; silvery on head and sides. Small specimens with a silvery stripe on sides that fades with age. Caudal fin yellowish to colorless. Snout sharply pointed. Maxilla rounded posteriorly, its tip well before preopercular margin. Anal fin moderately long-based, with 18–24 rays, its origin under rear of dorsal fin. Body deep over pelvic fins. HABITAT: Southern Gulf of Mexico and Caribbean Sea to southern Brazil. Occur in shallow coastal waters, along beaches, in lagoons, and in brackish water from surface to about 82 ft. BIOLOGY: Often form large schools.

Silver Anchovy - *Engraulis eurystole* (Swain & Meek, 1884)

FEATURES: Translucent blue green dorsally; silvery on head and jaws. Sides silvery or with broad, silver stripe that has a dark upper edge. Snout prominent and pointed. Maxilla bluntly tipped and extends to anterior margin of preopercle. Anal fin short-based with 13–15 rays and entirely posterior to dorsal-fin base. Body elongate. HABITAT: MA to Gulf of Mexico, and Venezuela to northern Brazil. Pelagic in shallow coastal and protected waters from near shore to about 200 ft. BIOLOGY: Feed on plankton and form large, compact schools. Of minor commercial value.

Atlantic Sabretooth Anchovy - *Lycengraulis grossidens* (Spix & Agassiz, 1829)

FEATURES: Bluish green or grayish blue to brownish dorsally. Silvery head and sides. Short arc of small dark spots usually on upper portion of opercle. Caudal fin often yellowish with dark margin. Snout short and blunt. Maxilla pointed at tip and reaches lower opercular margin. Lower jaw teeth enlarged, widely spaced, and canine-like. Anal fin long-based, with 26–28 rays, its origin under middle of dorsal fin. HABITAT: Belize to Brazil. Occur in coastal waters, lagoons, estuaries, river mouths, and in fresh water from shore to about 130 ft.

NOTE: There are at least six other poorly recorded Anchovies in the area.

Clupeidae - Herrings

Finescale Menhaden - *Brevoortia gunteri* Hildebrand, 1948

FEATURES: Grayish or dusky dorsally from eye level to tail. Lower sides and ventral area silvery. Single dark spot present behind upper opercular margin. No other spots present on sides. Dorsal and caudal fins dusky or yellowish. Head comparatively large. Mouth oblique. Upper jaw with a median notch. Eyes with adipose lids. Abdomen with a row of 27–30 scutes. Body deep, compressed. HABITAT: In the Gulf of Mexico from Chandeleur Sound, LA, to Gulf of Campeche, Mexico. Occur inshore along coast in shallow marine and brackish water. BIOLOGY: Spawn near shore and in bays during winter and early spring. Feed on plankton.

Anchovies

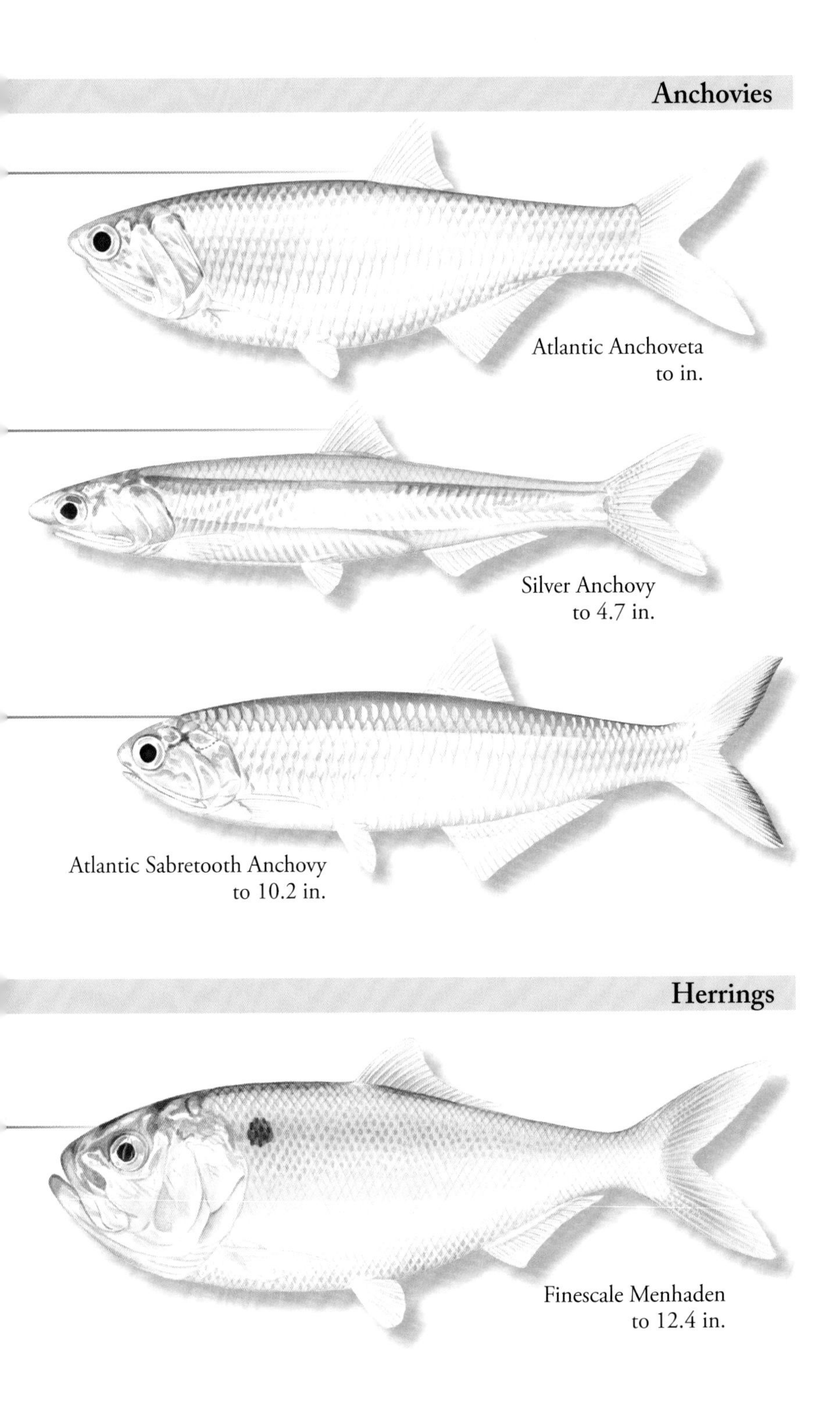

Atlantic Anchoveta
to in.

Silver Anchovy
to 4.7 in.

Atlantic Sabretooth Anchovy
to 10.2 in.

Herrings

Finescale Menhaden
to 12.4 in.

Gulf Menhaden - *Brevoortia patronus* Goode, 1878

FEATURES: Green blue to bluish gray dorsally. Sides brassy or silvery. Ventral area silvery. Large, dark spot behind upper opercular margin followed by single, double, or multiple rows of smaller, paler spots. Dorsal and caudal fins yellowish to brassy. Caudal fin with dusky margin. Head large. Mouth oblique. Upper jaw with a median notch. Eyes relatively large. Abdomen with a row of scutes. Body deep, compressed. HABITAT: In the Gulf of Mexico from Florida Bay to Gulf of Campeche, Mexico. Occur primarily in shallow marine waters along coast. BIOLOGY: Spawn near or off shore. Eggs hatch at sea and are carried inshore by currents. Caught commercially.

Threadfin Shad - *Dorosoma petenense* (Günther, 1867)

FEATURES: Olivaceous to bluish black with golden highlights dorsally. Sides and ventral area silvery. Large, dark spot behind upper opercular margin. Dorsal fin olivaceous with single long, trailing ray. Head relatively small; eyes large. Mouth small, slightly oblique. Upper jaw with a median notch. Abdomen with a row of scutes. Body deep, compressed. HABITAT: Gulf of Mexico from FL to N Guatemala. Pelagic, primarily in quiet fresh water. Also found in brackish to saltwater estuaries and bays. BIOLOGY: Anadromous and schooling. Primarily a filter feeder.

Round Herring - *Etrumeus sadina* (Mitchill, 1814)

FEATURES: Olive green dorsally. Sides and ventral area silvery. Head small. Eyes large. Dorsal fin tall, slightly forward of midbody. Single scute present before pelvic fins. Body elongate, cylindrical. HABITAT: Bay of Fundy to FL and Gulf of Mexico. Also Cuba and northern coast of South America. Usually occur over deep water off coast and along continental shelves and slopes. Pelagic. BIOLOGY: Migrate vertically in the water column. Spawning occurs at night. Females cast up to 19,000 eggs. NOTE: Previously known as *Etrumeus teres.*

False Pilchard - *Harengula clupeola* (Cuvier, 1829)

FEATURES: Blue green with faint streaks along scales dorsally and on upper sides. Silvery below. Faint orange to yellow spot followed by faint dark spot behind upper opercular margin. Fins colorless. Head and eyes large. Upper jaw lacks notch. Abdomen with row of scutes. HABITAT: FL Keys, Gulf of Mexico, and Caribbean Sea to northern Brazil. Pelagic in shallow coastal waters, estuaries, bays, and lagoons. Occur in turbid and clear water. BIOLOGY: Feed on plankton.

Redear Herring - *Harengula humeralis* (Cuvier, 1829)

FEATURES: Iridescent blue green dorsally with three or four narrow orange stripes along scales dorsally. Silvery below. Orange spot behind upper opercular margin. Dorsal-fin tip and caudal-fin margins dark. Head and eyes large. Upper jaw lacks notch. Abdomen with a row of scutes. HABITAT: Bermuda, S FL, northern and southern Gulf of Mexico, Bahamas, and Caribbean Sea. In clear coastal waters around reefs and over seagrass beds from surface to about 30 ft. BIOLOGY: Feed on plankton.

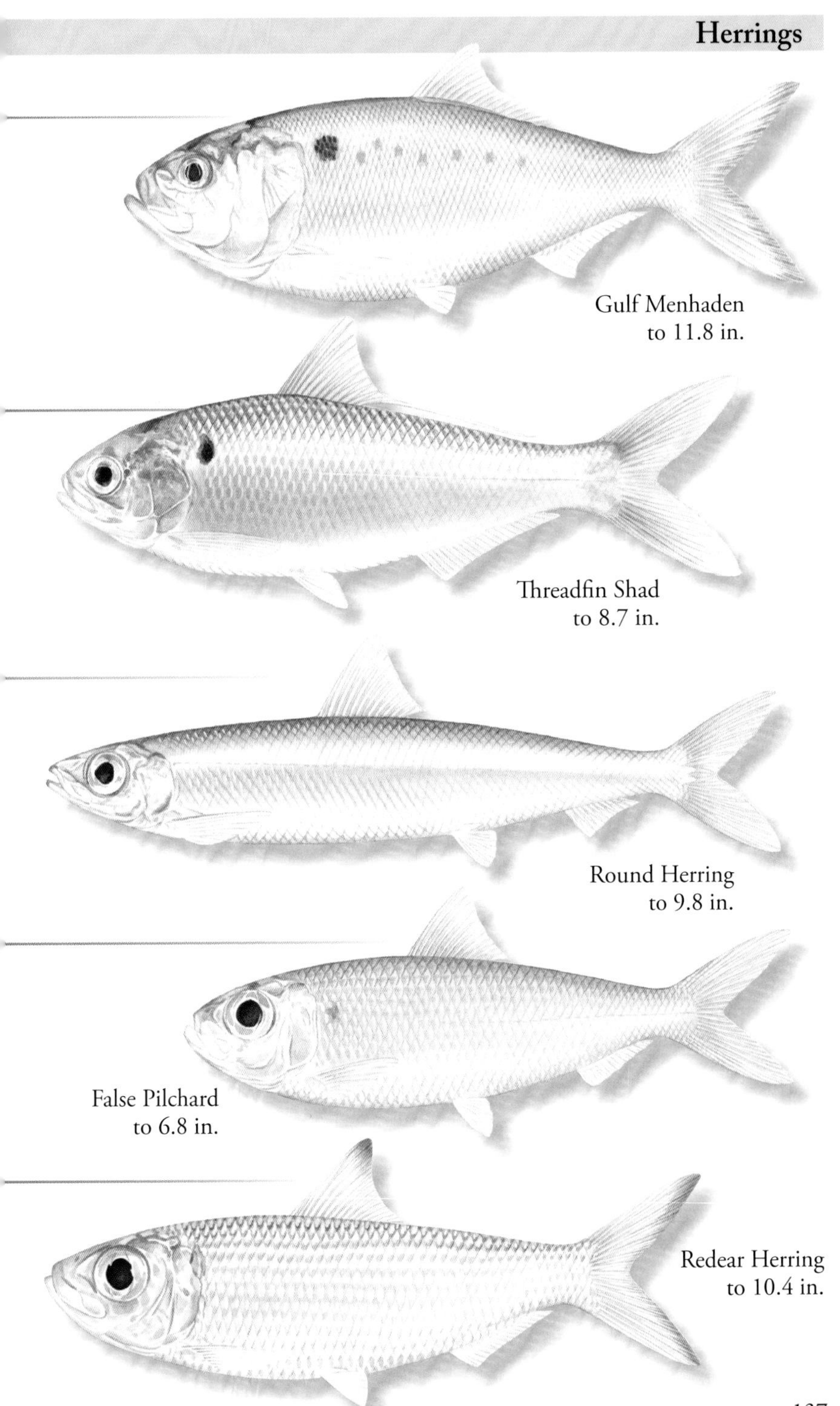
Gulf Menhaden
to 11.8 in.
Threadfin Shad
to 8.7 in.
Round Herring
to 9.8 in.
False Pilchard
to 6.8 in.
Redear Herring
to 10.4 in.

Scaled Sardine - *Harengula jaguana* Poey, 1865

FEATURES: Bluish gray, bluish black, or brownish dorsally with faint streaks along scales. Silvery below. Small, faint to conspicuous spot behind upper opercular margin. Caudal-fin margin often dusky. Head and eyes large. Upper jaw lacks a notch. Abdomen with a row of scutes. HABITAT: NJ to FL, Gulf of Mexico, Bahamas, and Caribbean Sea to Brazil. Pelagic or near sandy to muddy bottoms along coast. Also in bays, estuaries, and hypersaline lagoons. BIOLOGY: Schooling. Often targeted or caught as bycatch.

Dwarf Herring - *Jenkinsia lamprotaenia* (Gosse, 1851)

FEATURES: Pale yellowish green dorsally. Silvery white ventrally. Head silvery. Broad silvery stripe with bright blue margin on sides that is not narrow anteriorly. Small, dark spots on snout, back, and bases of dorsal and anal fins. Mouth oblique. Eyes large. Anal fin small, entirely posterior to dorsal fin. Body elongate. HABITAT: FL, Gulf of Mexico, Bermuda, and Caribbean Sea to Venezuela. Near surface, coastal, inshore, and in bays and estuaries. BIOLOGY: Form large, migratory schools. Feed on zooplankton.

Shortband Herring - *Jenkinsia stolifera* (Jordan & Gilbert, 1884)

FEATURES: Translucent green dorsally. Silvery white ventrally. Head silvery. Broad silvery stripe with bright blue margin on sides that is not narrow anteriorly. Small dark spots on scale margins. Mouth oblique. Eyes large. Anal fin small, entirely posterior to dorsal fin. Body elongate. HABITAT: FL Keys, Cayman Islands, Jamaica, Puerto Rico, British Virgin Islands to Tobago; coasts of Belize to Panama, and Venezuela to Trinidad. Occur in shallow inshore waters near surface.

Atlantic Piquitinga - *Lile piquitinga* (Schreiner & Miranda Ribeiro, 1903)

FEATURES: Greenish to blue green dorsally with iridescent highlights. Silvery on ventral area. Broad silvery stripe on sides about three-quarters of eye diameter. Dark to black spot on upper rear caudal peduncle. Mouth oblique. Upper jaw lacks median notch. Eyes large. Abdomen with a row of strong scutes. HABITAT: Southern Costa Rica to Brazil. Occur over shallow soft bottoms from estuaries and lagoons to protected coastal waters. Also enter fresh water.

Atlantic Thread Herring - *Opisthonema oglinum* (Lesueur, 1818)

FEATURES: Blue dorsally and on upper sides; may have dark spots forming lines along scales. Silvery below. Dark spot behind upper opercular margin. Spot may be followed by row of smaller spots. Dorsal and caudal fins tipped black. Mouth small, oblique. Last ray of dorsal fin long and trailing. Anal fin small. Abdomen with a row of scutes. Body deep, oval in profile. HABITAT: Gulf of Maine to FL, Gulf of Mexico, and Caribbean Sea to Brazil. Also Bermuda. Pelagic and coastal. BIOLOGY: Migratory and schooling.

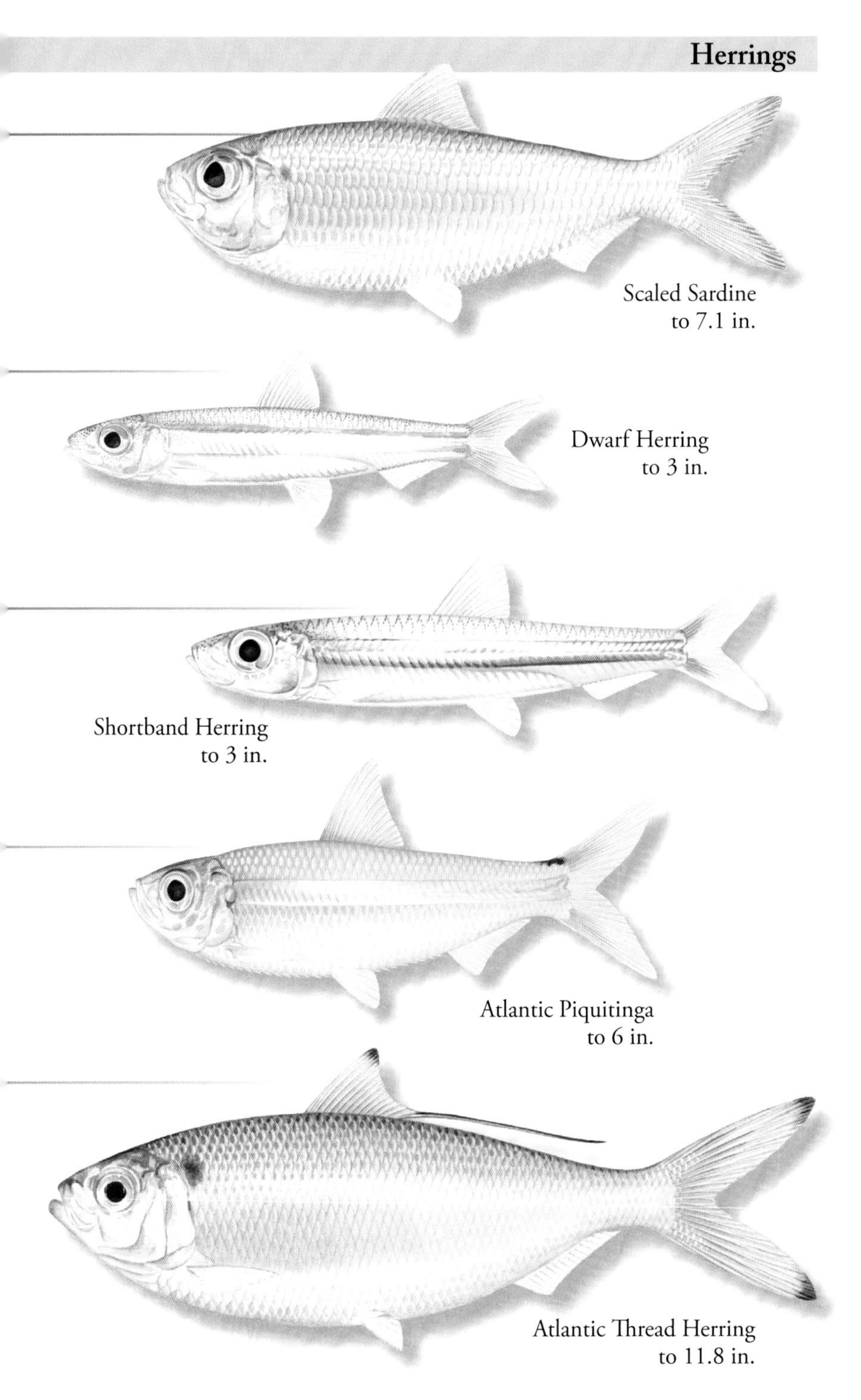
Scaled Sardine
to 7.1 in.
Dwarf Herring
to 3 in.
Shortband Herring
to 3 in.
Atlantic Piquitinga
to 6 in.
Atlantic Thread Herring
to 11.8 in.

Clupeidae - Herrings, *cont.*

Spanish Sardine - *Sardinella aurita* Valenciennes, 1847

FEATURES: Iridescent blue to green dorsally. Silvery below. Usually with dark spot on opercular margin. Faint, golden spot behind opercular margin followed by faint gold stripe on sides. Mouth oblique. Last two anal-fin rays elongate. Abdomen with a row of scutes. Body elongate. HABITAT: MA to FL, Gulf of Mexico, Bermuda, Bahamas, and Caribbean Sea to Argentina. Also along the eastern Atlantic coast and in the Mediterranean Sea. Coastal and pelagic in warm temperate to tropical marine waters from inshore to edge of continental shelves. BIOLOGY: Schooling and migratory.

ALSO IN THE AREA: *Jenkinsia majua*, *Jenkinsia parvula*, see p. 547.

Aspredinidae - Banjo Catfishes

Sevenbarbed Banjo - *Aspredinichthys filamentosus* (Valenciennes, 1840)

FEATURES: Shades of brown dorsally. May have paler blotches on tail. Anal fin dark anteriorly, pale posteriorly. Head and body broad and flattened, tapered to a whip-like tail. Eyes tiny. Pair of long barbels with a small accessory barbel over mouth. About seven pairs of short barbels under head and belly. Head with a bony shield. Dorsal fin with thread-like first ray. Pectoral fins with a strong serrated spine. HABITAT: Orinoco River to the Amazon River. Demersal over shallow muddy bottoms in estuaries and river mouths. BIOLOGY: Females brood eggs on abdomen.

Banded Banjo - *Platystacus cotylephorus* Bloch, 1794

FEATURES: Shades of brown to blackish dorsally. Often with irregular, pale, and diffuse or distinct mottling. Fins dark. Head and body broad and flattened, tapered to a whip-like tail. Eyes tiny. Pair of long, broad barbels over mouth joined to side of head by a membrane. Two pairs of short barbels under head and belly. Three to four low ridges along tail. Pectoral fins with a strong serrated spine. HABITAT: Orinoco River to the Amazon River. Demersal over shallow muddy bottoms in estuaries and river mouths. BIOLOGY: Females brood eggs on abdomen; each egg is attached to abdomen by a thread-like "cotylephore."

Auchenipteridae - Driftwood Catfishes

Cocosoda Catfish - *Pseudauchenipterus nodosus* (Bloch, 1794)

FEATURES: Color and pattern vary. Bluish gray to black dorsally. Creamy yellow to pinkish or white below. Caudal-fin margin dusky to black. Upper back with several rows of pale spots. Snout blunt, rounded. Eyes large, bulging. Two pairs of barbels under chin. Maxillary barbels thread-like and not reaching tip of pectoral spine. Head shield with a honeycomb pattern. Base of dorsal-fin spine enlarged in mature specimens. Lateral line pale, wavy, and branched at caudal-fin base. HABITAT: Gulf of Paria and Trinidad to N Brazil. Demersal over muddy bottoms in shallow brackish river mouths and low-salinity coastal waters.

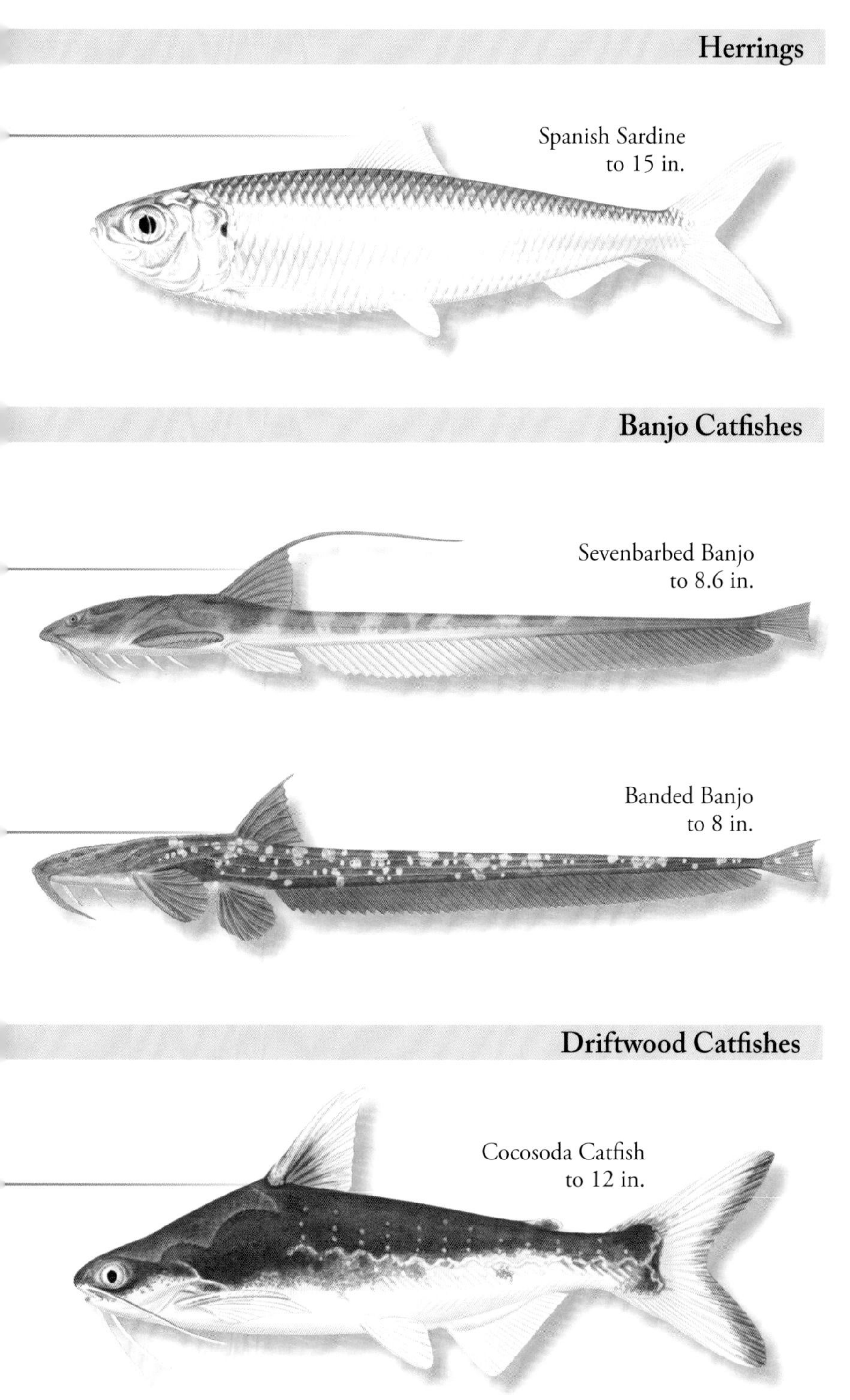
Herrings
Spanish Sardine
to 15 in.
Banjo Catfishes
Sevenbarbed Banjo
to 8.6 in.
Banded Banjo
to 8 in.
Driftwood Catfishes
Cocosoda Catfish
to 12 in.

Ariidae - Sea Catfishes

Hardhead Catfish - *Ariopsis felis* (Linnaeus, 1766)

FEATURES: Dark blue to brown dorsally. Whitish ventrally. Adipose fin blackish. Head rounded, slightly flattened. Narrow, fleshy groove along top of head. Fleshy groove between nostrils absent. Head shield rugous, extends to over eyes. Two pairs of barbels under mouth. Moderately long barbel on both sides of mouth that extends to pectoral fin. Dorsal and pectoral fins with a serrated spine. HABITAT: NC to FL and Gulf of Mexico to Yucatán. Along coast over muddy bottoms in turbid marine and brackish waters. Occasionally in fresh water. BIOLOGY: Prey on bottom-dwelling invertebrates. Spines are venomous.

Coco Sea Catfish - *Bagre bagre* (Linnaeus, 1766)

FEATURES: Silvery gray to bluish gray dorsally. Pale to white ventrally. Anal fin with large gray to black blotch anteriorly. Fleshy groove between nostrils absent. Profile slightly depressed at nape. Head shield smooth, with a crescent-shaped plate at dorsal-fin base. Single pair of barbels under mouth. Long barbel on both sides of mouth that is flap-like with a long filament that extends past pelvic fins. Head shield smooth. Dorsal and pectoral fins with a very long filament. HABITAT: Colombia to Brazil. Occur coastally and around river mouths from shore to about 165 ft. BIOLOGY: Prey on small fishes and invertebrates.

Gafftopsail Catfish - *Bagre marinus* (Mitchill, 1815)

FEATURES: Bluish gray to dark brown dorsally. White to pale ventrally. Fleshy groove between nostrils absent. Head slightly arched and flattened. Single pair of barbels under mouth. Long barbel on both sides of mouth that is flap-like with a filament that may extend to pelvic fins. Dorsal and pectoral fins long and trailing and with a serrated spine. HABITAT: MA to FL and Gulf of Mexico to Brazil. Occur along coast in marine and brackish waters to about 160 ft. Also in hypersaline lagoons. BIOLOGY: Feed on small fishes and invertebrates. Males mouth-brood eggs. Spines are venomous.

Thomas Sea Catfish- *Notarius grandicassis* (Valenciennes, 1840)

FEATURES: Shades of yellow brown to gray brown dorsally. May have a golden luster. Golden to white below. Snout roundly pointed, relatively long, and overhangs mouth. Fleshy groove between nostrils absent. Two pairs of barbels under mouth. Long barbel on both sides of mouth that reaches pectoral fin. Triangular and pointed bony plate present above pectoral fins. Lateral line forks before caudal-fin base. HABITAT: Occur in shallow coastal waters of South America from the Guajira Peninsula, Colombia, to Brazil. Found from shore to about 115 ft. Also in turbid estuaries.

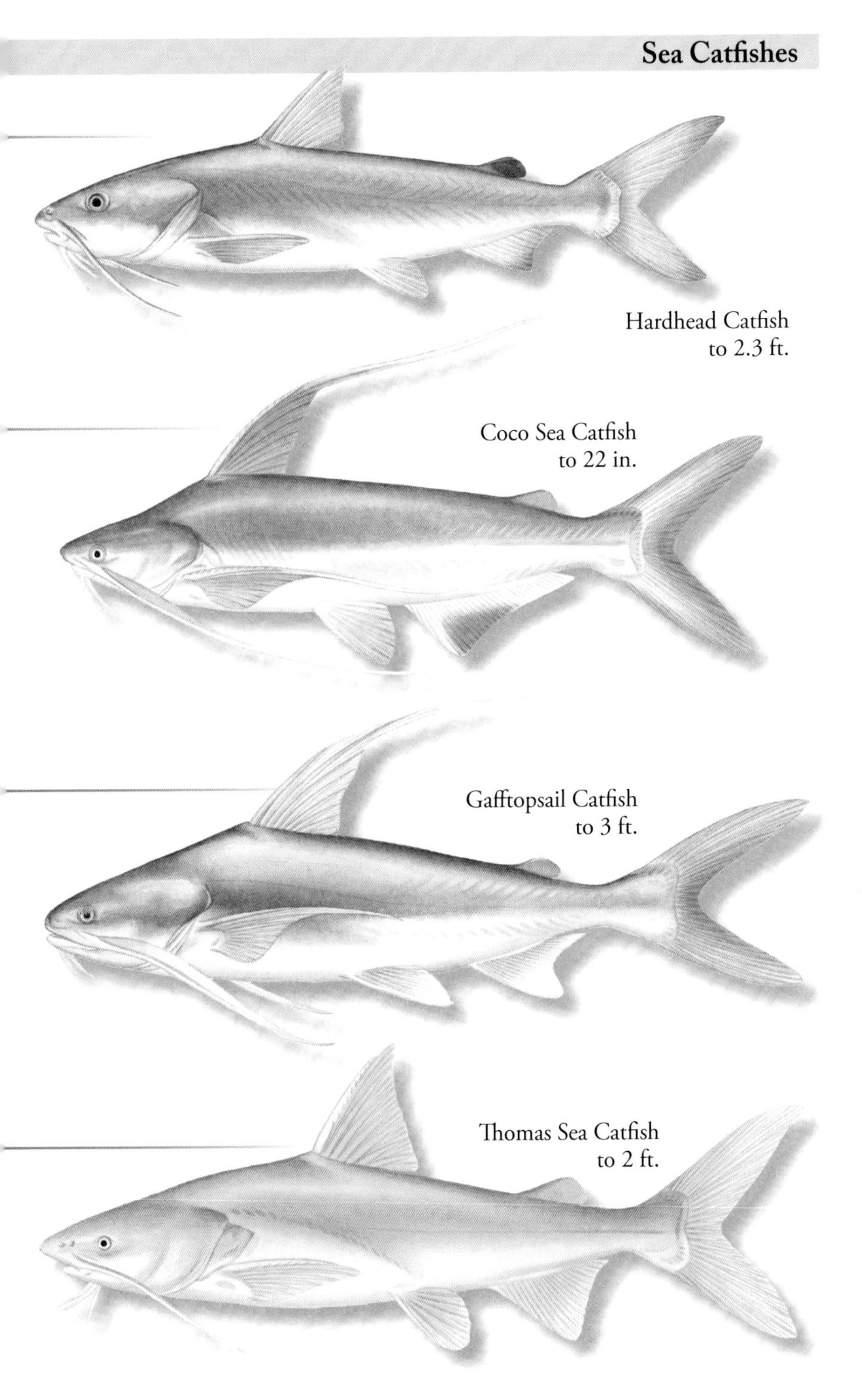
Hardhead Catfish
to 2.3 ft.
Coco Sea Catfish
to 22 in.
Gafftopsail Catfish
to 3 ft.
Thomas Sea Catfish
to 2 ft.

Pemecou Sea Catfish - *Sciades herzbergii* (Bloch, 1794)

FEATURES: Color varies. Shades of dark brown or gray brown to dark gray dorsally. Whitish below. Snout long, rounded, and overhangs mouth. Posterior nostrils connected by a shallow groove partially covered by skin. Two pairs of barbels under mouth. Long barbel on both sides of mouth reaches to or beyond pectoral fin. Head slightly flattened at top. Head shield very rough, extends to between eyes and has a short and crescent-shaped plate before first dorsal-fin spine. Lateral line forked before caudal-fin base. HABITAT: Colombia to Brazil. Demersal in estuaries, mangrove-lined lagoons, lower parts of rivers, and occasionally in shallow marine waters to about 65 ft.

Crucifix Sea Catfish - *Sciades proops* (Valenciennes, 1840)

FEATURES: Color varies. Shades of gray, brown, or blue dorsally. Paler to white below. Snout very broad, short, and nearly squared. Two pairs of barbels under mouth. Long barbel on both sides of mouth reaches pectoral fin. Head shield rugous, anterior tips extend to rear of eyes, central posterior margin with a spine that extends into a notch in the predorsal plate. Adipose fin large, its base about equal to anal-fin base. Lateral line forked before caudal-fin base. HABITAT: Colombia to Brazil. Primarily found in brackish estuaries and lagoons. Also in fresh water, hypersaline lagoons, and marine waters to about 65 ft.

ALSO IN THE AREA: *Cathorops higuchii*, *Cathorops wayuu*, see p. 547.

Argentinidae - Argentines or Herring Smelts

Striated Argentine - *Argentina striata* Goode & Bean, 1896

FEATURES: Translucent with a silvery stripe on sides and silvery reflections above and below. Snout bluntly pointed. Mouth small with upper jaws not meeting anteriorly. Lower limb of first gill arch with six or seven gill rakers. Eyes large. Dorsal fin followed by a small adipose fin. Pectoral fins low on body with 18–21 rays. Anal fin short-based with 11–15 rays. Body elongate and slender. HABITAT: Nova Scotia to FL, and Gulf of Mexico to Uruguay. Also off Cuba. Occur over soft bottoms from about 300 to 1,560 ft.

Pygmy Argentine - *Glossanodon pygmaeus* Cohen, 1958

FEATURES: Pale gray with diffuse dark saddles along back and a silvery stripe along sides. Snout pointed. Mouth small with upper jaws meeting or separated by a small space anteriorly. Lower limb of first gill arch with 21 to 23 gill rakers. Eyes large. First dorsal fin followed by a small adipose fin. Pectoral fins low on body with 12–14 rays. Anal fin short-based with 11–14 rays. Caudal peduncle depth equal to or greater than distance from anus to anal-fin origin. Body elongate and slender. HABITAT: SC to FL, Gulf of Mexico, and Caribbean Sea to Brazil. Occur over outer continental shelves from about 300 to 1,500 ft.

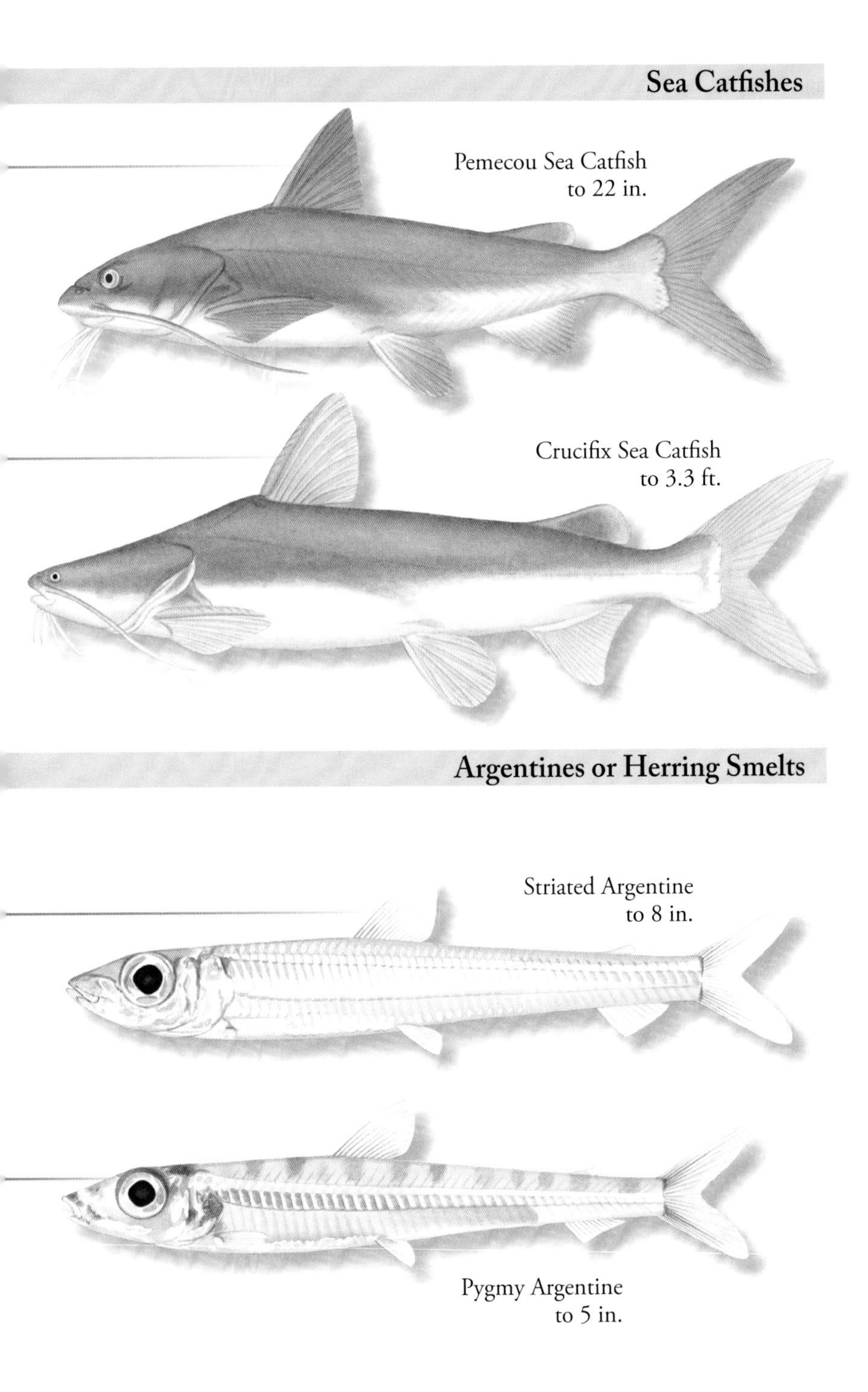
Sea Catfishes
Pemecou Sea Catfish
to 22 in.
Crucifix Sea Catfish
to 3.3 ft.
Argentines or Herring Smelts
Striated Argentine
to 8 in.
Pygmy Argentine
to 5 in.

Aulopidae - Flagfins

Yellowfin Aulopus - *Aulopus filamentosus* (Bloch, 1792)

FEATURES: Opalescent with yellowish blotches and dark saddles and blotches dorsally. Iridescent below. Males with yellow to reddish bands and blotches on fins and elongate filament on first dorsal fin. Fins of females similarly colored with a black blotch on tip of first dorsal fin. Snout bluntly pointed. Eyes and mouth large. First dorsal fin long-based. Pelvic fins broad. HABITAT: S FL, eastern Gulf of Mexico, Bermuda, and scattered in Caribbean Sea. Demersal over continental shelves and slopes from about 160 to 3,300 ft. Usually between 330 and 650 ft.

Synodontidae - Lizardfishes

Largescale Lizardfish - *Saurida brasiliensis* Norman, 1935

FEATURES: Brownish above with three vague darker saddles. Pale below. About six blotches along lateral line. May have a single band on dorsal fin. Snout length about equal to eye diameter. Lower jaw longer than upper jaw. Three scale rows between lateral line and dorsal-fin base. HABITAT: NC to FL Keys, Gulf of Mexico, western Caribbean Sea to Brazil. Demersal over continental shelves and slopes, from about 60 to 1,300 ft. Possibly off eastern Atlantic.

Smallscale Lizardfish - *Saurida caribbaea* Breder, 1927

FEATURES: Brownish to dusky with irregular dark blotches dorsally. Silvery on sides and below. Dorsal- and caudal-fin margins may be dark. Snout long and pointed. Lower jaw longer than upper jaw. Four scale rows between lateral line and dorsal-fin base. HABITAT: NC to FL Keys, Gulf of Mexico, western Caribbean Sea to Brazil. On or near bottom, offshore on continental shelves and slopes, from about 60 to 1,300 ft. Possibly off eastern Atlantic.

Shortjaw Lizardfish - *Saurida normani* Longley, 1935

FEATURES: Grayish with small dark blotches dorsally. Silvery below. Five to six dark blotches along lateral line. Pelvic fin with dark blotch. Snout short and blunt. Upper jaw longer than lower jaw. Four scale rows between lateral line and dorsal-fin base. HABITAT: NC to FL Keys, Gulf of Mexico, western Caribbean Sea to the Guianas. Also Bahamas and off Cuba. On sandy or muddy bottoms from about 160 to 1,800 ft.

Sharpnose Lizardfish - *Synodus bondi* Fowler, 1939

FEATURES: Brownish dorsally with six to eight vague marks along midline, often with pale centers. Back often with vague saddles. Dark blotch on upper portion of peduncle. Silvery below. Snout long, triangular, sharply pointed, and longer than eye diameter. Upper jaw shorter than lower jaw. Five or six scale rows between lateral line and dorsal-fin base. Dorsal-fin base longer than anal-fin base. HABITAT: Scattered in the central and lower Caribbean Sea to Brazil. Demersal over soft bottoms from shore to about 650 ft.

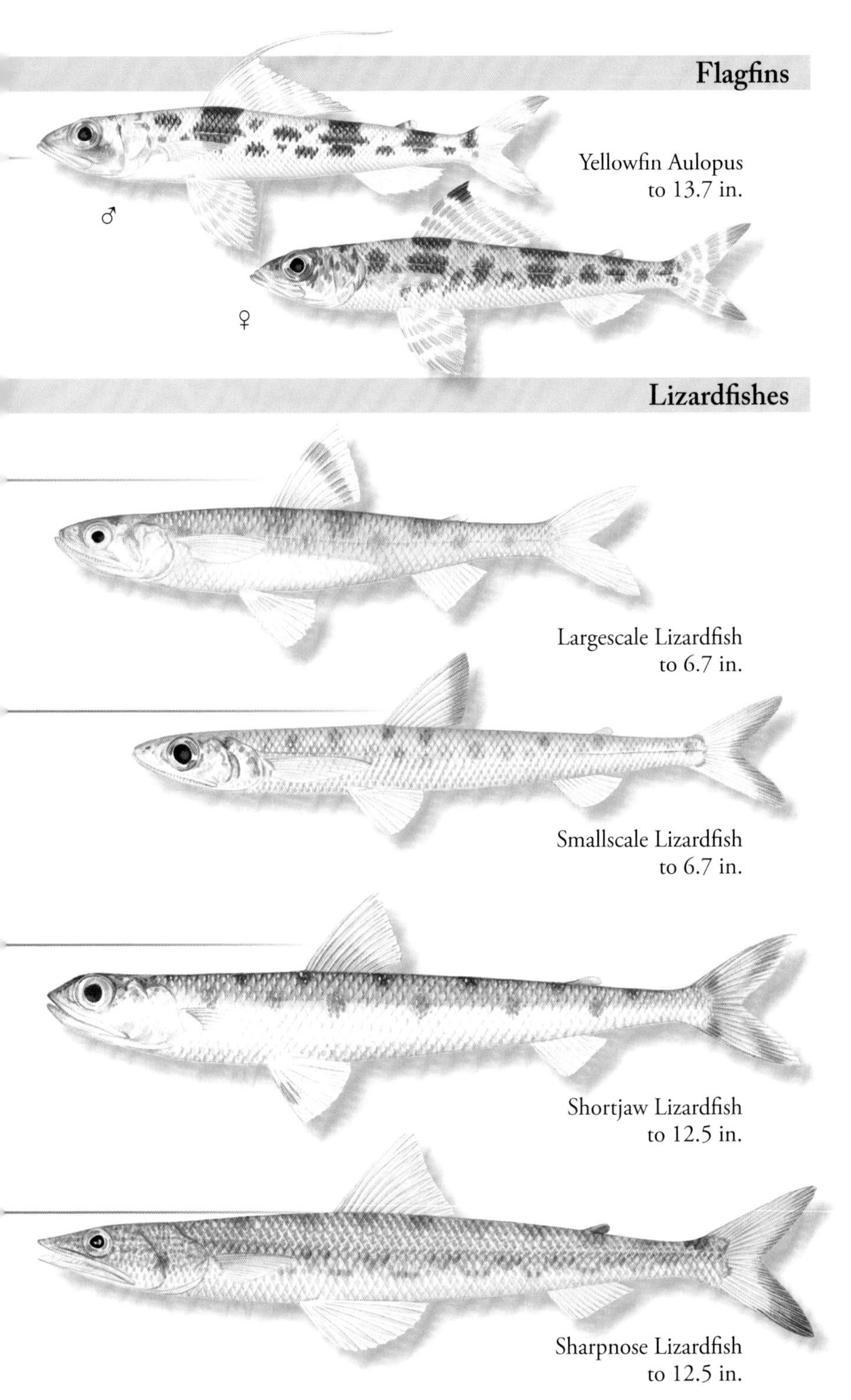
Flagfins
♂
♀
Yellowfin Aulopus
to 13.7 in.
Lizardfishes
Largescale Lizardfish
to 6.7 in.
Smallscale Lizardfish
to 6.7 in.
Shortjaw Lizardfish
to 12.5 in.
Sharpnose Lizardfish
to 12.5 in.

Inshore Lizardfish - *Synodus foetens* (Linnaeus, 1766)

FEATURES: Brownish, grayish, or pale to greenish dorsally. White ventrally. Throat yellowish. Adults with about eight diamond-shaped blotches along midline and smaller blotches on interspaces. Dorsal and caudal fins dusky. Dark spot on operculum absent. Snout conical, about as long as eye diameter. Five or six scale rows between lateral line and dorsal-fin base. Anal-fin base longer than dorsal-fin base. HABITAT: MA to FL, Gulf of Mexico, Caribbean Sea to northern Brazil. Also Bermuda. Near bottom. Along beaches and in bays, inlets, estuaries, and lagoons to about 600 ft.

Sand Diver - *Synodus intermedius* (Spix & Agassiz, 1829)

FEATURES: Color and pattern vary. Yellowish stripes through dark saddles dorsally and on sides. Head and jaws blotched and streaked. All usually with a dark spot at upper edge of gill opening. Young specimens more brightly colored. Snout pointed, longer than eye diameter. Upper jaw longer than lower jaw. Three scale rows between lateral line and dorsal-fin base. HABITAT: NC to FL, Bermuda, Gulf of Mexico, Caribbean Sea to Brazil. Demersal over sandy, rocky, or coralline bottoms to about 1,000 ft. Also in high-salinity lagoons.

Offshore Lizardfish - *Synodus poeyi* Jordan, 1887

FEATURES: Tawny dorsally; silvery below. About eight diamond-shaped marks along midline with a row of smaller blotches below. Ventral area silvery. Snout long, pointed, and equal to or longer than eye diameter. Lower jaw with a fleshy knob at posterior tip. Seven scale rows between lateral line and dorsal-fin base. HABITAT: NC to FL, Gulf of Mexico, and Caribbean Sea to Brazil. Demersal on sandy and muddy bottoms from about 88 to 1,050 ft. More common above 600 ft.

Bluestripe Lizardfish - *Synodus saurus* (Linnaeus, 1758)

FEATURES: Shades of pale ochre to reddish brown with darker flecks and numerous bluish white flecks and spots dorsally. Sides with several rows of bluish white wavy and nearly straight stripes bordered by brownish lines. White below. May display faint to dark saddles. Dark spot at upper opercular margin absent. Snout slightly longer than eye diameter. Three rows of scales between lateral line and dorsal-fin base. HABITAT: S FL, Bahamas, Leeward islands, and scattered in Caribbean Sea. Demersal over soft bottoms from shore to about 1,300 ft. Most common above 65 ft.

Red Lizardfish - *Synodus synodus* (Linnaeus, 1758)

FEATURES: Color varies. Broad, alternating pale and dark reddish to brownish or orangish saddles dorsally. Head and jaws blotched. Small, dark spot on snout tip. Snout conical. Upper jaw longer than lower jaw. Four to six scale rows between lateral line and dorsal-fin base. HABITAT: W FL, Gulf of Mexico, Bermuda, Bahamas, and Caribbean Sea to Uruguay. Demersal on rocky, sandy, muddy, or reef bottoms from near shore to about 295 ft.

Inshore Lizardfish
to 15.7 in.

Sand Diver
to 15.7 in.

Offshore Lizardfish
to 8 in.

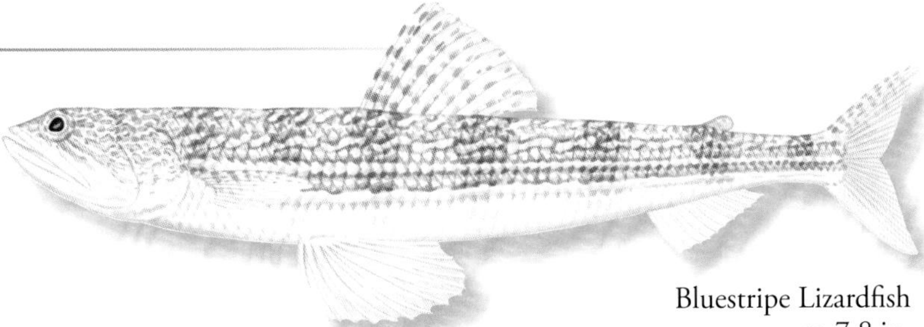

Bluestripe Lizardfish
to 7.8 in.

Red Lizardfish
to 11.8 in.

Synodontidae - Lizardfishes, *cont.*

Snakefish - *Trachinocephalus myops* (Forster, 1801)

FEATURES: Alternating yellowish and pearly blue stripes dorsally and on sides. Five to seven indistinct brown saddles along back. Head mottled. Dark spot at upper opercular margin. Snout very short. Mouth broad and gaping. Anal fin long-based. Body robust. HABITAT: Worldwide in warm seas. MA to FL, Gulf of Mexico, Caribbean Sea to Brazil. Demersal over sandy, shelly, muddy, or rocky bottoms to about 1,300 ft. Often around reefs. BIOLOGY: Bury themselves and swiftly ambush prey.

ALSO IN THE AREA: *Synodus macrostigmus*, see p. 547.

Chlorophthalmidae - Greeneyes

Shortnose Greeneye - *Chlorophthalmus agassizi* Bonaparte, 1840

FEATURES: Grayish with silvery to iridescent luster. Several diffuse dark bars and blotches on upper sides. Eyes reflective green, very large, and meet at top of head. Snout moderately long, upturned. Pupils teardrop-shaped. First dorsal fin tall; pectoral fins long. HABITAT: Nova Scotia to FL, Gulf of Mexico, Caribbean Sea to Suriname. Also eastern Atlantic and Mediterranean Sea. Demersal over mud and clay bottoms from about 165 to 3,300 ft.

Lampridae - Opahs

Opah - *Lampris guttatus* (Brünnich, 1788)

FEATURES: Iridescent bluish to greenish dorsally; silvery to pinkish on sides and below. Body covered in small silvery spots. All fins red. Mouth small. Dorsal fin tall and pointed anteriorly. Pectoral and pelvic fins long and pointed. Body robust, very deep, round in profile, and laterally compressed. HABITAT: Worldwide in tropical and temperate seas. Nova Scotia to Argentina and Gulf of Mexico. Pelagic in open water from surface to about 650 ft. BIOLOGY: Opahs are muscular and predatory. Swim by flapping pectoral fins. Recently discovered to generate and circulate heated blood through body. Mostly solitary. Feed on invertebrates and fishes.

Lophotidae - Crestfishes

Crestfish - *Lophotus lacepede* Giorna, 1809

FEATURES: Uniformly silvery. May have pale spots on sides. Fins reddish. First dorsal-fin ray elongate and erect, remainder of fin extends to tail. Prominent crest on head arches forward and upward. Body relatively deep and laterally compressed; tapers toward tail. Caudal and anal fins small. HABITAT: Western Atlantic, Mediterranean Sea, western Indian Ocean, and off southern Australia. In western Atlantic from FL, Gulf of Mexico, Bahamas, and Caribbean Sea to southern Brazil. Pelagic and oceanic to about 300 ft. Occasionally in inshore waters.

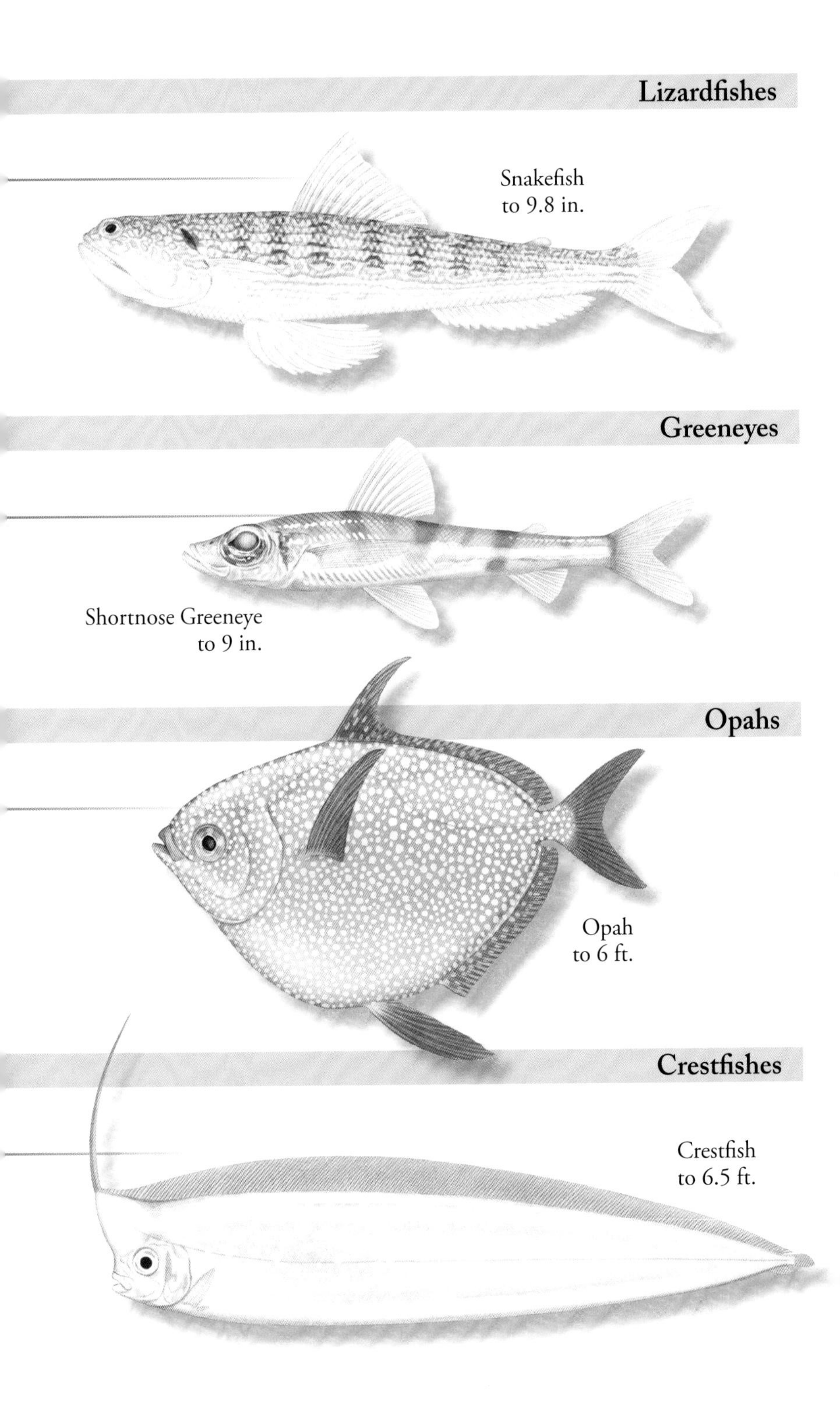
Lizardfishes
Snakefish
to 9.8 in.
Greeneyes
Shortnose Greeneye
to 9 in.
Opahs
Opah
to 6 ft.
Crestfishes
Crestfish
to 6.5 ft.

Regalecidae - Oarfishes

Oarfish - *Regalecus glesne* Ascanius, 1772

FEATURES: Silvery with variably spaced spots and wavy lines on sides. Fins reddish. Head angular. Jaws nearly vertical, highly protrusible. First 10–12 dorsal-fin rays elongate. Pelvic fins long and trailing. Body very long and laterally compressed. Tail blunt, with few to no caudal-fin rays. HABITAT: Worldwide in tropical to warm temperate seas. In western Atlantic from Greenland to Gulf of Mexico and eastern Caribbean Sea to northern Brazil. Pelagic from surface to about 3,300 ft. BIOLOGY: Use undulating fins and body to swim vertically in the water column. May be longest known bony fish.

Polymixiidae - Beardfishes

Beardfish - *Polymixia lowei* Günther, 1859

FEATURES: Shiny grayish blue to greenish blue dorsally. Sides silvery to silvery gray. Silvery below. Dorsal, caudal, and anal fins with back tips in males, dusky tips in females. Snout short, rounded. Eyes large. Two pinkish to pearly white barbels on lower jaw. Dorsal fin with 26–32 rays. Body comparatively slender. HABITAT: ME to Brazil, Gulf of Mexico, Caribbean Sea, and Bermuda. Usually near bottom over muddy bottoms of shelves and slopes from about 160 to 2,100 ft. Usually below 500 ft.

Stout Beardfish - *Polymixia nobilis* Lowe, 1836

FEATURES: Shiny grayish blue to greenish blue dorsally. Sides silvery to silvery gray. Silvery below. Dorsal, caudal, and anal fins with black tips. Snout short, rounded. Eyes large. Two pinkish to pearly white barbels on lower jaw. Dorsal fin with 34–37 rays. Body comparatively deep. HABITAT: FL, northern Gulf of Mexico, Bermuda, Bahamas, and Caribbean Sea. Occur over soft to semi-hard bottoms from about 230 to 2,600 ft.

Bregmacerotidae - Codlets

Antenna Codlet - *Bregmaceros atlanticus* Goode & Bean, 1886

FEATURES: Translucent grayish with silver on head and abdomen. May have small black flecks dorsally. First dorsal fin a single, erect ray on head. Second dorsal and anal fins tall anteriorly and low at middle. Pelvic fins jugular with filamentous anterior rays. HABITAT: Circumglobal in tropical to warm temperate seas. NJ to FL, Gulf of Mexico, Bermuda, Bahamas, and Caribbean Sea. Pelagic from about 165 to 650 ft.

Moridae - Deepsea Cods

Shortbeard Codling - *Laemonema barbatulum* Goode & Bean, 1883

FEATURES: Pearly white to pearly gray dorsally and below. Dorsal, caudal, and anal fins with black margins. Chin with a short barbel. Spiny dorsal fin short-based and tall and pointed. Pelvic fins thoracic with two moderately long and slender rays. Body deep anteriorly, very slender at peduncle. HABITAT: Nova Scotia to FL, Gulf of Mexico, Bahamas, and northern Brazil. Found on or near bottom over deep continental slopes.

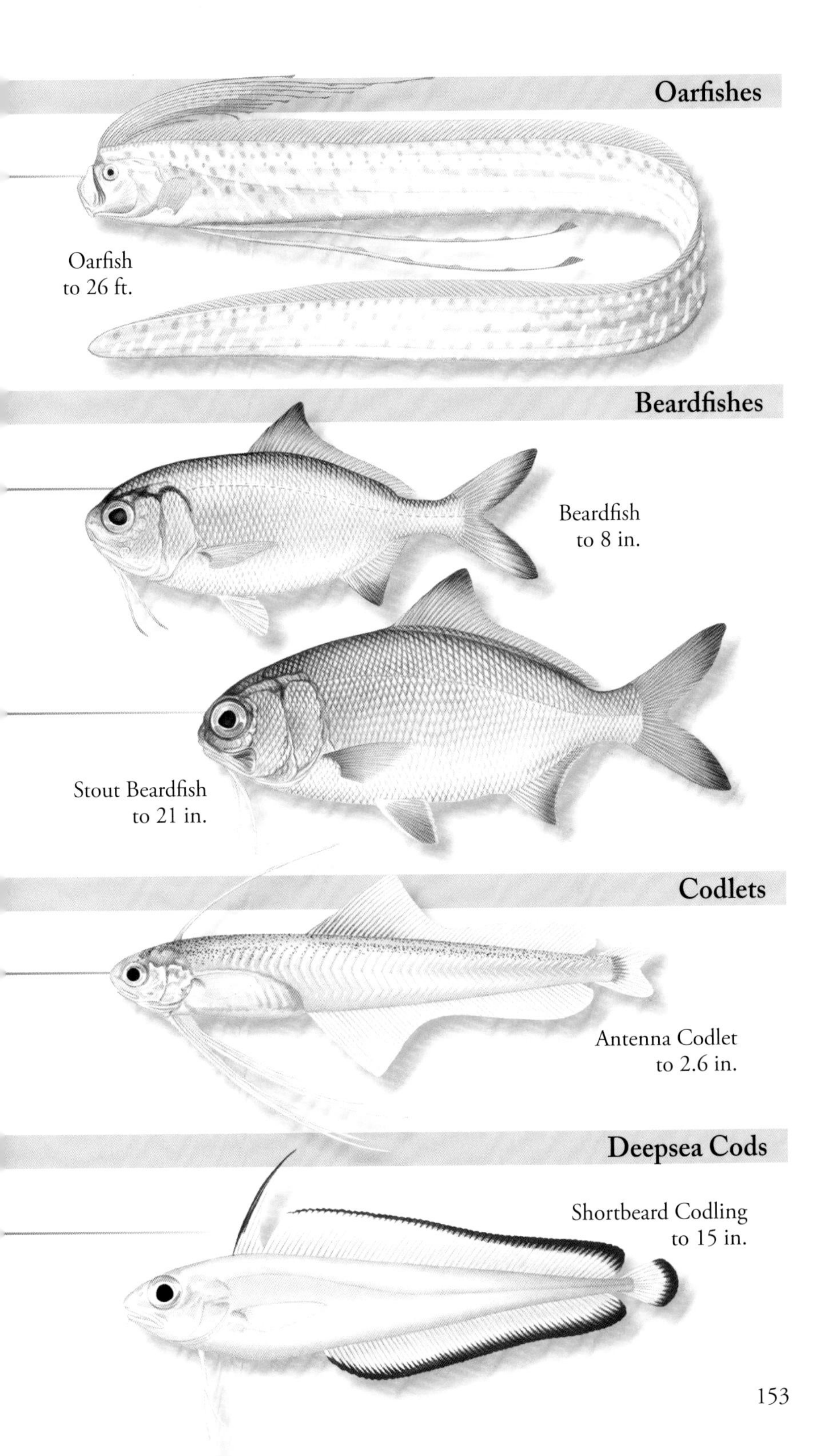
Oarfishes
Oarfish
to 26 ft.
Beardfishes
Beardfish
to 8 in.
Stout Beardfish
to 21 in.
Codlets
Antenna Codlet
to 2.6 in.
Deepsea Cods
Shortbeard Codling
to 15 in.

Moridae - Deepsea Cods, *cont.*

Metallic Codling - *Physiculus fulvus* Bean, 1884

FEATURES: Pinkish brown with a silvery tint dorsally and below. Bluish purple on abdomen. Dorsal, caudal, and anal fins with blackish margins. Chin with a moderately long barbel. Spiny dorsal fin short-based. Pelvic fins thoracic with two long and slender rays. Body robust anteriorly; very slender at peduncle. HABITAT: MA to FL, Gulf of Mexico, Bahamas, and Caribbean Sea to Brazil. Occur on or near bottom over deep continental shelves and upper slopes.

Merlucciidae - Merlucciid Hakes

Offshore Hake - *Merluccius albidus* (Mitchill, 1818)

FEATURES: Grayish dorsally, fading to silvery below. Snout depressed. Cheeks and operculums scaled. Second dorsal fin notched near midlength. Pectoral fins pointed at tips, reach to anus. Caudal-fin margin slightly concave. HABITAT: ME to FL, Gulf of Mexico, and Caribbean Sea to Brazil. Near bottom of continental shelves and slopes to about 3,800 ft. BIOLOGY: Females carry up to 340,000 eggs.

Steindachneriidae - Luminous Hake

Luminous Hake - *Steindachneria argentea* Goode & Bean, 1896

FEATURES: Overall silvery. Inside of mouth dark; abdomen blackish. First dorsal fin short-based with elongate first spine. Second dorsal fin long-based. Anal fin with a high and short-based anterior lobe, followed by a short and long-based portion. Body tapers to pointed tail. Caudal fin absent. HABITAT: Gulf of Mexico and Caribbean Sea. Occur over soft bottoms of continental shelves and upper slopes from 330 to 1,600 ft.

Phycidae - Phycid Hakes

Gulf Hake - *Urophycis cirrata* (Goode & Bean, 1896)

FEATURES: Pale brownish dorsally and on sides. Abdomen silvery. Sides may be spotted or blotched. Diffuse blotch on operculum. Dorsal and anal fins with dark margins. Chin barbel very small or absent. Pelvic-fin rays reach beyond anal-fin origin. HABITAT: E FL, Gulf of Mexico, and northern coast of South America. Along coast, over muddy bottoms from about 88 to 2,200 ft.

Spotted Hake - *Urophycis regia* (Walbaum, 1792)

FEATURES: Plain or blotchy brownish dorsally. Pale below. Rows of small dark spots, a vague blotch, and obscure to distinct oblique stripes on operculum. Lateral line dark with a series of pale spots. First dorsal fin with black spot and white margins. Second dorsal fin vaguely spotted; anal fin with dark margin. Chin barbel short. Pelvic-fin rays reach to or beyond anal fin origin. HABITAT: MA to FL and Gulf of Mexico. Absent from Cuba. On or near bottom from near shore to about 1,300 ft. BIOLOGY: Spawn offshore. Young found in estuaries.

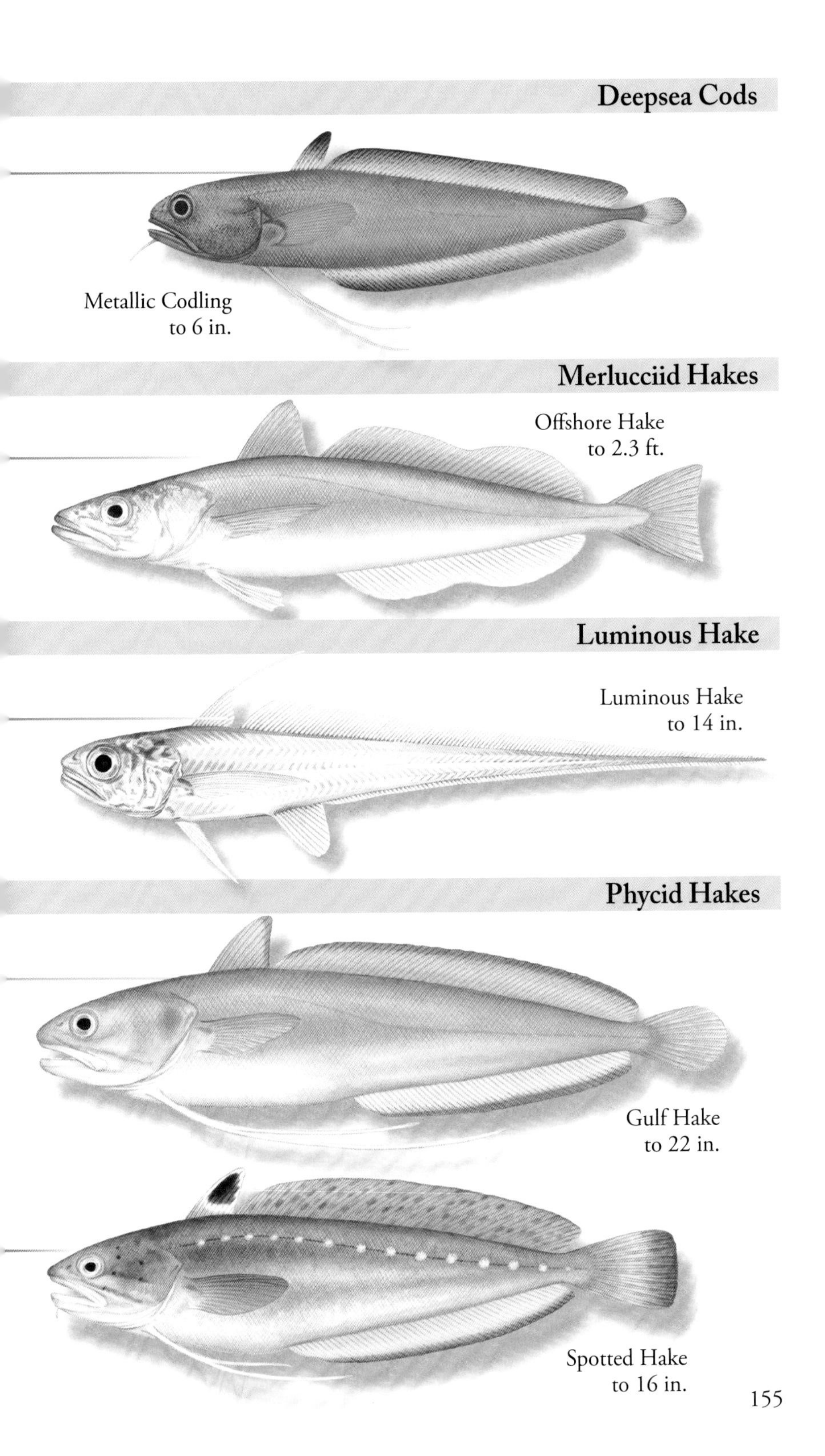
Deepsea Cods
Metallic Codling
to 6 in.
Merlucciid Hakes
Offshore Hake
to 2.3 ft.
Luminous Hake
Luminous Hake
to 14 in.
Phycid Hakes
Gulf Hake
to 22 in.
Spotted Hake
to 16 in.

Carapidae - Pearlfishes

Pearlfish - *Carapus bermudensis* (Jones, 1874)

FEATURES: Translucent. Silvery on head and abdomen. Vertebrae dark. Dark spots on sides. Anal fin longer and taller than dorsal fin. Body tapers to pointed tail. HABITAT: NC to Brazil, including Gulf of Mexico, Bermuda, Bahamas, and Caribbean Sea. Near bottom from shore to about 700 ft. BIOLOGY: Adults live commensally in sea cucumbers' intestines. Emerge at night to forage. SIMILAR SPECIES: Chain Pearlfish, *Echiodon dawsoni*.

Ophidiidae - Cusk-eels

Atlantic Bearded Brotula - *Brotula barbata* (Bloch & Schneider, 1801)

FEATURES: Brownish to reddish on most of body and fins. Juveniles and some adults speckled. Dorsal and anal fins may have white margins. Snout and lower jaw with three pairs of barbels. Pectoral fins broad. Body deep and stout, tapers to tail. HABITAT: FL, Gulf of Mexico, Caribbean Sea to South America. Also Bermuda. Absent from Bahamas. On sandy and muddy bottoms and on reefs from near shore to upper slopes. BIOLOGY: Females lay masses of gelatinous eggs.

Blackedge Cusk-eel - *Lepophidium brevibarbe* (Cuvier, 1829)

FEATURES: Tannish to grayish dorsally. Ventral area silvery. Body unmarked. Dorsal fin—and sometimes anal fin—with a dark margin. Snout with small spine at tip. Small pelvic fins below lower jaw. Body elongate, tapers to pointed tail. HABITAT: NC and Gulf of Mexico to Brazil, including Caribbean Sea. Demersal from shore to about 260 ft. BIOLOGY: Females lay floating, gelatinous masses of eggs. Taken as trawl bycatch.

Mottled Cusk-eel - *Lepophidium jeannae* Fowler, 1941

FEATURES: Tannish dorsally and on sides with irregular, darker splotches. Abdomen pale. Small dark spot on operculum. Dorsal fin with numerous spots at base and black saddles along margin. Anal fin with dark margin. Snout with small spine at tip. Small pelvic fins below lower jaw. Body elongate, tapers to pointed tail. HABITAT: NC to S FL and Gulf of Mexico to northern Yucatán Peninsula. Near bottom from about 59 to 295 ft. BIOLOGY: Females lay floating masses of eggs.

Upsilon Cusk-eel - *Lepophidium pheromystax* Robins, 1960

FEATURES: Yellowish brown to pale brown dorsally and on sides with two rows of large and small, irregularly spaced, darker spots on sides. Upper rows of spots merge onto dorsal-fin base. Abdomen pale. Dark "moustache" on upper jaw, and large dark spot on operculum. Dorsal fin with black saddles along margin. Anal fin with dark margin. Snout with small spine at tip. Small pelvic fins below lower jaw. HABITAT: Southern Gulf of Mexico and southern Caribbean Sea to Brazil. Also Puerto Rico. Demersal over soft bottoms from about 90 to 540 ft.

Pearlfishes

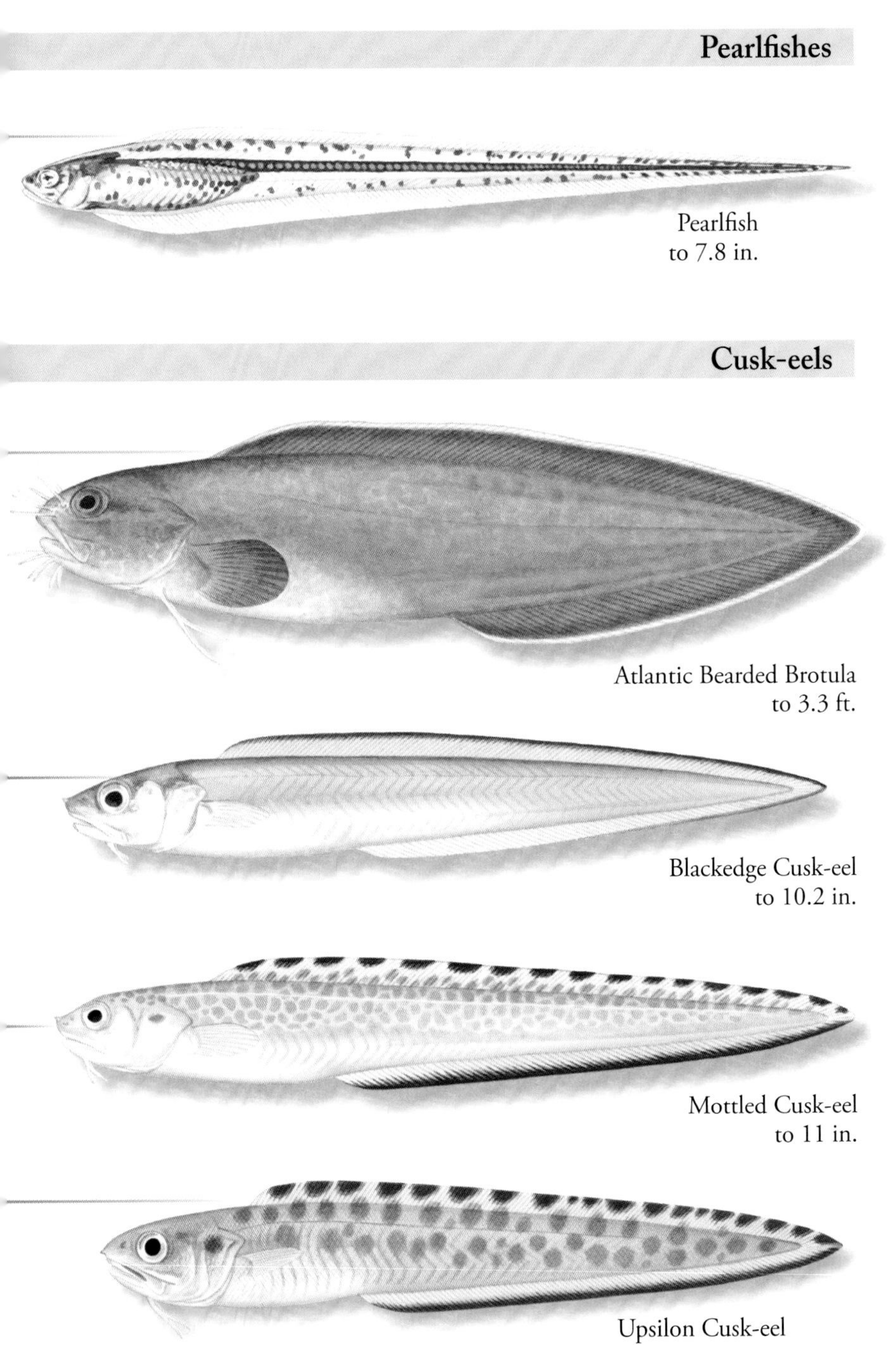

Pearlfish
to 7.8 in.

Cusk-eels

Atlantic Bearded Brotula
to 3.3 ft.

Blackedge Cusk-eel
to 10.2 in.

Mottled Cusk-eel
to 11 in.

Upsilon Cusk-eel
to 11 in.

Fawn Cusk-eel - *Lepophidium profundorum* (Gill, 1863)

FEATURES: Brownish yellow dorsally and on sides. Two rows of pearly, oval spots along upper sides. Dorsal and anal fins with dark margins posteriorly. Snout with small spine at tip. Small pelvic fins below lower jaw. Body elongate, tapers to pointed tail. HABITAT: ME to FL and eastern and southern Gulf of Mexico to northern South America. Over continental shelves from about 180 to 1,200 ft. BIOLOGY: More active at night. Preyed upon by Conger Eels.

Stripefin Brotula - *Neobythites marginatus* Goode & Bean, 1886

FEATURES: Pale yellowish brown. Two broken, narrow to wide, brown stripes from snout to posterior portion of body. Dorsal fin with dark inner stripe. Pelvic fins very slender, originate below preopercular margin. Body stout anteriorly, tapers to a pointed tail. HABITAT: Gulf of Mexico, Bahamas, and Caribbean Sea to northern South America. Also reported from NC to FL. Found near bottom from about 250 to 3,000 ft. BIOLOGY: Feed on crustaceans.

Bank Cusk-eel - *Ophidion holbrookii* Putnam, 1874

FEATURES: Uniformly brownish dorsally and on sides. Lower head and abdomen pale. Dorsal fin, and sometimes anal fin, with dark margins. Lateral line pale. Body otherwise unmarked. Pelvic fins below lower jaw. Head profile almost straight. Body stout, tapers to blunt tail. HABITAT: NC to FL and Gulf of Mexico to Brazil. More common in northeastern than northwestern Gulf of Mexico. Absent from Bahamas. On soft bottoms from shore to about 250 ft. BIOLOGY: Bank Cusk-eels are caught as bycatch of shrimp-trawling fisheries.

Sleeper Cusk-eel - *Otophidium dormitator* Böhlke & Robins, 1959

FEATURES: Nearly colorless. May have some golden to orange tint anteriorly. Body otherwise unmarked. Stout triangular spine under skin at snout tip. Pelvic fins below lower jaw. First pelvic-fin ray much shorter than second. Body elongate, tapers to pointed tail. HABITAT: S FL and Bahamas, and Lesser Antilles to Yucatán. Usually found over coral sand near reefs. Also reported over turtle-grass beds. Occur from shore to about 50 ft. BIOLOGY: Burrow in bottom sediment. Reported to be nocturnal.

Polka-dot Cusk-eel - *Otophidium omostigma* (Jordan & Gilbert, 1882)

FEATURES: Tannish with two to three irregular rows of brown spots and blotches on back and sides. Upper row forms saddles. Large spot just behind upper gill cover always darker than other spots. Dorsal fin with several black blotches along margin. Anal-fin base black. Eyes large. A spine is present on both the snout and operculum. Pelvic fins below lower jaw. Body stout anteriorly, tapers to pointed tail. HABITAT: NC to S FL, northern Gulf of Mexico, and Lesser Antilles. Occur on or near bottom from about 32 to 164 ft.

Fawn Cusk-eel
to 9 in.

Stripefin Brotula
to 8.7 in.

Bank Cusk-eel
to 11.8 in.

Sleeper Cusk-eel
to 3 in.

Polka-dot Cusk-eel
to 4 in.

Ophidiidae - Cusk-eels, *cont.*

Dusky Cusk-eel - *Parophidion schmidti* (Woods & Kanazawa, 1951)

FEATURES: Pale to dark dusky yellowish brown to greenish brown. Some silvery iridescence on operculum and abdomen. Fins colorless to yellowish. Small bony bump on snout. Partially hidden sharp spine on opercle. Pelvic fins below lower jaw. Body scales in a basket-weave pattern. HABITAT: S FL, Bahamas, and Caribbean Sea. Found in shallow coastal waters over seagrass beds from shore to about 26 ft.

Redfin Brotula - *Petrotyx sanguineus* (Meek & Hildebrand, 1928)

FEATURES: Shades of dark red. Small specimens paler. Dorsal and anal fins somewhat darker posteriorly. Snout blunt. Pelvic fins originate at gill isthmus. Head and body with embedded scales. Single lateral line on anterior portion of body. Body deep and robust anteriorly, tapers to pointed tail. HABITAT: S FL, Bahamas, and Caribbean Sea. Occur over shallow outer coral reef slopes from about 10 to 50 ft.

Bythitidae - Viviparous Brotulas

Reef-cave Brotula - *Grammonus claudei* (Torre y Huerta, 1930)

FEATURES: Dark blackish brown to dark gray. Fins blackish. Snout bluntly pointed. Rear margin of upper jaw broad, flattened. Preopercular margin with several small spines and an elongate lower corner. Pelvic fins originate at gill isthmus. Lateral line double; runs below dorsal profile and above ventral profile. Dorsal and anal fins confluent with caudal fin. HABITAT: FL Keys, Bermuda, Bahamas, and Caribbean Sea. Found in caverns of coral reefs and rocky shores. Uncommon. BIOLOGY: Females give birth to live young.

Cave Brotula - *Lucifuga spelaeotes* Cohen & Robins, 1970

FEATURES: Shades of brown or reddish brown to blackish. Dorsal and anal fins pale or with dark bases. Skin over mouth, eyes, and cavernous parts of head clear. Snout depressed; nape arched. Eyes small, vestigial. Rear margin of upper jaw expanded. Pelvic fins originate at gill isthmus. Lateral line in two separate parts: anterior part on upper sides, posterior part on midside. Dorsal and anal fins confluent with caudal fin. HABITAT: Bahamas. Occur on eight different islands in 12 locations. Occur in brackish and marine caves and sinkholes near coastal margins surface to about 90 ft. BIOLOGY: Live in partial to complete darkness. Females give birth to live young.

Black Brotula - *Stygnobrotula latebricola* Böhlke, 1957

FEATURES: Body and fins brownish black to black. Snout bluntly rounded and protruding. Rear margin of jaw narrow. Pelvic fins originate at gill isthmus. Lateral line interrupted, with upper and lower portions. Dorsal and anal fins confluent with caudal fin. HABITAT: Reported from S FL. Also Bahamas to Curaçao. Found over shallow rocky ledges and reefs. BIOLOGY: Hide in caves during the day. Females give birth to live young.

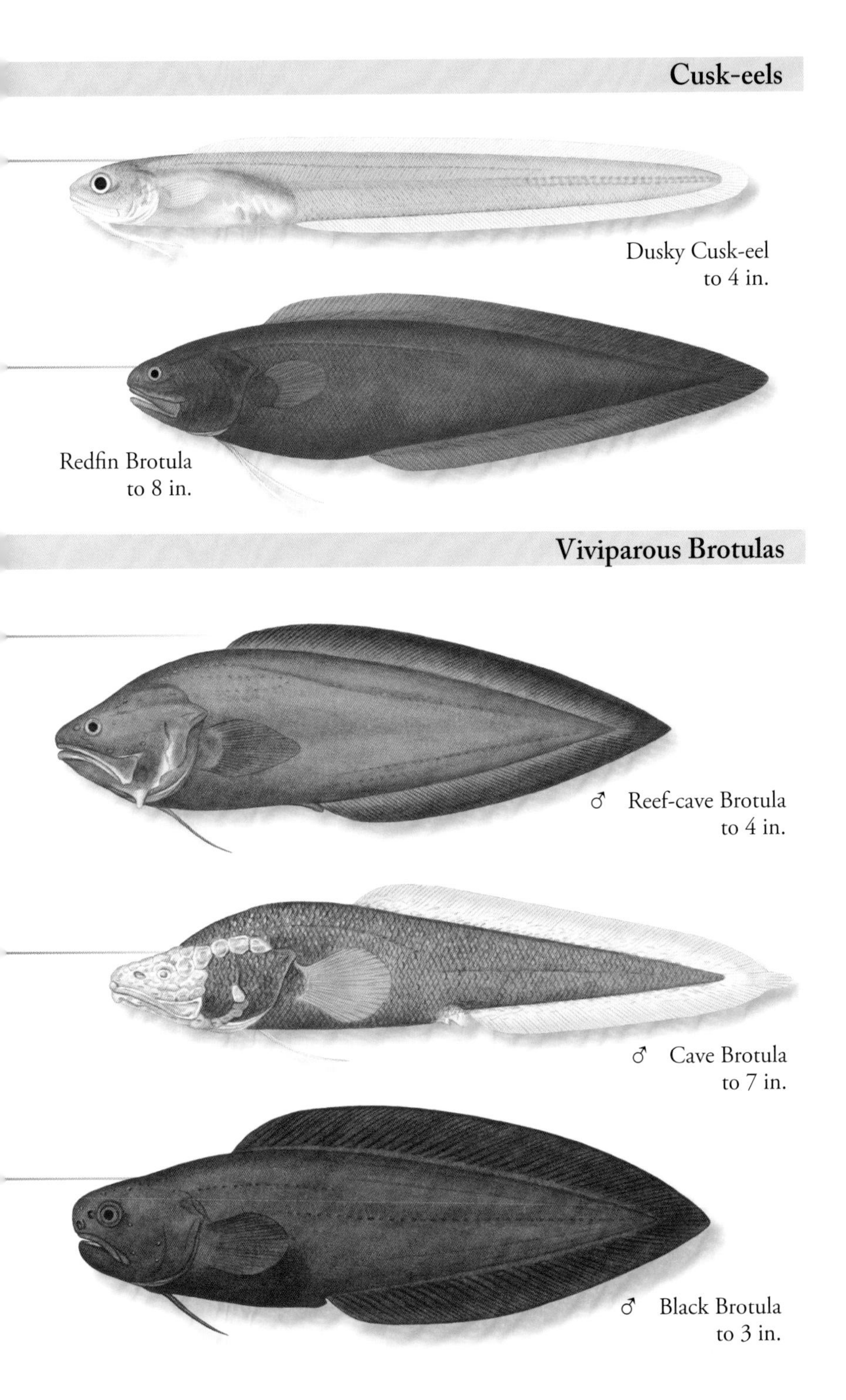
Cusk-eels
Dusky Cusk-eel
to 4 in.
Redfin Brotula
to 8 in.
Viviparous Brotulas
♂ Reef-cave Brotula
to 4 in.
♂ Cave Brotula
to 7 in.
♂ Black Brotula
to 3 in.

Dinematichthyidae - Brotulas

Boehlke's Brotula - *Ogilbia boehlkei* Møller, Schwarzhans & Nielsen, 2005

FEATURES: Shades of reddish brown to brown. Fin margins pale. Snout blunt. Rear margin of upper jaw expanded. Preopercle with a patch of 5–7 scales. Operculum with a broad spine with 2–4 points in mature specimens. Pelvic fins originate at gill isthmus. Lateral line complete, continuous but indistinct. Dorsal and anal fins separate from caudal fin. HABITAT: Bahamas and Puerto Rico. Found over reefs from shore to about 50 ft.

Key Brotula - *Ogilbia cayorum* Evermann & Kendall, 1898

FEATURES: Reddish brown to olive brown. Fins similarly colored with pale margins. Rear margin of upper jaw expanded posteriorly. Scales absent from opercle; four to six rows on cheek. Pelvic fins comparatively short, originate at gill isthmus. Dorsal fin with 66–76 rays. Dorsal and anal fins separate from caudal fin. Lateral line continuous and indistinct. Males possess two pairs of hardened claspers. HABITAT: S FL, Bermuda, and scattered in Gulf of Mexico and Cuba. Reef associated and over shallow grass and algae beds.

Williams' Brotula - *Ogilbia jeffwilliamsi* Møller, Schwarzhans & Nielsen, 2005

FEATURES: Pale yellowish brown. Dorsal and anal fins yellow to yellowish brown at bases, pale to colorless along margins. Rear margin of upper jaw expanded posteriorly with a knob at lower corner. Scales absent from opercle; six to nine rows on cheek. Pelvic fins originate at gill isthmus. Dorsal fin with 71–82 rays. Dorsal and anal fins separate from caudal fin. Lateral line continuous and indistinct. Males possess two pairs of hardened claspers. HABITAT: Puerto Rico, Lesser Antilles, and scattered in Caribbean Sea. Found over coral and rocky reefs from shore to about 50 ft.

Sabaj's Brotula - *Ogilbia sabaji* Møller, Schwarzhans & Nielsen, 2005

FEATURES: Yellowish with brownish flecks clustered on head, nape, and scales. Fins pale with dark flecks at base. Rear margin of upper jaw expanded posteriorly with a knob at lower corner. Scales absent from opercle; five to nine rows on cheek. Pelvic fins originate at gill isthmus. Dorsal fin with 66–72 rays. Dorsal and anal fins separate from caudal fin. Lateral line continuous and indistinct. Males with expanded claspers. HABITAT: S FL, Bahamas, and southeastern Caribbean Sea. Occur over coral reefs from shore to about 100 ft.

Suarez's Brotula - *Ogilbia suarezae* Møller, Schwarzhans & Nielsen, 2005

FEATURES: Overall brownish red to red. Fins paler. Rear margin of upper jaw expanded posteriorly with a knob at lower corner. Scales absent from opercle; five to seven broad rows on cheek. Pelvic fins long, reach anal fin, originate at gill isthmus. Dorsal fin with 66–82 rays. Dorsal and anal fins separate from caudal fin. Lateral line continuous and indistinct. Males with expanded claspers. HABITAT: S FL, Bahamas, and southeastern Caribbean Sea. Occur over coral reefs from shore to about 100 ft.

NOTE: There are at least 12 rare or poorly recorded Brotulas in the area.

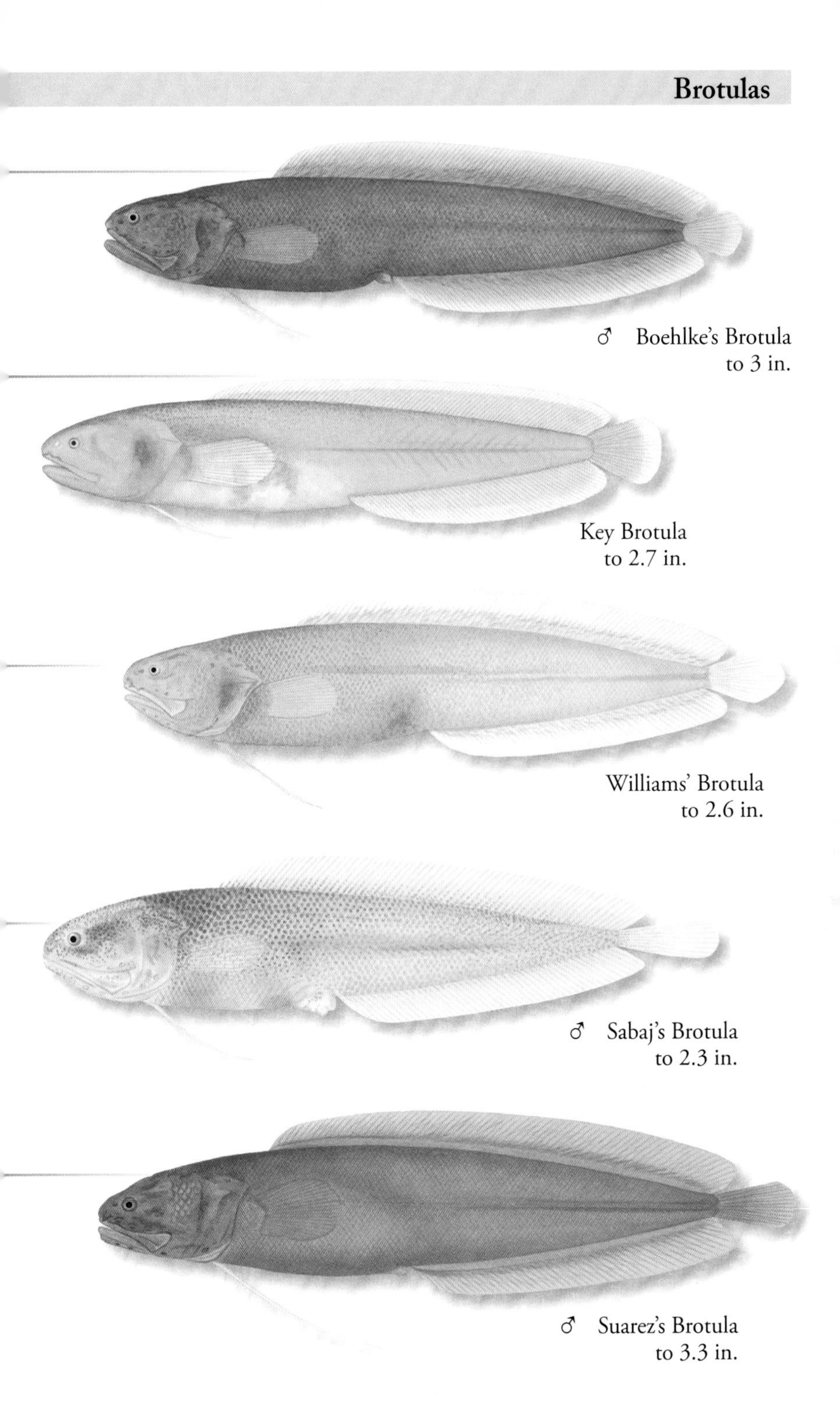

♂ Boehlke's Brotula
to 3 in.

Key Brotula
to 2.7 in.

Williams' Brotula
to 2.6 in.

♂ Sabaj's Brotula
to 2.3 in.

♂ Suarez's Brotula
to 3.3 in.

Batrachoididae - Toadfishes

Bocon Toadfish - *Amphichthys cryptocentrus* (Valenciennes, 1837)

FEATURES: Color and pattern vary. Shades of brown to yellowish brown. Head often with small, dark blotches and a dark band radiating below eyes. Body mottled; dorsal and anal fins banded. A prominent, branched tentacle present over eyes. Long branched tentacles around mouth. Opercle with two spines; preopercle with one spine. Second dorsal fin with 29 rays; anal fin with 23–25 rays. Lower lateral line with 27–34 pores. HABITAT: Panama to Brazil. Occur over sandy and rocky bottoms, and in caves and crevices from shore to about 230 ft. Sometimes enter lagoons. BIOLOGY: Feed on mollusks and crustaceans.

Large-eye Toadfish - *Batrachoides gilberti* Meek & Hildebrand, 1928

FEATURES: Adults nearly uniformly dark brown with vague paler bars and bands. Juveniles dark with distinct pale bars and banded dorsal, caudal, and anal fins. Lateral-line pores white. Head flattened. Thick-based barbels around mouth—may be simple or finely branched. Numerous small filaments on head. Eyes relatively large, without tentacles. Second dorsal fin with 24–26 rays; anal fin with 22-23 rays. Lower lateral line with 33–41 (usually more than 34) pores. Body covered with small, embedded scales. HABITAT: Yucatán Peninsula to Panama Canal zone. Demersal over soft bottoms from estuaries and coastal waters to about 100 ft. Also enter fresh water.

Cotuero Toadfish - *Batrachoides manglae* Cervigón, 1964

FEATURES: Color and pattern vary. Shades of ochre with dense dark brown mottling, spots, and marbling on the head. Sides with several irregular dark bars and blotches. Fins banded. Lower jaw and abdomen whitish. Head depressed. Numerous fringe-like barbels with fine filaments around mouth. Opercle with four spines. Second dorsal fin with 21–24 rays; anal fin with 19–21 rays. Body covered in small, embedded scales. HABITAT: Colombia to Venezuela. Demersal over shallow muddy bottoms. Abundant in mangrove-lined lagoons from shore to about 30 ft. BIOLOGY: Feed on gastropods and crustaceans.

Pacuma Toadfish - *Batrachoides surinamensis* (Bloch & Schneider, 1801)

FEATURES: Color and pattern vary. Shades of brown with several irregular darker bars on head and body. Abdomen pale to whitish. Fins variably spotted and banded. Others overall dark brown. Head depressed and very broad and squared anteriorly. Eyes small and without tentacles. Numerous barbels around mouth and sides of head. Second dorsal fin with 28–30 rays; anal fin with 25–27 rays. Lower lateral line with 48–63 pores. Body and top of head covered with minute, embedded scales. HABITAT: Honduras to Rio de Janeiro, Brazil. Found over muddy bottoms along the coast from estuaries to about 120 ft. offshore. BIOLOGY: Feed on small gastropods and crustaceans.

Bocon Toadfish
to 13.4 in.

Large-eye Toadfish
to 10.6 in.

Cotuero Toadfish
to 12 in.

Pacuma Toadfish
to 13.4 in.

Gulf Toadfish - *Opsanus beta* (Goode & Bean, 1880)

FEATURES: Color and pattern vary. Brownish to grayish with pale marbling and mottling. Posterior pale blotches often form rosettes. Fins banded or with spots forming bands. Fleshy tabs around mouth, on cheeks, and over eyes. Second dorsal fin with 24–25 rays, anal fin with 19–23 rays, pectoral fins with 18–19 rays. Scales absent. HABITAT: FL, Little Bahama Bank, Gulf of Mexico, Yucatán Peninsula to Belize, and Brazil. Demersal in shallow lagoons and bays and over seagrass beds and rocky areas. Also found in cans and empty shells from shore to about 20 ft.

Leopard Toadfish - *Opsanus pardus* (Goode & Bean, 1880)

FEATURES: Color somewhat variable. Pale brownish to yellowish with small to large dark spots that may blend together. Spots form irregular bands on fins. Fins may also have dark margins. Fleshy tabs around jaws, on cheeks, and over eyes. Second dorsal fin with 26 rays, anal fin with 19–22 rays, pectoral fins with 20–22 rays. Scales absent. HABITAT: North Carolina to Texas, northern Yucatán Peninsula, and northern Cuba. Demersal over rocky bottoms and reefs from shore to about 295 ft.

Scarecrow Toadfish - *Opsanus phobetron* Walters & Robins, 1961

FEATURES: Color and pattern vary. Tannish with large dark brown to blackish blotches and paler marbling between. Dark bands radiate from eyes. Fins strongly banded. Head moderately broad. Fleshy tabs around mouth, on head, and above eyes. Second dorsal fin with 23–25 rays, anal fin with 19–23 rays, pectoral fins with 17–18 rays. Scales absent. HABITAT: Bahamas, Cuba, and northern Yucatán Peninsula. Demersal. Found in empty shells and over muddy sand and seagrass from shore to about 40 ft.

Ocellated Toadfish - *Porichthys oculofrenum* Gilbert, 1968

FEATURES: Tannish to pale brownish dorsally with rows of brown spots on head and sides. Yellowish ventrally. Dorsal fin with rows of spots. Anal fin with dark margin. Pectoral fins with a dark ocellus at upper margin. Mouth upturned; eyes stalked. Two small spines precede second dorsal fin. Second dorsal fin with 29–32 rays; anal fin with 27–30 rays. Series of photophores on head, throat, and along lateral lines. HABITAT: Venezuela and northern Brazil. Demersal, primarily over soft bottoms from about 190 to 230 ft.

Few-rayed Toadfish - *Porichthys pauciradiatus* Caldwell & Caldwell, 1963

FEATURES: Tannish to greenish yellow with seven to eight brown saddles along back and one on head. Sides and portions of cheeks silvery. Lower body tannish to greenish yellow. Mouth upturned; eyes stalked. Two small spines precede second dorsal fin. Second dorsal fin with 29–32 rays; anal fin with 27–30 rays. Series of photophores on head, throat, and along lateral lines. HABITAT: Southern Belize to northern Brazil. Demersal, primarily over soft bottoms from shore to about 180 ft.

Gulf Toadfish
to 15 in.
Leopard Toadfish
to 15.3 in.
Scarecrow Toadfish
to 8.5 in.
Ocellated Toadfish
to 4.3 in.
Few-rayed Toadfish
to 3 in.

Atlantic Midshipman - *Porichthys plectrodon* Jordan & Gilbert, 1882

FEATURES: Metallic straw to dark brown or gray with dark spots dorsally. Larger spots along midbody lateral line. Lower sides silvery. Ventral area golden. Anal fin with dark margin. Series of photophores on head and along lateral lines. Two small spines precede second dorsal fin. Second dorsal fin with 33–39 rays; anal fin with 30–36 rays. HABITAT: VA to FL, and Gulf of Mexico to Argentina. On soft bottoms to about 840 ft. BIOLOGY: Atlantic Midshipman are one of few shallow-water fishes to possess photophores. Bioluminescence may be used during courtship.

Starry Toadfish - *Sanopus astrifer* (Robins & Starck, 1965)

FEATURES: Shades of brown to black with numerous, irregular, small white spots and flecks on head and body that are more dense anteriorly. Some spots form lines on head. Head broad and depressed. Thick and slender, long and pointed barbels line lower jaw and edge of operculum. Central one or two pairs branched. Upper lateral line with 36–41 pores. HABITAT: Belize. Demersal over reefs and rubble, and in crevices and recesses from about 10 to 100 ft. IUCN: Vulnerable.

Bearded Toadfish - *Sanopus barbatus* (Meek & Hildebrand, 1928)

FEATURES: Color varies. Marbled, mottled, and blotched in shades of ochre, brown, black, and white. Head broad and depressed. Dense rows of thick and complexly branched barbels line lower jaw and edge of operculum. Numerous fleshy tabs and filaments on head, body, and along lateral lines. Second dorsal fin with 31–34 rays. Scales absent from head and body. HABITAT: Belize to Panama. Demersal around shallow rocky and coral reefs from shore to about 65 ft.

Whitelined Toadfish - *Sanopus greenfieldorum* Collette, 1983

FEATURES: Brownish black to black with irregular white lines and spots radiating from eyes and circling head. Body peppered with white spots. Fins dark, may have narrow white margins. Head broad and depressed. Thick and slender, long and pointed barbels line lower jaw and edge of operculum. Central one or two pairs branched. Upper lateral line with 27–34 pores. HABITAT: Belize. Demersal around reefs, in crevices and recesses, and over associated sandy areas from 10 to 100 ft. IUCN: Vulnerable.

Reticulate Toadfish - *Sanopus reticulatus* Collette, 1983

FEATURES: Pale with brown with blackish reticulations covering head, body, and fins. Marks oblique on second dorsal and anal fins. Head broad and depressed. Branched barbels line lower jaw and edge of operculum. Small filaments on head. Second dorsal fin with 31–32 rays; anal fin with 25–26 rays. Scales absent from head and body. HABITAT: Northern Yucatán Peninsula. Demersal in caves and crevices of coral reefs from shore to about 30 ft. IUCN: Endangered.

Atlantic Midshipman
to 8.6 in.
Starry Toadfish
to 11 in.
Bearded Toadfish
to 16 in.
Whitelined Toadfish
to 12 in.
Reticulate Toadfish
to 10.2 in.

Splendid Toadfish - *Sanopus splendidus* Collette, Starck & Phillips, 1974

FEATURES: Color somewhat variable. Shades of pale gray to pale blue gray with dark gray to blackish thin and thick lines across head that become irregular, interconnected blotches on body. White under head. Inner portions of fins blackish, outer margins bright yellow. Head broad and depressed. Thick and slender barbels around mouth and edge of operculum. Small fleshy filaments on head. Second dorsal fin with 29–31 rays. HABITAT: Cozumel, Mexico, and Glover's Reef, Belize. In caves and crevices under coral heads from about 30 to 80 ft. IUCN: Endangered.

Caño Toadfish - *Thalassophryne maculosa* Günther, 1861

FEATURES: Color and pattern vary. Shades of pale tan with either distinct dark blotches, vague dark blotches, or irregular dark mottling. Dark spots or mottling radiate from eyes. Fins lack distinct white margins. Flattened fleshy tabs around mouth. Eyes elevated. Head large, flattened on top. Opercular spine and first dorsal-fin spines venomous. Single lateral line. Scales absent. HABITAT: Panama to Venezuela. Demersal over or buried in mud or sand from shore to about 650 ft.

Big-eyed Toadfish - *Thalassophryne megalops* Bean & Weed, 1910

FEATURES: Nearly uniformly tannish or with diffuse pale brown saddles and/or blotches on body. Vague to distinct dark blotches radiate from eyes. Fins lack distinct white margins. Small barbels around mouth. Eyes elevated. Head large and flattened on top. Opercular spine and first dorsal-fin spines venomous. Single lateral line. Scales absent. HABITAT: Panama to Venezuela. Demersal over or buried in muddy bottoms from about 255 to 600 ft.

Copper Joe Toadfish - *Thalassophryne nattereri* Steindachner, 1876

FEATURES: Shades of brown to grayish brown dorsally. Paler below. May have irregular dark blotches on head and vague saddles on body. Dorsal, anal, caudal, and pectoral fins with blackish inner margins and whitish outer margins. Small barbels around mouth. Eyes elevated. Head large and flattened on top. Opercular spine and first dorsal-fin spines venomous. Single lateral line. Scales absent. Body comparatively robust. HABITAT: Colombia to Brazil. Demersal over or buried in soft bottoms from about 40 to 240 ft.

Glover's Toadfish - *Vladichthys gloverensis* (Greenfield & Greenfield, 1973)

FEATURES: Yellowish to orange or whitish with orange brown to brown bands and blotches on head and four to five bars on body. Fins banded. Head broad. Fringed barbels around mouth and edge of operculum. Two cirri over eyes, the rear cirrus fringed and lap-like. Pores around head and along lateral lines with one to two skin flaps. Second dorsal fin with 20–21 rays. Scales absent. HABITAT: Known from barrier and oceanic reefs around Glover's Island, Belize. Demersal from about 50 to 100 ft. IUCN: Vulnerable.

ALSO IN THE AREA: *Opsanus dichrostomus, Sanopus johnsoni*, see p. 547.

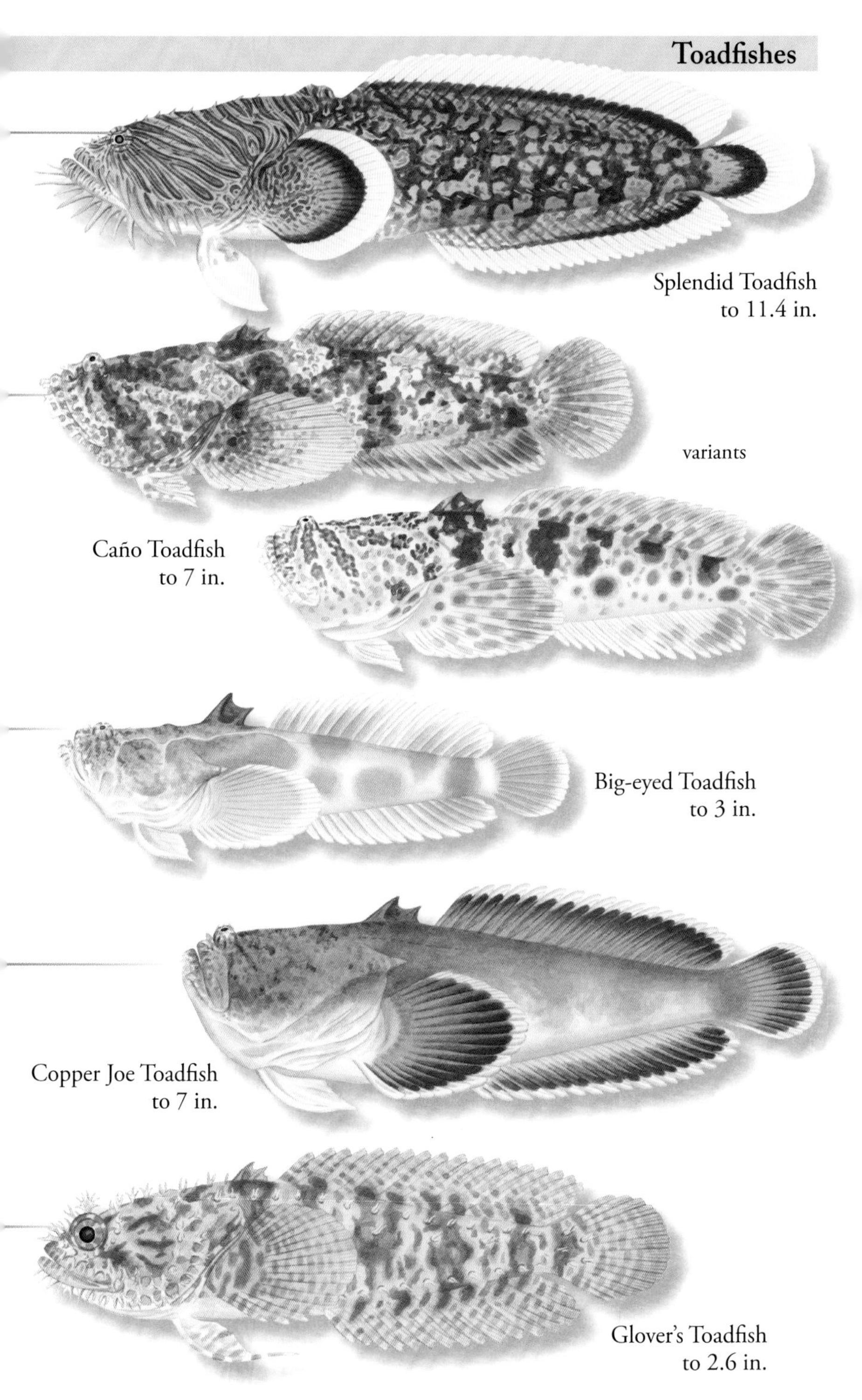

Splendid Toadfish
to 11.4 in.

Caño Toadfish
to 7 in.

Big-eyed Toadfish
to 3 in.

Copper Joe Toadfish
to 7 in.

Glover's Toadfish
to 2.6 in.

Lophiidae - Goosefishes

Reticulate Goosefish - *Lophiodes reticulatus* Caruso & Suttkus, 1979

FEATURES: Color highly and pattern highly variable. Pinkish, tannish, grayish, or brownish with a reticulating or chain-like pattern on body and fins. Reticulations may be very faint. Mouth very wide with numerous sharp teeth. First four dorsal-fin spines free and filamentous; the first is tall with a fleshy lure. Body broad, flattened, and plate-like. Branched, fleshy tabs around edge of mouth, body, and tail. HABITAT: NC to northern coast of South America. Also Puerto Rico and Virgin Islands. Demersal over sandy bottoms primarily on continental slopes from about 210 to 1,210 ft. BIOLOGY: Voracious. May eat prey as large as themselves. Use first spine as lure.

Antennariidae - Frogfishes

Island Frogfish - *Abantennarius bermudensis* (Schultz, 1957)

FEATURES: Color and pattern vary. Shades of tan, brownish, or yellowish, with irregular spots, mottling, and bands of pink, gray, or whitish. A large single ocellus under second dorsal-fin base. Well-defined spots absent from abdomen. Bands radiate from eyes. Lure short with a fleshy cluster of filaments and round, bulb-like appendages. Second dorsal-fin spine strongly curved and without membrane connected to back. Skin rough, with many fleshy filaments that are more clustered around mouth. HABITAT: Bermuda, Bahamas, and Caribbean Sea. Reef associated. Occur from about 10 to 100 ft. NOTE: Previously known as *Antennarius bermudensis*.

Longlure Frogfish - *Antennarius multiocellatus* (Valenciennes, 1837)

FEATURES: Color and pattern vary. Shades of yellow, orange, red, pink, white, tan, gray, or black; may be somewhat to highly mottled. May have whitish saddles on shoulder and peduncle. Variably sized pale to dark ocelli at base of dorsal and anal fins. Always with three ocelli on caudal fin. May have smaller ocelli on fins and body. Lure very long, about twice as long as second dorsal-fin spine. Esca is an irregular cluster of fringe. Second dorsal-fin spine connected to head by thin membrane. Anal fin with six or seven rays. HABITAT: FL Keys, Gulf of Mexico, Caribbean Sea to Brazil. Also Bermuda. Rarely north to NY. Common on Caribbean reefs. Occur from shore to about 370 ft. Prefer areas with sponges.

Dwarf Frogfish - *Antennarius pauciradiatus* Schultz, 1957

FEATURES: Yellowish, whitish, brown, or olivaceous to red. Some plainly patterned others heavily mottled and blotched. Small, vague, dark or pale spot at base of second dorsal fin. Lure shorter than second dorsal-fin spine. Esca is fleshy and complexly branched. Second dorsal-fin spine branched at tip, with a pocket to hold lure, and connected to head by membrane. Body may lack filaments or may be densely covered with filaments. Anal fin with seven or eight rays. HABITAT: Southern FL, Bermuda, Bahamas, and Caribbean Sea. Occur over patch reefs, sand, rubble, and seagrass beds from about 20 to 240 ft.

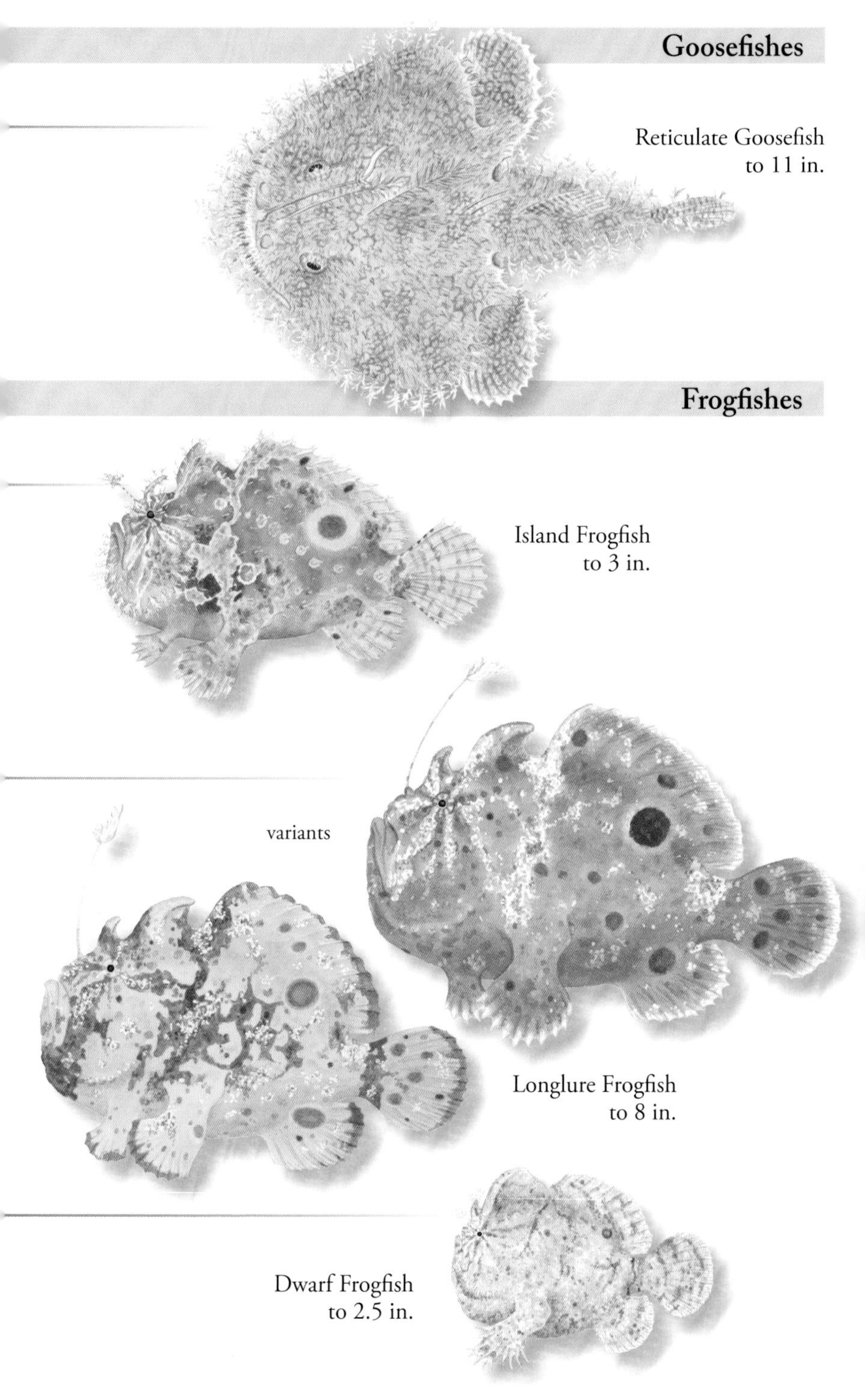
Goosefishes
Reticulate Goosefish
to 11 in.
Frogfishes
Island Frogfish
to 3 in.
variants
Longlure Frogfish
to 8 in.
Dwarf Frogfish
to 2.5 in.

Striated Frogfish - *Antennarius scaber* (Cuvier, 1817)

FEATURES: Color and pattern highly variable. Shades of brown, yellow, beige, red, or black. May be uniformly colored or densely to sparsely banded and/or spotted. Some are black with pale mottling. Body and fins may have few to numerous fleshy filaments. Lure about as long as second dorsal-fin spine and broad at base. Esca is fleshy, double or multi-branched, and worm-like. Second dorsal-fin spine erect, connected to head by a smooth membrane. Anal fin with six or seven rays. Skin rough, covered with small, close-set spicules. HABITAT: NY to FL, Gulf of Mexico, Bermuda, Bahamas, and Caribbean Sea to southern Brazil. Also in Central Indo-Pacific. Occur over reefs and sandy and muddy bottoms. NOTE: Previously known as *Antennarius striatus*.

Ocellated Frogfish - *Fowlerichthys ocellatus* (Bloch & Schneider, 1801)

FEATURES: Color and pattern highly variable. Brownish, yellowish, grayish, or purplish to reddish with variable darker and paler blotching. Small spots and smudges pepper head, body, and fins. Always with large, dark to pale ocelli at dorsal-fin base, on caudal fin, and on midbody. Lure short, about equal in length to second dorsal-fin spine. Esca is dense fleshy cluster. Second dorsal-fin spine connected to head by membrane. Skin may be densely covered with fine filaments. Anal fin with seven or eight rays. HABITAT: NC to FL, Gulf of Mexico, Bahamas, Bermuda, and western Caribbean Sea from Yucatán to Venezuela. Also around Puerto Rico. Occur over shelly, muddy, and sandy bottoms from shore to about 450 ft. Also found on reefs. BIOLOGY: Use lure to bait prey. NOTE: Previously known as *Antennarius ocellatus*.

Singlespot Frogfish - *Fowlerichthys radiosus* (Garman, 1896)

FEATURES: Color and pattern variable. Beige, pale brown to grayish brown. Single, dark ocellus under base of second dorsal fin. Body may have reticulating pattern. Dark lines radiate from eyes. Fins may be barred. Lure about as long as, or longer than, second dorsal-fin spine. Esca is small and simple. Second dorsal-fin spine nearly straight, connected to head by membrane. Anal fin with eight rays. HABITAT: NY to FL Keys, Gulf of Mexico, and Greater Antilles. On offshore banks and shelf waters from about 65 to 900 ft. NOTE: Previously known as *Antennarius radiosus*.

Sargassumfish - *Histrio histrio* (Linnaeus, 1758)

FEATURES: Yellowish, grayish, or olivaceous to brownish with irregular, darker and paler lines and spots. Few to many fleshy tabs on head, body, and fins. Lure short, about half as long as second dorsal-fin spine. Esca is branched and fleshy. Second dorsal-fin spine fleshy and erect, not connected to head by membrane. Pelvic fins long and flexible. Anal fin with six to eight rays. Skin smooth and without spicules. HABITAT: Circumglobal in warm seas. ME to FL, Gulf of Mexico, Bermuda, Bahamas, and Caribbean Sea to Uruguay. Occur from surface to about 20 ft. BIOLOGY: Usually found in floating *Sargassum* seaweed, but also found around mangroves and in seagrass beds. Well camouflaged. Use fins to cling and crawl. Lure used to bait prey, including juvenile Sargassumfish.

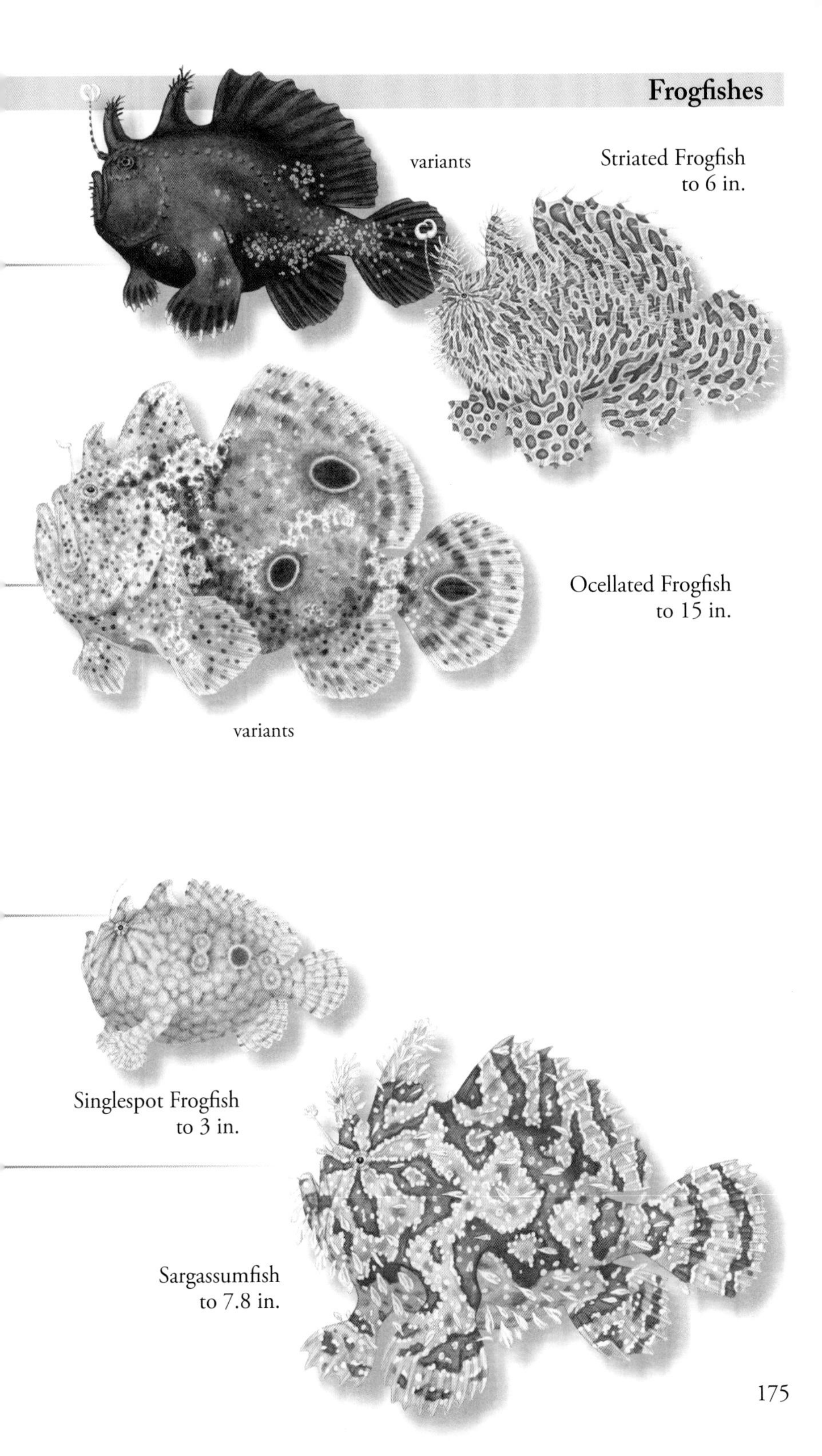
variants
Striated Frogfish
to 6 in.
Ocellated Frogfish
to 15 in.
variants
Singlespot Frogfish
to 3 in.
Sargassumfish
to 7.8 in.

Ogcocephalidae - Batfishes

Atlantic Batfish - *Dibranchus atlanticus* Peters, 1876

FEATURES: Uniformly brownish to reddish gray dorsally; slightly paler ventrally. Dorsal surface covered with tubercles. Outer margin of body with a row of multi-spined tubercles. Rear margins of disk with an elongate spine with multiple spinelets. Ventral surface completely covered in small, prickle-like and/or cone-shaped scales. Body depressed, rounded in outline. Tail long, tapered. HABITAT: RI to Brazil, Gulf of Mexico, and Lesser Antilles. Also eastern Atlantic. Demersal. Occur on soft bottoms from about 130 to 4,200 ft. BIOLOGY: Possess photophores as adults. Use lure to attract invertebrates.

Pancake Batfish - *Halieutichthys aculeatus* (Mitchill, 1818)

FEATURES: Pale tan, olive, or yellowish gray dorsally with variable darker network reticulations on dorsal surface. Pectoral and caudal fins banded brown and white with yellow margins. Dorsal surface covered with irregularly shaped tubercles. Fleshy tabs around margin of body. Ventral surface naked. Body flattened, almost circular in outline. Tail comparatively short. HABITAT: NC to FL, Gulf of Mexico, and Caribbean Sea to northern South America. Demersal over sandy bottoms from about 100 to 1,400 ft. BIOLOGY: May lie on the bottom partially covered with sand.

Caribbean Batfish - *Halieutichthys caribbaeus* Garman, 1896

FEATURES: Pale grayish with a fine network of irregular reticulations on dorsal surface. Tubercles are whitish. Pectoral and caudal fins white with reddish to reddish brown bands. Dorsal surface with irregularly shaped tubercles. Fleshy tabs around margin of body. Ventral surface naked. Body flattened, almost circular in outline. Tail comparatively short. HABITAT: Caribbean Sea to French Guyana. Demersal over sandy and muddy bottoms from about 230 to 900 ft.

Grotesque Batfish - *Malthopsis gnoma* Bradbury, 1998

FEATURES: Pale to brown with a pattern of dark brown reticulation covering dorsal surface. Ventral surface brown to purplish. Dorsal fin banded. Rostrum short and upturned. Body triangular in outline. Pectoral fins arm-like; tail long. Body covered with bucklers, spines, and prickles. HABITAT: Bahamas, Cuba, Puerto Rico, Virgin Islands, and southern Caribbean Sea. Demersal over sandy and muddy bottoms from about 300 to 1,560 ft.

Longnose Batfish - *Ogcocephalus corniger* Bradbury, 1980

FEATURES: Brown to gray with uniformly scattered, small pale spots dorsally. Spots may form reticulating pattern. Pale ventrally. Pectoral fins with broad, dark reddish brown to blackish margins and white tips. Caudal fin with broad dark reddish brown to blackish margin. Dorsal surface covered in tubercles and bucklers. Rostrum always long, protrudes straight from head. Pectoral fins with fleshy pads ventrally. HABITAT: NC to FL, Bahamas, and Greater Antilles. Also northeastern and southern Gulf of Mexico. Demersal. Found from about 95 to 750 ft.

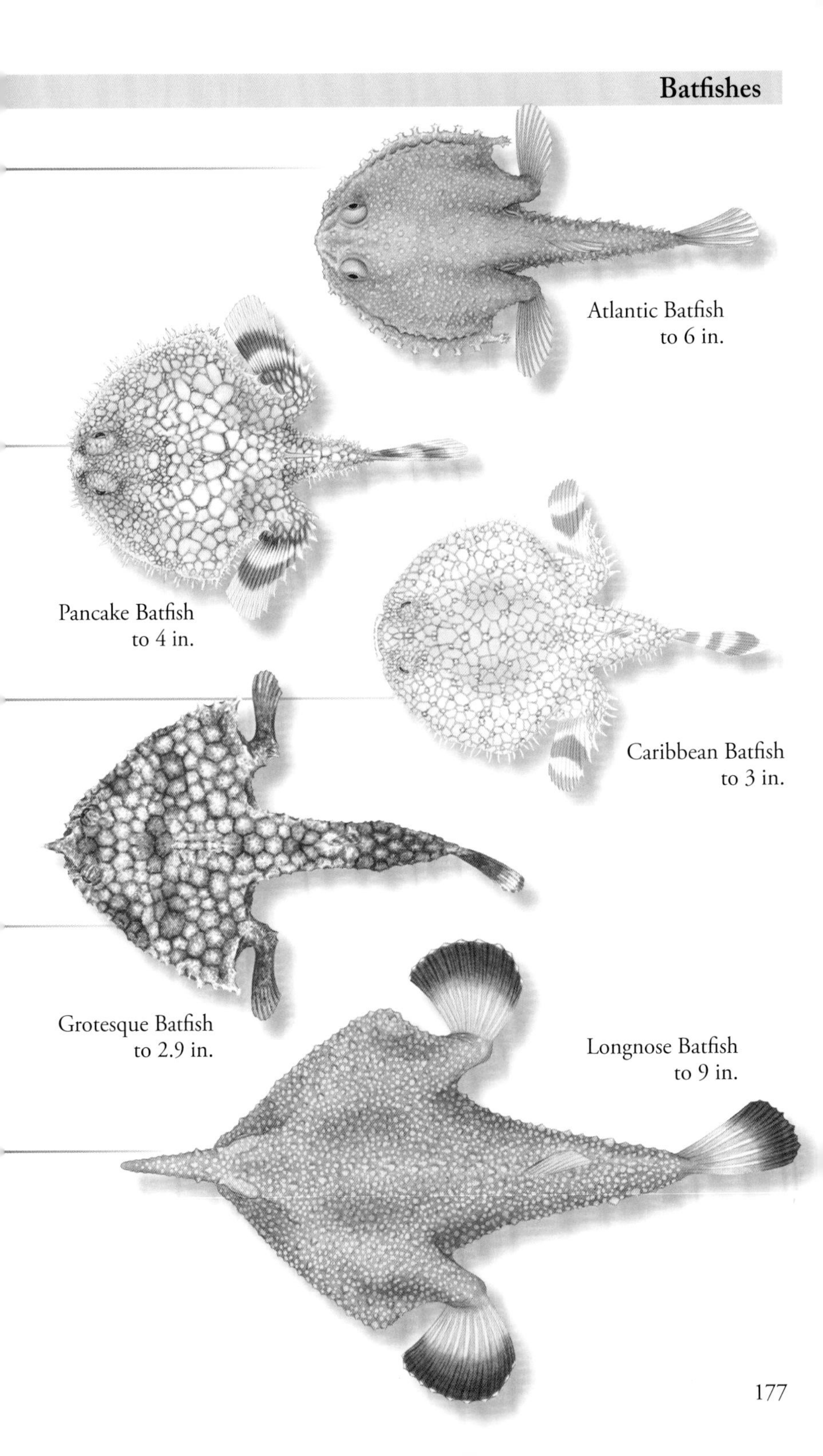

Atlantic Batfish
to 6 in.

Pancake Batfish
to 4 in.

Caribbean Batfish
to 3 in.

Grotesque Batfish
to 2.9 in.

Longnose Batfish
to 9 in.

Polka-dot Batfish - *Ogcocephalus cubifrons* (Richardson, 1836)

FEATURES: Tannish, brown, or reddish to grayish dorsally. Pale-ringed dark spots on shoulders and along sides. May have patches of yellow to orange dorsally. Pectoral fins always with wavy spots. Tubercles and bucklers scattered on dorsal surface. Fleshy tabs along margin of body and tail. Rostrum shortens with age. Pectoral fins lack fleshy pads ventrally. HABITAT: NC to FL, northeastern Gulf of Mexico, Yucatán, and Bahamas. Demersal. From shore to about 220 ft.

Shortnose Batfish - *Ogcocephalus nasutus* (Cuvier, 1829)

FEATURES: Tan to brownish dorsally with small clusters of spots on shoulders. May be somewhat uniformly colored or blotched. May also have spots under eyes and on lateral sides of tail. Pectoral fins dusky to almost black, darker toward margins. Caudal fin banded. Rostrum variable in length but relatively short and upturned. Body covered in close-set tubercles. Pectoral fins with fleshy pads ventrally. HABITAT: S FL, northern Gulf of Mexico, Bahamas, Caribbean Sea to mouth of Amazon River. Occur on bottom from shore to about 900 ft.

Roughback Batfish - *Ogcocephalus parvus* Longley & Hildebrand, 1940

FEATURES: Color and pattern highly variable. Tannish, pinkish, or yellowish to brownish dorsally. May be mottled or uniformly colored. Usually with patches of dark spots at shoulders. Pectoral fins pale to white at base with broad dark margins and white tips. Caudal fin colorless or with dark margin. Dorsal surface covered in irregular tubercles and bucklers. May also be covered dorsally with numerous irregular fleshy tabs. Rostrum finger-shaped to cone-shaped, variable in length. Pectoral fins with fleshy pads ventrally. HABITAT: NC to FL, E Gulf of Mexico, Caribbean Sea to Brazil. Reef associated. On bottom from about 95 to 410 ft.

Dwarf Batfish - *Ogcocephalus pumilus* Bradbury, 1980

FEATURES: Tannish with areas of lavender or pink and clusters of dark spots dorsally. Others described as blue gray. Inner margins of pectoral fins dark orange, outer margins dark. Ventral surface pale, suffused with orange. Rostrum long and horizontal to body. Dorsal surface covered with close-set tubercles and bucklers. Pectoral fins lack fleshy pads ventrally. HABITAT: S Bahamas, Puerto Rico, Lesser Antilles, and southern Caribbean Sea. Occur over soft bottoms from about 115 to 1,150 ft.

Tricorn Batfish - *Zalieutes mcgintyi* (Fowler, 1952)

FEATURES: Uniformly pale to dark brown, or olive brown dorsally. Pectoral and caudal fins pale and unmarked. Tubercles and bucklers cover dorsal surface of body and are paler than dorsal color. Rostrum short with prominent, recurved horns on either side. Pectoral fins lack fleshy pads ventrally. Tail long, tapered at end. HABITAT: FL, Gulf of Mexico, Caribbean Sea to northern South America. Found on bottom from about 295 to 590 ft.

ALSO IN THE AREA: *Ogcocephalus notatus*, see p. 547.

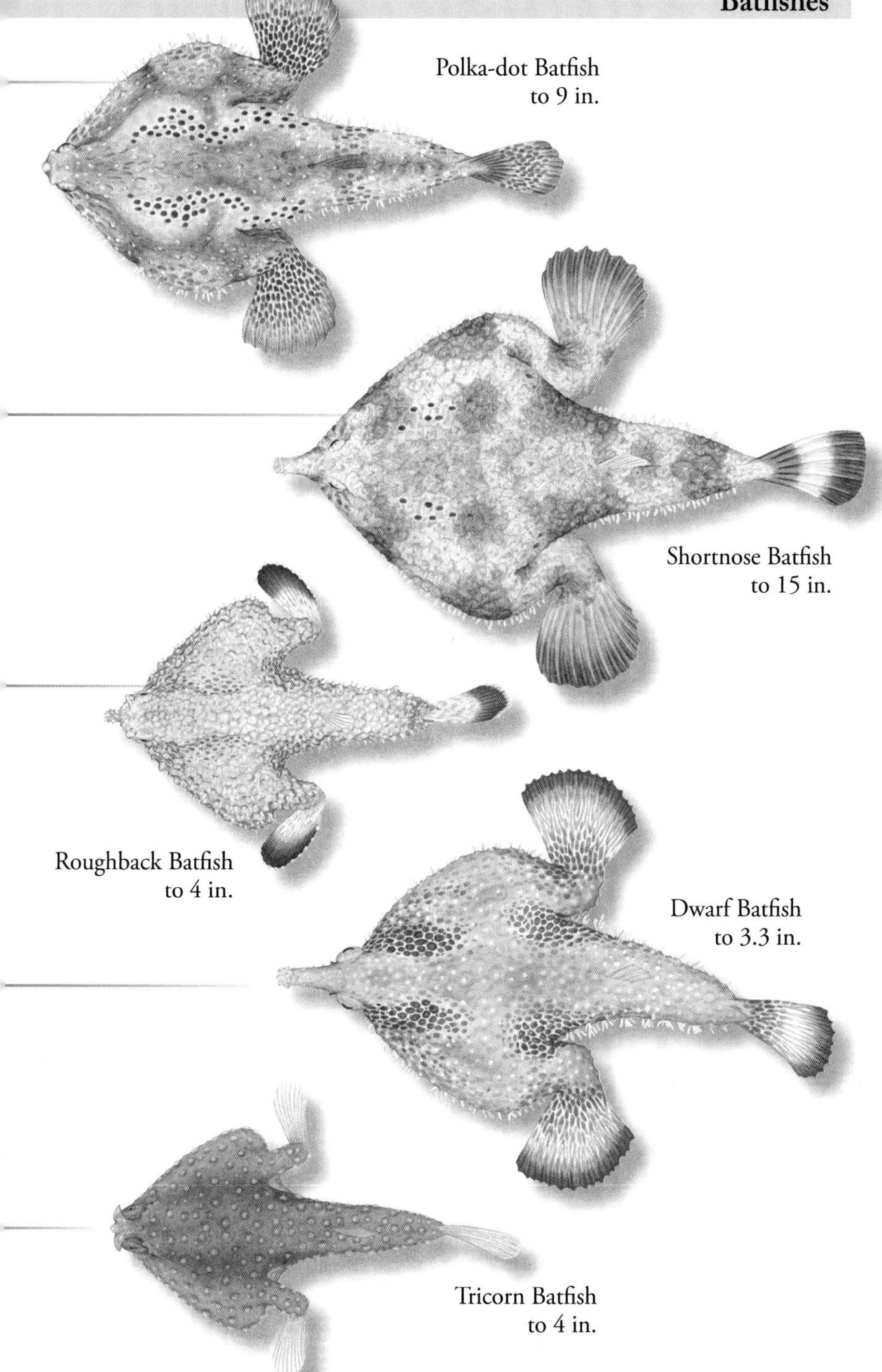
Polka-dot Batfish
to 9 in.
Shortnose Batfish
to 15 in.
Roughback Batfish
to 4 in.
Dwarf Batfish
to 3.3 in.
Tricorn Batfish
to 4 in.

Mugilidae - Mullets

Mountain Mullet - *Dajaus monticola* (Bancroft, 1834)

FEATURES: Brownish dorsally; whitish below. May have silvery band on sides. Dark spot at bases of pectoral and caudal fins. First dorsal fin yellowish with dark areas. Second dorsal and caudal fins yellowish with dusky margins. Adipose eyelids absent. Nape profile convex. Second dorsal and anal fins scaled on anterior base. HABITAT: NC, FL, Gulf of Mexico to Venezuela. Also Bahamas and Greater Antilles. Inshore and in brackish and fresh waters with sea access to about 20 ft. BIOLOGY: Adults are anadromous. Spawn in lower rivers and at sea. NOTE: Previously known as *Agonostomus monticola*.

Bobo Mullet - *Joturus pichardi* Poey, 1860

FEATURES: Grayish green dorsally, golden to silvery on sides, silvery white below. Fins dark with pale margins. Scales with dark inner and outer margins. Juveniles with dark bands on dorsal and caudal fins and a dark blotch on anal fin. Snout rounded, overhangs mouth. Adipose eyelids absent. Eyes moderate in juveniles, small in adults. Head deep and broad. HABITAT: Southwestern Gulf of Mexico to Colombia. Also Bahamas and Greater Antilles. Adults primarily in freshwater streams but probably spawn in lagoons or in the sea. Occur from shore to about 20 ft.

Striped Mullet - *Mugil cephalus* Linnaeus, 1758

FEATURES: Grayish olive to grayish brown dorsally. Sides silvery. Abdomen whitish to pale yellow. Dark spots on scales form stripes along sides. Dorsal and caudal fins dusky. Pectoral fins with dark spot at base. Second dorsal and anal fins with small scales on anterior base. Second dorsal-fin origin over anal-fin origin. HABITAT: Worldwide in tropical to warm temperate coastal, brackish, and fresh waters. Nova Scotia to FL; Gulf of Mexico to Brazil. Found near bottom to about 390 ft. BIOLOGY: Form large schools. Catadromous. Feed on plankton and detritus.

White Mullet - *Mugil curema* Valenciennes, 1836

FEATURES: Bluish green to olive dorsally. Sides and abdomen silvery to whitish. Yellowish to orangish blotch at upper edge of opercle. Pectoral fins with dark spot at base. Caudal fin with yellowish base and blackish margin. Second dorsal and anal fins almost entirely scaled. Anal fin with nine rays. HABITAT: MA to FL, Gulf of Mexico, and Caribbean Sea to Brazil. Also Bermuda. In tropical to warm temperate, coastal, and brackish waters from shore to about 100 ft. Occasionally in fresh water. BIOLOGY: Schooling and catadromous; sometimes enter rivers. Feed on plankton and detritus.

Dwarf Mullet - *Mugil curvidens* Valenciennes, 1836

FEATURES: Bluish gray dorsally, silvery on sides, silvery white below. Some gold to orange may be present on iris, on upper opercular margin, and on pectoral fins. Small dark blue spot on upper pectoral-fin origin. Second dorsal and anal fins fully scaled in adults. Anal fin with eight (rarely seven) rays. HABITAT: Bermuda, Bahamas, Antilles to Rio de Janeiro, Brazil. Also Ascension Island. Occur in shallow coastal waters from shore to about 100 ft. Adults over sand; juveniles in rock pools.

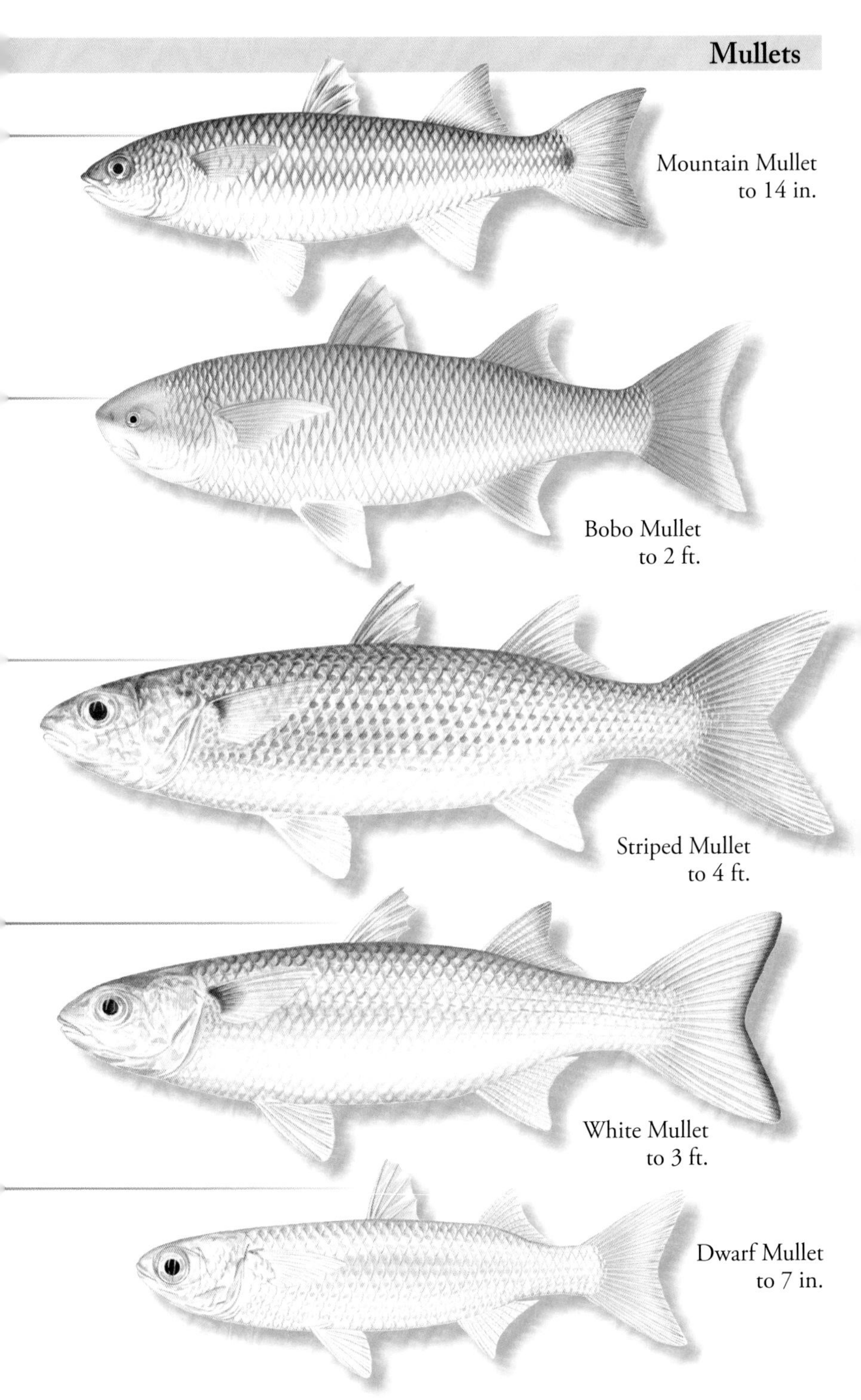
Mountain Mullet
to 14 in.
Bobo Mullet
to 2 ft.
Striped Mullet
to 4 ft.
White Mullet
to 3 ft.
Dwarf Mullet
to 7 in.

Hospe Mullet - *Mugil hospes* Jordan & Culver, 1895

FEATURES: Olivaceous dorsally; silvery on sides and below. Dorsal and caudal fins dusky. Pelvic and anal fins white. Small black blotch on upper pectoral-fin base. Snout pointed and compressed. Head triangular in profile. Pectoral fins long, reach first dorsal-fin origin. Anal fin with nine (rarely eight) rays. HABITAT: Colombia to Brazil. Also reported from coastal waters around Belize City. Found in inshore waters over sandy and muddy bottoms from shore to about 60 ft. May enter river mouths. Also in eastern Pacific from Mexico to Ecuador.

Parassi Mullet - *Mugil incilis* Hancock, 1830

FEATURES: Olivaceous to bluish gray dorsally; silvery on sides and below. Blackish blotch on pectoral-fin base. Dorsal and anal fins dusky to grayish. Second dorsal fin darker at tip. Caudal fin darker at margin. Anal and pelvic fins white. Second dorsal and anal fins well scaled. Anal fin with nine rays in adults. HABITAT: Haiti, Dominican Republic, and Panama to southeastern Brazil. Occur in shallow inshore waters and estuaries from shore to about 80 ft. May enter fresh water.

Liza - *Mugil liza* Valenciennes, 1836

FEATURES: Dusky bluish dorsally. Sides and abdomen silvery to white. Indistinct, dark stripes follow scale rows on upper sides. Pelvic fins pale or yellowish; other fins dusky. Head slender, triangular in profile. Second dorsal and anal fins with small scales on anterior base. Anal fin with eight rays. Caudal fin deeply forked. HABITAT: Southern FL, Bahamas, and Antilles to northern coast of South America. Also Bermuda. Occur in inshore marine and brackish waters from shore to about 400 ft. BIOLOGY: May occasionally enter fresh water. Schooling. Catadromous. Feed on algae and detritus.

Longtail Mullet - *Mugil longicauda* Guitart Manday & Alvarez-Lajonchere, 1976

FEATURES: Silvery gray dorsally. Silvery on sides and below. Opercle with some golden yellow. Pectoral, dorsal, and caudal fins dark; pelvic and anal fins white. Body comparatively long and slender. Caudal peduncle comparatively long. Second dorsal and anal fins covered with scales. HABITAT: Northern coast of Cuba. Found in shallow coastal waters from shore to about 30 ft.

Redeye Mullet - *Mugil rubrioculus* Harrison, Nirchio, Oliveira, Ron & Gaviria, 2007

FEATURES: Olivaceous to bluish dorsally; silvery on sides and below. Scales lack distinct dark spots. Irises red to orange. Small golden blotch may be present on upper portion of operculum. Small black spot present on upper pectoral-fin base. Second dorsal fin grayish, darker at tip. Second dorsal and anal fins fully scaled. Anal fin with 9 rays (10 in juveniles). HABITAT: Southern FL, Haiti, and Dominican Republic to northern coast of South America. Also Bermuda. Occur around mangroves, in estuaries, and in brackish lagoons from shore to about 30 ft. Adults do not enter fresh water.

Hospe Mullet
to 14 in.

Parassi Mullet
to 15.7 in.

Liza
to 3.3 ft.

Longtail Mullet
to 15.7 in.

Redeye Mullet
to 13 in.

Mugilidae - Mullets, *cont.*

Fantail Mullet - *Mugil trichodon* Poey, 1875

FEATURES: Dusky olive dorsally with bluish reflections. Silvery to whitish below. Pectoral fins with dark bluish spot at base. Dorsal and caudal fins dusky. Caudal fin with dusky margin. Second dorsal and anal fins scaled. Anal fin with eight rays (very rarely nine). HABITAT: Southern FL, Gulf of Mexico, and Caribbean Sea to NE Brazil. Also Bermuda. Occur in inshore marine and brackish waters and in river mouths from shore to about 80 ft. BIOLOGY: Prefer clear waters.

Atherinopsidae - New World Silversides

Robust Silverside - *Atherinella brasiliensis* (Quoy & Gaimard, 1825)

FEATURES: Translucent olivaceous dorsally, paler below. Others described as pale yellow above and below. Silvery on cheeks and abdomen. Silvery stripe on sides about as wide as iris diameter with dark upper edge. Pectoral fins comparatively long, reach to rear or tips of pelvic fins. First dorsal-fin origin over second to fifth anal-fin ray. Anal fin with 18–20 rays. HABITAT: Colombia to Brazil. Found along shallow beaches, in estuaries, and around mangroves from shore to about 30 ft.

Chargres Silverside - *Atherinella chagresi* (Meek & Hildebrand, 1914)

FEATURES: Greenish to yellowish green dorsally with dark cross-hatching on upper scales. Pale below. Silvery on cheeks and abdomen. Silvery stripe on sides about half eye diameter with dark upper edge. Pectoral fins reach to middle of pelvic fins. First dorsal-fin origin over third to sixth anal-fin ray. Anal fin long-based. HABITAT: Nicaragua to Panama. Primarily in fresh water but also along the coast in estuaries from shore to about 30 ft.

Miller's Silverside - *Atherinella milleri* (Bussing, 1979)

FEATURES: Translucent with dark speckles on dorsal scales. Silvery on cheeks and abdomen. Silvery stripe on sides narrow. Dark specks in two or three rows on upper portion of operculum and extend to pectoral-fin base. Pectoral fins large, extend past pelvic-fin tips. First dorsal-fin origin over seventh to ninth anal-fin rays. HABITAT: Northern Lagoon, Belize, to Westfalia, Costa Rica. Occur in shallow estuaries, lagoons, and around brackish water mangroves from shore to about 30 ft.

Querimana Silverside - *Melanorhinus microps* (Poey, 1860)

FEATURES: Dark bluish black on back; upper sides shades of blue. Silvery on sides and below. Silvery stripe on sides absent. Head profile rounded. Pectoral fins high on body, above eye midline. First dorsal fin with six to nine spines. Body deep and laterally compressed. Dorsal fin deeply forked. HABITAT: Bahamas and Greater and Lesser Antilles to coastal islands of northern Brazil. Also off Panama. Pelagic in coastal waters from surf zone to about 30 ft.

NOTE: There are at least eight other rare New World Silversides in the area.

Mullets

Fantail Mullet
to 9.8 in.

New World Silversides

Robust Silverside
to 6.3 in.

Chargres Silverside
to 4.5 in.

Miller's Silverside
to 5.5 in.

Querimana Silverside
to 3.6 in.

Atherinidae - Old World Silversides

Reef Silverside - *Atherina harringtonensis* Goode, 1877

FEATURES: Translucent greenish dorsally. Silvery below. Broad silver stripe with black margin bordered by iridescent green. Black flecks on dorsal scales only. Caudal-fin tips dusky to black. First dorsal fin with five to seven spines. HABITAT: S FL, E and S Gulf of Mexico, Bahamas, Antilles, and Bermuda to northern South America. Occur in coastal and offshore waters from shore to about 30 ft. BIOLOGY: Form large, dense schools.

Hardhead Silverside - *Atherinomorus stipes* (Müller & Troschel, 1848)

FEATURES: Translucent bluish green dorsally. Pale below. Silver stripe with black margin on sides. Scales flecked with black. Outer caudal-fin lobes blackish in large specimens. Head and eyes large. First dorsal fin with four to six spines. HABITAT: S FL, E and S Gulf of Mexico, Bahamas, and Caribbean Sea to Brazil. Found in coastal waters from shore to about 30 ft. BIOLOGY: Schooling and planktivorous.

Exocoetidae - Flyingfishes

Margined Flyingfish - *Cheilopogon cyanopterus* (Valenciennes, 1847)

FEATURES: Dark iridescent blue dorsally. Silvery white below. Pectoral fins bluish black with a thin, pale margin. Dorsal fin grayish with a large black blotch along central margin; blotch may cover entire fin in smaller specimens. Caudal fin dark gray to black. Pelvic and anal fins transparent white. Lower jaw protrudes slightly. Pectoral fins extend past middle of dorsal-fin base. HABITAT: NJ to Brazil, including Gulf of Mexico and Caribbean Sea. Also Bermuda and Indo-West Pacific. Found along coast to about 400 miles offshore from surface to about 20 ft.

Bandwing Flyingfish - *Cheilopogon exsiliens* (Linnaeus, 1771)

FEATURES: Dark iridescent blue dorsally. Silvery white below. Pectoral fin dark bluish black with a pale inner band that is broader ventrally. Dorsal fin grayish with a large, black blotch. Upper lobe of caudal fin pale, lower lobe blackish. Pelvic fins grayish; may be blackish at margin. Anal fin small and transparent. Lower jaw protrudes slightly. Pectoral fins extend to rear or past dorsal fin. HABITAT: Oceanic in temperate to tropical Atlantic. NJ to Brazil, including Gulf of Mexico, northern Caribbean Sea, and off Bermuda from surface to about 20 ft. May also enter clear, coastal waters.

Spotfin Flyingfish - *Cheilopogon furcatus* (Mitchill, 1815)

FEATURES: Dark iridescent blue dorsally. Silvery white below. Pectoral fins gray to blue gray with a clear band across center and a broad pale margin. Dorsal fin gray. Caudal fin gray to black. Pelvic fins with some gray at base and distally. Anal fin small and transparent. Lower jaw slightly protrudes. Pectoral fins reach past anal-fin base. HABITAT: Circumglobal in temperate to tropical seas. Oceanic from ME to Brazil, including Gulf of Mexico, Caribbean Sea, and off Bermuda from surface to about 20 ft. BIOLOGY: Feed on zooplankton and small fishes.

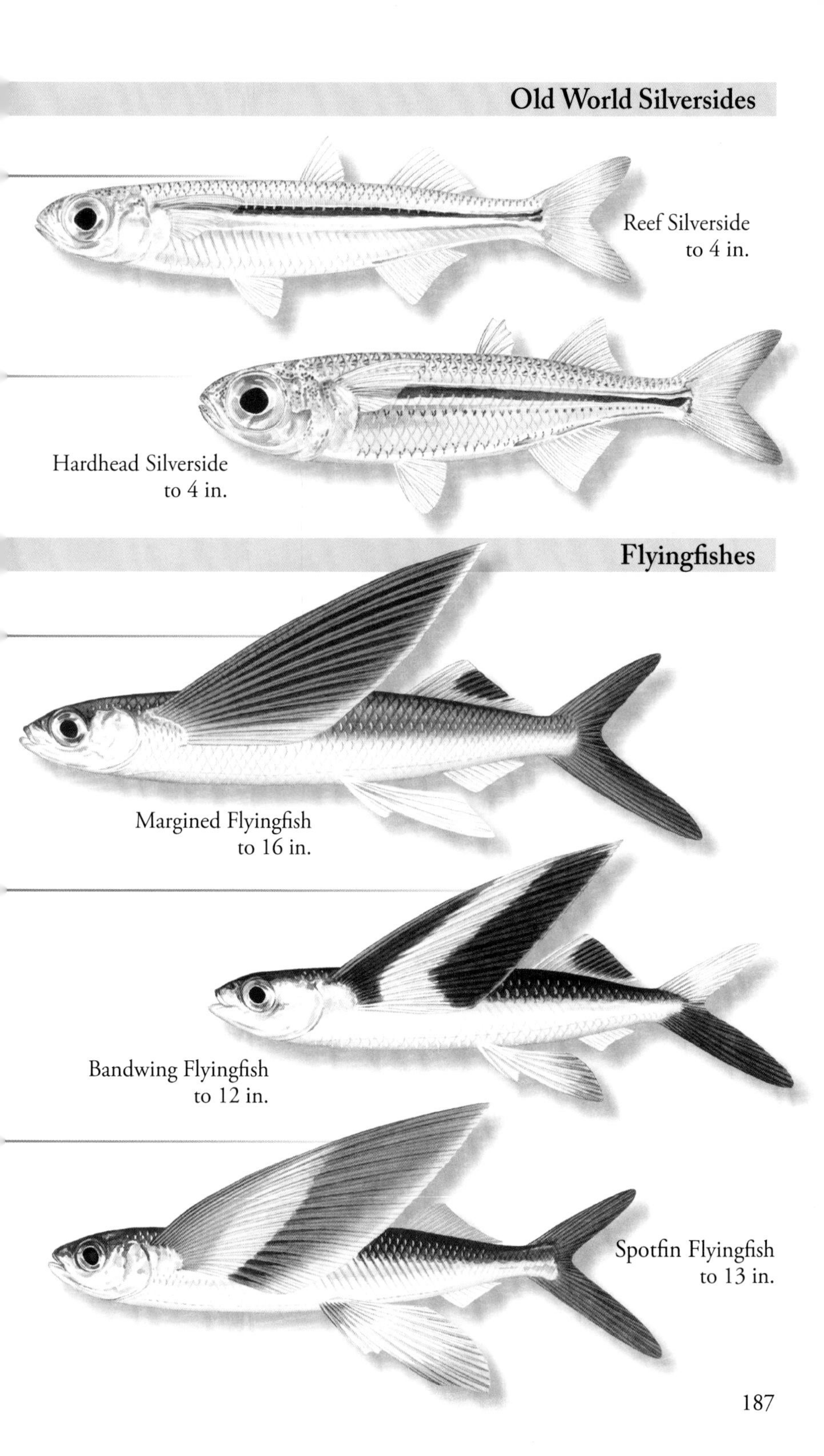
Old World Silversides
Reef Silverside
to 4 in.
Hardhead Silverside
to 4 in.
Flyingfishes
Margined Flyingfish
to 16 in.
Bandwing Flyingfish
to 12 in.
Spotfin Flyingfish
to 13 in.

Blotchwing Flyingfish - *Cheilopogon heterurus* (Rafinesque, 1810)

FEATURES: Dark iridescent blackish blue dorsally. Silvery white below. Pectoral fins coppery to grayish with pale inner band and pale margins. Dorsal fin translucent and unmarked. Caudal fin grayish. Lower jaw slightly protrudes. Pelvic fins white, may have some copper at base. Pectoral fins extend to rear dorsal-fin tip. HABITAT: Occur around Bermuda and in eastern Atlantic and Mediterranean Sea. Also in eastern Pacific. Occur from nearshore waters to well offshore from surface to about 20 ft.

Atlantic Flyingfish - *Cheilopogon melanurus* (Valenciennes, 1847)

FEATURES: Dark iridescent bluish dorsally. Silvery white below. Pectoral fins grayish with pale inner band and a thin, pale margin. Dorsal and caudal fins grayish. Anal and pelvic fins transparent white. Pectoral fins reach beyond middle of dorsal-fin base. Lower jaw slightly protrudes. HABITAT: MA to FL, Gulf of Mexico, Bermuda, and Caribbean Sea to Brazil. Occur over continental shelves to about 400 miles offshore from surface to about 20 ft. BIOLOGY: Feed on zooplankton. Leap and glide across the surface.

Clearwing Flyingfish - *Cypselurus comatus* (Mitchill, 1815)

FEATURES: Iridescent bluish dorsally. Silvery white below. Pectoral fins uniform pale gray to gray. Dorsal and caudal fins grayish. Pelvic fins transparent white, grayish at base. Snout comparatively pointed. Lower jaw slightly shorter than upper. Pectoral fins reach to end of dorsal-fin base. Anal fin originates under fourth to sixth dorsal-fin ray. HABITAT: NY to central Brazil. Also in central Gulf of Mexico and Caribbean Sea. Occur over continental shelves from surface to about 20 ft.

Oceanic Two-wing Flyingfish - *Exocoetus obtusirostris* Günther, 1866

FEATURES: Iridescent bluish dorsally. Silvery white below. Pectoral fins brownish gray with a broad, pale margin. Dorsal and caudal fins grayish. Pelvic and anal fins transparent white. Pectoral fins reach past dorsal-fin base. Pelvic fins small, originate forward of midbody line. Anal fin originates slightly anterior to dorsal-fin origin. HABITAT: NJ to FL, Gulf of Mexico, and Caribbean Sea to Brazil. Also Bermuda. Oceanic in warm waters. Found near surface to about 20 ft. BIOLOGY: Feed on zooplankton. Preyed upon by predatory fishes, squids, and sea birds.

Tropical Two-wing Flyingfish - *Exocoetus volitans* Linnaeus, 1758

FEATURES: Dark iridescent blue dorsally. Silvery white below. Pectoral fins grayish. Pelvic and anal fins transparent white. Caudal fin gray. Pectoral fins reach past dorsal fin. Pelvic fin small, forward of midbody line. Anal fin originates under first to third dorsal-fin ray. HABITAT: Worldwide in tropical seas. NJ to FL, eastern Gulf of Mexico, Caribbean Sea to S Brazil. Rare in Gulf of Mexico. Occur in offshore waters from surface to about 20 ft. Absent inshore. BIOLOGY: Feed mainly on copepods. Leap and glide across surface. Spawn intermittently. Eggs are pelagic.

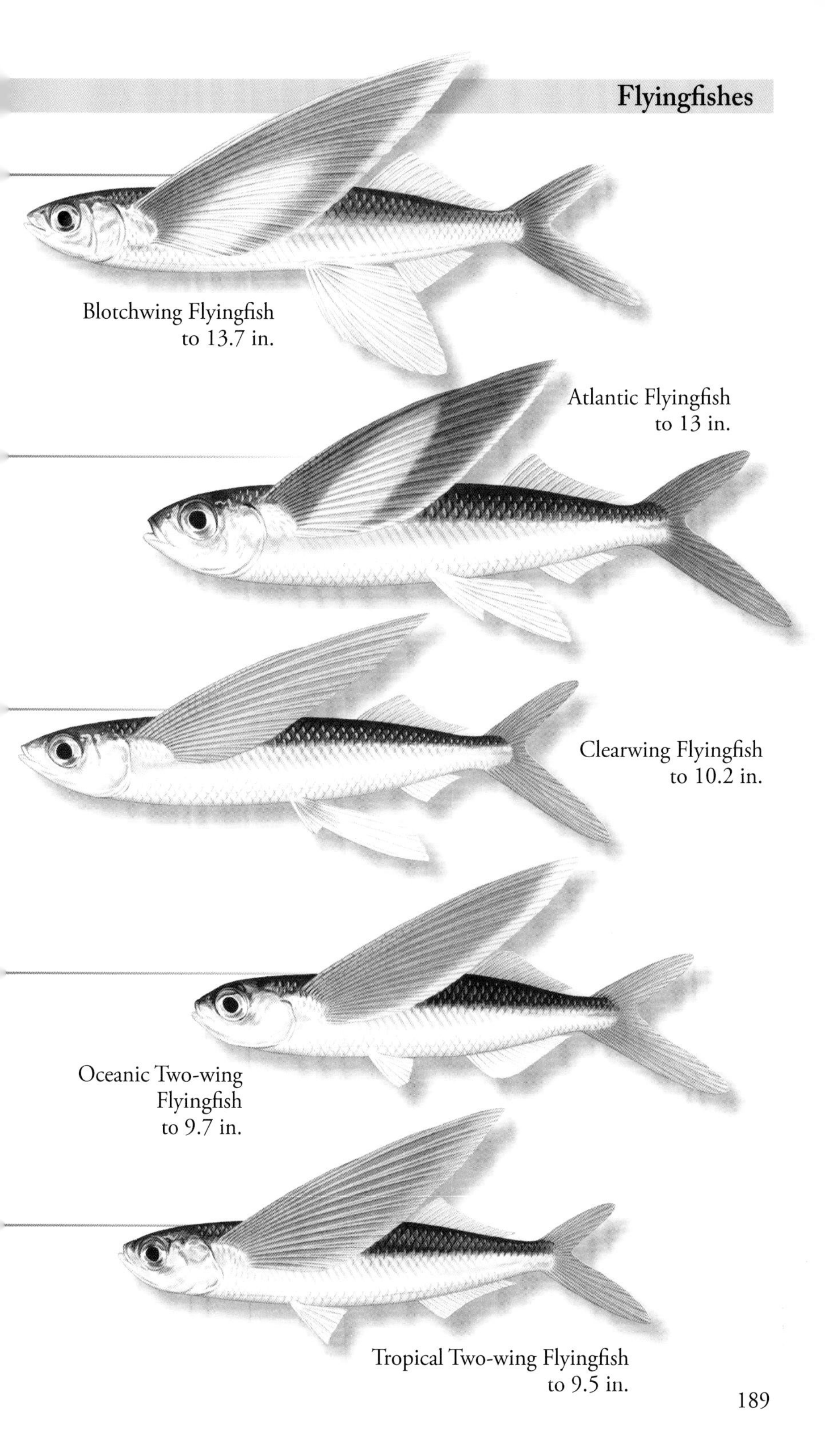
Blotchwing Flyingfish
to 13.7 in.
Atlantic Flyingfish
to 13 in.
Clearwing Flyingfish
to 10.2 in.
Oceanic Two-wing
Flyingfish
to 9.7 in.
Tropical Two-wing Flyingfish
to 9.5 in.

Fourwing Flyingfish - *Hirundichthys affinis* (Günther, 1866)

FEATURES: Dark iridescent bluish dorsally. Silvery white below. Pectoral fins gray with pale inner band and thin, pale margin. Caudal fin grayish. Pelvic and anal fins transparent white. Pectoral fins extend beyond end of dorsal-fin base. Anal-fin origin slightly anterior or posterior to dorsal-fin origin. HABITAT: VA to FL, Bermuda, Gulf of Mexico, Caribbean Sea to Brazil. Found over continental shelves from surface to about 70 ft. BIOLOGY: Preyed upon by dolphinfishes. Sought commercially in the Caribbean.

Blackwing Flyingfish - *Hirundichthys rondeletii* (Valenciennes, 1847)

FEATURES: Dark iridescent bluish dorsally. Silvery white below. Pectoral fin bluish black with thin, pale margin. Dorsal and caudal fins grayish. Anal fin transparent whitish. Pelvic fins usually with large, black blotch. Pectoral fins reach rear dorsal-fin base. Anal-fin origin slightly anterior or posterior to dorsal-fin origin. HABITAT: Nearly worldwide in temperate seas. MA to FL, Gulf of Mexico, Bahamas, Caribbean Sea to Brazil. Also Bermuda. Found in open oceanic waters from surface to about 70 ft.; also inshore. Juveniles in bays. BIOLOGY: Leap and glide for long distances across the surface. Feed on zooplankton.

Mirrorwing Flyingfish - *Hirundichthys speculiger* (Valenciennes, 1847)

FEATURES: Dark iridescent bluish dorsally. Silvery white below. Pectoral fin variably colored and patterned. Pale gray to dark gray with a short to broad clear band at lower center and a pale margin. Dorsal and caudal fins shades of gray. Anal and pelvic fins usually colorless; some with dark areas on pelvic fins. Pectoral fins reach past anal-fin base. Anal-fin origin slightly anterior or posterior to dorsal-fin origin. HABITAT: Circumtropical in warm seas. NJ to northern Argentina including Gulf of Mexico and Caribbean Sea. Found in oceanic waters from surface to about 70 ft. BIOLOGY: Feed on zooplankton. Eggs are laid on floating objects.

Sailfin Flyingfish - *Parexocoetus hillianus* (Gosse, 1851)

FEATURES: Iridescent greenish blue dorsally. Silvery below. Dorsal fin tall and expanded with large, black blotch. Pectoral fins transparent. Other fins pale to transparent. Upper jaw protrusible. Pectoral fins reach to about middle of dorsal fin. HABITAT: MA to N Brazil. Gulf of Mexico, Bermuda, and Caribbean Sea. Occur in inshore and offshore waters from surface to about 70 ft. BIOLOGY: Prey on crustacean plankton. They are preyed upon by other fishes and sea birds. NOTE: Previously known as *Parexocoetus brachypterus.*

ALSO IN THE AREA: *Prognichthys glaphyrae, Prognichthys occidentalis*, see p. 547.

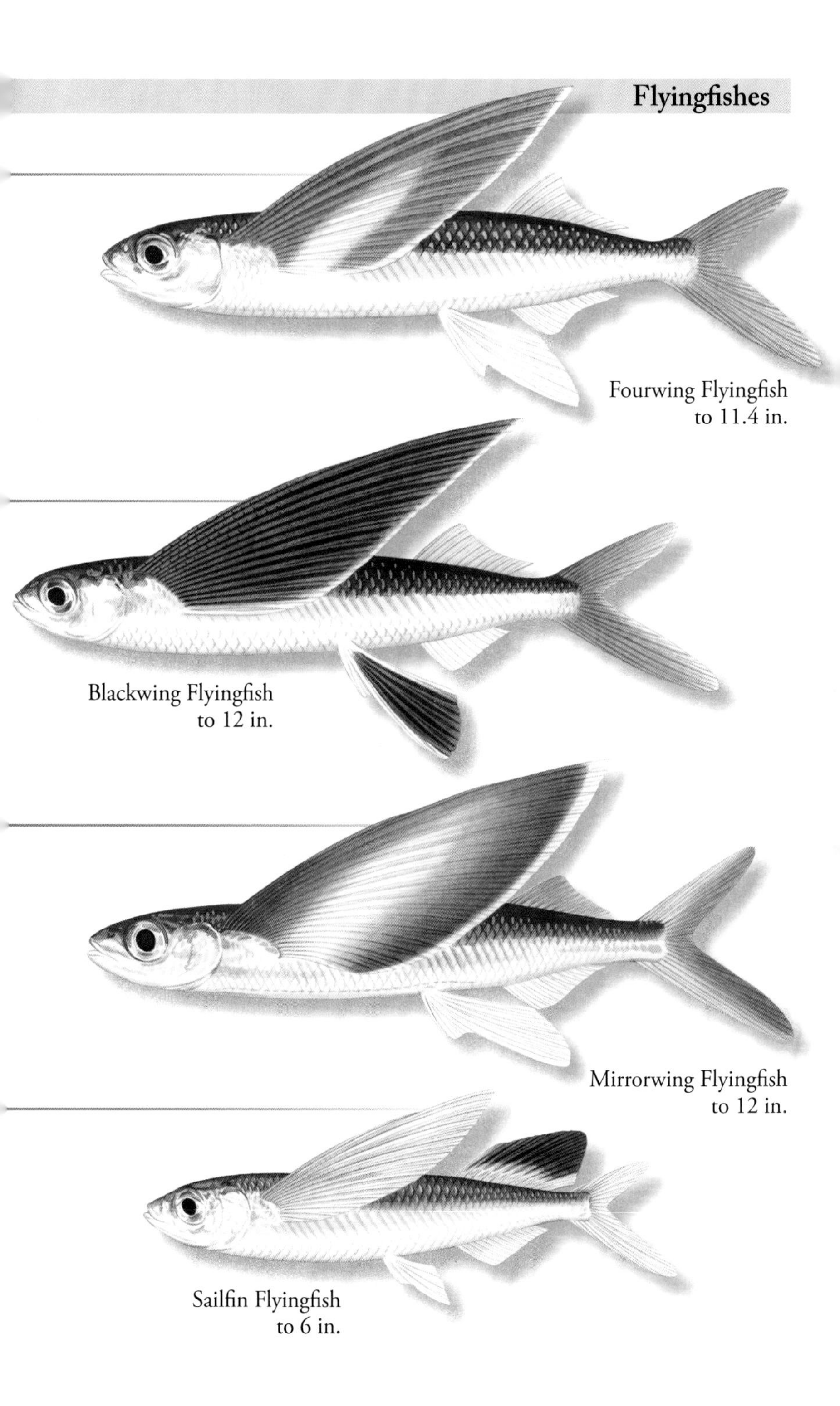

Fourwing Flyingfish
to 11.4 in.

Blackwing Flyingfish
to 12 in.

Mirrorwing Flyingfish
to 12 in.

Sailfin Flyingfish
to 6 in.

Hemiramphidae - Halfbeaks

Hardhead Halfbeak - *Chriodorus atherinoides* Goode & Bean, 1882

FEATURES: Translucent olive or tan dorsally; silvery below. Darker spots on scales form streaks. Silver stripe on sides. Lower jaw short, rounded. Caudal fin symmetrical. HABITAT: S FL, Bahamas, Cuba, Yucatán, and Belize. In clear nearshore waters and in brackish lakes or fresh water to about 20 ft.

Flying Halfbeak - *Euleptorhamphus velox* Poey, 1868

FEATURES: Olivaceous dorsally. Ventral area silvery. Lower jaw very long and pointed. Pectoral fins long. Caudal fin deeply forked. Body very long and ribbon-like. HABITAT: RI to FL, Gulf of Mexico, Caribbean Sea to Brazil. Also Bermuda and in eastern Atlantic. Pelagic. Found along coast and offshore to about 20 ft. BIOLOGY: May leap and glide for short distances.

Balao - *Hemiramphus balao* Lesueur, 1821

FEATURES: Bluish dorsally. Silvery white below. Both upper and lower caudal-fin lobes bluish violet. Lower jaw long and dark with red tip. Caudal fin deeply forked. HABITAT: NY to FL, Gulf of Mexico, Bahamas, and Caribbean Sea to Brazil. Found at the surface of inshore waters to about 20 ft. BIOLOGY: Form large schools. Feed primarily on planktonic invertebrates. Sought as bait and foodfish.

Ballyhoo - *Hemiramphus brasiliensis* (Linnaeus, 1758)

FEATURES: Dark bluish green dorsally. Silvery white below. Upper lobe of caudal fin yellowish orange. Lower jaw long and dark with red tip. Caudal fin deeply forked. HABITAT: MA to FL, Gulf of Mexico, Bahamas, and Caribbean Sea to Brazil. Occur at the surface of inshore waters to about 20 ft. BIOLOGY: Form large schools. Feed on seagrasses and invertebrates. Sought as bait. Often caught with Balao.

Bermuda Silverstripe Halfbeak - *Hyporhamphus collettei* Banford, 2010

FEATURES: Translucent greenish dorsally. Silvery below. Silver stripe on sides. Lower jaw tip red. Caudal-fin margin dark. Lower jaw long and with red tip. First gill arch with 26–32 rakers. Pectoral fins short. Only bases of dorsal and anal fins scaled. Caudal fin emarginate. HABITAT: Bermuda. Occur inshore and at surface to about 30 ft. BIOLOGY: Form large schools. Mainly caught as bait.

Slender Halfbeak - *Hyporhamphus roberti* (Valenciennes, 1847)

FEATURES: Translucent greenish dorsally with three black lines. Silvery below. Silver stripe on sides bordered above by narrow dark lines. Caudal-fin margin dark. Lower jaw long and with red tip. Usually more than 38 gill rakers on first arch. Pectoral fins short. Scales usually absent from dorsal and anal fins, or with only a few anteriorly. Caudal fin emarginate. HABITAT: Southwestern Gulf of Mexico to southern Brazil. Absent from northern and eastern Caribbean islands. Found in estuaries and river mouths to about 100 ft.

Hardhead halfbeak
to 6.5 in.

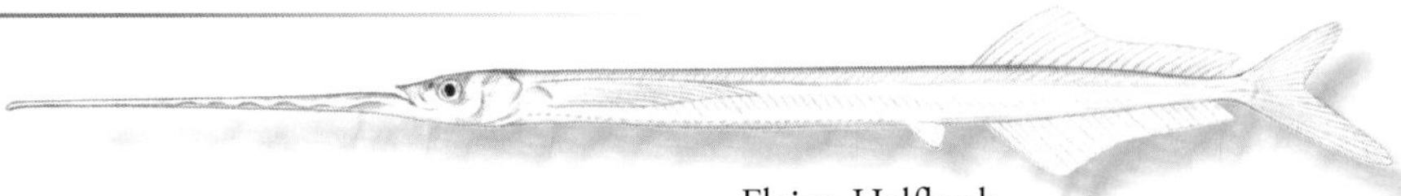

Flying Halfbeak
to 11 in.

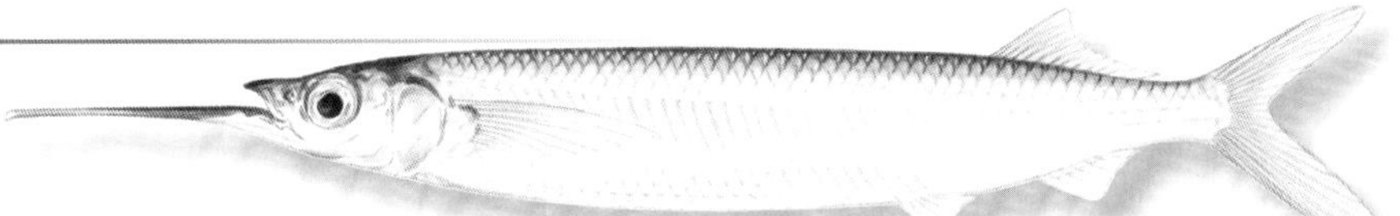

Balao
to 13.8 in.

Ballyhoo
to 13.8 in.

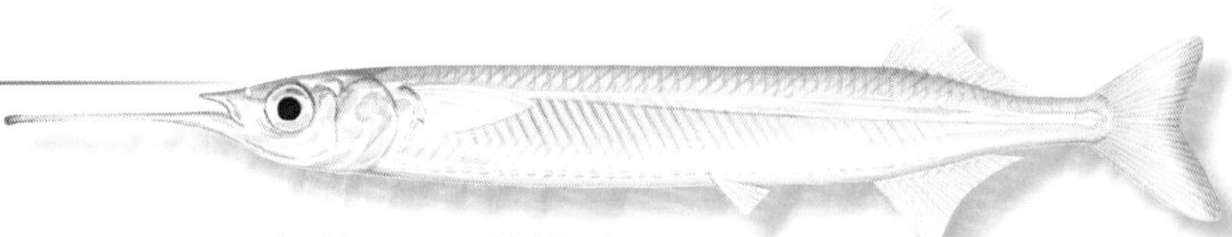

Bermuda Silverstripe Halfbeak
to 10 in.

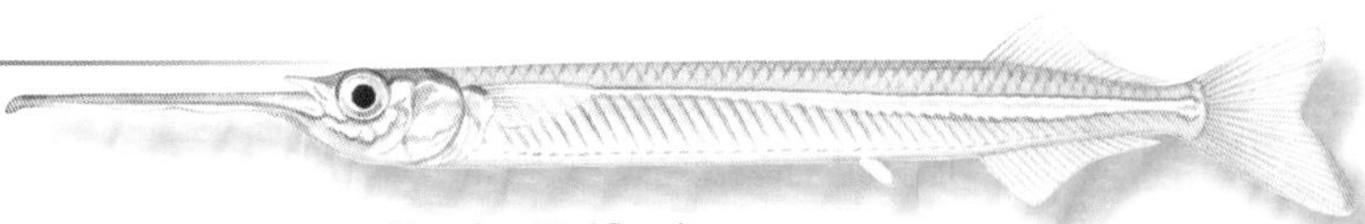

Slender Halfbeak
to 8.3 in.

Hemiramphidae - Halfbeaks, *cont.*

Atlantic Silverstripe Halfbeak - *Hyporhamphus unifasciatus* (Ranzani, 1841)

FEATURES: Transparent greenish dorsally. Silvery below. Silvery stripe on sides. Three thin black lines on back from head to dorsal fin. Lower jaw long and with a red tip. Caudal-fin margin blackish. First gill arch with 27–35 rakers. Dorsal and anal fins covered with scales. Caudal fin emarginate. HABITAT: SW FL, Bahamas, Caribbean Sea, Veracruz, and Bermuda. Occur inshore and at surface to about 100 ft. Also enter estuaries.

Smallwing Flyingfish - *Oxyporhamphus similis*, Bruun, 1935

FEATURES: Greenish blue dorsally. Silvery on sides and below. Lower jaw slightly elongate in juveniles, short and slightly projecting in adults. Pectoral fins long, but do not reach pelvic-fin origin. Pelvic fins small, triangular. Caudal fin deeply forked. HABITAT: Circumtropical. NC to FL, Gulf of Mexico, and Caribbean Sea to the equator. Also Bermuda. Marine, pelagic, and near surface to about 20 ft. Found inshore and offshore.

ALSO IN THE AREA: *Hemiramphus bermudensis*, *Hyporhamphus meeki*, see p. 547.

Belonidae - Needlefishes

Flat Needlefish - *Ablennes hians* (Valenciennes, 1846)

FEATURES: Bluish dorsally; silvery below. Dark bars on sides. Dorsal fin with expanded, black, posterior lobe. Dorsal and anal fins with a tall, falcate anterior lobe. Caudal keels absent. Body laterally compressed. HABITAT: Worldwide in tropical and subtropical seas. MA to FL, Gulf of Mexico, Bahamas, Caribbean Sea to Brazil. Also Bermuda. Common offshore to about 40 ft. BIOLOGY: Feed mainly on fishes.

Keeltail Needlefish - *Platybelone argalus* (Lesueur, 1821)

FEATURES: Greenish dorsally; silvery below. Narrow stripe on sides. Fins transparent. Lower jaw up to 20% longer than upper jaw. Well-developed, bluish black caudal keels present. Body round in cross-section. HABITAT: Worldwide in tropical and warm temperate seas. VA to FL, Gulf of Mexico, Bahamas, and Caribbean Sea to Trinidad. Occur from surface to about 20 ft. BIOLOGY: Feed on small fishes. Attracted to lights.

Atlantic Needlefish - *Strongylura marina* (Walbaum, 1792)

FEATURES: Bluish green dorsally; silvery below. Black area behind eyes. Blue stripe on sides. Caudal keels absent. Body oval in cross-section. HABITAT: MA to FL; Gulf of Mexico to Brazil. Occur coastally and in brackish and fresh water to about 20 ft.

Redfin Needlefish - *Strongylura notata* (Poey, 1860)

FEATURES: Bluish green dorsally; silvery below. Black bar on preopercular margin. Dorsal, caudal, and anal fins with areas of red to orange. Caudal keels absent. HABITAT: FL, Bahamas, Antilles, and coast of Central America. Occur inshore and in bays and estuaries to about 20 ft.

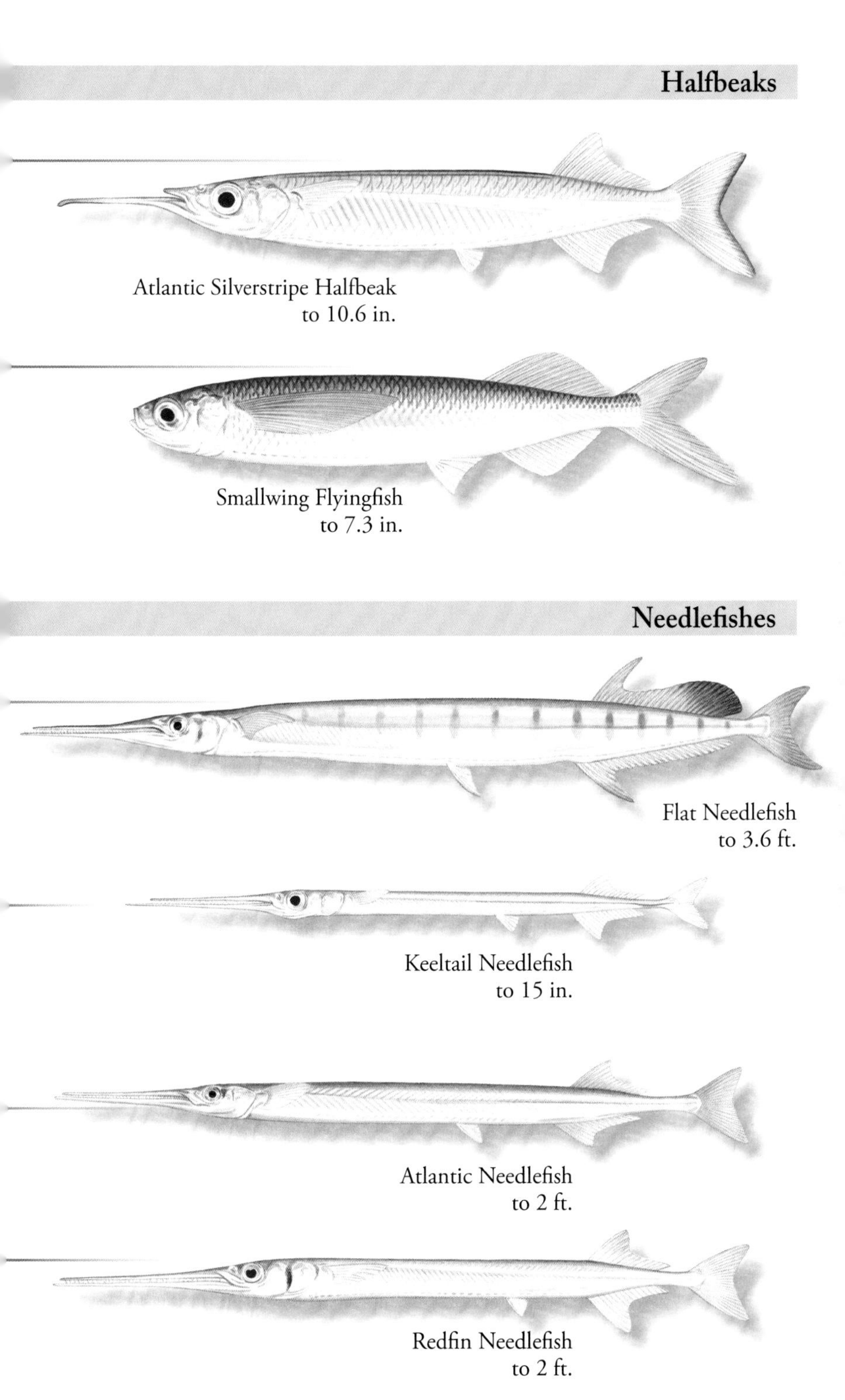
Halfbeaks
Atlantic Silverstripe Halfbeak
to 10.6 in.
Smallwing Flyingfish
to 7.3 in.
Needlefishes
Flat Needlefish
to 3.6 ft.
Keeltail Needlefish
to 15 in.
Atlantic Needlefish
to 2 ft.
Redfin Needlefish
to 2 ft.

Belonidae - Needlefishes, *cont.*

Timucu - *Strongylura timucu* (Walbaum, 1792)

FEATURES: Bluish green to dusky green above; silvery below. Dusky stripe on sides. Dark pigment in front of and behind eyes. Caudal keels absent. Body oval in cross-section. Predorsal scales comparatively large, number 120 to 185. HABITAT: FL, southern Gulf of Mexico, Bahamas, and Caribbean Sea to Brazil. In coastal waters, lagoons, and estuaries to about 20 ft. BIOLOGY: Feed mainly on fishes. Young hide in seagrasses.

Atlantic Agujon - *Tylosurus acus* (Lacepède, 1803)

FEATURES: Dark bluish dorsally; silvery below. Dark blue stripe on sides. Posterior dorsal-fin lobe blackish; elevated in juveniles. Small, blackish caudal keels present. Beak about twice the length of the head. HABITAT: Circumglobal in tropical to warm temperate seas. MA to FL, Bermuda, Gulf of Mexico, Bahamas, and Caribbean Sea to Brazil. Occur coastally and offshore to about 30 ft. BIOLOGY: Feed on fishes.

Houndfish - *Tylosurus crocodilus* (Péron & Lesueur, 1812)

FEATURES: Dark bluish green to greenish dorsally; silvery below. Blue stripe on sides. Juveniles with black, elevated posterior dorsal-fin lobe. Small, black caudal keels present. Beak about one and a half times length of head. HABITAT: Circumglobal in warm seas. NC to FL, Bermuda, Bahamas, and Caribbean Sea to Brazil. Occur along the coast in near surface waters to about 30 ft. BIOLOGY: Prey on fishes.

Rivulidae - New World Rivulines

Giant Rivulus - *Anablepsoides hartii* (Boulenger, 1890)

FEATURES: Color and pattern somewhat variable. Shades of brown dorsally; sides iridescent blue to white; abdomen pearly white. Reddish brown spots on scales form six to eight connected and broken stripes on sides. Dorsal fin with reddish brown spots. Male caudal fin with yellow on upper and lower caudal-fin margins; anal fin white basally and yellow orange distally with reddish brown spots. Females with a white patch followed by a black blotch on upper caudal peduncle; anal fin yellowish with dark spots. Head flattened on top. Upper jaw protrusible. HABITAT: Gulf of Venezuela to Trinidad and Tobago. Found in very shallow coastal brackish and fresh waters.

Mangrove Rivulus - *Kryptolebias marmoratus* (Poey, 1880)

FEATURES: Females mottled and blotched in shades of brown or tan to olive with a dark blotch above pectoral fin and a white-ringed black ocellus on upper caudal peduncle. Males are generally more orange with black fin margins that fade with age. HABITAT: E FL to Tampa Bay, FL, Bahamas, and West Indies to Brazil. Found in poorly oxygenated brackish marshes and swamps. BIOLOGY: Typically hermaphroditic and can self-fertilize and lay fertilized eggs. Males are rare. Feed on invertebrates.

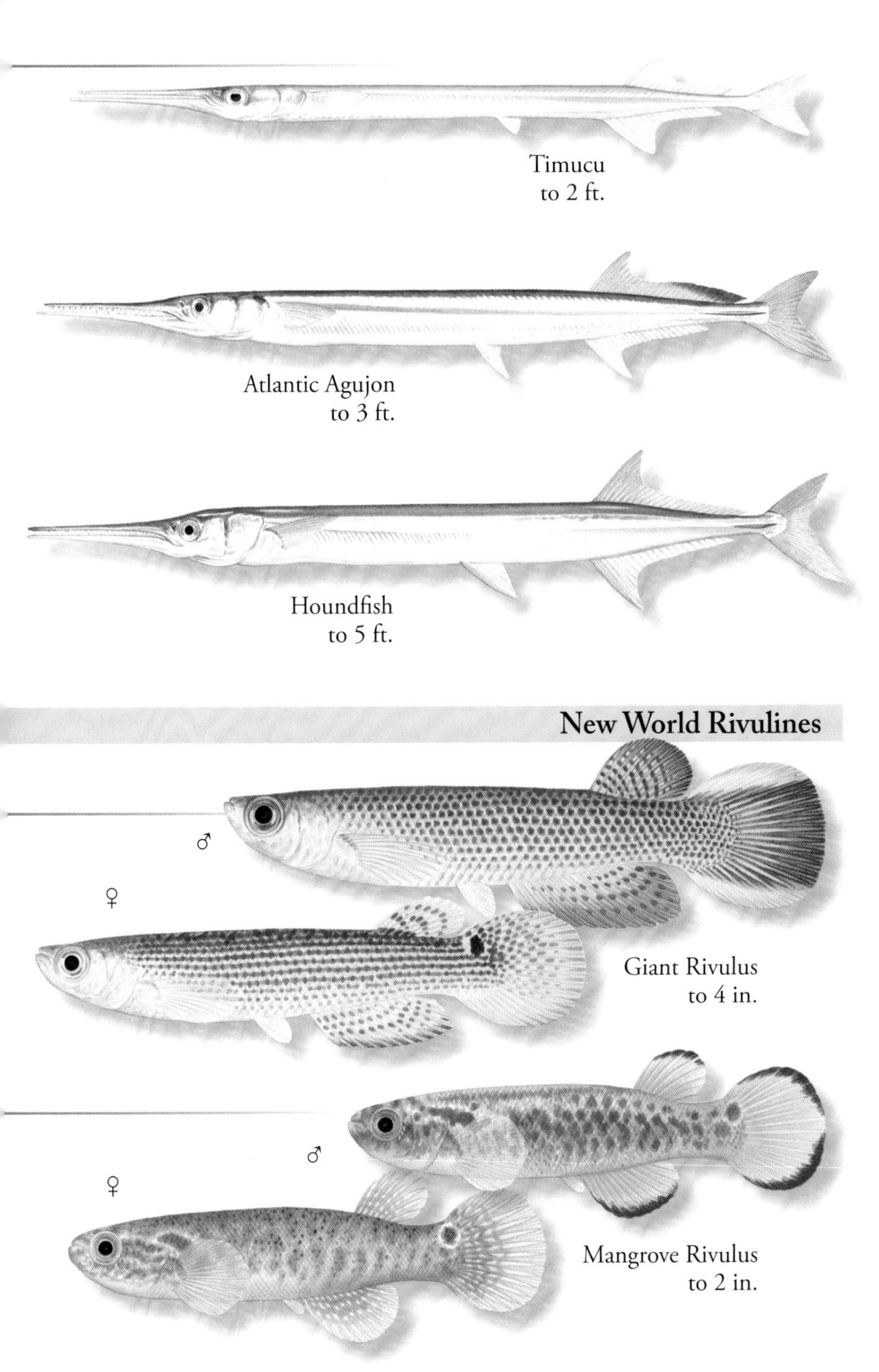
Needlefishes
Timucu
to 2 ft.
Atlantic Agujon
to 3 ft.
Houndfish
to 5 ft.
New World Rivulines
♂
♀
Giant Rivulus
to 4 in.
♂
♀
Mangrove Rivulus
to 2 in.

Fundulidae - Topminnows

Mummichog - *Fundulus heteroclitus* (Linnaeus, 1766)

FEATURES: Females: olivaceous above with pale abdomen; may have dusky bars; fins unmarked. Males: olivaceous above; pale below with yellow tinge; sides spotted, with up to 15 pale bars; fins spotted, with yellowish margins; dark spot usually present on dorsal fin. Both with convex snout profile and eight pores on lower jaw. Caudal-fin margin broadly rounded. HABITAT: Labrador to NE FL. Also Bermuda. In coastal bays, marshes, and channels and over seagrass flats. NOTE: *Fundulus bermudae* is a synonym of *F. heteroclitus*.

NOTE: There are five other rare Topminnows in the area.

Cyprinodontidae - Pupfishes

Sheepshead Minnow - *Cyprinodon variegatus* Lacepède, 1803

FEATURES: Olive or gray to brown dorsally and on sides with alternating, irregular bars. May be very pale. Females: one or two ocelli on dorsal fin. Males: blue highlights dorsally; margin of caudal fin blackish. Both with convex snout profile and deep body. Caudal fin of both broad with nearly straight margin. HABITAT: Cape Cod to FL, Gulf of Mexico, Bahamas, Cuba, Jamaica to Venezuela. In shallow, coastal, vegetated, fresh, brackish, and marine waters. BIOLOGY: Male Sheepshead Minnows defend territories. Females may have multiple spawns with up to 140 eggs.

Yucatán Flagfish - *Garmanella pulchra* Hubbs, 1936

FEATURES: Females white to tan with silvery reflections and six to seven dusky gray bars that become more narrow and numerous with age. Males tan to orange with silvery spots on dorsal scales and vague dark saddles; lower sides silvery to pinkish with many broken and wavy brownish lines and spots; fins orange with rows of brown spots basally. Both sexes with a blackish oblique bar under eyes and a black blotch on sides above pectoral fins. Dorsal fin comparatively long-based. HABITAT: Western Yucatán Peninsula to Belize. Found in shallow pools and coastal lagoons in brackish and fresh waters.

NOTE: There are five other fresh and brackish water Pupfishes in the area.

Anablepidae - Four-eyed Fishes

Finescale Four-eyes - *Anableps microlepis* Müller & Troschel, 1844

FEATURES: Shades of greenish gray to bluish or brownish gray dorsally; silvery white below. Sides with five to six gray stripes on sides; upper and lower stripes create an inverted C-shape at peduncle. Snout short, mouth protrusible; head flattened. Eyes elevated, with horizontally divided pupils. Body with 76–83 lateral scales. Pectoral fins broad; dorsal fin located posteriorly. Anal fin modified in males, used for internal fertilization. HABITAT: Trinidad and the Orinoco Delta to the Amazon Delta. Occur near or at surface in shallow waters around mangroves, in river mouths, and in fresh water.

ALSO IN THE AREA: *Anableps anableps*, see p. 547.

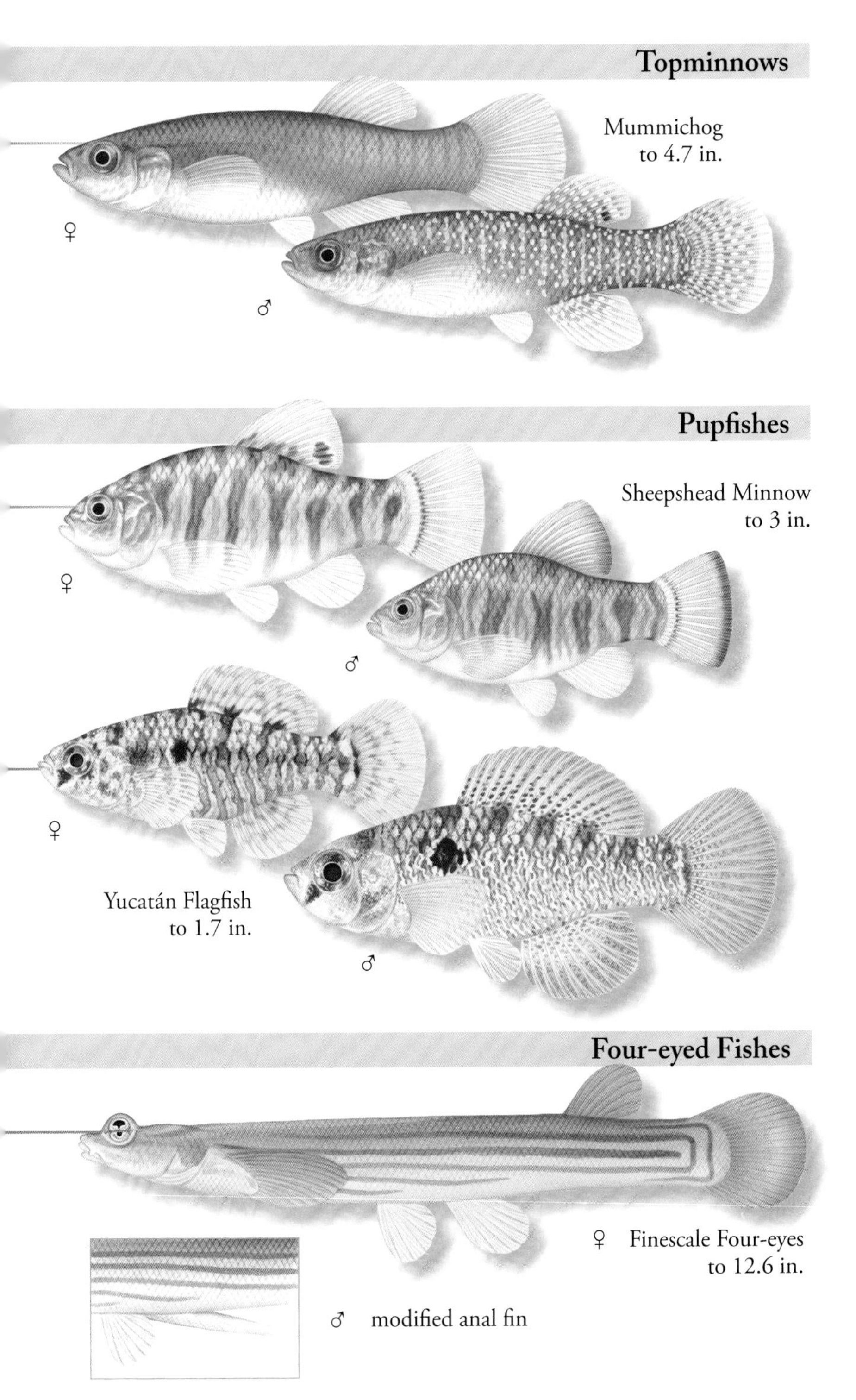
Topminnows
Mummichog
to 4.7 in.
♀
♂
Pupfishes
Sheepshead Minnow
to 3 in.
♀
♂
♀
Yucatán Flagfish
to 1.7 in.
♂
Four-eyed Fishes
♀ Finescale Four-eyes
to 12.6 in.
♂ modified anal fin

Sleek Mosquitofish - *Gambusia luma* Rosen & Bailey, 1963

FEATURES: Translucent pale olive dorsally; pale to white below. Abdomen silvery. Oblique dark bar under eyes. Dorsal and caudal fins with faint to distinct dark spots. Dark line runs along ventral portion of caudal peduncle. Body otherwise unmarked. Bodies of both sexes moderately robust. Anal-fin base anterior to dorsal-fin origin. Male gonopodium curved down at tip and lacks a fleshy palp. HABITAT: Belize to Honduras. Found in very shallow fresh and brackish waters.

Bahama Gambusia - *Gambusia manni* Hubbs, 1927

FEATURES: Shades of brownish olive dorsally and on sides. Abdomen paler. Oblique dark bar below eyes. Gravid females with a dark spot on abdomen. Female dorsal fin yellowish to pale orange with several rows of dark spots. Male dorsal fin orange to yellow with two (sometimes one or three) rows of dark spots. Caudal fin with one or two rows of dark spots. Anal-fin insertion below dorsal-fin origin. Bodies of both sexes moderately robust. Male gonopodium curved upward at tip and lacks a fleshy palp. HABITAT: Bahamas. Occur in shallow fresh to brackish waters, including tide pools and around mangroves.

Cuban Gambusia - *Gambusia punctata* Poey, 1854

FEATURES: Translucent tannish to greenish yellow dorsally. Abdomen pale to white. Dark bar below eyes absent. Spots form rows along scales on upper body. Dorsal fin with dusky to blackish margin and a row of dark spots at base. Caudal fin with small to minute dark spots. Anal-fin insertion below dorsal-fin origin. Bodies of both sexes moderately robust. Male gonopodium curved upward at tip and lacks a fleshy palp. HABITAT: FL Keys, Cuba, and coast of Belize. Primarily in shallow fresh water and streams; also in brackish waters.

Caribbean Gambusia - *Gambusia puncticulata* Poey, 1854

FEATURES: Shades of olive, pale olive to brownish green. Abdomen pale to silvery white. Body with few to many dark spots posteriorly. Dark bar below eyes present or absent. A faint to distinct stripe extends from behind eyes and fades behind pectoral fins. Dorsal and caudal fins with rows of dark spots in all except subspecies around Bahamas. Male gonopodium curved upward at tip and lacks a fleshy palp. HABITAT: Bahamas, Cuba, Jamaica, Cayman Islands, and tip of Yucatán Peninsula. Occur in fresh water and coastal brackish and marine waters.

Mangrove Gambusia - *Gambusia rhizophorae* Rivas, 1969

FEATURES: Tannish, grayish to olivaceous. Distinct blackish spots follow scale rows on sides. Dorsal fin yellowish with dark spots. Females: may have a gravid spot on abdomen; body deep; larger than males. Males: anal fin elongate at tip; body slender. HABITAT: S FL and Cuba. Primarily occur around Red Mangroves in marine and hypersaline waters. Also found somewhat inland in brackish water with White and Black Mangroves and in fresh water. BIOLOGY: Tolerant of high water temperatures. Feed on spiders and ants.

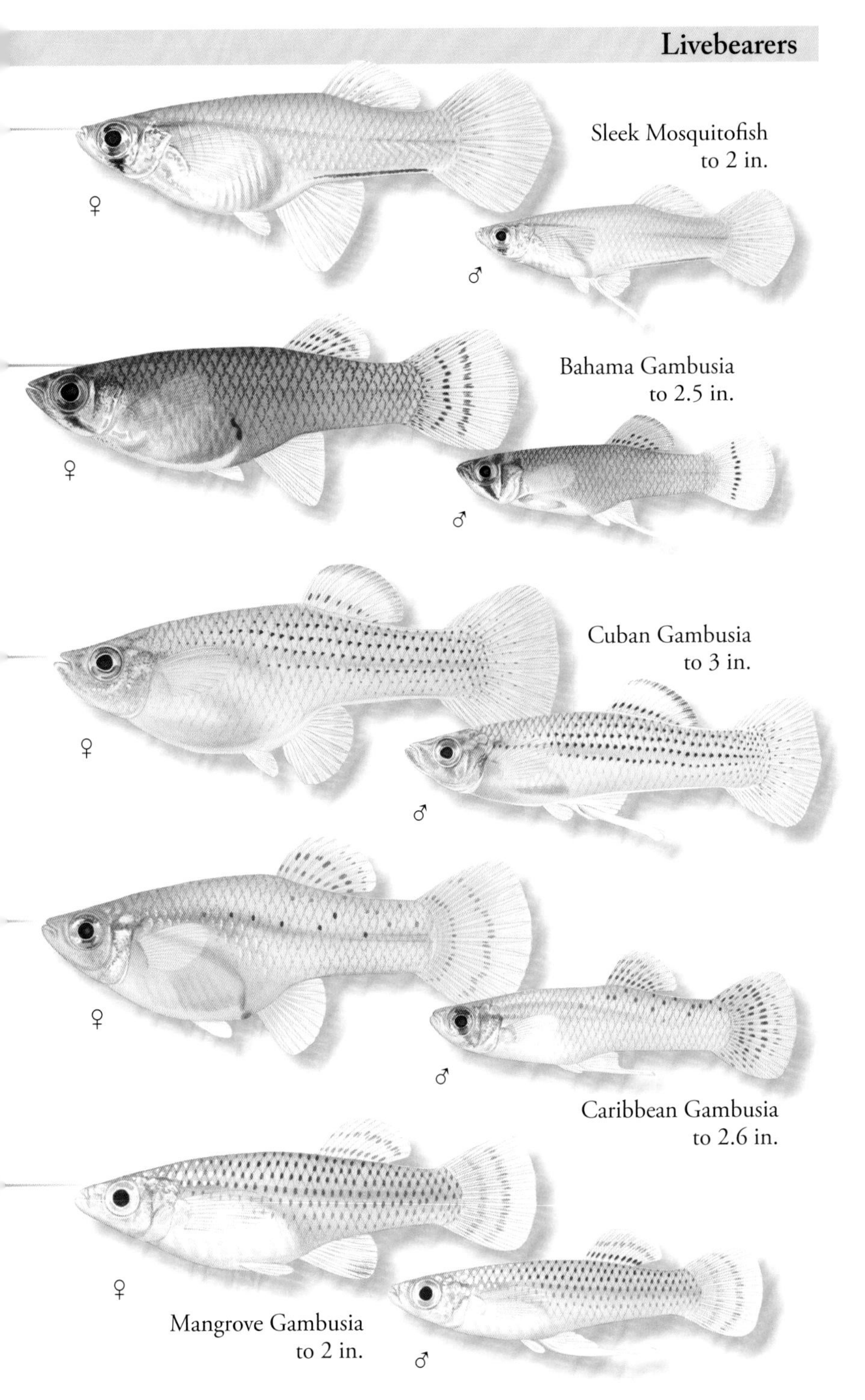
Sleek Mosquitofish
to 2 in.
♀
♂
Bahama Gambusia
to 2.5 in.
♀
♂
Cuban Gambusia
to 3 in.
♀
♂
♀
♂
Caribbean Gambusia
to 2.6 in.
♀
Mangrove Gambusia
to 2 in.
♂

Molly - *Poecilia gillii* (Kner, 1863)

FEATURES: Color and pattern variable. Shades of tan to grayish with silvery reflections and rows of orange to yellowish spots along scale rows. Females with few to many small dark spots on dorsal and/or caudal fins. Males with a large blackish area and black spots at base, black spots on caudal fin, and yellow to orange or reddish margins. Dorsal fin expanded and sometimes very long in males, short in females. Anal-fin origin under or a little anterior to dorsal-fin origin. Body robust; caudal peduncle comparatively deep. HABITAT: Guatemala to Panama. Occur primarily in fresh water; large adults also in shallow coastal brackish waters.

Sailfin molly - *Poecilia latipinna* (Lesueur, 1821)

FEATURES: Olivaceous to brownish dorsally. Pale below. Scales form diamond-shaped pattern. Spots on scales form stripes along sides. Females: dorsal fin low, banded; peduncle moderately deep. Males: dorsal fin tall, sail-like, and banded with a yellow to orange margin; caudal peduncle deep. HABITAT: NC to FL Keys; in Gulf of Mexico from SW FL to Yucatán. In weedy and quiet fresh, brackish, and marine waters. Prefer clear waters. BIOLOGY: Feed on algae, detritus, insects, and crustaceans. Tolerant of polluted waters. Southern populations are larger and more colorful than northern populations. Color variants are popular in the aquarium trade.

Shortfin Molly - *Poecilia mexicana* Steindachner, 1863

FEATURES: Color variable. Females (not shown): body color similar to males; fins transparent; caudal peduncle moderately deep. Males: olivaceous or brownish to grayish dorsally with orangish spots on scales; dorsal and caudal fins may be blackish and often with yellowish to orange margins; membranes may be spotted or blotched; caudal peduncle deep. Both with short-based dorsal fin. HABITAT: S TX to Colombia. Also southern islands of the West Indies. Possibly also FL. Occur in fresh to salt water, mostly in standing or slow-moving currents.

Mangrove Molly - *Poecilia orri* Fowler, 1943

FEATURES: Shades of dusky green to olive dorsally; sides and abdomen silvery bluish. Others described as bluish black dorsally. Faint orange spots form rows along scales. Males with a dark blotch at pectoral-fin base. Dorsal fin red with large black and white blotches at base and reddish caudal fins with black spots or dorsal and caudal fins yellowish orange with many black spots. Females with some black and white on dorsal fin. Anal-fin origin under or slightly anterior to dorsal-fin origin. HABITAT: Eastern Yucatán to Honduras. Occur in full salt water around mangroves, in lagoons, and along sandy beaches from shore to about 10 ft. Also in brackish to full fresh water.

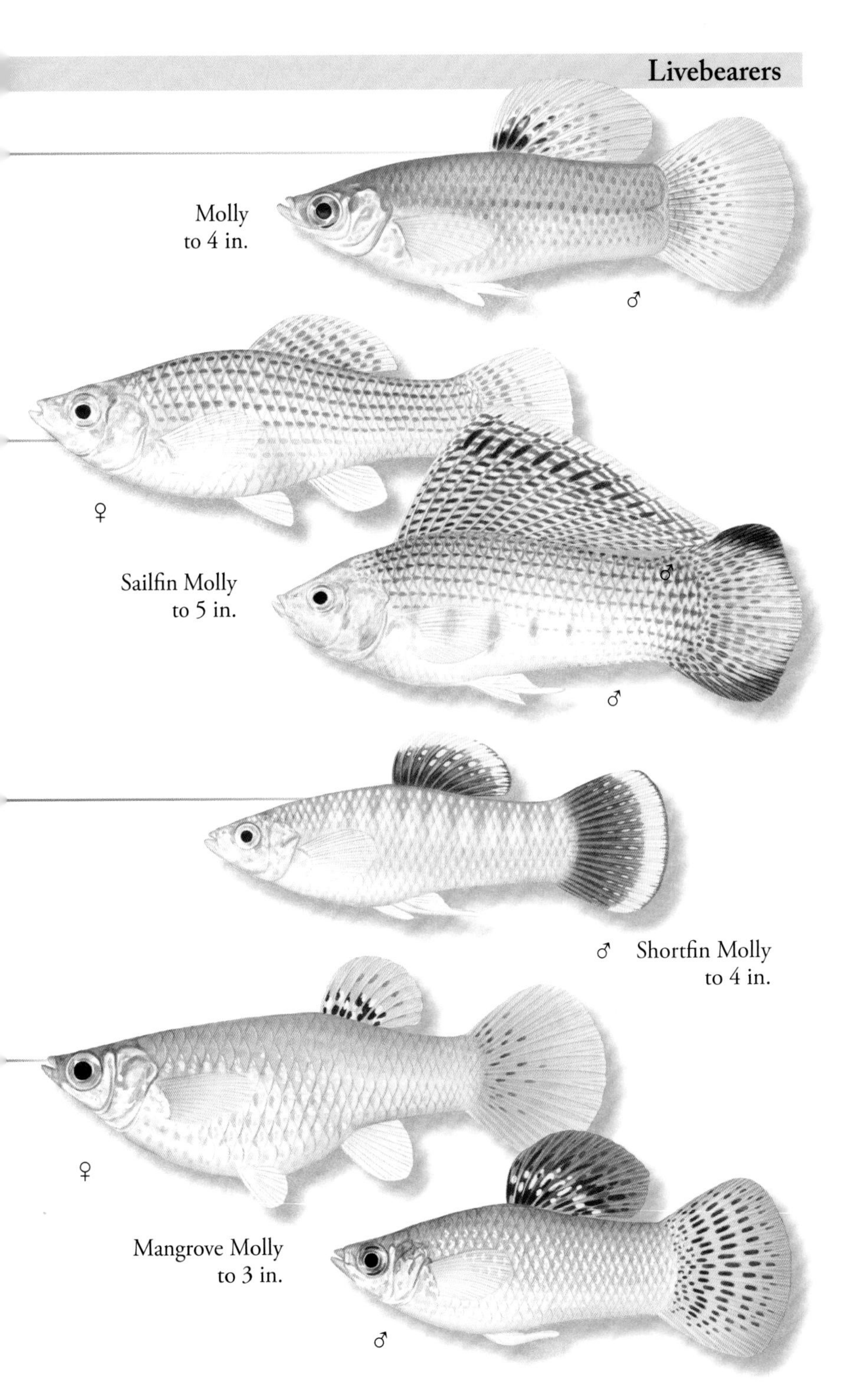
Molly
to 4 in.
♂
♀
Sailfin Molly
to 5 in.
♂
♂
♂ Shortfin Molly
to 4 in.
♀
Mangrove Molly
to 3 in.
♂

Guppy - *Poecilia reticulata* Peters, 1859

FEATURES: Female shades of grayish green to dusky; sides of head and abdomen silvery white; variably colored gravid spot on abdomen; fins colorless or with small dark scattered flecks. Male color and pattern highly variable; translucent grayish green to dusky with black, golden orange, and/or iridescent blue blotches, stripes, reticulations and/or spots on body and caudal fin. Dorsal fin moderate to tall; caudal fin rounded to bilobed or sword-like in males. HABITAT: Venezuela to Guyana and northern Brazil. Widely introduced in other regions. Occur in shallow, warm, and quiet vegetated streams, ponds, canals, ditches, and in brackish waters. BIOLOGY: Believed to be responsible for the decline of native fishes in Nevada, Wyoming, and Hawaii by the transport of parasites and by eating native fish eggs.

Island Molly - *Poecilia vandepolli* van Lidth de Jeude, 1887

FEATURES: Female shades of translucent pale yellowish to tannish orange with iridescent highlights; cheeks and abdomen silvery; indistinct spots on upper scale rows; dorsal and caudal fins with rows of dark spots. Male shades of translucent yellowish green to yellowish with iridescent highlights; some with spots along scale rows; cheeks silvery; abdomen silvery to golden; dorsal fin yellow to orange with a black blotch at base and black spots and/or margin above; caudal fin yellowish to orange with rows of black spots. All with deep caudal peduncle. HABITAT: Coast of Venezuela, and introduced around St. Martin. Occur in shallow fresh and brackish coastal waters.

Cuban Limia - *Poecilia vittata* Guichenot, 1853

FEATURES: Grayish to grayish green dorsally. Paler to silvery white on abdomen. Body and dorsal and caudal fins with large and small, irregular and scattered, black and orange blotches. Blotches may be sparse to very dense. Scales on upper body with faint to distinct dark margins. Dorsal fin more densely blotched in males. HABITAT: Cuba. Found along shallow coastal brackish waters and around mangroves. Also in freshwater lakes and streams opening to the ocean. BIOLOGY: Popular in the aquarium trade. NOTE: Previously known as *Limia vittata*.

Southern Molly - *Poecilia vivipara* Bloch & Schneider, 1801

FEATURES: Shades of tan to olivaceous or grayish dorsally. Silvery on cheeks and abdomen. Males may have some orange on sides. Both sexes usually with a very faint to distinct black blotch on upper sides that may be bordered anteriorly and posteriorly by silvery white. Dorsal fin with pale to dark orange and black spots near base. May also have faint to distinct black marks at upper and lower caudal-fin base or small spots on caudal-fin rays. HABITAT: Venezuela to Argentina. Introduced to Martinique and Puerto Rico. Occur in shallow salt, brackish, and hypersaline lagoons. Rarely in fresh water.

NOTE: There are 10 other primarily brackish or marine-water Livebearers in the area.

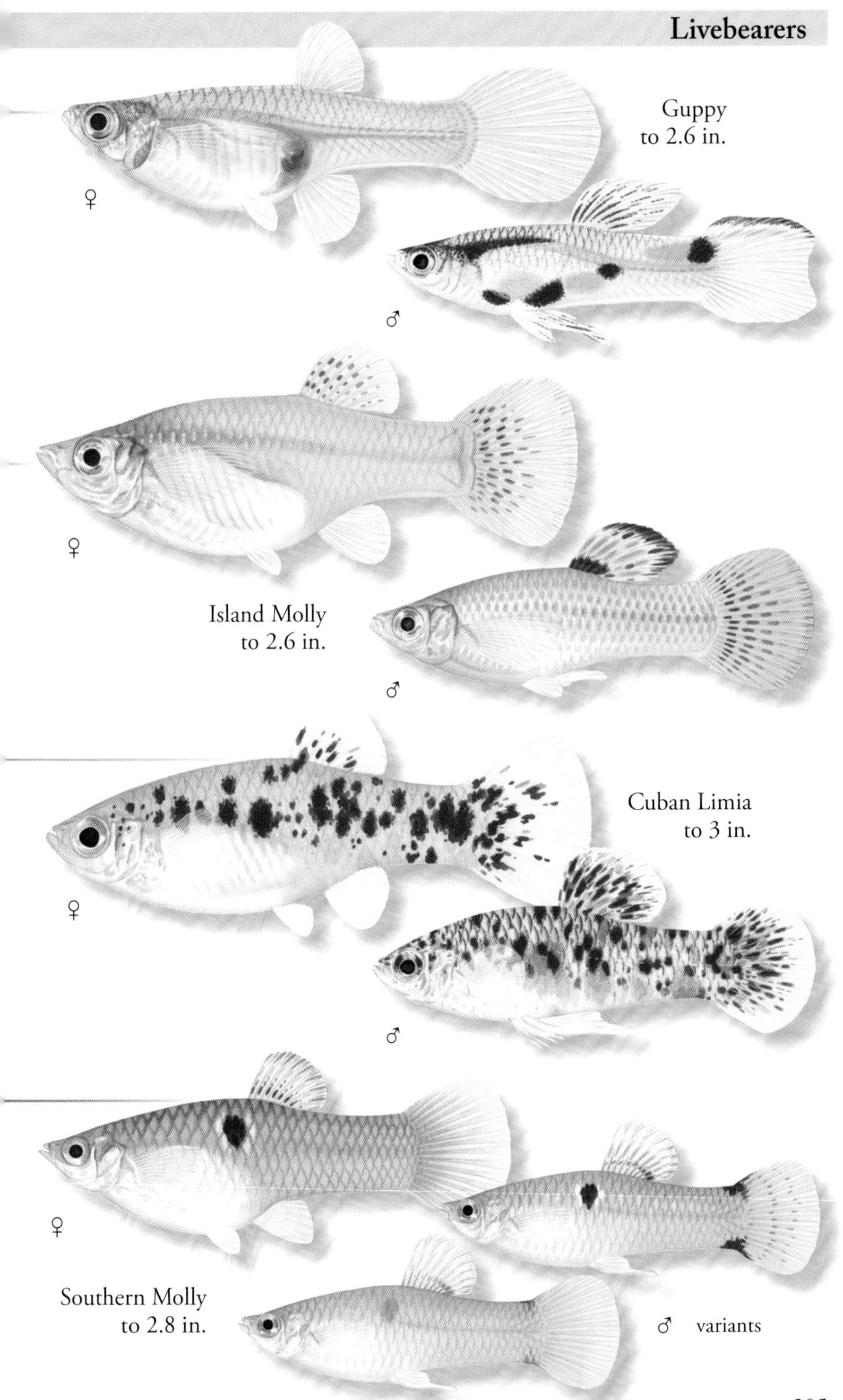
Guppy
to 2.6 in.
♀
♂
♀
Island Molly
to 2.6 in.
♂
Cuban Limia
to 3 in.
♀
♂
♀
Southern Molly
to 2.8 in.
♂ variants

Anomalopidae - Flashlight Fishes

Atlantic Flashlight Fish - *Kryptophanaron alfredi* (Silvester & Fowler, 1926)

FEATURES: Body and fin rays shades of black. Large white light organ under eyes visible when dark lower lid is retracted. Lateral-line scales whitish. Cheeks and abdomen may be washed with dark blue. Snout short and steep. Eyes very large. HABITAT: Scattered around Bahamas, Greater Antilles, and Caribbean Sea. Occur near bottom in caves and crevices from about 65 to 650 ft. Emerge at night to feed in the water column.

Trachichthyidae - Roughies

Big Roughy - *Gephyroberyx darwinii* (Johnson, 1866)

FEATURES: Reddish dorsally. Silvery red to pink below. Low ridges around eyes. Opercles with large spines at lower and upper corners. Abdomen with row of heavy scutes. Body deep, compressed. HABITAT: DE to Panama, northern Gulf of Mexico, and western Caribbean Sea. On or near bottom of slopes from about 240 to 2,000 ft. Young near coast.

Silver Roughy - *Hoplostethus mediterraneus* Cuvier, 1829

FEATURES: Pinkish dorsally. Sides and ventral area silvery grayish to blackish. All fins reddish. Snout short. Inside of mouth black. Low ridges and spines on head. Lateral-line scales enlarged. Row of 8–12 scutes along abdomen. Body deep, compressed. HABITAT: NJ to FL, Gulf of Mexico, Bermuda, and Greater Antilles. Also in Indian and western Pacific oceans. Occur near bottom in tropical to warm temperate waters from about 1,000 to 4,800 ft. BIOLOGY: Feed on crustaceans. Caught commercially.

Berycidae - Alfonsinos

Splendid Alfonsino - *Beryx splendens* Lowe, 1834

FEATURES: Pinkish red dorsally. Silvery below. Fins pinkish. Eyes large. Opercle with two spines. Pectoral fin erect. Single, midbody dorsal fin. Caudal fin deeply forked. HABITAT: Scattered worldwide in warm seas. ME to Brazil. Near bottom from about 650 to 3,300 ft. BIOLOGY: Form dense schools. Feed on fishes, crustaceans, and cephalopods.

Holocentridae - Squirrelfishes

Spinycheek Soldierfish - *Corniger spinosus* Agassiz, 1831

FEATURES: Head and body uniformly colored in shades of deep red. All fins red. Mouth and eyes large. Three long spines below each eye. Snout, cheeks, and opercles very spiny. Scales serrated. Body deep. Soft dorsal, caudal, and anal fins comparatively small. HABITAT: SC, FL, NE Gulf of Mexico, eastern Caribbean Sea, and Brazil. Also in the eastern Atlantic at St. Helena. Over deep rocky slopes and reefs from about 150 to 1,000 ft.

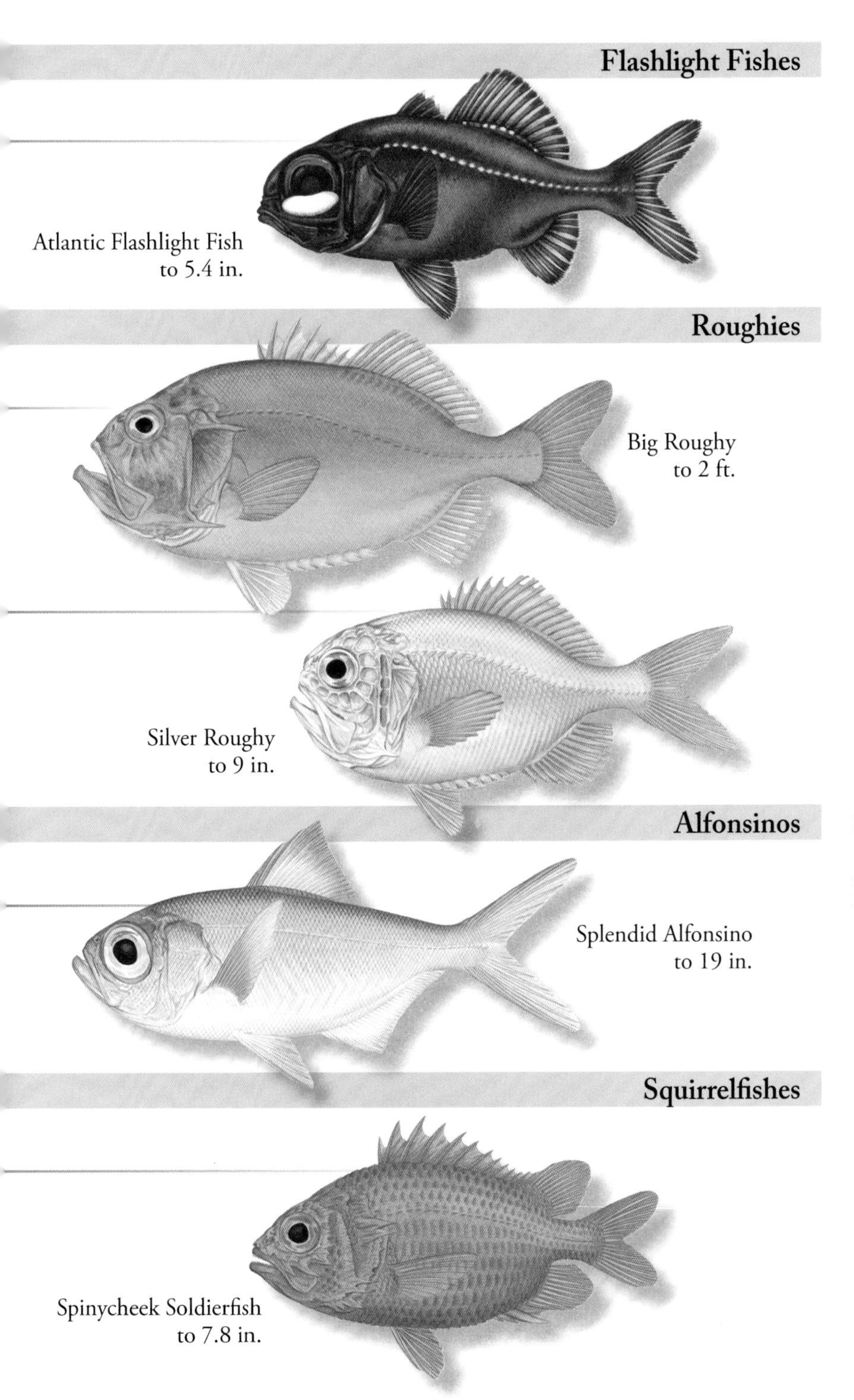
Flashlight Fishes
Atlantic Flashlight Fish
to 5.4 in.
Roughies
Big Roughy
to 2 ft.
Silver Roughy
to 9 in.
Alfonsinos
Splendid Alfonsino
to 19 in.
Squirrelfishes
Spinycheek Soldierfish
to 7.8 in.

Squirrelfish - *Holocentrus adscensionis* (Osbeck, 1765)

FEATURES: Body with alternating reddish and silvery stripes. May have pale blotches or bars. Snout red. White, diagonal streak from jaw to corner of preopercle. Spiny dorsal fin yellowish, other fins pale pinkish. Jaws extend to rear margin of pupil. Soft dorsal fin and upper lobe of caudal fin elongate. HABITAT: VA to FL, Bahamas, Caribbean Sea to Brazil. Also Bermuda and portions of Gulf of Mexico. Over shallow inshore coral reefs to deeper offshore waters to about 295 ft. BIOLOGY: Nocturnal. Seek shelter in crevices and under ledges during the day.

Longspine Squirrelfish - *Holocentrus rufus* (Walbaum, 1792)

FEATURES: Reddish dorsally with alternating dark and pale stripes. Usually with irregular, pale bars on sides. Top of head deep red. Eyes dark red. White stripe from jaw to preopercular corner and along preopercular margin. Spiny dorsal fin with a white blotch at upper margin of each membrane. Other fins pinkish. Jaws short. HABITAT: NC to FL, Bahamas, Caribbean Sea to Central and South America. Also Bermuda and portions of the Gulf of Mexico. Over shallow reefs to about 105 ft. BIOLOGY: Hide in crevices by day, feed over seagrass beds at night.

Blackbar Soldierfish - *Myripristis jacobus* Cuvier, 1829

FEATURES: Reddish dorsally. Ventral area pale to silvery. Broad, dark bar from top of opercle to rear pectoral-fin base. Spiny dorsal fin red with white at tips and between spines. Anterior margins of soft dorsal, anal, pelvic, and caudal fins white. Other fin portions reddish. Body and fins may appear pale to grayish. Eyes and mouth large. Small spine on opercle. HABITAT: NC to FL, Bahamas, Caribbean Sea to Brazil. Also Bermuda and portions of Gulf of Mexico. Over shallow coral reefs to deeper waters to about 295 ft. BIOLOGY: Nocturnal. Form aggregations.

Longjaw Squirrelfish - *Neoniphon marianus* (Cuvier, 1829)

FEATURES: Alternating reddish and yellowish stripes dorsally and on sides. Ventral area silvery. Upper head and snout reddish. White bar on preopercle and yellowish bar on opercle. Spiny dorsal fin yellow with white on margins and membranes. Other fins reddish to pinkish. Lower jaw protrudes. Upper jaw extends to below center of pupil. HABITAT: FL Keys, W Gulf of Mexico, Bahamas, Antilles to Trinidad and coast of Central America. Over patch reefs from surface to about 200 ft., more commonly below 50 ft. NOTE: Previously known as *Holocentrus marianus*.

Dusky Squirrelfish - *Neoniphon vexillarium* (Poey, 1860)

FEATURES: Alternating broad dusky red and whitish stripes dorsally and on sides. Silvery white below. Head dusky red above with dusky red band from lower eye to corner of preopercle. Body may have small, dark specks. Dorsal fin red with white areas around spines. Other fins reddish or yellowish to clear. HABITAT: FL, Gulf of Mexico, Bahamas, Caribbean Sea to northern South America. Also Bermuda. Occur over inshore reefs to about 65 ft. BIOLOGY: Nocturnal. Feed on invertebrates and small fishes. NOTE: Previously known as *Sargocentron vexillarium*.

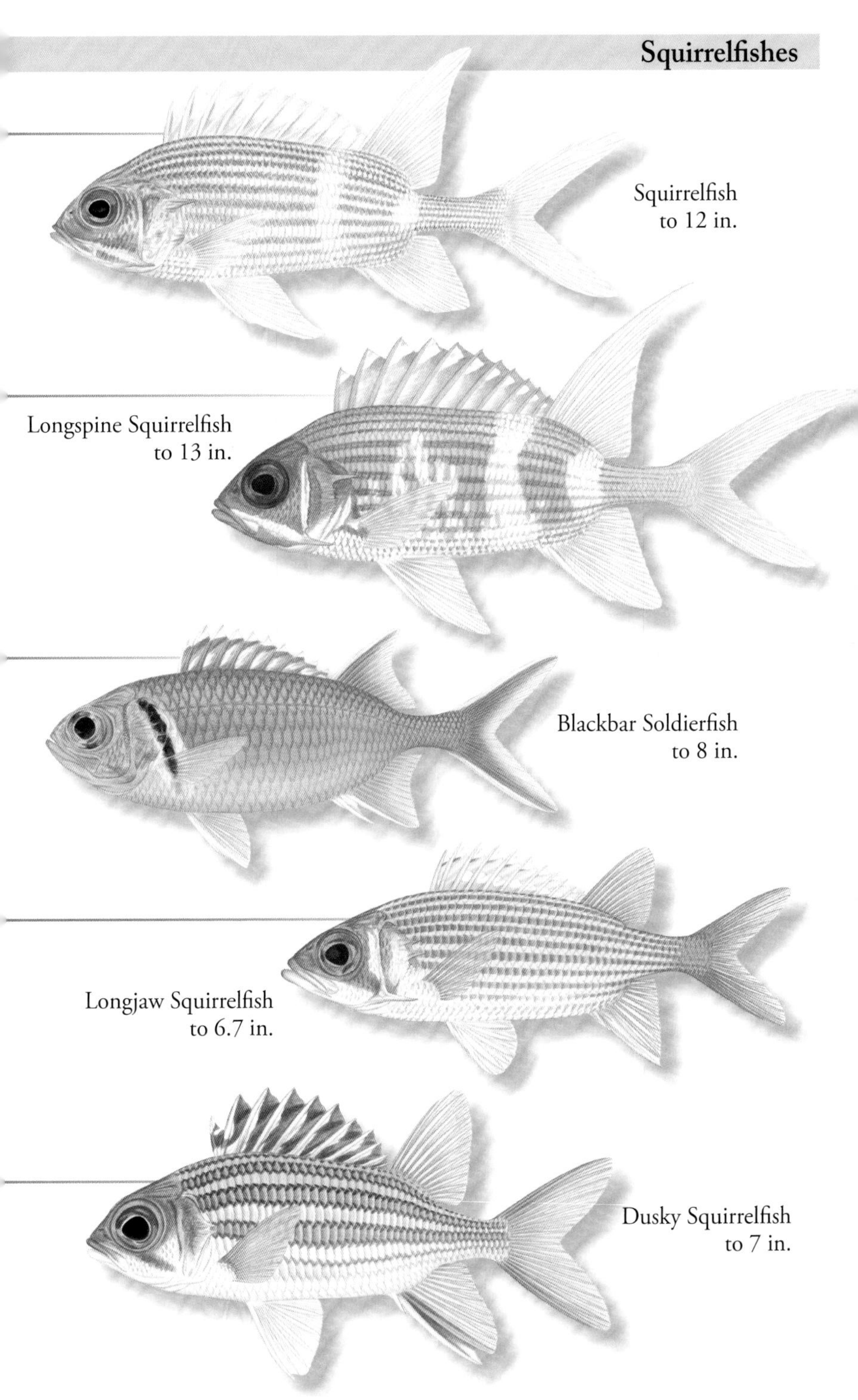
Squirrelfish
to 12 in.
Longspine Squirrelfish
to 13 in.
Blackbar Soldierfish
to 8 in.
Longjaw Squirrelfish
to 6.7 in.
Dusky Squirrelfish
to 7 in.

Bigeye Soldierfish - *Ostichthys trachypoma* (Günther, 1859)

FEATURES: Reddish with indistinct pale or bright silver stripes dorsally and on sides. May appear very red or very silvery. Mouth large. Upper jaw extends beyond posterior margin of eyes. Preopercle serrated. Opercle with single spine at upper margin. Lobes of soft dorsal, caudal, and anal fins angular. Body very deep. HABITAT: NY to FL, Gulf of Mexico, and Caribbean Sea to Brazil. Usually near bottom between about 125 and 1,600 ft. BIOLOGY: Nocturnal. Feed at night; seek shelter during the day.

Cardinal Soldierfish - *Plectrypops retrospinis* (Guichenot, 1853)

FEATURES: Uniformly red. May be paler below. Spiny dorsal fin red; membranes pale at tips. Other fins reddish. Irises may be red, yellow, or silvery. Spines under and behind eyes overlap upper jaw. Opercles serrated and spiny. Body relatively deep. HABITAT: S FL, Bahamas, Caribbean Sea to Brazil. In Gulf of Mexico from the Flower Garden Banks. Also Bermuda. Occur around reefs from about 10 to 800 ft. BIOLOGY: Reclusive. Hide in crevices during the day. May swim upside down.

Deepwater Squirrelfish - *Sargocentron bullisi* (Woods, 1955)

FEATURES: Orange yellow to red dorsally and on sides with narrow white stripes bordered above and below by brownish to black stripes. Head reddish orange above, white below, with reddish band from lower eye to preopercular corner. Dorsal fin yellow with white areas at base and a black spot between first and second spines. Other fins reddish yellow to clear. HABITAT: NC to FL, Gulf of Mexico, Bermuda, Bahamas, and Caribbean Sea to Brazil. Occur over offshore reefs from about 120 to 390 ft. BIOLOGY: Nocturnal. Seek deep crevices during the day; forage at night.

Reef Squirrelfish - *Sargocentron coruscum* (Poey, 1860)

FEATURES: Alternating red and white stripes dorsally and on sides. Head red above, white below, with red band from lower eye to preopercular corner. Dorsal fin red with white at tips and at base. Black splotch between first and third or first and fourth dorsal-fin spines. Other fins reddish to clear. HABITAT: S FL, Bahamas, and Caribbean Sea. Also Bermuda. Occur over reefs and adjacent areas from surface to about 100 ft. BIOLOGY: Nocturnal. Hide in reef crevices during the day, forage over sandy areas at night.

Saddle Squirrelfish - *Sargocentron poco* (Woods, 1965)

FEATURES: Alternating red and white stripes on sides, or appearing almost uniformly red. Head shades of red. Dark red to blackish saddle under soft dorsal fin, and dark red to blackish blotch on peduncle. Saddle and blotch may be separated by a white area. Black blotch between first and third dorsal-fin spines. Outer margins of other fins may be red. HABITAT: SW FL, N Gulf of Mexico, Bermuda, Bahamas, and Caribbean Sea. Occur over reefs from near surface to about 500 ft.

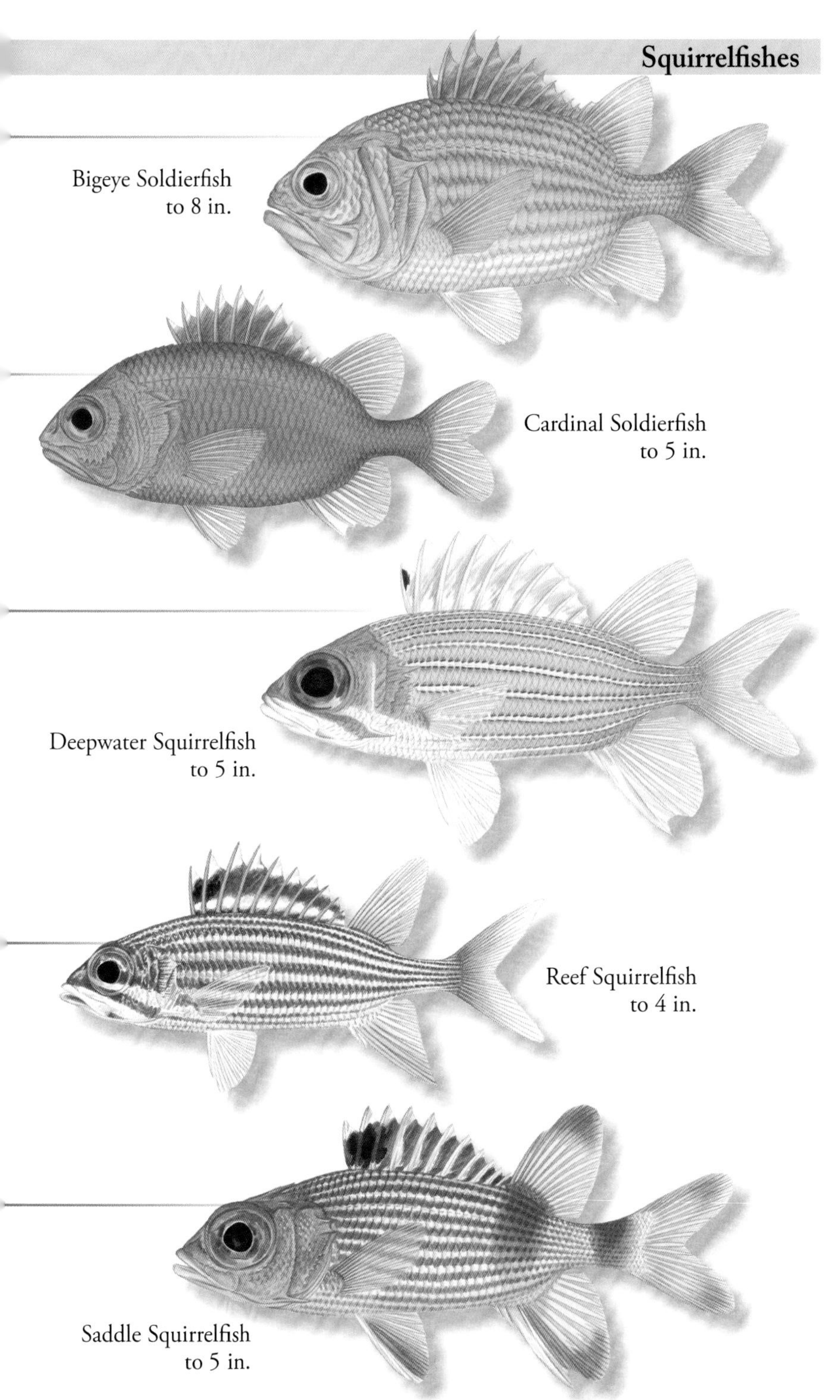
Bigeye Soldierfish
to 8 in.
Cardinal Soldierfish
to 5 in.
Deepwater Squirrelfish
to 5 in.
Reef Squirrelfish
to 4 in.
Saddle Squirrelfish
to 5 in.

Parazenidae - Smooth Dories

Red Dory - *Cyttopsis rosea* (Lowe, 1843)

FEATURES: Head and body silvery red. Head, eyes, and mouth large. Head profile convex. Pelvic fins large and with blackish membranes. Spiny scutes absent. Body deep, compressed. HABITAT: In warm waters of Atlantic, Indian, and Indo-West Pacific oceans. In western Atlantic from Canada, SC, N Gulf of Mexico, portions of Caribbean Sea. Near bottom from about 800 to 2,000 ft.

ALSO IN THE AREA: *Parazen pacificus*, see p. 547.

Grammicolepididae - Tinselfishes

Thorny Tinselfish - *Grammicolepis brachiusculus* Poey, 1873

FEATURES: Silvery overall. Head, snout short. Eyes large; mouth very small. Caudal fin very long. Scales tall and narrow. Body deep and compressed. Juveniles diamond-shaped in profile with irregular bars and blotches on sides, dark caudal-fin margin, pronounced thorns on sides, and elongate first dorsal- and anal-fin spines. HABITAT: Eastern and western Atlantic, and northwest Pacific to western Indian oceans. In western Atlantic from Georges Bank to FL, Gulf of Mexico, Bermuda, Bahamas, and Caribbean Sea to central Brazil. Found in midwater and over bottoms from about 980 to 3,000 ft.

Spotted Tinselfish - *Xenolepidichthys dalgleishi* Gilchrist, 1922

FEATURES: Silvery with faint to distinct spots on sides. Dorsal profile and caudal-fin margin dark. Head, snout short. Eyes large; mouth very small. Scales tall and narrow. Body very deep, compressed, and diamond-shaped. First dorsal- and anal-fin rays very elongate in juveniles; relatively shorter with age. HABITAT: Eastern and western Atlantic, western Pacific, and eastern Indian oceans. In western Atlantic from Nova Scotia to FL, Gulf of Mexico to French Guiana. Also Bermuda. Occur near bottom from about 300 to 3,000 ft.

Zeidae - Dories

Buckler Dory - *Zenopsis conchifera* (Lowe, 1852)

FEATURES: Overall silvery with a vague, dark blotch on sides. Juveniles spotted on body and fins. Head and mouth large. Head profile concave. Anterior dorsal-fin spines tall with trailing filaments; filaments very long in juveniles. Juvenile pectoral fins broad. Scales absent. Bucklers present along bases of dorsal and anal fins. Body deep, compressed. HABITAT: Eastern and western Atlantic and western Indian oceans. In western Atlantic from Nova Scotia to FL, Gulf of Mexico to Brazil. Over bottoms and in midwater from about 320 to 1,300 ft. BIOLOGY: Form small schools. Feed on fishes.

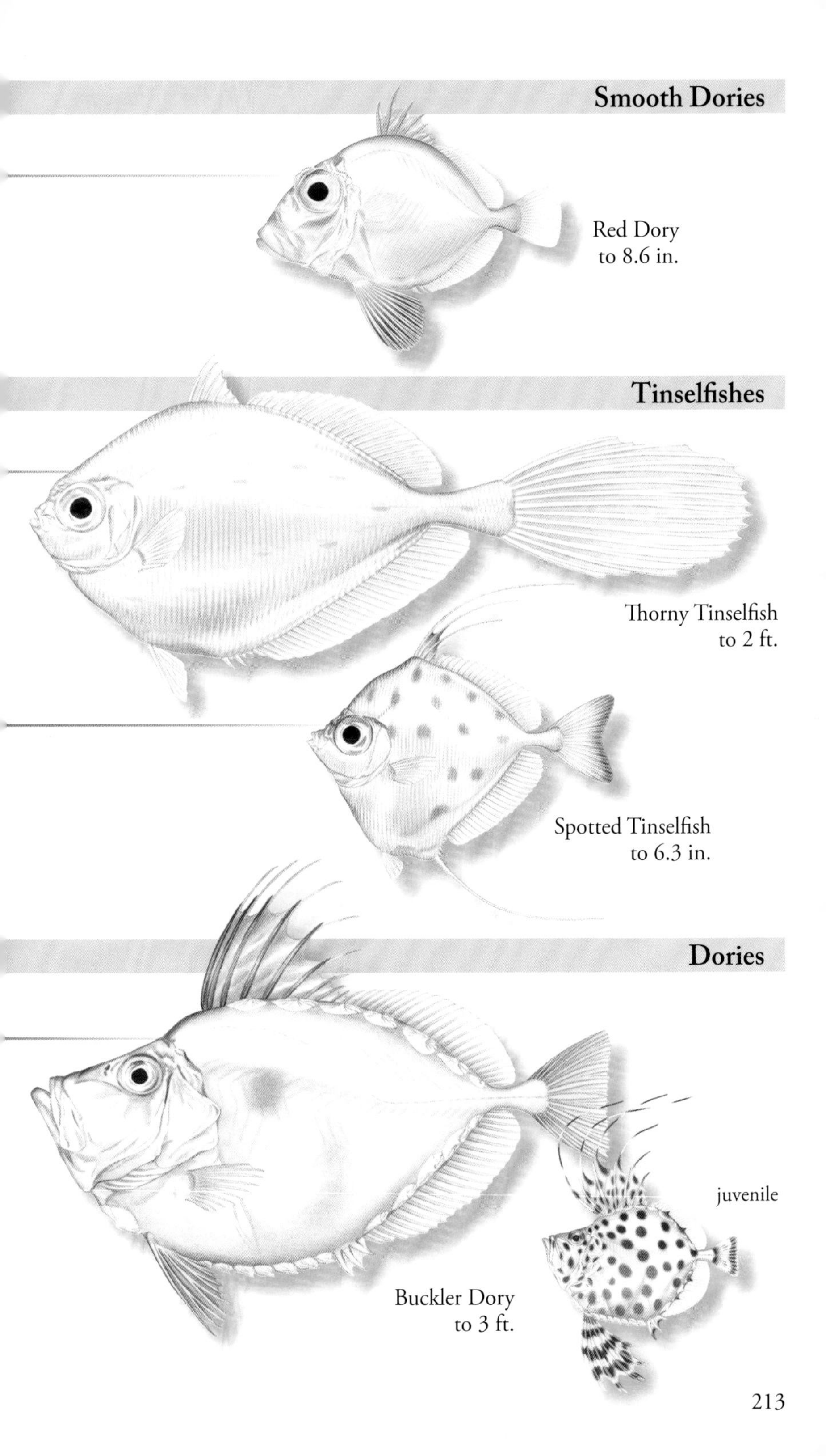
Smooth Dories
Red Dory
to 8.6 in.
Tinselfishes
Thorny Tinselfish
to 2 ft.
Spotted Tinselfish
to 6.3 in.
Dories
juvenile
Buckler Dory
to 3 ft.

Syngnathidae - Pipefishes and Seahorses

Pipehorse - *Amphelikturus dendriticus* (Barbour, 1905)

FEATURES: Color and pattern highly variable. Shades of greenish to brownish or red with dark to pale blotches. May be uniformly colored. Slender, elongate, with head bent slightly downward. Small caudal fin at end of tail. Dermal flaps on head and body may be simple or highly branched. HABITAT: New Brunswick to FL, Bermuda, N Gulf of Mexico, and Bahamas to Brazil. Demersal in shallow to oceanic waters from near surface to about 100 ft. NOTE: Previously known as *Acentronura dendritica*.

Lined Seahorse - *Hippocampus erectus* Perry, 1810

FEATURES: Color and pattern highly variable. Pale and dark dots on neck and along back. Pale and dark lines on head and body. Snout shorter than distance from eye to opercular margin. Body may be covered in fleshy filaments. HABITAT: Nova Scotia (rare) to FL, Gulf of Mexico, Bermuda, Bahamas, and Caribbean Sea to Venezuela. In vegetated waters to about 240 ft. IUCN: Vulnerable.

Longsnout Seahorse - *Hippocampus reidi* Ginsburg, 1933

FEATURES: Color and pattern highly variable. Small dark spots on head and body. May have paler or darker blotches. Snout usually longer than distance from eye to opercular margin. HABITAT: FL to N Gulf of Mexico, Bermuda, Bahamas, and Caribbean Sea to Venezuela. Over seagrass beds, around pilings, and on reefs to about 50 ft. IUCN: Near Threatened.

Dwarf Seahorse - *Hippocampus zosterae* Jordan & Gilbert, 1882

FEATURES: Usually shades of tan to brown. May be plain or with pale and dark mottling. Snout short. Head large relative to body. May have thin fleshy tabs on body. HABITAT: E FL, Gulf of Mexico, Bermuda, and Bahamas. In shallow brackish to marine waters over seagrass flats to about 70 ft. Occasionally in floating weeds. BIOLOGY: Feed on crustaceans. Males and females are monogamous during their breeding periods. Males brood eggs in pouch.

Insular Pipefish - *Anarchopterus tectus* (Dawson, 1978)

FEATURES: Tannish to brownish. Body with either darker spots on body rings or fairly evenly spaced, diffuse dark bars on body rings. May have dark band from eye to snout. Snout short; head profile slightly concave. Anal fin absent. Trunk with 17–18 rings. HABITAT: FL Keys, eastern Gulf of Mexico, Bahamas to northern South America. Reported on turtle-grass beds from near shore to about 85 ft.

Pugnose Pipefish - *Bryx dunckeri* (Metzelaar, 1919)

FEATURES: Reddish brown to pale. May be uniformly mottled or may have narrow to broad pale bars. Snout very short. Anal fin absent. Trunk with 16–17 rings. HABITAT: NC to FL, Bermuda, Caribbean Sea to northern South America. In shallow estuaries and over seagrass beds and reefs to about 230 ft. Also found with *Sargassum* seaweed.

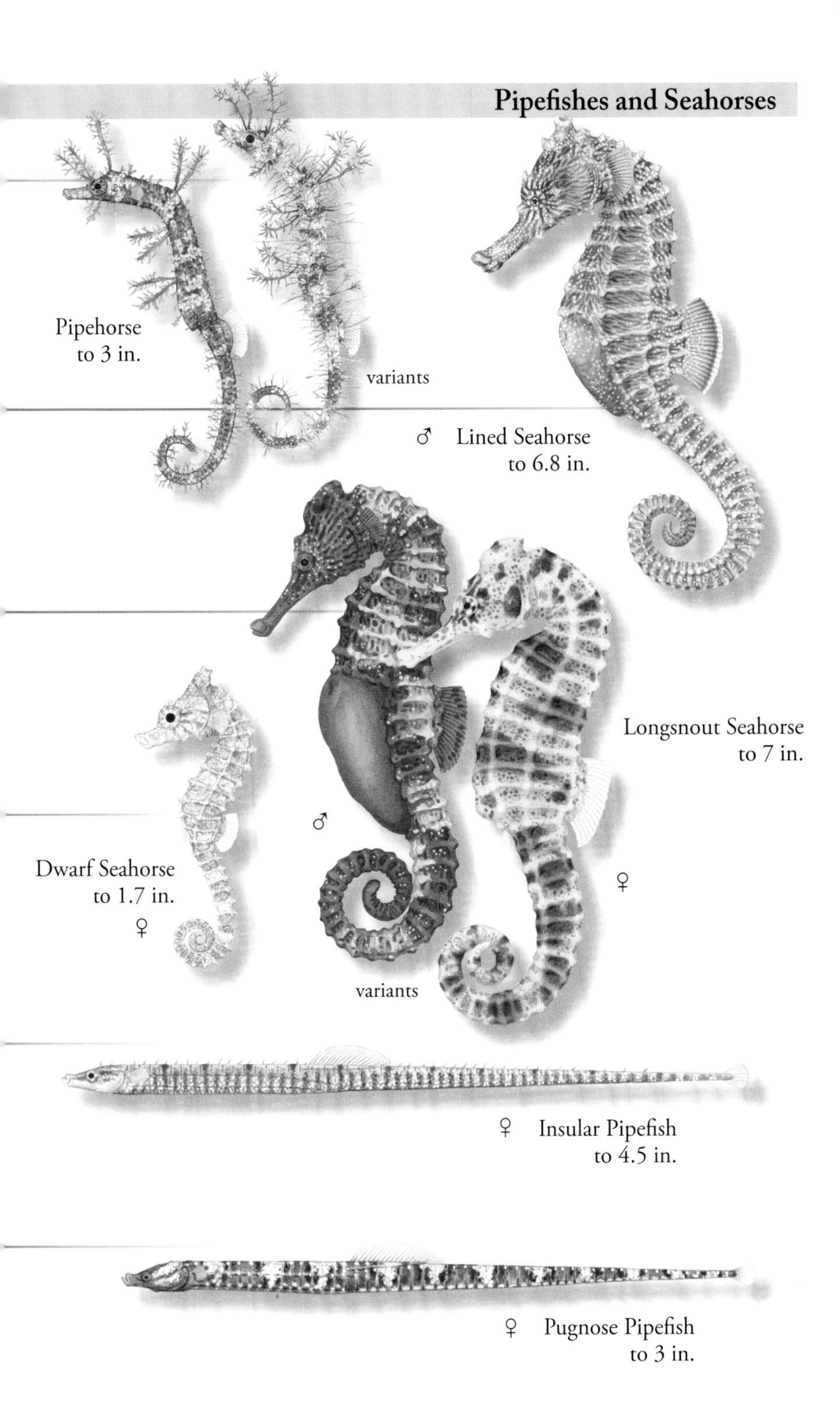
Pipehorse
to 3 in.
variants
♂ Lined Seahorse
to 6.8 in.
Longsnout Seahorse
to 7 in.
♂
♀
variants
Dwarf Seahorse
to 1.7 in.
♀
♀ Insular Pipefish
to 4.5 in.
♀ Pugnose Pipefish
to 3 in.

Ocellated Pipefish - *Bryx randalli* (Herald, 1965)

FEATURES: Grayish brown to brown with pale ocellated spots that form bars from head to tail. Top of snout pale. Lower snout and cheeks with narrow pale lines. HABITAT: Western Caribbean Sea, Haiti, Dominican Republic, and Virgin Islands to Trinidad and Tobago. Demersal in shallow water over rocks and reefs, around algae and mangroves, and in estuaries to about 100 ft.

Whitenose Pipefish - *Cosmocampus albirostris* (Kaup, 1856)

FEATURES: Grayish to brownish or whitish with broad, diffuse bars on body. Snout long, always whitish. Head may be darker than body. Body may or may not have dermal flaps. HABITAT: SC to FL, Gulf of Mexico, Bahamas, and Antilles to northern South America. Over shallow weedy or reef bottoms to about 180 ft.

Crested Pipefish - *Cosmocampus brachycephalus* (Poey, 1868)

FEATURES: Brownish with darker bars and mottling. Underside of head banded. Snout short. Prominent bony crest on head. Body short, stout. HABITAT: FL, Bahamas, and Antilles to Venezuela. Over shallow, coastal seagrass flats to about 180 ft.

Shortfin Pipefish - *Cosmocampus elucens* (Poey, 1868)

FEATURES: Color variable. Blackish, purplish, brownish to pinkish. Two bands radiate from eyes. Evenly spaced pale bars over length of body. May or may not have dermal flaps on head and body. Trunk with 16–18 rings. HABITAT: NJ to FL, eastern Gulf of Mexico, and Caribbean Sea. Also Bermuda. Occur to about 260 ft.

Banded Pipefish - *Halicampus crinitus* (Jenyns, 1842)

FEATURES: Variably barred in distinct shades of yellow and brick red to purplish brown. Bars usually form rings around body. Snout short. HABITAT: S FL, Bahamas to Brazil. Also Bermuda. Reef associated from shore to about 70 ft. BIOLOGY: Some authors believe there are two distinct species. OTHER NAMES: Harlequin Pipefish, *Micrognathus ensenadae*, *Micrognathus crinitus*.

Opossum Pipefish - *Microphis lineatus* (Kaup, 1856)

FEATURES: Brownish with a series of red blotches on each trunk ring. Silvery stripe on sides. Lower snout with red and black bars. Caudal fin red with black stripe. Juveniles less colorful. Trunk with 16–20 rings. HABITAT: NJ to FL, Gulf of Mexico, and Caribbean Sea to Brazil. Occur in shallow fresh to salt water to about 30 ft.

Freshwater Pipefish - *Pseudophallus mindii* (Meek & Hildebrand, 1923)

FEATURES: Observed specimens shades of dark brown or blackish brown to greenish brown. Others described as pale brown above, greenish below. Dark stripe on snout. Snout short, upturned. HABITAT: Greater Antilles and southern Caribbean Sea. Primarily in fresh water, but also in estuaries and very shallow coastal waters. Juveniles have been recorded at sea.

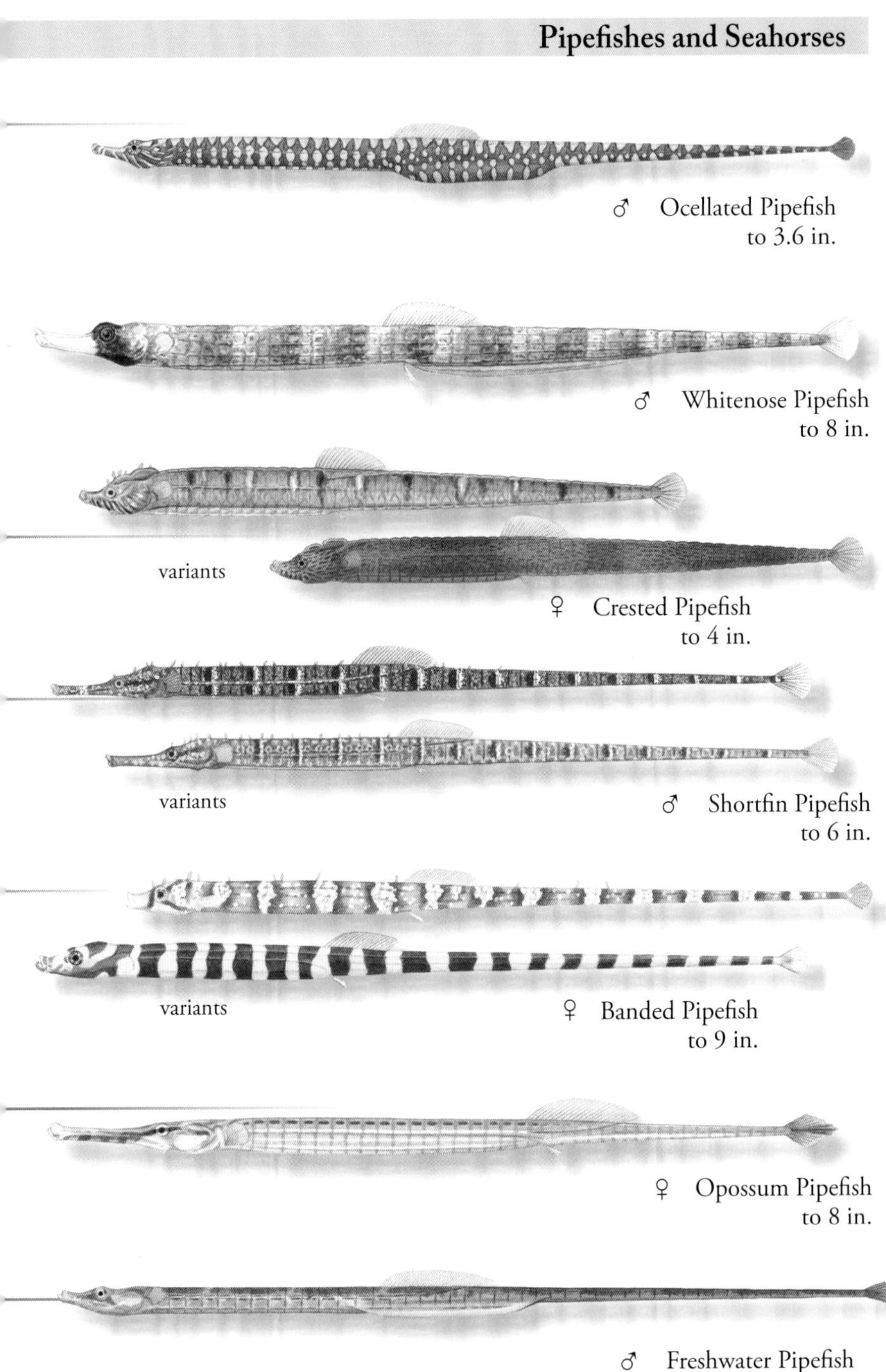
♂ Ocellated Pipefish
to 3.6 in.
♂ Whitenose Pipefish
to 8 in.
variants
♀ Crested Pipefish
to 4 in.
variants
♂ Shortfin Pipefish
to 6 in.
variants
♀ Banded Pipefish
to 9 in.
♀ Opossum Pipefish
to 8 in.
♂ Freshwater Pipefish
to 6.3 in.

Caribbean Pipefish - *Syngnathus caribbaeus* Dawson, 1979

FEATURES: Color and pattern varies. Greenish orange to greenish brown with pale spots and dark streaks, or brown with eight or nine diffuse pale bars and scattered pale spots. Snout long and slender. Low ridge on opercle. Trunk with 31–35 rings. HABITAT: Cuba and Caribbean Sea. Occur coastally and in estuaries over muddy and weedy bottoms from shore to about 20 ft.

Antillean Pipefish - *Syngnathus dawsoni* (Herald, 1969)

FEATURES: Whitish to tannish dorsally. Dark brown edged by black on sides. Abdomen pale with narrow dark bars. Dark band radiates below eyes. Snout long and slender. Trunk with 33–36 rings. Body comparatively very slender. HABITAT: Puerto Rico and Lesser Antilles. Demersal in bays over weedy bottoms from shore to about 25 ft.

Dusky Pipefish - *Syngnathus floridae* (Jordan & Gilbert, 1882)

FEATURES: Greenish, brownish to whitish. May be mottled or blotched dorsally. Bars absent. Dorsal fin with 26–35 rays (usually 32). Trunk with 16–19 rings. HABITAT: VA to FL, northern Gulf of Mexico, Bahamas, western Caribbean Sea. Also Bermuda. Over shallow seagrass beds and in estuaries and channels to about 70 ft.

Chain Pipefish - *Syngnathus louisianae* Günther, 1870

FEATURES: Tannish to whitish or brownish. Row of dark, diamond-shaped marks along lower sides. Dark stripe from snout to pectoral fin. Body ridges rather low. Trunk with 19–21 rings. HABITAT: NJ to S FL and Gulf of Mexico from surface to about 125 ft. Found near shore over seagrass beds and around marsh grasses and offshore with *Sargassum* seaweed rafts.

Sargassum Pipefish - *Syngnathus pelagicus* Linnaeus, 1758

FEATURES: Pale tan to brown with pale and dark bars on body. Snout with dark stripe. Dark bands may radiate from eyes. Dorsal fin banded. Head with low ridge. Body ridges rather low. Trunk with 15–18 rings. HABITAT: Nova Scotia to FL, Bermuda, Gulf of Mexico, Bahamas, Caribbean Sea to Colombia. Primarily oceanic. Found from near surface to about 240 ft. Associated with *Sargassum* seaweed.

Gulf Pipefish - *Syngnathus scovelli* (Evermann & Kendall, 1896)

FEATURES: Dark brownish, olive brown to tannish. Males and females with Y-shaped silvery white bars on sides; bars may be more distinct in females. Dorsal fin banded. Caudal fin rounded. Females with keeled abdomen. Male trunk almost square in cross-section. Trunk with 16–17 rings. HABITAT: NE FL, Gulf of Mexico to Brazil. Occur over seagrass beds and in marshes, vegetated streams, and river shorelines to about 20 ft.

ALSO IN THE AREA: *Penetopteryx nanus*, *Syngnathus makaxi*, *Syngnathus springeri*, see p. 547.

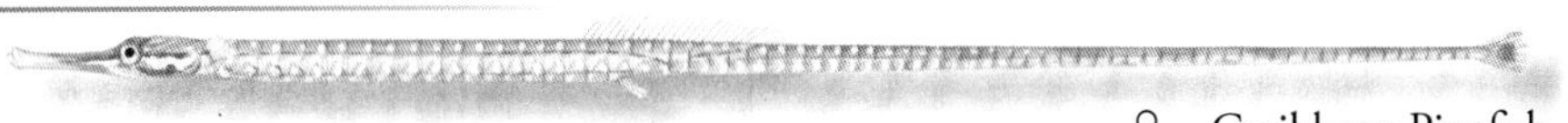

♀ Caribbean Pipefish
to 12 in.

♀ Antillean Pipefish
to 7 in.

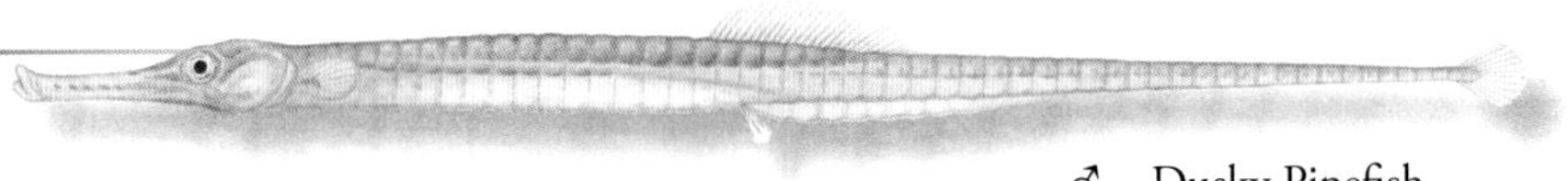

♂ Dusky Pipefish
to 10 in.

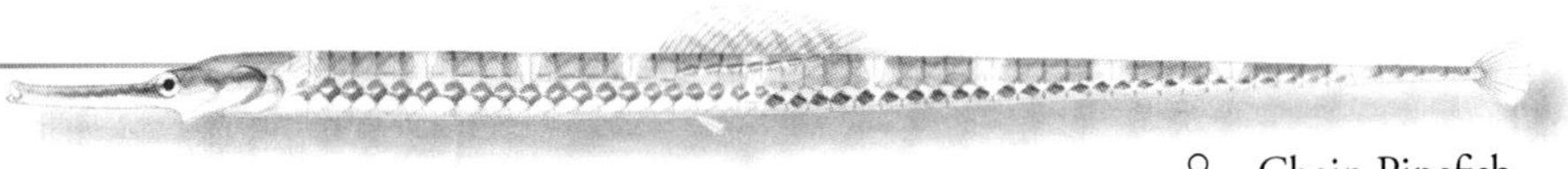

♀ Chain Pipefish
to 15 in.

♀ Sargassum Pipefish
to 8 in.

♀

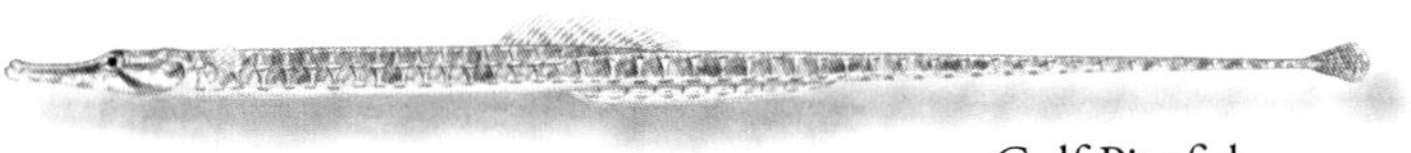

♂

Gulf Pipefish
to 7 in.

Aulostomidae - Trumpetfishes

Atlantic Trumpetfish - *Aulostomus maculatus* Valenciennes, 1841

FEATURES: Color and pattern highly variable. Brownish, olivaceous, or yellowish to reddish with blackish spots and whitish lines on head and body. May have faint bars. Second dorsal and anal fins with dark bands or spots. Caudal fin with one or two black spots or blotches. Lower jaw with short barbel. Body elongate, slightly compressed. First dorsal-fin spines short, separate. HABITAT: S FL, Gulf of Mexico, Bahamas, and Caribbean Sea to Brazil. Also Bermuda. Over reefs and among weeds in clear, shallow water. BIOLOGY: Often hover and drift vertically among coral branches. Dart toward and suck in prey. Feed on small crustaceans and fishes.

Fistulariidae - Cornetfishes

Red Cornetfish - *Fistularia petimba* Lacepède, 1803

FEATURES: Reddish to orange brown dorsally. Ventral area silvery. May have faint, dark bars. Snout, head, and body slender, elongate. First dorsal fin absent. Caudal fin forked with long, trailing filament. HABITAT: Circumtropical. MA to FL, Gulf of Mexico, and Caribbean Sea to Brazil. Found over coastal soft bottoms and reefs from about 30 to 650 ft.

Bluespotted Cornetfish - *Fistularia tabacaria* Linnaeus, 1758

FEATURES: Brownish dorsally; pale ventrally. Pale blue spots and lines along snout and body. May have faint bars. Snout, head, and body elongate. First dorsal fin absent. Caudal fin forked with long, trailing filament. HABITAT: Nova Scotia to FL, Gulf of Mexico, and Caribbean Sea to Brazil. Also Bermuda. Over seagrass beds and reefs. BIOLOGY: Feed on fishes.

Macroramphosidae - Snipefishes

Slender Snipefish - *Macroramphosus gracilis* (Lowe, 1839)

FEATURES: Bluish to grayish or pinkish with silvery tint. Snout long, tube-like. Body laterally compressed, comparatively shallow. Second dorsal-fin spine serrated and shorter than snout length. HABITAT: Worldwide in tropical seas. In western Atlantic from FL, northern Gulf of Mexico, and Bahamas to Cuba. Adults near bottom from about 656 to 985 ft. Juveniles pelagic. NOTE: Some authors consider this species a junior synonym of *Macroramphosus scolopax*.

Longspine Snipefish - *Macroramphosus scolopax* (Linnaeus, 1758)

FEATURES: Reddish to orange dorsally. Sides and abdomen silvery pink. Snout long, tube-like. Body laterally compressed, comparatively deep. Second dorsal-fin spine serrated and about as long as snout. HABITAT: Gulf of Maine to Argentina. Also Gulf of Mexico and Bermuda. Occur over sandy lower continental shelves. Adults near bottom from about 164 to 1,150 ft. Juveniles pelagic. BIOLOGY: Feed on plankton and small fishes. Schooling.

Trumpetfishes

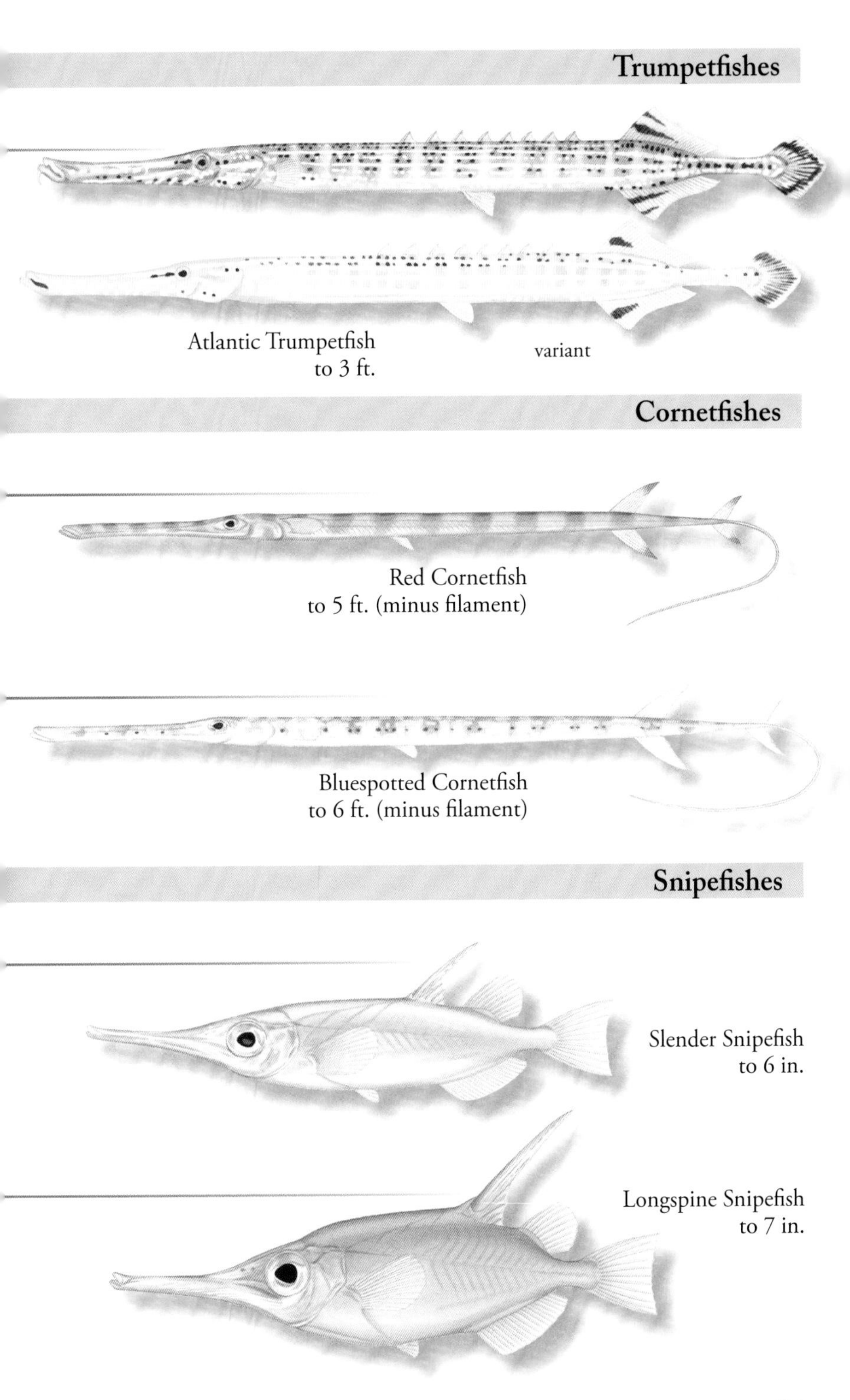

Dactylopteridae - Flying Gurnards

Flying Gurnard - *Dactylopterus volitans* (Linnaeus, 1758)

FEATURES: Typically brownish. Bright blue spots cover head, body, and pectoral fins. Head short. Body elongate. Bony shield and long, keeled spine extend from head to first dorsal fin. Opercles spiny. Pectoral fins greatly expandable. HABITAT: MA to FL, Gulf of Mexico, Bahamas, and Caribbean Sea to Argentina. Also Bermuda and eastern Atlantic. Bottom-dwelling in sandy and muddy coastal waters to about 330 ft. BIOLOGY: "Walk" across the bottom using pelvic and pectoral rays. Spread pectoral fins when alarmed. Feed on crustaceans, mollusks, and fishes.

Scorpaenidae - Scorpionfishes

Blackbelly Rosefish - *Helicolenus dactylopterus* (Delaroche, 1809)

FEATURES: Reddish dorsally. Ventral area pinkish to white. Usually with six diffuse to distinct darker bars on upper body. Bars below soft dorsal fin form a V-shape. Cheeks may have dark grayish cast. May have dark blotch at rear of spiny dorsal fin. Pectoral fins with 19 rays; upper margin squarish; tips of lower rays free. HABITAT: Nova Scotia to FL. Scattered in Gulf of Mexico, around Bahamas, and along northern coast of South America. Over soft bottoms and continental shelves from about 650 to 2,000 ft. BIOLOGY: Feed on crustaceans and fishes.

Spotwing Scorpionfish - *Neomerinthe beanorum* (Evermann & Marsh, 1900)

FEATURES: Rosy red on head and back with irregular brownish marks. Abdomen white with some rose. First dorsal fin pale with large brownish spots near margin. Soft dorsal, upper caudal, and upper pectoral fins spotted. Pair of small cirri present over eyes. Pectoral fins with 16–18 rays, upper three to nine rays branched in adults. HABITAT: Bahamas, Greater Antilles, and Belize to Curaçao. Demersal over hard bottoms from about 300 to 1,200 ft.

Spinycheek Scorpionfish - *Neomerinthe hemingwayi* Fowler, 1935

FEATURES: Dark reddish with dark brown spots and mottling dorsally and on sides. All fins spotted except pelvic fins. Always with three spots on posterior portion of lateral line. Snout moderately long. Pectoral fins with 17 rays; third to seventh or eighth rays branched. HABITAT: NJ to S FL, and northern and southern Gulf of Mexico. Found near bottom over middle to outer continental shelf from about 100 to 750 ft. BIOLOGY: Spines are venomous.

Short-tube Scorpionfish - *Phenacoscorpius nebris* Eschmeyer, 1965

FEATURES: Pinkish with pale blotches and irregular dark red saddle under spiny dorsal fin, and dark red bars under soft dorsal fin and on peduncle. Dark red bands radiate from eyes. Dark reddish black blotch on spiny dorsal fin between fifth and ninth spines. Simple cirri over eyes and above upper jaw. Pectoral fins with 15–17 rays, upper rays branched. HABITAT: N Gulf of Mexico, N Cuba, Puerto Rico, Colombia, and Venezuela. Demersal over soft bottoms from about 200 to 1,500 ft.

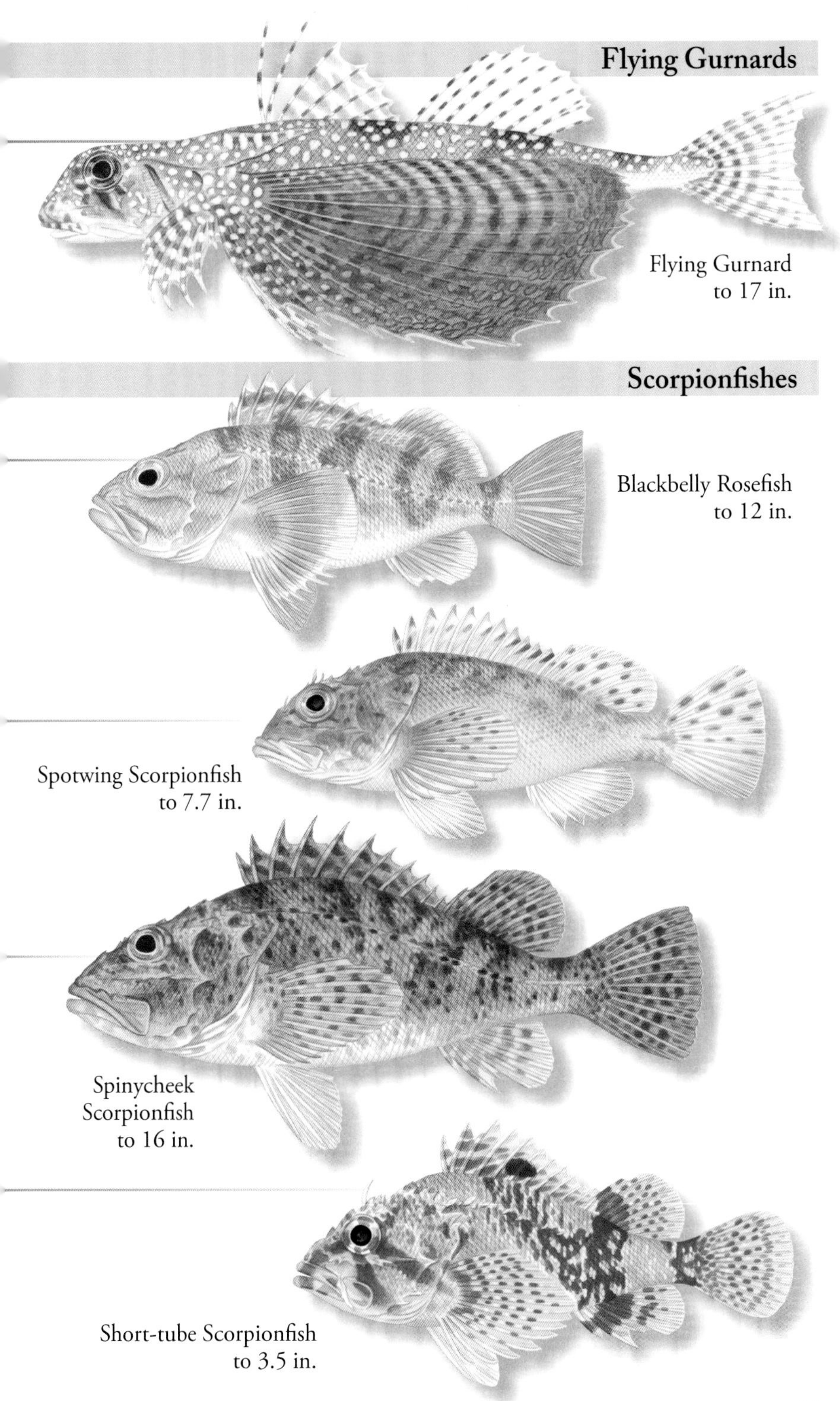
Flying Gurnards
Flying Gurnard
to 17 in.
Scorpionfishes
Blackbelly Rosefish
to 12 in.
Spotwing Scorpionfish
to 7.7 in.
Spinycheek
Scorpionfish
to 16 in.
Short-tube Scorpionfish
to 3.5 in.

Longsnout Scorpionfish - *Pontinus castor* Poey, 1860

FEATURES: Mottled and blotched in shades of reddish orange, yellow, and silvery white. Vague reddish orange and white saddles along back. May have few to many scattered dark brown spots. Fins heavily spotted. Snout long and pointed; mouth large. Head large and spiny. Fleshy tabs present on head and around mouth. Pectoral fins broad, with 15–18 unbranched rays. HABITAT: FL, Bermuda, Bahamas, and Caribbean Sea to Venezuela. Demersal over rocky, soft, and reef bottoms from about 105 to 1,800 ft.

Longspine Scorpionfish - *Pontinus longispinis* Goode & Bean, 1896

FEATURES: Reddish to pinkish dorsally; pale below. Dusky red spots or blotches on upper sides and along lateral line. Soft dorsal, pectoral, and caudal fins spotted. Fleshy tentacle over eyes tall, small, or absent. Third dorsal-fin spine elongate in larger specimens. Pectoral fins with 17–18 unbranched rays. HABITAT: SC to FL, Gulf of Mexico, and scattered along coasts of Central and South America. Occur over soft or semi-hard bottoms from about 260 to 1,400 ft. BIOLOGY: Feed on invertebrates and fishes. May be venomous.

Spinythroat Scorpionfish - *Pontinus nematophthalmus* (Günther, 1860)

FEATURES: Blotched and mottled in shades of red to dark orange with four vague saddles along back. White on chest and peduncle. Dorsal, caudal, and pectoral fins spotted; anal fin banded. Snout short and blunt. Head large and spiny. Cirri over eyes may be very long. Pectoral fins with 15–17 unbranched rays. HABITAT: GA to FL, Bahamas, and Caribbean Sea. Demersal over rocky, shelly, and coral bottoms from about 270 to 1,300 ft.

Highfin Scorpionfish - *Pontinus rathbuni* Goode & Bean, 1896

FEATURES: Blotched and mottled in shades of red, orange, or white. Pale band on peduncle. Small specimens with saddles along back. Irregular blotches along lateral line. Fins spotted and banded. Fleshy tentacle over each eye. Spiny dorsal fin tall with third spine taller than others. Pectoral fins with 17 unbranched rays. HABITAT: VA to FL, and NE Gulf of Mexico. Scattered from Yucatán to Brazil. Occur near bottom from about 240 to 1,200 ft.

Red Lionfish - *Pterois volitans* (Linnaeus, 1758)

FEATURES: Head and body with alternating reddish brown and whitish bars. Spiny dorsal fin and pectoral fins banded. Other fins transparent with reddish spots. Fleshy tabs above eyes and around mouth. Dorsal and pectoral fins typically expanded and fan-like. HABITAT: Native to the Pacific and Indian oceans. In the western Atlantic from NY to Bahamas, Caribbean Sea, and Bermuda. Range expanding. Typically occur over nearshore and offshore coral and rocky reefs from near surface to about 800 ft. Also reported from around dock pilings. BIOLOGY: Believed to have been introduced to Florida waters in the 1990s. Stalk, corner, and engulf prey in one swift gulp. Spines are highly venomous. May be aggressive when threatened.

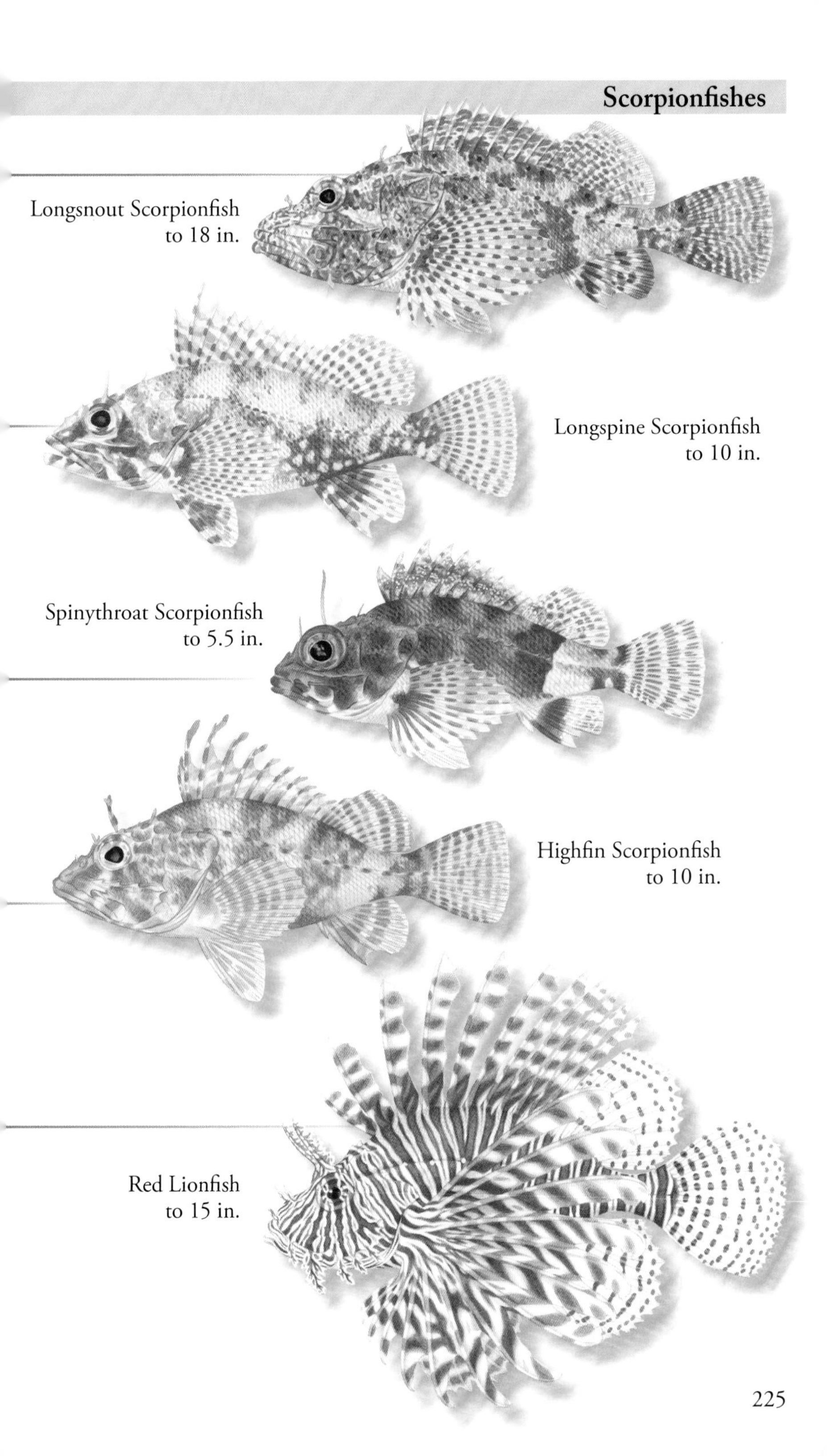
Longsnout Scorpionfish
to 18 in.
Longspine Scorpionfish
to 10 in.
Spinythroat Scorpionfish
to 5.5 in.
Highfin Scorpionfish
to 10 in.
Red Lionfish
to 15 in.

Longfin Scorpionfish - *Scorpaena agassizii* Good & Bean, 1896

FEATURES: Shades of bright red with paler and darker spots and blotches. May be pale with dense mottling. Dark blotch on spiny dorsal fin absent. Dorsal fin mottled, caudal fin usually banded, and pectoral fins spotted. Eyes very large. Pectoral fins very broad and fan-like, reaching past anal-fin base, with 18–20 rays; uppermost rays branched. HABITAT: NC to FL, Gulf of Mexico, and southern Caribbean Sea to Brazil. Demersal over soft bottoms of continental shelves from about 150 to 980 ft. BIOLOGY: Feed on crustaceans and mollusks.

Coral Scorpionfish - *Scorpaena albifimbria* Evermann & Marsh, 1900

FEATURES: Body and head mottled in shades of red, pink, and brown. Wide, diffuse dark bar behind head. Anal fin dark red at base. Pelvic fins with dark outer margin. Eyes large with a pair of short to long cirri. Many fleshy tabs on head and body. Pectoral fins with 19–21 rays; uppermost and twelfth to fifteenth rays unbranched. HABITAT: S FL, Bahamas, and scattered through Caribbean Sea. Also Bermuda. Occur inshore over reefs from about 3 to 120 ft.

Goosehead Scorpionfish - *Scorpaena bergii* Evermann & Marsh, 1900

FEATURES: Speckled and barred in shades of dark red or reddish brown to brown. Dark, irregular spot on spiny dorsal fin between third, fourth, or fifth spines to seventh or eighth spines. Pelvic fins with dusky margin. Anal fin with three dark bands. Caudal fin with three dark bands. Fleshy tabs over eyes moderate in size. Well-developed pit behind eyes in adults. Pectoral fins with 16–17 rays; some upper rays branched. HABITAT: Scattered from NY to FL, Bahamas, Caribbean Sea to Brazil. Occur in clear water over turtle grass, coral reefs, and rocky and sandy bottoms.

Shortfin Scorpionfish - *Scorpaena brachyptera* Eschmeyer, 1965

FEATURES: May be sparsely to densely mottled and blotched with pale pink, yellow, or white. Irises yellow. Underside of pectoral fins with a bright yellow area. Spiny dorsal fin with a large dark blotch between third and seventh spines. Other fins spotted and banded. Anal fin dark red anteriorly. Eyes large. Adults with a shallow pit behind eyes. Pectoral fins comparatively short, not reaching first anal-fin spine, and with 19 or 20 rays; some upper rays branched in adults. HABITAT: S FL, and southern Caribbean Sea. Demersal over rocky, rubble, and sponge bottoms from 150 to 500 ft.

Barbfish - *Scorpaena brasiliensis* Cuvier, 1829

FEATURES: Shades of red, brown, and yellow. Always with two dark blotches behind head and small dark spots at pectoral-fin base. Usually with dark spots on abdomen. Caudal fin with two dark bands. Other fins banded. Fleshy tabs on eyes well developed. Pectoral fins with 18–20 rays. HABITAT: VA to FL, Gulf of Mexico, and scattered through Caribbean Sea to Brazil. Possibly Bermuda. Over shallow, soft bottoms to about 165 ft. Sometimes on coral reefs. BIOLOGY: Possess potent venomous spines.

Longfin Scorpionfish
to 6 in.
Coral Scorpionfish
to 3 in.
Goosehead Scorpionfish
to 4 in.
Shortfin Scorpionfish
to 3 in.
Barbfish
to 9 in.

Hunchback Scorpionfish - *Scorpaena dispar* Longley & Hildebrand, 1940

FEATURES: Mottled and blotched in shades of red, orange, and pink. Often with one or two dark blotches behind head, and oblique bands on body. Lateral line spotted. Fins with spots forming bands. Eyes moderate with a slender cirrus above. Adults with a well-developed pit behind eyes. Pectoral fins with 17–19 rays; a few upper rays branched in adults. HABITAT: SC to FL, Gulf of Mexico to Brazil. Demersal over rocky and coral reefs from about 120 to 2,260 ft.

Plumed Scorpionfish - *Scorpaena grandicornis* Cuvier, 1829

FEATURES: Mottled and barred in shades of brown to reddish. Small white spots on lower portion of operculum, on chest, and at pectoral-fin base. Underside of pectoral-fin base dark with white specks. Fins irregularly banded and mottled. Large, well-developed fleshy tabs over eyes. Pectoral fins with 18–19 rays; upper and lower rays unbranched. HABITAT: S FL, Bahamas, and scattered in Caribbean Sea. Possibly Bermuda. Over sandy bottoms and seagrass beds. Also in channels and bays. BIOLOGY: Lie motionless. Reported to be preyed upon by sharks.

Mushroom Scorpionfish - *Scorpaena inermis* Cuvier, 1829

FEATURES: Color variable. Mottled, speckled, and barred in shades of red to brown. Tiny, dark spots at pectoral-fin base. Soft dorsal, anal, and pectoral fins with dark band at margin. Caudal fin with two dark bands. Eyes with small, inverted, mushroom-shaped cirri over pupils. Pectoral fins with 19–21 rays. HABITAT: GA to S FL, Bahamas, and scattered through Caribbean Sea to Venezuela. In clear waters over sandy and grassy bottoms and coral reefs from shore to about 240 ft. BIOLOGY: Common on shrimp grounds. Feed on crustaceans and fishes.

Smoothcheek Scorpionfish - *Scorpaena isthmensis* Meek & Hildebrand, 1928

FEATURES: Heavily mottled and blotched in shades of brown to red with some green. Irises yellow. Spiny dorsal fin with a large dark blotch between third and seventh spines. Pelvic fins pink to red at base. Anal fin with a large reddish blotch. Large branched cirrus above each eye. Bony ridge below eyes lacks spines. Pectoral fins with 18-19 rays; some upper rays branched in adults. HABITAT: NC to FL, Caribbean Sea to Brazil. Demersal over rocky and muddy bottoms with some sand and rubble from near shore to about 330 ft.

Spotted Scorpionfish - *Scorpaena plumieri* Bloch, 1789

FEATURES: Color, mottling highly variable in shades of red to brown with greens and/or yellows. Broad, pale bar on caudal peduncle. Fins banded and blotched. Underside of pectoral-fin base black with large, white spots. Head and body may have numerous fleshy tabs. Pectoral fins with 18–21 rays. HABITAT: MA to FL, Gulf of Mexico, Bermuda, Bahamas, and Caribbean Sea. More common in southern waters. Over shallow, rocky, and reef areas. Also around pilings, jetties, and oil platforms. BIOLOGY: Lie motionless in wait for prey. Spines are venomous.

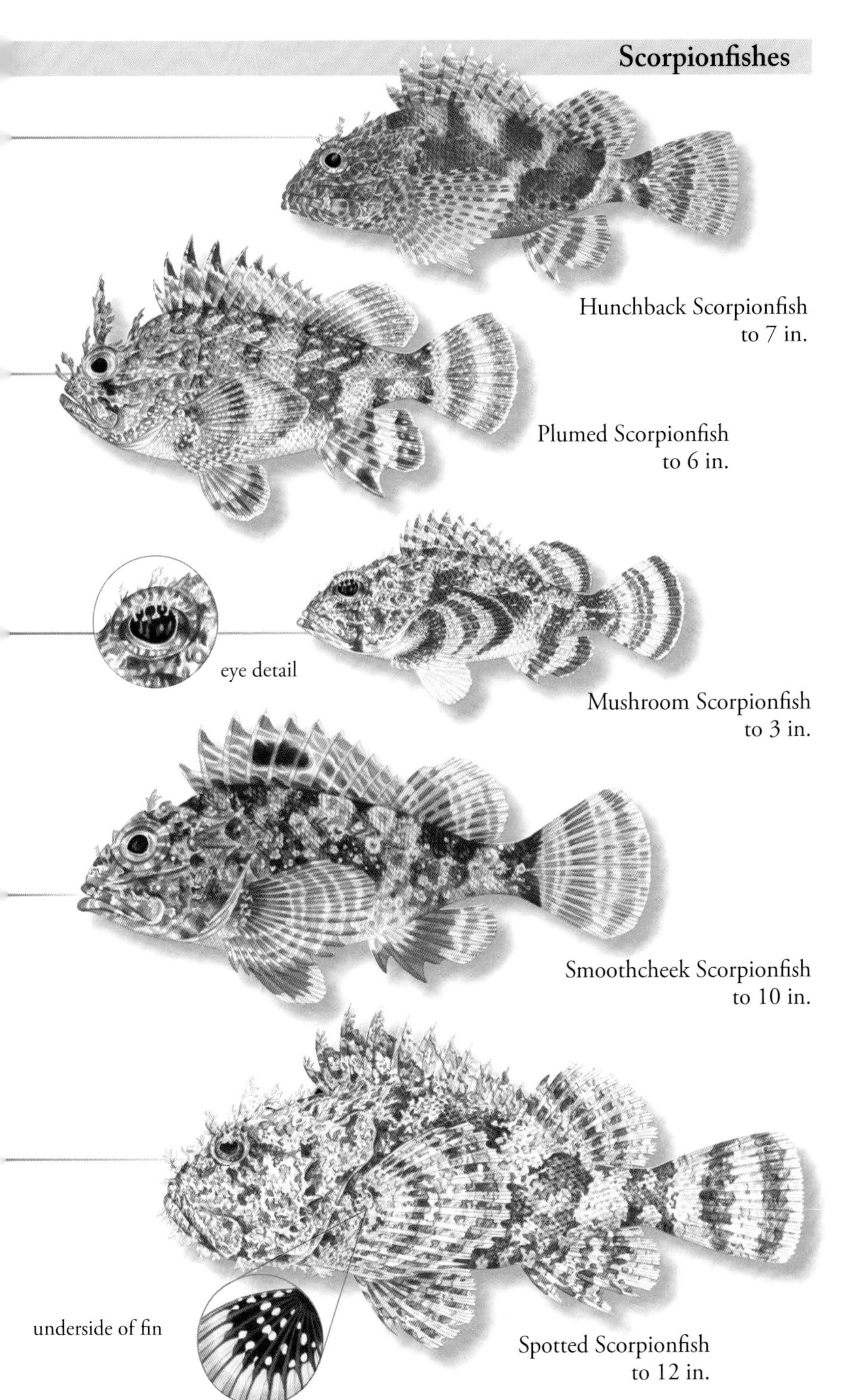
Hunchback Scorpionfish
to 7 in.
Plumed Scorpionfish
to 6 in.
eye detail
Mushroom Scorpionfish
to 3 in.
Smoothcheek Scorpionfish
to 10 in.
underside of fin
Spotted Scorpionfish
to 12 in.

Scorpaenidae - Scorpionfishes, *cont.*

Reef Scorpionfish - *Scorpaenodes caribbaeus* Meek & Hildebrand, 1928

FEATURES: Densely spotted and blotched in shades of red to brown. May have wide, whitish area on forebody. Spots on fins form bands. Spiny dorsal fin with a dark blotch posteriorly. Caudal peduncle spotted in large specimens. Two rows of spines present below eyes. Pectoral fins with 18–20 rays. HABITAT: FL, Bahamas, and Caribbean Sea. Also Bermuda. Demersal in clear inshore waters over coral or rocky areas. From near shore to about 60 ft. BIOLOGY: Hide in protective crevices. Spines are venomous.

Deepreef Scorpionfish - *Scorpaenodes tredecimspinosus* (Metzelaar, 1919)

FEATURES: Mottled reddish or brownish; predominantly reddish at deeper depths. Fins either spotted or mostly transparent. A black blotch is present at posterior portion of spiny dorsal fin. Single row of spines present below eyes. Pectoral fins long, pointed at tip, with 16–18 rays. HABITAT: NC to FL, Bahamas, and Caribbean Sea. Over coral reefs and rocky bottoms from about 26 to 255 ft. BIOLOGY: Spines are venomous.

ALSO IN THE AREA: *Pontinus helena, Scorpaena calcarata, Scorpaena elachys,* see p. 547.

Triglidae - Searobins

Shortfin Searobin - *Bellator brachychir* (Regan, 1914)

FEATURES: Mottled and blotched in shades of red to reddish brown dorsally. White below. Pectoral fins with a large ocellated black blotch. Spiny dorsal fin may have a brown to blackish blotch between fourth and fifth spines. Snout short and steep with short triangular projections. Pectoral fins short; first free ray longer than joined rays. HABITAT: NC to FL, Gulf of Mexico, and Caribbean Sea to Uruguay. Demersal over sandy, muddy, and gravel bottoms from about 90 to 1,200 ft. Usually from about 450 to 900 ft.

Streamer Searobin - *Bellator egretta* (Goode & Bean, 1896)

FEATURES: Mottled, blotched, and spotted in shades of orange, brown, and red dorsally. White below. Spiny dorsal fin mottled. Soft dorsal fin with alternating white, red, and yellow stripes. Upper caudal fin with red-bordered yellow spots. Pectoral fins with dark brown spots. Snout short and steep with very short projections. Pectoral fins short; first free ray shorter than joined rays. First dorsal-fin spine with a long filament in males. HABITAT: NC to FL, northern and southern Gulf of Mexico, and Caribbean Sea to Brazil. Demersal over sandy and muddy bottoms from about 130 to 760 ft.

Horned Searobin - *Bellator militaris* (Goode & Bean, 1896)

FEATURES: Reddish brown to brownish dorsally. Pale below. Two to three yellow stripes on lower sides. Spiny dorsal fin with yellowish spots. Soft dorsal and caudal fins with alternating white, red, and yellow stripes. Pectoral fins with blackish spots along uppermost rays. Mature males with filamentous first and second dorsal-fin rays. HABITAT: NC to FL, and Gulf of Mexico to Colombia. Occur over outer continental shelves from about 65 to 700 ft. BIOLOGY: Feed on crustaceans and fishes.

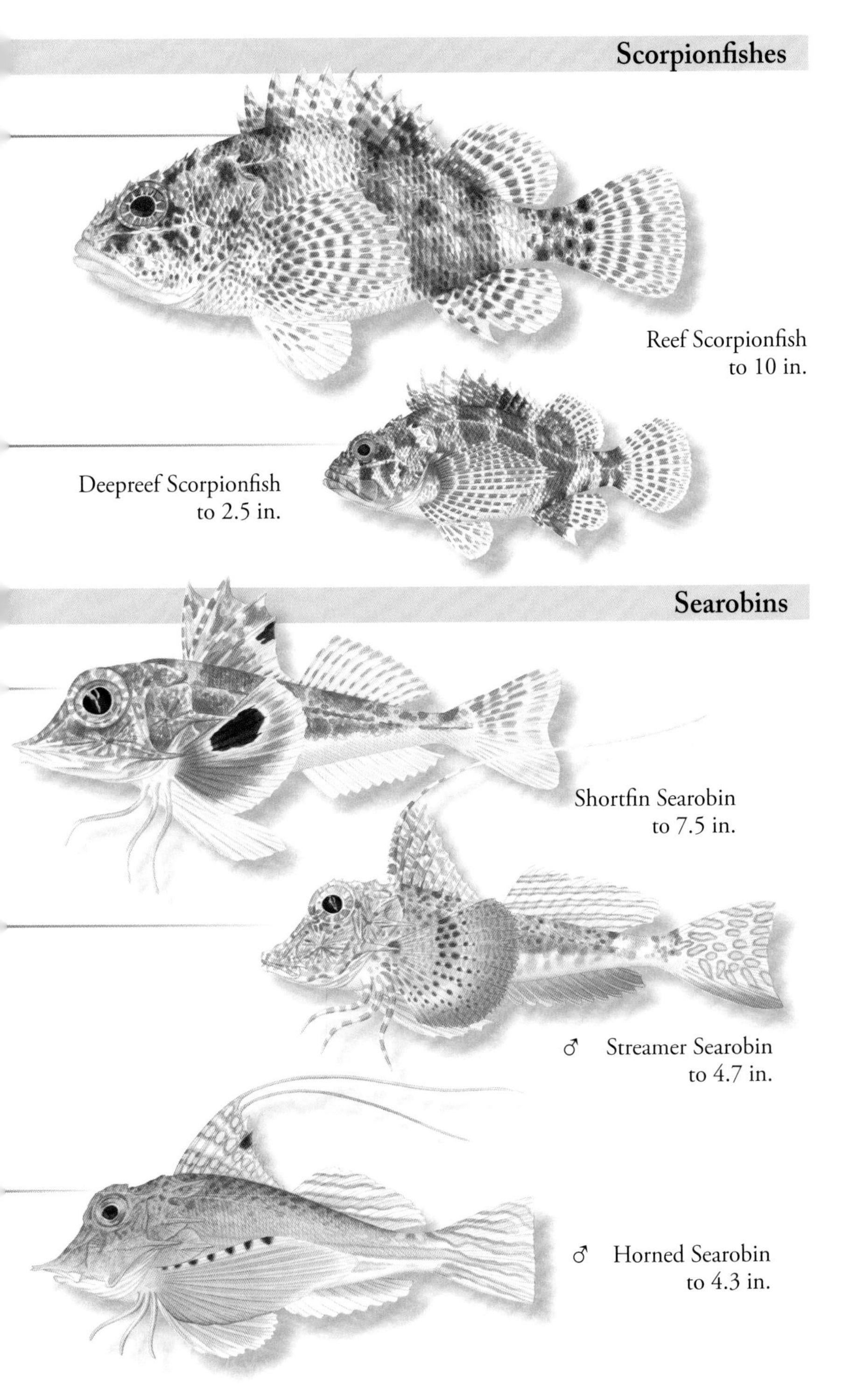
Scorpionfishes
Reef Scorpionfish
to 10 in.
Deepreef Scorpionfish
to 2.5 in.
Searobins
Shortfin Searobin
to 7.5 in.
♂ Streamer Searobin
to 4.7 in.
♂ Horned Searobin
to 4.3 in.

Spiny Searobin - *Prionotus alatus* Goode & Bean, 1883

FEATURES: Shades of brown with darker spots dorsally. Spiny dorsal fin with black ocellated spot between fourth and fifth spines. Pectoral fins with dark bands; middle posterior margins concave; lower rays elongate. Head very spiny. HABITAT: VA to eastern Gulf of Mexico. Also Bahamas and off Yucatán Peninsula. Occur over shelly, fragmented, and sedimentary bottoms from about 180 to 1,500 ft. BIOLOGY: Feed on worms, crustaceans, and fishes. May hybridize with Mexican Searobin, *Prionotus paralatus,* where ranges overlap.

Bean's Searobin - *Prionotus beanii* Goode, 1896

FEATURES: Greenish gray brown with reddish brown to orange brown slots and blotches dorsally. Spiny dorsal fin with black ocellated spot between fourth and fifth spines. Soft dorsal and pectoral fins with rows of reddish brown spots. Caudal fin banded. Snout moderately steep. Pectoral fins with straight to concave posterior margins; extend to middle of anal-fin base. HABITAT: Honduras to Brazil. Demersal over muddy sand bottoms from about 180 to 900 ft. Usually below 420 ft.

Bandtail Searobin - *Prionotus ophryas* Jordan & Swain, 1885

FEATURES: Variably mottled and spotted in shades of black, brown, red, and gold. Spiny dorsal fin with irregular spots forming bands; distinct spot absent. Second dorsal fin with a dark blotch on anterior portion. Upper pectoral-fin rays with pale and dark bands; lower portion dark with darker spots. Caudal fin dark at base, with two distinct dark bands. A single filament over the nostrils and a fringed tentacle over the eyes are present. Body robust. HABITAT: NC to FL, Bahamas, and Gulf of Mexico to Venezuela. Occur over soft bottoms from about 60 to 210 ft.

Mexican Searobin - *Prionotus paralatus* Ginsburg, 1950

FEATURES: Pinkish gray with reddish spots and blotches dorsally. Pinkish white below. Spiny dorsal fin with a black, ocellated spot between fourth and fifth spines. Pectoral fins banded in red, black, pink, and white. Caudal fin banded. Pectoral fins with straight to concave posterior margins; extend to middle of anal-fin base. HABITAT: W FL and Gulf of Mexico to northern Yucatán Peninsula. Demersal over sandy and muddy bottoms from about 30 to 900 ft. Usually below 90 ft.

Bluewing Searobin - *Prionotus punctatus* (Bloch, 1793)

FEATURES: Grayish with numerous reddish brown spots and blotches dorsally. White below. Spiny dorsal fin blotched; lacks distinct spot in adults. Pectoral fins dark blackish brown to olivaceous with darker and paler spots above and bright blue lower margin. Ventral fins white to pink. Caudal fin with spots forming bands. HABITAT: S Gulf of Mexico and Caribbean Sea to Argentina. Demersal over sandy and muddy bottoms from about 20 to 620 ft. Usually below 100 ft.

Spiny Searobin
to 8 in.

Bean's Searobin
to 6 in.

Bandtail Searobin
to 8 in.

Mexican Searobin
to 8 in.

Bluewing Searobin
to 8 in.

Bluespotted Searobin - *Prionotus roseus* Jordan & Evermann, 1887

FEATURES: Brownish with darker blotches dorsally. Whitish below. First dorsal fin with a dark spot between fourth and fifth spines. Pectoral fins dusky to dark brownish with blue spots on membranes. Caudal fin with two dark bands. HABITAT: NC to FL, Gulf of Mexico, Bahamas, Caribbean Sea to Brazil. On or near soft bottoms from about 30 to 600 ft. BIOLOGY: Feed on invertebrates and fishes. Spawn from April to September.

Blackwing Searobin - *Prionotus rubio* Jordan, 1886

FEATURES: Brownish to grayish brown dorsally with irregular darker spots and blotches. First dorsal fin with dark spot between third and fourth spines. Spot fades with age. Pectoral fins dark with diffuse spotting and a bright blue lower margin. HABITAT: NC to FL, Cuba, and the Gulf of Mexico from FL to TX. On or near bottom from shore to about 700 ft.; more commonly between about 30 and 180 ft. BIOLOGY: Feed on invertebrates, fishes, and worms.

Leopard Searobin - *Prionotus scitulus* Jordan & Gilbert, 1882

FEATURES: Tannish with dark brownish spots dorsally and on sides. Darker, diffuse bands may be present. Pectoral, dorsal, and caudal fins spotted. First dorsal fin with two blackish spots: one between the first and second spines, a second between the fourth and fifth spines. Body comparatively elongate. HABITAT: VA to FL, and Gulf of Mexico to Venezuela. Usually found near bottom in shallow bays and from shore to about 150 ft. BIOLOGY: Feed on a variety of invertebrates. Spawn in early fall and spring. Prefer high-salinity waters.

Shortwing Searobin - *Prionotus stearnsi* Jordan & Swain, 1885

FEATURES: Silvery black to gray dorsally. May have vague blotches. Ventral area silvery. Pectoral fins and caudal-fin margin blackish. First dorsal fin lacks distinct spot. Small, bony knob at tip of lower jaw. Pectoral fins short. HABITAT: NC to FL, and Gulf of Mexico to French Guiana. Found offshore over soft and semi-hard bottoms from about 120 to 360 ft. BIOLOGY: Feed on crustaceans and small fishes. Spawn during winter and spring.

Bighead Searobin - *Prionotus tribulus* Cuvier, 1829

FEATURES: Grayish brown dorsally with darker, oblique bars and pale flecks. Pale ventrally. First dorsal fin with a dark blotch between fourth and fifth spines. Pectoral fins dark with irregular pale cross bands. Caudal fin with a single broad, dark bar. Head comparatively large and elevated. HABITAT: NY to FL, and Gulf of Mexico to Yucatán Peninsula. Demersal over mud-sand bottoms from about 10 to 600 ft. Usually above 200 ft. BIOLOGY: Feed on crustaceans, small fishes, and worms. Spawn from fall to spring.

ALSO THE AREA: *Bellator ribeiroi*, *Prionotus martis*, see p. 547.

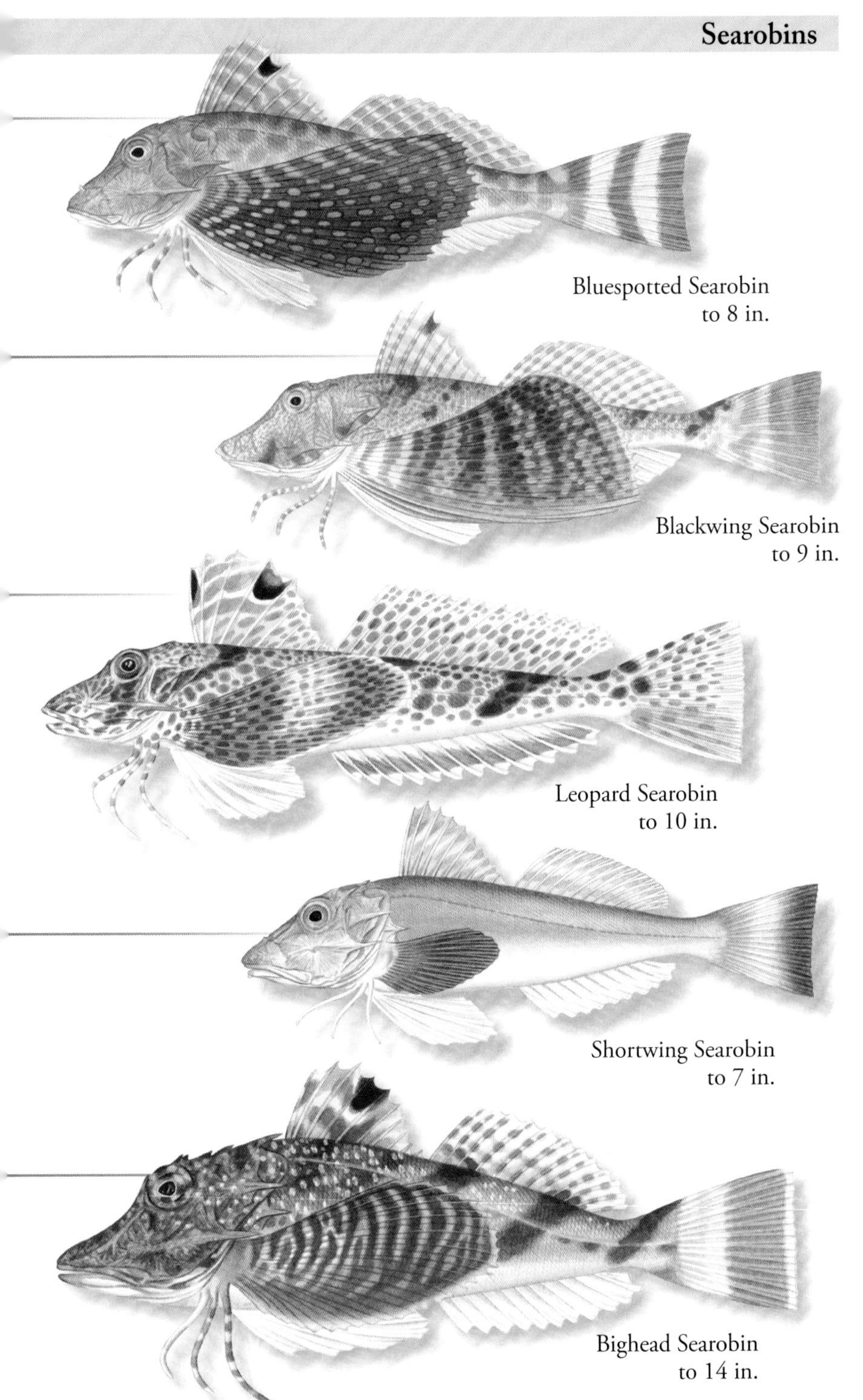

Bluespotted Searobin
to 8 in.

Blackwing Searobin
to 9 in.

Leopard Searobin
to 10 in.

Shortwing Searobin
to 7 in.

Bighead Searobin
to 14 in.

Peristediidae - Armored Searobins

Flathead Armored Searobin - *Peristedion brevirostre* (Günther, 1860)

FEATURES: Color and pattern vary. Dorsal surface grayish with patches of orange to mottled and blotched in orange and yellow. White below. Dorsal, caudal, and anal fins with red to orange margins. Rostral projections short. Sides of head with a relatively wide bony rim. Lower lip with a pair of long, branched barbels. Chin with seven clusters of 32 to 42 barbels. HABITAT: E FL to LA, southern Bahamas, and Caribbean Sea. Demersal over sandy, muddy, and hard bottoms from about 180 to 1,800 ft.

Slender Searobin - *Peristedion gracile* Goode & Bean, 1896

FEATURES: Reddish to straw-colored dorsally. May have wide dark bars along back. Silvery white below. Dark stripe through middle of first and second dorsal fins. Caudal fin slightly forked with dark margin. Rostral projections very long and slender. Sides of head with narrow bony rim. Series of short and somewhat elongate barbels on chin. HABITAT: VA to FL, and Gulf of Mexico to Yucatán Peninsula. Demersal over continental shelves and slopes from about 240 to 1,500 ft.

Widehead Armored Searobin - *Peristedion longispatha* Goode & Bean, 1886

FEATURES: Dark red dorsally to dark red with numerous blackish wavy lines dorsally. White below. Dorsal and caudal fins with black margins. Rostral projections long. Head very wide, sides with a right-angled bony rim. Chin with two long, filamentous barbels and six clusters of single and branched barbels. HABITAT: FL to northern Gulf of Mexico, Bahamas, and Caribbean Sea. Demersal over sandy and muddy bottoms from about 330 to 2,500 ft.

Armored Searobin - *Peristedion miniatum* Goode, 1880

FEATURES: Reddish dorsally. White below. Margins of dorsal fins blackish red. Rostrum with short projections. Sides of head with spiny, wing-like projections. Several short and two long filamentous barbels on chin. HABITAT: Canada and Georges Bank to FL and Gulf of Mexico to Brazil. Demersal over outer sandy and muddy continental shelves and slopes from about 200 to 3,000 ft. BIOLOGY: Feed on crustaceans. Preyed upon by Spiny Dogfish.

Rimspine Searobin - *Peristedion thompsoni* Fowler, 1952

FEATURES: Yellowish orange to red dorsally. Pale below. Dorsal fins unmarked. Caudal fin with dark margin. Rostrum with short projections; each bears a small spine on base. Sides of head with relatively wide rim. Lips and chin with many short, clustered barbels. HABITAT: NC to FL Keys, northern Gulf of Mexico, and Honduras to Brazil. Demersal over sandy and muddy continental shelves and slopes from about 370 to 1,500 ft. BIOLOGY: Use free pectoral rays to "walk" along bottom and barbels to locate prey.

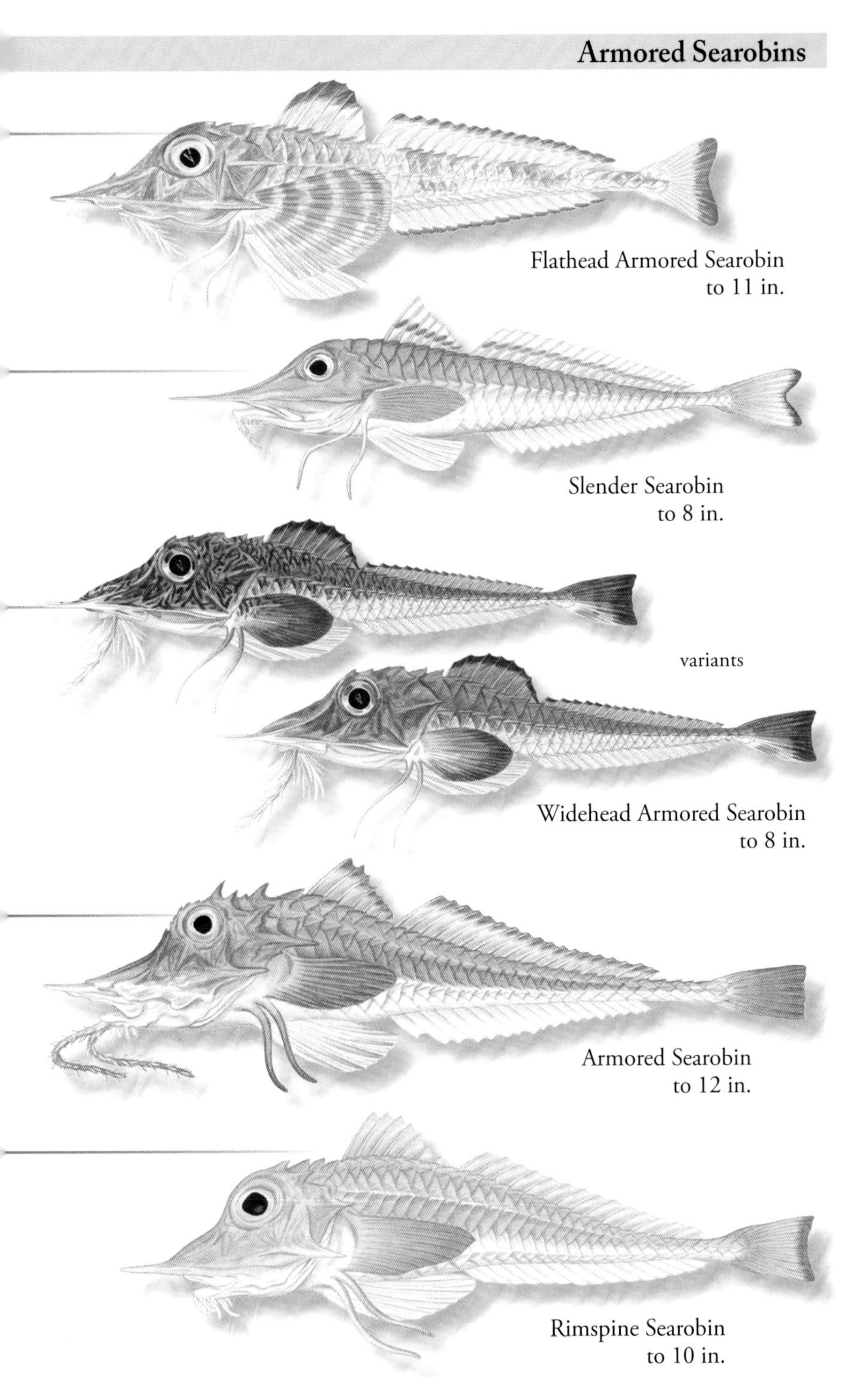

Flathead Armored Searobin
to 11 in.

Slender Searobin
to 8 in.

Widehead Armored Searobin
to 8 in.

Armored Searobin
to 12 in.

Rimspine Searobin
to 10 in.

Centropomidae - Snooks

Swordspine Snook - *Centropomus ensiferus* Poey, 1860

FEATURES: Yellow brown to brownish green above; silvery on sides and below. Lateral line dark. Fins dusky. Pelvic fins may have dark blotch. Pectoral fins reach to about anus. Pelvic fins reach to or past anus. Second anal-fin spine extends to or beyond caudal-fin base. Usually with 9 or 10 scales between lateral line and soft dorsal-fin origin. HABITAT: S FL, southern Gulf of Mexico to Brazil, Greater and Lesser Antilles. Along coast and in estuaries, lagoons, and fresh water. Prefer lower salinities. BIOLOGY: Feed on fishes and crustaceans.

Largescale Fat Snook - *Centropomus mexicanus* Bocourt, 1868

FEATURES: Brownish green to yellow brown above; silvery on sides and below. Lateral line black. Fins dusky. Dark blotch on pelvic-fin tips. Pectoral about as long as pelvic fins. Pelvic fins reach to or past anus. Second anal-fin spine curved, not reaching caudal-fin base. Usually with 11–14 scales between lateral line and soft dorsal-fin origin. HABITAT: E FL, western Gulf of Mexico, and Caribbean Sea to Brazil. Occur along the coast in bays, estuaries, and full salt water from shore to about 50 ft. Also found in streams.

Smallscale Fat Snook - *Centropomus parallelus* Poey, 1860

FEATURES: Brownish green to yellow brown above; silvery on sides and below. Lateral line dark. Fins dusky. Pectoral fins do not reach anus. Pelvic fins reach to or past anus. Second anal-fin spine may reach caudal-fin base. Body comparatively deep; 9–16 scales between lateral line and soft dorsal-fin origin. HABITAT: S FL, southern Gulf of Mexico, Greater and Lesser Antilles to Brazil. Along coast and in estuaries, lagoons, and fresh water. Prefer low salinities.

Tarpon Snook - *Centropomus pectinatus* Poey, 1860

FEATURES: Brownish green to yellow brown above; silvery on sides and below. Lateral line dark. Fins dusky. Dark blotch on pelvic-fin tips. Pectoral fins distinctly shorter than pelvic fins. Pelvic fins reach to or past anus. Second anal-fin spine slightly outcurved at tip, not reaching caudal-fin base. Usually with 10–12 scales between lateral line and soft dorsal-fin origin. HABITAT: S FL, southern Gulf of Mexico, Greater and Lesser Antilles to Brazil. Also in Pacific from Mexico to Colombia. Along coast and in estuaries, lagoons, and fresh water. Prefer low salinities.

Mexican Snook - *Centropomus poeyi* Chávez, 1961

FEATURES: Brownish green to yellow brown above; silvery on sides and below. Lateral line dark. Fins dusky. Pectoral fins about as long as pelvic fins. Pelvic fins not reaching anus in adults. Second anal-fin spine slightly curved, not reaching caudal-fin base. Usually with 11 or 12 scales between lateral line and soft dorsal-fin base. HABITAT: SW Gulf of Mexico to Belize. Occur along coast in bays, estuaries, and lagoons. Enter fresh water. Prefer low salinities. BIOLOGY: Feed on fishes and crustaceans.

Swordspine Snook
to 14 in.

Largescale Fat Snook
to 17 in.

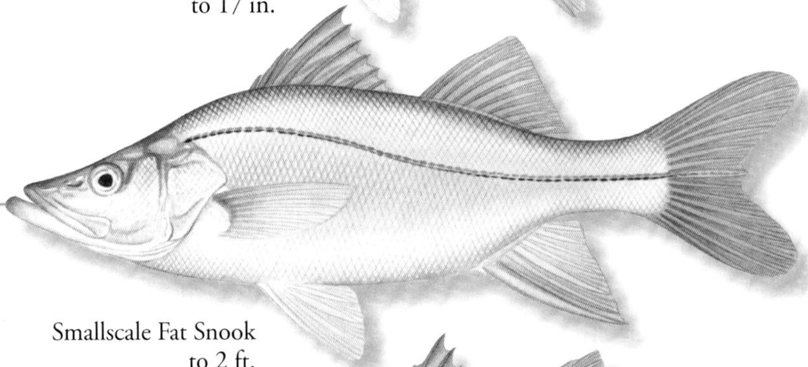

Smallscale Fat Snook
to 2 ft.

Tarpon Snook
to 20 in.

Mexican Snook
to 3 ft.

Common Snook - *Centropomus undecimalis* (Bloch, 1792)

FEATURES: Yellow brown to brownish green above; silvery on sides and below. Lateral line dark. Fins dusky to yellowish. Pelvic fins not reaching anus. Second anal-fin spine not reaching caudal-fin base. Body comparatively elongate; 10–14 scales between lateral line and soft dorsal-fin origin. HABITAT: NC to FL, Gulf of Mexico, Greater and Lesser Antilles to Brazil. Along coast and in estuaries, marshes, lagoons, mangroves, river mouths, and freshwater streams to about 65 ft. BIOLOGY: Intolerant of cool water temperatures. Feed on fishes and crustaceans. Congregate in river mouths during summer spawning. Sought commercially and for sport.

Acropomatidae - Lanternbellies

Blackmouth Bass - *Synagrops bellus* (Goode & Bean, 1896)

FEATURES: Pearly white to pearly blue with large black saddles along back and large patches of black blotches on sides. Spiny dorsal fin black distally, pale at base. Caudal fin pale with black upper and lower margins. Dorsal fins separate; fin spines lack serrations. Opercle with one or two flat points. HABITAT: Canada to FL, Gulf of Mexico. Bermuda, Bahamas, and Caribbean Sea to Brazil. Found near or on soft and hard bottoms from about 200 to 3,300 ft. Usually below 330 ft.

Symphysanodontidae - Slopefishes

Slope Bass - *Symphysanodon berryi* Anderson, 1970

FEATURES: Shades of pink, orange, or red dorsally with iridescent highlights. Silvery below. Fins mostly translucent. Eye diameter larger than snout length. Opercle with two spines. Most of head and jaws scaled. Dorsal fin continuous and slightly notched. Pelvic fins extend to anus in females; first pelvic soft ray long and thread-like in males. Caudal fin deeply forked with thread-like filaments at lobe tips in males. Soft dorsal and anal fins with basal scaly sheath. Body comparatively long. HABITAT: NC to FL, Bermuda, Bahamas, and Caribbean Sea to northern Brazil. Also northern Yucatán and Quintana Roo. Occur near bottoms over deep reefs from about 330 to 1,560 ft.

Insular Bunquelovely - *Symphysanodon octoactinus* Anderson, 1970

FEATURES: Shades of orange to reddish orange with iridescent highlights. A narrow to broad iridescent purplish band present from opercle to peduncle. Dorsal fin translucent pinkish yellow along outer margin. Anal and pelvic fins translucent pinkish yellow. Upper caudal-fin lobe yellow; lower lobe orange to yellow orange. Eye diameter larger than snout length. Opercle with two spines. Most of head and jaws scaled. Dorsal fin continuous, lacks notch. Soft dorsal and anal fins with a basal scaly sheath. HABITAT: Bahamas, Puerto Rico and adjacent islands, and western and southern Caribbean Sea. Occur near bottom over deep reefs from about 500 to 2,100 ft.

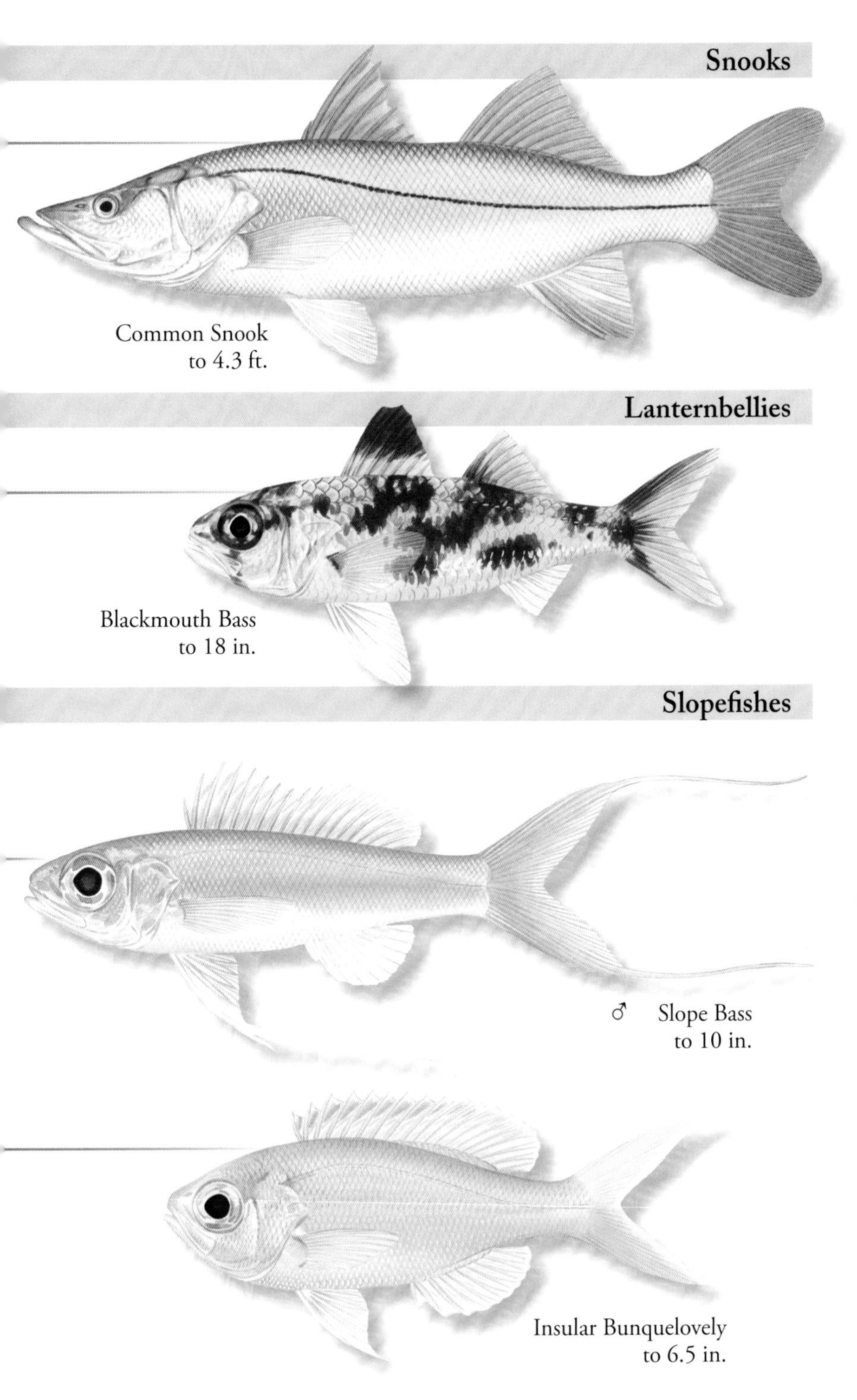
Snooks
Common Snook
to 4.3 ft.
Lanternbellies
Blackmouth Bass
to 18 in.
Slopefishes
♂ Slope Bass
to 10 in.
Insular Bunquelovely
to 6.5 in.

Polyprionidae - Wreckfishes

Wreckfish - *Polyprion americanus* (Bloch & Schneider, 1801)

FEATURES: Shades of dark brownish to dark grayish dorsally; paler below. Juveniles blackish with whitish mottling. Lower jaw protrudes. Bony knob over eyes and a distinct ridge at nape. Upper operculum with a horizontal ridge. HABITAT: Newfoundland to FL, Bermuda, and northwestern Bahamas. Also in eastern Atlantic, Mediterranean, southwest Pacific, and southern Indian Ocean. Adults near bottom over rocky slopes and seamounts and around caves and wrecks in temperate waters from about 160 to 2,600 ft. Juveniles pelagic and under floating objects. BIOLOGY: Spawn in summer. Feed on fishes and crustaceans. Sought commercially and for sport. Vulnerable to overfishing.

Epinephelidae - Groupers

Mutton Hamlet - *Alphestes afer* (Bloch, 1793)

FEATURES: Variably colored. Shades of red, brown, or gray to olive with darker, irregular bars and paler spots and blotches. Small, orange spots on cheeks, sides, and abdomen. Usually with a dark band from eyes to dorsal-fin origin. Fins spotted and mottled. Snout very short. Head profile somewhat steep, nearly straight. HABITAT: NC to FL, Bermuda, Bahamas, and Caribbean Sea to Brazil. Also over seagrass beds and around rocks and sponges from shore to about 115 ft. BIOLOGY: Sedentary during day; predatory at night. Will hide in crevices or lie partially covered in sand to avoid detection. Feed on crustaceans.

Graysby - *Cephalopholis cruentata* (Lacepède, 1802)

FEATURES: Body and fins grayish, bluish, or brownish with small, evenly spaced orange brown to reddish spots. Four black or white spots along back under dorsal-fin base. Dorsal, caudal, and anal fins with reddish inner margin. May pale or darken in color or become blotched while resting. HABITAT: NC to FL, Gulf of Mexico, Bermuda, Bahamas, and Caribbean Sea to Brazil. Found over seagrass beds and coral reefs from shore to about 560 ft. BIOLOGY: Feed at dusk and dawn; stay near hiding places during the day.

Coney - *Cephalopholis fulva* (Linnaeus, 1758)

FEATURES: Three color phases: uniform red (deep water); bicolored (shallow water); yellow (shallow to deep water). Uniform and bicolored phases with have small, dark-circled, pale blue spots on body. Yellow form with scattered, small blue spots on head and often on back. All with two dark spots on lower lip and two dark spots on upper portion of peduncle. Night pattern is usually pale with irregular bars. HABITAT: SC to FL, Gulf of Mexico, Bermuda, Bahamas, and Caribbean Sea to Brazil. Around coral reefs and rocky bottoms in clear water to about 150 ft. BIOLOGY: Hermaphroditic. May hide in caves or under ledges during day. Feed on crustaceans and small fishes. Occasionally follow Moray Eels in search of prey.

Wreckfishes
Wreckfish
to 6.5 ft.
juvenile
Groupers
Mutton Hamlet
to 13 in.
Graysby
to 13 in.
Coney
to 15.4 in.

Atlantic Creolefish - *Cephalopholis furcifer* (Valenciennes, 1828)

FEATURES: Brownish red or purplish to grayish dorsally; creamy to salmon below. Three pale or dark spots below dorsal fin. Sometimes with spots on posterior portion of lateral line. Body may display pale mottling. All with a dark red blotch at upper pectoral-fin base. Spiny dorsal-fin membranes with yellowish spots. Caudal fin deeply forked. HABITAT: NC to FL, Gulf of Mexico, Bermuda, Bahamas, and Caribbean Sea to Brazil. Also south Atlantic islands. Occur over hard bottoms and coral reefs from about 33 to 210 ft. BIOLOGY: Feed on zooplankton.

Marbled Grouper - *Dermatolepis inermis* (Valenciennes, 1833)

FEATURES: Pattern highly variable. Brownish to grayish with pale blotches and mottling. Dark spots scattered between blotches. Margins of fins usually pale. Juveniles black to dark brown with irregular white to tan blotches and rounded caudal fin. Head profile steep, nearly straight. Body deep. HABITAT: NC to FL, Gulf of Mexico, Bahamas, and Caribbean Sea to Brazil. Around ledges, over reefs, in caves and crevices, and around oil rigs from about 70 to 820 ft. BIOLOGY: Solitary and shy. Sought as sportfish.

Rock Hind - *Epinephelus adscensionis* (Osbeck, 1765)

FEATURES: Body and fins greenish to buff with numerous reddish brown spots and scattered pale blotches. Three to five dark blotches below dorsal-fin base. Single dark blotch at top of caudal peduncle. Dark spots along margin of caudal fin may form a band. Juveniles with larger and fewer dark spots. HABITAT: MA (rare) to FL, Gulf of Mexico, and Caribbean Sea. Also Bermuda and eastern Atlantic. Occur around rocky bottoms, reefs, jetties, and oil rigs from about 6 to 330 ft. BIOLOGY: Feed on crabs and fishes.

Red Hind - *Epinephelus guttatus* (Linnaeus, 1758)

FEATURES: Buff or greenish white to pale reddish brown with reddish brown spots dorsally and bright red spots ventrally. Spiny dorsal fin with yellow tips. Soft dorsal, caudal, and anal fins with a broad blackish inner margin and a narrow white margin. Juveniles with fewer and larger spots. HABITAT: NC to FL, Gulf of Mexico, Bahamas, and Caribbean Sea to Brazil. Also Bermuda. Found over rocky bottoms and reefs from about 6 to 330 ft. BIOLOGY: Hermaphroditic. Females may carry up to 3 million eggs. Feed on octopods, crustaceans, and fishes.

Atlantic Goliath Grouper - *Epinephelus itajara* (Lichtenstein, 1822)

FEATURES: Brownish, brownish yellow to greenish with small dark and pale spots. Smaller specimens with broken bars on body. Bars often radiate from eyes. Fins variably banded and spotted. Head broad, flattened between comparatively small eyes. Spiny dorsal fin low. Caudal fin rounded. Body robust. HABITAT: FL, Gulf of Mexico, Bahamas, and Caribbean Sea to Brazil. In shallow waters, around wrecks, and in mangrove swamps, bays, and harbors to about 100 ft. BIOLOGY: Territorial and slow growing. May live to 37 years old. Females may carry up to 5 million eggs. Protected in the United States. OTHER NAME: Jewfish. IUCN: Vulnerable.

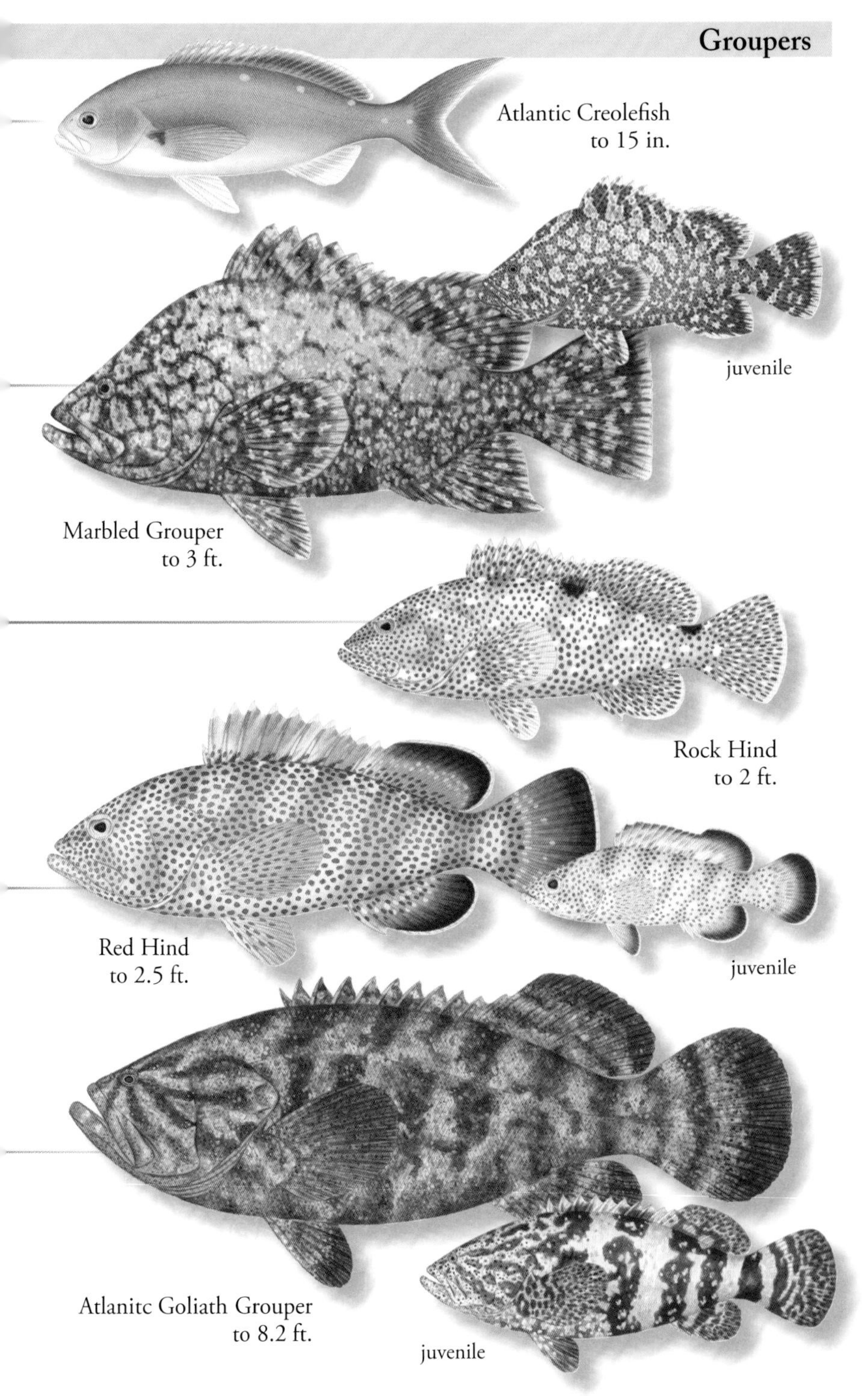
Atlantic Creolefish
to 15 in.
juvenile
Marbled Grouper
to 3 ft.
Rock Hind
to 2 ft.
Red Hind
to 2.5 ft.
juvenile
Atlanitc Goliath Grouper
to 8.2 ft.
juvenile

Red Grouper - *Epinephelus morio* (Valenciennes, 1828)

FEATURES: Reddish brown with diffuse, pale blotches and spots. Bars often radiate from eyes. May appear uniform brown with pale spots. Margins on soft dorsal, caudal, and anal fins dark. Inside of mouth bright reddish orange. Spiny dorsal fin with a tall second spine and nearly straight margin. HABITAT: NC to FL, Gulf of Mexico, Bermuda, Bahamas, Caribbean Sea to Brazil. Around reefs, in crevices, under ledges, and over sandy or mud bottoms to about 950 ft. BIOLOGY: Hermaphroditic. Females carry up to 5 million eggs. IUCN: Vulnerable.

Nassau Grouper - *Epinephelus striatus* (Bloch, 1792)

FEATURES: Reddish brown with five buff colored bars on body. May have pale blotches scattered on head and body. Specimens from deep water are pinkish to reddish ventrally. Diagonal band runs from upper jaw to dorsal-fin origin. Small dark spots around eyes. Margin of spiny dorsal fin yellowish. Black saddle on caudal peduncle. Color may change fairly rapidly. HABITAT: SC to FL, NW Gulf of Mexico to Yucatán, Bermuda, Bahamas, and Caribbean Sea to Brazil. Occur from shore to about 300 ft. Adults are found over coral reefs, juveniles over seagrass beds. BIOLOGY: Form large, complex spawning aggregations. IUCN: Critically Endangered.

Spanish Flag - *Gonioplectrus hispanus* (Cuvier, 1828)

FEATURES: Body covered in alternating yellow and salmon-colored stripes. Yellow bar runs from snout through eye and along base of dorsal fins. Cheeks with yellow spots and lines. Large, deep red spot on anterior portion of anal fin. White blotch on abdomen. HABITAT: NC to FL, Gulf of Mexico, Cuba, Jamaica, and Venezuela to Brazil. Occur over rocky bottoms from about 200 to 1,200 ft.

Speckled Hind - *Hyporthodus drummondhayi* (Goode & Bean, 1878)

FEATURES: Adults shades of brown with a dense covering of small, pearly speckles on body and fins. Juveniles yellowish with numerous pearly spots on body and fins. Caudal fin with nearly straight margin. Body relatively deep. HABITAT: NC to FL Keys; Gulf of Mexico to Quintana Roo. Also Bermuda. Occur over rocky bottoms from about 80 to 600 ft. BIOLOGY: May live to 25 years. Females carry up to 2 million eggs. Feed on fishes and invertebrates. NOTE: Previously known as *Epinephelus drummondhayi*.

Yellowedge Grouper - *Hyporthodus flavolimbatus* (Poey, 1865)

FEATURES: Yellowish tan to grayish brown dorsally; pale below. Thin, pale blue line from eye to preopercular corner. Adults may be uniformly colored or may display several rows of white spots. Juveniles with white spots and a black saddle on caudal peduncle. Margins of dorsal, caudal, and pectoral fins yellow. HABITAT: NC to FL, Gulf of Mexico to Brazil. Over rocky, sandy, or muddy bottoms from about 200 to 1,200 ft. BIOLOGY: Hermaphroditic. May live to about 20 years. Feed on invertebrates and fishes. NOTE: Previously known as *Epinephelus flavolimbatus*. IUCN: Vulnerable.

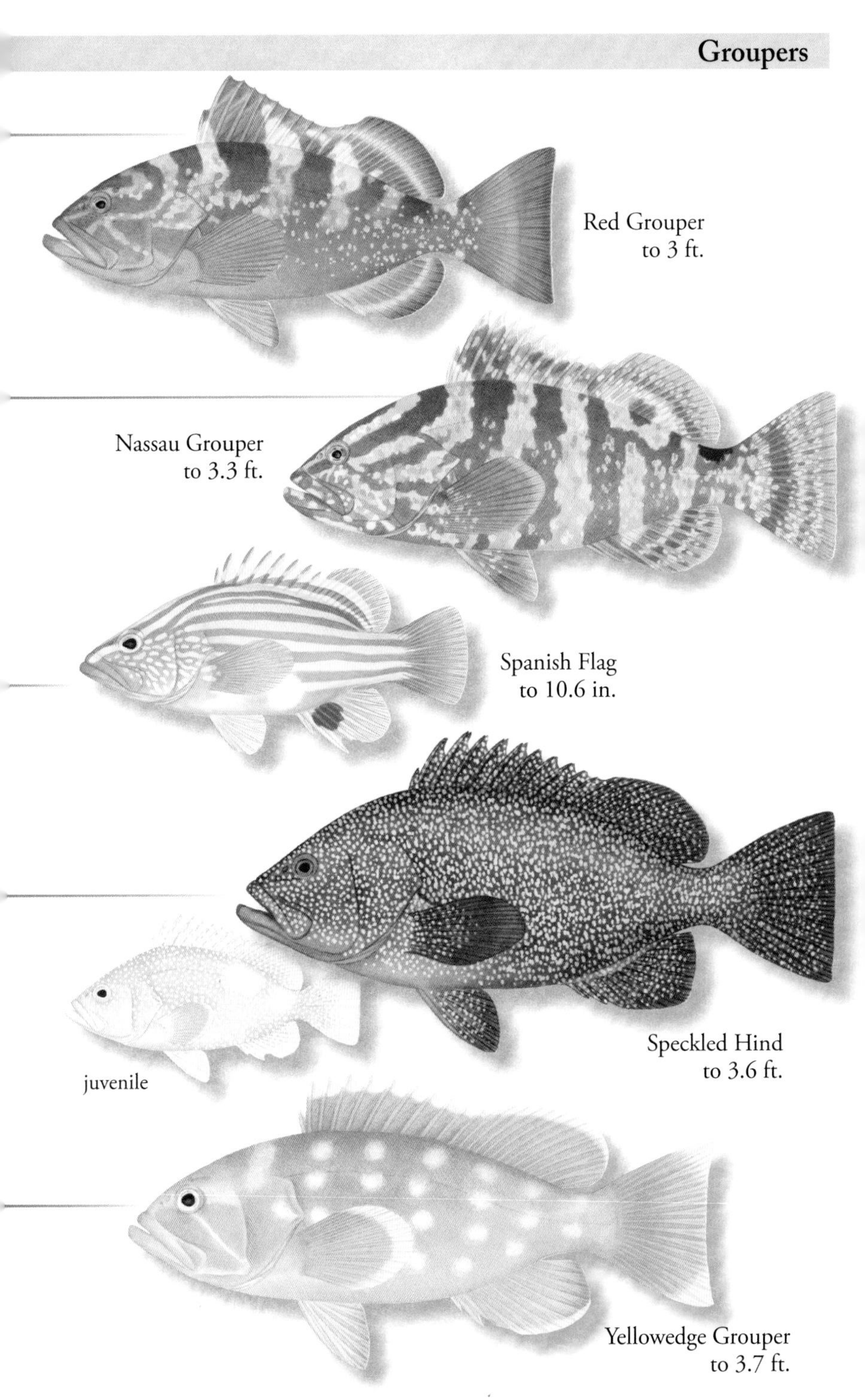
Red Grouper
to 3 ft.
Nassau Grouper
to 3.3 ft.
Spanish Flag
to 10.6 in.
Speckled Hind
to 3.6 ft.
juvenile
Yellowedge Grouper
to 3.7 ft.

Misty Grouper - *Hyporthodus mystacinus* (Poey, 1852)

FEATURES: Brownish with eight or nine darker bars and paler, diffuse spots. Bands radiate from eyes. Last two bands on peduncle are darker than others and fuse to appear as one. Dark "moustache" just above jaws. Fin margins pale. Juveniles with more contrasting body bars and pale caudal fin. HABITAT: NC to FL, eastern and western Gulf of Mexico, Bermuda, Bahamas, and Caribbean Sea. Found over rocky reefs from about 100 to 1,600 ft. NOTE: Previously known as *Epinephelus mystacinus*.

Warsaw Grouper - *Hyporthodus nigritus* (Holbrook, 1855)

FEATURES: Dark reddish brown to almost black dorsally, fading to dull reddish gray below. Juveniles with yellowish caudal fin and scattered whitish spots on body. Spiny dorsal fin with 10 spines; second spine distinctly elongate. Pelvic-fin origin anterior to lower pectoral-fin base. HABITAT: MA to FL, Gulf of Mexico, Bermuda, Greater Antilles, and Venezuela to Brazil. Adults over rough, rocky bottoms from about 180 to 1,700 ft. Juveniles may be found around reefs and jetties. NOTE: Previously known as *Epinephelus nigritus*. IUCN: Near Threatened.

Snowy Grouper - *Hyporthodus niveatus* (Valenciennes, 1828)

FEATURES: Large specimens uniformly dark brown to gray brown. Spiny dorsal-fin margin blackish. Other fin margins may be pale blue. Intermediate specimens with regularly spaced white spots. Small juveniles also with a black saddle on peduncle and yellowish pectoral and caudal fins. Pelvic-fin origin below or posterior to lower pectoral-fin base. Upper opercular margin distinctly arched. HABITAT: MA to Cuba; Gulf of Mexico to Brazil. Bermuda, Great Bahama Bank, and Bimini Islands. Adults over rocky bottoms from about 30 to 1,300 ft. BIOLOGY: Feed on crustaceans and fishes. May live to 27 years. NOTE: Previously *Epinephelus niveatus*. IUCN: Vulnerable.

Bladefin Bass - *Jeboehlkia gladifer* (Poey, 1860)

FEATURES: Color varies. Shades of white to pink or red. All with a broad red band through eyes, a broad red bar from soft dorsal to anal fin, and a large red area on caudal fin. Fin margins white. Second dorsal-fin spine tall and erect. Pelvic fins long and pointed. Soft dorsal-, caudal-, and anal-fin margins rounded. HABITAT: SC, northern Gulf of Mexico, and western, eastern, and southern Caribbean Sea. Occur over deep reefs from about 330 to 1,300 ft.

Eyestripe Basslet - *Liopropoma aberrans* (Poey, 1860)

FEATURES: Yellow dorsally with some scattered orange red spots and mottling. Orange red to pink below. Top of head lavender, pink, or orange red. Yellow stripe from snout to opercular margin. Lower cheeks with irregular yellow spots. Fins yellow, may have lavender or orange red margins. Tips of second dorsal, caudal, and anal fins pointed. HABITAT: FL Keys, Flower Garden Reef, Bahamas, Greater and Lesser Antilles, Quintana Roo to Belize, and Venezuela to Brazil. Reef associated from about 300 to 750 ft.

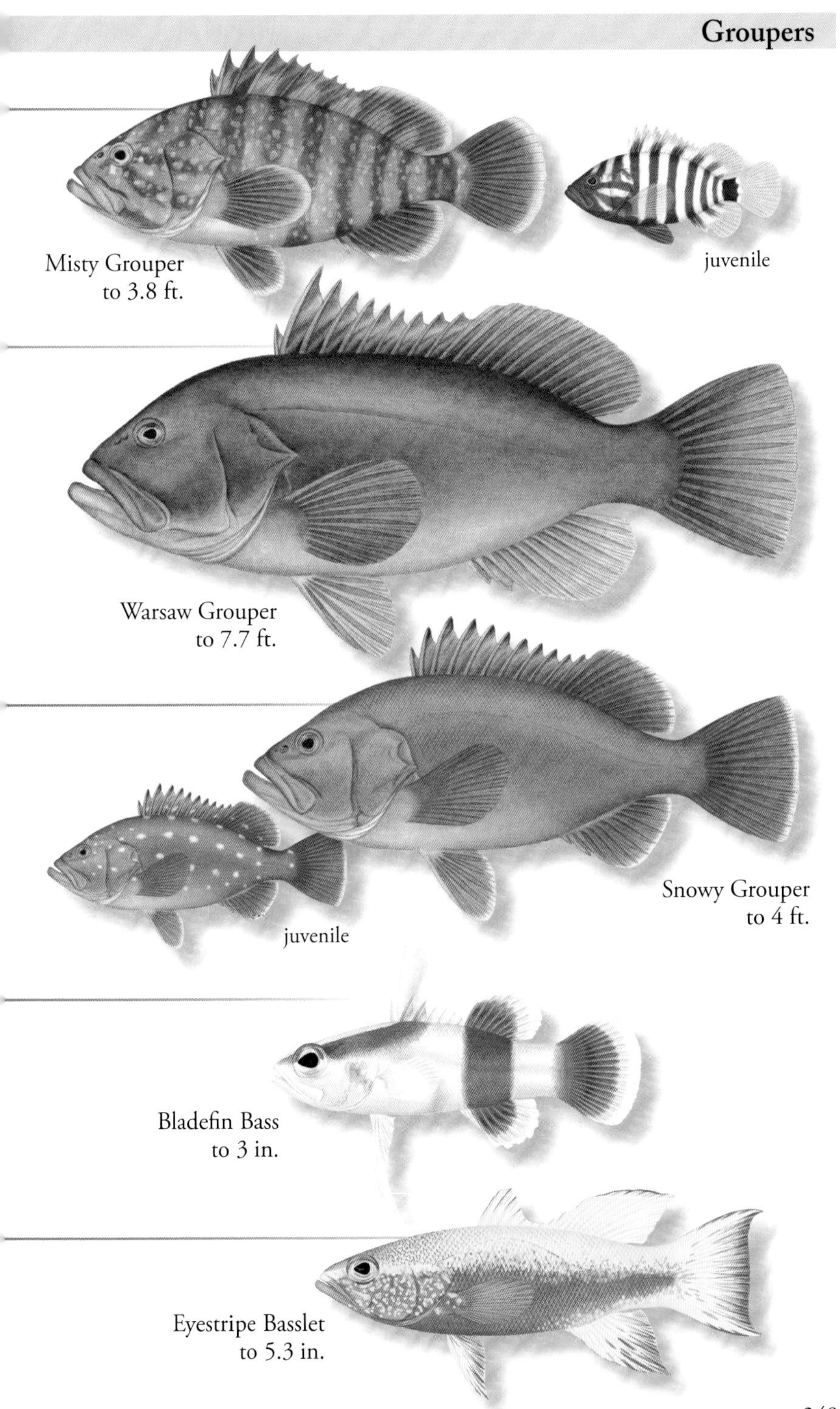
Misty Grouper
to 3.8 ft.
juvenile
Warsaw Grouper
to 7.7 ft.
juvenile
Snowy Grouper
to 4 ft.
Bladefin Bass
to 3 in.
Eyestripe Basslet
to 5.3 in.

Candy Basslet - *Liopropoma carmabi* (Randall, 1963)

FEATURES: Yellow with five blue to purplish stripes on body. Each stripe is bordered above and below by a red stripe. Dorsal fin with a large black spot bordered by a blue ring. Caudal fin with two separate black spots bordered by blue rings. Anal fin lacks black spot. HABITAT: FL Keys, Bahamas, southern Gulf of Mexico, and Caribbean Sea. Reef associated over coral, sand, and coral rubble bottoms from about 50 to 230 ft. BIOLOGY: Reclusive and territorial.

Cave Basslet - *Liopropoma mowbrayi* Woods & Kanazawa, 1951

FEATURES: Body and fins shades of red. Yellowish stripe from snout tip to eyes. Dorsal and anal fins with a small black spot and bluish white margin at tips. Spot on anal fin may be small to absent. Caudal fin with bluish white margin and black inner band. HABITAT: S FL, Bermuda, Bahamas, and Caribbean Sea. Reef associated from about 98 to 200 ft. BIOLOGY: Solitary.

Yellow-spotted Basslet - *Liopropoma olneyi* Baldwin & Johnson, 2014

FEATURES: Yellow with brighter yellow spots dorsally. Pinkish to white below with yellow spots. Top of head pink to red. Bright yellow stripe from snout to opercular margin. Dorsal, caudal, and anal fins yellow, may have pink to red margins. Tips of second dorsal, caudal, and anal fins bluntly rounded. HABITAT: Known from Bonaire, Curaçao, Dominica, and St. Eustatius. May also occur north of Cuba. Reef associated from about 435 to 630 ft.

Peppermint Basslet - *Liopropoma rubre* Poey, 1861

FEATURES: Yellowish with five reddish to reddish brown stripes on body. Each stripe bordered above and below by a pinkish to reddish stripe. Dorsal and anal fins with a large black spot and white margin. Caudal fin with white margin and two spots that may merge at center. HABITAT: S FL, Bahamas, Caribbean Sea to Venezuela. Also Bermuda and portions of the Gulf of Mexico. Reef associated in deep recesses from about 10 to 150 ft. BIOLOGY: Solitary and secretive. OTHER NAME: Swissguard Basslet.

Tail-spot Basslet - *Liopropoma santi* Baldwin & Robertson, 2104

FEATURES: Color and pattern somewhat variable. Top of head lavender. Upper lip bright yellow. Irises yellow. Back bright yellow with scattered orange red spots. Sides with orange red spots along scale rows. Whitish ventrally with bright white chevrons. Dorsal fins bright yellow distally. Caudal fin bright yellow on upper and lower lobes; lower-lobe tip with a reddish black spot. Ventral fins white. Top of snout slightly to prominently humped in adults. Lower rear corner of upper jaw with a ventral projection. HABITAT: Curaçao. Found over rocky ledges and slopes and in small caves and crevices from about 600 to 900 ft.

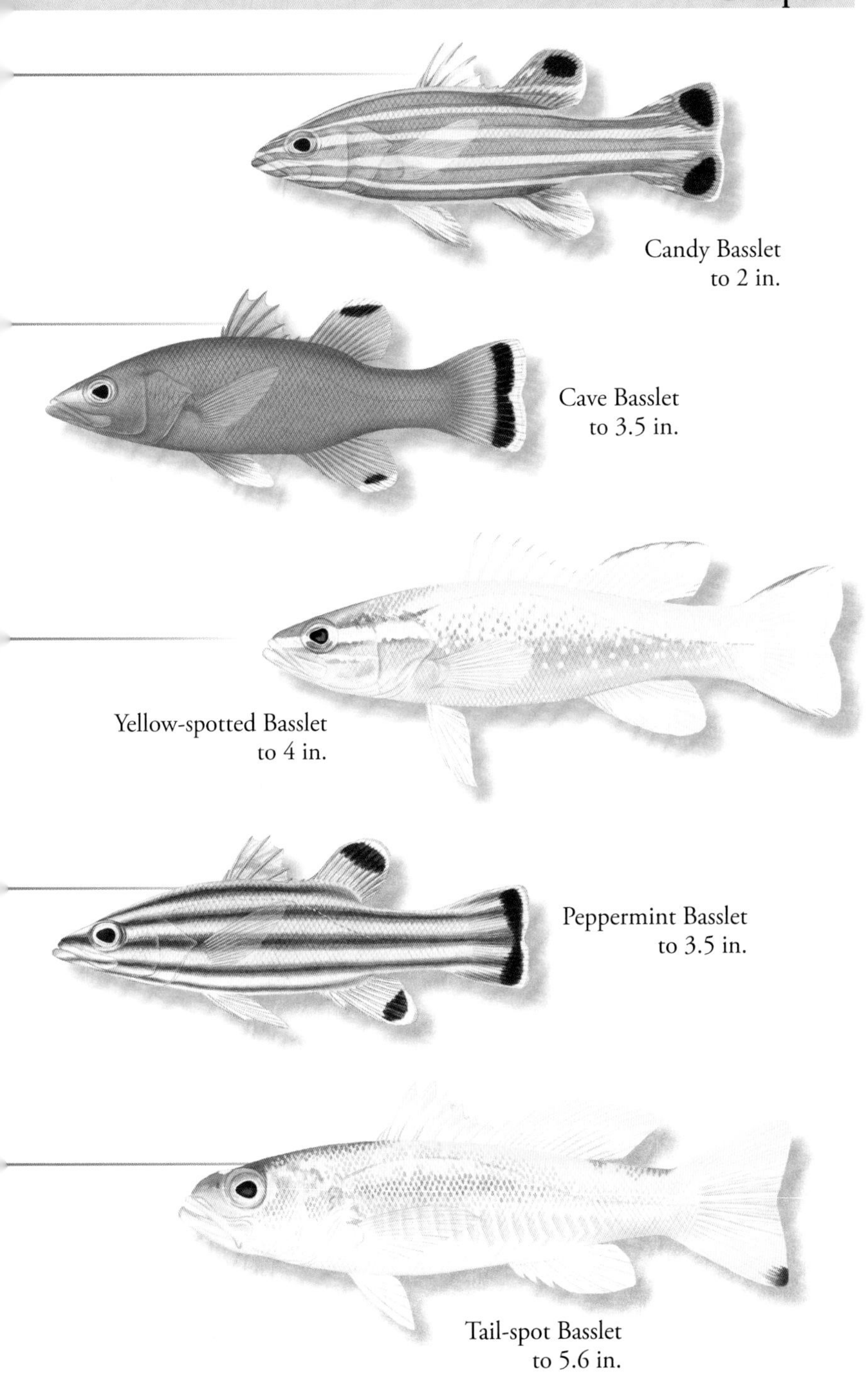

Candy Basslet
to 2 in.

Cave Basslet
to 3.5 in.

Yellow-spotted Basslet
to 4 in.

Peppermint Basslet
to 3.5 in.

Tail-spot Basslet
to 5.6 in.

Western Comb Grouper - *Mycteroperca acutirostris* (Valenciennes, 1828)

FEATURES: Grayish brown with paler blotches and spots on head and body. Largest individuals mostly uniformly colored. Three to four brown stripes radiate from eyes to opercular margin. Another brown stripe radiates from upper jaw to lower opercular margin. Ventral stripes become wavy on chest. Fins dark with irregular pale spots and streaks. Trailing margin of anal fin pointed in adults. Caudal-fin margin rounded in juveniles, becoming concave in adults. HABITAT: NW Gulf of Mexico, Greater and Lesser Antilles, and northern South America to Brazil. Adults occur over rocky bottoms from shore to about 82 ft. Juveniles occur around turtle-grass beds, mangroves, and shallow reefs.

Black Grouper - *Mycteroperca bonaci* (Poey, 1860)

FEATURES: Color and pattern vary. Head and body grayish to dark brown with dark grayish, brownish, or reddish spots that blend into streaks and rectangular, chain-like patterns. Dark streaks radiate from eyes. May appear nearly uniformly dark. Soft dorsal, caudal, and anal fins with broad, dark inner margins and narrow, whitish outer margins. Pectoral fins usually with yellowish to orange margin. Juveniles similarly patterned. Corner of preopercle evenly rounded, lacks notch. HABITAT: SC to FL, Gulf of Mexico, Bermuda, Bahamas, and Caribbean Sea to Brazil. Found over rocky bottoms and coral reefs and around jetties to about 330 ft. Juveniles found around mangroves. BIOLOGY: Hermaphroditic. IUCN: Near Threatened.

Venezuelan Grouper - *Mycteroperca cidi* Cervigón, 1966

FEATURES: Body brownish gray to pale brown with irregular darker spots and vermiculations. Dark spots on snout, and dark streaks radiating from eyes. Dorsal, caudal, and anal fins brownish dark inner margins and narrow, bluish outer margins. Abdomen pale. Preopercle serrated with a distinct lobe at lower corner. Second dorsal and anal fins bluntly pointed with a trailing margin. HABITAT: Colombia to Venezuela. Also reported from Jamaica. Juveniles occur over soft coral beds, adults over reefs from about 260 to 520 ft.

Yellowmouth Grouper - *Mycteroperca interstitialis* (Poey, 1860)

FEATURES: Pale brownish gray with close-set, brown spots on head and body that may coalesce into irregular dark blotches. Some may have faint bars or may be uniformly colored. Mouth and spiny dorsal-fin margin always yellowish. Pectoral fins with dark rays, pale membranes, and whitish margin. Pelvic fins comparatively short. Caudal-fin margin evenly serrated. Juveniles tricolored: pale dorsally, black to brown along midsides, white below; become blotched with age. Preopercle with a distinct lobe at lower corner. HABITAT: FL, Gulf of Mexico, Bahamas, Antilles to Brazil. Over coral reefs and rocky bottoms to about 490 ft. BIOLOGY: Feed primarily on fishes. Sought commercially. IUCN: Vulnerable.

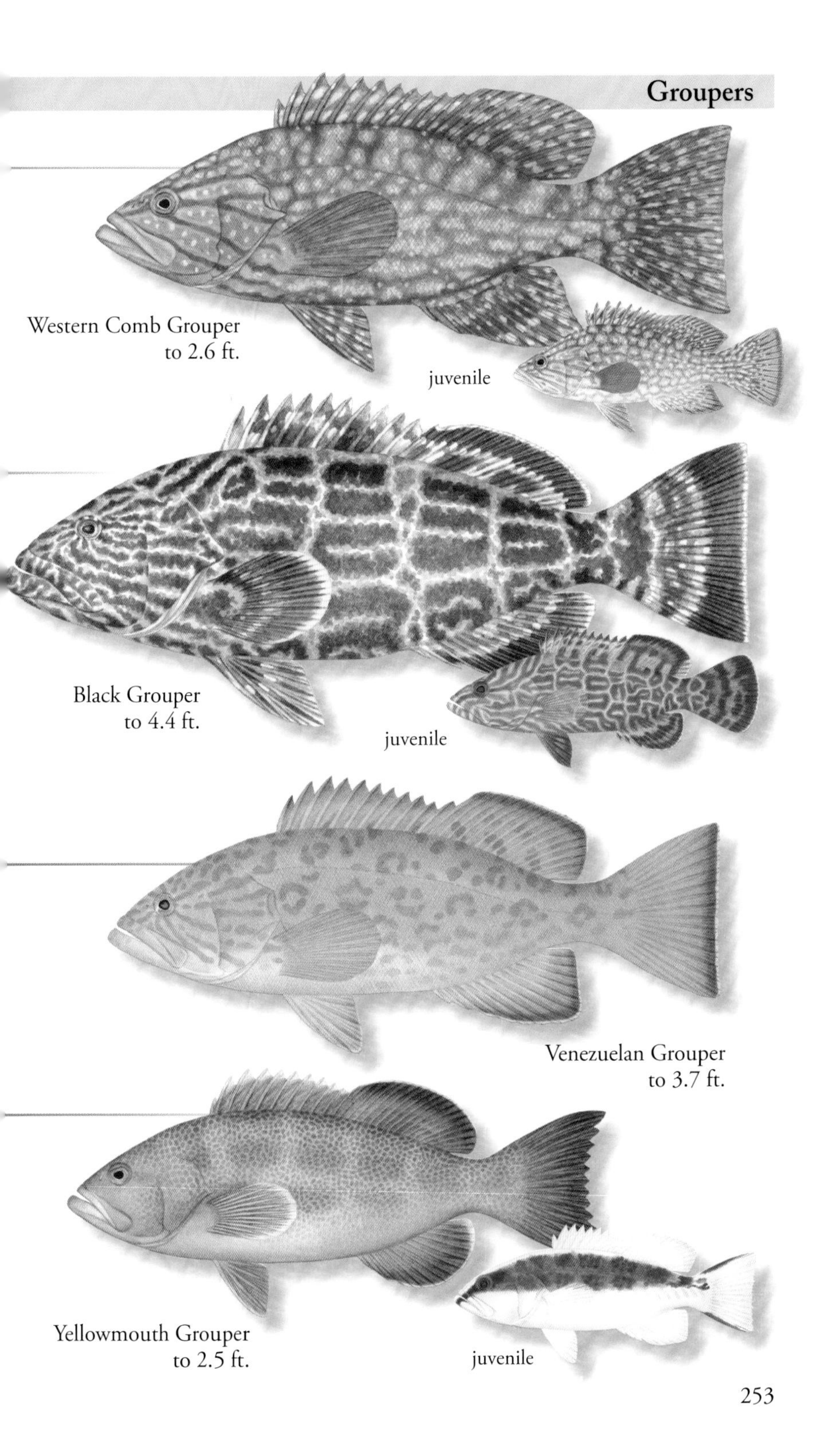
Western Comb Grouper
to 2.6 ft.
juvenile
Black Grouper
to 4.4 ft.
juvenile
Venezuelan Grouper
to 3.7 ft.
Yellowmouth Grouper
to 2.5 ft.
juvenile

Gag - *Mycteroperca microlepis* (Goode & Bean, 1879)

FEATURES: Females and juveniles grayish to greenish gray with darker vermiculations that may blend into saddles or bars. Dark lines radiate from eyes. Large males with two patterns: pale gray with posterior dark vermiculations and dark ventral area or with posterior vermiculations, blackish snout, and dark ventral and posterior areas. Soft dorsal- and anal-fin margins always rounded and often with a narrow blue margin. HABITAT: NC to FL, and Gulf of Mexico to Yucatán. Also Cuba. Juveniles north to MA. Also Bermuda. Adults offshore over rocky or reef bottoms to about 500 ft. Juveniles in estuaries and over seagrass beds. BIOLOGY: Hermaphroditic. Adults solitary or in large groups. Form spawning aggregations. Sought recreationally. IUCN: Vulnerable.

Scamp - *Mycteroperca phenax* Jordan & Swain, 1884

FEATURES: Usually grayish brown with small, close-set, darker spots. Spots may form clusters and overlie diffuse vermiculations. Larger specimens pale anteriorly, dark posteriorly, with dark vermiculations. Corners of mouth yellowish. Pectoral fins with dark inner margin, pale outer margin. Juveniles with large brownish spots that form irregular blotches and vermiculations. Caudal fin concave, unevenly serrated. HABITAT: NC to FL, Gulf of Mexico, northern South America. Over rocky and live bottoms from about 33 to 330 ft. Juveniles occur around jetties and mangroves. BIOLOGY: Feed on octopods, crustaceans, and fishes. May live to 21 years old.

Tiger Grouper - *Mycteroperca tigris* (Valenciennes, 1833)

FEATURES: Color and pattern vary. Head and body with close-set brownish or reddish to blackish spots over whitish or grayish to greenish background. Spots may be small to large and cluster or blend to form oblique bars on upper sides. Ventral area marbled. Fins with irregular spots and bands. Pectoral fin usually with yellowish margin. Juveniles yellow with blackish midbody stripe that develops into oblique bars. HABITAT: S FL, Bahamas, Antilles to Brazil. In Gulf of Mexico from FL; also TX to Yucatán. Also Bermuda. Over reefs and rocky bottoms from about 33 to 130 ft. BIOLOGY: Hermaphroditic. Ambush prey.

Yellowfin Grouper - *Mycteroperca venenosa* (Linnaeus, 1758)

FEATURES: Color and pattern vary. Greenish gray in shallow water; reddish in deeper water. Both with oblong, darker blotches and small blackish spots outlined in red dorsally and small red spots ventrally. Both phases with black inner margins on dorsal, caudal, anal, and pelvic fins. Pectoral fins always with yellow margin. Juveniles with large reddish and blackish spots that merge into irregular blotches and stripes. HABITAT: NC, FL, Gulf of Mexico, Bahamas, and Antilles to Brazil. Also Honduras, Nicaragua, and Bermuda. Around reefs and rocky and muddy bottoms from about 6 to 450 ft. Juveniles over seagrass beds. BIOLOGY: Feed on squids and fishes. Flesh may be toxic. IUCN: Endangered in the Gulf of Mexico. Near Threatened globally.

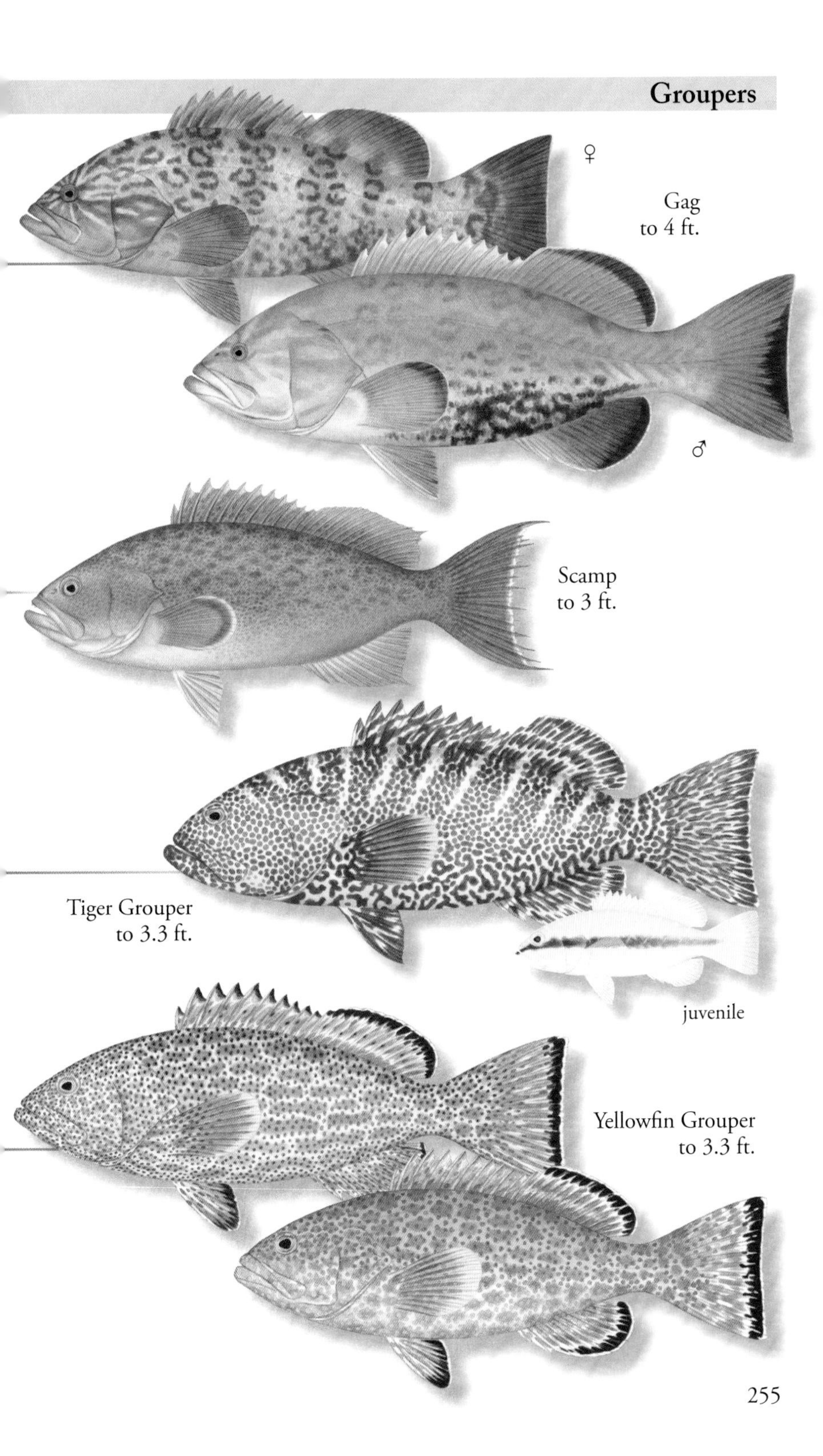

♀
Gag
to 4 ft.
♂
Scamp
to 3 ft.
Tiger Grouper
to 3.3 ft.
juvenile
Yellowfin Grouper
to 3.3 ft.

Reef Bass - *Pseudogramma gregoryi* (Breder, 1927)

FEATURES: Body and fins brownish red with diffuse, paler blotches. Head brownish to greenish with wide, dark-edged, whitish bars radiating from eyes. Opercle with large, black ocellated spot. Each eye with a small, fleshy tentacle. Spiny dorsal fin low. HABITAT: S FL, Bermuda, and Caribbean Sea. Occur around hard corals and coral rubble from about 3 to 200 ft. BIOLOGY: Secretive. Reported to flare gills at rivals. Feed on crustaceans and worms.

Freckled Soapfish - *Rypticus bistrispinus* (Mitchill, 1818)

FEATURES: Brownish above; yellowish to creamy below. A pale stripe runs along profile from snout to dorsal-fin origin. Body and unpaired fins covered with close-set, tiny spots. Spots are more dense dorsally. Pattern may be uniform or broken by underlying, diffuse, pale blotches. Juveniles with a dark stripe on sides. Dorsal fin with two spines. HABITAT: E FL, eastern and southern Gulf of Mexico, Bahamas, and Antilles to Venezuela. Near bottom over sandy areas from near shore to about 260 ft. BIOLOGY: Solitary. Hide in shells and burrows during the day; feed on crustaceans at night. Skin is covered in toxic mucus.

Large-spotted Soapfish - *Rypticus bornoi* Beebe & Tee-Van, 1928

FEATURES: Tannish gray to brown with large brown spots on head and anterior body. A pale stripe above a brown stripe runs along profile from snout to dorsal-fin origin. Dorsal fin with two (sometimes three) spines. HABITAT: Bahamas, Haiti, and Dominican Republic, and off Belize, Honduras, and Colombia. Found over silty marl and dead and living coral from near shore to about 230 ft.

Slope Soapfish - *Rypticus carpenteri* Baldwin & Weigt, 2012

FEATURES: Yellowish brown. A pale stripe runs along profile from snout to dorsal-fin origin. Head with blackish spots with vague pale rings that are smaller than the pupil. Body with larger blackish spots with vague pale rings. Smaller blackish spots on lower portions of fins. Pectoral fins yellowish. Dorsal fin with three to four spines, usually four. HABITAT: S FL, Bermuda, Bahamas, and Caribbean Sea. Found over steep slopes and walls of rocks and coral from near shore to about 130 ft., usually between 20 and 100 ft.

Whitespotted Soapfish - *Rypticus maculatus* Holbrook, 1855

FEATURES: Brownish with scattered white spots on upper sides of body. Spots may merge. A pale stripe runs along profile from snout to dorsal-fin origin. Lower portion of head and abdomen creamy. Dorsal fin with two to three spines. HABITAT: NC to FL and Gulf of Mexico. Occur near bottom around rocky areas, coral reefs, jetties, and pilings from near shore to about 300 ft. BIOLOGY: Secretive and nocturnal. Reported to lie on their side pressed against rocks or under ledges. Feed on crustaceans and fishes.

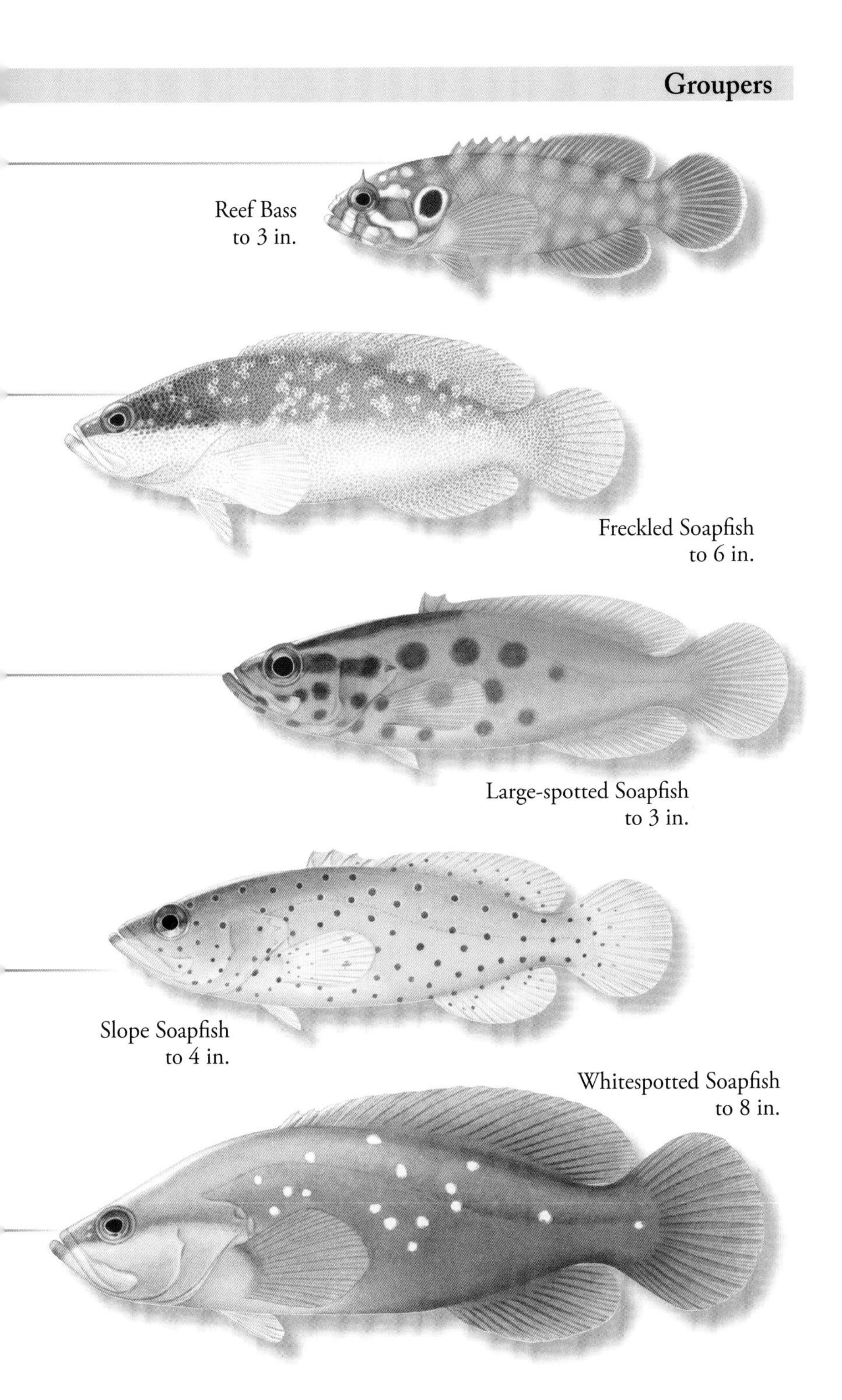
Reef Bass
to 3 in.
Freckled Soapfish
to 6 in.
Large-spotted Soapfish
to 3 in.
Slope Soapfish
to 4 in.
Whitespotted Soapfish
to 8 in.

Epinephelidae - Groupers, *cont.*

Randall's Soapfish - *Rypticus randalli* Courtenay, 1967

FEATURES: Brown to reddish brown. Lower head and abdomen creamy to pale brown. Fin margins may be darker brown; rear soft dorsal and anal fin may have a pale blotch. Distinct spots or mottling absent. Juveniles with a pale stripe along profile from snout to dorsal-fin origin. Dorsal fin with two to three spines, usually two. HABITAT: Southwest Gulf of Mexico and Caribbean Sea. Found over soft bottoms in brackish water and hypersaline lagoons from shore to about 50 ft.

Greater Soapfish - *Rypticus saponaceus* (Bloch & Schneider, 1801)

FEATURES: Dark bluish black to brownish with dense scattering of small, irregular, pale flecks and blotches on head, body, and fins. Lower jaw and abdomen pale. Profile of head and nape may be pale. Juveniles pale gray with dark markings that form reticulations. Dorsal fin with three spines. HABITAT: NC to FL, Gulf of Mexico, Bermuda, Bahamas, and Caribbean Sea to Brazil. Found around soft bottoms and hard bottoms from about 3,200 ft. BIOLOGY: Nocturnal and secretive. Often lie motionless on side. Skin is covered in toxic mucus. Feed on crustaceans and fishes.

Spotted Soapfish - *Rypticus subbifrenatus* Gill, 1861

FEATURES: Brownish to olivaceous with fairly evenly spaced dark ocellated spots on head, body, and fins. Spots behind head round or elongate and at least one the same size or larger than pupil. Pectoral and vertical fins tan to dark brown. Abdomen pale. Older specimens with fewer spots posteriorly. Head profile with pale stripe. HABITAT: S FL, southern Gulf of Mexico, Bermuda, Bahamas, and Caribbean Sea. Found around rocky areas and reefs to about 70 ft. BIOLOGY: Secretive and solitary. Adults prefer clear water. Skin is covered in toxic mucus.

Serranidae - Sea Basses

Yellowfin Bass - *Anthias nicholsi* Firth, 1933

FEATURES: Reddish lavender with yellow patches dorsally. Irregular yellow stripes on sides. Silvery below. Yellow stripe from snout to pectoral fin and from eye to opercular margin. Fins bright yellow with reddish to lavender markings and margins. Eyes large. Spiny dorsal fin with short, trailing filaments. Caudal fin deeply forked. HABITAT: Nova Scotia to FL, NE Gulf of Mexico, Nicaragua, and Guyana to Brazil. Found near bottom from about 180 to 1,400 ft. BIOLOGY: Hermaphroditic. Spawn in the spring.

Streamer Bass - *Baldwinella aureorubens* (Longley, 1935)

FEATURES: Reddish orange dorsally; silvery on sides and below. Dorsal fins and caudal-fin lobes yellowish. Pectoral fins pinkish. Eyes large, mostly yellow with red. Snout short; mouth oblique. Spiny dorsal fin with small trailing tabs or filaments. Caudal fin deeply forked. HABITAT: NJ to FL, and Gulf of Mexico to Yucatán. Also Colombia to Suriname. Reported over semi-hard bottoms from about 300 to 2,000 ft. NOTE: Previously known as *Hemanthias aureorubens.*

Groupers
Randall's Soapfish
to 8.5 in.
Greater Soapfish
to 13 in.
Spotted Soapfish
to 7 in.
Sea Basses
Yellowfin Bass
to 10 in.
Streamer Bass
to 12 in.

Red Barbier - *Baldwinella vivanus* (Jordan & Swain, 1885)

FEATURES: Reddish dorsally; pale to silvery below. Two yellow stripes on each side of head. Yellowish areas on sides. Males with long dorsal-fin filaments, yellow anal fin, and red pelvic fins. Females with relatively short dorsal-fin filaments, mottled anal fin, and pink pelvic fins. Caudal fin deeply forked in both. HABITAT: NJ to FL, Gulf of Mexico, Bahamas, and southern Caribbean Sea to Brazil. Found from about 65 to 1,400 ft. BIOLOGY: Hermaphroditic. Form large, fast-moving schools. NOTE: Previously known as *Hemanthias vivanus*.

Pugnose Bass - *Bullisichthys caribbaeus* Rivas, 1971

FEATURES: Color and pattern vary. Shades of reddish orange with irregular whitish saddles dorsally. Sides orange red with silvery white blotches that merge onto silvery white below. A dark brownish band runs through eyes. A large blackish blotch on anterior dorsal fin, and a small black ocellated spot on upper peduncle. Dorsal and caudal fins yellow with white flecks. Snout very short and indented in front of eye. Lower jaw protrudes. HABITAT: Bahamas and Caribbean Sea. Occur over deep reefs from about 265 to 1,800 ft.

Twospot Bass - *Centropristis fuscula* Poey, 1861

FEATURES: Color and pattern vary. Tan to pinkish dorsally; silvery white on lower head and abdomen. Orange to yellow bar below eyes. Sides with six to seven vague to distinct solid or blotchy bars that may be intersected by wavy stripes and blotches. A large dark blotch on midside, and another large dark blotch at caudal-fin base. Preopercle finely serrated, with an evenly rounded lower edge. HABITAT: SC to S FL, N Cuba, Puerto Rico, and Lesser Antilles to Curaçao. Occur over hard bottoms from about 230 to 850 ft. NOTE: Some authors classify this species as *Serranus fuscula*.

Bank Sea Bass - *Centropristis ocyurus* (Jordan & Evermann, 1887)

FEATURES: Cream to yellowish white dorsally with six or seven dark bars on upper sides. Each bar merges with a dark patch at lateral line. Head yellowish with pale blue lines. Dorsal and caudal fins with blackish spots. Spiny dorsal fin with short, trailing flaps. Caudal fin rounded in juveniles, trilobed or double-concave in adults. HABITAT: Cape Hatteras to FL Keys; Gulf of Mexico to Yucatán. Occur over hard bottoms and sandy-shell bottoms near reefs from about 36 to 330 ft. BIOLOGY: Hermaphroditic. Feed on invertebrates and fishes.

Rock Sea Bass - *Centropristis philadelphica* (Linnaeus, 1758)

FEATURES: Brownish to olive gray dorsally. About seven dark oblique bars on upper sides. Midbody bar merges with a dark spot at base of dorsal fin. Head with blue and rusty lines. Dorsal and caudal fins spotted. Spiny dorsal fin with trailing flaps. Caudal fin rounded in juveniles, trilobed or double-concave in adults. HABITAT: VA to FL and in the Gulf of Mexico to Yucatán. Found over soft mud and sandy bottoms from about 30 to 560 ft. BIOLOGY: Hermaphroditic. Females mature in one year, then become male in second year. May live to four years of age.

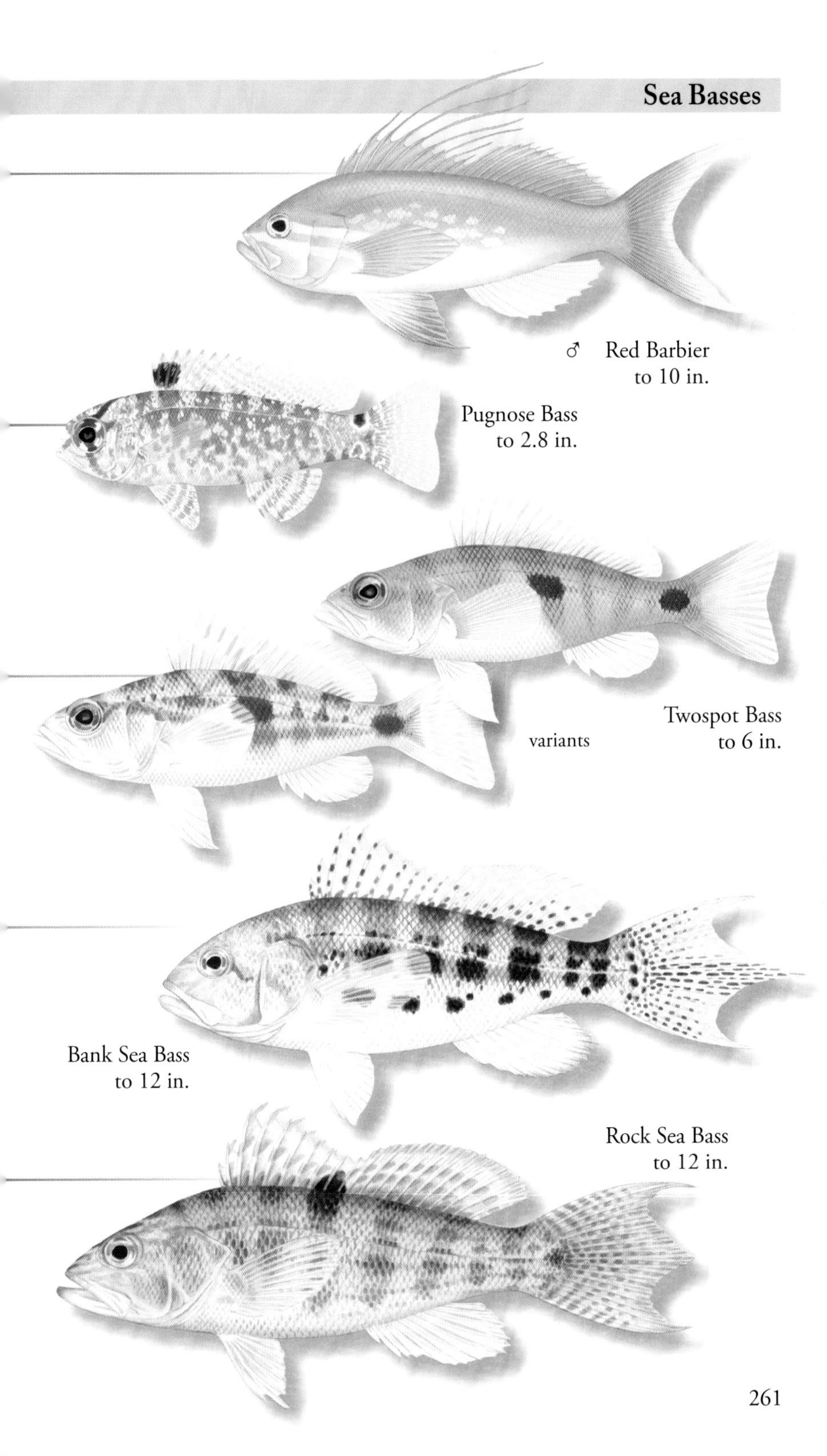
♂ Red Barbier
to 10 in.
Pugnose Bass
to 2.8 in.
Twospot Bass
to 6 in.
variants
Bank Sea Bass
to 12 in.
Rock Sea Bass
to 12 in.

Threadnose Bass - *Choranthias tenuis* (Nichols, 1920)

FEATURES: Rosy to orange red dorsally. Pale below. Dorsal and anal fins spotted. Nostrils each with a thread-like filament. Eyes with yellow irises. Dorsal fin spines with short, trailing filaments. Caudal fin with pointed lobes and red tips. HABITAT: NC to FL Keys, northern Gulf of Mexico, Yucatán, Bermuda, Bahamas, and eastern Caribbean Sea. Found near bottom from about 180 to 3,000 ft. Most occur above 490 ft. BIOLOGY: Planktivorous and form large schools. NOTE: Previously known as *Anthias tenuis*.

Dwarf Sand Perch - *Diplectrum bivittatum* (Valenciennes, 1828)

FEATURES: Pale greenish gray above; pearly below. Two brownish stripes on upper sides. Stripes are continuous in juveniles, become broken and intersected by dark saddles in adults. Snout and cheeks with pale blue lines. Dorsal fins with orange margins and spots. Caudal fin with irregular orange bands and elongate upper ray. Two small dark spots on upper caudal peduncle. HABITAT: FL, Gulf of Mexico, and Caribbean Sea to Brazil. Absent from Bahamas. Occur over muddy silt bottoms from near shore to about 390 ft.

Sand Perch - *Diplectrum formosum* (Linnaeus, 1766)

FEATURES: Shades of brown dorsally, fading to white below. Narrow, wavy blue lines on snout and head merge with narrow blue stripes on body. Dark midlateral stripe may run from head to a dark spot at caudal-fin base. Five to seven diffuse dark bars on sides. Dorsal and caudal fins tannish with blue lines. Caudal fin usually forked, with upper lobe longer than lower lobe. HABITAT: VA to FL, Gulf of Mexico, and from Colombia to Brazil. Also Bermuda. Rare in Bahamas and Antilles. Found over sandy to shelly bottoms from shore to about 240 ft.

Aquavina - *Diplectrum radiale* (Quoy & Gaimard, 1824)

FEATURES: Shades of gray to tan dorsally and on sides with darker, irregular oblique and broken bars overlaid with dark brown to blackish blotches that loosely form rows. Ventral area silvery to golden with irregular dark blotches on lower flanks. Irregular dark brown to black blotch at caudal-fin base. Dorsal and caudal fins with rows of reddish, orange, or brownish spots. HABITAT: Cuba, Greater and Lesser Antilles, and Nicaragua to southern Brazil. Found over inshore seagrass flats and sandy to rubble bottoms from about 30 to 260 ft. Usually below 65 ft.

Longtail Bass - *Hemanthias leptus* (Ginsburg, 1952)

FEATURES: Rosy to reddish dorsally. Silvery below. May have indistinct yellowish markings on sides. Yellow stripe below eye from snout to opercular margin. All fins except pectorals yellow with red at base. Larger specimens with an elongate filament on third dorsal-fin spine. Dorsal and anal fins with trailing filaments. Pelvic fins long, trailing. HABITAT: NC to FL, northern and southern Gulf of Mexico, Venezuela to Suriname. Over hard bottoms from about 100 to 2,000 ft.

Threadnose Bass
to 4.3 in.

Dwarf Sand Perch
to 10 in.

Sand Perch
to 12 in.

Aquavina
to 10 in.

Longtail Bass
to 18 in.

Yellowbelly Hamlet - *Hypoplectrus aberrans* Poey, 1868

FEATURES: Color and pattern vary. Upper head, back, and spiny dorsal dark brownish or blue brown to gray brown; yellow below. Dark area may extend to midline or to belly. May have irregular blue spots and thin bars on head. Sometimes with a small black spot on upper tail base. Occasionally with dark stripes along the top and bottom edges of the tail fin. Rarely with a large dark spot on the side of the snout. HABITAT: S FL, S Gulf of Mexico, Bahamas, and Caribbean Sea. Reef associated from near shore to about 50 ft.

Yellowtail Hamlet - *Hypoplectrus chlorurus* (Cuvier, 1828)

FEATURES: Body uniformly dark brownish to bluish black. Caudal fin pale to yellow. Pectoral fins usually colorless, occasionally yellowish. HABITAT: S FL (rare; observations may be erroneous) and Antilles to Venezuela. Reef associated; near bottom from about 10 to 75 ft. BIOLOGY: Hermaphroditic and solitary. Believed to mimic Yellowtail Damselfish, *Microspathodon chrysurus.*

Blue Hamlet - *Hypoplectrus gemma* Goode & Bean, 1882

FEATURES: Body and fins iridescent blue to bluish black, with upper and lower margins of caudal fin dark blue to black. Color may pale at night. Body relatively deep, compressed. Dorsal fin broad, rounded. Caudal fin concave. HABITAT: S FL, FL Keys, and Bahamas. Reef associated in shallow water. BIOLOGY: Hermaphroditic. Thought to mimic Blue Chromis, *Chromis cyanea.*

Golden Hamlet - *Hypoplectrus gummigutta* (Poey, 1851)

FEATURES: Body and fins bright yellow. Blackish blotch on snout and jaws surrounded by bright blue lines. May also have blue lines behind eyes and at base of opercle. Dorsal, anal, and pelvic fins may have thin blue margins. HABITAT: S FL, Bahamas, Greater Antilles, Nicaragua, Honduras. Reef associated to about 150 ft. BIOLOGY: Observed hybridizing with Yellowbelly Hamlet, *H. aberrans.*

Shy Hamlet - *Hypoplectrus guttavarius* (Poey, 1852)

FEATURES: Head, abdomen, and all fins yellow. Body from nape to caudal-fin base dark bluish black to purplish black. Blackish blotch with bright blue border on snout. May have blue lines around eyes and to corner of preopercle. Body relatively deep and compressed. Dorsal fin rounded. Caudal fin concave. HABITAT: S FL, Bahamas, and Antilles. Reef associated in shallow water. BIOLOGY: Hermaphroditic. Thought to mimic Rock Beauty, *Holacanthus tricolor.*

Indigo Hamlet - *Hypoplectrus indigo* (Poey, 1851)

FEATURES: Head and body with alternating deep blue and whitish bars. Blue bar below spiny dorsal fin widest. Pelvic fins deep blue, others pale blue. Body relatively deep and compressed. Dorsal fin broad, rounded. HABITAT: S FL, Bahamas, Greater Antilles. Also off portions of central and northern South America. Reef associated. Occur in shallow water to about 145 ft. BIOLOGY: Hermaphroditic. Found singly or in pairs.

Yellowbelly Hamlet
to 5 in.

Yellowtail Hamlet
to 5 in.

Blue Hamlet
to 5 in.

Golden Hamlet
to 5 in.

Shy Hamlet
to 5 in.

Indigo Hamlet
to 5.5 in.

Bicolor Hamlet - *Hypoplectrus maculiferus* Poey, 1871

FEATURES: Upper head, back, and spiny dorsal fin shades of brownish to purplish and often infused with blue. Remainder of body and fins yellow. Fins may have narrow blue margins. Blackish blotch with bright blue border on snout. Blackish saddle on upper portion of peduncle. HABITAT: Bahamas, Greater Antilles, and scattered of northern Yucatán and in Caribbean Sea. Occur over reefs from about 10 to 40 ft.

Belizian Blue Hamlet - *Hypoplectrus maya* Lobel, 2011

FEATURES: Overall iridescent indigo blue; somewhat darker dorsally. Pectoral and caudal fins unmarked. HABITAT: Belize. Occur from South Water Marine Reserve area at Wee Wee Caye southward and throughout the Pelican Cays and to the Sapodilla Cays at the southern most margin of the Meso-American Barrier Reef. Occur in mangrove cays and over coral reefs from about 10 to 50 ft. IUCN: Endangered.

Black Hamlet - *Hypoplectrus nigricans* (Poey, 1852)

FEATURES: Overall brownish black, grayish, bluish black, or purplish black. Some with blue lines on head and snout and a bluish to purplish wash on snout. Pectoral fins black on black-bodied specimens, colorless on gray-bodied specimens. HABITAT: S FL, southern Gulf of Mexico, Bahamas, and Caribbean Sea. Occur over hard and soft coral in shallow, tropical water. BIOLOGY: Hermaphroditic. Found singly or in pairs. May defend feeding territories. Thought to mimic Dusky Damselfish, *Stegastes adustus*.

Masked Hamlet - *Hypoplectrus providencianus* Acero P. & Garzón-Ferreira, 1994

FEATURES: Creamy white to tan with a broad black band bordered by blue lines below eyes. Pectoral fins entirely black or black dorsally. Caudal fin black. Pelvic fins deep blue. Tips of first few dorsal-fin membranes may be black. May also have black blotches on rear portion of dorsal fin and/or on peduncle. HABITAT: Southern Bahamas and scattered in Caribbean Sea. Occur over and around reefs from about 20 to 50 ft.

Barred Hamlet - *Hypoplectrus puella* (Cuvier, 1828)

FEATURES: Tan to yellowish with brownish bars below eyes, nape, and on head and body. Bar below spiny dorsal fin is widest. Bright blue lines and spots border bar below eyes and are followed by blue lines that extend onto chest. Color may pale or darken. HABITAT: FL, Gulf of Mexico, Bahamas, and Caribbean Sea to Venezuela. Also Bermuda. Reef associated. Prefer murky water to about 100 ft. BIOLOGY: Hermaphroditic. Pairs spawn at same site on consecutive nights. May hybridize with other Hamlets. Most abundant Hamlet in the area.

Bicolor Hamlet
to 5 in.
Belizian Blue Hamlet
to 5 in.
Black Hamlet
to 5 in.
Masked Hamlet
to 5 in.
Barred Hamlet
to 5 in.

Tan Hamlet - *Hypoplectrus randallorum* Lobel, 2011

FEATURES: Color and pattern vary. Uniformly pale brown to yellowish brown, or with a purplish blue wash dorsally. Bluish wash around mouth. May be unmarked or may have a black blotch with a dark or bright blue border on snout, a small black spot on upper pectoral-fin base, and a faint or distinct black blotch on upper portion of peduncle. HABITAT: FL Keys, SE Gulf of Mexico, Caribbean Sea. Reef associated from near shore to about 50 ft.

Butter Hamlet - *Hypoplectrus unicolor* (Walbaum, 1792)

FEATURES: Pale yellow to creamy with whitish or bluish highlights. Blue ring around eyes. Snout may have black blotch and may have numerous lines and spots on head. Small to large black blotch present on caudal peduncle that may form a ring around peduncle. HABITAT: S FL, N and S Gulf of Mexico, Bahamas, Caribbean Sea. Reef associated from near shore to about 115 ft. BIOLOGY: May hybridize with Barred Hamlet.

Vieja Parrot Rock-bass - *Paralabrax dewegeri* (Metzelaar, 1919)

FEATURES: Head and upper body grayish with numerous irregular orange brown spots. Lower body creamy white with about six wide blackish bars. Large black blotch at pectoral-fin base. Dorsal, caudal, and ventral fins spotted. Lower preopercular margin rounded and finely serrated. Opercle with three spines, the uppermost inconspicuous. Third and fourth dorsal-fin spines elongate and sharply pointed. HABITAT: Venezuela to northern Brazil. Occur over soft coral beds, rubble areas, and rocky reefs from about 30 to 150 ft.

Splitfin Bass - *Parasphyraenops incisus* (Colin, 1978)

FEATURES: White to pinkish with three irregular brown to reddish stripes on sides that are peppered with white spots. May also display wide pinkish bars on sides. Head yellowish to greenish. Upper and lower caudal-fin margins yellowish. A small to large black blotch or band on spiny dorsal fin that is bordered by white. Dorsal fins deeply notched. Spiny dorsal fin tall and pointed. HABITAT: NC to FL, Bahamas, Greater Antilles, and coast of Colombia. Form aggregations over steep reef slopes and dropoffs from about 100 to 650 ft.

Apricot Bass - *Plectranthias garrupellus* Robins & Starck, 1961

FEATURES: Reddish to reddish orange on anterior back and spiny dorsal fin, bordered below by broken white blotches or a white stripe. Sides shades of yellow to white with dark scale margins. Lower head and chest white. Spiny dorsal fin tall and pointed. HABITAT: NC to FL, eastern Gulf of Mexico, Bahamas, and Caribbean Sea. Found near rocky and rubble bottoms from about 42 to 1,200 ft. BIOLOGY: Feed on crustaceans. Believed to be hermaphroditic.

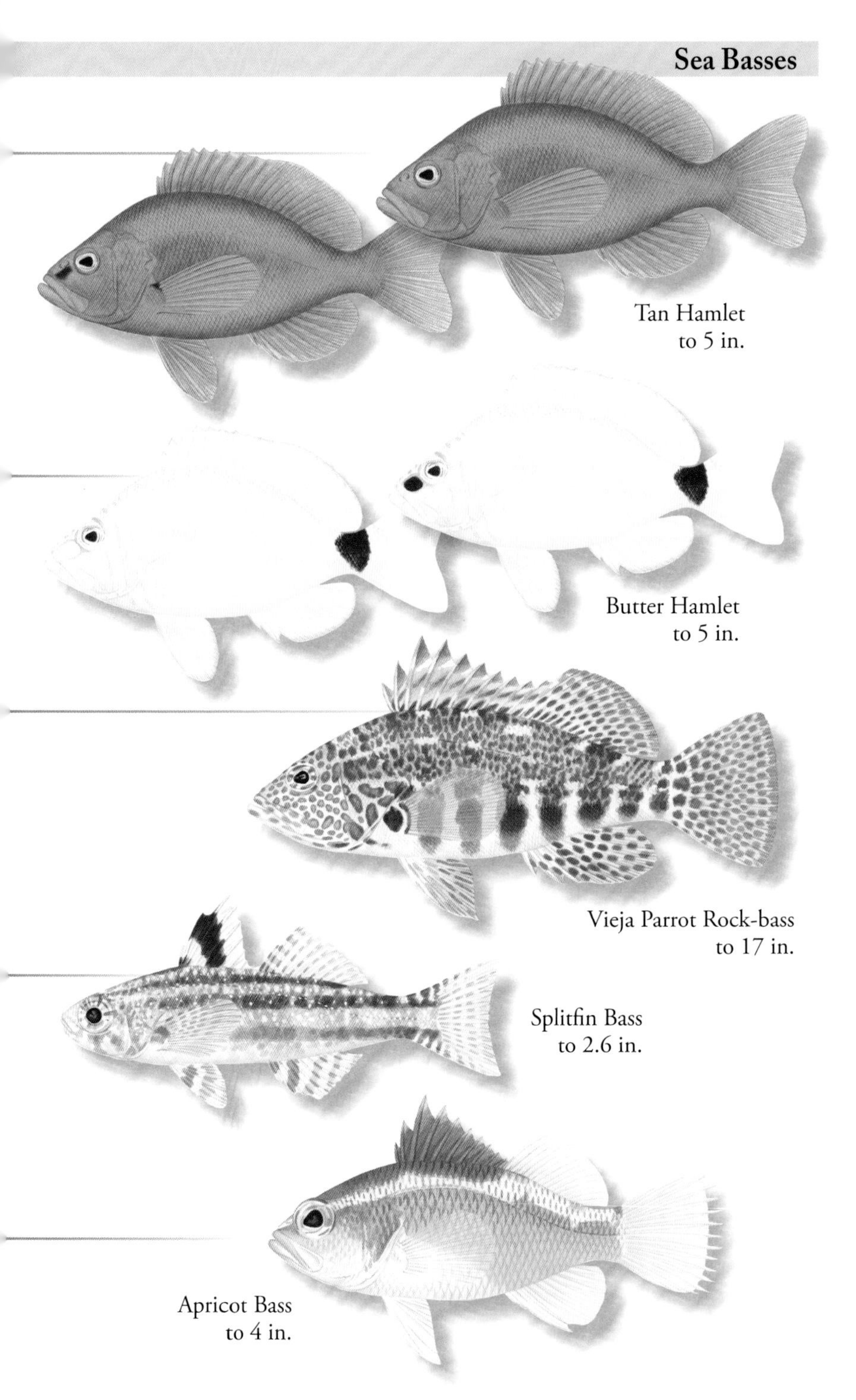
Tan Hamlet
to 5 in.
Butter Hamlet
to 5 in.
Vieja Parrot Rock-bass
to 17 in.
Splitfin Bass
to 2.6 in.
Apricot Bass
to 4 in.

Roughtongue Bass - *Pronotogrammus martinicensis* (Guichenot, 1868)

FEATURES: Rosy to orange red. A pale yellow or greenish yellow band runs from snout to lower opercular margin. May also have two other bars extending from eyes. Irregular yellowish bars and blotches under spiny dorsal fin that may merge into a bar or V-shape. Fins rosy to orange red. Third dorsal-fin spine tall. Caudal fin may have elongate upper and lower rays. HABITAT: NC to FL, Gulf of Mexico, Bermuda, Bahamas, and Caribbean Sea to Brazil. Found near bottom, often around soft corals, from about 180 to 750 ft. BIOLOGY: Hermaphroditic. Feed on crustaceans. NOTE: Previously known as *Holanthias martinicensis.*

School Bass - *Schultzea beta* (Hildebrand, 1940)

FEATURES: Rusty orange dorsally turning orange below with three rows of irregular pale to white blotches that form a chain-like pattern dorsally and on sides. Snout brown. Reddish bar below each eye. Caudal fin yellow, sometimes with a dark, chevron-shaped mark. Mouth very protrusible, lacks teeth. HABITAT: NC to FL, eastern and southern Gulf of Mexico, Bahamas, Antilles to northern South America. Reef associated from about 70 to 550 ft. BIOLOGY: Form small schools close to bottom while feeding on plankton. NOTE: Previously known as *Serranus beta.*

Pygmy Sea Bass - *Serraniculus pumilio* Ginsburg, 1952

FEATURES: Buff colored with irregular, brownish bars on sides and peduncle. Abdomen whitish. Reddish bars radiate from eyes. Small, black spots below dorsal-fin base. Dark blotch at rear of spiny dorsal fin. Lateral line with row of dark brown dots. HABITAT: NC to FL, Gulf of Mexico, Greater Antilles to Venezuela. Occur around reefs, seagrass beds, and sandy and shelly bottoms from near shore to about 540 ft. BIOLOGY: Hermaphroditic. Feed on crustaceans.

Orangeback Bass - *Serranus annularis* (Günther, 1880)

FEATURES: Color slightly variable. Broad, reddish to blackish saddle below spiny dorsal fin. Yellowish saddles below soft dorsal fin and on peduncle. Lower sides with yellowish bars and spots. Two rectangular-shaped, black-outlined, orange markings behind eyes. HABITAT: S FL, E and N Gulf of Mexico, Bahamas, and Antilles to northern South America. Also Bermuda. Reef associated from about 33 to 230 ft. Also reported over sandy, rocky, and rubble areas. BIOLOGY: Hermaphroditic. Usually occur in pairs. Reported staying close to sheltering crevices.

Blackear Bass - *Serranus atrobranchus* (Cuvier, 1829)

FEATURES: Pearly whitish with six to eight tannish and brownish bars dorsally and on sides. Midbody bar extends from spiny dorsal-fin margin to abdomen; bar may be dark to faint. Yellowish patch below midbody bar may be bright to absent. Opercle with a black blotch on inner surface that is visible from the outside. Fins pale to yellowish. HABITAT: FL and northern Gulf of Mexico to Brazil. Occur near bottom from about 33 to 900 ft. BIOLOGY: Hermaphroditic.

Roughtongue Bass
to 8 in.

School Bass
to 4 in.

Pygmy Sea Bass
to 3 in.

Orangeback Bass
to 2.5 in.

Blackear Bass
to 3.5 in.

Lantern Bass - *Serranus baldwini* (Evermann & Marsh, 1899)

FEATURES: Color and pattern vary with water depth; colors brighter in deeper waters. Upper sides with rows of reddish to blackish blotches forming stripes. Always with irregular, yellow midbody stripe. Lower sides with four dark, oval blotches, each with a yellowish to reddish bar below. Always four small black spots at caudal-fin base. HABITAT: S FL, Bahamas, Antilles to Brazil. Occur near bottom over rocky and rubble areas from about 3 to 260 ft. Also over seagrass beds. Juveniles may be found in abandoned conch shells. BIOLOGY: Hermaphroditic and territorial. Feed on shrimps and fishes.

Snow Bass - *Serranus chionaraia* Robins & Starck, 1961

FEATURES: Whitish with broad, brown stripes radiating from eyes that merge with rows of brown blotches on body. Lower cheeks and chest with brown blotches. Large white area on abdomen behind pectoral-fin base. Midsection of caudal peduncle usually lacks blotches. Brownish bar at caudal-fin base. Upper and lower caudal-fin margins with row of brown spots. HABITAT: SE FL, FL Keys, Puerto Rico, Honduras. Near bottom around reefs and rubble bottoms from about 150 to 295 ft.

Twinspot Bass - *Serranus flaviventris* (Cuvier, 1829)

FEATURES: Color and pattern somewhat variable. Whitish to tan with orange brown bands that run from snout and onto back. Orange red to brownish spots scattered on body and fins. Dorsal fin with a large blackish blotch anteriorly followed by several smaller blackish blotches. Caudal-fin base with two black ocellated spots. Large bright white patch on mid-abdomen. Preopercle serrated. HABITAT: Caribbean Sea to Brazil. Occur over reefs, rocky areas with sand and rubble, and soft bottoms from near shore to about 1,300 ft.

Crosshatch Bass - *Serranus luciopercanus* Poey, 1852

FEATURES: Shades of pale gray to white dorsally; white on sides and below. Dark brown to black blotches form broken stripes from mouth to lower caudal fin, from snout to upper caudal fin, and along back. Spiny dorsal fin with a large black oblique band. Snout long and pointed. Spiny dorsal fin tall and pointed; membranes deeply incised. HABITAT: Southern Gulf of Mexico to Honduras, Bahamas, Greater and Lesser Antilles to Curaçao. Occur over reefs from about 200 to 980 ft.

Saddle Bass - *Serranus notospilus* Longley, 1935

FEATURES: Color and pattern vary. Whitish with a brown oblique band through eyes and interconnecting brown bars on body, some merge onto dorsal fin; abdomen with a broad, bright white bar followed by a narrow brown to black bar; rear half of caudal fin orange. Others more white with brown blotches on body, some of which extend onto lower dorsal fin; narrow silvery bar on abdomen; caudal-fin tips orange. HABITAT: GA to N Gulf of Mexico, N Yucatán, Bahamas, and Caribbean Sea. Occur over soft bottoms from about 20 to 750 ft. Usually below 260 ft.

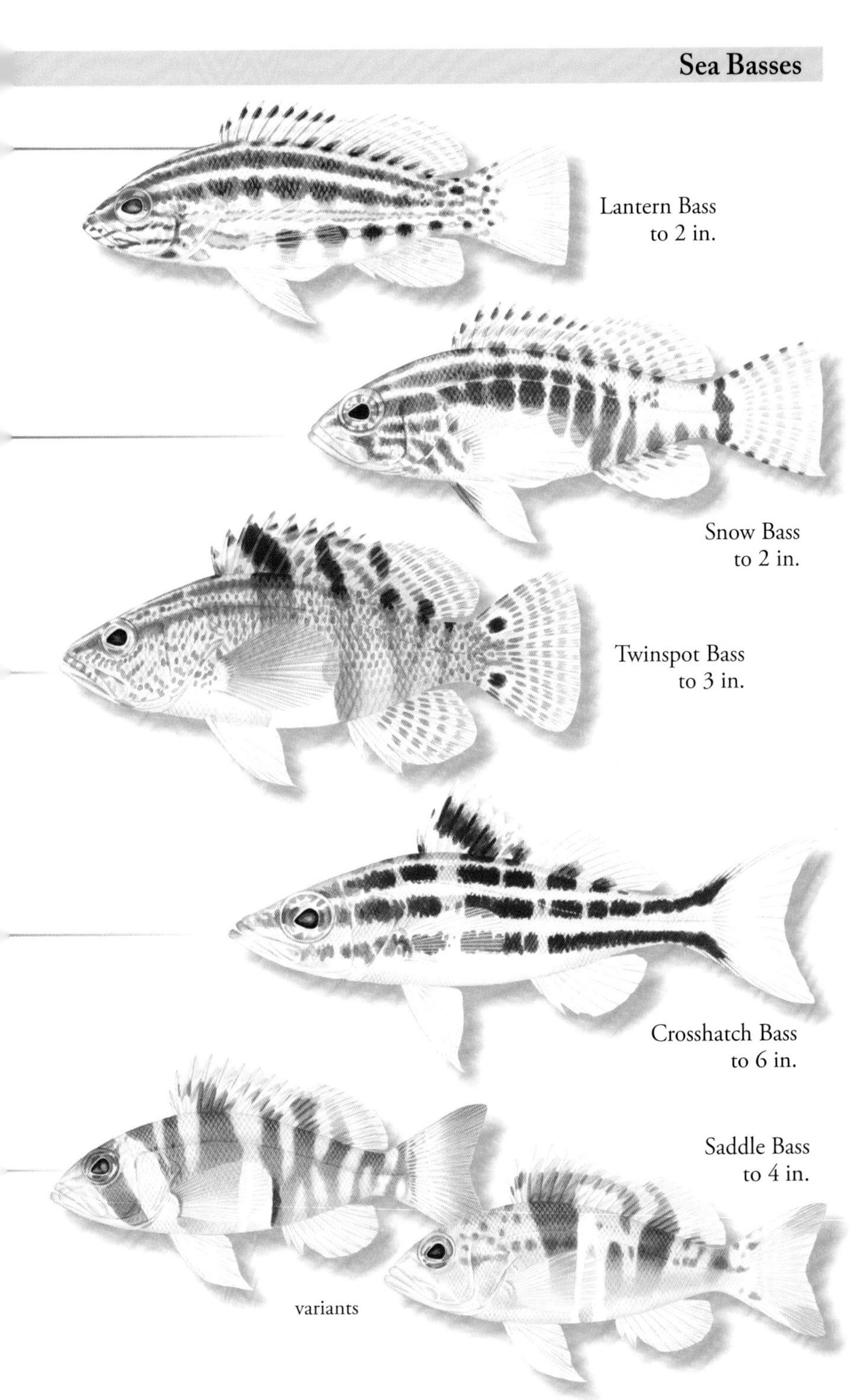
Lantern Bass
to 2 in.
Snow Bass
to 2 in.
Twinspot Bass
to 3 in.
Crosshatch Bass
to 6 in.
Saddle Bass
to 4 in.
variants

Tattler - *Serranus phoebe* Poey, 1851

FEATURES: Whitish to tannish with a brassy to brownish bar below eyes. A brassy to brownish bar runs from anterior margin of spiny dorsal fin to abdomen and is followed by a silvery white bar on abdomen above anus. Dark stripe on sides from midbody to caudal-fin base. May also have indistinct bars at nape and on posterior sides. All markings except silvery white bar fade with age. HABITAT: SC, FL, Gulf of Mexico, Greater Antilles to northern South America. Also Bermuda. Around reefs and rocky areas from about 88 to 590 ft. BIOLOGY: Hermaphroditic and solitary. Feed primarily on shrimps.

Belted Sandfish - *Serranus subligarius* (Cope, 1870)

FEATURES: Head with irregular, reddish brown spots and a dark band through eye. Scales outlined in reddish brown. Large black blotch on anterior portion of soft dorsal fin merges with dark bar below. Abdomen abruptly silvery white. Fins banded. Snout pointed; head profile sloping. HABITAT: NC to FL, and in Gulf of Mexico from FL to Veracruz. Rare in FL Keys. Found near bottom over rocky and mixed bottoms, around jetties and outcroppings from about 3 to 60 ft.; possibly deeper. BIOLOGY: Hermaphroditic. Territorial. Prefer turbid water.

Tobaccofish - *Serranus tabacarius* (Cuvier, 1829)

FEATURES: Whitish with orange brown midbody stripe or with orange brown from midline to abdomen. Alternating brown to grayish saddles on back. Saddles may form U-shapes. Caudal fin with a dark brown stripe on upper and lower lobes, often appearing as a V-shape. HABITAT: GA to FL, Bahamas, Antilles to Brazil. Also E Gulf of Mexico and Bermuda. Reef associated. Occur from about 10 to 200 ft. BIOLOGY: May follow Goatfishes in search of prey. Hermaphroditic.

Harlequin Bass - *Serranus tigrinus* (Bloch, 1790)

FEATURES: Whitish with irregular grayish to black bars on head and body that are intersected by horizontal rows of pale to dark spots that together form a loose cross-hatch pattern. Sides may have yellow cast. Dorsal, caudal, and anal fins spotted. Spiny dorsal fin may have black blotch between third and fifth spines. Snout pointed. HABITAT: S FL, Bahamas, Antilles to Venezuela. Also Bermuda. Over reefs and around scattered rock or coral from about 3 to 120 ft.

Chalk Bass - *Serranus tortugarum* Longley, 1935

FEATURES: Alternating reddish brown and bluish white bars dorsally. Reddish bars may become pinkish below. Sides with bluish to lavender sheen. Ventral area whitish. Caudal fin may be bluish white, reddish brown, or slightly banded. HABITAT: S FL, SE Gulf of Mexico, Bahamas, Antilles to Venezuela. Reef associated from about 40 to 1,300 ft. BIOLOGY: Often in small groups.

NOTE: There are seven other rare Sea Basses in the area.

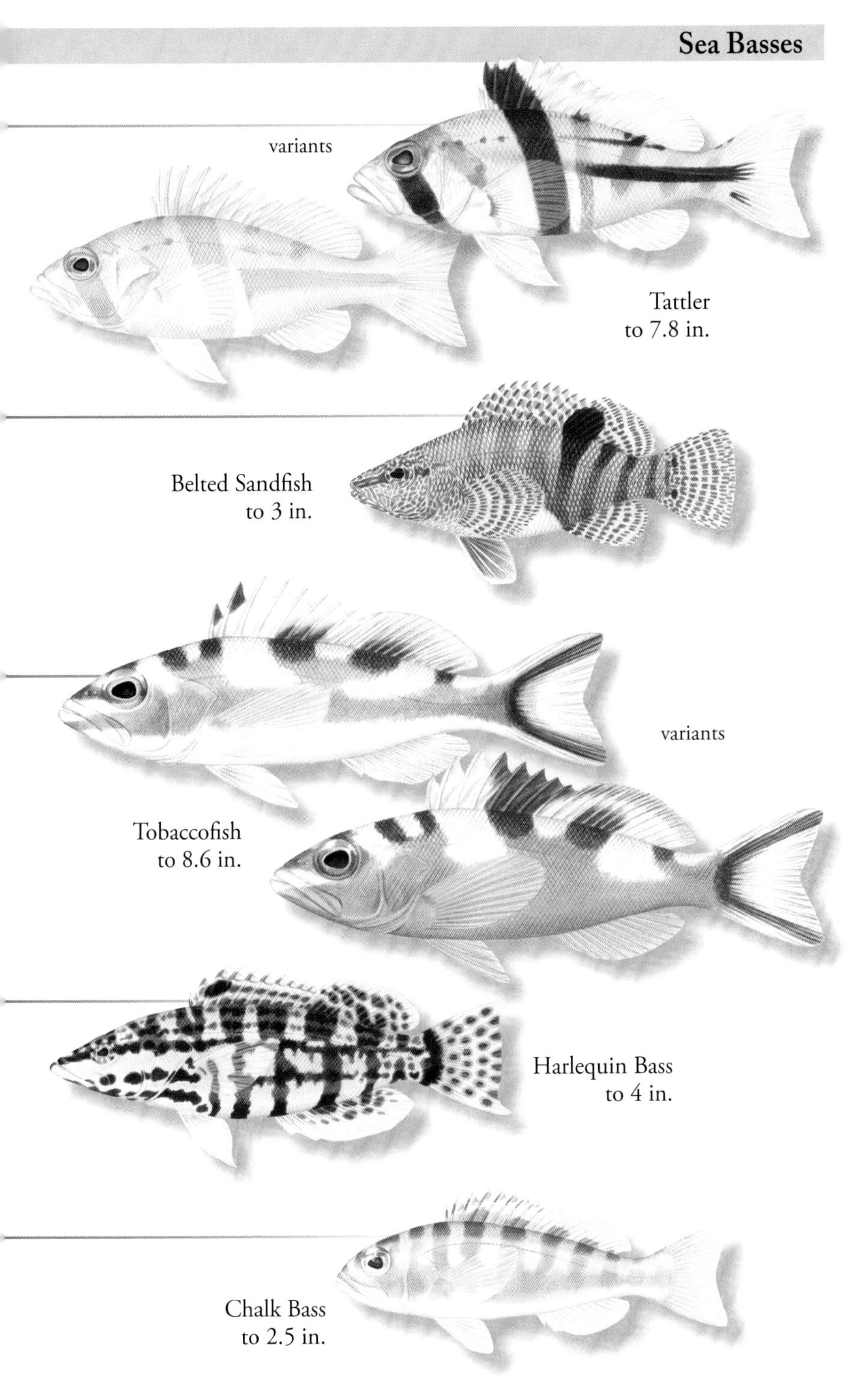
variants
Tattler
to 7.8 in.
Belted Sandfish
to 3 in.
variants
Tobaccofish
to 8.6 in.
Harlequin Bass
to 4 in.
Chalk Bass
to 2.5 in.

Grammatidae - Basslets

Cuban Gramma - *Gramma dejongi* Victor & Randall, 2010

FEATURES: Shades of golden yellow on head, body, and dorsal, caudal, and anal fins. Anterior dorsal-fin margin magenta with a large black blotch on first few spines. Breast and pelvic fins magenta with bright blue on second pelvic-fin ray. Anal fin may have a narrow magenta margin. Pelvic fins sometimes reach past anal-fin base. HABITAT: Southern Cuba with a few strays found around the Cayman Islands. Found over coral reef dropoffs from about 30 to 100 ft.

Yellowcheek Basslet - *Gramma linki* Starck & Colin, 1978

FEATURES: Color varies. Shades of dark blue, bluish green, or purplish blue to grayish. All with yellowish wavy lines radiating from upper jaw to opercular margin. May have yellowish wash on head. Sides with minute to moderate yellowish spots. Dorsal-fin margin blue. Pelvic fins sometimes reach past anal-fin base. Caudal-fin margin concave to deeply forked. HABITAT: Bahamas and Caribbean Sea. Found over rocky and coral reefs from about 65 to 425 ft.

Fairy Basslet - *Gramma loreto* Poey, 1868

FEATURES: Deep bluish purple anteriorly and yellow posteriorly with delineation highly variable. Rusty bar from lower jaw tip through eye. Yellowish bar from lower eye to opercle. Black spot on anterior portion of spiny dorsal fin. Caudal-fin margin either rounded or slightly concave. HABITAT: FL Keys, Bahamas, Antilles, Quintana Roo to Venezuela. Reef associated. Around caves and ledges from about 3 to 200 ft. BIOLOGY: Often swim upside down. Males care for nests. Apparently introduced to FL Keys.

Blackcap Basslet - *Gramma melacara* Böhlke & Randall, 1963

FEATURES: Color and pattern vary. Shades of bright magenta to deep purple with a solid to wavy black "cap" that extends from snout tip to dorsal-fin origin. Orange to yellowish or bluish lines radiate from eyes. Dorsal fin with a black wedge-shaped stripe that narrows toward rear tip. Sides may be unmarked or with orange to yellow spots. Pelvic and anal fins may be streaked with blue, green, or pink. Pelvic fins sometimes reach past anal-fin margin. HABITAT: Bahamas and Caribbean Sea. Occur over rocky and coral reefs from about 50 to 440 ft.

Blue-spotted Basslet - *Lipogramma barrettorum* Baldwin, Nonaka & Robertson, 2018

FEATURES: Shades of greenish to brownish yellow with seven to nine faint bars on rear body. A very narrow bright blue line extends from lower lip to nape along dorsal profile. A large black spot with a blue ring soft on dorsal-fin base that extends onto back. Dorsal, caudal, and anal fins yellow with bluish spots and lines. Pelvic fins blue with yellow spots. Pelvic fins long, extend to or past anal-fin base. HABITAT: Curaçao. Found over elevated rocky bottoms with cracks or holes from about 400 to 530 ft.

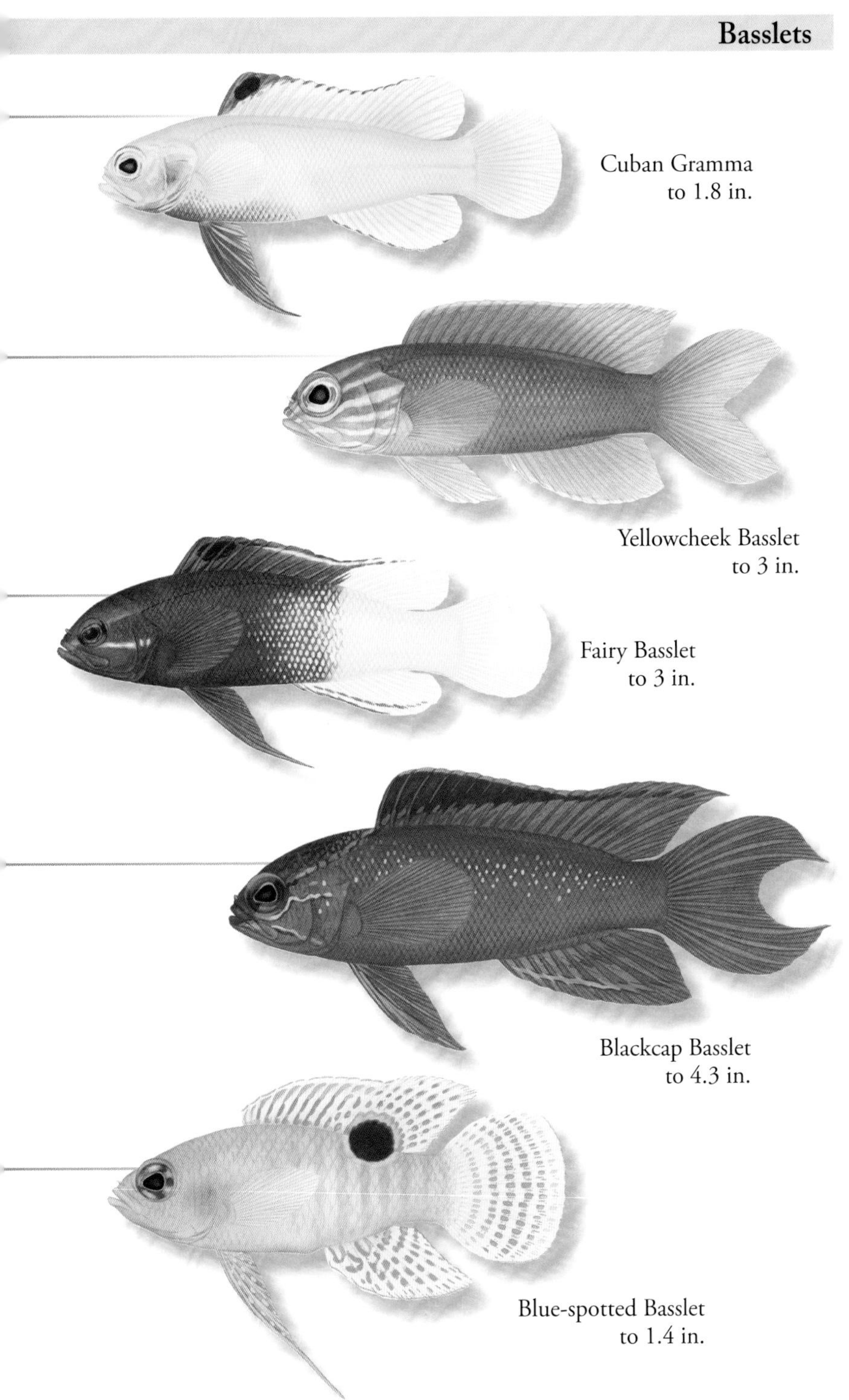
Cuban Gramma
to 1.8 in.
Yellowcheek Basslet
to 3 in.
Fairy Basslet
to 3 in.
Blackcap Basslet
to 4.3 in.
Blue-spotted Basslet
to 1.4 in.

Banded Basslet - *Lipogramma evides* Robins & Colin, 1979

FEATURES: Pale pinkish to tannish white. A blackish bar extends through eyes, a dark brown bar present under first few dorsal-fin spines, and a pale to dark brown bar under a large black blotch on soft dorsal-fin base. Dorsal and anal fins bluish with yellow inner margin and yellow spots. Caudal fin yellow with pale blue spots. Caudal-fin margin concave. Pelvic fins reach to or past anal-fin base. HABITAT: NE Yucatán to Curaçao, and northern Lesser Antilles. Occur over deep reefs from about 450 to 1,060 ft.

Yellow Basslet - *Lipogramma flavescens* Gilmore & Jones, 1988

FEATURES: Lemon yellow to canary yellow dorsally; pinkish cast on cheeks and chest. Brownish wedge-shaped bar below eyes. Dorsal fin with yellow inner margin and rows of yellow spots. Caudal fin yellow with darker yellow spots. Anal fin with blackish margin, yellow inner margin, and yellow spots on membranes. Pelvic fins bluish with yellow spots. A large black ocellated spot on soft dorsal-fin base that extends onto back. Pelvic fins reach to or past anal-fin base. HABITAT: Bahamas and Belize to Honduras. Found over deep reefs from about 660 to 930 ft.

Yellow-banded Basslet - *Lipogramma haberorum* Baldwin, Nonaka & Robertson, 2016

FEATURES: Dusky yellow on cheeks and along back. A broad dusky yellow bar below a large, black ocellated spot at soft dorsal-fin base. Brownish bar through eyes, and a brownish bar below first few dorsal-fin spines. Dorsal, caudal, and anal fins bluish gray with yellow spots; spots elongate on anterior portion of dorsal fin. Pelvic fins whitish with a few spots. Pelvic fins extend to or past anal-fin base. HABITAT: Curaçao. Occur over deep reefs from about 500 to 760 ft.

Blue-backed Basslet - *Lipogramma idabeli* Tornabene, Robertson & Baldwin, 2018

FEATURES: Bright purplish blue on top of head and along back. Irises bright blue. Pinkish on cheeks. Remainder of body pale brownish yellow. A large, irregular, blue-ringed black and brown spot on anterior back and a small blue or black ocellated spot at rear soft dorsal-fin base. Dorsal and anal fins bluish with yellow spots. Caudal fin with a concave margin. HABITAT: Roatan. Found over vertical rock faces of deep reefs from about 400 to 540 ft.

Bicolor Basslet - *Lipogramma klayi* Randall, 1963

FEATURES: Shades of lavender on head and anterior back, becoming yellow with scattered lavender spots. Yellow posteriorly and on caudal fin. Spiny dorsal fin lavender with yellow spots. Soft dorsal, anal, and pelvic fins yellow with lavender spots. Eyes purple and blue. Pelvic fins extend to or past anal-fin base. Caudal fin emarginate or with trailing rays. HABITAT: Bahamas, Dominican Republic, Puerto Rico and surrounding islands, off Belize, and off coast of Venezuela. Occur over rocky and coral reefs from about 115 to 475 ft.

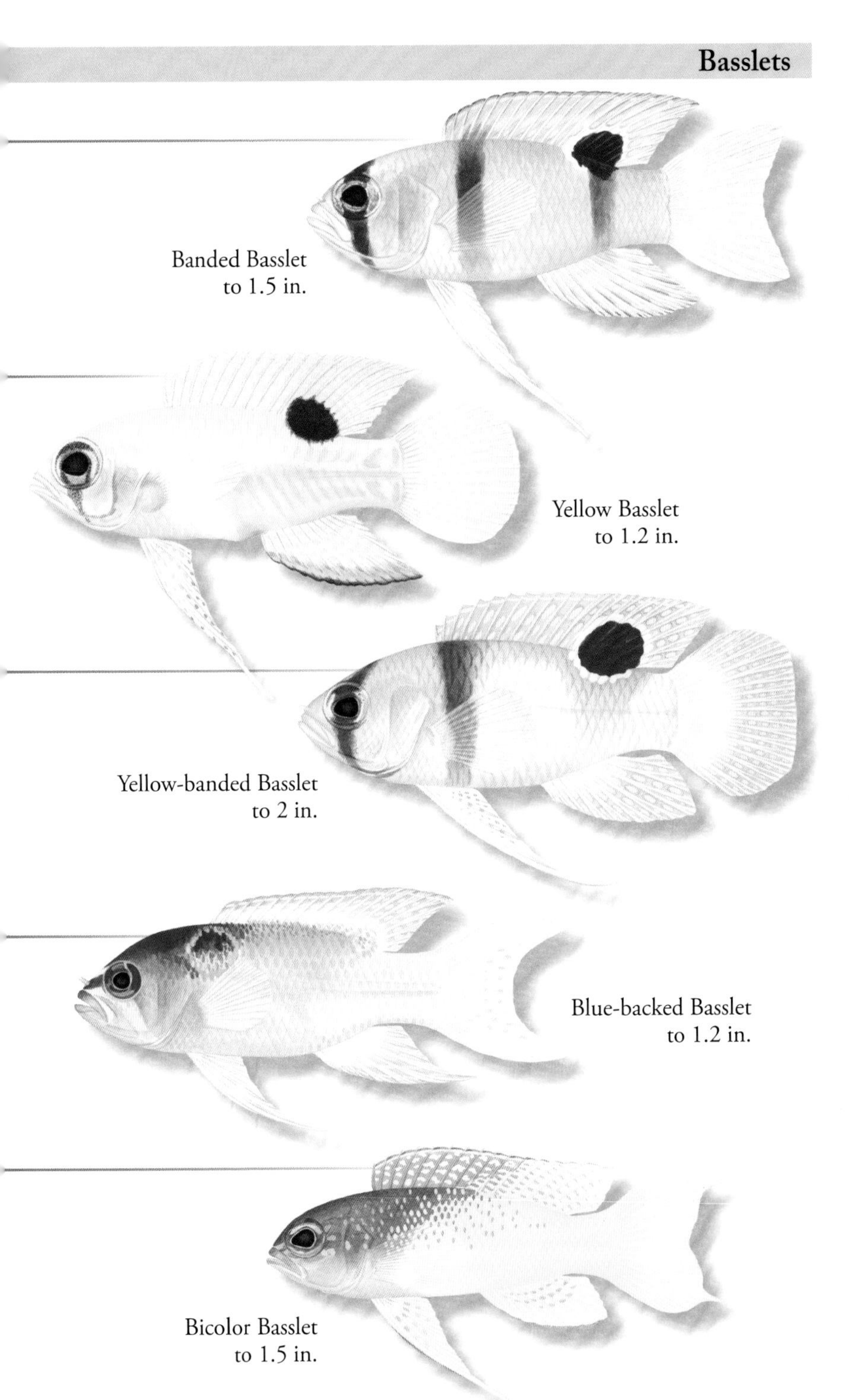
Banded Basslet
to 1.5 in.
Yellow Basslet
to 1.2 in.
Yellow-banded Basslet
to 2 in.
Blue-backed Basslet
to 1.2 in.
Bicolor Basslet
to 1.5 in.

Hourglass Basslet - *Lipogramma levinsoni* Baldwin, Nonaka & Robertson, 2016

FEATURES: White with faint gray on nape. A pale to dark brown bar extends through eyes. A second, nearly straight to hourglass-shaped brown bar extends from anterior dorsal fin to abdomen. A third bar extends from rear dorsal fin to rear anal fin. Third bar is complete, or may appear as an hourglass, or as two separate blotches. Dorsal and anal fins with orange inner margins. Caudal and pelvic fins white. HABITAT: Bahamas, Jamaica, St. Eustatius and Saba, Honduras, Dominica, and Curaçao. Occur over deep reefs, rocky rubble on gradual slopes, and patches of cobble from about 300 to 610 ft.

Royal Basslet - *Lipogramma regia* Robins & Colin, 1979

FEATURES: Alternating bluish and yellow to dusky orange stripes from snout to nape. Lower cheeks pinkish. Body with alternating dusky orange to yellow bars that extend onto fins. A broad white bar present behind pectoral fins. Large black ocellated spot on rear dorsal-fin base that extends onto upper back. Irises bright yellow, bordered with blue. HABITAT: Off the Mississippi Delta, Bahamas, NW Cuba, Puerto Rico, St. Eustatius and adjacent islands, and Honduras. Occur over steep and sloping outer reef faces and rocky bottoms from about 150 to 380 ft.

Maori Basslet - *Lipogramma schrieri* Baldwin, Nonaka & Robertson, 2018

FEATURES: Shades of tan to pale brown with white wavy markings on top of snout and head that become a white band to dorsal-fin origin. Two whitish bars below eyes. Body with seven or eight dark brown bars. Dorsal, caudal, and anal fins shades of blue with yellow to orange lines and spots and blue margins. Inner caudal fin may be entirely yellow to orange. Large black ocellated spot present at rear soft dorsal-fin base. Pelvic fins blue. Dorsal, caudal, and anal fins fan-like. HABITAT: Flower Garden Banks, northern Gulf of Mexico, and off Curaçao. Demersal over deep reefs with elevated rocky bottoms with cracks and holes from about 460 to 680 ft.

Threeline Basslet - *Lipogramma trilineatum* Randall, 1963

FEATURES: Yellowish with scales outlined in greenish blue. Bright blue stripe from tip of upper jaw to dorsal-fin origin. Bright blue line from top of eyes to below spiny dorsal fin. May have short blue line below eyes. Cheeks pinkish to purplish. Dorsal fin yellowish with blue margin. Anal and pelvic fins with yellowish spots. HABITAT: SE FL, FL Keys, in the Gulf of Mexico from the Flower Garden Banks, Bahamas, and scattered in Caribbean Sea. Found near bottom around coral and rocky ledges and walls to about 310 ft.

ALSO IN THE AREA: *Lipogramma anabantoides*, *Lipogramma rosea*, see p. 547.

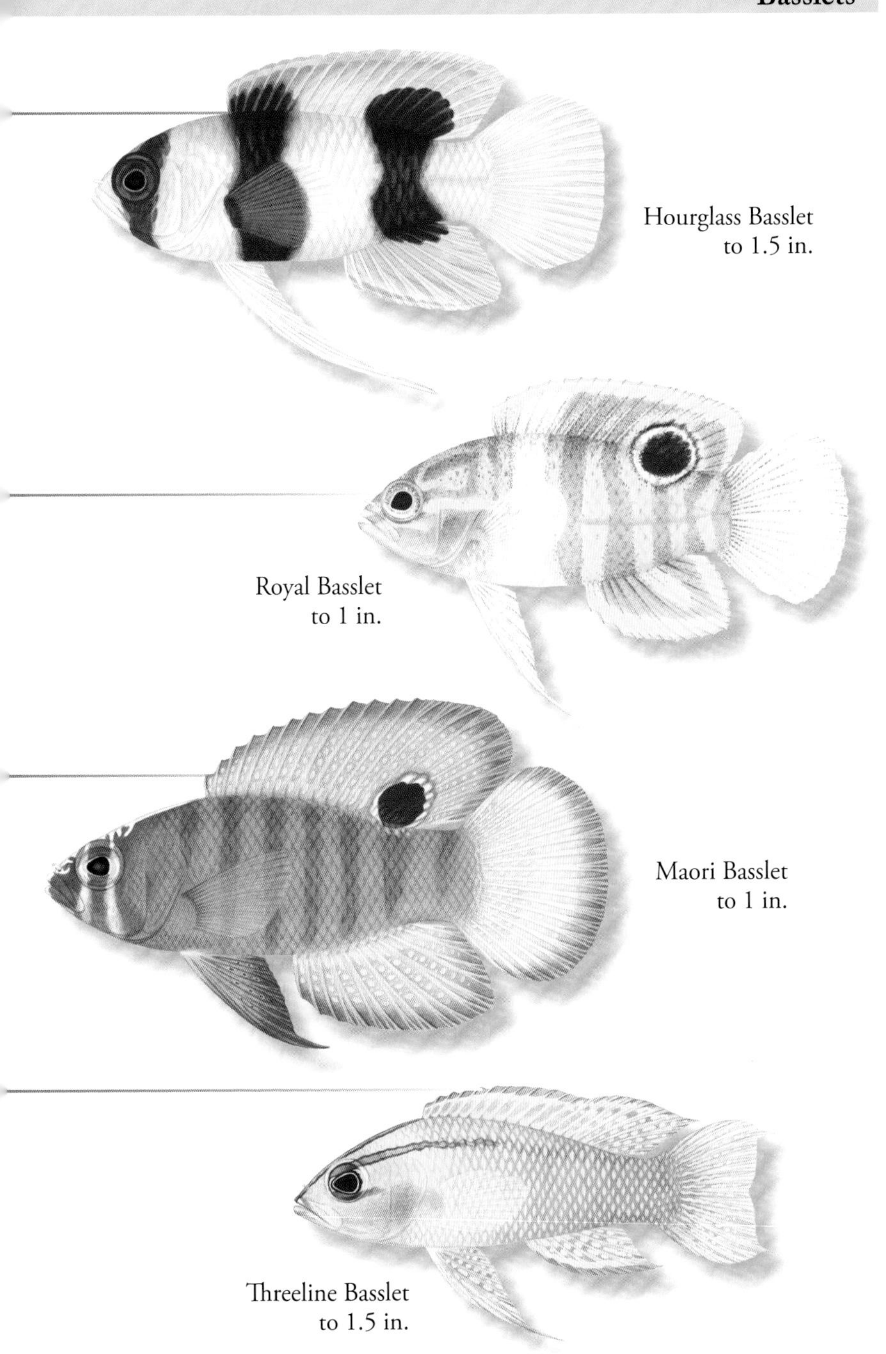

Hourglass Basslet
to 1.5 in.

Royal Basslet
to 1 in.

Maori Basslet
to 1 in.

Threeline Basslet
to 1.5 in.

Opistognathidae - Jawfishes

Higman's Jawfish - *Lonchopisthus higmani* Mead, 1959

FEATURES: Brownish orange with whitish bars anteriorly and scattered spots posteriorly. Opercle with a large white-ringed black spot. Dorsal, anal, and pelvic fins yellowish basally, grayish black distally, with pale spots. Eyes very large. Upper jaw expanded at rear. Dorsal, caudal, anal, and pelvic fins pointed at tips. Body deep anteriorly. HABITAT: Belize to French Guiana. Occur over muddy, sandy, and gravel bottoms from about 130 to 330 ft. Usually above 230 ft. BIOLOGY: Build and maintain burrows. Males mouth-brood eggs until hatching.

Swordtail Jawfish - *Lonchopisthus micrognathus* (Poey, 1860)

FEATURES: Tan with narrow, pale blue bars anteriorly and spots posteriorly. Dorsal and anal fins with blue margins. Pelvic fins bluish. Caudal fin with bluish upper and lower margins and elongate, trailing rays. HABITAT: S FL, Gulf of Mexico, and Caribbean Sea to Suriname. Over relatively uniform silty or muddy bottoms to about 280 ft. BIOLOGY: Usually colonial. Build and maintain burrows that may reach about 14 in. in depth. May live commensally with the Smoothwrist Soft Crab, *Chasmocarcinus cylindricus*. Feed on zooplankton.

Yellowhead Jawfish - *Opistognathus aurifrons* (Jordan & Thompson, 1905)

FEATURES: Head bright to pale yellow fading to pearly bluish to tannish posteriorly. Membranes under mouth may be unmarked, with single or multiple pairs of black spots, or a pair of black lines. Dorsal fin with narrow blue margin. Body otherwise unmarked. Caudal fin rounded to slightly pointed. HABITAT: S FL, Gulf of Mexico, Bahamas, and Caribbean Sea. Over sand and coral rubble around reefs from about 3 to 200 ft. BIOLOGY: Use their mouths to excavate and maintain burrows. Males mouth-brood eggs. Popular in aquarium trade.

Yellow Jawfish - *Opistognathus gilberti* Böhlke, 1976

FEATURES: Females pearly grayish to bluish with yellow bars and yellow and blue blotches on head; dorsal and anal fins with broad, yellow inner ban; caudal fin mostly yellow. Males may be uniform shades of gray, becoming dark posteriorly, or mottled yellow; yellow and blue blotches on head; dorsal fin with a small to large black blotch on fourth, fifth, or sixth spines; caudal fin with a large dark to black inner blotch. Upper jaw expanded posteriorly. HABITAT: Bahamas and western and northern Caribbean Sea. Demersal on rubble bottoms or outer reef slopes from about 90 to 180 ft. BIOLOGY: Build and maintain burrows.

Moustache Jawfish - *Opistognathus lonchurus* Jordan & Gilbert, 1882

FEATURES: Olive gray dorsally; tannish on sides. Abdomen whitish. Snout dark. Jaws bluish. Dusky orange moustache above upper jaw. Irregular blue lines on sides. Dorsal and anal fins with blue banding. All fins except pectoral fins with blue margins. HABITAT: SC to TX, NW Cuba, Hispaniola, Puerto Rico, Honduran Bay Islands, and from the Gulf of Venezuela to Campos, Brazil. On sandy, rubble bottoms around reefs to about 300 ft. BIOLOGY: Build and maintain burrows.

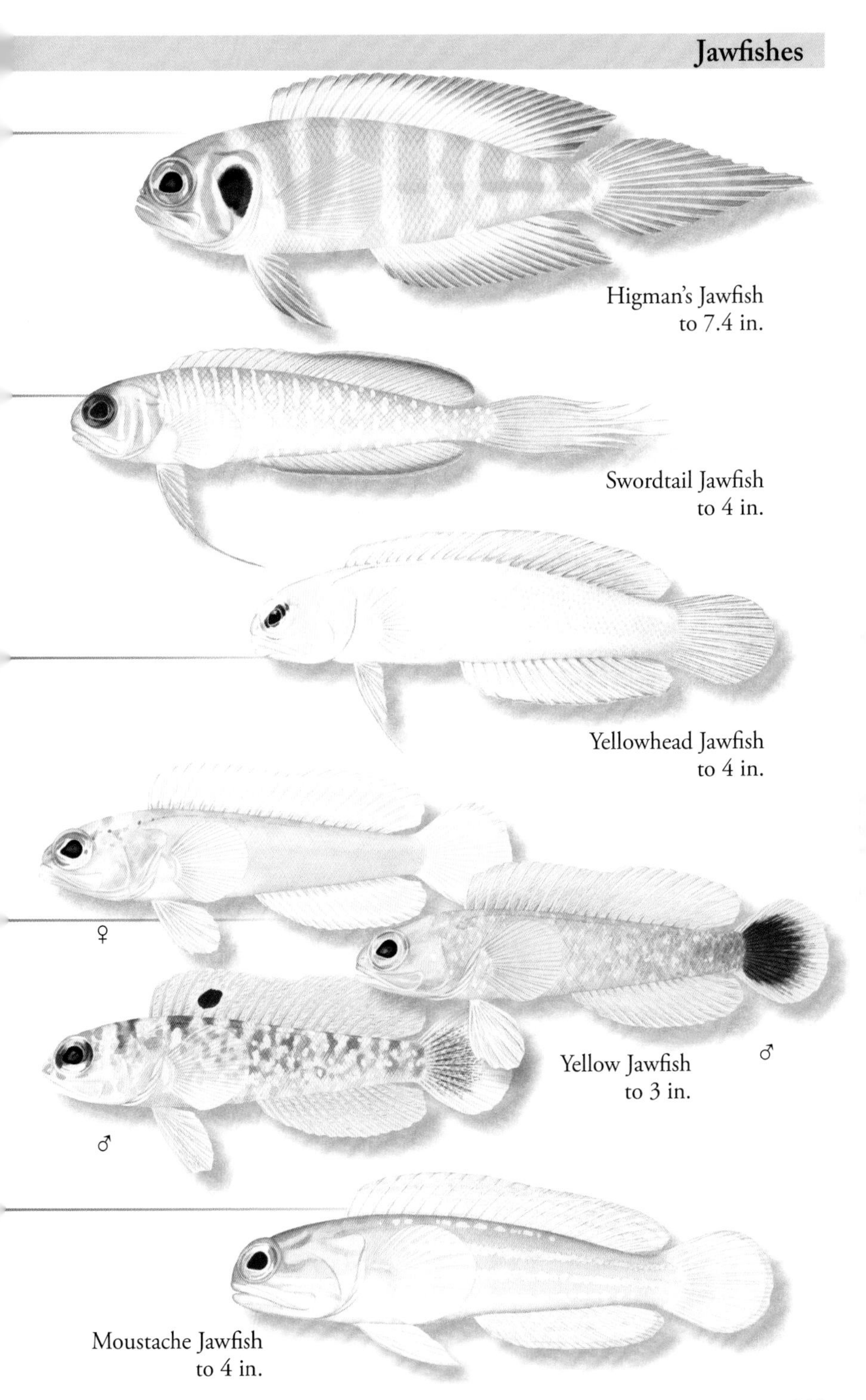
Higman's Jawfish
to 7.4 in.
Swordtail Jawfish
to 4 in.
Yellowhead Jawfish
to 4 in.
♀
Yellow Jawfish
to 3 in.
♂
♂
Moustache Jawfish
to 4 in.

Banded Jawfish - *Opistognathus macrognathus* Poey, 1860

FEATURES: Body brownish with irregular, pale spots and mottling. Four to six brown blotches at dorsal-fin base. Blotches may merge with brown bars below. Dorsal fin yellowish with irregular, pale bands. Dark blotch between sixth and ninth dorsal-fin spines may be present. Males with elongate and pointed upper jaw that has two blackish bands in inner surface. Female upper jaw squared to rounded posteriorly, and with one, short dark band on inner surface. HABITAT: S FL to northern South America. Absent from Jamaica. Over sand, rubble, and rocky bottoms around reefs from near surface to about 140 ft. BIOLOGY: Build and maintain burrows that they line with small stones. Apparently monogamous.

Mottled Jawfish - *Opistognathus maxillosus* Poey, 1860

FEATURES: Head and body mottled in shades of brownish to grayish brown. Large pale blotches form rows below dorsal-fin base and along ventral area. Two large, pale blotches at caudal-fin base. Dorsal fin brownish with pale spotting and five to six dark blotches along base; first blotch may be indistinct. Blotch between sixth and eighth spines largest and darkest. Rear jaw margin similar in male and female. HABITAT: S FL, Yucatán Peninsula, Bahamas, Greater and Lesser Antilles to Venezuela. Demersal over sandy, rocky, and coral rubble bottoms from about 3 to 25 ft. BIOLOGY: Construct elaborate burrows lined with coral debris.

Spotfin Jawfish - *Opistognathus robinsi* Smith-Vaniz, 1997

FEATURES: Head brownish with paler spots and mottling. Body brownish with pale blotches arranged in uneven rows on sides. Dorsal fin brownish to yellowish brown with bands of pale spots. Prominent ocellated black spot between third and seventh dorsal-fin spines. Males with elongated and bluntly pointed upper jaw that has one blackish band on inner surface. HABITAT: SC to FL, Bahamas, and northern Gulf of Mexico. On sandy to rubble bottoms from near shore to about 150 ft. Reported in estuaries and lagoons. BIOLOGY: Construct and maintain burrows. Males mouth-brood eggs until hatching.

Curaçao Jawfish - *Opistognathus schrieri* Smith-Vaniz, 2017

FEATURES: Brownish gray with brown and white mottling on head and five to six irregular and wide brownish bars on body. Upper jaw bluish white posteriorly with a blackish blotch. A large white patch present at pectoral-fin base. Dorsal fin yellowish with bands of brown and white spots, and a large black ocellated spot on first to sixth spines. Caudal fin yellowish with rows of brownish spots. Anal fin yellowish distally, whitish at base. Inner lining of rear upper and lower jaws brownish black. Rear edge of upper jaw squared. HABITAT: Roatan, Honduras, and Curaçao. Demersal over sandy and rubble bottoms from about 330 to 750 ft. BIOLOGY: Construct and maintain burrows. Observed in colonies.

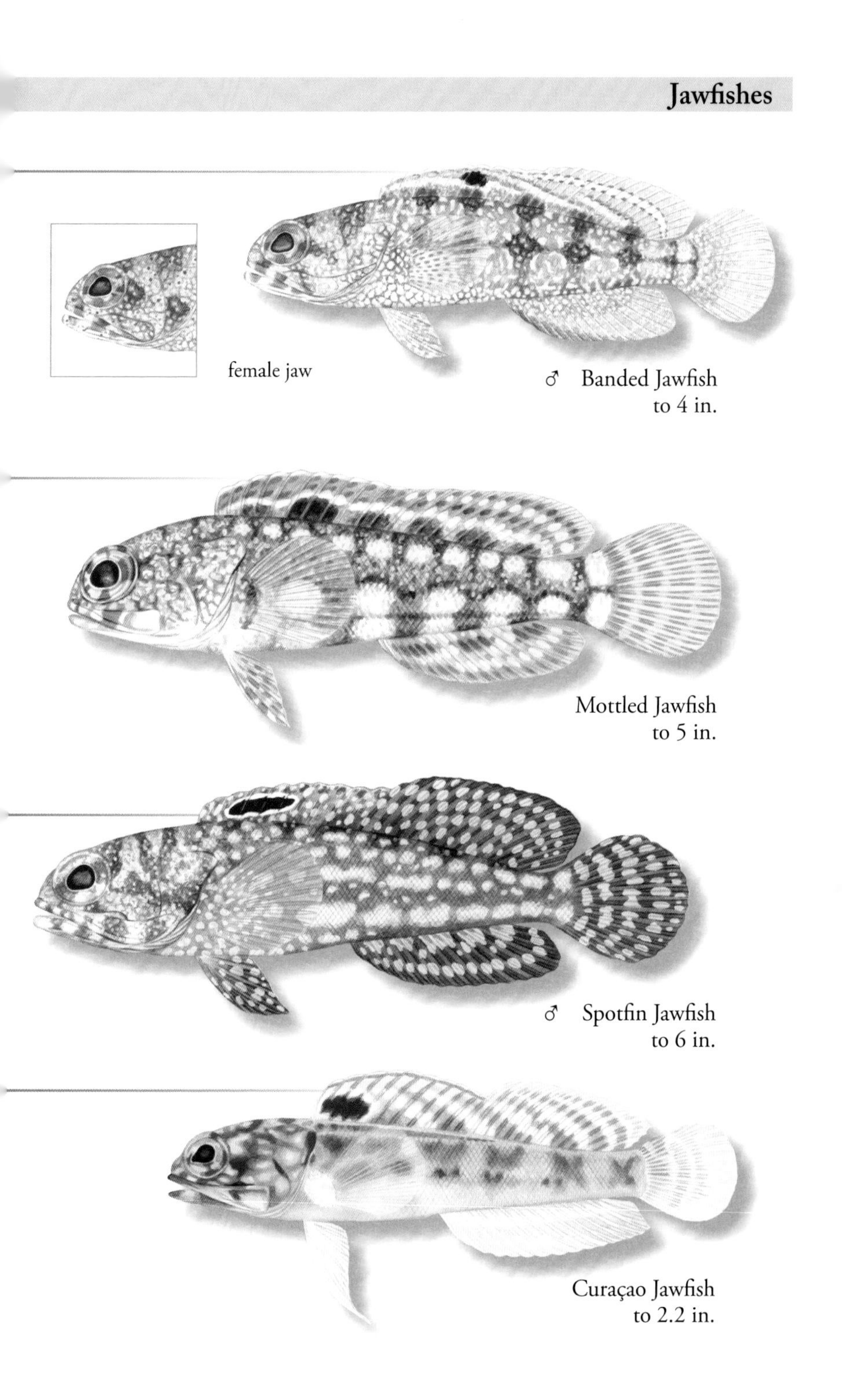

female jaw

♂ Banded Jawfish
to 4 in.

Mottled Jawfish
to 5 in.

♂ Spotfin Jawfish
to 6 in.

Curaçao Jawfish
to 2.2 in.

Opistognathidae - Jawfishes, *cont.*

Dark-spotted Jawfish - *Opistognathus signatus* Smith-Vaniz, 1997

FEATURES: Shades of tan to brown with scattered whitish and dark brown flecks and spots on sides. Head darkly mottled. Dorsal fin tannish with bands of blackish spots. Large black ocellated spot between third and seventh spines. Caudal fin blackish with two bands of whitish spots. Anal fin black with white spots at base. Males with elongated and pointed upper jaw with one blackish band on inner surface. HABITAT: Nicaragua to Venezuela. Demersal over sandy to rubble bottoms from about 90 to 220 ft.

Dusky Jawfish - *Opistognathus whitehursti* (Longley, 1927)

FEATURES: Head mottled brown. Body mottled dark brown with obscure dark bars reaching dorsal-fin base. Row of close-set, irregular white blotches along dorsal profile. Usually with pale blue to green blotch between second and fourth dorsal-fin spines; blotch may be obscure to absent. Soft dorsal fin yellowish with pale and dark banding. HABITAT: S FL, SW Gulf of Mexico, Bahamas, and Caribbean Sea. Occur over sandy and rubble bottoms, often around edges of turtle-grass beds from about 3 to 156 ft. BIOLOGY: Construct and maintain burrows. May live in groups.

ALSO IN THE AREA: *Opistognathus megalepis*, see p. 547.

Priacanthidae - Bigeyes

Bulleye - *Cookeolus japonicus* (Cuvier, 1829)

FEATURES: Uniformly deep red or fading to pinkish or silvery below. Dorsal, caudal, and anal fins red with blackish margins. Pelvic fins with blackish membranes. Eyes large with red irises. Dorsal and anal fins elongate posteriorly. Pelvic fins expanded with inner rays attached to abdomen by membrane—become shorter with age. HABITAT: Circumglobal in tropical to temperate seas. In western Atlantic from VA to Argentina. Juveniles recorded to Nova Scotia. Over hard bottoms from about 200 to 1,300 ft. BIOLOGY: Feed on crustaceans and fishes. May live to nine years or more. Often confused with Glasseye Snapper, *Heteropriacanthus cruentatus*. OTHER NAME: Longfin Bulleye.

Glasseye Snapper - *Heteropriacanthus cruentatus* (Lacepède, 1801)

FEATURES: Uniformly red or orange red to pink, or with pale to silvery bars and blotches. Soft dorsal, caudal, and anal fins faintly spotted. Spiny dorsal- and caudal-fin margins sometimes dusky. Eyes large with red irises. Pelvic fins comparatively short, attached to abdomen by membrane. HABITAT: Circumglobal in tropical to temperate seas. In western Atlantic from NJ (rare) to Argentina and Gulf of Mexico. Occur as deep as 1,400 ft.—more commonly over shallow reefs. BIOLOGY: Secretive and nocturnal. Hide in holes and crevices and under ledges during the day. Feed on a variety of invertebrates. Taken by hook-and-line, by spearing, and in traps. NOTE: Previously known as *Priacanthus cruentatus* and *Cookeolus boops*.

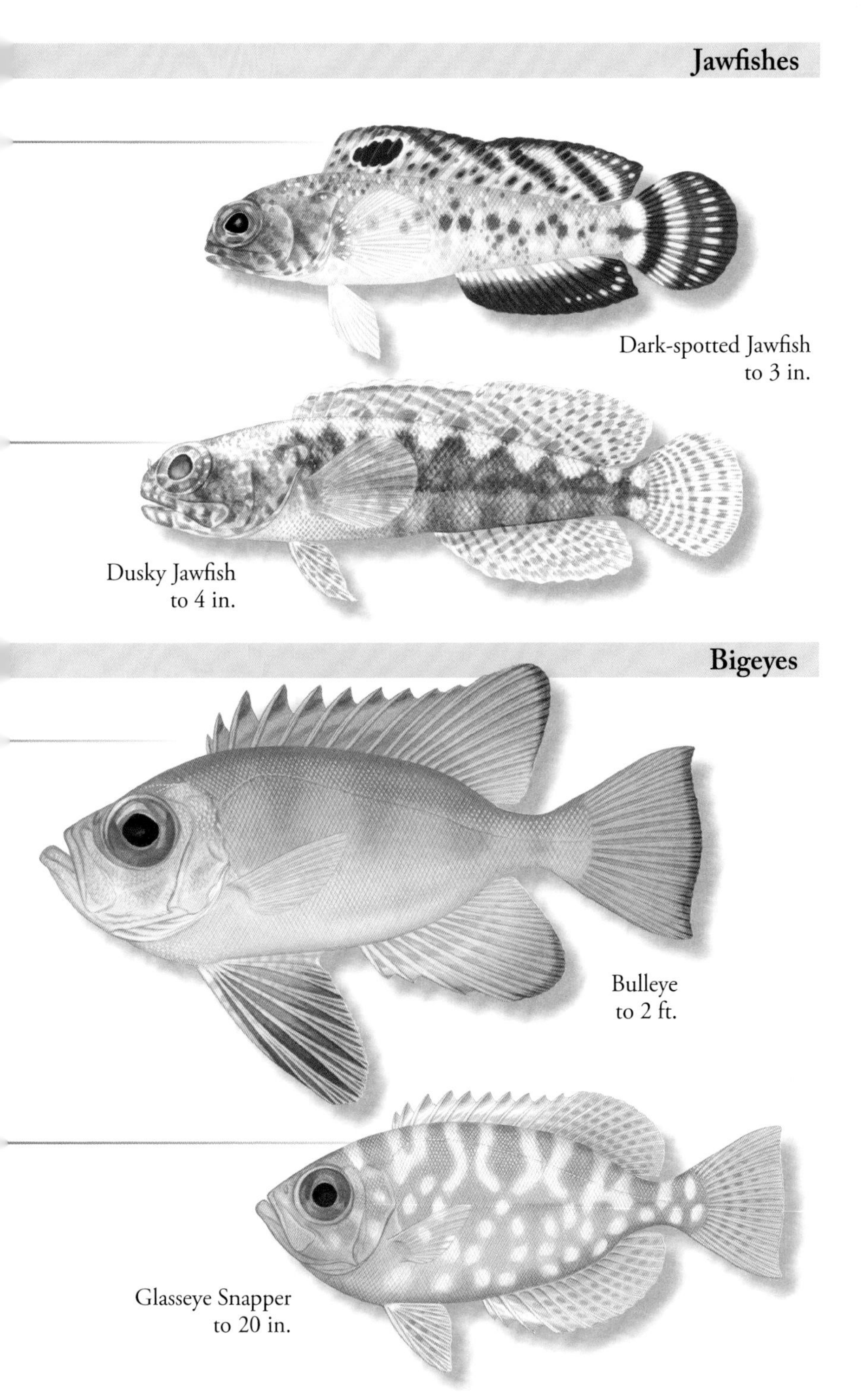
Dark-spotted Jawfish
to 3 in.
Dusky Jawfish
to 4 in.
Bigeyes
Bulleye
to 2 ft.
Glasseye Snapper
to 20 in.

Priacanthidae - Bigeyes, *cont.*

Bigeye - *Priacanthus arenatus* Cuvier, 1829

FEATURES: Uniformly red, but may change to silvery with broad bars. Row of small, dark red spots sometimes along lateral line. Dorsal and anal fins with faint spots. Pelvic fins with blackish membranes and small black spot at base. Spiny dorsal-fin margin comparatively straight. Caudal fin margin concave. HABITAT: On both sides of Atlantic in tropical to subtropical waters. In western Atlantic from NC to FL, Gulf of Mexico, Bahamas, Caribbean Sea to Argentina. Also Bermuda. Over reefs and rocky bottoms from about 65 to 820 ft.

Short Bigeye - *Pristigenys alta* (Gill, 1862)

FEATURES: Body red to salmon-colored, often with diffuse, paler bars dorsally. Soft dorsal, caudal, and anal fins pale with black margins. Pelvic fins red with black margins. Juveniles mottled in red to salmon with banded fins. Eyes large with red irises. Inner rays of pelvic fins attached to abdomen. Scales comparatively large. Body comparatively deep. HABITAT: MA to FL, Gulf of Mexico, Bahamas, and Caribbean Sea to Venezuela. Also Bermuda. Juveniles to ME. Found over hard bottoms from about 16 to 980 ft. BIOLOGY: Secretive, hide in recesses.

Apogonidae - Cardinalfishes

Bridle Cardinalfish - *Apogon aurolineatus* (Mowbray, 1927)

FEATURES: Translucent salmon or pinkish to golden with enlarged melanophores on body. Body and fins otherwise unmarked. May have two streaks behind eyes. HABITAT: S FL, northern and southern Gulf of Mexico, Bahamas to northern South America. Over seagrass beds and around coral reefs. From about 3 to 245 ft. BIOLOGY: Form congregations and seek shelter among sea anemones. Feed on zooplankton. Form pairs during spawning. Males mouth-brood eggs.

Barred Cardinalfish - *Apogon binotatus* (Poey, 1867)

FEATURES: Pinkish to pale red and somewhat translucent. May pale or darken or have iridescent highlights. Always with one blackish bar from rear second dorsal-fin base to rear anal-fin base and one blackish bar on caudal peduncle near caudal-fin base. HABITAT: SE FL, eastern Gulf of Mexico, Bahamas, Antilles to Venezuela. Reef associated from shore to about 200 ft. BIOLOGY: Nocturnal. Reported to form aggregations in hiding places during the day.

Deepwater Cardinalfish - *Apogon gouldi* Smith-Vaniz, 1977

FEATURES: Red to reddish orange on anterior portion of body. Posterior portion of body pale to whitish with a broad dark red to blackish saddle or bar under second dorsal fin and a second dark red to blackish bar on caudal peduncle. Tips of dorsal, caudal, and anal fins reddish to blackish. HABITAT: NC, Bermuda, Bahamas, Chinchorro Bank (Mexico) to Belize and Curaçao. Found over calcareous vertical walls from about 180 to 860 ft.

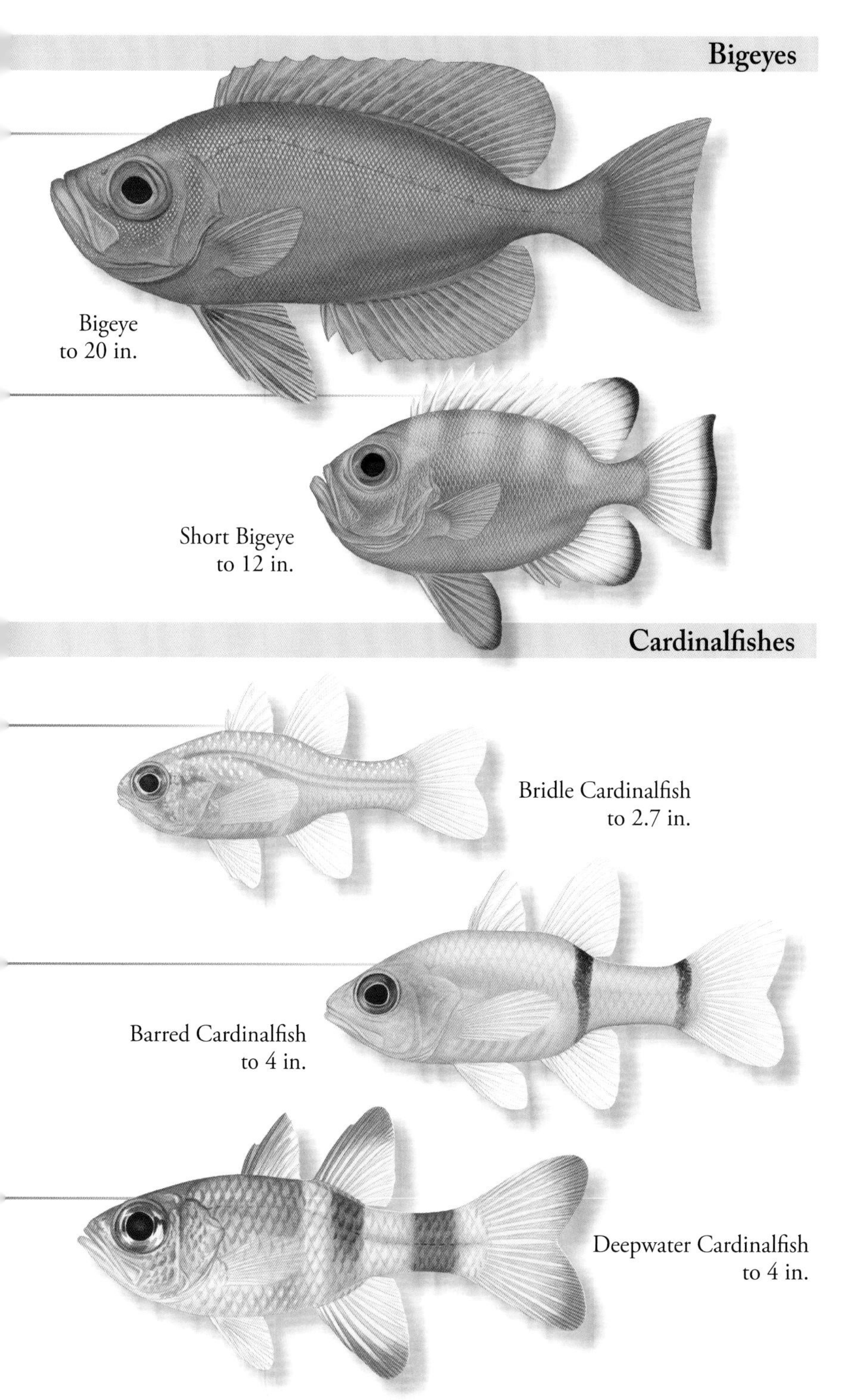
Bigeyes
Bigeye
to 20 in.
Short Bigeye
to 12 in.
Cardinalfishes
Bridle Cardinalfish
to 2.7 in.
Barred Cardinalfish
to 4 in.
Deepwater Cardinalfish
to 4 in.

Whitestar Cardinalfish - *Apogon lachneri* Böhlke, 1959

FEATURES: Reddish with small black spot followed by small white spot on upper caudal peduncle just behind second dorsal-fin base. Black speckles cluster on outer margins of dorsal and anal fins. Scales may appear to have iridescent flecks. HABITAT: S FL, southern Gulf of Mexico, Bahamas, Antilles to Belize. Reef associated in clear waters from about 20 to 230 ft. BIOLOGY: Secretive and nocturnal. Remain hidden during the day.

Slendertail Cardinalfish - *Apogon leptocaulus* Gilbert, 1972

FEATURES: Pale peach to deep red with three dark bars on body. First bar under first dorsal fin; second bar from second dorsal-fin base to anal-fin base; third bar circles end of caudal peduncle. Anterior portions of second dorsal and anal fins dark. Upper and lower lobes of caudal fin dark. Snout pointed. Caudal peduncle long and slender. HABITAT: Boca Raton, FL, Bahamas, Jamaica, and Belize to Colombia. Occur in caves and crevices over coral reefs and rocky slopes from about 60 to 100 ft. May be more common in deeper water.

Flamefish - *Apogon maculatus* (Poey, 1860)

FEATURES: Deep orange red to dusky red with black flecks that cluster to form a bar through eye to opercular margin. All with a white stripe above and below pupil. Black blotch below second dorsal-fin base. Usually with faint to dark, bar-like saddle on caudal peduncle. May also display a white bar anterior to blotch on peduncle. HABITAT: MA to FL, Gulf of Mexico, Bermuda, Bahamas, and Caribbean Sea. Around reefs, sea walls, and pilings from near shore to about 330 ft. BIOLOGY: Hide in or near reef caves and crevices during the day. Feed at night.

Dwarf Cardinalfish - *Apogon mosavi* Dale, 1977

FEATURES: Translucent pinkish orange with iridescent highlights on cheeks and abdomen. Head with dense orange spotting and scattered blackish spots. Body covered with tiny orange dots and flecks. Cluster of orange or orange and black dots form an oval or rectangular bar on rear caudal peduncle. Fins nearly colorless. HABITAT: Bahamas, Haiti, Jamaica, Belize, and Panama. Occur over reefs and rocks from about 40 to 130 ft. May associate with anemones.

Mimic Cardinalfish - *Apogon phenax* Böhlke & Randall, 1968

FEATURES: Deep orange red to red with iridescent highlights. At night, highlights become turquoise blue. Diffuse, triangular-shaped blackish bar below rear second dorsal-fin base. Wide, blackish saddle or bar on caudal peduncle near caudal-fin base. HABITAT: S FL, Bahamas, and Caribbean Sea. Occur around reefs and rocky areas with little sand from about 10 to 165 ft. BIOLOGY: Males mouth-brood eggs. Solitary, secretive. Reported hiding in crevices during the day, foraging at night.

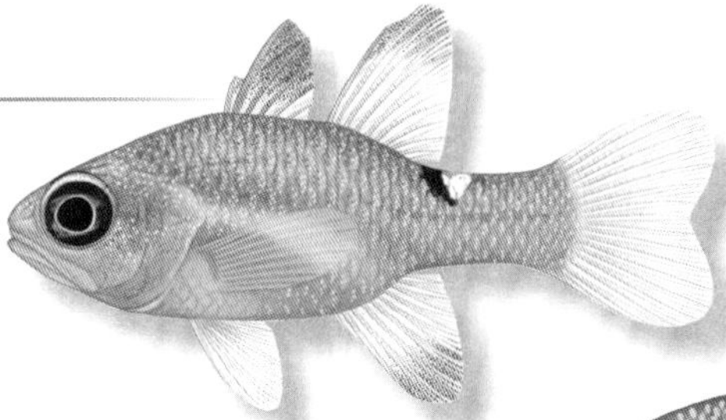

Whitestar Cardinalfish
to 2.6 in.

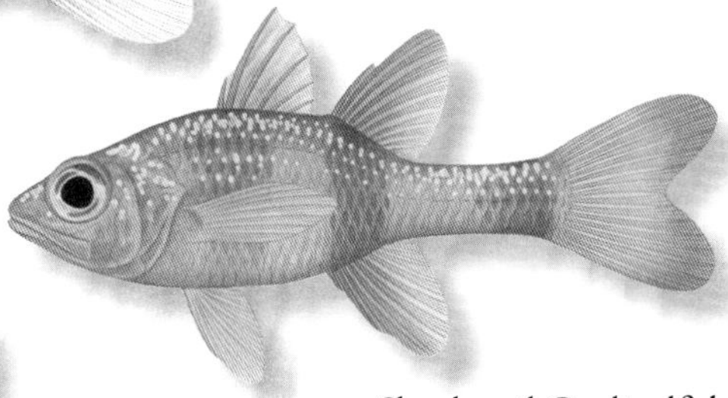

Slendertail Cardinalfish
to 2.5 in.

Flamefish
to 4.4 in.

variants

Dwarf Cardinalfish
to 1.7 in.

Mimic Cardinalfish
to 3.2 in.

Broadsaddle Cardinalfish - *Apogon pillionatus* Böhlke & Randall, 1968

FEATURES: Reddish and somewhat transparent with some luster. Diffuse, blackish bar below rear of second dorsal fin is followed by a white bar and a broad, diffuse blackish saddle. A small, white bar at caudal-fin base may be present or absent. HABITAT: S FL, Bahamas, and Caribbean Sea. Found over reefs and rocky areas from about 50 to 295 ft. Reported around reef faces, slopes, and walls. BIOLOGY: Secretive. Hide in reef caves and crevices by day, emerge at night to feed on plankton.

Pale Cardinalfish - *Apogon planifrons* Longley & Hildebrand, 1940

FEATURES: Iridescent whitish or pale pink to reddish orange. Narrow reddish to blackish bar from rear of second dorsal-fin base to rear of anal-fin base. Large, reddish to blackish spot or wide bar on caudal peduncle near caudal-fin base. Bars and spots may be very faint to almost absent. Body color becomes pearly at night. HABITAT: S FL, Bermuda, Bahamas, eastern Gulf of Mexico, and Caribbean Sea. Reef associated from about 10 to 100 ft. BIOLOGY: Reclusive and solitary. Reported leaving crevices and caves at night to forage on zooplankton. Form pairs during spawning. Males mouth-brood eggs.

Twospot Cardinalfish - *Apogon pseudomaculatus* Longley, 1932

FEATURES: Deep orange red with faint to dark blotch on opercle. Black spot on back below rear second dorsal fin. Black spot on upper caudal peduncle near caudal-fin base. Tips of dorsal, caudal, and anal fins may be reddish black. Eyes with a whitish stripe above and below pupil. Juveniles translucent. HABITAT: MA (rare) to FL, Bermuda, Bahamas, Gulf of Mexico, and Caribbean Sea. Over reefs and hard bottoms, around harbors, pilings, and sea walls from near shore to about 1,320 ft. BIOLOGY: Nocturnal. Males brood eggs in mouth. Feed on plankton.

Sawcheek Cardinalfish - *Apogon quadrisquamatus* Longley, 1934

FEATURES: Translucent reddish orange, or reddish to brownish orange, with darker melanophores covering body. Small to large diffuse, dark spot on caudal peduncle at caudal-fin base. Fins yellowish. Dorsal fin often with a central pinkish to orange bar. HABITAT: GA to FL, Bahamas, Gulf of Mexico, and Caribbean Sea. Occur over reefs, seagrass flats, sand, and rubble bottoms from about 4 to 245 ft. Also associated with sea urchins, anemones, and tube sponges. Feed at night. NOTE: This species is one of a multi-species complex.

Striped Cardinalfish - *Apogon robbyi* Gilbert & Tyler, 1997

FEATURES: Dark reddish orange to translucent pinkish. Wide, dark stripes in shades of orange to brown run from head to caudal peduncle. A faint to distinct large spot present on caudal peduncle at caudal-fin base. Fins yellowish. HABITAT: SE FL and eastern and southeastern Caribbean Sea. Also reported from northern and eastern Brazil. Occur over seagrass beds and rocky, rubble, and sandy bottoms, and on reefs from about 3 to 240 ft. Also associated with sponges and sea anemones. Around sea urchins, anemones, and tube sponges.

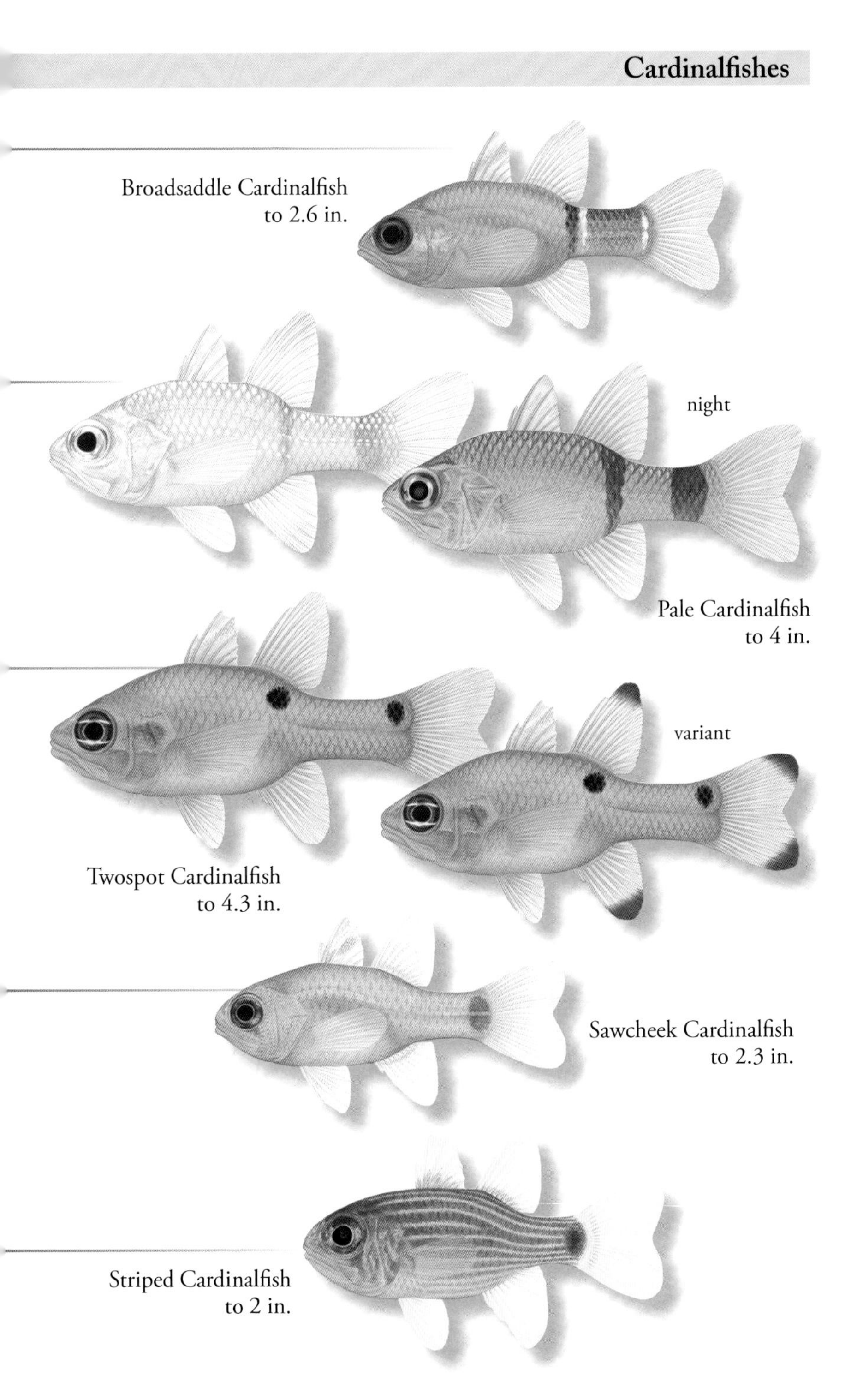
Broadsaddle Cardinalfish
to 2.6 in.
night
Pale Cardinalfish
to 4 in.
variant
Twospot Cardinalfish
to 4.3 in.
Sawcheek Cardinalfish
to 2.3 in.
Striped Cardinalfish
to 2 in.

Roughlip Cardinalfish - *Apogon robinsi* Böhlke & Randall, 1968

FEATURES: Shades of pink to bright red with iridescent highlights. Dark red to black bar under rear second dorsal fin and a wider dark bar on caudal peduncle. Space between dark bars may be identical to body color or iridescent pearly white. Tips of dorsal, anal, and caudal fins may be dark red. Teeth in upper jaw extend outside jawbone. HABITAT: Bermuda, Bahamas, and Caribbean Sea. Found over patch reefs and rocky bottoms from about 20 to about 500 ft. BIOLOGY: Feed on plankton at night.

Belted Cardinalfish - *Apogon townsendi* (Breder, 1927)

FEATURES: Reddish yellow to pinkish and somewhat translucent. Narrow dark bar from posterior base of second dorsal fin to anal-fin base. Narrow dark bar on caudal peduncle, and a broken bar at caudal-fin base. Area between bars on peduncle darkens at night to form a ringed blotch. HABITAT: S FL, eastern and western Gulf of Mexico, Bahamas, Antilles to northern South America. Also Bermuda. Around coral and rocky areas from about 10 to 180 ft. BIOLOGY: Nocturnal. Seek shelter among sea urchin spines.

Bronze Cardinalfish - *Astrapogon alutus* (Jordan & Gilbert, 1882)

FEATURES: Head and abdomen silvery; body brownish. Dark bars radiate from eyes. Small brown flecks form larger spots on head and on scales. Fins brownish to bronze and densely covered in darker flecks. Pelvic fins with bronze hue; may have blackish margins. Pelvic fins do not reach past anterior third of pelvic-fin base. HABITAT: NC to Venezuela and eastern Gulf of Mexico. Reported from Bahamas. Occur around reefs and seagrass beds in shallow water. Also occur in the mantle cavity of the West Indian Fighting Conch, *Strombus pugilis.*

Blackfin Cardinalfish - *Astrapogon puncticulatus* (Poey, 1867)

FEATURES: Head and abdomen silvery. Dark bars radiate from eyes. Small brown flecks form larger spots on head and on scales. Fins finely to densely peppered with brown flecks. Margins of second dorsal and anal fins lack pigment. Pelvic fins reach between anterior and middle third of anal-fin base. HABITAT: S FL, eastern Gulf of Mexico, Bahamas to Brazil. Also Bermuda. Found around reefs, over seagrass beds, and in empty shells to about 26 ft. BIOLOGY: Form pairs during spawning. Males mouth-brood eggs.

Conchfish - *Astrapogon stellatus* (Cope, 1867)

FEATURES: Head and abdomen silvery. Dark bars radiate from eyes. Small brown flecks form larger spots on head and on scales. Fins brownish and densely peppered with small brown flecks. Pelvic fins greatly expanded, reach to rear third of pelvic-fin base. HABITAT: S FL, eastern and southern Gulf of Mexico, Bahamas, Antilles to Venezuela. Also Bermuda. Reef associated from about 3 to 130 ft. BIOLOGY: Live commensally in the mantle cavity of the Queen Conch, *Strombus gigas*, and in the Stiff Penshell, *Atrina rigida*. Nocturnal.

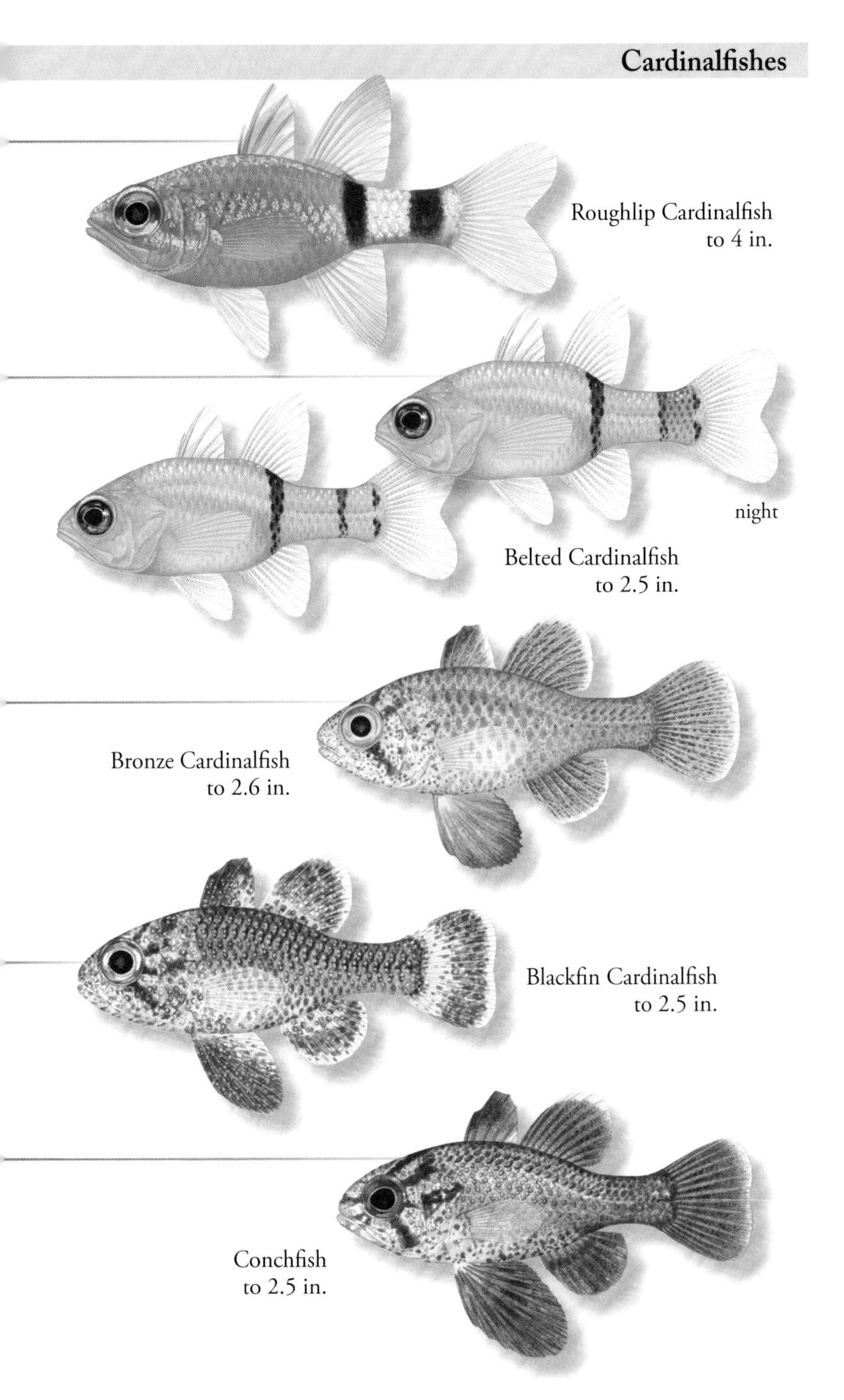
Roughlip Cardinalfish
to 4 in.
night
Belted Cardinalfish
to 2.5 in.
Bronze Cardinalfish
to 2.6 in.
Blackfin Cardinalfish
to 2.5 in.
Conchfish
to 2.5 in.

Bigtooth Cardinalfish - *Paroncheilus affinis* (Poey, 1875)

FEATURES: Translucent pale pink or salmon to bronze. Always with dusky to blackish stripe from snout tip through eyes. Caudal fin may have dusky margin. Otherwise unmarked. HABITAT: S FL and FL Keys, Gulf of Mexico, Bahamas to Suriname. Also Bermuda. Reported from the eastern central Atlantic. Reef associated around hollow coral heads, caves, and overhangs. Typically from about 65 to 295 ft.; reported to 984 ft. BIOLOGY: Spawn in pairs. Males mouth-brood eggs.

Freckled Cardinalfish - *Phaeoptyx conklini* (Silvester, 1915)

FEATURES: Translucent, usually with pinkish cast. Body densely covered with dark flecks that concentrate on scales. Flecks form a bar below eyes and a blotch near caudal-fin base. Second dorsal and anal fins always with dark band just above base. Second and third dorsal-fin spines about equal in length. Eyes very large. HABITAT: S FL, northern and southern Gulf of Mexico, Bahamas to Venezuela. Also Bermuda. Over shallow coral, rocky, and rubble bottoms. Often in empty shells or containers.

Dusky Cardinalfish - *Phaeoptyx pigmentaria* (Poey, 1860)

FEATURES: Translucent with brownish to reddish cast. Body with dark flecks and spots concentrated on scales. Flecks form a bar below eyes and a blotch near caudal-fin base. Second dorsal and anal fins unmarked. Third dorsal-fin spine usually taller than second spine. HABITAT: S FL, northern and southern Gulf of Mexico, Antilles to Brazil. Also Bermuda. On coral, seagrass, or shelly bottoms from near shore to about 140 ft. BIOLOGY: Court and pair during spawning. Males mouth-brood eggs.

Sponge Cardinalfish - *Phaeoptyx xenus* (Böhlke & Randall, 1968)

FEATURES: Translucent orangish to lavender brown. Head with yellow cast. Small brownish spots and flecks on head and body. Flecks form a bar below eyes and a blotch near caudal-fin base. Second dorsal and anal fins yellowish, with or without narrow, faint to dark band at base. Second dorsal-fin spine slightly taller than third spine. HABITAT: S FL, eastern and northern Gulf of Mexico, Bahamas to Venezuela. Associated with cylindrical sponges on rocky and coral bottoms. BIOLOGY: Share tube-shaped sponge cavities with Gobies and Brittle Stars. Feed at night.

Oddscale Cardinalfish - *Zapogon evermanni* (Jordan & Snyder, 1904)

FEATURES: Shades of red to reddish orange, often with three broad, vague bars on rear body. A dark brown to blackish stripe runs from snout to opercular margin; stripe may be broken. A black spot followed by a white spot present under rear anal-fin base. Snout long and bluntly pointed. HABITAT: Bahamas and Caribbean Sea. Also in the Central Indo-Pacific and eastern Atlantic. Demersal over calcareous walls and in deep recesses of caves on out reef slopes from about 10 to 820 ft. BIOLOGY: Hides inside caves during the day; emerges at night to feed. Often swims upside down on cave ceilings.

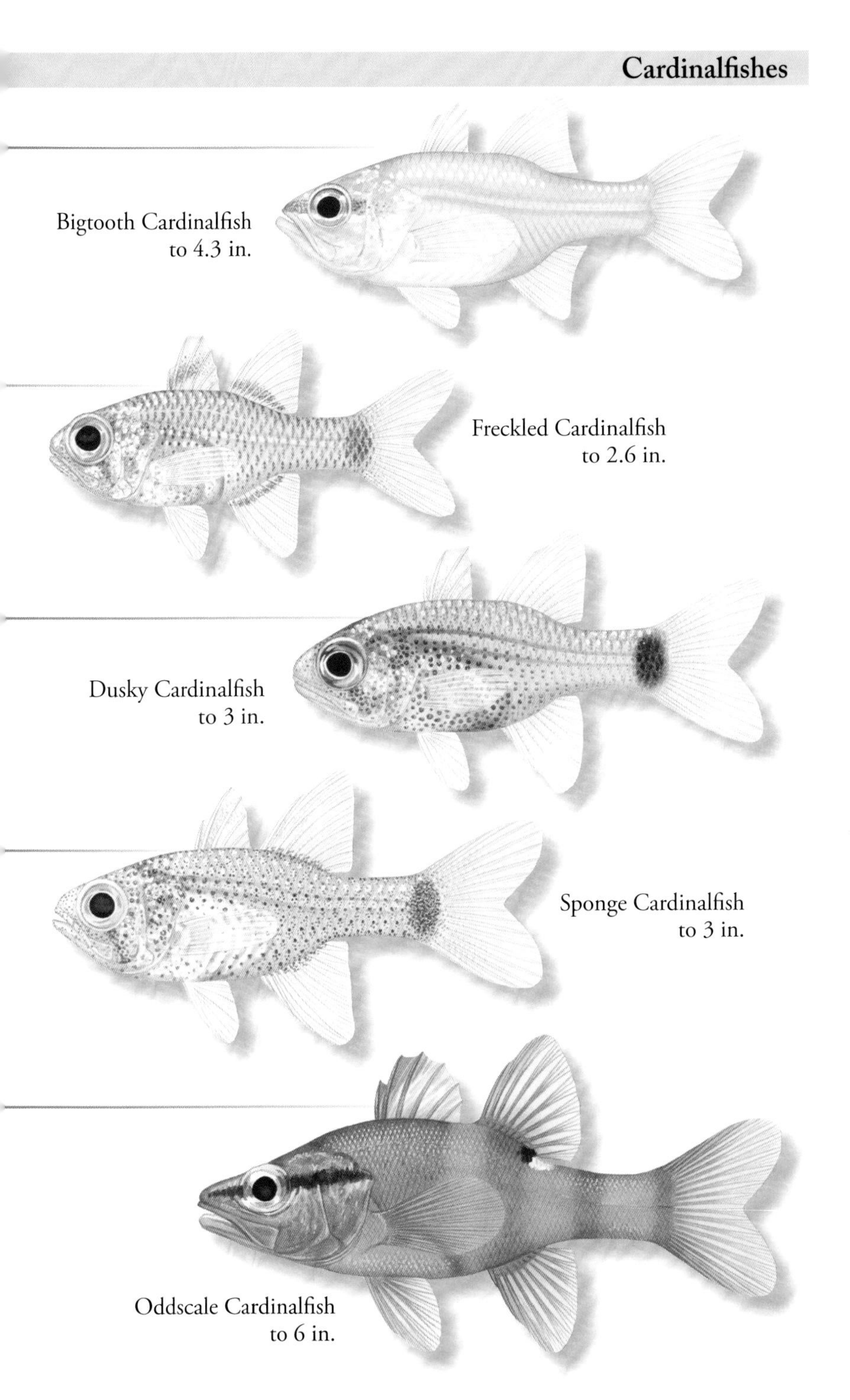
Bigtooth Cardinalfish
to 4.3 in.
Freckled Cardinalfish
to 2.6 in.
Dusky Cardinalfish
to 3 in.
Sponge Cardinalfish
to 3 in.
Oddscale Cardinalfish
to 6 in.

Goldface Tilefish - *Caulolatilus chrysops* (Valenciennes, 1833)

FEATURES: Pale bluish violet with yellow cast dorsally. Silvery to pearly white below. Bright yellow crescent below eye. Faint blue streak below yellow streak. Irises golden. Black spot just above pectoral-fin base. Membranes of soft dorsal and caudal fins with pale yellow spots. HABITAT: NC to FL, Gulf of Mexico, and Greater Antilles to Brazil. Bottom-dwelling, usually over rubble bottoms from 295 to about 630 ft. BIOLOGY: Feed on invertebrates and small fishes.

Blackline Tilefish - *Caulolatilus cyanops* Poey, 1866

FEATURES: Violet to bluish with yellowish sheen and reticulations dorsally. Silvery to white below. Predorsal ridge yellow. May have dark stripe below dorsal-fin base. Spiny portion of dorsal fin yellowish. Upper and lower caudal-fin lobes yellowish. Black spot above pectoral-fin base. HABITAT: NC to FL, Gulf of Mexico, Dominican Republic to Saba Bank, and Nicaragua and Colombia to Guyana. Demersal over sand, mud, and rubble from about 150 to 1,625 ft.

Yellow Barred Tilefish - *Caulolatilus williamsi* Dooley & Berry, 1977

FEATURES: Dusky dorsally; white below. Sides with 17–20 yellowish bars that become short posteriorly and overlay a yellowish stripe along upper midline that extends onto caudal fin. Lower caudal-fin lobe with a large yellow blotch. Silvery band from mouth to eyes. HABITAT: Bahamas and northeastern Caribbean Sea. Found over outer slopes of coral reef ledges, and over sandy and muddy bottoms from about 410 to 1,000 ft.

Tilefish - *Lopholatilus chamaeleonticeps* Goode & Bean, 1879

FEATURES: Bluish green to bluish gray dorsally with small, close-set, irregular yellow spots. Fading to milky white ventrally. Juveniles with fewer, larger yellow spots. Yellowish predorsal flap present. Caudal fin with irregular yellow spots and lines. HABITAT: Nova Scotia to FL, northern and southern Gulf of Mexico, and Venezuela to Suriname. Near bottom on soft bottoms, usually from 265 to 1,770 ft. BIOLOGY: Construct burrows in bottom substrate. Feed on a variety of invertebrates and fishes. Sought commercially. IUCN: Endangered.

Sand Tilefish - *Malacanthus plumieri* (Bloch, 1786)

FEATURES: Blue gray dorsally; pearly white below. Irregular pale blue lines on head. Dorsal and anal fins with yellow margins. Caudal fin with yellow upper and lower margins and a dusky patch on upper lobe. Body elongate. HABITAT: Ocracoke Island Inlet, NC, to FL, Gulf of Mexico, Bermuda, Bahamas, and Caribbean Sea to southern Brazil. Demersal over sand and rubble from about 33 to 500 ft. BIOLOGY: Usually in pairs. Construct and live in burrows.

ALSO IN THE AREA: *Caulolatilus bermudensis*, *Caulolatilus guppyi*, *Caulolatilus intermedius*, *Caulolatilus microps*, see p. 547.

Goldface Tilefish
to 21 in.

Blackline Tilefish
to 14 in.

Yellow Barred Tilefish
to 2 ft.

Tilefish
to 3.5 ft.

Sand Tilefish
to 2 ft.

Pomatomidae - Bluefish

Bluefish - *Pomatomus saltatrix* (Linnaeus, 1766)

FEATURES: Greenish to greenish blue dorsally; silvery below. Dark blotch at pectoral-fin base. Dorsal and caudal fins olivaceous. Body moderately elongate and compressed. Dorsal profile moderately convex. Lower jaw protrudes. Teeth sharp and numerous. HABITAT: Worldwide in eight major populations. In western Atlantic from Nova Scotia to FL, Gulf of Mexico to Yucatán and Cuba, Colombia to Argentina, and Bermuda. Occur coastally over continental shelves. BIOLOGY: Swift and voracious. Adults hunt in loose groups, juveniles in schools, often mangling prey. Migrate north in summer, south in winter. Sought commercially and for sport.

Coryphaenidae - Dolphinfishes

Pompano Dolphinfish - *Coryphaena equiselis* Linnaeus, 1758

FEATURES: Bright green blue dorsally; silvery below with golden highlights. Small dark spots scattered on sides. Dorsal fin tall, long-based. Caudal fin deeply forked. Body comparatively deep; dorsal and ventral profiles comparatively convex. Dorsal fin with 52–59 rays. Pectoral fins about half of head length. Anal fin convex, lacks deep anterior notch. HABITAT: Worldwide in tropical and warm temperate seas. Usually oceanic, may enter coastal waters. Associated with flotsam and *Sargassum* seaweed. BIOLOGY: School with and often misidentified as Dolphinfish.

Dolphinfish - *Coryphaena hippurus* Linnaeus, 1758

FEATURES: Bright blue green above; golden to silvery below. Small dark spots on sides. Pectoral fins yellowish. Juveniles with bars that extend into dorsal and anal fins. Dorsal fin tall, long-based. Caudal fin deeply forked. Body comparatively shallow; dorsal and anal profiles comparatively straight. Dorsal fin with 58–66 rays. Pectoral fins more than half of head length. Anal fin notched anteriorly. Males head profile angular; females head profile rounded. HABITAT: Worldwide in tropical to warm temperate seas. Usually offshore and oceanic, may enter coastal waters. Associated with flotsam and *Sargassum* seaweed. BIOLOGY: Form small schools. Highly migratory. Feed on fishes, crustaceans, and squids. OTHER NAME: Mahi-mahi.

Rachycentridae - Cobia

Cobia - *Rachycentron canadum* (Linnaeus, 1766)

FEATURES: Dark brown dorsally and on sides; whitish below. Two bright whitish stripes on sides fade with age. Dorsal, caudal, and pectoral fins dark brown. Pelvic and anal fins dusky. Snout broad; head compressed. First dorsal fin with seven to nine separate spines. Caudal fin tall, forked. HABITAT: Circumglobal in tropical to warm temperate seas, except eastern Pacific. MA to FL, Gulf of Mexico, Bermuda, Bahamas, and Caribbean Sea to Brazil. Pelagic around reefs, over rocky bottoms, and in estuaries from near surface to about 4,000 ft. BIOLOGY: Grow rapidly and may reach eight years of age. Feed on invertebrates and fishes. Sought commercially and for sport.

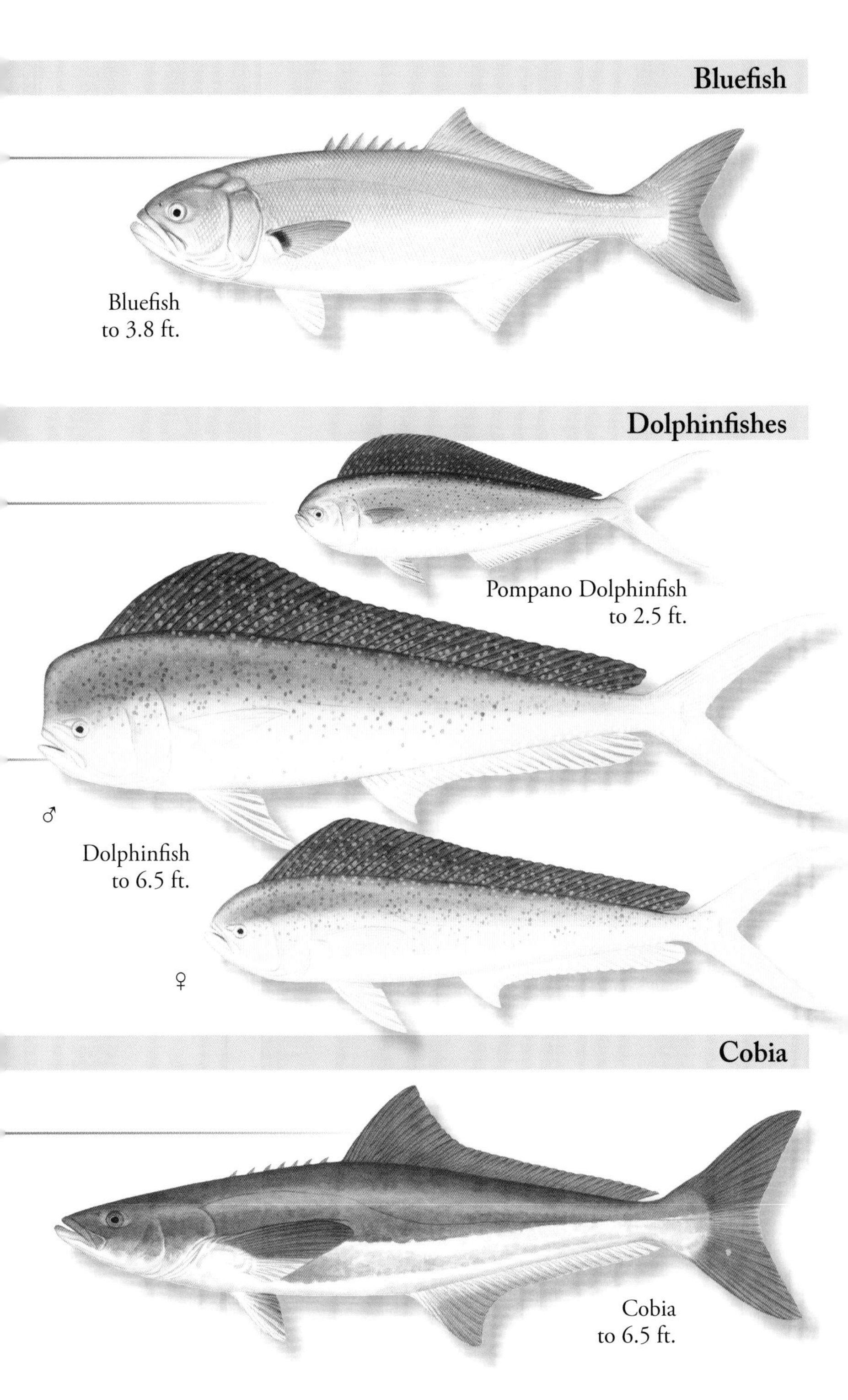
Bluefish
Bluefish
to 3.8 ft.
Dolphinfishes
Pompano Dolphinfish
to 2.5 ft.
♂
Dolphinfish
to 6.5 ft.
♀
Cobia
Cobia
to 6.5 ft.

Echeneidae - Remoras

Sharksucker - *Echeneis naucrates* Linnaeus, 1758

FEATURES: Dark bluish to brownish gray with a dark stripe running from mouth to caudal-fin base that is bordered above and below by whitish stripes. White stripes may be obscure in larger specimens. Dorsal, caudal, and anal fins dark with narrow, pale to white margins. Cephalic disk with 21 to 28 laminae (usually 23). Caudal fin trilobed or with a nearly concave margin. HABITAT: Circumglobal in tropical to warm temperate seas. From Nova Scotia to Uruguay, including Gulf of Mexico and Caribbean Sea. BIOLOGY: Attach to a wide variety of hosts, including sharks, rays, bony fishes, sea turtles, and marine mammals. Also free-swimming.

Whitefin Sharksucker - *Echeneis neucratoides* Zuiew, 1789

FEATURES: Brownish gray to blackish with a dark stripe running from tip of snout to caudal-fin base that is bordered above and below by whitish stripe. Stripes may be obscure in some specimens. Dorsal, caudal, and anal fins with broad whitish margins. Disk with 18 to 23 laminae (usually 21). Caudal-fin margin slightly trilobed to truncate. HABITAT: Western Atlantic only. MA to FL, Gulf of Mexico, Bermuda, Bahamas, and Caribbean Sea. BIOLOGY: Attach to a wide variety of hosts, including sharks, rays, bony fishes, sea turtles, and whales.

Slender Suckerfish - *Phtheirichthys lineatus* (Menzies, 1791)

FEATURES: Dark brownish gray to black with a darker stripe running from mouth to caudal-fin base that is bordered above and below by whitish stripes. All fins with pale to white margins. Disk comparatively short with 9 to 11 laminae (usually 10). Head comparatively short; body very long. Caudal-fin margin slightly trilobed to rounded. HABITAT: Circumglobal in tropical to subtropical seas. NY to Brazil, including Gulf of Mexico and Caribbean Sea. BIOLOGY: Usually associated with barracudas but also with other large bony fishes, sharks, and sea turtles.

White Suckerfish - *Remora albescens* (Temminck & Schlegel, 1850)

FEATURES: Usually whitish. May be pale tannish or grayish. Cephalic disk wide and short with 12 to 14 laminae and almost reaching pectoral-fin tips. Dorsal-, anal-, and pectoral-fin tips rounded. Body comparatively robust. HABITAT: Circumtropical in tropical to warm temperate seas. NC to Brazil, including Gulf of Mexico and Caribbean Sea; may occur farther north. BIOLOGY: Usually found attached to Giant Manta and Black Marlin, with a few accounts of sharks as hosts. NOTE: Previously known as *Remorina albescens.*

Whalesucker - *Remora australis* (Bennett, 1840)

FEATURES: Uniform dark slate blue, bluish white, dark blue, grayish white, or slate gray to brown or almost black. Margins of dorsal, anal, and caudal fins may be white. Cephalic disk very large, about half as long as body with 24 to 28 laminae (usually 26). HABITAT: Circumtropical in tropical to warm temperate seas. NC to Brazil, including Gulf of Mexico and Caribbean; may occur farther north. BIOLOGY: Attach to whales, dolphins, and porpoises.

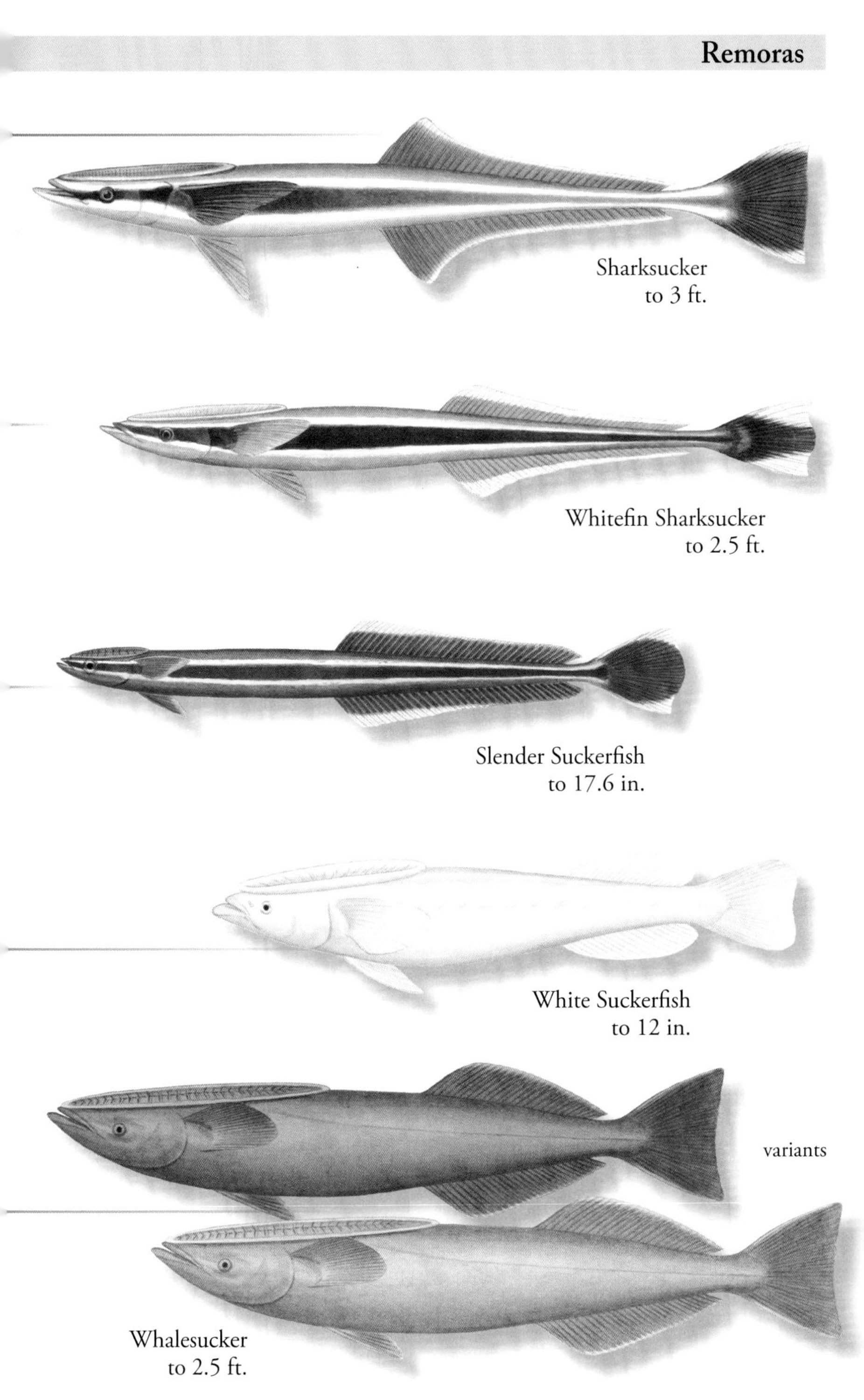
Sharksucker
to 3 ft.
Whitefin Sharksucker
to 2.5 ft.
Slender Suckerfish
to 17.6 in.
White Suckerfish
to 12 in.
variants
Whalesucker
to 2.5 ft.

Echeneidae - Remoras, *cont.*

Spearfish Remora - *Remora brachyptera* (Lowe, 1839)

FEATURES: Color corresponds to host color. May be brownish or bluish brown to whitish. Dark specimens with pale dorsal- and anal-fin margins. Disk comparatively short with 15 to 18 laminae; does not reach past pectoral-fin tips. Pectoral fins rounded. Caudal-fin margin slightly concave. Body comparatively robust. HABITAT: Circumglobal in tropical to warm temperate seas. Nova Scotia to Brazil, including Gulf of Mexico and Caribbean Sea. BIOLOGY: Attach to billfishes, also to sharks and molas.

Marlinsucker - *Remora osteochir* (Cuvier, 1829)

FEATURES: Brownish gray to grayish or blackish. Cephalic disk large with 15 to 19 laminae and reaching well past pectoral-fin tips. Fin margins rounded. Caudal peduncle comparatively narrow. Caudal-fin margin truncate to concave. HABITAT: Circumglobal in tropical to warm temperate seas. MA to Brazil, including Bermuda, Bahamas, Gulf of Mexico, and Caribbean Sea. BIOLOGY: Attach to billfishes, particularly White Marlin and Sailfish.

Remora - *Remora remora* (Linnaeus, 1758)

FEATURES: Overall brownish gray to sooty or blackish. Mottled or uniformly colored. Cephalic disk with 16 to 20 laminae and not reaching past pectoral-fin tips. Pectoral-fin tips blunt. Caudal-fin margin deeply concave. HABITAT: Circumglobal in tropical to warm temperate seas. Nova Scotia to Argentina, including Gulf of Mexico and Caribbean Sea. BIOLOGY: Usually attach to sharks, but also other large fishes, sea turtles, and ships. Sometimes free-swimming.

Carangidae - Jacks and Pompanos

African Pompano - *Alectis ciliaris* (Bloch, 1787)

FEATURES: Bluish dorsally; silvery below. May have obscure, dark blotch on upper opercular margin. Grayish blue chevrons on body fade with age. Head profile strongly arched over eyes. Body profile rhomboid. Dorsal and anal fins with pointed lobes in adults, filamentous in juveniles. HABITAT: Circumglobal in tropical to warm temperate seas. MA to FL, Gulf of Mexico, Bermuda, Bahamas, and Caribbean Sea to Brazil. Adults near bottom; juveniles pelagic. BIOLOGY: Solitary, strong swimmers. Feed mainly on fishes and squid.

Yellow Jack - *Caranx bartholomaei* Cuvier, 1833

FEATURES: Silvery to golden or bronze. May display spots, bands, or bars. Dorsal profile and most fins golden. Body otherwise unmarked. Upper and lower profiles almost identical in shape. Straight portion of lateral line with 25–28 scutes and begins posterior to anal-fin origin. HABITAT: MA to FL, Gulf of Mexico, Bahamas, Caribbean Sea to Brazil. Also Bermuda. Reef associated over continental shelves and slopes from surface to about 164 ft. Juveniles pelagic.

Remoras

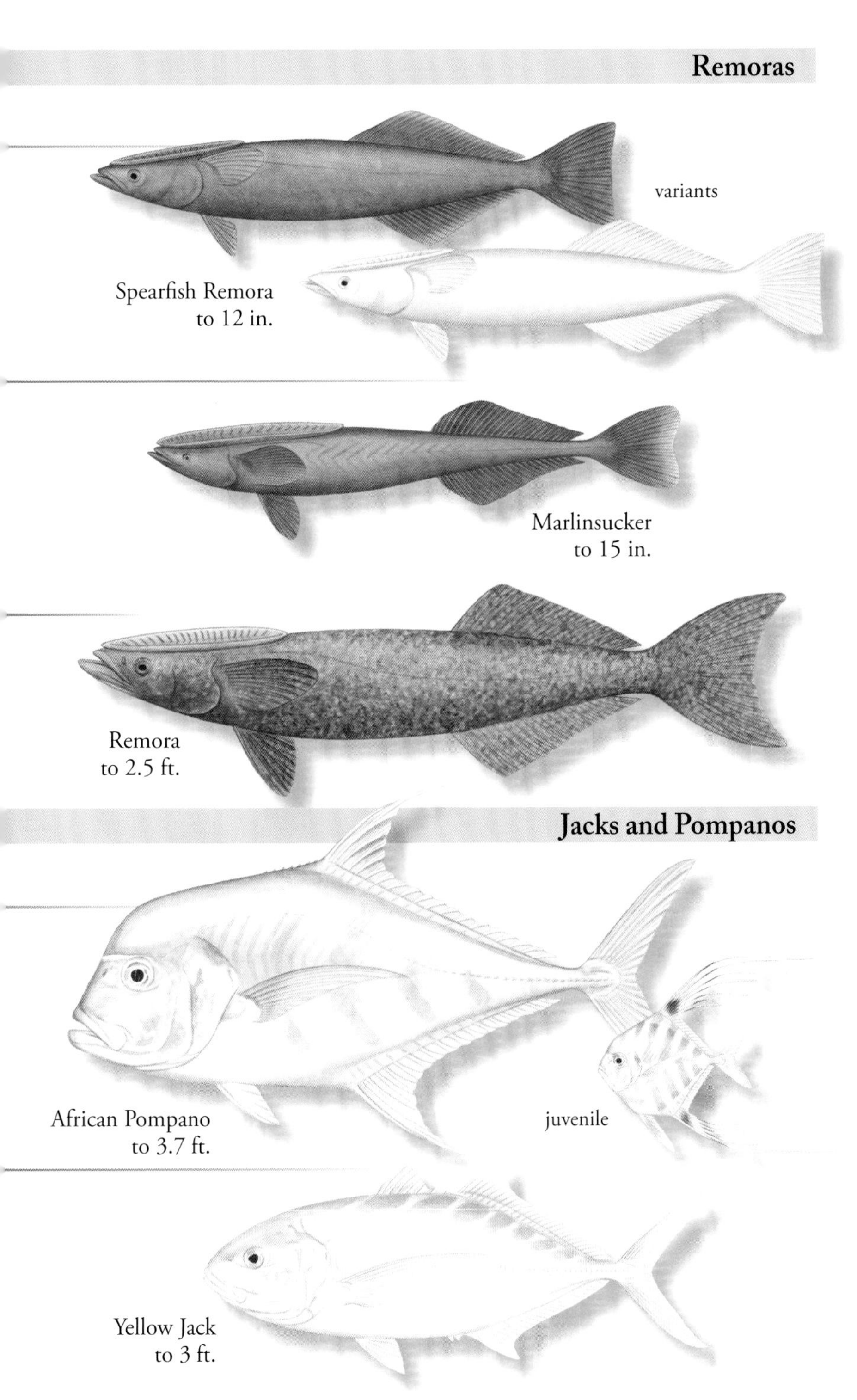
variants
Spearfish Remora
to 12 in.
Marlinsucker
to 15 in.
Remora
to 2.5 ft.
Jacks and Pompanos
African Pompano
to 3.7 ft.
juvenile
Yellow Jack
to 3 ft.

Blue Runner - *Caranx crysos* (Mitchill, 1815)

FEATURES: Shades of blue to green dorsally with metallic sheen. Silvery to golden below. May display distinct bars on sides. Breeding males blackish. Small black spot on opercular margin. Tips of upper and lower caudal-fin lobes black. Upper and lower body profiles almost identical in shape. Straight portion of lateral line begins anterior to anal-fin origin. HABITAT: Nova Scotia to FL, Gulf of Mexico, Bahamas, and Caribbean Sea to Brazil. Also Bermuda. Over outer continental shelves from surface to about 330 ft. Juveniles associated with *Sargassum* seaweed. BIOLOGY: Schooling. Feed on fishes and invertebrates.

Crevalle Jack - *Caranx hippos* (Linnaeus, 1766)

FEATURES: Greenish or bluish to bluish black above. Silvery to golden below. Small black spot on opercular margin. Black blotch on lower pectoral-fin rays. Jaws extend to below or just beyond rear margin of eyes. Head profile steeply convex. HABITAT: Nova Scotia to FL, Gulf of Mexico, Bahamas, and Greater Antilles to Uruguay. Also Bermuda and eastern Atlantic. In brackish to marine waters from surface to about 1,100 ft. Larger specimens may be solitary and in deeper water. May ascend rivers. Juveniles pelagic, around *Sargassum* seaweed. BIOLOGY: Form moderate to large schools. Prized gamefish.

Horse-eye Jack - *Caranx latus* Agassiz, 1831

FEATURES: Metallic bluish to bluish gray dorsally; silvery to golden below. Small black spot at upper opercular margin. Posterior scutes silvery to blackish. Upper dorsal-fin lobe may be blackish. Caudal fin yellowish; upper margin may be blackish. Eyes comparatively large. Jaws extend to under rear margin of eyes. HABITAT: NJ to FL, Gulf of Mexico, Bermuda, Bahamas, and Caribbean Sea to Brazil. Pelagic. Inshore, offshore, and along sandy beaches. Also enter brackish water and rivers. From surface to about 460 ft. BIOLOGY: Form small schools. Feed on fishes and invertebrates.

Black Jack - *Caranx lugubris* Poey, 1860

FEATURES: Olive gray to brownish gray or blackish. May be paler below or uniformly colored. May also be densely speckled. A small black spot present on upper opercular margin. Dorsal, anal, and pelvic fins and rear margin of caudal fin dark gray to black. Lateral line scutes blackish. Head profile angular. Jaws extend to under middle of eyes. Second dorsal- and anal-fin lobes comparatively tall and pointed. HABITAT: Circumglobal in tropical marine waters. In western Atlantic from Gulf of Mexico, Bahamas, Antilles to Brazil. Also Bermuda. Found offshore or off islands from about 78 to 1,000 ft. BIOLOGY: Pelagic in clear ocean water. Occasionally form schools. Feed at night.

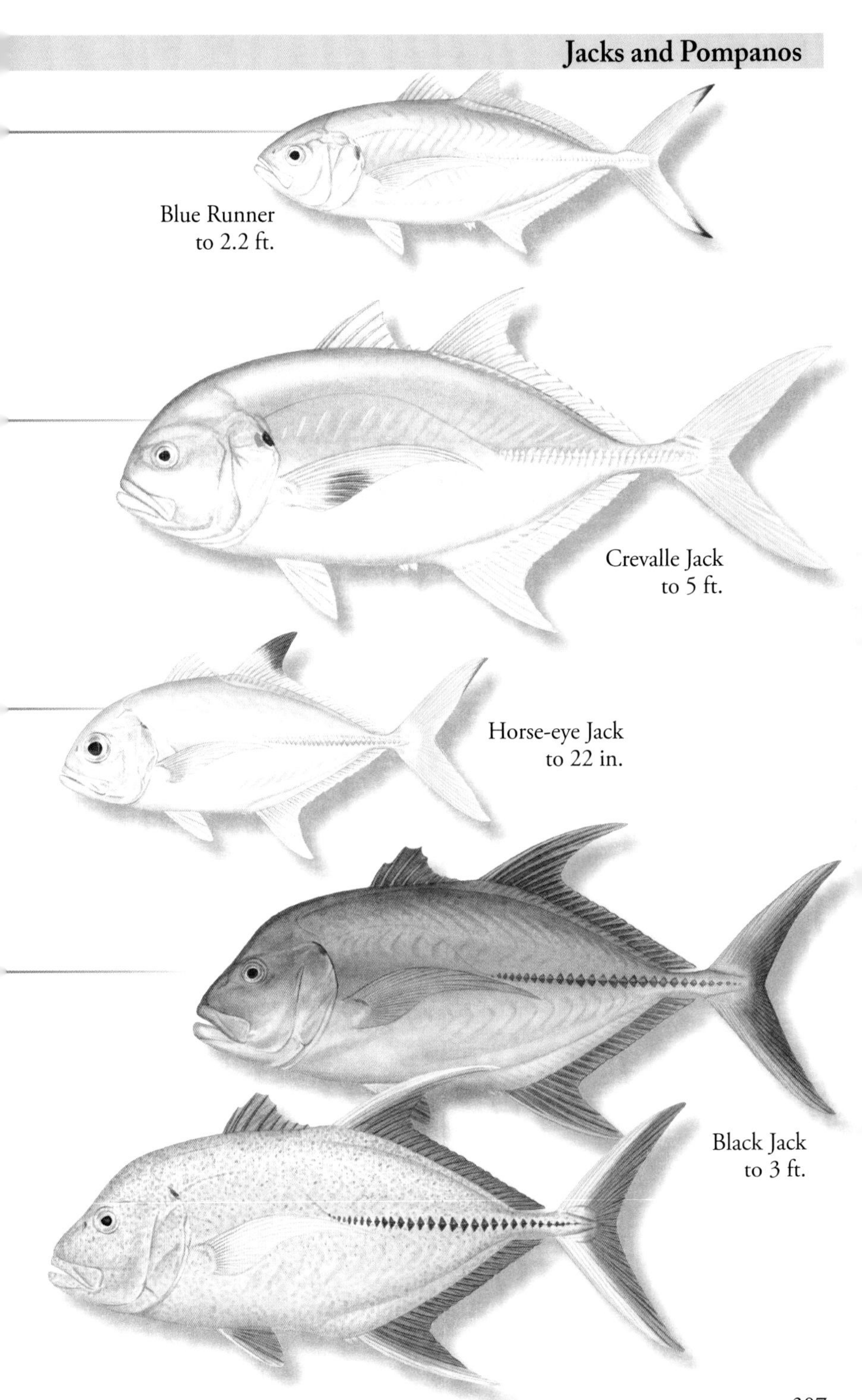
Blue Runner
to 2.2 ft.
Crevalle Jack
to 5 ft.
Horse-eye Jack
to 22 in.
Black Jack
to 3 ft.

Bar Jack - *Caranx ruber* (Bloch, 1793)

FEATURES: Pale grayish blue above; silvery below. Blackish band overlying iridescent bluish band runs along dorsal profile, through caudal peduncle, and onto lower caudal-fin lobe. Body and fins may turn entirely blackish, particularly while hunting. HABITAT: NJ to FL, Gulf of Mexico, Bermuda, Bahamas, and Caribbean Sea to Venezuela. In clear, shallow waters around coastal reefs and seagrass bottoms. Juveniles associate with *Sargassum* seaweed rafts. BIOLOGY: Form small to large schools; occasionally solitary. Make grunting sounds when distressed. Flesh may be toxic.

Atlantic Bumper - *Chloroscombrus chrysurus* (Linnaeus, 1766)

FEATURES: Metallic blue dorsally; silvery below. Black blotch on upper caudal peduncle. Caudal fin yellowish, others with yellowish tint. Body compressed and with ventral profile more convex than dorsal profile. HABITAT: MA to FL, Gulf of Mexico, Bermuda, Bahamas, and Caribbean Sea to Uruguay. Also eastern Atlantic. Occur in shallow coastal marine and estuarine waters. Also around mangrove-lined lagoons. Juveniles may occur well offshore and are associated with jellyfish. BIOLOGY: Schooling. Feed on fishes, cephalopods, zooplankton, and detritus. Grunt when in distress.

Mackerel Scad - *Decapterus macarellus* (Cuvier, 1833)

FEATURES: Bluish black to metallic blue dorsally; may have greenish cast posteriorly. Silvery to white ventrally. Black spot on upper opercular margin. Silvery to bluish stripe on midline. Caudal fin yellowish green to yellowish. Single, separate finlet behind dorsal and anal fins. HABITAT: Circumtropical in clear, open tropical and warm temperate seas. Gulf of Maine to Brazil, including Gulf of Mexico and Caribbean Sea. BIOLOGY: Schooling. Feed on planktonic invertebrates.

Round Scad - *Decapterus punctatus* (Cuvier, 1829)

FEATURES: Bluish to greenish dorsally; silvery to whitish below. Small black spot on opercular margin. Yellowish to olivaceous stripe on body midline. Up to 14 small, black spots on curved portion of lateral line. Single, separate finlet behind dorsal and anal fins. HABITAT: MA to FL, Gulf of Mexico, Bermuda, Bahamas, and Caribbean Sea to Brazil. Also Bermuda and eastern Atlantic. Occur primarily from midwater to near bottom to about 295 ft. Juveniles pelagic and near surface. BIOLOGY: Schooling. Feed on planktonic invertebrates.

Redtail Scad - *Decapterus tabl* Berry, 1968

FEATURES: Bluish black to metallic blue dorsally; white ventrally. Small black spot on upper opercular margin. Caudal fin red. Tips of anterior second dorsal-fin rays reddish. Single, separate finlet behind dorsal and anal fins. HABITAT: Circumglobal in tropical to warm temperate seas. NC to LA, Bermuda, Bahamas, NE Yucatán, NW Caribbean Sea to Uruguay. Also Indian and western and central Pacific oceans. Found in midwater to near bottom from about 490 to 650 ft. BIOLOGY: Form schools and feed on planktonic invertebrates. Used as bait.

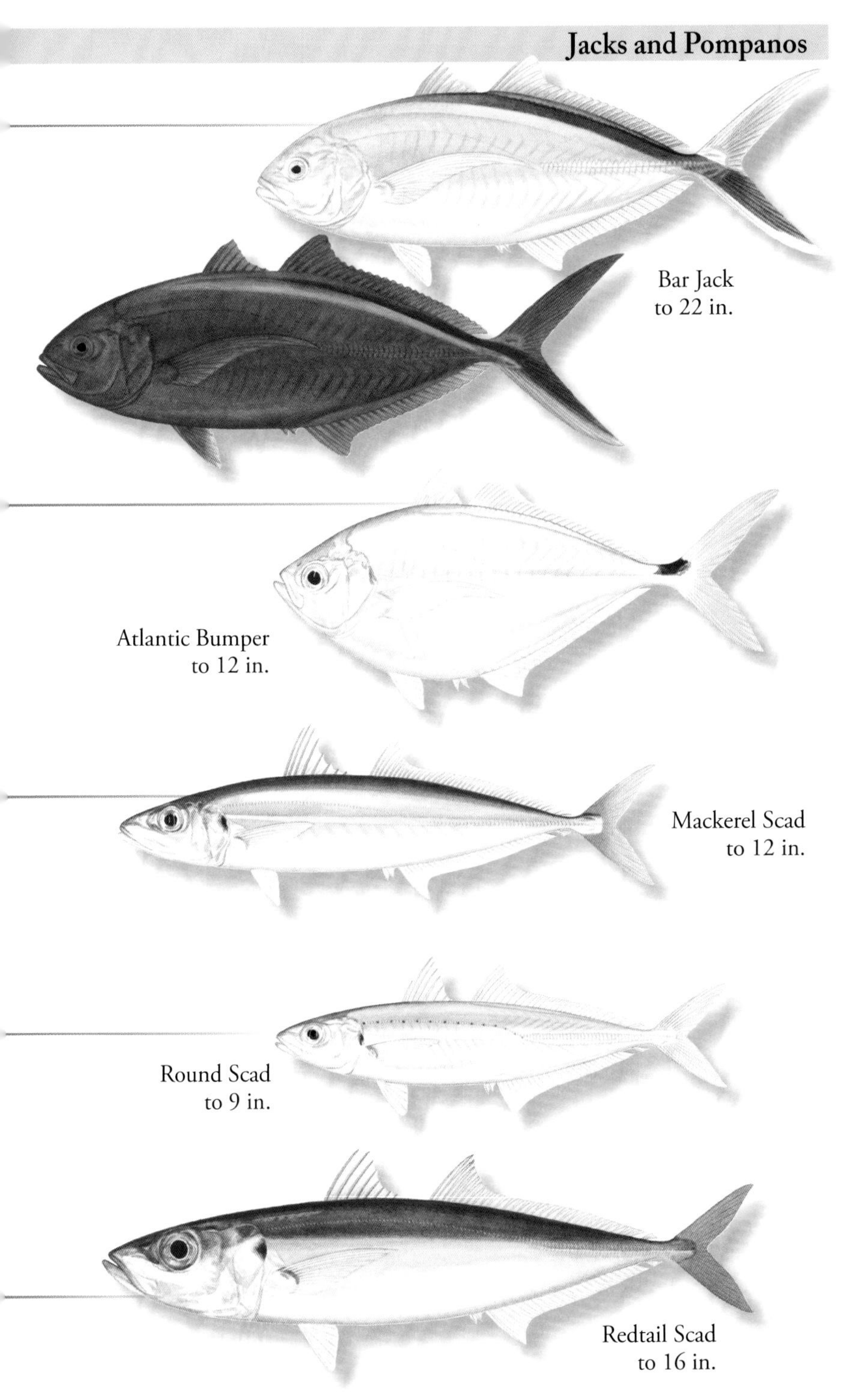
Bar Jack
to 22 in.
Atlantic Bumper
to 12 in.
Mackerel Scad
to 12 in.
Round Scad
to 9 in.
Redtail Scad
to 16 in.

Rainbow Runner - *Elagatis bipinnulata* (Quoy & Gaimard, 1825)

FEATURES: Dark blue or olive blue to dark green dorsally, fading to greenish or yellowish on sides. White ventrally. Two narrow bluish stripes run from snout to caudal peduncle. Juveniles pale green dorsally and on sides and white below; two blue stripes run from snout to caudal peduncle. Two separate rays form finlets behind dorsal and anal fins. Dorsal and ventral profiles similarly shaped. Mouth small. HABITAT: Circumglobal in tropical to warm temperate seas. MA to Brazil, including Gulf of Mexico and Caribbean Sea. Pelagic. Found near surface over reefs and offshore. BIOLOGY: May form large schools. Feed on fishes, invertebrates.

Golden Trevally - *Gnathodon speciosus* (Forsskål, 1775)

FEATURES: Adults silvery with a golden sheen and a faint grayish bar through eyes and alternating moderate and narrow grayish bars on sides. Juveniles golden yellow with a blackish band through eyes and alternating moderate and narrow blackish bars on sides. Snout slightly humped. Pectoral fins longer than head. HABITAT: Eastern Pacific, Indo-Pacific, and Indian oceans and the Red Sea. Recorded in Caribbean Sea from the locks of the Panama Canal and off southern Cuba.

Bluntnose Jack - *Hemicaranx amblyrhynchus* (Cuvier, 1833)

FEATURES: Bluish green dorsally; silvery below. Large, blackish blotch on opercle. Dorsal-fin margins blackish. Caudal fin yellowish with blackish tip on upper lobe. Body deep and strongly compressed. Mouth small; snout blunt. HABITAT: NC to FL; Gulf of Mexico to northern South America. Also Cuba. Reported rare along eastern coast of United States. Coastal; near bottom of marine and brackish waters to about 160 ft. BIOLOGY: Usually solitary. Juveniles associated with jellyfish.

Pilotfish - *Naucrates ductor* (Linnaeus, 1758)

FEATURES: Bluish silver with bluish black bars on body. Three bars extend into second dorsal fin. Caudal-fin lobes with white tips. Band from eyes to dorsal-fin origin absent. Eyes large. First dorsal fin with small, separate spines. Caudal peduncle with lateral keels. HABITAT: Circumglobal in tropical to warm temperate seas. Nova Scotia to Argentina, including Gulf of Mexico and Caribbean Sea. Pelagic. BIOLOGY: Adults follow large sharks, rays, and sea turtles. Juveniles associate with seaweed and jellyfish. Feed on ectoparasites, host's prey scraps, and fishes.

Maracaibo Leatherjack - *Oligoplites palometa* (Cuvier & Valenciennes, 1832)

FEATURES: Dorsal profile with a narrow band of blue green. Sides and abdomen silvery, may have washes of golden yellow. Second dorsal-fin lobe blackish. Caudal fin golden to amber. Head profile nearly straight. Ventral profile of lower jaw weakly convex. First dorsal fin as four separate spines. Posterior dorsal- and anal-fin rays as semi-detached finlets. HABITAT: Guatemala to southern Brazil. Pelagic in fresh and brackish coastal waters. Also in muddy marine waters from shore to about 150 ft. BIOLOGY: Feed on fishes and crustaceans.

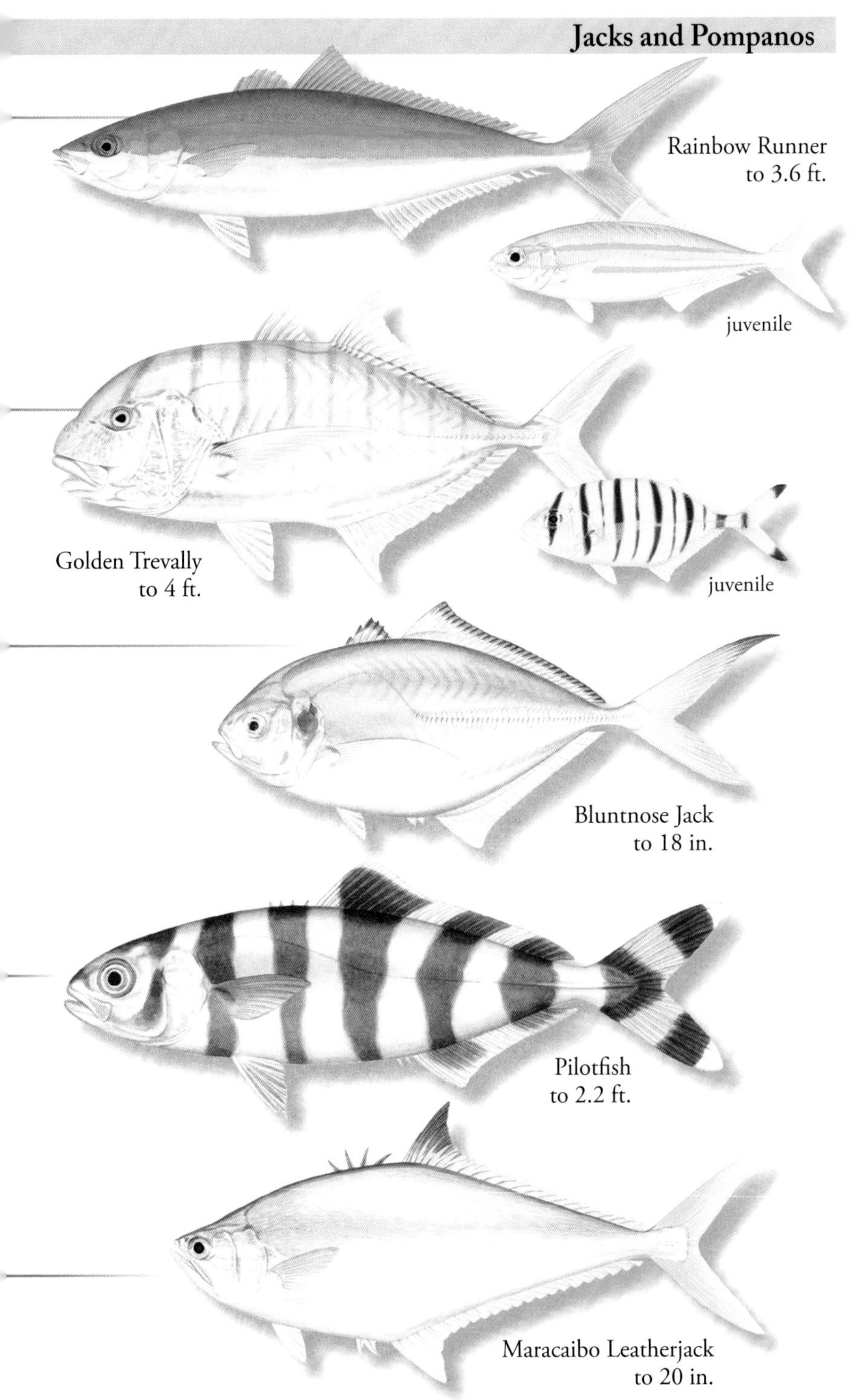
Rainbow Runner
to 3.6 ft.
juvenile
Golden Trevally
to 4 ft.
juvenile
Bluntnose Jack
to 18 in.
Pilotfish
to 2.2 ft.
Maracaibo Leatherjack
to 20 in.

Castin Leatherjack - *Oligoplites saliens* (Bloch, 1793)

FEATURES: Dorsal profile with a narrow band of blue green. Sides and abdomen silvery with some areas of golden. Lower anterior portion of second dorsal fin dark. Caudal fin golden to amber. Head profile nearly straight. Ventral profile of lower jaw strongly convex. First dorsal fin as four separate spines. Posterior dorsal- and anal-fin rays as semi-detached finlets. HABITAT: Honduras to southern Brazil. Pelagic in brackish to marine waters from shore to about 60 ft. BIOLOGY: Feed on scales of other fishes.

Leatherjack - *Oligoplites saurus* (Bloch & Schneider, 1801)

FEATURES: Metallic bluish to greenish dorsally; silvery to white below. Caudal fin yellowish to colorless. Head profile nearly straight. Ventral profile of lower jaw weakly convex. First dorsal fin as four to six separate spines. Posterior dorsal- and anal-fin rays as semi-detached finlets. HABITAT: MA to FL, Gulf of Mexico, and Caribbean Sea to Brazil. Also in eastern Pacific. Inshore along sandy beaches, bays, and inlets. More often in turbid water. BIOLOGY: Form large schools. Juveniles feed on ectoparasites of other fishes. Dorsal- and anal-fin spines venomous.

White Trevally - *Pseudocaranx dentex* (Bloch & Schneider, 1801)

FEATURES: Pale greenish blue dorsally, fading to silvery below. Yellowish stripe on sides along midline and at bases of dorsal and anal fins. Black spot at opercular margin. Fins yellowish. Dorsal and ventral body profiles similar. Jaws protrusible; upper jaw not reaching anterior margin of eyes. HABITAT: In subtropical waters of Atlantic, Pacific, and Indian oceans from about 260 to 650 ft. In western Atlantic from NC to FL, Bermuda, and Brazil. BIOLOGY: Schooling. Filter and suck invertebrates from the sea bed. Sometimes placed in the genus *Caranx*.

Bigeye Scad - *Selar crumenophthalmus* (Bloch, 1793)

FEATURES: Metallic bluish to bluish green dorsally. Silvery below. May have yellowish stripe from opercular margin to upper caudal peduncle. Small black blotch on notched opercular margin. Eyes very large. Lateral line with large pored scales anteriorly, scutes posteriorly. HABITAT: Worldwide in tropical to warm temperate seas. Nova Scotia to FL, Gulf of Mexico, Bahamas, and Caribbean Sea to Brazil. In shallow coastal waters to about 550 ft. BIOLOGY: Schooling. Planktivorous.

Caribbean Moonfish - *Selene brownii* (Cuvier, 1816)

FEATURES: Silvery white with metallic bluish sheen. Faint dark spot on upper opercular margin. Dorsal, caudal, and pectoral fins may have dusky to yellowish cast; other fins colorless. Small juveniles with dark blotch over straight portion of lateral line. Body very deep and compressed. Anterior profile of head steeply sloping, concave in front of eyes, humped at top. Lower jaw protrudes. HABITAT: Caribbean Sea to Brazil. Usually demersal over continental shelves from near shore to about 180 ft. Young occur offshore near surface; juveniles may occur in bays and river mouths.

Castin Leatherjack
to 20 in.

Leatherjack
to 12 in.

White Trevally
to 18 in.

Bigeye Scad
to 12 in.

Caribbean Moonfish
to 11 in.

Atlantic Moonfish - *Selene setapinnis* (Mitchill, 1815)

FEATURES: Silvery white with metallic bluish sheen. Faint dark blotch on opercle and upper caudal peduncle. Caudal fin may have dusky to yellowish cast; other fins colorless. Juveniles with dark blotch on straight portion of lateral line. Body very deep and compressed. Anterior head profile steeply sloping, rounded at top. Lower jaw protrudes. HABITAT: Nova Scotia to FL; Gulf of Mexico to Argentina. Absent from Bahamas and Antilles. Adults near bottom from inshore to about 180 ft. Juveniles pelagic, near surface. BIOLOGY: Feed on fishes and crustaceans.

Lookdown - *Selene vomer* (Linnaeus, 1758)

FEATURES: Silvery white with metallic bluish to yellowish sheen. May have silvery bars on body. Body very deep and compressed. Anterior profile of head very steep, angular at top. First dorsal-fin spines elongated in juveniles, reduced in adults. Second dorsal- and anal-fin lobes elongate. Pelvic fins elongate in juveniles, very small in adults. HABITAT: ME to FL, Gulf of Mexico, Bermuda, Bahamas, and Caribbean Sea to Uruguay. Occur near bottom of shallow, coastal waters to about 170 ft. Juveniles in estuaries and off beaches. BIOLOGY: Feed in schools.

Greater Amberjack - *Seriola dumerili* (Risso, 1810)

FEATURES: Shades of brown to olivaceous dorsally, often with pinkish luster on sides. Silvery below. Usually with dark band from eyes to first dorsal-fin origin. Faint amber stripe on sides from eyes to caudal fin. Upper jaw broad and rounded posteriorly; reaches to about middle of pupils. Body elongate, comparatively shallow. HABITAT: Circumglobal in tropical to warm temperate seas. Nova Scotia to Brazil, including Bermuda, Bahamas, Gulf of Mexico, and Caribbean Sea. Near bottom over continental shelves and slopes and around reefs and rocky ledges from about 60 to 235 ft.

Lesser Amberjack - *Seriola fasciata* (Bloch, 1793)

FEATURES: Dark dusky pinkish to dusky violet dorsally, fading to silvery below. Faint band from eyes to nape. Often with faint amber stripe on sides from eyes to caudal fin. Upper jaw moderately broad posteriorly, reaching to anterior margin of pupil. Eyes comparatively large. HABITAT: MA to FL, northern Gulf of Mexico, Greater Antilles, Bermuda, and Venezuela. Also eastern Atlantic and Mediterranean Sea. On or near bottom from about 180 to 490 ft.

Almaco Jack - *Seriola rivoliana* Valenciennes, 1833

FEATURES: Brown or olivaceous to bluish green dorsally; paler below. Dark band runs from eyes to first dorsal-fin origin. Faint amber stripe on sides from eyes to caudal fin. Upper jaw angular, very broad posteriorly; reaches below anterior margin of pupil. Second dorsal- and anal-fin lobes comparatively tall and pointed. Body comparatively deep. HABITAT: Circumglobal in tropical to warm temperate seas. MA to Argentina, including Bermuda, Bahamas, Gulf of Mexico, and Caribbean Sea. Pelagic in the water column from near surface to about 1,100 ft.

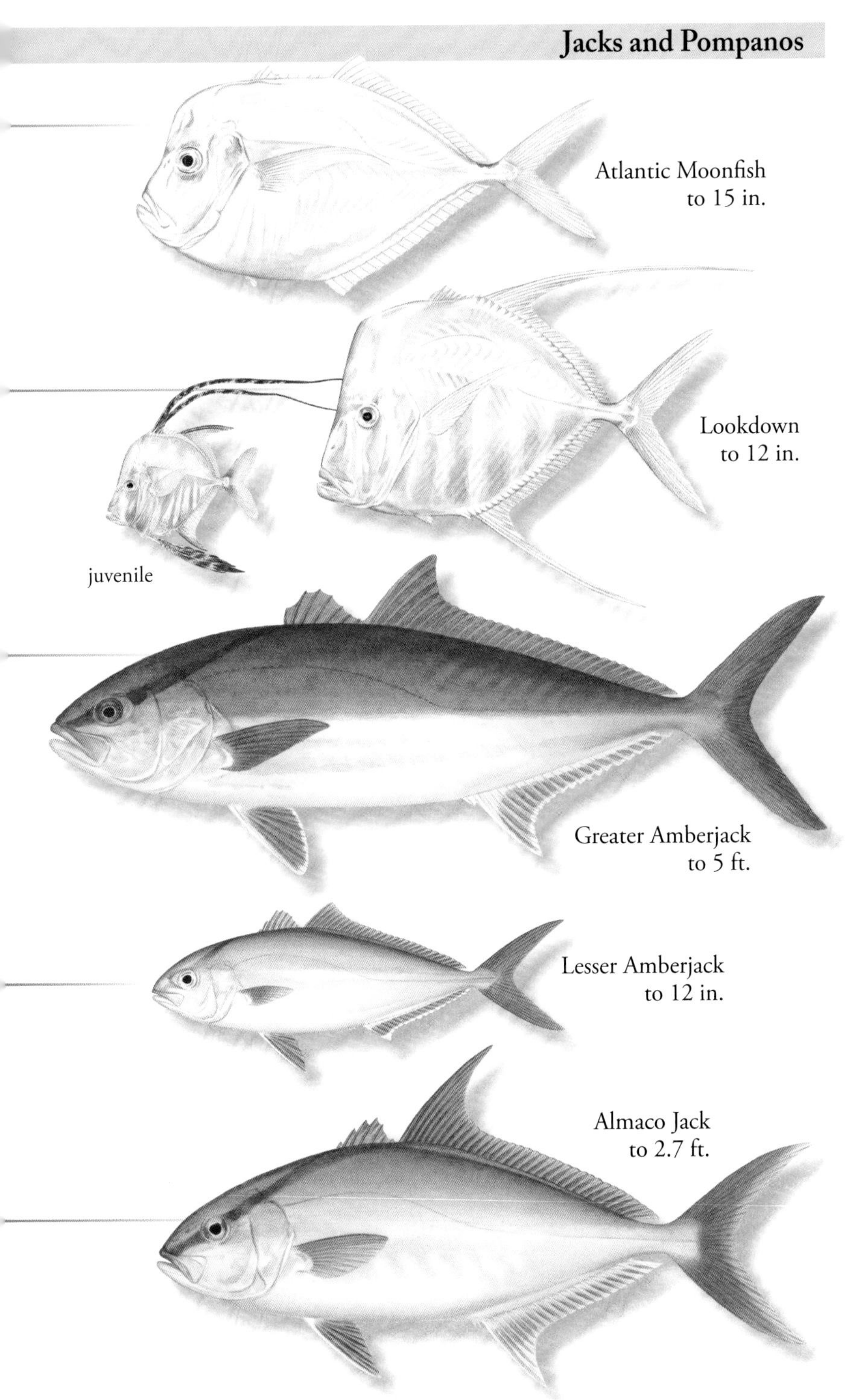
Atlantic Moonfish
to 15 in.
Lookdown
to 12 in.
juvenile
Greater Amberjack
to 5 ft.
Lesser Amberjack
to 12 in.
Almaco Jack
to 2.7 ft.

Banded Rudderfish - *Seriola zonata* (Mitchill, 1815)

FEATURES: Dark grayish to greenish blue dorsally; silvery white below. May have a dark band from eyes to first dorsal fin and an amber stripe on sides from eyes to caudal peduncle. Fins dusky. Second dorsal and caudal fins with pale margins. Juveniles with six bars on sides and dark band from eyes to first dorsal-fin origin. Upper jaw reaches to about rear margin of eyes in adults. Body elongate, comparatively shallow. Snout pointed. HABITAT: MA to FL, Gulf of Mexico, Bahamas, and Colombia to Brazil. Pelagic or near bottom over continental shelves. Juveniles with drifting plants, jellyfish, and larger fishes.

Florida Pompano - *Trachinotus carolinus* (Linnaeus, 1766)

FEATURES: Silvery with metallic bluish to greenish sheen. Abdomen silvery to golden. Anal- and caudal-fin lobes yellowish. First dorsal-fin spines small, separate. Second dorsal fin with 22–27 rays. Head profile somewhat rounded; snout blunt. Body deep, compressed. HABITAT: MA to FL, Gulf of Mexico, and Caribbean Sea to Brazil. Absent from Bahamas. Along sandy beaches and in brackish bays and inlets from shore to about 130 ft. Adults are pelagic; juveniles occur in beach surf zone. BIOLOGY: Juveniles form large schools, chase prey up onto beach faces. Sought commercially and for sport.

Cayenne Pompano - *Trachinotus cayennensis* Cuvier, 1832

FEATURES: Grayish to bluish dorsally; silvery to golden below. Body unmarked. First dorsal-fin ray blackish. Caudal fin yellowish with blackish upper and lower margins. Anal-fin lobe yellow. Snout very short and blunt. Mouth comparatively small; head comparatively short. First dorsal-fin spines small, separate. Body deep and compressed. HABITAT: Venezuela to northern Brazil. Occur along sandy and muddy coastal bottoms to about 200 ft. Also in estuaries. Young in shallow water.

Permit - *Trachinotus falcatus* (Linnaeus, 1758)

FEATURES: Silvery with metallic bluish to greenish sheen. Abdomen silvery, often with golden areas. Usually with a large gray to black smudge on sides. Anal- and caudal-fin lobes blackish. Head profile and snout rounded. First dorsal-fin spines small, separate. Body robust. HABITAT: MA to FL, Gulf of Mexico, Bermuda, Bahamas, and Caribbean Sea to Brazil. In coastal waters to about 118 ft. Adults pelagic or near bottom in channels and over seagrass flats, reefs, or mud bottoms. Juveniles occur in beach surf.

Palometa - *Trachinotus goodei* Jordan & Evermann, 1896

FEATURES: Silvery with metallic blue sheen dorsally. Abdomen may have golden tint. Usually with narrow bars on upper sides. Second dorsal- and anal-fin lobes elongate, blackish. Upper and lower margins of caudal fin blackish. Snout short and blunt. Head comparatively small. First dorsal-fin spines small, separate. Body deep, compressed. HABITAT: MA to FL, Gulf of Mexico, Bermuda, Bahamas, and Caribbean Sea to Brazil. Coastal. In beach surf zone and around reefs and rocky areas to about 120 ft.

Banded Rudderfish
to 2 ft.

juvenile

Florida Pompano
to 2 ft.

Cayenne Pompano
to 22 in.

Permit
to 2.6 ft.

Palometa
to 20 in.

Rough Scad - *Trachurus lathami* Nichols, 1920

FEATURES: Pale to dark blue or bluish green dorsally. Silvery to whitish below. Black spot on opercular margin. First dorsal-fin spines and anterior portion of second dorsal fin dusky. Caudal fin colorless to dusky at margin. Well-developed scutes along entire lateral line; anterior scutes may be somewhat overgrown by body scales. HABITAT: Gulf of Maine to FL, Gulf of Mexico, southern Caribbean Sea to southern Brazil. Usually occur coastally and near bottom to about 295 ft.

Cottonmouth Jack - *Uraspis secunda* (Poey, 1860)

FEATURES: Grayish or brownish gray to bluish black or dusky. Juveniles and specimens up to about 12 in. with six or seven bars on body. Bars fade with age. Vertical fins dark. Tongue and floor and roof of mouth white to creamy. Body oval in shape, laterally compressed. HABITAT: Circumglobal in warm seas. NJ to FL, Gulf of Mexico, Bermuda, Bahamas, and Caribbean Sea to Brazil. Oceanic; in water column from surface to bottoms of about 120 ft. BIOLOGY: Solitary or in small schools. May grunt when in distress.

Bramidae - Pomfrets

Atlantic Pomfret - *Brama brama* (Bonnaterre, 1788)

FEATURES: Grayish silver. Margins of dorsal, caudal, and anal fins blackish. Pectoral-fin margins transparent. Lower jaw protrudes. Area between eyes prominently arched. Vertical fins scaled, rigid. Dorsal-fin lobe moderately tall. Caudal-fin lobes similarly shaped. Anal fin usually with 31 rays. HABITAT: Circumglobal in tropical to warm temperate seas. Scattered from Nova Scotia to Argentina. Oceanic from surface to about 3,300 ft. BIOLOGY: Form small schools. Migrations follow water temperatures. Feed on small fishes and invertebrates.

Caribbean Pomfret - *Brama caribbea* Mead, 1972

FEATURES: Dark brown to coppery. Membranes of dorsal and anal fins black. Inside of mouth black. Lower jaw protrudes. Vertical fins scaled, rigid. Upper lobe of caudal fin considerably longer than lower lobe. Anal fin low. Body comparatively deep. HABITAT: NC to Brazil, including Bermuda, Bahamas, and Caribbean Sea. Pelagic. From surface to about 1,300 ft.

Lowfin Pomfret - *Brama dussumieri* Cuvier, 1831

FEATURES: Grayish silvery to dark gray silver. Dorsal fin dark. Pectoral and pelvic fins translucent. Vertical fins scaled, relatively rigid. Dorsal-fin lobe comparatively low. Upper caudal-fin lobe longer than lower lobe in adults; considerably longer in juveniles. Area between eyes arched. Anal fin usually with 26–28 rays; 27–32 rays reported in Atlantic specimens. HABITAT: Circumglobal in tropical seas. NC to Brazil, including Bermuda, Bahamas, Gulf of Mexico, and Caribbean Sea. Pelagic. From surface to about 650 ft.

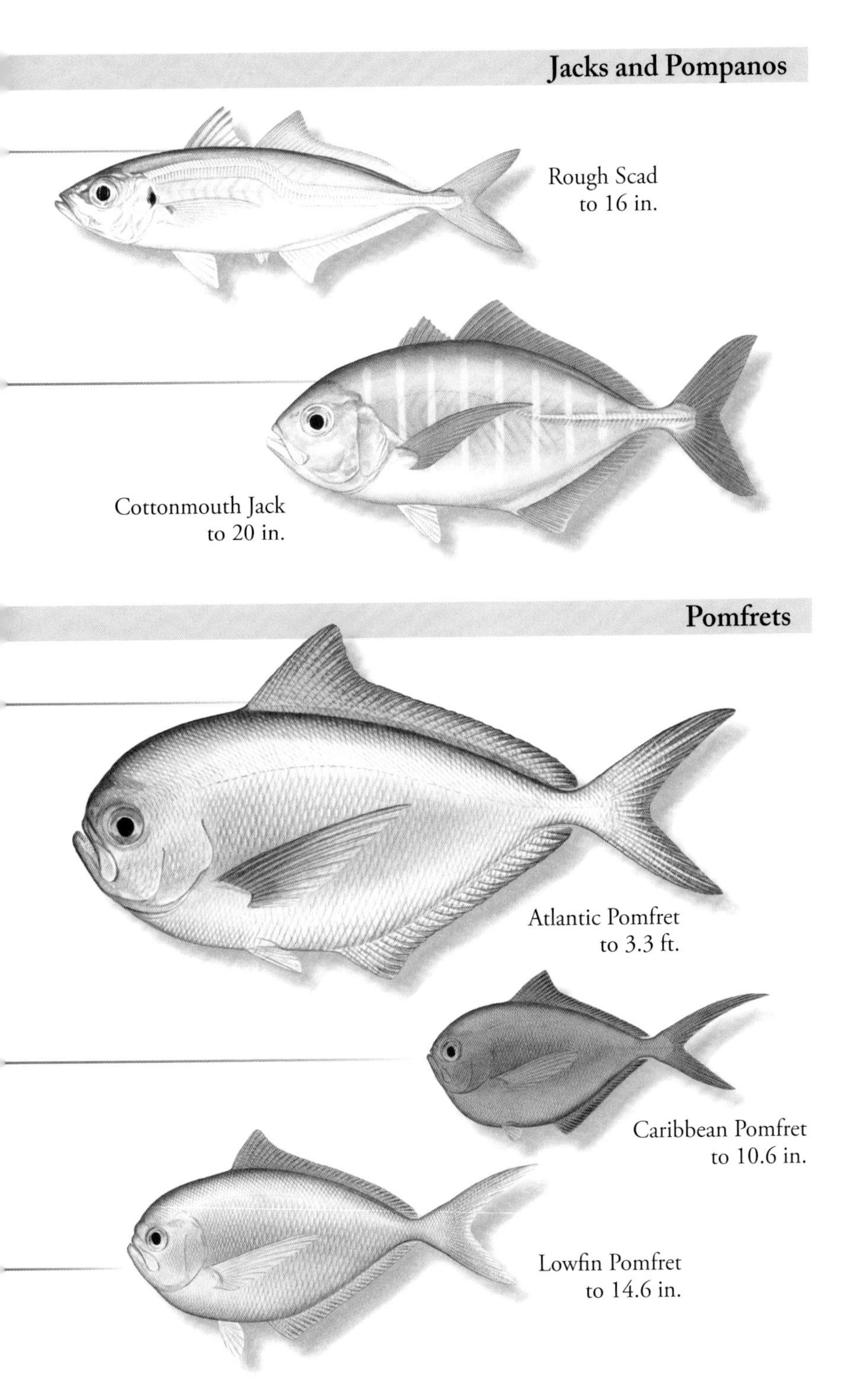
Jacks and Pompanos
Rough Scad
to 16 in.
Cottonmouth Jack
to 20 in.
Pomfrets
Atlantic Pomfret
to 3.3 ft.
Caribbean Pomfret
to 10.6 in.
Lowfin Pomfret
to 14.6 in.

Bramidae - Pomfrets, *cont.*

Atlantic Fanfish - *Pterycombus brama* Fries, 1837

FEATURES: Dark gray with silvery sheen. Dorsal and anal fins dark. Eyes large. Lower jaw protrudes. Dorsal fin originates over posterior margin of eyes in adults and over pectoral fins in juveniles. Dorsal and anal fins are tall, sail-like, and scaleless; they depress into grooves along dorsal and ventral profiles. Dorsal and anal fin lengths vary with age and conditions. Scales on body may or may not be spined. Lateral line absent in adults. HABITAT: Newfoundland to FL, Gulf of Mexico, Bermuda, Bahamas, and Caribbean Sea. Also eastern Atlantic and Mediterranean Sea. Offshore, pelagic. Occur from about 80 to 1,000 ft.

Bigscale Pomfret - *Taractichthys longipinnis* (Lowe, 1843)

FEATURES: Body and fins blackish with silvery to coppery sheen. Caudal-fin margin, pelvic-fin tips, and lower pectoral-fin margins pale. Lower jaw protrudes. Dorsal- and anal-fin lobes scaled, stiff, long, and pointed. Caudal fin crescent-shaped. Body deep, somewhat compressed. Posterior scales keeled, forming horizontal rows. Lateral line indistinct. HABITAT: Nova Scotia to FL, Gulf of Mexico, Bermuda, Bahamas, and Caribbean Sea to Brazil. Also eastern Atlantic. Oceanic; pelagic. BIOLOGY: Bigscale Pomfret are highly migratory.

ALSO IN THE AREA: *Pteraclis carolinus*, *Taractes rubescens*, see p. 547.

Emmelichthyidae - Rovers

Red Rover - *Emmelichthys ruber* (Trunov, 1976)

FEATURES: Crimson red dorsally; silvery below. Others may appear uniformly red. Pectoral and caudal fins red; other fins pinkish. Eyes large with orange irises. Upper jaw protrusible. Lower preopercular corner bluntly rounded. Dorsal fins separated by a distinct gap. Second dorsal and anal fins with a scaly sheath at base. Body comparatively round in cross-section. HABITAT: Eastern and western Atlantic. In western Atlantic from Bermuda, Bahamas, Jamaica, and eastern Gulf of Mexico. Pelagic. Occur between about 330 and 650 ft. Juveniles found near surface.

Crimson Rover - *Erythrocles monodi* Poll & Cadenat, 1954

FEATURES: Dark crimson red dorsally to about midline; pinkish to silvery below. Pectoral and caudal fins red; other fins pinkish. Eyes large. Upper jaw highly protrusible. Mouth toothless or with a few minute teeth. Lower preopercular corner bluntly rounded. Dorsal fins not separated by a gap. Second dorsal and anal fins with a scaly sheath at base. Body comparatively deep. HABITAT: SC, northern Gulf of Mexico, Bahamas, St. Lucia, Colombia, and Venezuela. Also eastern Atlantic. Found near sand and mud bottoms from about 330 to 1,000 ft. BIOLOGY: Schooling. Little is known of their life history. Taken as bycatch. OTHER NAME: Atlantic Rubyfish.

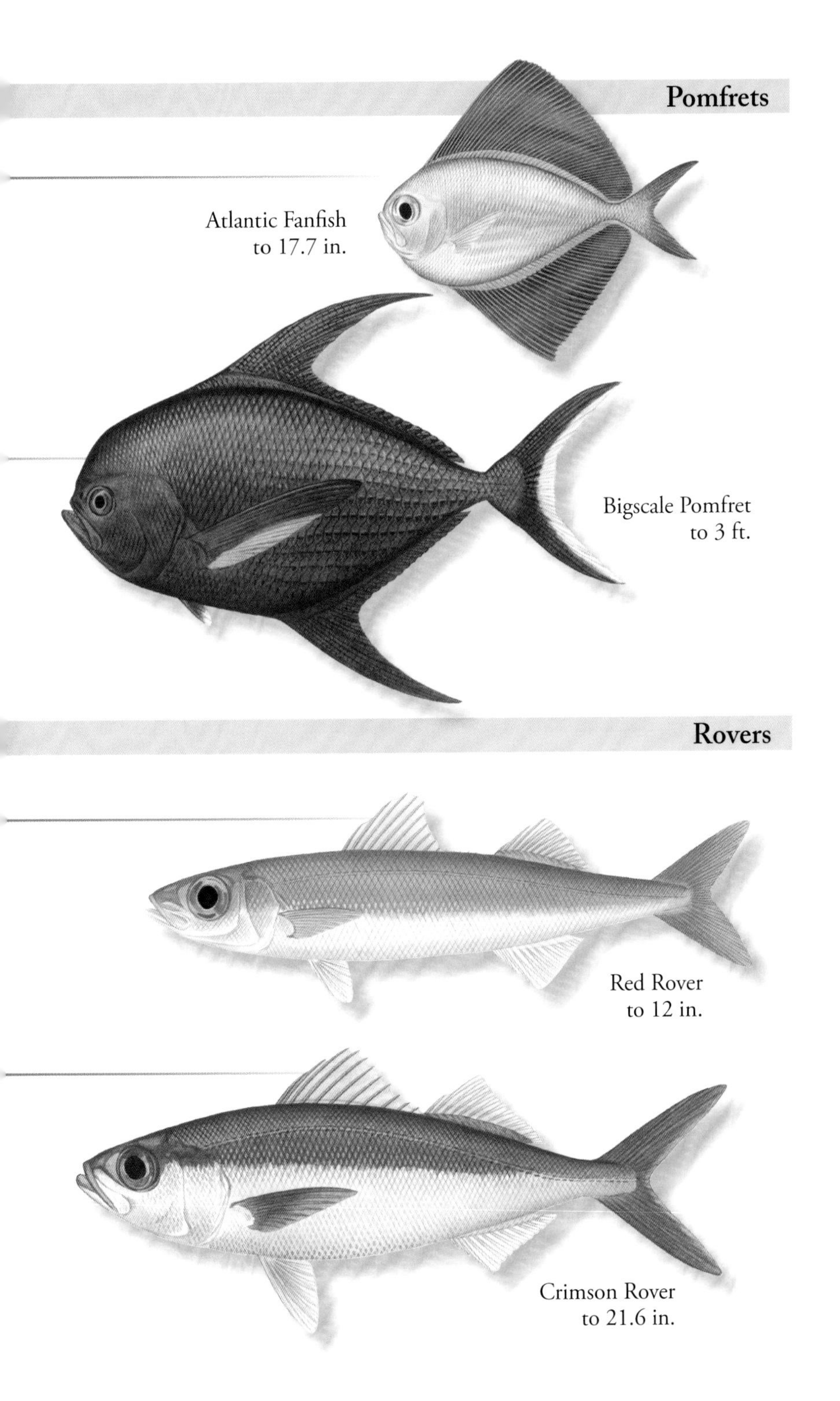
Pomfrets
Atlantic Fanfish
to 17.7 in.
Bigscale Pomfret
to 3 ft.
Rovers
Red Rover
to 12 in.
Crimson Rover
to 21.6 in.

Black Snapper - *Apsilus dentatus* Guichenot, 1853

FEATURES: Body and fins brownish violet to brownish black. Outer margin of caudal fin pale. Irises bronze with dark outer ring. Small juveniles deep blue. Last rays of dorsal and anal fins shorter than next-to-last ray. Caudal fin forked to almost straight. HABITAT: FL Keys, Gulf of Mexico. Yucatán, Bahamas, and Caribbean Sea. Occur over rocky bottoms and along steep dropoffs from about 40 to 800 ft. BIOLOGY: Feed on fishes, cephalopods, and tunicates. Juveniles believed to mimic Blue Chromis, *Chromis cyanea*.

Queen Snapper - *Etelis oculatus* (Valenciennes, 1828)

FEATURES: Dark red to pinkish red above midline. Pinkish to silvery below midline. Spiny dorsal and caudal fins reddish; other fins pinkish. Dorsal fin deeply notched. Last rays of soft dorsal and anal fins long. Lengths of caudal-fin lobes about equal in smaller specimens; upper lobe considerably longer than lower in larger specimens. HABITAT: NC to FL, Gulf of Mexico, Bermuda, Bahamas, and Caribbean Sea to Brazil. Over rocky bottoms between about 440 and 1,500 ft. BIOLOGY: Feed on squids, crustaceans, and small fishes.

Mutton Snapper - *Lutjanus analis* (Cuvier, 1828)

FEATURES: Olive with reddish tinge dorsally. Sides reddish; ventral area whitish. Sides may be uniformly colored or with pale bars. Blue lines below and behind eyes. Small, black spot present on upper sides. Caudal fin with thin, black margin. Anal fin angular. HABITAT: MA to FL, Gulf of Mexico, Bermuda, Bahamas, Caribbean Sea to Brazil. Found over sandy bottoms, in bays and estuaries, around mangroves, and over coral reefs from near shore to about 500 ft. BIOLOGY: Solitary. Form large spawning aggregations. Feed on invertebrates and fishes. IUCN: Near Threatened.

Schoolmaster - *Lutjanus apodus* (Walbaum, 1792)

FEATURES: Reddish brown to olive gray dorsally. Sides and ventral area with reddish tinge. Eight pale bars on sides that fade with age. Fins yellow, or yellow green to pale orange. Usually with solid or broken blue line under eyes. Large canine teeth in upper jaw. Anal fin rounded. HABITAT: MA (rare) to FL, Gulf of Mexico, Bermuda, Bahamas, and Caribbean Sea to Brazil. Found in coastal waters over a variety of bottoms from shore to about 500 ft. Juveniles enter brackish waters. BIOLOGY: Form aggregations during the day. Feed nocturnally on invertebrates and fishes.

Blackfin Snapper - *Lutjanus buccanella* (Cuvier, 1828)

FEATURES: Deep red dorsally. Pale reddish to silvery ventrally. Irises yellow to orange. Fins orangish to yellowish. Always with blackish blotch at pectoral-fin base. Base of soft dorsal fin dark. Smaller specimens with yellowish upper caudal peduncle and caudal fin. Anal fin rounded. HABITAT: NC to FL, Gulf of Mexico, Bermuda, Bahamas, and Caribbean Sea to southern Brazil. Adults over sand and rocky bottoms, dropoffs, and ledges from about 260 to 800 ft. Juveniles from about 100 to 170 ft.

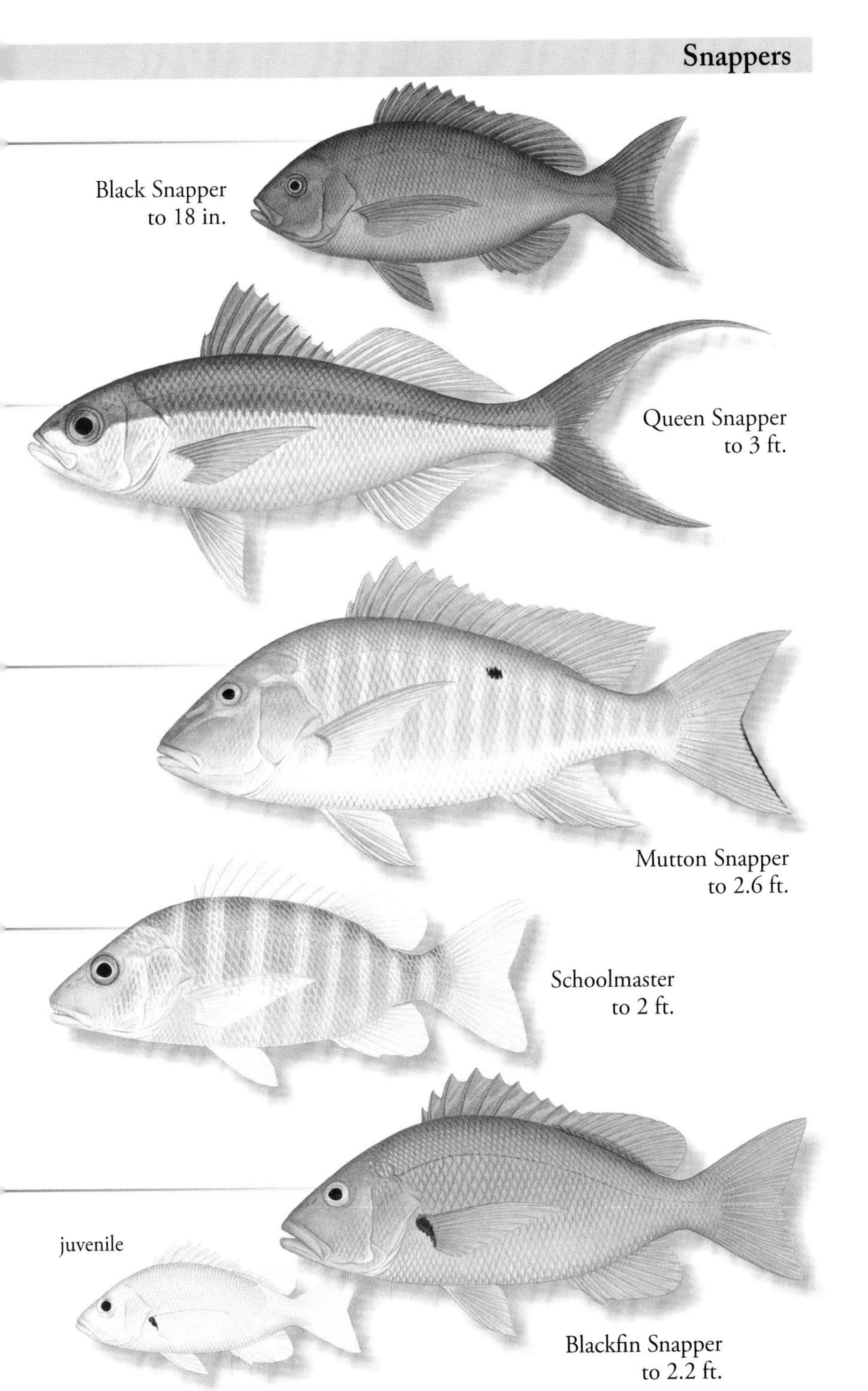
Black Snapper
to 18 in.
Queen Snapper
to 3 ft.
Mutton Snapper
to 2.6 ft.
Schoolmaster
to 2 ft.
juvenile
Blackfin Snapper
to 2.2 ft.

Red Snapper - *Lutjanus campechanus* (Poey, 1860)

FEATURES: Scarlet red to brick red dorsally and on sides. Pinkish below. Irises red. Caudal fin usually with narrow black margin. Small specimens have black blotch on posterior upper sides that fades with age. Anal fin angular with elongate inner rays. Roof of mouth with an anchor-shaped tooth patch. Lateral line with 47–49 scales. HABITAT: MA (rare) to FL, Bahamas (possibly), and Gulf of Mexico. Adults over rocky bottoms and reefs from about 30 to 650 ft. Juveniles in shallow waters over sandy and muddy bottoms. BIOLOGY: Feed on invertebrates and fishes. Heavily fished in the United States. IUCN: Vulnerable.

Cubera Snapper - *Lutjanus cyanopterus* (Cuvier, 1828)

FEATURES: Dark to pale reddish gray with silvery reflections dorsally; paler below. Pale bars on upper body fade with age. Pectoral fins grayish to translucent. Body comparatively shallow. Teeth in upper and lower jaws equally developed. Anal fin rounded. HABITAT: Nova Scotia to FL, Gulf of Mexico, Bermuda, Bahamas, and Caribbean Sea to Brazil. Adults over rocky bottoms, ledges, and reefs to about 400 ft. Juveniles around mangroves. BIOLOGY: Largest snapper in the area. Form spawning aggregations. Feed on crustaceans and fishes. IUCN: Vulnerable.

Gray Snapper - *Lutjanus griseus* (Linnaeus, 1758)

FEATURES: Dark olive or grayish green to grayish dorsally. Sides and ventral area paler with a reddish cast. Centers of scales on sides reddish. Juveniles with a blue line below eyes, a dark bar through eyes, and pale bars on sides that fade with age. HABITAT: MA to FL, Gulf of Mexico, Bermuda, Bahamas, and Caribbean Sea to Venezuela. Around mangroves, rocky areas, coral reefs, estuaries, tidal creeks, and river mouths. From shore to about 590 ft. Young may enter fresh water. BIOLOGY: Feed primarily at night on a variety of invertebrates and fishes. Spawning takes place during full-moon phases. OTHER NAME: Mangrove Snapper.

Dog Snapper - *Lutjanus jocu* (Bloch & Schneider, 1801)

FEATURES: Olive brown with bronze cast dorsally. Reddish with coppery cast below. Upper sides may have pale bars. Usually with a whitish bar below eyes to corner of jaws. May have a series of blue spots or a blue line below eyes to opercular margin. Anal fin rounded. HABITAT: MA (rare) to FL, Gulf of Mexico, Bahamas, and Caribbean Sea to southern Brazil. Adults over coral reefs from shore to about 600 ft. Juveniles in coastal waters, estuaries, and occasionally fresh water. BIOLOGY: Solitary and territorial.

Mahogany Snapper - *Lutjanus mahogoni* (Cuvier, 1828)

FEATURES: Olive or grayish dorsally. Sides silvery with reddish cast. Abdomen silvery. Dorsal and caudal fins reddish to yellowish with reddish to dusky margins. Dark blotch may be present on lateral line below soft dorsal fin. HABITAT: NC to FL, Gulf of Mexico, Bahamas, Caribbean Sea to Venezuela. In clear, shallow waters over rocky, sandy, and grassy bottoms. Also near coral reefs. Occur from shore to about 700 ft. BIOLOGY: Schooling. Feed at night on cephalopods, crustaceans, and fishes.

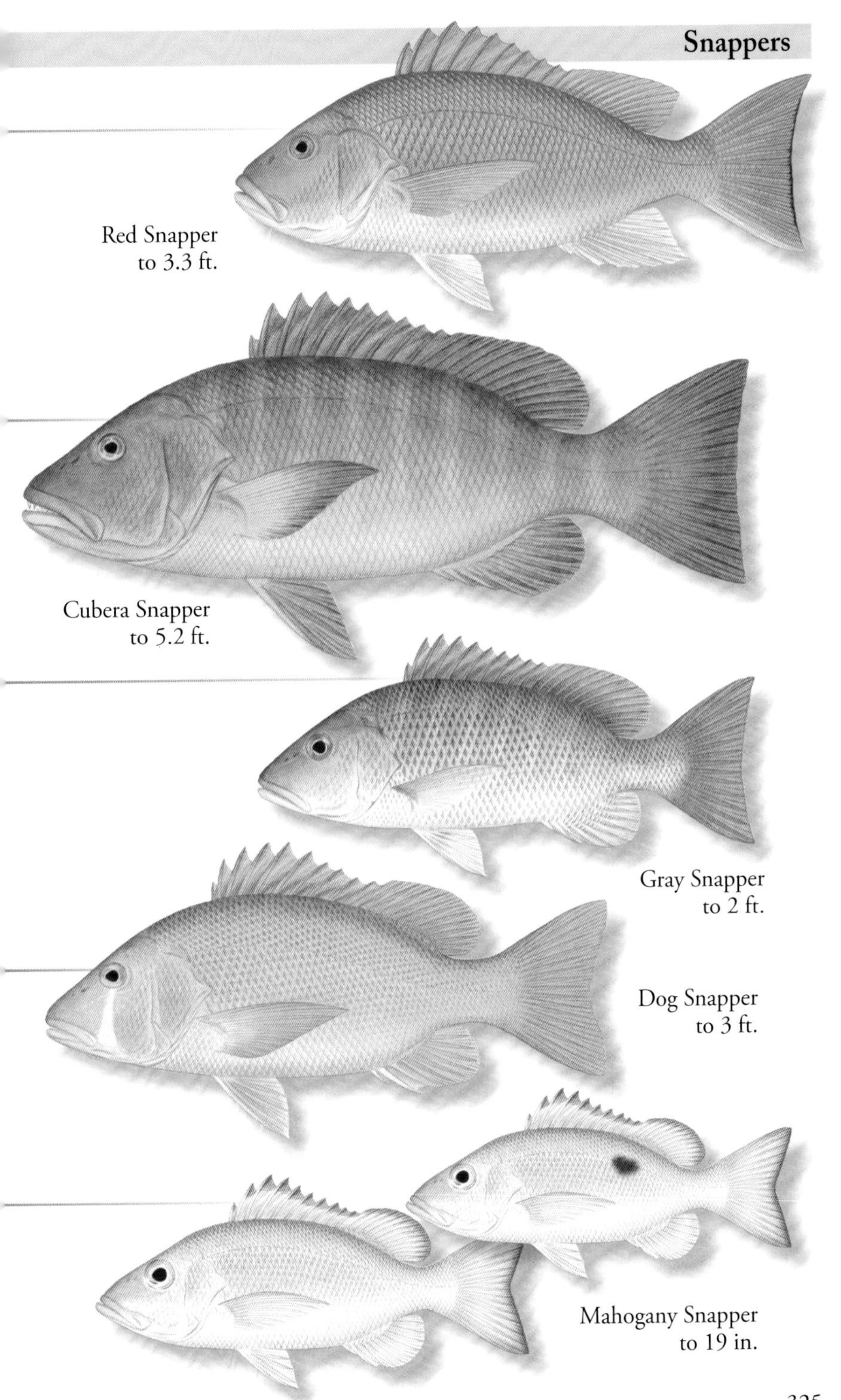
Red Snapper
to 3.3 ft.
Cubera Snapper
to 5.2 ft.
Gray Snapper
to 2 ft.
Dog Snapper
to 3 ft.
Mahogany Snapper
to 19 in.

Caribbean Red Snapper - *Lutjanus purpureus* (Poey, 1866)

FEATURES: Scarlet to brick red dorsally, fading to silvery pink below. Shallow water specimens pink. Irises red. Edges of fins may be narrowly black. Small specimens have black blotch on posterior upper sides that fades with age. Anal fin angular with elongate inner rays. Roof of mouth with an anchor-shaped tooth patch. Lateral line with 50–51 scales. HABITAT: Caribbean Sea to eastern Brazil. Found over rocky bottoms from about 100 to 1,115 ft., usually from 230 to 400 ft. Juveniles occur over sandy and muddy bottoms. BIOLOGY: Feed on fishes and invertebrates.

Lane Snapper - *Lutjanus synagris* (Linnaeus, 1758)

FEATURES: Silvery pink to reddish dorsally. Silvery below. Narrow, yellow stripes on head and sides. May display pale bars. Dark spot below anterior portion of soft dorsal fin; spot may be faint or absent. Dorsal fin with yellow margin. Caudal fin rosy with narrow dark margin. Anal fin rounded. HABITAT: NC to FL, Gulf of Mexico, Bermuda, Bahamas, and Caribbean Sea to southern Brazil. Occur in shallow coastal waters over a variety of bottoms. Primarily around coral reefs and vegetated sandy bottoms. Occur from shore to about 1,000 ft. BIOLOGY: Feed on fishes and invertebrates.

Silk Snapper - *Lutjanus vivanus* (Cuvier, 1828)

FEATURES: Rosy to pinkish dorsally. Pinkish to silvery below. Irises bright yellow. Juveniles with dark spot below soft dorsal fin. Indistinct, narrow, yellowish stripes on sides. May display pale bars. Fins rosy to yellowish. Caudal fin with narrow, dark reddish margin. Anal fin angular. HABITAT: NC to FL, Gulf of Mexico, Bermuda, Bahamas, Caribbean Sea to Brazil. Over sandy, gravel, and coral bottoms from about 300 to 650 ft. Juveniles found in shallower waters. BIOLOGY: Feed on fishes and a variety of invertebrates.

Yellowtail Snapper - *Ocyurus chrysurus* (Bloch, 1791)

FEATURES: Color and pattern vary. Grayish, bluish, pinkish, or reddish dorsally with irregular yellowish spots. Bright to dusky yellow or greenish stripe from snout to caudal fin. Stripe becomes broader posteriorly, merges with yellow caudal fin. Stripe may be complete or broken. Ventral area silvery white to pinkish. Dorsal fin similarly colored to spots and stripe. Anal fin broadly rounded. HABITAT: MA to FL, Gulf of Mexico, Bermuda, Bahamas, and Caribbean Sea to southern Brazil. Coastal and around reefs and hard bottoms, and in water column from shore to about 540 ft. BIOLOGY: Feed on fishes and invertebrates, primarily at night. Prized foodfish.

Wenchman - *Pristipomoides aquilonaris* (Goode & Bean, 1896)

FEATURES: Reddish to pinkish dorsally. Pinkish to silvery below. Irises yellow orange. Pectoral-fin base and margins of dorsal and caudal fins yellowish. Eyes comparatively large. Scales absent on dorsal and anal fin bases. Last dorsal and anal fin rays elongate. Body comparatively deep. HABITAT: NC to FL, Gulf of Mexico, Bahamas, and Caribbean Sea to southern Brazil. Found over hard bottoms, including natural and artificial reefs from about 80 to 2,130 ft. BIOLOGY: Feed on fishes.

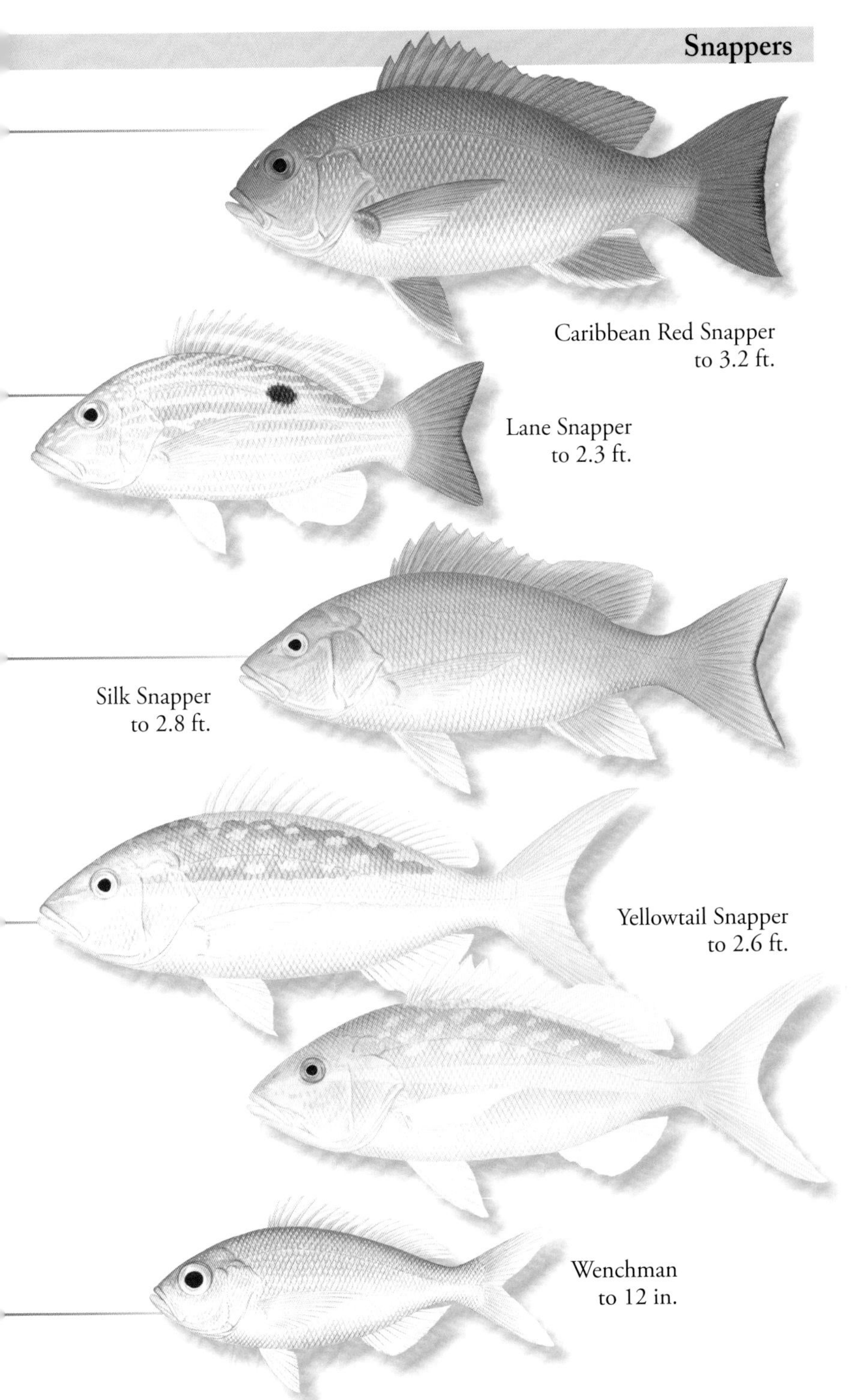
Caribbean Red Snapper
to 3.2 ft.
Lane Snapper
to 2.3 ft.
Silk Snapper
to 2.8 ft.
Yellowtail Snapper
to 2.6 ft.
Wenchman
to 12 in.

Lutjanidae - Snappers, *cont.*

Slender Wenchman - *Pristipomoides freemani* Anderson, 1966

FEATURES: Orange to dark red dorsally. Pinkish to silvery pink below. Irises pale yellow. Dorsal fin pinkish to translucent with yellow outer portion. Upper caudal-fin lobe yellow; lower lobe pinkish to translucent. Eyes comparatively large. Scales absent on dorsal- and anal-fin bases. Last dorsal- and anal-fin rays elongate. Body comparatively shallow. HABITAT: Scattered records from NC to FL; southern Caribbean Sea to Uruguay. Occur from midwater to bottom at depths of about 285 to 720 ft.

Cardinal Snapper - *Pristipomoides macrophthalmus* (Müller & Troschel, 1848)

FEATURES: Shades of red to pink dorsally, fading to silvery or silvery pink below. Dorsal fin pinkish with yellow. Caudal fin reddish to orange. Scales absent on dorsal- and anal-fin bases. Last dorsal- and anal-fin rays elongate. Eyes moderately large. Body comparatively deep. HABITAT: Bermuda, Bahamas, western Gulf of Mexico, and Caribbean Sea. Occur near edges of continental slopes over soft and semi-hard bottoms from about 330 to 2,000 ft.

Vermilion Snapper - *Rhomboplites aurorubens* (Cuvier, 1829)

FEATURES: Deep red above lateral line; pale pinkish to silver below. Fine, bluish lines follow scales above lateral line. Sides with fine, oblique yellowish lines. Irises silvery red. Dorsal fin reddish with orange margin. Caudal fin red. Roof of mouth with a rhomboid tooth patch. Anal fin broadly rounded. HABITAT: NC to FL, Gulf of Mexico, Bermuda, Bahamas, Caribbean Sea to southern Brazil. Over rocky or soft bottoms of edges of continental shelves from about 80 to 1,300 ft.

Lobotidae - Tripletails

Atlantic Tripletail - *Lobotes surinamensis* (Bloch, 1790)

FEATURES: Variably mottled, flecked, and blotched in shades of brown to olive. Dark bands radiate from eyes. May display pale bars or blotches. Pectoral fins translucent. Juveniles often yellowish, but may be brown. Head profile steep, concave above eyes. Snout short; mouth large with a slightly protrusible upper jaw. Spiny dorsal fin incised. Soft dorsal, caudal, and anal fins broad, rounded, and overlapping. Body deep and compressed. HABITAT: Tropical to warm temperate waters of the Atlantic, Indian, and Indo-West Pacific oceans. In western Atlantic from Nova Scotia to Argentina, including Gulf of Mexico and Caribbean Sea. Occur in nearshore and offshore waters, usually near surface. Often associated with flotsam and *Sargassum* seaweed. BIOLOGY: Sluggish, often floating on sides at the surface. Juveniles may mimic fallen mangrove leaves.

Snappers

Tripletails

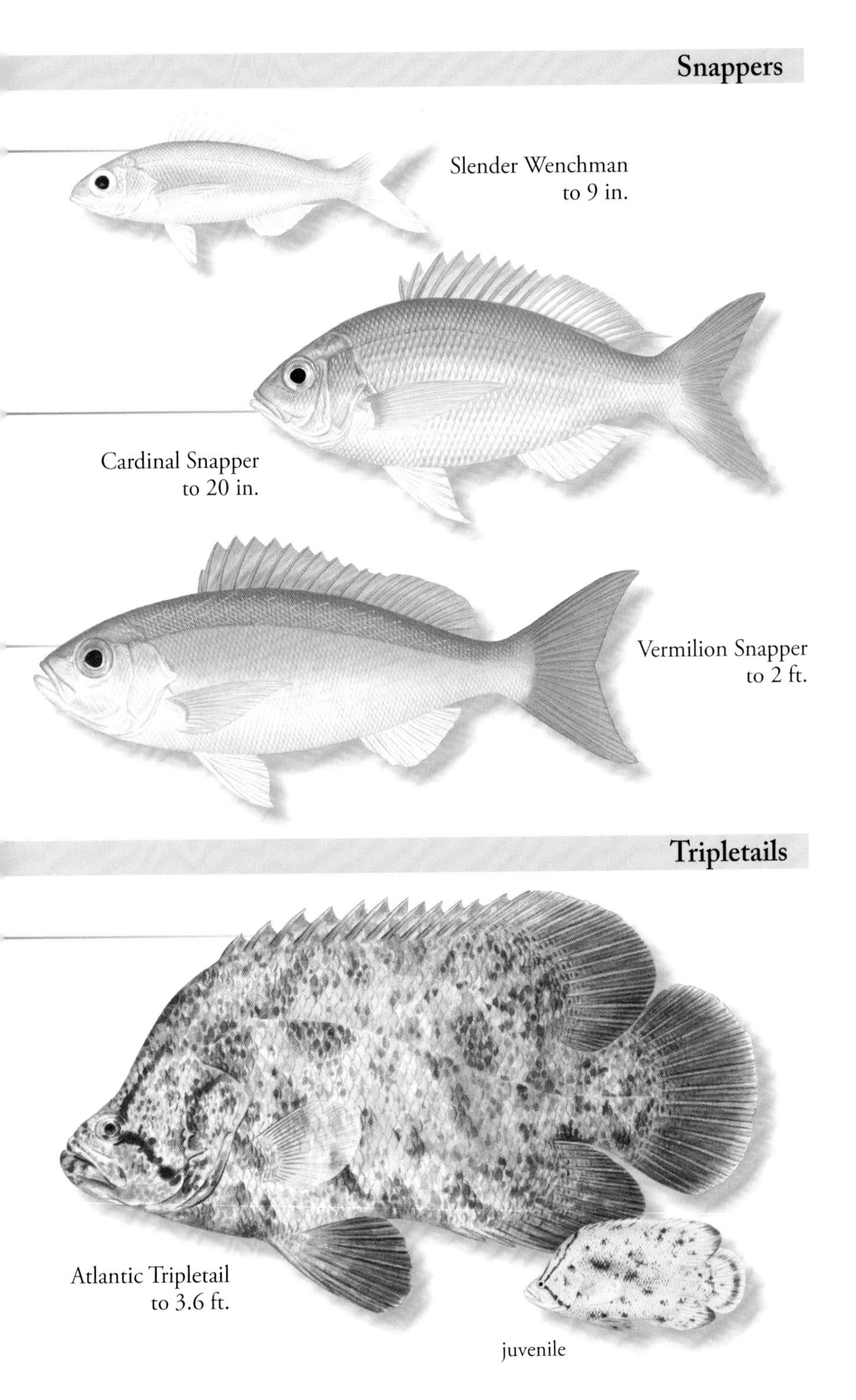

Gerreidae - Mojarras

Irish Pompano - *Diapterus auratus* Ranzani, 1842

FEATURES: Silvery; somewhat darker dorsally. Smaller specimens with three dark bars on sides. Spiny dorsal fin with dusky margin. Anal and pelvic fins yellowish. Lower preopercular margin serrated. Anal fin lobed with three spines and eight rays; second spine strong. Body rhomboid in shape, deep, and compressed. HABITAT: FL, western Gulf of Mexico, Caribbean Sea to southern Brazil. Occur in shallow coastal waters, including estuaries, over seagrass beds, around mangroves, and in lagoons and bays. Commonly enter fresh water.

Rhombic Pompano - *Diapterus rhombeus* (Cuvier, 1829)

FEATURES: Uniformly silvery. Spiny dorsal-fin margin dusky. Anal and pelvic fins yellowish. Preopercular margin serrated. Anal fin lobed with two spines and nine rays; second spine strong. Body rhomboid in shape, deep, and compressed. HABITAT: Western Gulf of Mexico, Caribbean Sea to southern Brazil. Occur in shallow, sandy bays and estuaries, and in mangrove lagoons. May enter fresh water. Juveniles common in brackish water and hypersaline lagoons.

Spotfin Mojarra - *Eucinostomus argenteus* Baird & Girard, 1855

FEATURES: Silvery. Smaller specimens with dark, oblique bars and spots that fade with age. Spiny dorsal fin with dusky margin. Tips of caudal fin may be dusky. Anal fin low, fairly straight-edged. Scaleless area between eyes surrounded by scales. Body moderately slender. HABITAT: NJ (rare) to FL, Gulf of Mexico, Bermuda, Bahamas, Antilles, northern South America to southern Brazil. Over sandy and shelly bottoms, and around reefs from near shore to about 200 ft. Occasionally in inlets and estuaries. Juveniles occur in mangrove lagoons. BIOLOGY: Feed on benthic invertebrates.

Silver Jenny - *Eucinostomus gula* (Quoy & Gaimard, 1824)

FEATURES: Silvery; somewhat darker dorsally. Smaller specimens with about seven oblique bars and blotches on upper sides that fade with age. Spiny dorsal fin with dusky margin. Scaleless area between eyes surrounded by scales. Body moderately deep. HABITAT: MA (rare) to FL, Gulf of Mexico, Bermuda, Bahamas, and Caribbean Sea to Argentina. Primarily over shallow seagrass beds, but also over open sandy bottoms. Occur from shore to about 230 ft. Also in estuaries and hypersaline areas. Rarely in fresh water. BIOLOGY: Feed on benthic invertebrates.

Tidewater Mojarra - *Eucinostomus harengulus* Goode & Bean, 1879

FEATURES: Silvery with oblique dark bars and spots on upper sides; markings more evident in smaller specimen. Second and third bars form Y-shape. Spiny dorsal fin with dusky to blackish margin. Scaleless area between eyes open anteriorly. Pectoral fins lack scales. Body moderately slender. HABITAT: VA to FL, Gulf of Mexico, Bermuda, Bahamas, and Caribbean Sea to southern Brazil. In protected estuaries, over seagrass beds and sandy and muddy bottoms, and around mangroves from near shore to about 30 ft. Commonly enter freshwater tributaries.

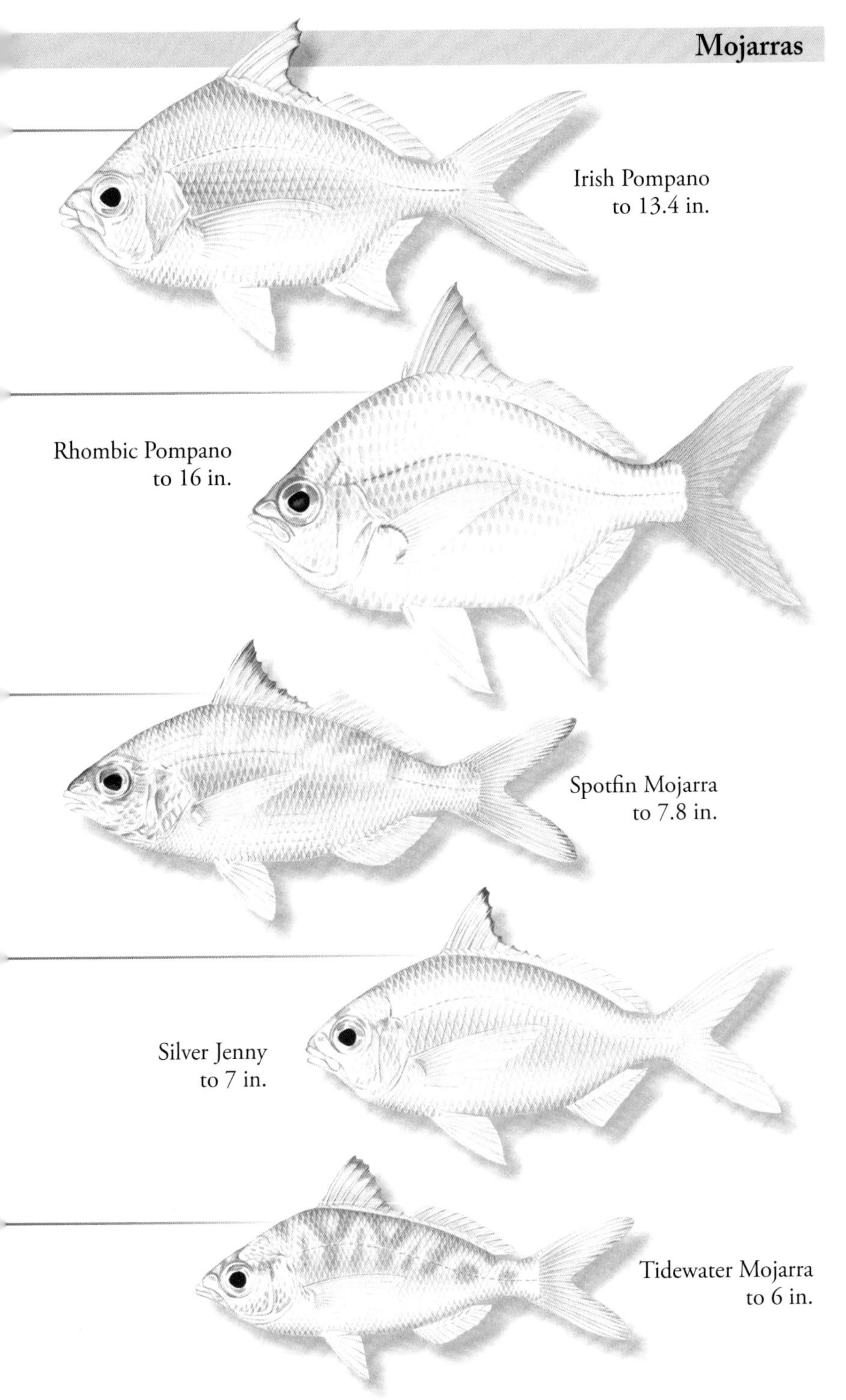
Irish Pompano
to 13.4 in.
Rhombic Pompano
to 16 in.
Spotfin Mojarra
to 7.8 in.
Silver Jenny
to 7 in.
Tidewater Mojarra
to 6 in.

Bigeye Mojarra - *Eucinostomus havana* (Nichols, 1912)

FEATURES: Silvery with darker, oblique bars on upper sides. Spiny dorsal fin with a black upper margin or with a wedge-shaped black blotch on upper portion. Eyes very large. Pectoral fins completely scaled in adults. Body moderately slender. HABITAT: Eastern FL, Bermuda, Bahamas, and Caribbean Sea. Also off Port St. Joe, FL, Corpus Christi, TX, and Yucatán Peninsula. Coastal. Over shallow sandy, muddy, and vegetated bottoms from near shore to about 30 ft. Also around mangroves and sandy beaches. Do not enter estuaries. BIOLOGY: Feed on benthic invertebrates.

Slender Mojarra - *Eucinostomus jonesii* (Günther, 1879)

FEATURES: Silvery greenish dorsally; silvery below. Smaller specimens with darker, oblique bars that fade with age. Distinct V-shaped mark on snout. Spiny dorsal fin with dusky margin. Scaleless area between eyes open anteriorly. Pectoral fins lack scales. Body slender. HABITAT: FL, Gulf of Mexico, Bermuda, Bahamas, and Caribbean Sea to northeastern Brazil. Over high-energy sandy and grassy bottoms, in inlets, and along beaches from near shore to about 150 ft. Do not enter brackish water or estuaries.

Mottled Mojarra - *Eucinostomus lefroyi* (Goode, 1874)

FEATURES: Silvery with about seven to nine dusky, oblique, wavy bars on upper sides. Bars may be indistinct to nearly absent. Spiny dorsal fin may be colorless or may have dusky margin. Scaleless area between eyes open anteriorly. Pectoral fins lack scales. Anal fin with two spines (compared to three spines in other *Eucinostomus*). HABITAT: NC (rare) to FL, Gulf of Mexico, Bermuda, Bahamas, all Antilles, and Caribbean Sea to northeastern Brazil from near shore to about 150 ft. Occur along high-energy sandy beaches and inlets, and around reefs. Do not enter brackish water or estuaries.

Flagfin Mojarra - *Eucinostomus melanopterus* (Bleeker, 1863)

FEATURES: Silvery; somewhat darker dorsally. Body unmarked. Dorsal fin with a black outer margin and a distinct white inner band. Scaleless area between eyes open anteriorly. Pectoral fins lack scales. HABITAT: SC (rare) to FL, Gulf of Mexico, Bahamas, and Caribbean Sea to central Brazil. Also eastern Atlantic. Absent from Bermuda. In shallow coastal waters, inlets, and lagoons over sandy, muddy, and shelly bottoms from near shore to about 80 ft. Occasionally in fresh water.

Maracaibo Mojarra - *Eugerres awlae* Schultz, 1949

FEATURES: Silvery gray dorsally, fading to silver below with brownish black or gray spots that form stripes along scale rows. Eyes with a yellowish cast. Pelvic and anal fins pale yellow; other fins dusky. Spiny dorsal fin very tall and pointed with second spine thick and reaching fifth or sixth soft dorsal ray when depressed. Anal fin tall and pointed with second spine reaching caudal-fin base when depressed. HABITAT: Southern Gulf of Mexico and southern Caribbean Sea to Venezuela. Also around Puerto Rico and adjacent islands. Occur in shallow coastal waters with grassy and muddy bottoms, in lagoons, around mangroves, and in river mouths from near shore to about 100 ft.

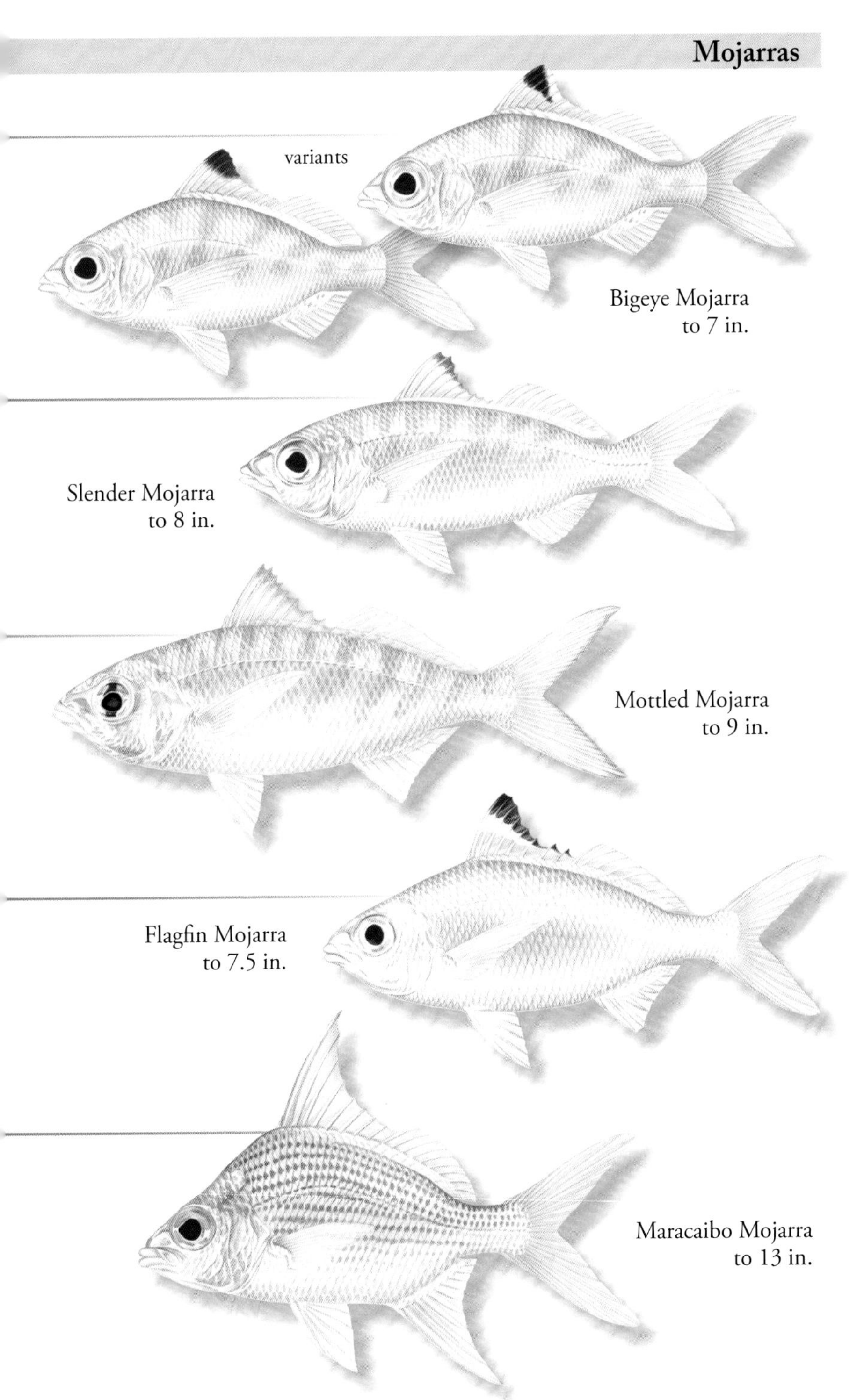
variants
Bigeye Mojarra
to 7 in.
Slender Mojarra
to 8 in.
Mottled Mojarra
to 9 in.
Flagfin Mojarra
to 7.5 in.
Maracaibo Mojarra
to 13 in.

Gerreidae - Mojarras, *cont.*

Brazilian Mojarra - *Eugerres brasilianus* (Cuvier, 1830)

FEATURES: Dark silver dorsally; silvery below. Grayish spots form lines along scale rows on upper sides. Spiny dorsal fin pointed, with second spine not reaching soft dorsal-fin base when depressed. Second anal-fin spine strong, but not reaching caudal-fin base when depressed. HABITAT: Cuba, Puerto Rico and adjacent islands, and SW Gulf of Mexico to Brazil. Occur over soft bottoms in shallow coastal estuaries, in lagoons, and around mangroves from near shore to about 100 ft. Tolerates fresh water.

Striped Mojarra - *Eugerres plumieri* (Cuvier, 1830)

FEATURES: Silvery with brownish, greenish, or bluish tinge dorsally. Silvery below. Dark brown to blackish spots form rows along scale rows on upper sides. Second dorsal-fin spine reaches to anterior soft dorsal-fin base. Second anal-fin spine strong, reaches caudal-fin base when depressed. HABITAT: SC to W FL, W Gulf of Mexico to Brazil including Caribbean Sea. Occur in shallow coastal waters, mangrove-lined creeks, and lagoons from near shore to about 100 ft. Often in fresh water.

Yellowfin Mojarra - *Gerres cinereus* (Walbaum, 1792)

FEATURES: Silvery to tannish with darker bars or rows of spots on upper body. Pelvic fins yellowish. Anal fin may have yellowish tint. Dorsal fin unmarked. Dorsal profile over eyes slightly concave. HABITAT: FL, Gulf of Mexico, Bermuda, Bahamas, and Caribbean Sea to Brazil. Also in eastern Pacific. In shallow coastal waters along sandy beaches, around mangroves, over seagrass beds and coral reefs, and in bays from near shore to about 100 ft. Enter brackish and fresh water.

Haemulidae - Grunts

Black Margate - *Anisotremus surinamensis* (Bloch, 1791)

FEATURES: Dusky silver with blackish to brownish spots on scales. Broad, black to dark brown bar on lower midbody. All fins blackish to dark brown. Juvenile with two dark stripes on sides. Body deep, compressed. Head profile sloping. HABITAT: FL, southern Gulf of Mexico, Bahamas, and Caribbean Sea to southern Brazil. Over shallow coral reefs and rocky areas from shore to about 130 ft. BIOLOGY: Feed on echinoderms, crustaceans, and small fishes. May form spawning aggregations.

Porkfish - *Anisotremus virginicus* (Linnaeus, 1758)

FEATURES: Adults silvery white and alternating yellow stripes. Two black bands on head: the first from nape to mouth; the second from anterior dorsal-fin base to gill margin. All fins yellow. Juveniles with two dark stripes on sides and a large black blotch on peduncle. Body deep, compressed. Head profile steeply sloping. HABITAT: FL, southern Gulf of Mexico, Bahamas, and Caribbean Sea to Brazil. Bahamas. Over coral reefs and hard bottoms from shore to about to 165 ft. BIOLOGY: Adults feed on invertebrates, juveniles on ectoparasites of other fishes.

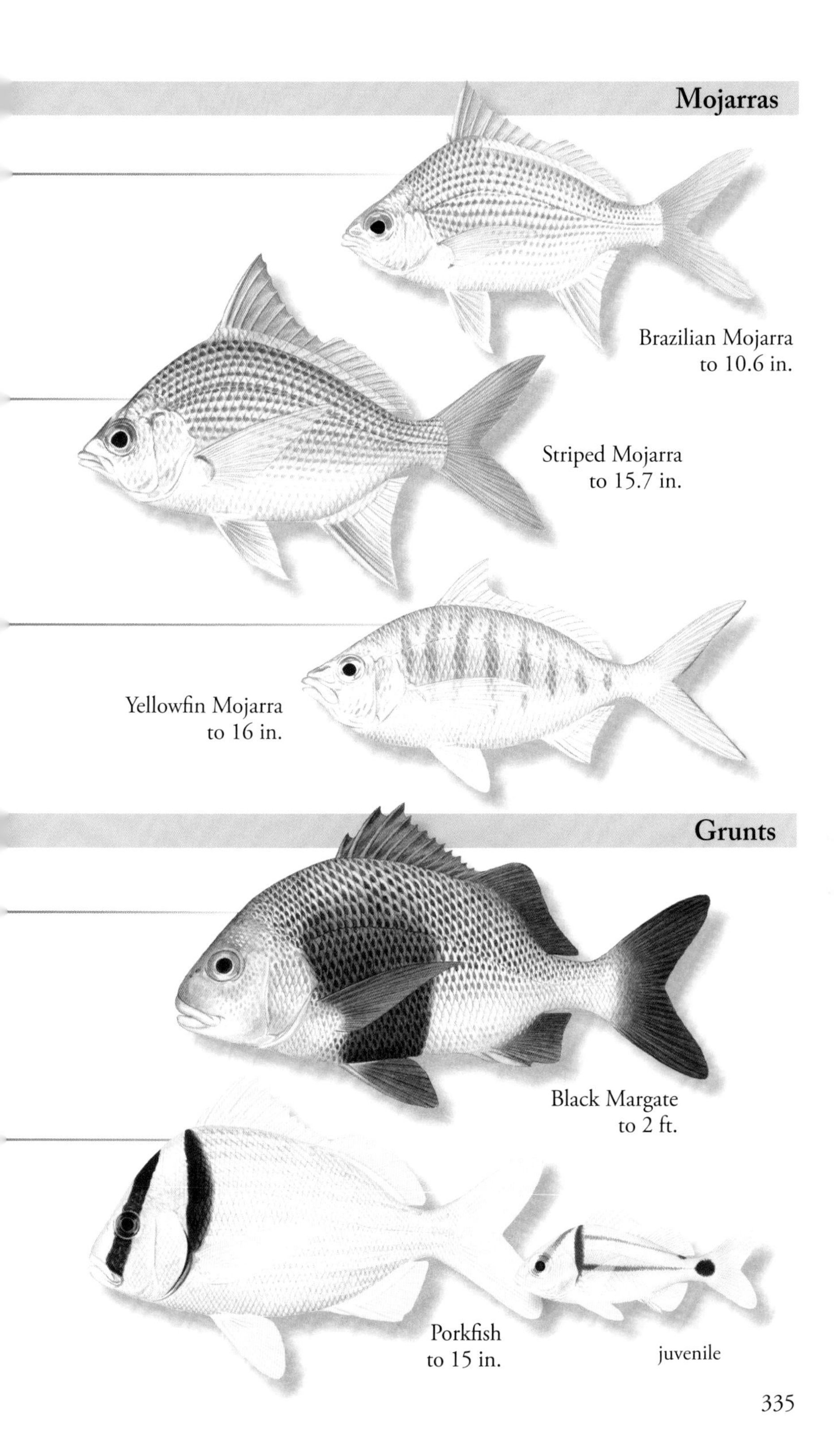
Mojarras
Brazilian Mojarra
to 10.6 in.
Striped Mojarra
to 15.7 in.
Yellowfin Mojarra
to 16 in.
Grunts
Black Margate
to 2 ft.
Porkfish
to 15 in.
juvenile

Smallmouth Grunt - *Brachygenys chrysargyreum* (Günther, 1859)

FEATURES: Adults silvery white with bright yellow stripes on head and sides. All fins except pectoral fins yellow. Juveniles silvery with three blackish stripes on sides, the lowermost ending in an angular blotch at caudal-fin base. Scales below lateral line form uniformly horizontal rows. Body oblong, somewhat elongate, and compressed. Head profile evenly convex. HABITAT: S FL, southern Gulf of Mexico, Bahamas, and Caribbean Sea to Brazil. Adults over coral reefs. Juveniles over coral reefs, hard bottoms, and seagrass beds to about 130 ft. BIOLOGY: Feed on small crustaceans and plankton. NOTE: Previously known as *Haemulon chrysargyreum*.

Barred Grunt - *Conodon nobilis* (Linnaeus, 1758)

FEATURES: Brownish dorsally, becoming pale below. Upper body with brown bars and tan stripes that form checkerboard pattern. All fins with some yellow. Preopercle serrated, with two enlarged spines at lower corner. HABITAT: E FL, Gulf of Mexico, and Caribbean Sea to Brazil. Absent from Cuba. In shallow, turbid waters over sandy shores and soft bottoms to about 330 ft. Also in bays and estuaries. BIOLOGY: Feed primarily at night on crustaceans and small fishes.

Bonnetmouth - *Emmelichthyops atlanticus* Schultz, 1945

FEATURES: Metallic greenish to yellowish gray dorsally. Silvery on sides; white along abdomen. Three brownish stripes on upper sides. Mouth highly protrusible. First dorsal fin widely separated from second dorsal fin. HABITAT: FL Keys, Bermuda, Bahamas, and Caribbean Sea. Scattered in Gulf of Mexico. Found around coral and patch reefs and over sandy bottoms to about 300 ft. BIOLOGY: Swift and schooling. Feed on zooplankton and small fishes. NOTE: Previously in the Family Inermiidae.

Torroto Grunt - *Genyatremus luteus* (Bloch, 1790)

FEATURES: Grayish silver dorsally; silvery below. May have a golden cast. Lower caudal-fin lobe, anal fin, and pelvic fins yellowish. Mouth small, nearly horizontal. Spiny dorsal fin tall, incised, with fifth spine tallest. Second anal-fin spine strong. HABITAT: Gulf of Venezuela to southern Brazil. Recorded from Trinidad. Primarily in brackish estuaries and adjacent waters over sandy and muddy bottoms. Sometimes found in marine waters. BIOLOGY: Feed on crustaceans. NOTE: Previously *G. cavifrons*.

Margate - *Haemulon album* Cuvier, 1830

FEATURES: Pearly gray dorsally; silvery below. Scales on upper body with dark margins. Soft dorsal and caudal fins dark gray. Inside of mouth pale red. Body oblong, compressed. Profile of head almost straight. HABITAT: NC to FL, southern Gulf of Mexico, Bermuda, Bahamas, and Caribbean Sea to Brazil. Occur over coral reefs, hard bottoms, and nearby soft bottoms from about 65 to 200 ft. BIOLOGY: Feed on a variety of invertebrates and small fishes. OTHER NAME: White Margate.

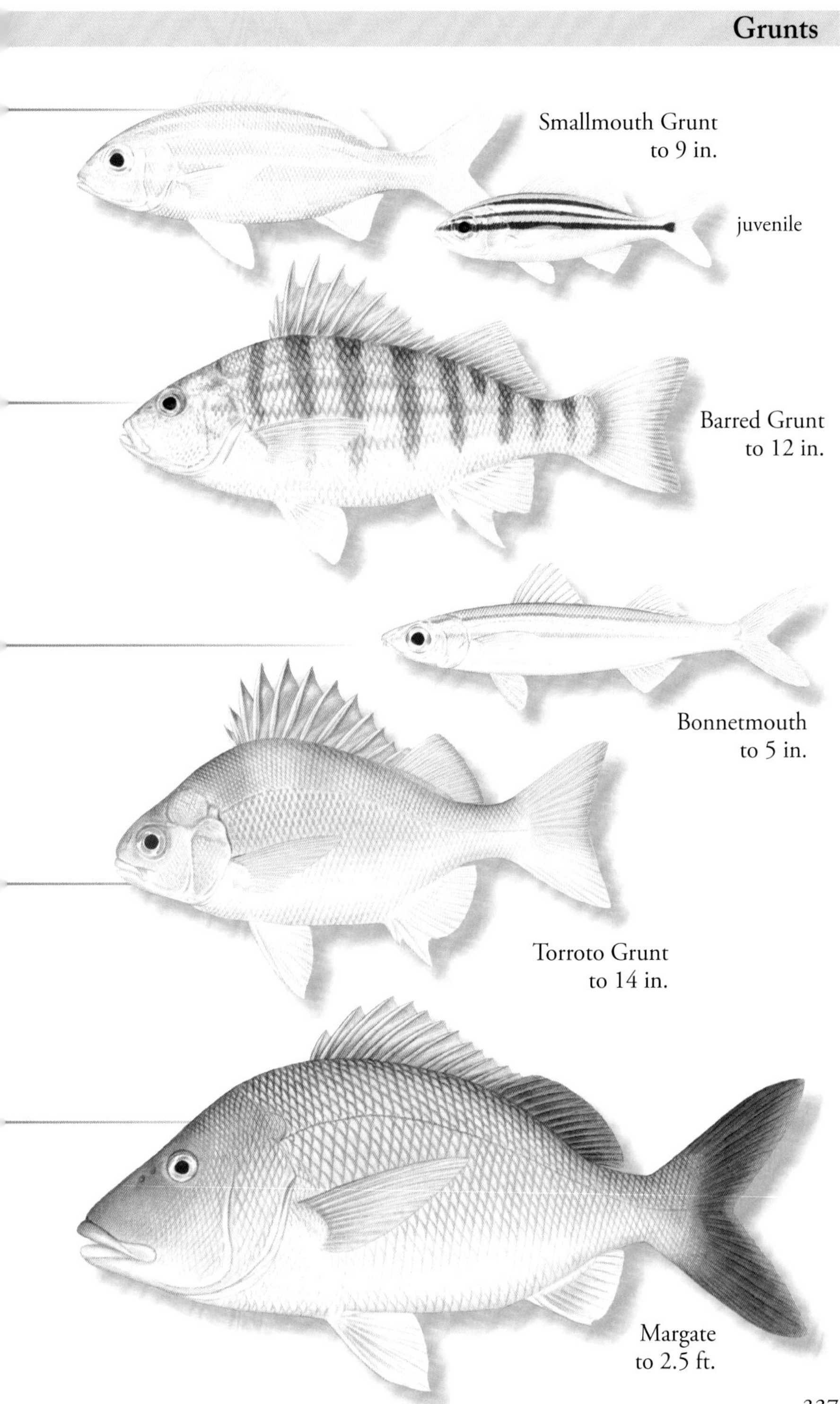
Smallmouth Grunt
to 9 in.
juvenile
Barred Grunt
to 12 in.
Bonnetmouth
to 5 in.
Torroto Grunt
to 14 in.
Margate
to 2.5 ft.

Latin Grunt - *Haemulon atlanticus* Carvalho, Marceniuk, Oliveira & Wosiacki, 2021

FEATURES: Adults silvery gray with bright white spots on scale centers that form oblique lines above lateral line. Scales become horizontal ventrally. Dark lines comparatively narrow. Dorsal and anal fins grayish, caudal fin with a blackish spot or wedge at base. Juveniles with two brownish to grayish stripes on upper sides, and a black blotch on peduncle. HABITAT: Eastern Yucatán to southern Brazil. Occur over sand, rubble, rocky reefs, and large corals from near shore to about 164 ft.

Tomtate - *Haemulon aurolineatum* Cuvier, 1830

FEATURES: Adults silvery white to tannish with two dark yellowish stripes: first from nape to rear soft dorsal-fin base; second from snout to caudal peduncle. May also have narrow, faint yellow stripes along scale rows. Black blotch often present at base of caudal fin. Juveniles silvery with a brown stripe on upper sides and on midside and a squarish black blotch on peduncle. HABITAT: Chesapeake Bay to FL, Gulf of Mexico, Bermuda, Bahamas, and Caribbean Sea to Brazil. Occur over a variety of natural and artificial bottoms from shore to about 130 ft. BIOLOGY: Form large schools. Feed on invertebrates, small fishes, and algae.

Black Grunt - *Haemulon bonariense* Cuvier, 1830

FEATURES: Silvery with large black spots that merge to form comparatively large, irregular wavy lines along scale rows on upper sides. Rows become horizontal ventrally. Dorsal fin grayish; caudal fin dark gray to black. Soft dorsal and anal fins densely scaled Second and third anal-fin spines similar in length. HABITAT: S FL, southern Gulf of Mexico, and Caribbean Sea to northern Brazil. Occur over soft and hard bottoms, and around reefs from shore to about 160 ft.

Bronzestripe Grunt - *Haemulon boschmae* (Metzelaar, 1919)

FEATURES: Silvery with bronze spots that merge to form irregular wavy lines along scale rows. Lateral line bronze. Usually with a black blotch at caudal-fin base. Small juvenile with four bronze stripes on upper sides. Scales on lower sides in nearly horizontal rows. Body elongate. HABITAT: Two separate populations: one in southern Gulf of Mexico, another from Colombia to Guyana. Occur over sandy bottoms adjacent to reefs from near shore to about 295 ft.

Caesar Grunt - *Haemulon carbonarium* Poey, 1860

FEATURES: Adults silvery gray with bronze stripes on head and body. Abdomen may be dusky gray to black. Spiny dorsal fin with bronze membranes. Soft dorsal, caudal, and anal fins dark gray to blackish. Inside of mouth red. Juveniles with three broad, brownish stripes on sides, bronze spots on scales, and a squarish spot on peduncle. HABITAT: FL, southern Gulf of Mexico, Bermuda, Bahamas, and Caribbean Sea to Brazil. Found over coral reefs and hard bottoms from shore to about 100 ft. BIOLOGY: Feed at night on a variety of invertebrates.

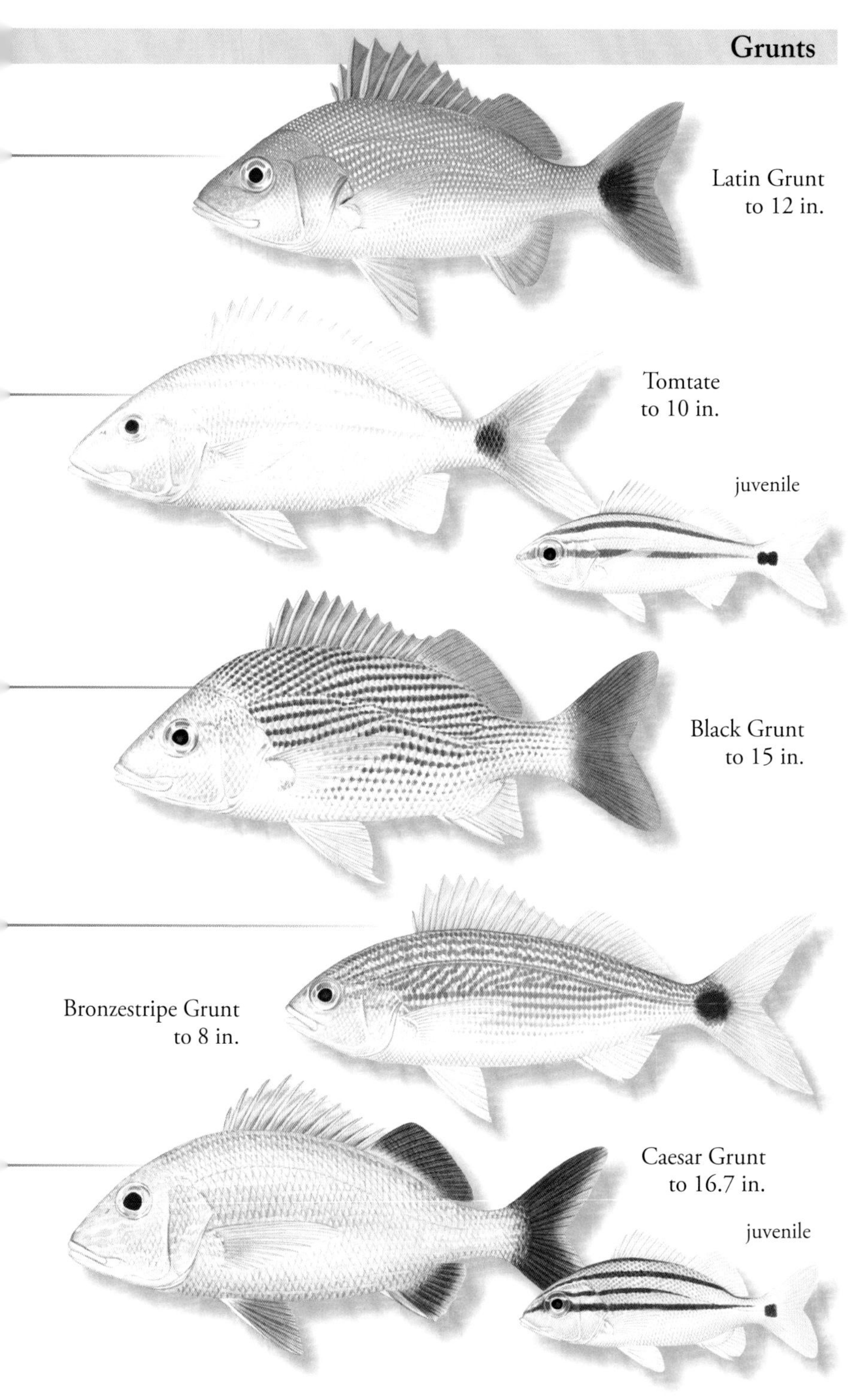
Latin Grunt
to 12 in.
Tomtate
to 10 in.
juvenile
Black Grunt
to 15 in.
Bronzestripe Grunt
to 8 in.
Caesar Grunt
to 16.7 in.
juvenile

French Grunt - *Haemulon flavolineatum* (Desmarest, 1823)

FEATURES: Adults with alternating silvery white and yellow to bronze stripes that follow scales rows. Fins yellowish to bronze. Inside of mouth is reddish. Juveniles with two broad brownish stripes on sides, a short brown stripe behind eyes, and oblique yellow spots along scale rows, and a dark brown blotch on peduncle. Scales below lateral line comparatively tall and in oblique rows. HABITAT: SC to FL, southern Gulf of Mexico, Bermuda, Bahamas, and Caribbean Sea. Occur around a variety of soft and hard bottoms from shore to about 130 ft. BIOLOGY: Schooling. Feed at night on crustaceans.

Spanish Grunt - *Haemulon macrostoma* Günther, 1859

FEATURES: Body silvery gray with dark grayish stripes on sides. Dorsal profile and caudal peduncle yellowish. Ventral area may be grayish to blackish. Pectoral fins and outer margins of dorsal, caudal, and anal fins yellowish. Juveniles yellow dorsally; silvery below with three black stripes on sides and a squarish black blotch on peduncle. HABITAT: FL, Bermuda, Bahamas, and Caribbean Sea to Brazil. Scattered in Gulf of Mexico. Occur in clear water over coral reefs and hard bottoms to about 130 ft. BIOLOGY: Feed on crustaceans and echinoderms. Rarely schooling. NOTE: Previously known as *Haemulon macrostomum*.

Cottonwick - *Haemulon melanurum* (Linnaeus, 1758)

FEATURES: Body pearly white to silvery white with bronze to yellow stripes on head and sides. Black band runs from nape along dorsal profile and lower dorsal fins to caudal fin, where it forms a V-shape along the inner rays. Outer portions of dorsal fins pearly to colorless. Inside of mouth pale red. Juveniles silvery with three blackish stripes on sides; middle stripe joins lower stripe; lower stripe ends at caudal-fin base. HABITAT: FL, northern and southern Gulf of Mexico, Bermuda, Bahamas, and Caribbean Sea to Brazil. Found in clear water over coral reefs, hard bottoms, and adjacent seagrass beds from shore to about 130 ft.

Sailors Choice - *Haemulon parra* (Desmarest, 1823)

FEATURES: Adults pearly gray with large blackish spots on scales that form oblique rows. Spiny dorsal, anal, pelvic, and pectoral fins grayish. Soft dorsal and caudal fins dark gray to blackish. Outer margins of irises yellowish. Juveniles silvery with three blackish stripes on sides and a squarish black blotch on peduncle. HABITAT: FL, Bahamas, and Caribbean Sea to southern Brazil. Scattered in Gulf of Mexico. Occur over reefs, hard bottoms, and other structure from shore to about 130 ft. Juveniles over seagrass beds and hard bottoms. BIOLOGY: Schooling. Feed at night on invertebrates and small fishes.

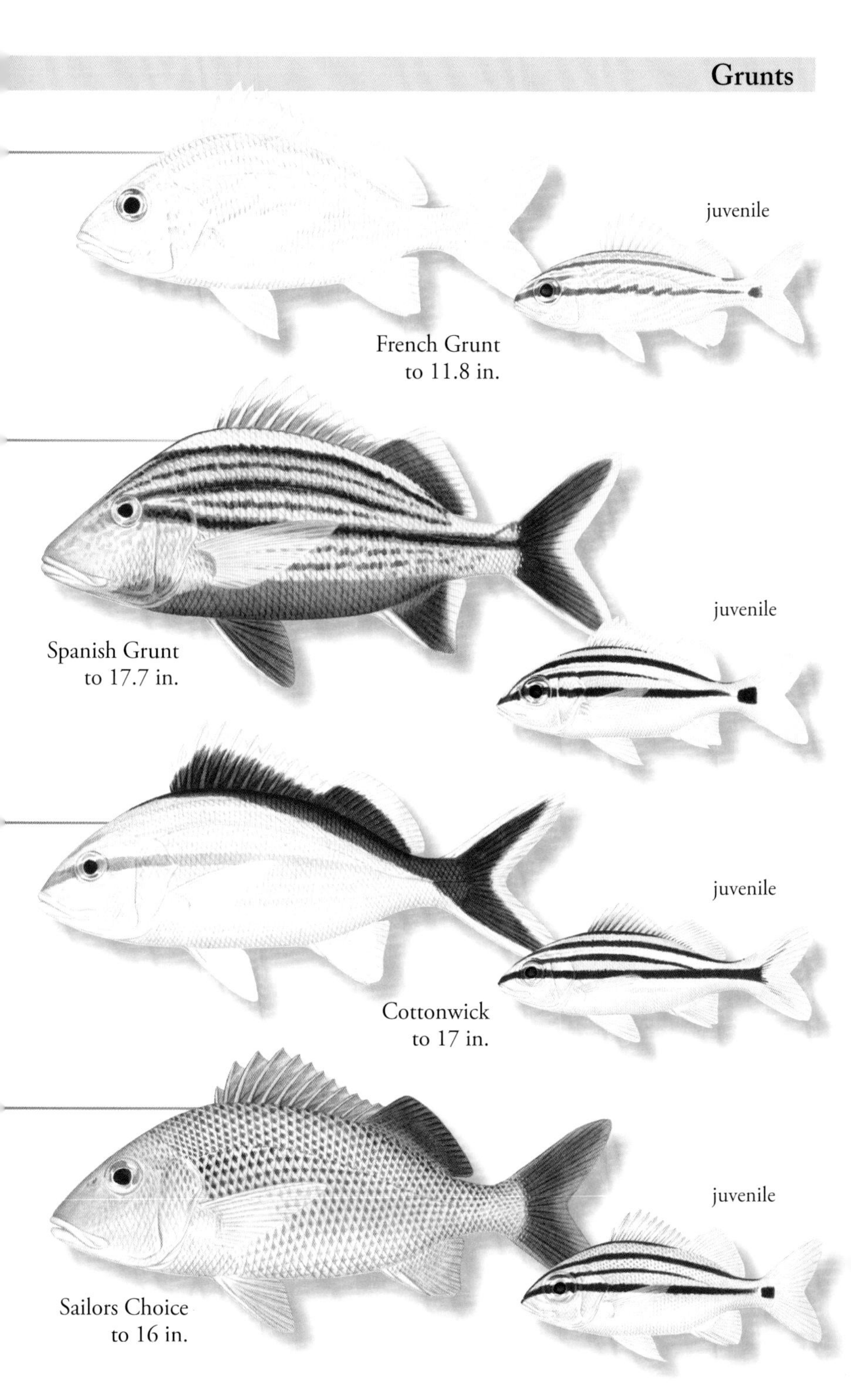
juvenile
French Grunt
to 11.8 in.
juvenile
Spanish Grunt
to 17.7 in.
juvenile
Cottonwick
to 17 in.
juvenile
Sailors Choice
to 16 in.

White Grunt - *Haemulon plumierii* (Lacepède, 1801)

FEATURES: Color varies. Adults pale yellowish to grayish with narrow, blue, wavy lines on head and above pectoral fins. Scales with dark centers and silvery blue margins; lines and scales appear blue and yellow under water. Inside of mouth bright red. Juveniles with alternating white and brownish stripes dorsally, yellow and blue stripes below, and a black blotch on caudal-fin base. May also display irregular dark blotches. HABITAT: Chesapeake Bay to FL, Gulf of Mexico, Bermuda, Bahamas, and Caribbean Sea to southern Brazil. Adults over a variety of bottoms and reefs from shore to about 130 ft. Juveniles over very shallow hard and vegetated bottoms.

Bluestriped Grunt - *Haemulon sciurus* (Shaw, 1803)

FEATURES: Adults with alternating yellow to bronze and bright blue narrow wavy lines and spots on head and body. Bright blue bordered above and below by darker blue. Inner portions of soft dorsal and caudal fins blackish to brownish black. Anal, pelvic, and pectoral fins yellow to pale. Inside of mouth bright red. Juveniles with fewer and more separated blue lines and a black spot on caudal-fin base. HABITAT: SC to FL, southern Gulf of Mexico, Bermuda, Bahamas, Caribbean Sea. Found over a variety of bottoms, structures, and reefs from shore to about 130 ft.

Striped Grunt - *Haemulon striatum* (Linnaeus, 1758)

FEATURES: Grayish white dorsally, shading to silvery below, with five distinct brownish to yellowish stripes on head and upper sides. Snout yellowish. Fins colorless to chalky. Juveniles with two brown stripes on upper sides. Scales below lateral line in oblique rows. HABITAT: GA to FL, Gulf of Mexico, Bahamas, and Caribbean Sea to Brazil. Occur over outer reefs from about 40 to 330 ft. BIOLOGY: Feed on plankton and small crustaceans.

Boga - *Haemulon vittatum* (Poey, 1860)

FEATURES: Metallic bluish to greenish dorsally. Silvery on sides; white along abdomen. A broad yellowish green stripe runs from snout to caudal-fin base. Several broken greenish brown stripes on upper sides. Caudal fin with purplish cast. Mouth highly protrusible. Spiny and soft dorsal fins continuous, deeply notched. HABITAT: NC (rare) to FL, Gulf of Mexico, Bermuda, Bahamas, and Caribbean Sea. Occur in open water and along coast to about 300 ft. NOTE: Previously *Inermia vittata* in the Family Inermiidae.

Roughneck Grunt - *Haemulopsis corvinaeformis* (Steindachner, 1868)

FEATURES: Olivaceous dorsally with silvery reflections. Silvery on sides and below. Scales above lateral line with dark centers; scales on sides with dark spots that form stripes. Dark blotch behind upper opercular margin. May have several vague saddles and bars on sides. Soft dorsal and caudal fins with blackish margins. Second anal-fin spine comparatively short. HABITAT: Caribbean Sea to Brazil. Demersal over muddy and sandy bottoms and low relief hard and rubble bottoms from shore to about 165 ft.

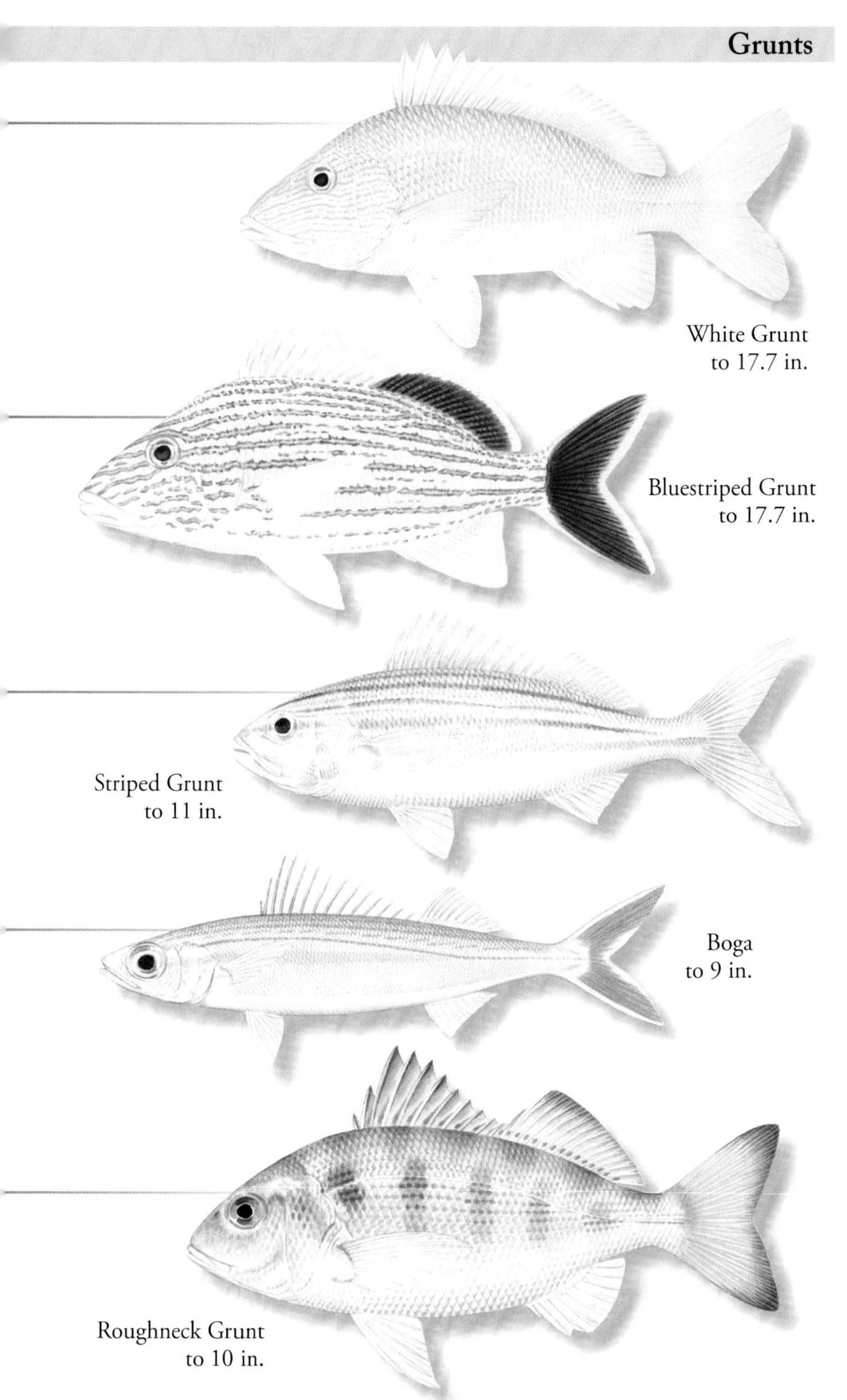
White Grunt
to 17.7 in.
Bluestriped Grunt
to 17.7 in.
Striped Grunt
to 11 in.
Boga
to 9 in.
Roughneck Grunt
to 10 in.

Pigfish - *Orthopristis chrysoptera* (Linnaeus, 1766)

FEATURES: Iridescent greenish gray dorsally. Silvery on sides; pearly white below. Dark orange spots and irregular lines on head and body. Inside of mouth orange. May display irregular dark bars, saddles, and blotches. Fins dusky to yellowish; may have dusky margins. Upper caudal-fin lobe slightly longer than lower lobe. HABITAT: NY to FL, Gulf of Mexico, Bermuda, Bahamas, and Cuba. Demersal over soft bottoms from near shore to about 65 ft. Often in bays and estuaries, sometimes over reefs.

Corocoro Grunt - *Orthopristis rubra* (Cuvier, 1830)

FEATURES: Silvery gray dorsally; silvery on sides and below. Irregular brownish spots form wavy lines on upper head and body. Also with a faint brownish stripe from snout through eyes. May also display two pale brown stripes and brown bars on upper body. Dorsal fins with brownish spots along base. HABITAT: Colombia to northeastern Brazil. Also reported to southern Brazil. Demersal over sandy and muddy bottoms, and low relief hard bottoms from shore to about 230 ft. Also in estuaries. BIOLOGY: Feed on crustaceans, mollusks, worms, and small fishes.

Brownstriped Grunt - *Paranisotremus moricandi* (Ranzani, 1842)

FEATURES: Anterior head shades of brown with two silvery crescents below eyes. Ventral gill area silvery. A large blackish blotch present on rear opercular margin. Body golden with wide, pale and dark brown stripes that are bordered above and below by narrow silvery stripes. A single or broken brown saddle on upper peduncle, and a dark brown spot at caudal-fin base. Pelvic fins dark brown. HABITAT: Costa Rica to northern Brazil. Found over coral and rocky bottoms from near shore to about 40 ft.

Burro Grunt - *Rhonciscus crocro* (Cuvier, 1830)

FEATURES: Shades of olivaceous to greenish gray dorsally and on sides. Abdomen silvery white. Sides with dusky spots on scale centers. Some with several irregular broken stripes on sides. Dorsal and caudal fins dusky olivaceous. Second anal-fin spine long and very strong. HABITAT: FL, western and southern Gulf of Mexico, and Caribbean Sea to southern Brazil. Found over shallow soft and vegetated bottoms in brackish and full salt water to about 65 ft. Also in river mouths and coastal streams.

Juvenile Grunts

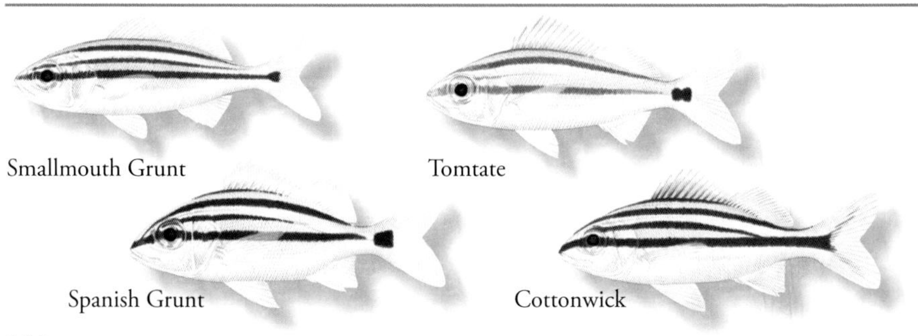

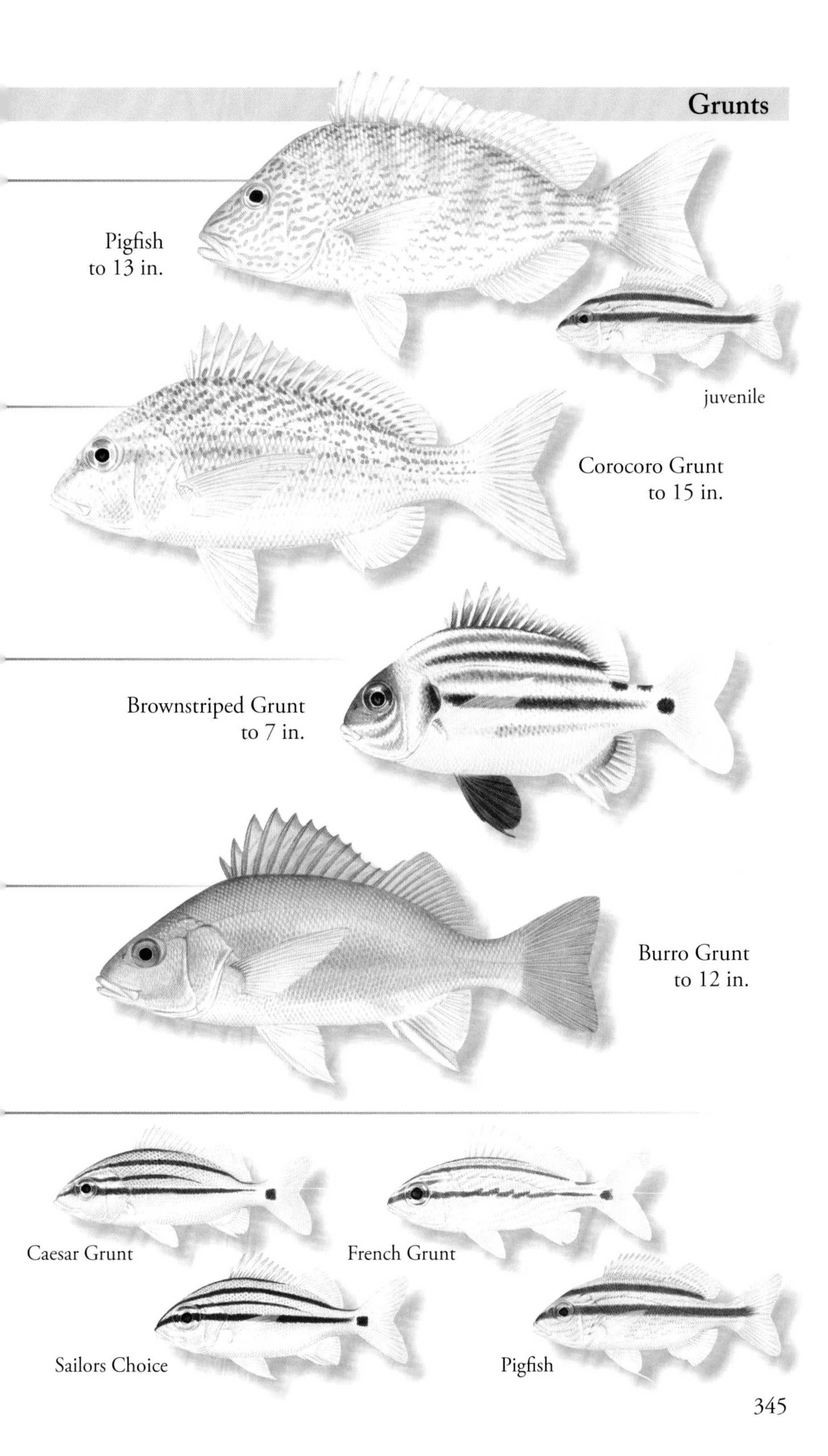
Pigfish
to 13 in.
juvenile
Corocoro Grunt
to 15 in.
Brownstriped Grunt
to 7 in.
Burro Grunt
to 12 in.
Caesar Grunt
French Grunt
Sailors Choice
Pigfish

Sparidae - Porgies

Sheepshead - *Archosargus probatocephalus* (Walbaum, 1792)

FEATURES: Grayish to brownish dorsally, fading to silvery below. May have golden cast. Five to six blackish to dark brown bars on body. Head and fins grayish to dusky. Anterior teeth are incisor-like. Upper and lower lips grooved. Pectoral fins long. Caudal fin forked. HABITAT: Nova Scotia to FL; Gulf of Mexico to Venezuela. Occur along coast, in estuaries around muddy and rocky bottoms, and over hard substrates to about 40 ft. Also around pilings. Enter brackish water. BIOLOGY: Feed on mollusks and crustaceans.

Sea Bream - *Archosargus rhomboidalis* (Linnaeus, 1758)

FEATURES: Silvery gray to olivaceous with golden yellow stripes on head and body. Blackish spot about as large as eye on lateral line near origin. Dorsal-fin margins blackish. Anal and pelvic fins yellowish. Anterior teeth incisor-like. Upper and lower lips grooved. HABITAT: NJ to FL, Gulf of Mexico, Caribbean Sea to southern Brazil. Over muddy and vegetated bottoms and around mangroves. BIOLOGY: Feed on mollusks and crustaceans.

Jolthead Porgy - *Calamus bajonado* (Bloch & Schneider, 1801)

FEATURES: Silvery to pale brownish. Scale centers silvery to pale with brassy edges. Scaleless portion of head brownish. Snout with horizontal silvery stripe. Corners of mouth always dark orange. Silvery blue lines above and below eyes. Juveniles with dark bars on body and caudal fin. Profile of snout forms between 43° to 55° angle with midline. HABITAT: RI (rare) to FL, northern and southern Gulf of Mexico, Bahamas, and Antilles to Brazil. Also Bermuda. Occur in coastal waters to about 150 ft. BIOLOGY: Feed on sea urchins, mollusks, and crabs.

Saucereye Porgy - *Calamus calamus* (Valenciennes, 1830)

FEATURES: Silvery with brassy to golden highlights. May display dark blotches. Scaleless portion of head silvery with golden to brassy areas. Corners of mouth silvery to yellowish. Blue line along lower margin of eyes. Small blue spot at upper pectoral-fin base. Profile of snout forms between 60° to 65° angle with midline. HABITAT: NC to FL, Gulf of Mexico, Bermuda, Bahamas, and Caribbean Sea to southeastern Brazil. Reported to be rare to absent from W FL and in western Gulf of Mexico. Found along coastal waters to about 250 ft. Adults occur around coral reefs, juveniles over seagrass beds. BIOLOGY: Can quickly change color and pattern.

Whitebone Porgy - *Calamus leucosteus* Jordan & Gilbert, 1885

FEATURES: Bluish silver on head and body. May show dark blotches or about five dark bars on sides. Snout purplish gray. Narrow blue lines above and below eyes. Fins dusky. Spot at upper pectoral-fin base always absent. Pectoral fins comparatively long, reaching beyond anal-fin origin. HABITAT: NC to FL Keys, Gulf of Mexico to Yucatán. Commonly found over sponge and coral, and also sandy bottoms from about 33 to 330 ft., usually above 100 ft.

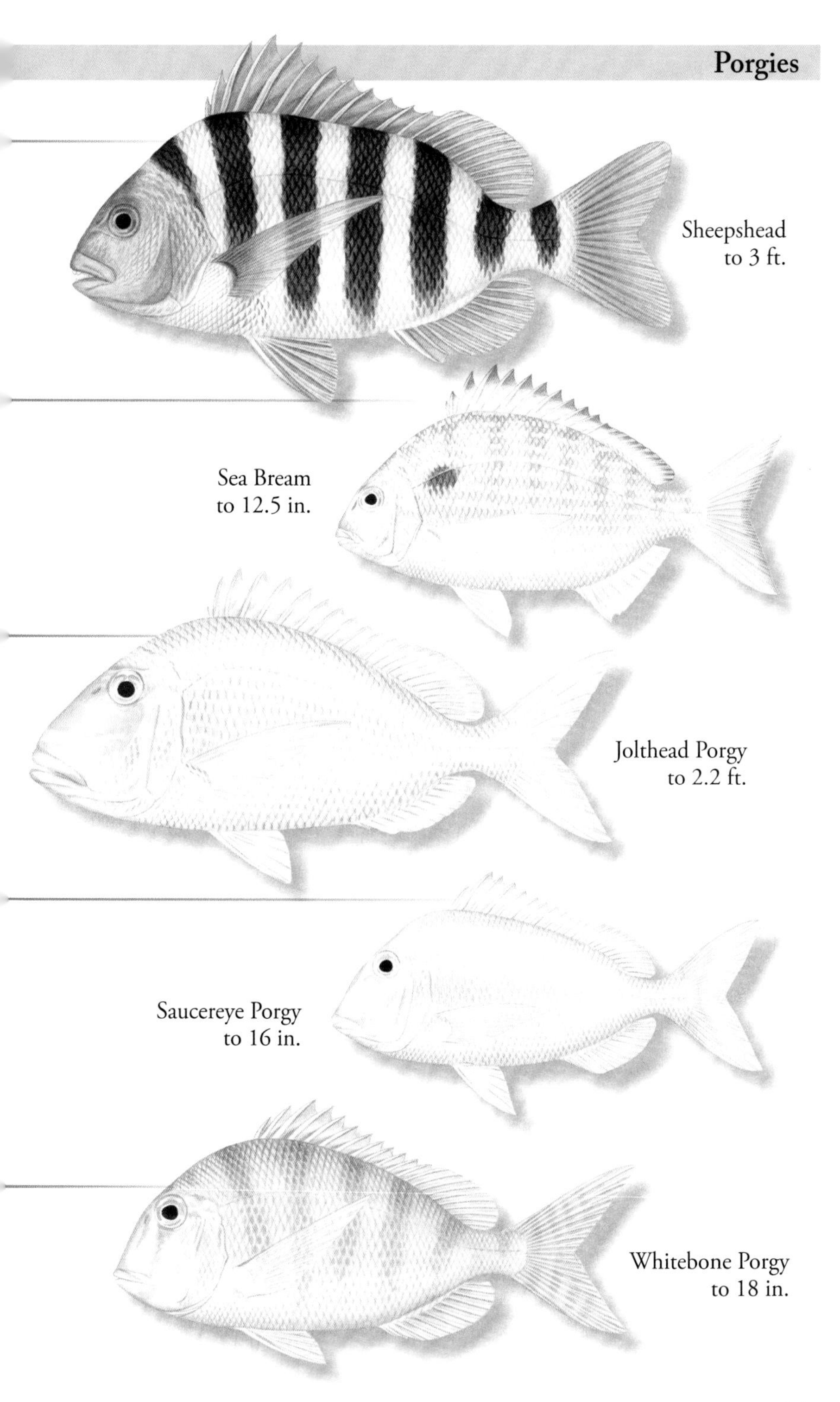
Sheepshead
to 3 ft.
Sea Bream
to 12.5 in.
Jolthead Porgy
to 2.2 ft.
Saucereye Porgy
to 16 in.
Whitebone Porgy
to 18 in.

Knobbed Porgy - *Calamus nodosus* Randall & Caldwell, 1966

FEATURES: Rosy silver. Scale centers iridescent bluish. Snout purplish with yellowish to bronze spots. Narrow blue line under eyes. A diffuse blue spot often present on upper pectoral-fin base. Snout steeply sloping, forms between 57° to 65° angle with midline. Bony knob present above nostrils; well developed in large adults. Nape humped in large adults. HABITAT: NC to FL, and Gulf of Mexico to Yucatán. Occur over wrecks, reefs, and rocky bottoms from about 30 to 290 ft.

Sheepshead Porgy - *Calamus penna* (Valenciennes, 1830)

FEATURES: Silvery with iridescent reflections. May display about seven dark bars on sides. Cheeks silvery with a wash of yellowish brown. A brownish bar runs from below eyes to corner of mouth. Sometimes with a faint blue line below eye. Always with a small black spot on upper pectoral-fin base. Upper lip evenly divided by a lengthwise groove. Head profile evenly convex. Pectoral fins reach to above anal fin origin. HABITAT: NC to FL, Bahamas, Gulf of Mexico, and Caribbean Sea to southern Brazil. Found over hard to semi-hard bottoms in clear water from about 10 to 285 ft.

Pluma Porgy - *Calamus pennatula* Guichenot, 1868

FEATURES: Overall silvery. May display some golden areas on top of head, on nape, and on snout. May also display diffuse body bars. All with a rectangular bright blue blotch behind eye to edge of operculum. Blue lines below, above, and in front of eyes. Faint to distinct blue lines on snout. Corners of mouth yellow. Scales may shine with a reflective blue dot. Upper head profile forms between 32° to 40° angle with midline. HABITAT: FL, southern Gulf of Mexico, Bahamas, and Caribbean Sea to southern Brazil. Occur over reefs, rocky bottoms and adjacent sandy areas, and seagrass beds from about 10 to 280 ft.

Littlehead Porgy - *Calamus proridens* Jordan & Gilbert, 1884

FEATURES: Iridescent silvery with yellowish highlights. Scale centers blue. Snout yellowish with narrow, blue wavy lines. Blue lines below and behind each eye. Small blue patch at upper opercular margin. Corners of mouth pale yellow. Head profile steeply sloping, especially in adults. Profile of snout forms between 57° to 64° angle with midline. HABITAT: FL, western and southern Gulf of Mexico, Bahamas, and Greater Antilles. Found over natural and artificial reefs, around offshore platforms, and live bottoms from near surface to about 195 ft.

Bermuda Porgy - *Diplodus bermudensis* Caldwell, 1965

FEATURES: Overall silvery. About nine faint and narrow bars on upper sides that fade with age. Large black blotch on caudal peduncle. Opercular membrane blackish. Small black spot on upper pectoral-fin base. Lateral line with 62–67 scales. HABITAT: Bermuda. Found over shallow rocky and coral bottoms in clear water from shore to about 80 ft.

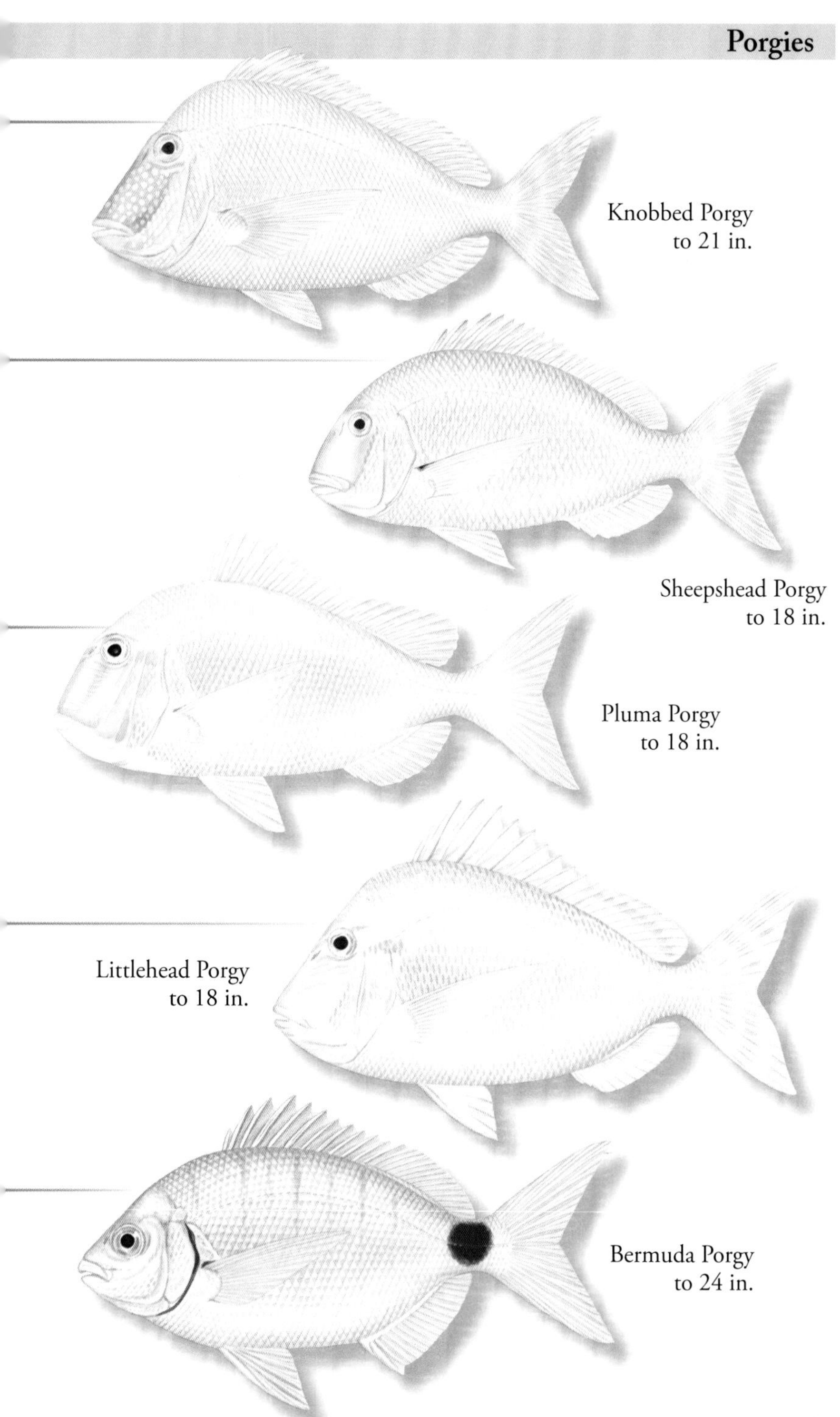
Knobbed Porgy
to 21 in.
Sheepshead Porgy
to 18 in.
Pluma Porgy
to 18 in.
Littlehead Porgy
to 18 in.
Bermuda Porgy
to 24 in.

Silver Porgy - *Diplodus caudimacula* (Poey, 1860)

FEATURES: Silvery with pale yellow reflections. Opercular membrane blackish. About nine faint, narrow bars on upper body. Bars fade with age. Black blotch on upper caudal peduncle. Fins with yellowish cast. Lateral line with 56–65 scales. HABITAT: NC to FL, Bahamas, southern Gulf of Mexico to southern Caribbean Sea. Occur in clear, shallow coastal waters over rocky and coral bottoms. NOTE: Previously known as *Diplodus argenteus*, now a separate species from Brazil to Argentina.

Spottail Pinfish - *Diplodus holbrookii* (Bean, 1878)

FEATURES: Iridescent golden dorsally and on sides. Silvery below. Opercular membrane black. About nine faint, narrow bars on upper sides. Bars fade with age. Large black saddle on upper caudal peduncle. Dorsal-fin membranes with brownish cast. Pectoral, caudal, and anal fins with golden cast. Lateral line with 50–61 scales. HABITAT: Chesapeake Bay to northwestern Gulf of Mexico, and northern coast of Cuba. Occur in shallow coastal waters over reefs and in bays and harbors. Prefer vegetated bottoms. BIOLOGY: Feed on a variety of invertebrates.

Pinfish - *Lagodon rhomboides* (Linnaeus, 1766)

FEATURES: Body with alternating iridescent bluish and yellowish to bronze stripes that intersect grayish bars. Stripes and bars form checkerboard pattern. Black spot on lateral-line origin. Dorsal fin with yellowish bands. Anal fin with yellowish inner band. Body oval and laterally compressed, with upper and lower profiles similarly shaped. HABITAT: MA (rare) to FL, Gulf of Mexico to Belize, northern Cuba, Bermuda, and Bahamas. Occur near bottom in a variety of shallow coastal habitats, including bays, estuaries, and canals. Often around vegetated and hard bottoms. May enter fresh water. BIOLOGY: Feed on a variety of plants, small fishes, and invertebrates.

Red Porgy - *Pagrus pagrus* (Linnaeus, 1758)

FEATURES: Iridescent pinkish to reddish dorsally, becoming silvery below. Dorsal, caudal, and pectoral fins with pinkish to reddish cast. May display reddish bars. Opercular margin dark. Body oblong and laterally compressed. Head profile gently sloping. HABITAT: NY to FL, Gulf of Mexico, and southern Caribbean Sea. Also eastern Atlantic and Mediterranean Sea. Found near bottom over rocky and hard sand bottoms from about 30 to 260 ft.

Longspine Porgy - *Stenotomus caprinus* Jordan & Gilbert, 1882

FEATURES: Silvery with olivaceous cast dorsally. Juveniles may have faint, dark, narrow bars. Body otherwise unmarked. Teeth in front of jaws incisor-like. Head profile almost straight, steeply sloping. First two dorsal-fin spines very short. Third through fifth spines very tall. Body deeply oblong in profile. HABITAT: DE to FL and Gulf of Mexico to Yucatán Peninsula. Found over muddy bottoms from about 15 to 390 ft. BIOLOGY: Feed on a variety of invertebrates, detritus, and small fishes.

ALSO IN THE AREA: *Calamus campechanus*, *Calamus cervigoni*, see p. 547.

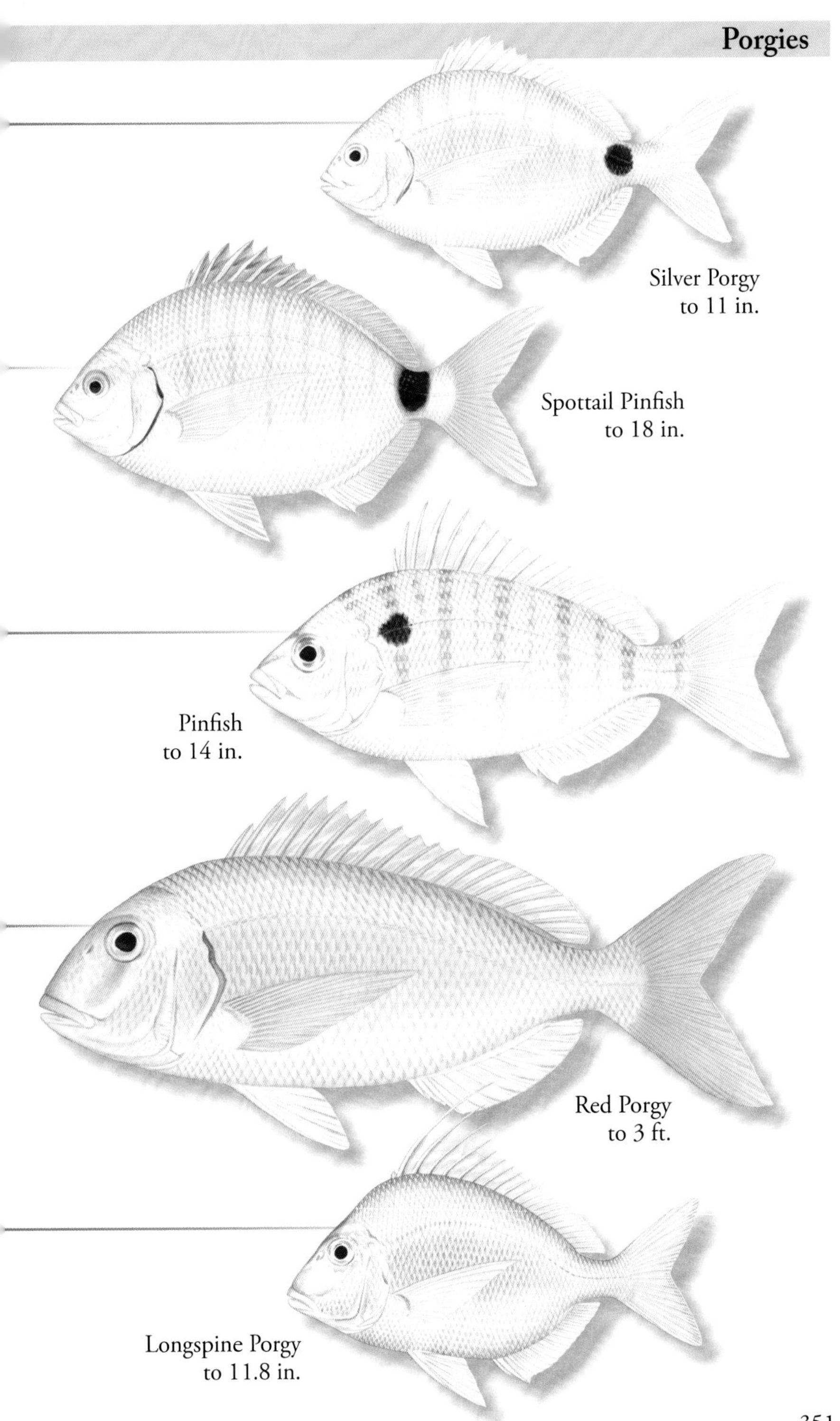
Silver Porgy
to 11 in.
Spottail Pinfish
to 18 in.
Pinfish
to 14 in.
Red Porgy
to 3 ft.
Longspine Porgy
to 11.8 in.

Polynemidae - Threadfins

Littlescale Threadfin - *Polydactylus oligodon* (Günther, 1860)

FEATURES: Dusky silver dorsally; whitish ventrally. Dorsal and caudal fins dark, blackish at margins. Anal and pelvic fins with dark inner portions. Pectoral fins black. Lower seven (sometimes eight) pectoral-fin rays separate, long, and filamentous. Scales comparatively small. HABITAT: SE FL, Bahamas, Yucatán, and northern and southern Caribbean Sea to southern Brazil. Occur near bottom and close to shore along exposed sandy beaches.

Barbu - *Polydactylus virginicus* (Linnaeus, 1758)

FEATURES: Silvery olive to blue gray dorsally; whitish ventrally. Dorsal and caudal fins dusky to yellowish with blackish margins. Pelvic and anal fins with pale margins. Pectoral fins pale with dark inner area. Lower seven pectoral-fin rays separate, long, and filamentous. HABITAT: NJ (rare) to FL, Bermuda, Bahamas, southern Gulf of Mexico, and Caribbean Sea to eastern Brazil. Occur near bottom along sandy and muddy flats and beaches and around mangroves. Also found in estuaries and river mouths.

ALSO IN THE AREA: *Polydactylus octonemus*, see p. 547.

Sciaenidae - Drums

Ground Croaker - *Bairdiella ronchus* (Cuvier, 1830)

FEATURES: Silvery gray dorsally; silvery on sides and below. Faint streaks along sides. Pectoral and pelvic fins yellowish. Anal fin dusky anteriorly. Snout bluntly rounded. Preopercle serrated with a few strong spines at corner. Second anal-fin spine long and stout, and as long as first ray. HABITAT: Western Gulf of Mexico and Caribbean Sea to Guyana. Occur along shallow sandy and muddy bottoms from about 10 to 130 ft. Also in brackish water.

Blue Croaker - *Corvula batabana* (Poey, 1860)

FEATURES: Bluish gray with silvery to brassy highlights. Dark spots scattered on upper sides; form rows along scales on sides. Preopercle finely serrated at corner. Soft dorsal and anal fins scaled at base. Caudal-fin margin rounded. HABITAT: S FL, Bay of Campeche, and in Caribbean Sea from Cuba to St. Barthélemy. One record off Trinidad. Occur in clear, highly saline waters over vegetated mud flats and over coral reefs from about 30 to 100 ft. BIOLOGY: Feed mainly on crustaceans. NOTE: Previously placed in the genus *Bairdiella*.

Striped Croaker - *Corvula sanctaeluciae* Jordan, 1890

FEATURES: Grayish with silvery highlights dorsally. Silvery below. Brownish spots follow scales along back and on sides. Rows on back are straight anteriorly, oblique under dorsal fins, and straight posteriorly. Fins pale yellowish with small dark flecks. Soft dorsal and anal fins scaled at base. HABITAT: E FL (rare), Bay of Campeche, and Caribbean Sea. Over muddy and sandy bottoms of inshore waters. Juveniles around rocky areas. BIOLOGY: Feed mainly on shrimps.

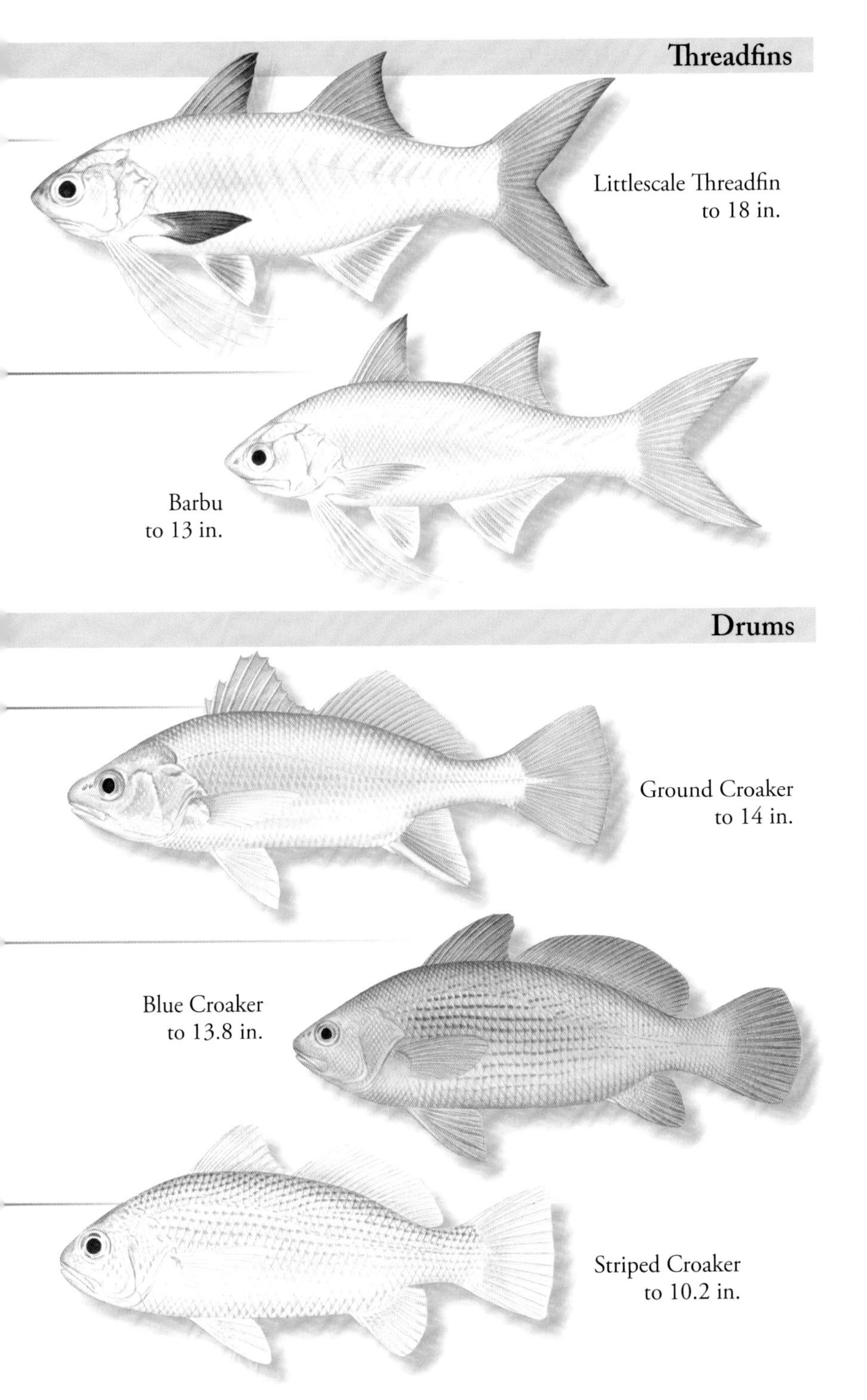

Threadfins
Littlescale Threadfin
to 18 in.
Barbu
to 13 in.
Drums
Ground Croaker
to 14 in.
Blue Croaker
to 13.8 in.
Striped Croaker
to 10.2 in.

Barbel Drum - *Ctenosciaena gracilicirrhus* (Metzelaar, 1919)

FEATURES: Silvery gray dorsally; silvery on sides and below. Small dark spot at pectoral-fin base. Others described with a large black spot at pectoral-fin base. Fins pale dusky to yellowish. Upper spiny dorsal fin dusky. Snout bluntly pointed and slightly overhangs mouth. Chin with a slender barbel, its length about half diameter of eye. Preopercular margin finely serrated. HABITAT: Guatemala to southern Brazil. Demersal over sandy and muddy bottoms from about 30 to 425 ft.

Acoupa Weakfish - *Cynoscion acoupa* (Lacepède, 1801)

FEATURES: Color varies. Dusky dorsally, fading to uniformly silver on sides and below, or silver on sides with a pale to deep golden wash below. Small to moderately large dark spot at upper pectoral-fin base. Dorsal fins dusky to amber; ventral fins pale yellow to amber. Inside of mouth yellow to orange. Mouth large and oblique with one or two large canine-like teeth at tip of upper jaw. Soft dorsal fin unscaled. Caudal-fin margin angular. HABITAT: Belize to Brazil. Demersal over sandy and muddy bottoms from about 30 to 260 ft. Caught as deep as 425 ft.

Jamaica Weakfish - *Cynoscion jamaicensis* (Vaillant & Bocourt, 1883)

FEATURES: Grayish dorsally; silvery on sides and below. Faint dark spots from lines along scale rows above lateral line. Upper pectoral-fin rays dark. Anal and pelvic fins yellowish. Lower half of soft dorsal fin scaled. Mouth large and oblique with one or two large canine-like teeth at tip of upper jaw. Caudal-fin margin straight to trilobed. Body comparatively deep. HABITAT: Haiti, Dominican Republic, Puerto Rico, and Honduras to southern Brazil. Occur over sandy and muddy bottoms in coastal waters from shore to about 390 ft. Juveniles in estuaries.

Smooth Weakfish - *Cynoscion leiarchus* (Cuvier, 1830)

FEATURES: Iridescent grayish blue with tiny dark spots along scales dorsally. Silvery on sides and below. Soft dorsal and caudal fins with dark margins. Anal and pelvic fins pale yellow to colorless. Mouth large and oblique with one or two large canine-like teeth at tip of upper jaw. One or two rows of scales at soft dorsal-fin base. Caudal-fin margin trilobed to concave in adults. Scales smooth and comparatively small. HABITAT: Colombia to southeast Brazil. Occur over sandy bottoms in coastal waters from shore to about 164 ft. Common in surf zone and estuaries.

Spotted Seatrout - *Cynoscion nebulosus* (Cuvier, 1830)

FEATURES: Silvery gray with iridescent reflections dorsally. Silvery on sides and below. Posterior upper sides with round black spots. Dorsal and caudal fins spotted. Mouth large with lower jaw protruding. Pair of large canine-like teeth in upper jaw tip. Juveniles with angular caudal fin; adults with straight to slightly concave caudal-fin margin. HABITAT: NY to FL, and entire Gulf of Mexico. Absent from Cuba. In coastal waters over sandy bottoms and seagrass beds, around rocks, and in marshes, tide pools, and estuaries from near shore to about 60 ft.

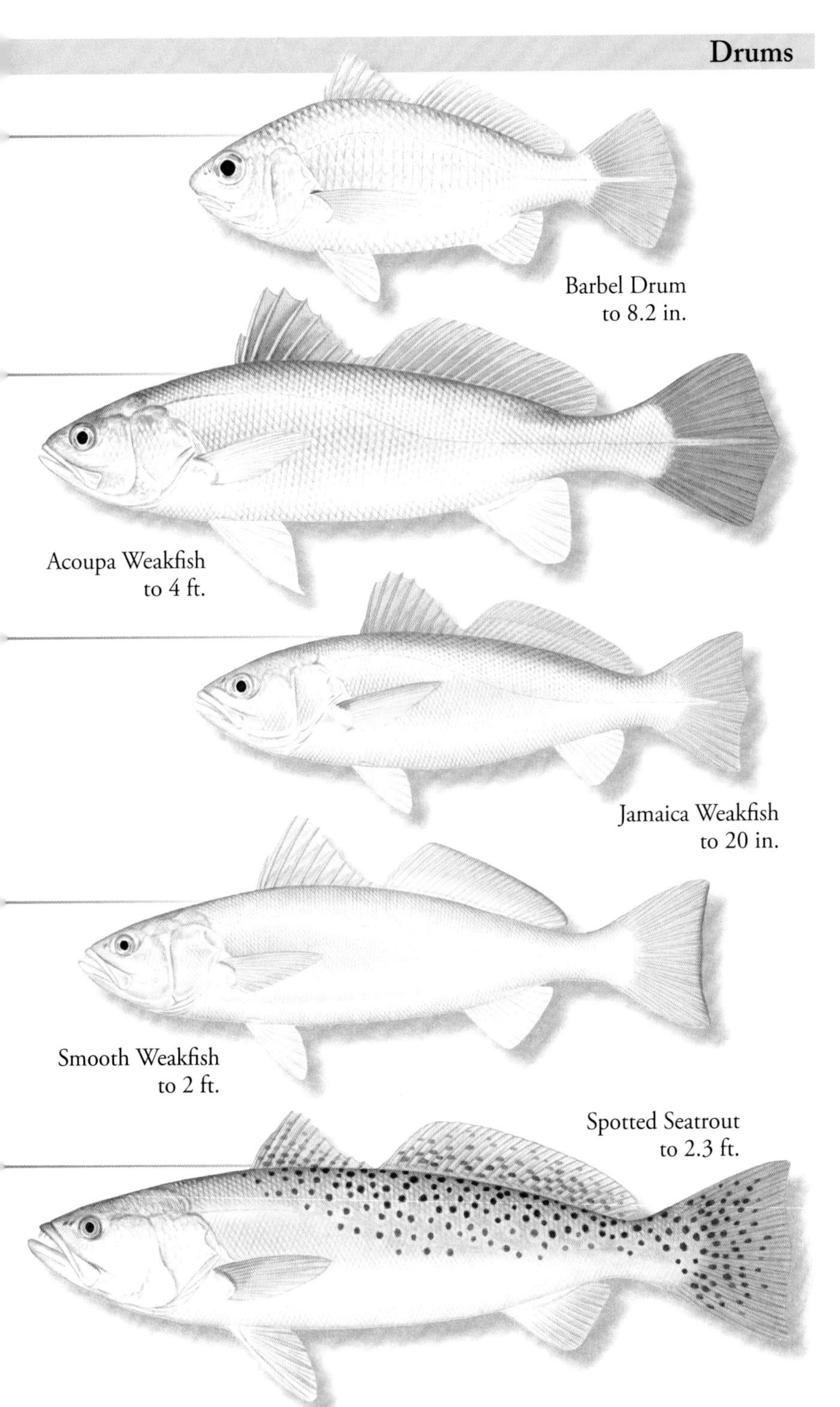
Barbel Drum
to 8.2 in.
Acoupa Weakfish
to 4 ft.
Jamaica Weakfish
to 20 in.
Smooth Weakfish
to 2 ft.
Spotted Seatrout
to 2.3 ft.

Green Weakfish - *Cynoscion virescens* (Cuvier, 1830)

FEATURES: Pale coppery with iridescent highlights dorsally. Silvery white below. Inside of mouth orange red. Dorsal fins dusky; caudal fin dark. Pectoral, anal, and pelvic fins yellowish to orange. Mouth large and oblique with one or two large canine-like teeth at tip of upper jaw. Pectoral fins slightly longer than pelvic fins. Body very long with a narrow caudal peduncle. Scales very small; lateral-line scales larger. HABITAT: Honduras to southern Brazil. Found near bottom over sandy and muddy bottoms along the coast near river mouth. Juveniles in estuaries.

Spotted Drum - *Eques punctatus* Bloch & Schneider, 1801

FEATURES: Whitish with wide, dark brown bands on head and body: one through eyes; a second from nape to pelvic fins; the third from first dorsal fin to caudal-fin base with narrow bands above and below. Dorsal, caudal, and anal fins dark brown with white spots. Pectoral fins dark brown. Head profile almost straight. Caudal fin bluntly pointed. Juveniles with dark spot on snout. HABITAT: S FL, Gulf of Mexico, Bahamas, and Caribbean Sea to Brazil. Found primarily over coral reefs. BIOLOGY: Secretive and solitary. Feed on soft corals and invertebrates.

Jackknife-fish - *Equetus lanceolatus* (Linnaeus, 1758)

FEATURES: Pearly tan with iridescent reflections. Three white-edged, blackish bands on body: one through eyes; a second from nape through pelvic fins; the third from first dorsal fin along body to tip of caudal fin. Pectoral fins pale to colorless. Head profile nearly straight. Caudal fin pointed. Juveniles with black streak on snout. HABITAT: SC to FL, Gulf of Mexico, Bahamas, and Caribbean Sea to southern Brazil. Also Bermuda. Occur over coastal sandy and muddy bottoms. Also around reefs and hard bottoms to about 196 ft. BIOLOGY: Feed on invertebrates.

Shorthead Drum - *Larimus breviceps* Cuvier, 1830

FEATURES: Silvery gray with iridescent highlights dorsally; silvery on sides and below. A small dark spot present at upper pectoral-fin base. Lower caudal-fin lobe, anal, and pelvic fins yellowish. Snout short, blunt. Mouth very large and oblique. Lower jaw protrudes. Caudal-fin margin angular. Body deep and compressed. HABITAT: Greater Antilles and southern Caribbean Sea to southern Brazil. Occur over muddy and sandy bottoms from near shore to about 195 ft. BIOLOGY: Feed mainly on small shrimps. Form spawning aggregations along beaches near estuaries.

King Weakfish - *Macrodon ancylodon* (Bloch & Schneider, 1801)

FEATURES: Silvery gray dorsally; silvery to pale yellow below. Upper pectoral fin, lower caudal fin, anal and pelvic fins yellowish. Mouth large and oblique. Teeth in front of mouth very long, curved and with arrow-like tips. Soft dorsal and anal fins entirely scaled. Caudal-fin margin concave above and below. Scales small and smooth, those on lateral line larger. HABITAT: Colombia to Bahia Blanca, Argentina. Occur over muddy and sandy bottoms in coastal waters to about 230 ft. Juveniles in estuaries and lagoons. BIOLOGY: Feed on shrimps and small fishes.

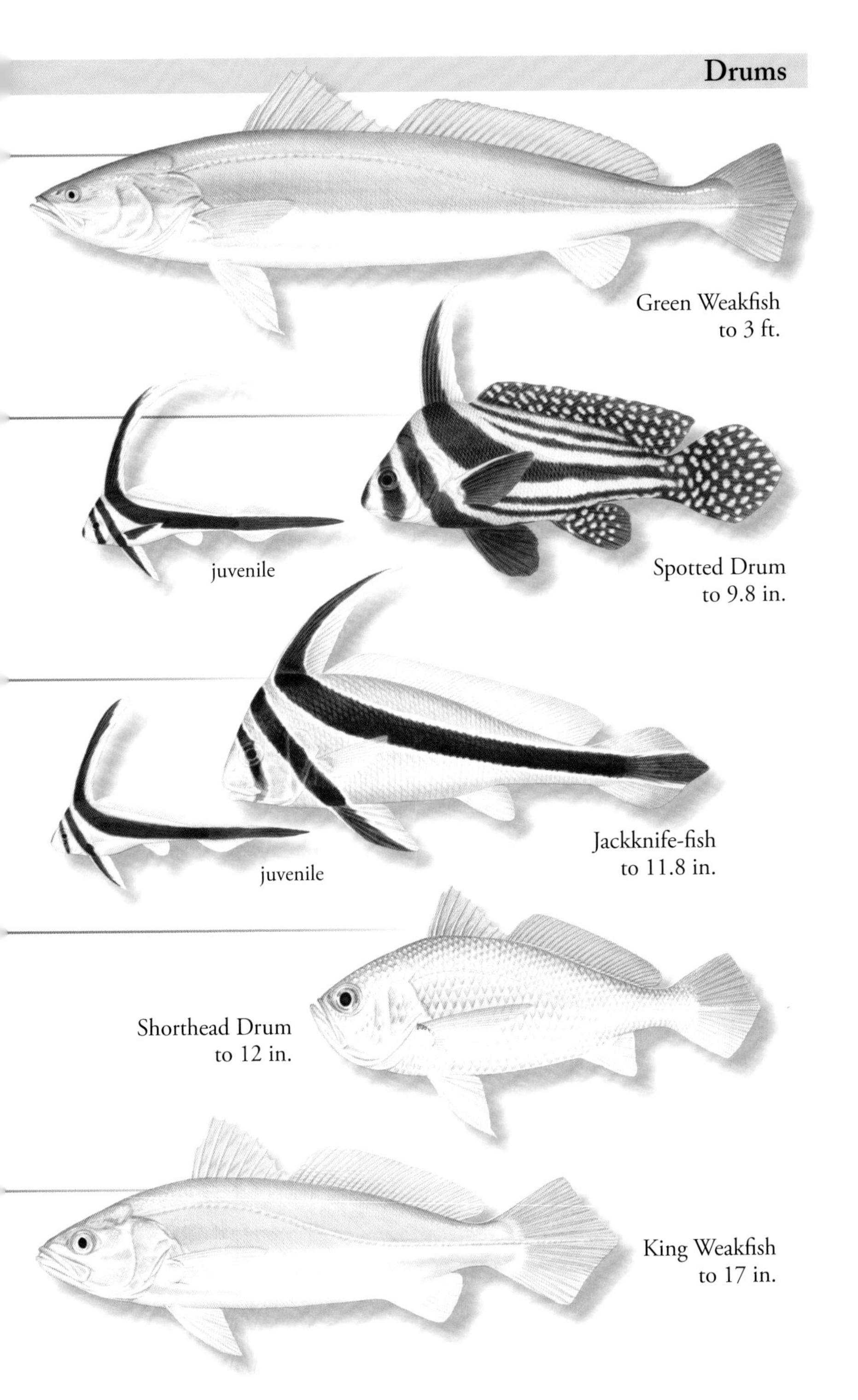

Green Weakfish
to 3 ft.

Spotted Drum
to 9.8 in.

Jackknife-fish
to 11.8 in.

Shorthead Drum
to 12 in.

King Weakfish
to 17 in.

Southern Kingfish - *Menticirrhus americanus* (Linnaeus, 1758)

FEATURES: Silvery gray to golden brown dorsally and on sides. Ventral area white. Intensity of color varies. Often with seven or eight faint to dark brownish bands on upper body; when present, first two bands form V-shape. Mouth small, nearly horizontal, with a fleshy barbel on lower lip. Ventral profile nearly straight. Abdomen flattened. HABITAT: MA to FL, Gulf of Mexico, and Caribbean Sea to southern Uruguay. Over sandy mud to hard sand bottoms in shallow coastal waters. Also in surf and estuaries. Juveniles often in brackish water.

Gulf Kingfish - *Menticirrhus littoralis* (Holbrook, 1847)

FEATURES: Pale silvery gray dorsally and on sides. Ventral area white. First dorsal-fin tip dusky. Upper caudal-fin tip usually black. Body otherwise unmarked. Mouth small, nearly horizontal, with a fleshy barbel on lower lip. Ventral profile nearly straight. Abdomen flattened. HABITAT: VA to FL, Gulf of Mexico, western Caribbean Sea to Brazil. Reported as far north as MA. In surf and over sandy and sandy mud bottoms. Occasionally in estuaries. BIOLOGY: Feed on bottom-dwelling invertebrates.

Whitemouth Croaker - *Micropogonias furnieri* (Desmarest, 1823)

FEATURES: Silvery with iridescent reflections and brown spots that form oblique and wavy lines along scale rows dorsally and on sides. Spots may also form several bars on midsides. Ventral area silvery to golden. Anal and pelvic fins with some yellow. Three or four pairs of tiny barbels along inner sides of lower lips. Preopercular margin serrated with two to three sharp spines at corner. HABITAT: Southern Gulf of Mexico and Caribbean Sea to Argentina. Found over muddy bottoms of coastal waters from shore to about 400 ft. Also in Estuaries.

Smalleye Croaker - *Nebris microps* Cuvier, 1830

FEATURES: Shades of golden to golden orange, or silvery with five to six brownish saddles along back. Pectoral fins orange with black lower margin. Anal and pelvic fins yellow to orange, may have black tips. Mouth large; eyes very small. Head large; body robust. Lateral-line scales enlarged. Caudal-fin margin trilobed. HABITAT: Colombia to Guyana. Found coastally over sandy and muddy bottoms from shore to about 164 ft. Also enters estuaries.

Reef Croaker - *Odontoscion dentex* (Cuvier, 1830)

FEATURES: Silvery gray to brownish gray with scattered brownish spots on scales that may form rows. Large black blotch at pectoral-fin base. Eyes large, dark. Snout short, blunt. Mouth large with pair of canine teeth in lower jaw tip. Chin lacks barbels. Caudal-fin margin straight to slightly rounded. Dorsal profile evenly convex, similar to ventral profile. HABITAT: FL, Gulf of Mexico, and Caribbean Sea to southern Brazil. In shallow coastal waters primarily over reefs and also over sandy mud bottoms. BIOLOGY: Active at night; shelters in caves during the day. Feed on shrimps, small fishes.

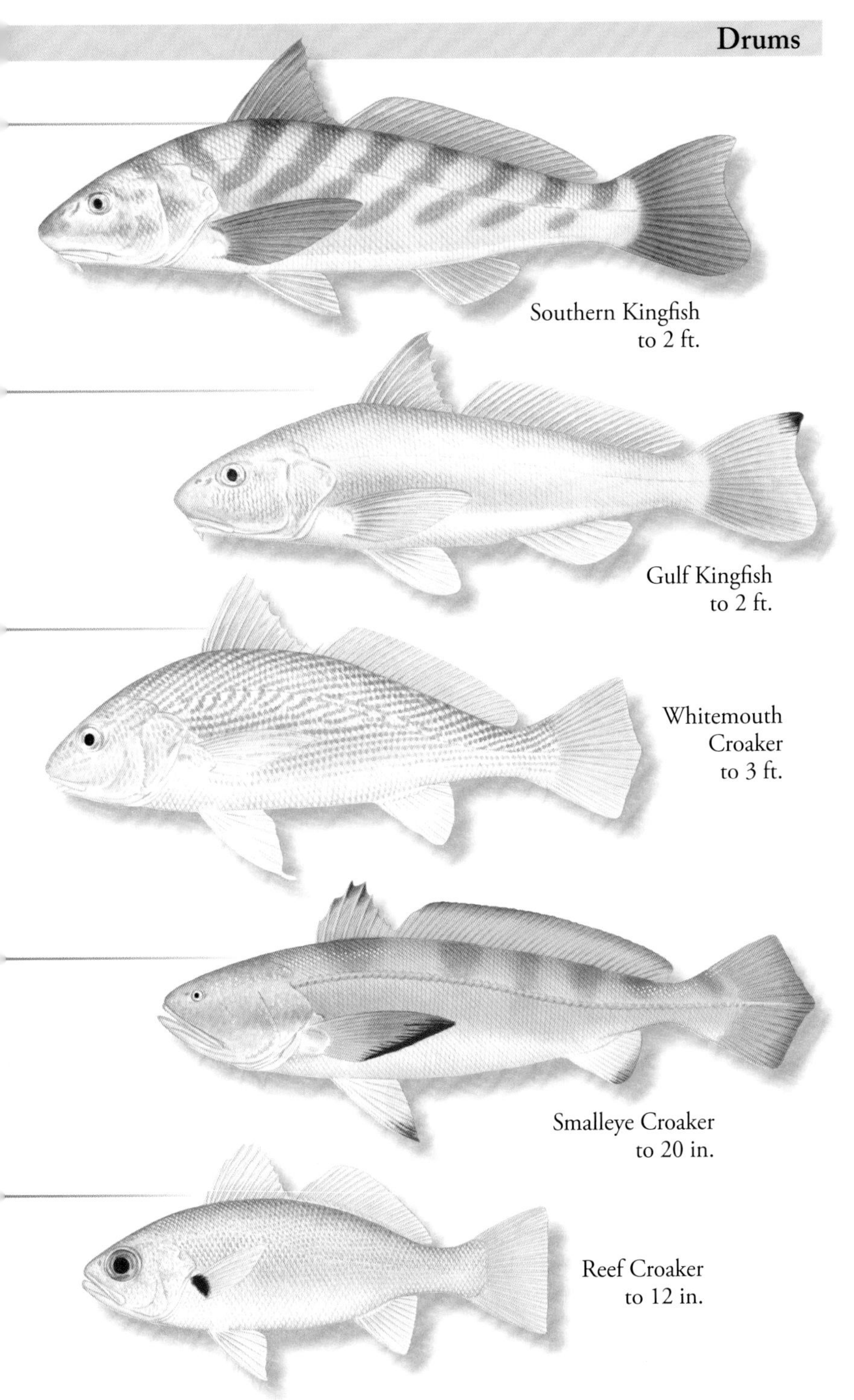

Southern Kingfish
to 2 ft.

Gulf Kingfish
to 2 ft.

Whitemouth
Croaker
to 3 ft.

Smalleye Croaker
to 20 in.

Reef Croaker
to 12 in.

Spotted Croaker - *Ophioscion punctatissimus* Meek & Hildebrand, 1925

FEATURES: Dark gray dorsally, fading to silvery gray on sides and silvery below. Scales with dark margins and iridescent centers. Fins grayish. Snout overhangs mouth. Chin lacks barbels. Second anal-fin spine long and stout. Caudal-fin margin angular. HABITAT: Belize to northeastern Brazil. Reported from Puerto Rico and adjacent islands and to southern Brazil. Occur over sandy mud bottoms, and commonly along beaches from shore to about 165 ft. BIOLOGY: Feed on bottom-dwelling worms and crustaceans.

Banded Croaker - *Paralonchurus brasiliensis* (Steindachner, 1875)

FEATURES: Color and pattern vary. Shades of silvery gay to yellowish dorsally and on sides. Silvery white to golden below. Large black blotch present behind upper portion of opercle. Seven to nine black to gray bars present on upper body that may appear as alternating wide and narrow. Snout rounded, overhangs nearly horizontal; mouth small. Nape humped. Caudal fin asymmetrical. HABITAT: Honduras to southern Brazil. Occur over muddy bottoms from near shore to about 165 ft.

High-hat - *Pareques acuminatus* (Bloch & Schneider, 1801)

FEATURES: Head and body with alternating, irregular, pearly white and dark brown to grayish stripes. First dorsal fin with dark brown inner band; other fins dark brown. Mouth small, nearly horizontal. Dorsal profile arched; ventral profile nearly straight. Juveniles with two spots on snout, dark upper pectoral fins, and tall, trailing dorsal fin. HABITAT: VA to FL, Bahamas, Gulf of Mexico, and Caribbean Sea to southern Brazil. In clear, coastal waters over sandy and muddy bottoms and coral reefs to about 200 ft. BIOLOGY: Feed on bottom-dwelling invertebrates.

Blackbar Drum - *Pareques iwamotoi* Miller & Woods, 1988

FEATURES: Pearly white to grayish with one dark brown bar through eyes. A second broad, dark brown bar from dorsal to pelvic fins merges with a brown horizontal stripe that extends to caudal-fin margin. Broad pale brown area below soft dorsal fin. Mouth small, nearly horizontal. Dorsal profile arched; ventral profile nearly straight. Juveniles similarly patterned in white and black, but may lack pale brown area. HABITAT: NC to FL, Gulf of Mexico, and southern Caribbean Sea to Brazil. Over sandy mud, reefs, and hard bottoms from about 120 to 600 ft. BIOLOGY: Feed on gastropods.

Cubbyu - *Pareques umbrosus* (Jordan & Eigenmann, 1889)

FEATURES: Brown to grayish brown with alternating narrow and very thin dark stripes on sides. Stripes merge on head to form V-shapes. All fins dark. Mouth small, nearly horizontal. Ventral profile nearly straight. Juveniles with oval-shaped mark between eyes, colorless pectoral fins, and tall, but not trailing, dorsal fin. HABITAT: NC to FL, Gulf of Mexico, Bermuda, scattered in Caribbean Sea, and Venezuela to French Brazil. Found in shallow coastal waters over sandy mud, hard bottoms, and reefs from about 10 to 400 ft.

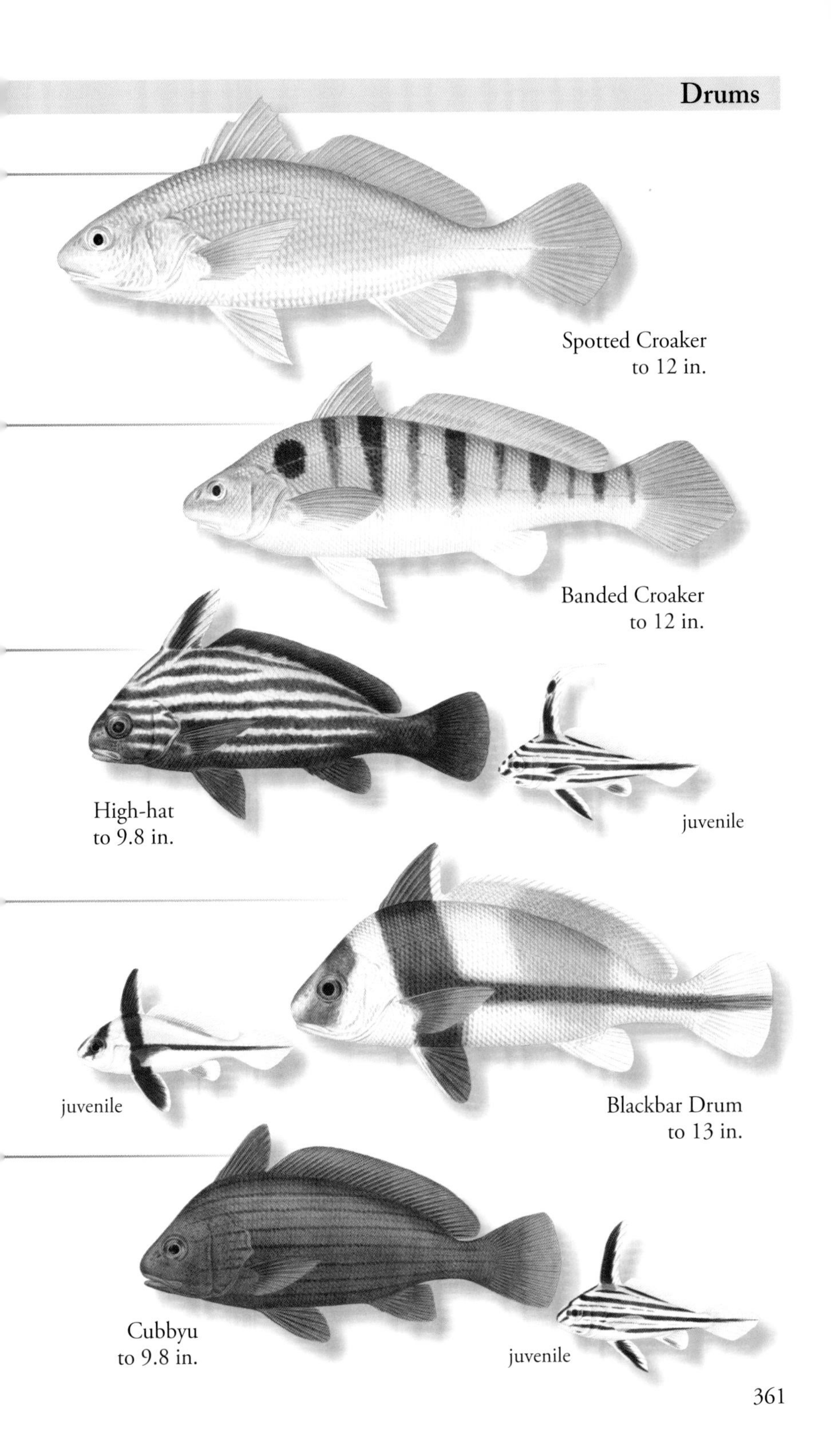

Spotted Croaker
to 12 in.

Banded Croaker
to 12 in.

High-hat
to 9.8 in.

juvenile

juvenile

Blackbar Drum
to 13 in.

Cubbyu
to 9.8 in.

juvenile

Black Drum - *Pogonias cromis* (Linnaeus, 1766)

FEATURES: Silvery bronze or golden to almost black dorsally and on sides. Ventral area silvery. Juveniles silvery gray with four to five black bars on sides. Bars fade with age. Mouth small, nearly horizontal. Chin with 10–13 pairs of barbels. Dorsal profile arched. Ventral profile nearly straight. Dorsal fin deeply notched. Caudal-fin margin nearly straight. HABITAT: Nova Scotia to FL, and Gulf of Mexico to Bay of Campeche. Occur over coastal sandy and mud bottoms, near river mouths, and in estuaries. Also in surf. BIOLOGY: Feed on bottom-dwelling organisms.

Chao Stardrum - *Stellifer chaoi* Aguilera, Solano & Valdez, 1983

FEATURES: Iridescent brownish too grayish dorsally; silvery on sides and below. Dorsal fins dusky; spiny dorsal fin may be darker at margin. Pectoral, anal, and pelvic fins brownish to blackish. Snout rounded; top of head spongy and slightly concave. Chin barbels absent. Preopercular corner with three to four prominent spines. Second anal-fin spine long and stout. Caudal fin margin angular. HABITAT: Bocas del Toro, Panama, to the Gulf of Venezuela. Occur over sandy and muddy bottoms of shallow coastal waters and estuaries.

Star Drum - *Stellifer lanceolatus* (Holbrook, 1855)

FEATURES: Silvery grayish to silvery olive dorsally. Fins dusky to pale. Spiny dorsal fin with dark margin. Snout bluntly rounded. Mouth moderately large, oblique. Preopercular corner with four to six strong spines. Snout rounded; top of head spongy and slightly concave. Caudal fin bluntly pointed. HABITAT: VA to FL, Gulf of Mexico, and Yucatán to Belize. Absent from Cuba. Found in coastal waters over hard sandy mud bottoms from shore to about 65 ft. Also in estuaries. BIOLOGY: Feed on small crustaceans.

Southern Stardrum - *Stellifer stellifer* (Bloch, 1790)

FEATURES: Silvery gray dorsally; silvery to golden below. Upper third of spiny dorsal fin dusky, dark at tip. Pectoral, anal, and pelvic fins yellowish. Snout very short and blunt; top of head spongy. Mouth large, oblique, and extends below rear eye margin. Preopercular corner with three strong spines, sometimes with a smaller spine above. Caudal fin bluntly pointed. HABITAT: Honduras to southeastern Brazil. Also Puerto Rico. Found over sandy and muddy bottoms in warm inshore waters to about 115 ft.

Sand Drum - *Umbrina coroides* Cuvier, 1830

FEATURES: Silvery gray dorsally and on sides. Silvery below. Centers of scales dark, forming faint, oblique stripes. Upper sides with eight or nine dark to faint bars. Fins dusky. Snout bluntly rounded. Mouth small. Small barbel on chin. HABITAT: VA to FL, western Gulf of Mexico, Bahamas, and Caribbean Sea to northeastern Brazil. Occur along shallow beaches, over muddy bottoms, and in estuaries. Sometimes over reefs.

NOTE: There are 23 other poorly recorded Drums in the area.

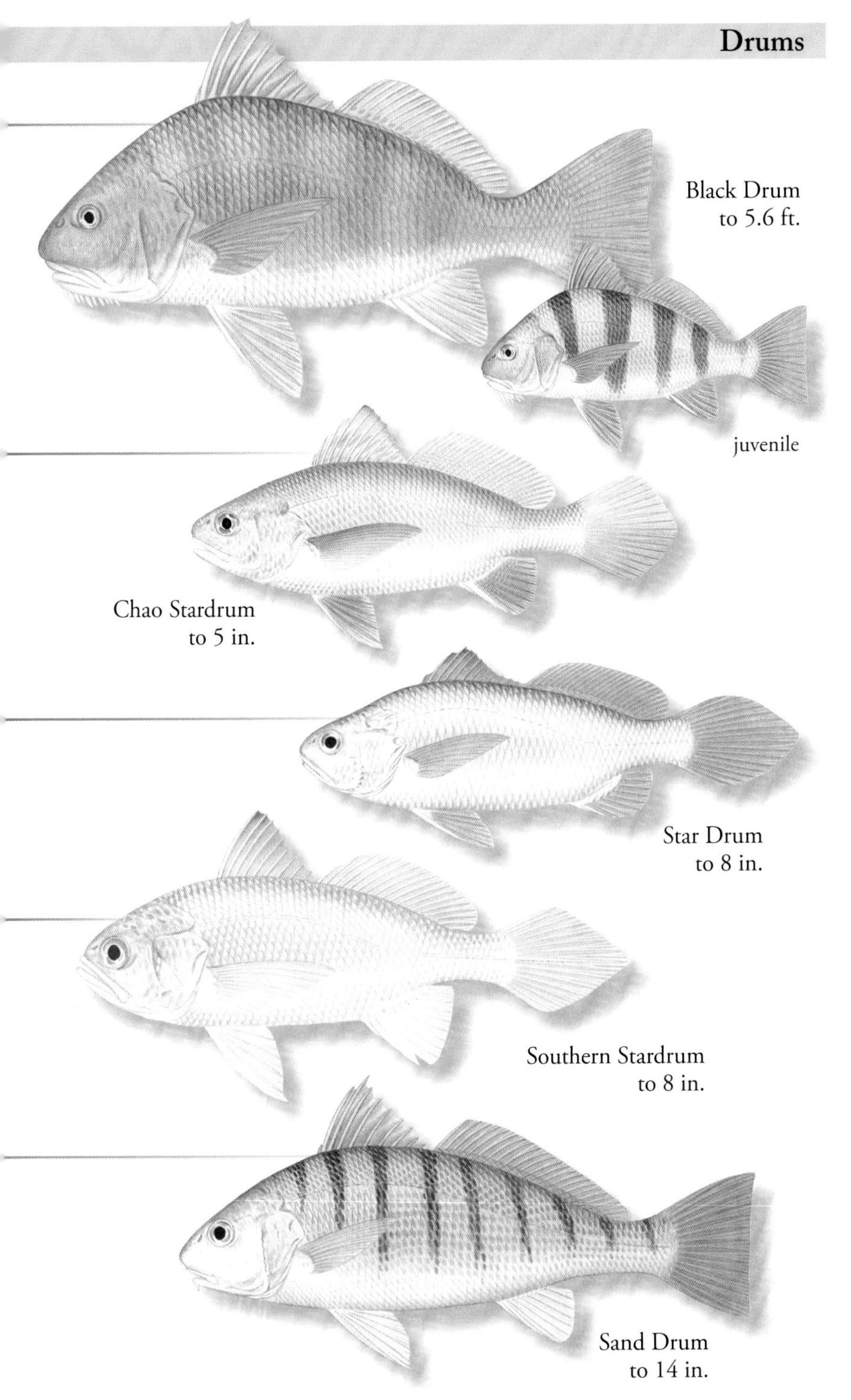
Black Drum
to 5.6 ft.
juvenile
Chao Stardrum
to 5 in.
Star Drum
to 8 in.
Southern Stardrum
to 8 in.
Sand Drum
to 14 in.

Yellow Goatfish - *Mulloidichthys martinicus* (Cuvier, 1829)

FEATURES: Silvery olive gray to silvery tan dorsally. Ventral area silvery white. A bright yellow stripe runs from eyes, along sides, and merges with yellow caudal fin. Dorsal fins and tips of anal and pelvic fins yellow. Can turn nearly uniformly pinkish, or may display dark red blotches on body at night. Snout gently sloping. Two long barbels on chin. Ventral profile nearly straight. HABITAT: S FL, Gulf of Mexico, Bermuda, Bahamas, and Caribbean Sea to southern Brazil. Also found around islands off the African coast. Found along shallow coasts over and around reefs and seagrass beds. BIOLOGY: Usually feed at night. Form large schools.

Red Goatfish - *Mullus auratus* Jordan & Gilbert, 1882

FEATURES: Reddish dorsally; silvery below. Pinkish to yellowish stripes follow scales dorsally. A pale to dark red stripe runs from eyes to caudal fin. First dorsal fin pale, with blackish red stripe above a yellow to orangish stripe that are separated by white stripes. Second dorsal fin banded. Caudal fin reddish to spotted, lacks distinct bands. Snout steeply sloping. Two long barbels on chin. HABITAT: MA to FL, Gulf of Mexico, Cuba, and southern Caribbean Sea to Guyana. Demersal over coastal muddy and silty bottoms from about 30 to 300 ft. More common in oceanic and continental shelf waters. BIOLOGY: Use barbels to locate bottom-dwelling invertebrates.

Spotted Goatfish - *Pseudupeneus maculatus* (Bloch, 1793)

FEATURES: Pearly pink to yellow with white spots along scales dorsally. Silvery to pinkish or yellow below. Three large, blackish red blotches present on sides. Night pattern reddish with interconnected reddish to pinkish blotches. Daytime resting pattern similar to night pattern. Snout comparatively long; snout profile almost straight. Two barbels on chin. HABITAT: NJ to FL, Gulf of Mexico, Bermuda, Bahamas, Caribbean Sea to southern Brazil. Demersal along shallow coastal waters over sandy and rocky bottoms near reefs. Juveniles found on seagrass beds. Rarely found below 130 ft. BIOLOGY: Feed on bottom-dwelling invertebrates. Form groups.

Dwarf Goatfish - *Upeneus parvus* Poey, 1852

FEATURES: Salmon pink to reddish dorsally with a continuous to broken yellowish stripe upper side behind eyes. Ventral area pearly white. Dorsal and caudal fins distinctly banded. Lower caudal-fin bands dark. May display bright orange blotches while resting. Snout gently sloping. Two barbels on chin. HABITAT: NC to FL, Gulf of Mexico, and Caribbean Sea to southern Brazil. Demersal over sandy, silty, and muddy bottoms from about 130 to 330 ft. BIOLOGY: Feed on bottom-dwelling invertebrates. Preyed upon by Inshore Lizardfish.

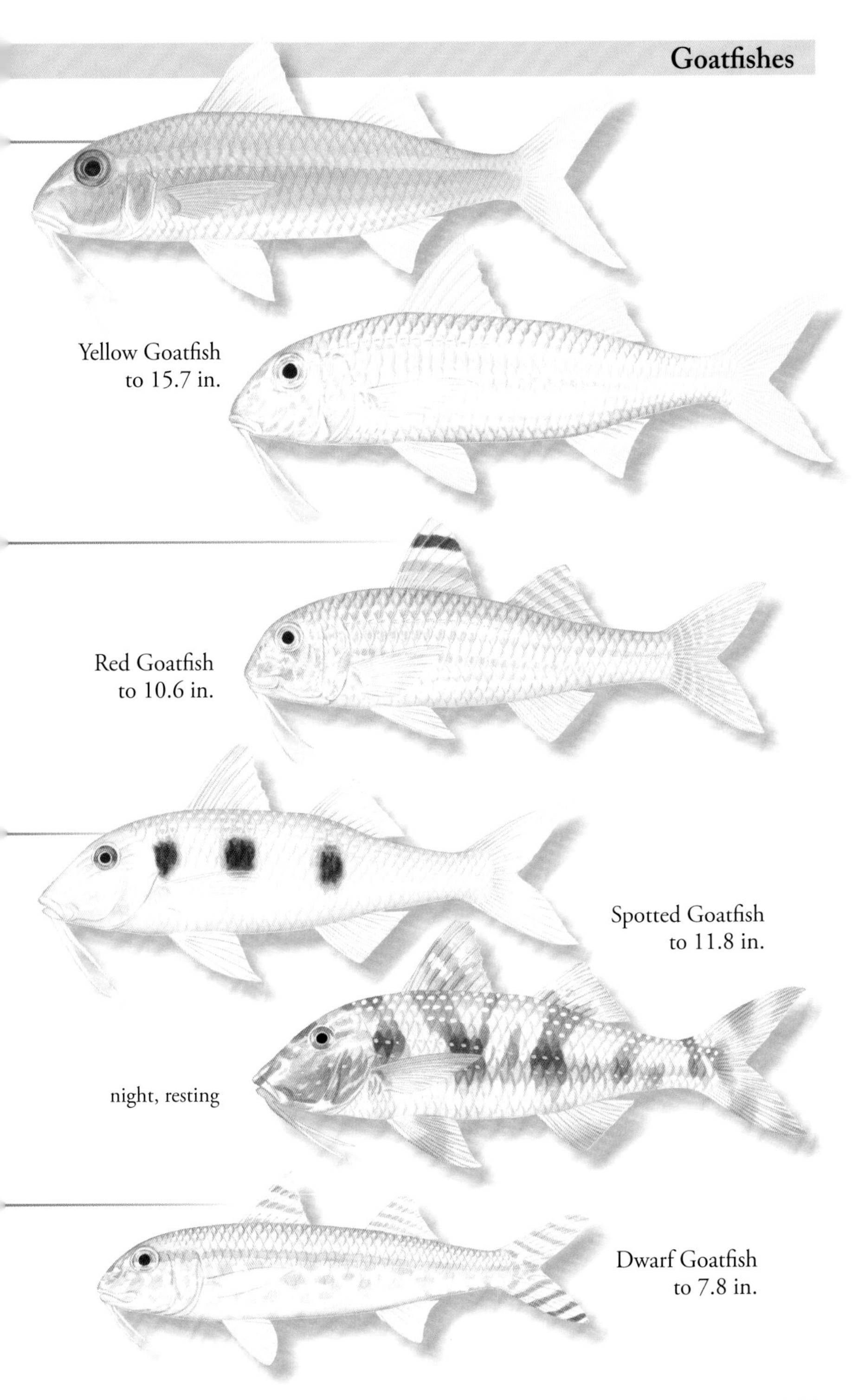
Yellow Goatfish
to 15.7 in.
Red Goatfish
to 10.6 in.
Spotted Goatfish
to 11.8 in.
night, resting
Dwarf Goatfish
to 7.8 in.

Pempheridae - Sweepers

Shortfin Sweeper - *Pempheris poeyi* Bean, 1885

FEATURES: Shades of very pale to dark green with silvery to golden highlights dorsally. Silvery below. Dark band at anal-fin base absent. Mouth large, oblique. Eyes very large. Single dorsal fin at midbody line. Anal-fin base comparatively short. Posterior profile nearly horizontal; ventral profile deeply convex. HABITAT: Bermuda, Bahamas, Cuba, and Caribbean Sea. Occur in caves and crevices of coral reefs from near shore to about 80 ft. BIOLOGY: Nocturnal.

Glassy Sweeper - *Pempheris schomburgkii* Müller & Troschel, 1848

FEATURES: Dark coppery with iridescent and greenish highlights. Dark band at anal-fin base. Mouth large, oblique. Eyes very large. Single dorsal fin at midbody line. Posterior profile nearly horizontal; ventral profile deeply convex. HABITAT: S FL and Dry Tortugas, Bahamas, and Antilles to Brazil. Also Bermuda. In crevices and caves of coral reefs during the day. In water column at night.

Kyphosidae - Sea Chubs

Darkfin Sea Chub - *Kyphosus bigibbus* Lacèpede, 1801

FEATURES: Shades of gray to brownish with iridescent highlights dorsally; silvery below. Usually two gray streaks below eyes. Opercular margin black. Dark blotch behind pectoral fins. Faint stripes follow scale rows. Fins dark gray; soft dorsal and anal fins with blackish margins. Head humped in front of eyes. Soft dorsal fin with 11–12 rays. Anal fin with 10–12 rays. HABITAT: Circumglobal in tropical seas. In western Atlantic from Bermuda, Grand Cayman, Belize, and San Blas, Panama. Pelagic over shallow rock and coral reefs to about 80 ft.

Topsail Sea Chub - *Kyphosus cinerascens* (Forsskål, 1775)

FEATURES: Shades of gray dorsally, fading to silvery below. May appear very pale to almost uniformly dark. Two grayish stripes below eyes. Opercular margin blackish. Grayish to brownish stripes follow scale rows. Dorsal profile in from of eyes humped. Soft dorsal and anal fins comparatively high. Soft dorsal fin with 12–13 rays (usually 12). Anal fin with 11–13 rays (usually 11). HABITAT: S FL, northern and southern Gulf of Mexico, Bahamas, and Caribbean Sea. Also Indo-Pacific. Occur over coral reefs and rocky bottoms from shore to about 130 ft.

Bermuda Chub - *Kyphosus sectatrix* (Linnaeus, 1758)

FEATURES: Silvery bluish gray with yellowish stripes following scales on sides. Two yellowish stripes on sides of head. May display pale spots on head and body. Opercular membrane blackish. Dorsal profile in front of eyes humped. Soft dorsal fin with 11–13 (usually 11) rays. Anal fin with 10–11 (usually 11) rays. HABITAT: MA to FL, Gulf of Mexico, Bermuda, Bahamas, Caribbean Sea to Brazil. Also eastern Atlantic and Mediterranean. In shallow water over seagrass beds and sandy and rocky bottoms and around coral reefs. Sometimes offshore. Young often associated with *Sargassum* seaweed. NOTE: Previously known as *Kyphosus sectator*.

Sweepers

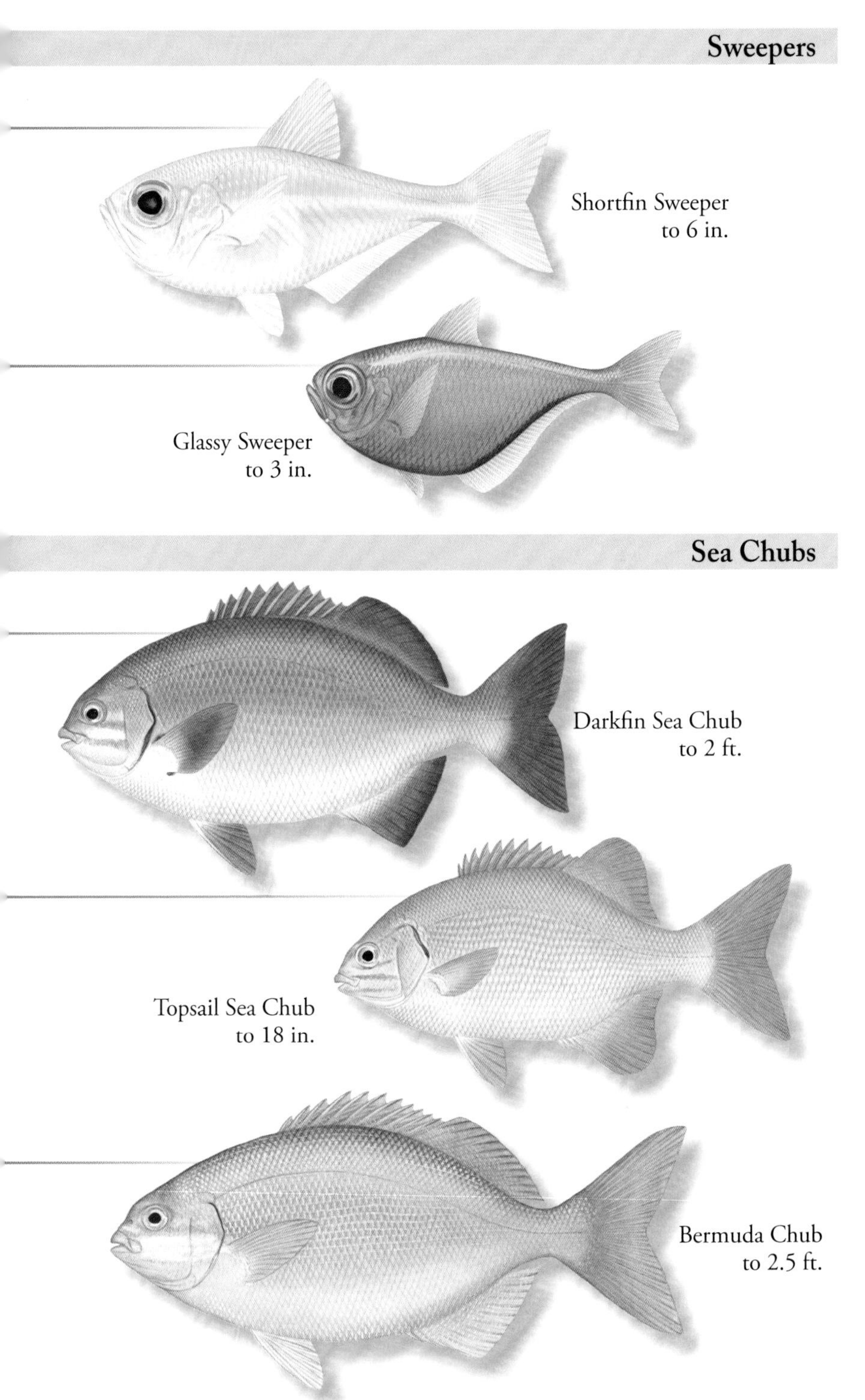

Shortfin Sweeper
to 6 in.
Glassy Sweeper
to 3 in.
Sea Chubs
Darkfin Sea Chub
to 2 ft.
Topsail Sea Chub
to 18 in.
Bermuda Chub
to 2.5 ft.

Kyphosidae - Sea Chubs, *cont.*

Yellow Chub - *Kyphosus vaigiensis* (Quoy & Gaimard, 1825)

FEATURES: Silvery bluish gray with brassy stripes following scales on sides. Two brassy stripes on sides of head. Opercular membrane gray. Dorsal profile in front of eyes slightly convex. Soft dorsal fin with 13–15 (usually 14) rays. Anal fin with 12–13 (usually 13) rays. HABITAT: MA to FL, Gulf of Mexico, Bahamas, Caribbean Sea to Brazil. Also Bermuda and eastern Atlantic. Usually occur offshore and associated with flotsam and *Sargassum* seaweed. Also in shallow water over hard bottoms and reefs. NOTE: Previously known as *Kyphosus incisor.*

Chaetodontidae - Butterflyfishes

Foureye Butterflyfish - *Chaetodon capistratus* Linnaeus, 1758

FEATURES: Head and ventral area yellowish. Grayish band through each eye from nape to lower opercle. Body pale gray to pale yellow with dark, oblique dashes that follow scales. Large, black ocellated spot on posterior body. Juveniles with black band through eyes, two broad bars on body, and two ocellated spots posteriorly; upper spot fades with age. HABITAT: MA (rare) to FL, Gulf of Mexico, Bermuda, Bahamas, and Caribbean Sea. Found over shallow coastal reefs and rocky areas. Juveniles occur over seagrass beds. BIOLOGY: Feed on small, bottom-dwelling gorgonians, tunicates, and worms. Found singly or in pairs. Popular in the aquarium trade.

Spotfin Butterflyfish - *Chaetodon ocellatus* Bloch, 1787

FEATURES: Body and most of head silvery white. Black band through eyes. Yellow areas on snout, and yellow band from opercular margin to pectoral-fin base. Fins and posterior portion of body yellow. A black blotch at middle base of soft dorsal fin appears during night, becomes faint or disappears during day. A small black spot at rear tip of soft dorsal fin always present. Juveniles silvery white on head and body; black band through each eye; black blotch at middle base of soft dorsal fin blends with black band on posterior body. HABITAT: MA (rare) to FL, Gulf of Mexico, Bermuda, Bahamas, and Caribbean Sea to southern Brazil. Occur over shallow coastal reefs from shore to about 100 ft. BIOLOGY: Usually occur in pairs. Feed on bottom-dwelling invertebrates. Popular in aquarium trade.

Reef Butterflyfish - *Chaetodon sedentarius* Poey, 1860

FEATURES: Yellowish dorsally, fading to silvery white below. Broad black band from nape, through eyes, to lower opercular margin. Several faint, oblique, brownish bars on sides. A broad blackish band extends from rear of soft dorsal fin, through caudal peduncle, to rear of anal fin; upper portion may be very dark to faint. Caudal fin yellow, lacks banding. Juveniles similar to adults but with diffuse, dark spot on rear soft dorsal fin. HABITAT: NC to FL, Gulf of Mexico, Bermuda, Bahamas, and Caribbean Sea to southern Brazil. Found over reefs from about 50 to 130 ft. Recorded to 330 ft. BIOLOGY: Feed on bottom-dwelling invertebrates.

Sea Chubs

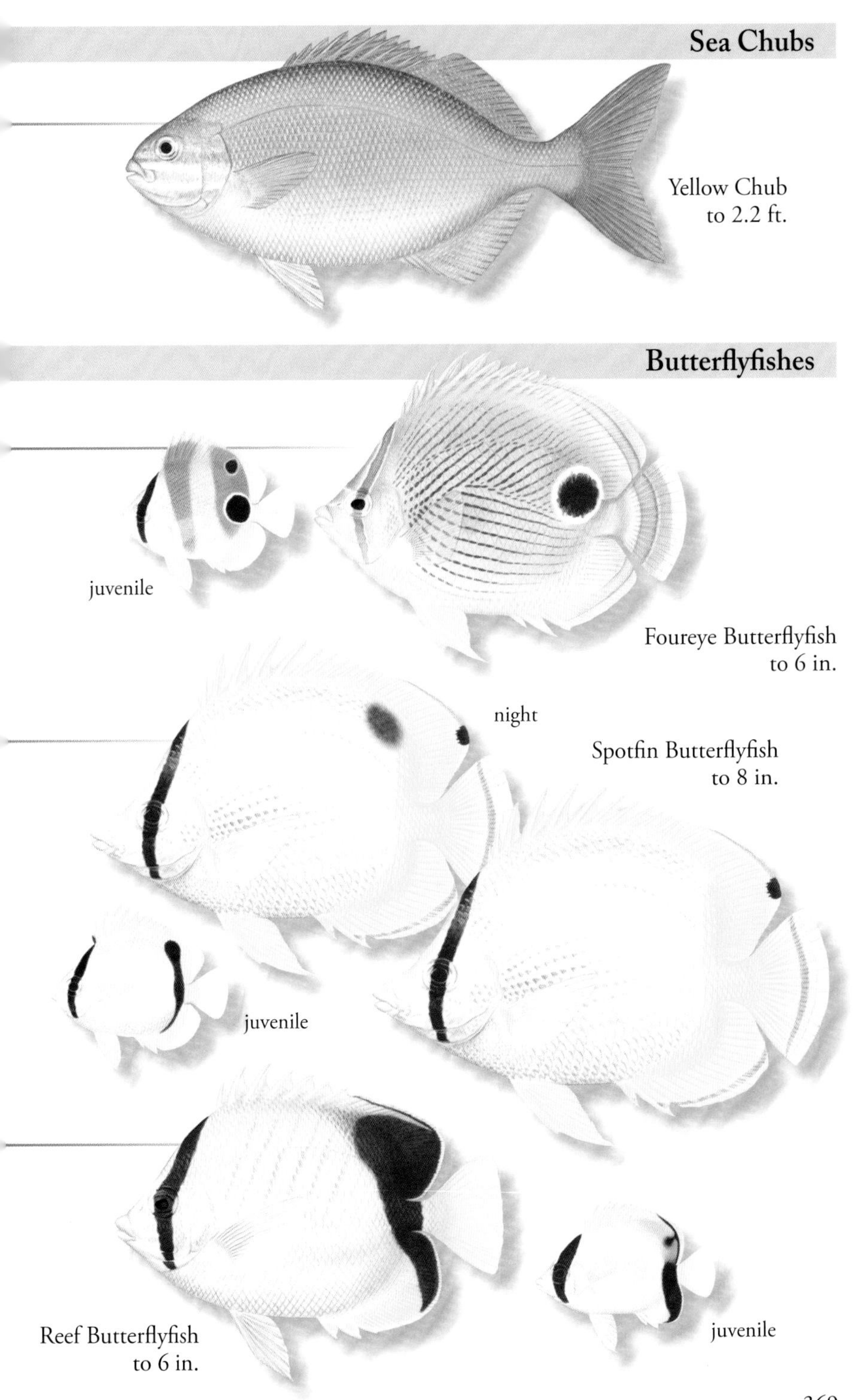

Yellow Chub
to 2.2 ft.

Butterflyfishes

Foureye Butterflyfish
to 6 in.

Spotfin Butterflyfish
to 8 in.

Reef Butterflyfish
to 6 in.

Banded Butterflyfish - *Chaetodon striatus* Linnaeus, 1758

FEATURES: Whitish with broad dark brown to grayish bars on body. The first bar is from nape through eye to breast. The second is from spiny dorsal fin to abdomen. The third is from anterior portion of soft dorsal fin to middle base of anal fin. Sides with dark, oblique lines that follow scale rows. Soft dorsal and anal fins with two broad dark bands. Caudal fin with dark inner band. Juveniles similar to adults but with a large ocellated spot on anterior soft dorsal fin; sides lack oblique lines. HABITAT: MA (rare) to FL, Gulf of Mexico, Bermuda, Bahamas, and Caribbean Sea to southern Brazil. Also recorded from eastern Atlantic. Found over shallow coastal reefs and rocky areas. Juveniles occur over seagrass beds. BIOLOGY: Feed on coral polyps, worms, crustaceans, and mollusk eggs. Popular in aquarium trade.

Longsnout Butterflyfish - *Prognathodes aculeatus* (Poey, 1860)

FEATURES: Dark yellow dorsally, fading to silvery below. A yellow band extends from eyes to dorsal-fin origin. Spiny dorsal fin, posterior portion of back, and part of soft dorsal fin dark brownish. A narrow yellow band bisects soft dorsal fin. Part of caudal peduncle, inner anal fin, and pelvic fins yellow. Remainder of fins colorless. Snout long, pointed. Dorsal-fin spines tall and incised. HABITAT: S FL, Gulf of Mexico, Bermuda, Bahamas, and Caribbean Sea. Also recorded from NC. Over moderate to deep reefs and rocky areas usually from about 50 to 300 ft. BIOLOGY: Use slender mouth to pick invertebrates from small crevices and pick flesh of urchins from between spines. May form schools.

Bank Butterflyfish - *Prognathodes aya* (Jordan, 1886)

FEATURES: Body silvery white to silvery tan. Dark brown stripe along snout profile. Dark blackish brown band from dorsal-fin origin, through eyes, to corner of mouth. Broad, dark blackish brown band from about fourth dorsal-fin spine to anal-fin base. Yellowish band from about the fifth dorsal-fin spine through caudal peduncle. Pelvic fins yellow. Snout long, pointed. Dorsal-fin spines very tall and incised. HABITAT: NC (rare) to FL and Gulf of Mexico to Campeche Banks. Found around rocky slopes between about 65 and 650 ft. BIOLOGY: Feed on small, bottom-dwelling invertebrates.

Guyana Butterflyfish - *Prognathodes guyanensis* (Durand, 1960)

FEATURES: Body pale yellow with darker yellow stripes along scales. Dark brown stripe along snout profile. Dark blackish brown band from dorsal-fin origin, through eyes, to corner of mouth. Broad, dark blackish brown band from about fourth dorsal-fin spine to anal-fin base. Another blackish brown band from posterior spiny dorsal fin to caudal peduncle. Pelvic fins brownish. Snout long, pointed. Dorsal-fin spines very tall and incised. HABITAT: Bahamas, Greater and Lesser Antilles, northwestern Caribbean Sea, and Venezuela to southern Brazil. Occur over rocky and reef slopes from about 200 to 990 ft. BIOLOGY: Feed on bottom-dwelling invertebrates.

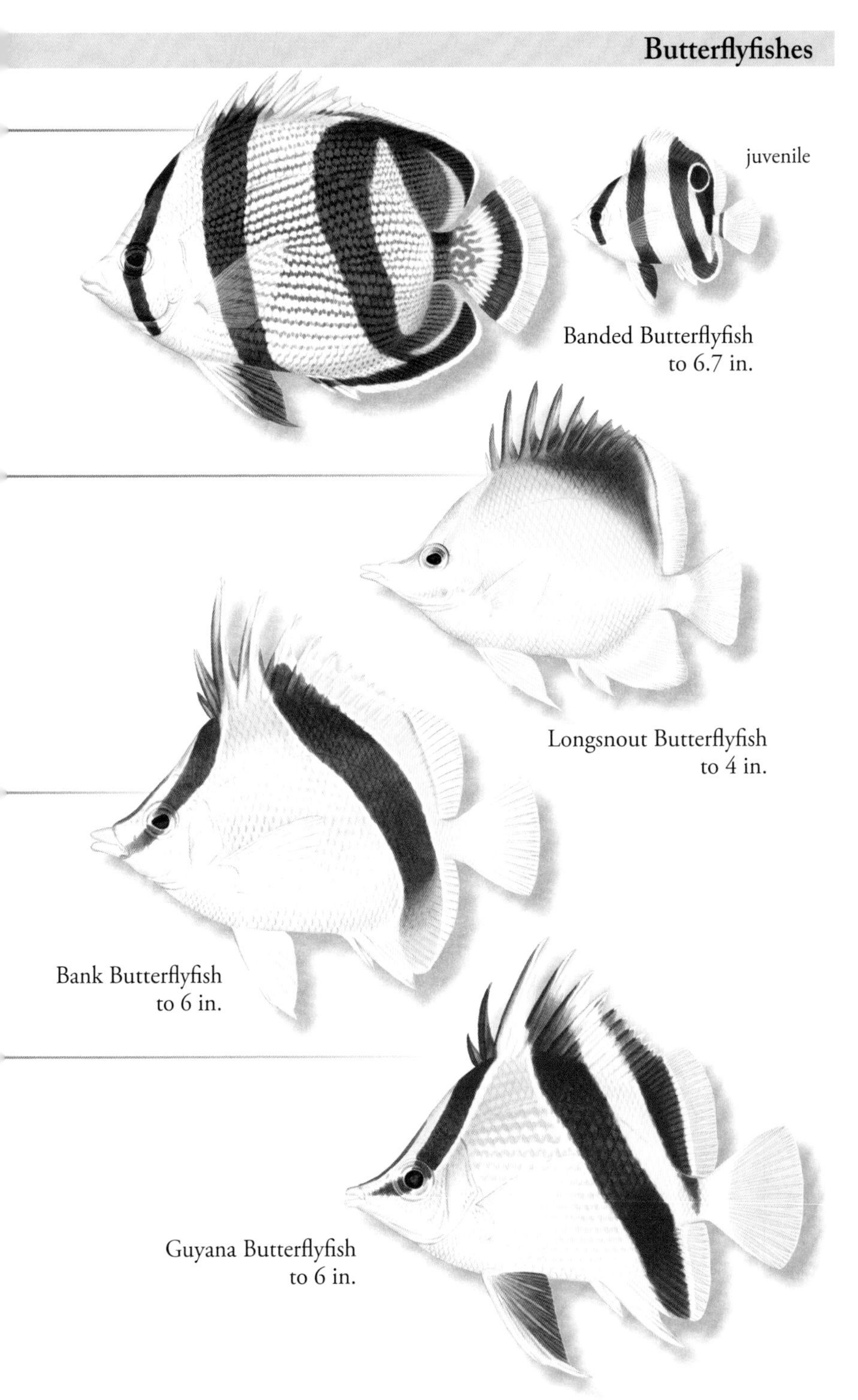

Banded Butterflyfish
to 6.7 in.

Longsnout Butterflyfish
to 4 in.

Bank Butterflyfish
to 6 in.

Guyana Butterflyfish
to 6 in.

Cherubfish - *Centropyge argi* Woods & Kanazawa, 1951

FEATURES: Lower head and chest yellow. Top of head and rest of body dark indigo blue. Eyes with a blue ring. Dark spots on scales. Dorsal and anal fins with two blackish inner bands and pale blue margins. Three spines on cheek posterior to jaws. Preopercle with long spine at corner. Body somewhat deep, compressed. HABITAT: GA to FL, Gulf of Mexico, Bermuda, Bahamas, Caribbean Sea to Guyana. Also recorded from NC. Around coastal rocky and reef areas from near shore to about 560 ft. BIOLOGY: Territorial and somewhat secretive. Feed on algae and invertebrates.

Flameback Angelfish - *Centropyge aurantonotus* Burgess, 1974

FEATURES: Upper head, back, and most of dorsal fin shades of yellow to orange yellow. Remainder of body and fins dark indigo blue. Eyes with a blue ring. Dark spots on scales. Caudal, anal, and pelvic fins with blackish inner band and pale blue margins. Two spines on cheek posterior to jaws. Preopercle with long spine at corner. Body somewhat deep, compressed. HABITAT: Lesser Antilles, and Venezuelan offshore islands to southern Brazil. Occur over coral reefs and rocky rubble from about 50 to 980 ft. Usually above 80 ft. BIOLOGY: Territorial. Feed on algae and sponges.

Blue Angelfish - *Holacanthus bermudensis* Goode, 1876

FEATURES: Tannish green to tannish blue on most of body and fins. Nape and chest pale blue. Pectoral fins with inner yellow band. Dorsal and anal fins with inner yellow margin, outer blue margin. Caudal fin with yellow margin. Preopercular margin spiny, with a prominent spine at corner. Juveniles dark greenish blue with pale blue bars on sides that are nearly vertical; yellow snout and chest; dark band through eyes, bordered by pale blue bands. HABITAT: NJ (rare) to FL, Bermuda, Bahamas, Gulf of Mexico to Yucatán. Rare in northern and central Caribbean Sea. Occur around coral reefs from near shore to about 200 ft. Also around oil platforms. BIOLOGY: Feed on bottom-dwelling invertebrates. Known to hybridize with Queen Angelfish. The resulting fish is known as the Townsend Angelfish and shares several visual attributes with each.

Queen Angelfish - *Holacanthus ciliaris* (Linnaeus, 1758)

FEATURES: Body greenish to bluish with yellow-margined scales. Nape with dark blue ocellated blotch. Chest blue. Pectoral fins and caudal fin yellow. Dorsal and anal fins fade to yellowish orange near tips; margins blue. Preopercular margin spiny, with a prominent spine at corner. Juveniles dark greenish blue with pale blue bars on sides that are slightly curved; yellow snout and chest; dark band through eyes bordered by pale blue bands. HABITAT: NC (rare) to FL, Gulf of Mexico, Bermuda, Bahamas, and Caribbean Sea to southern Brazil. Found around shallow coral reefs and hard structures from near shore to about 196 ft. BIOLOGY: Feed on bottom-dwelling invertebrates. Juveniles pick ectoparasites from other fishes. Hybridizes with Blue Angelfish—see above.

Cherubfish
to 2 in.
Flameback Angelfish
to 2 in.
juvenile
Blue Angelfish
to 15 in.
Queen Angelfish
to 10 in.
juvenile

Pomacanthidae - Angelfishes, *cont.*

Rock Beauty - *Holacanthus tricolor* (Bloch, 1795)

FEATURES: Nape, head, and anterior portion of body yellow. Jaws bluish to blackish. Sides of body and most of dorsal and anal fins dark blackish blue. Posterior margins of dorsal and anal fins yellow. Other fins yellow. Preopercular margin spiny, with prominent spine at corner. Juveniles yellow with dark spot bordered by blue ring on posterior upper sides. Intermediate specimens yellow with dark area on rear upper portion of body. Both with yellow jaws. HABITAT: NC (rare) to FL, Gulf of Mexico, Bermuda, Bahamas, and Caribbean Sea to southern Brazil. Found over coral reefs, rocky bottoms, and hard structures from about 3 to 295 ft. Juveniles associated with stinging corals. BIOLOGY: Feed on sponges.

Gray Angelfish - *Pomacanthus arcuatus* (Linnaeus, 1758)

FEATURES: Body and fins gray to brownish gray. Jaws and chin whitish. Head pale gray; chest dark gray. Scales with dark centers, pale edges. Preopercular margin spiny, with a prominent spine at corner. Middle soft rays of the dorsal and anal fins are long and trailing. Juveniles blackish with yellow bars on head and body; yellow bar extending along forehead splits over jaws to form an upside-down cross shape; caudal fin with a yellow band at base and broken yellowish to whitish margin. HABITAT: NY (rare) to FL, Gulf of Mexico, Bermuda, Bahamas, and Caribbean Sea to southern Brazil. Found over shallow coral reefs and around hard structures. BIOLOGY: Feed on a variety of invertebrates and algae. Juveniles pick ectoparasites from other fishes.

French Angelfish - *Pomacanthus paru* (Bloch, 1787)

FEATURES: Body and fins blackish; head dark bluish gray. Jaws and chin whitish. Eyes circled with yellow. Opercular margin, pectoral-fin base, and scale margins yellow. Middle soft rays of dorsal and anal fins long and trailing. Preopercular margin spiny, with a prominent spine at corner. Juveniles blackish with yellow bars on head and body; yellow bar extending along forehead splits over jaws to form an upside-down Y-shape; caudal fin with a yellow border that forms a ring. HABITAT: NY (rare) to FL, Bermuda, Bahamas, Antilles, and Caribbean Sea to southern Brazil. Also Ascension Island in eastern Atlantic. Over shallow reefs and hard structures. BIOLOGY: Feed on invertebrates and algae.

Cirrhitidae - Hawkfishes

Redspotted Hawkfish - *Amblycirrhitus pinos* (Mowbray, 1927)

FEATURES: Pale tannish to pale olivaceous. Five broad brownish to olivaceous bars on body, anterior four extend into dorsal fins. Upper portion of fourth bar and bar on caudal peduncle are blackish. Bright orange to red spots scattered on head, forebody, and dorsal fin. Dorsal-fin spine tips with tufts of cirri. HABITAT: S FL, Gulf of Mexico, Bermuda, Bahamas, and Caribbean Sea to southern Brazil. On reefs and hard substrates to about 147 ft. Also in areas of strong currents. BIOLOGY: Feed on small invertebrates. Use free pectoral-fin rays to move over substrate.

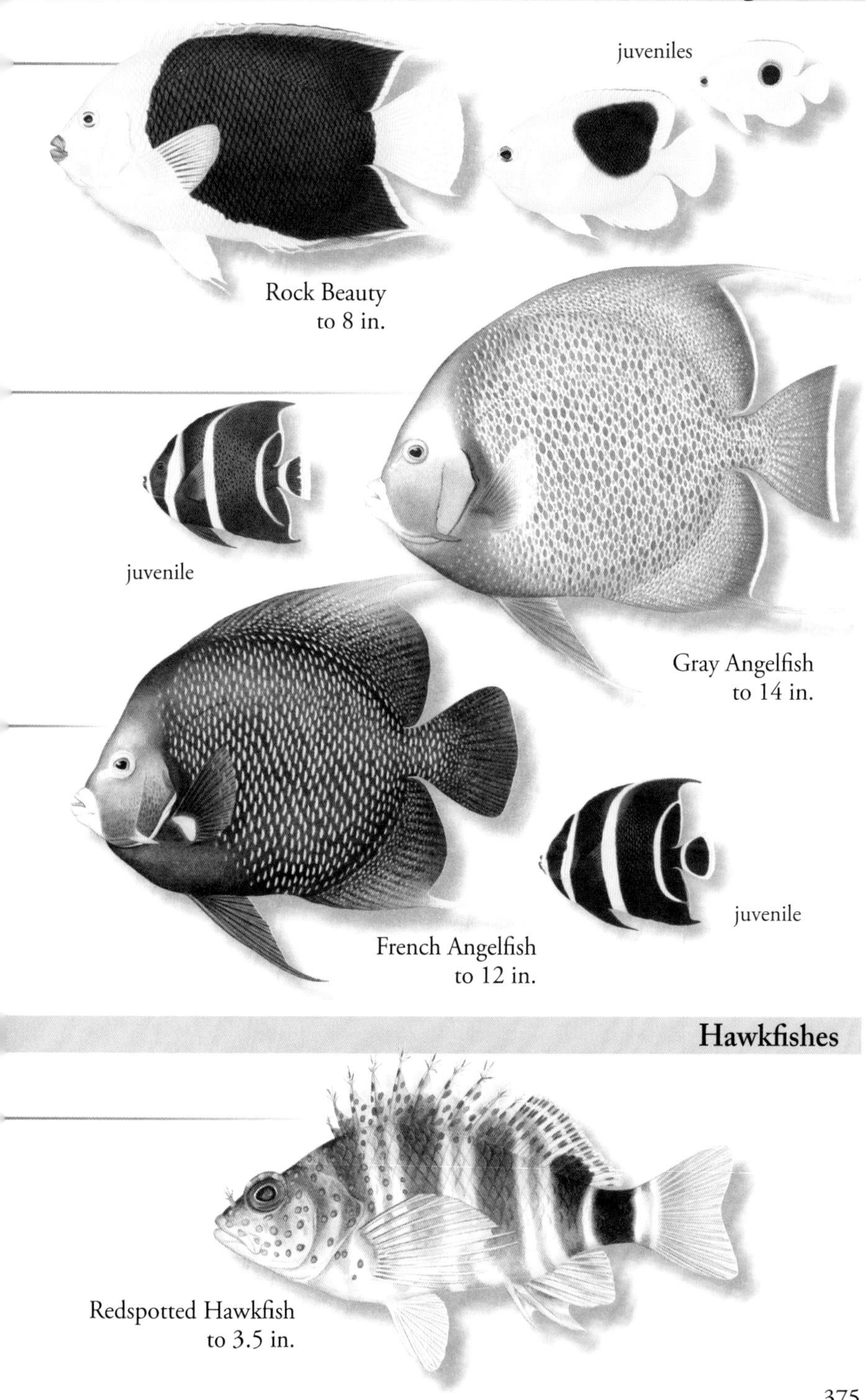
juveniles
Rock Beauty
to 8 in.
juvenile
Gray Angelfish
to 14 in.
juvenile
French Angelfish
to 12 in.
Hawkfishes
Redspotted Hawkfish
to 3.5 in.

Cichlidae - Cichlids

Mayan Cichlid - *Mayaheros urophthalmus* (Günther, 1862)

FEATURES: Color highly variable. Pale to dark olivaceous, or brown to orange or brick red dorsally. Tan to pink or yellow ventrally. All with six to eight (usually seven) dark green or dark bluish to blackish bars on upper sides. Bars may be bordered with narrow bluish lines or spots. A black ocellus present on upper caudal-fin base. Scales with pale centers. Jaws protrusible. Spiny dorsal fin tall with incised membranes. Soft dorsal- and anal-fin tips pointed. HABITAT: S FL, and the Coatzacoalcos River basin, Mexico, to the Prinzapolka River, Nicaragua, including Isla Mujeres and Isla Contoy. Occur in shallow rivers, streams, ponds, and wetlands, and around mangroves, in coastal lagoons, sinkholes, and caves.

Cuban Cichlid - *Nandopsis tetracanthus* (Valenciennes, 1831)

FEATURES: Color and pattern highly variable. Shades of white, gray, or tan with numerous blackish blotches, wavy lines, and bands on head and back that overlay vague bars on sides. Ventral area irregularly spotted. Scales with dark margins. Dorsal, caudal, and anal fins spotted. Snout bluntly pointed. Spiny dorsal fin low; soft dorsal and anal fins tall and pointed at tip. Large individuals develop a large bulge on head. Lateral line in two sections. HABITAT: Cuba and Isla de la Juventud (Isle of Pines). Primarily in fresh water, but also in shallow brackish estuaries and around mangroves.

Mozambique Tilapia - *Oreochromis mossambicus* (Peters, 1852)

FEATURES: Females and non-breeding males dark olivaceous to blue gray dorsally. Paler below. Scales with dark centers. Several dark saddles and blotches on sides. Margins of dorsal, caudal, and anal fins red. Breeding males nearly black with white under head. Scale edges may be pale. Margins of dorsal and caudal fins bright red. Dorsal-, caudal-, and anal-fin membranes with pale spots that form bands. Lips in mature males enlarged. Snout upturned. Spiny dorsal fin membranes incised. Soft dorsal and anal fin tips pointed, more so in large males. Lateral line in two sections. HABITAT: FL, Gulf of Mexico, and scattered in Caribbean Sea. Occur in coastal brackish, salt, and hypersaline waters from shore to about 30 ft. Also in fresh water. NOTE: Introduced into the area from southeastern Africa.

Spotted Cichlid - *Vieja maculicauda* (Regan, 1905)

FEATURES: Color and pattern highly variable. Shades of pale gray to pale tan. Lower cheeks and breast pink to deep red. Irregular black spots merge to form a large bar or blotch on midsides. May also have a blotch on caudal peduncle. Dorsal and anal fins with irregular black spots anteriorly, red margins posteriorly. Caudal fin mostly red in breeding males. Eyes small. Head gently sloping in small specimens, humped in large males. Spiny dorsal fin low; soft dorsal and anal fins tall and pointed at tip. Lateral line in two sections. HABITAT: Chachalaca River, Mexico, to Chagres River, Panama. Occur in lakes and slow-moving rivers over muddy and sandy bottoms, often around vegetation. Commonly found in shallow brackish and salt water.

NOTE: There are five other primarily freshwater but sometimes brackish-water Cichlids in the area.

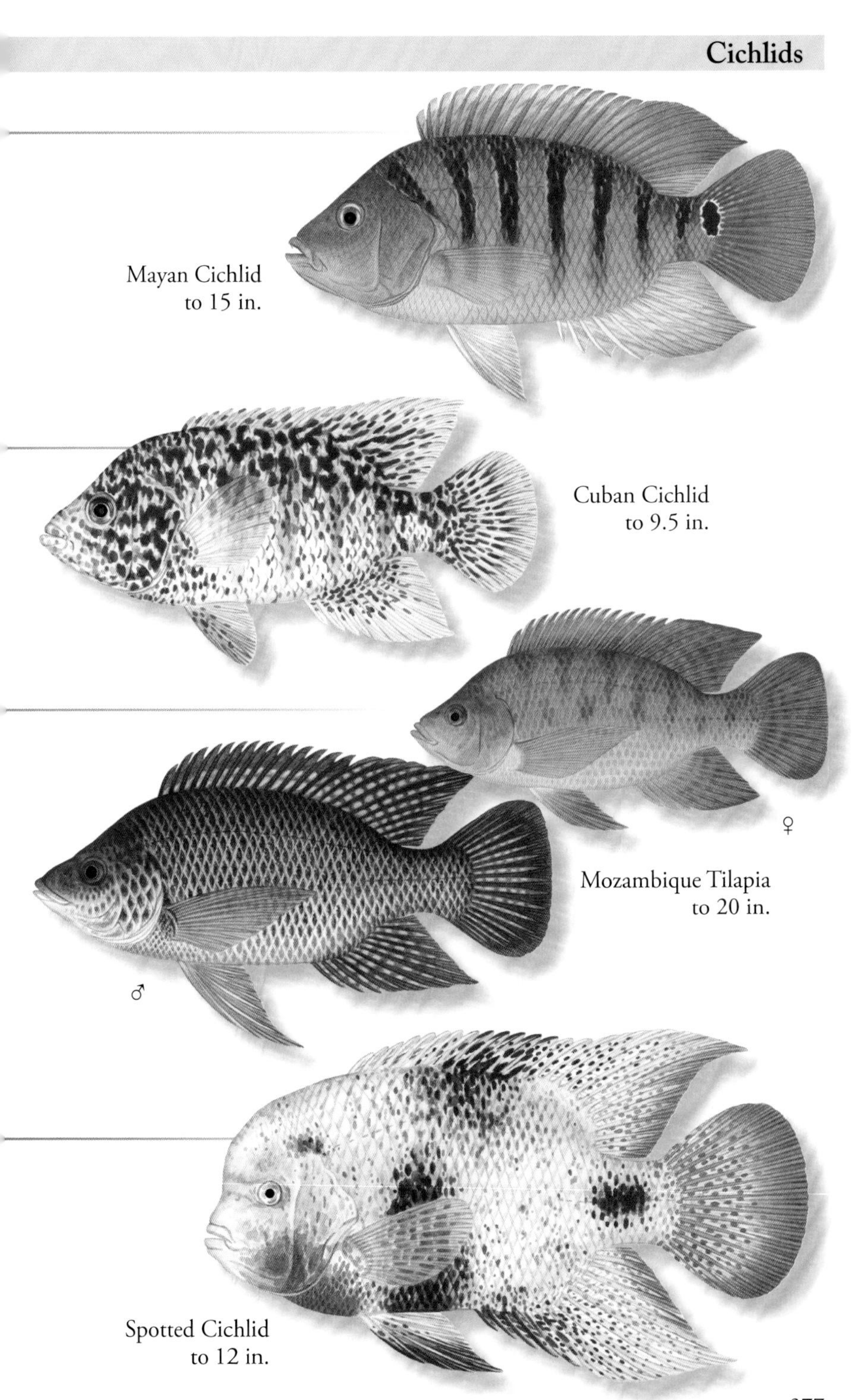

Mayan Cichlid
to 15 in.

Cuban Cichlid
to 9.5 in.

Mozambique Tilapia
to 20 in.

Spotted Cichlid
to 12 in.

Pomacentridae - Damselfishes

Sergeant Major - *Abudefduf saxatilis* (Linnaeus, 1758)

FEATURES: Upper head and nape bluish green. Body usually bright yellow dorsally, bluish white below, with five black bars on sides. Juveniles similarly colored and patterned. Adult males turn dark blue during spawning and while guarding eggs. Body deep, laterally compressed. HABITAT: RI to FL, Gulf of Mexico, Bermuda, Bahamas, and Caribbean Sea to Brazil. Also eastern Atlantic. In shallow water around jetties and over coral and rocky reefs to about 45 ft. Juveniles associated with *Sargassum* seaweed, often far offshore. BIOLOGY: Feed on a variety of invertebrates.

Night Sergeant - *Abudefduf taurus* (Müller & Troschel, 1848)

FEATURES: Yellowish tan above; paler below. Five wide brownish bars on sides extend onto dorsal fins. May have sixth bar on caudal peduncle. Bars may be complete or incomplete. Small dark spot at upper pectoral-fin base. Juveniles similarly colored and patterned. HABITAT: S FL, southern Gulf of Mexico, Bahamas, and Caribbean Sea. Also eastern Atlantic. Occur over shallow and turbulent rocky inshore reefs. Prefer wave-cut limestone shorelines and rocky ledges and tide pools with surf. Occasionally in low-salinity waters. BIOLOGY: Feed mainly on algae.

Blue Chromis - *Azurina cyanea* (Poey, 1860)

FEATURES: Head and body bright blue. Dorsal portion of head and body blackish. All fins blue. Dorsal and anal fins with blackish margins. Upper and lower caudal-fin margins blackish. Body moderately elongate; caudal fin deeply forked. HABITAT: NC to FL, Gulf of Mexico, Bermuda, Bahamas, and Caribbean Sea. Associated with shallow to moderately deep coral reefs and slopes. BIOLOGY: Maintain territories on the reef face. School with other species in the water column. Feed in aggregations on passing zooplankton. NOTE: Previously known as *Chromis cyanea*.

Brown Chromis - *Azurina multilineata* (Guichenot, 1853)

FEATURES: Olive brown to grayish brown dorsally, fading to whitish below. Dark spot at upper pectoral-fin base. Usually with white spot under rear dorsal-fin base. Outer margins of dorsal fin and tips of caudal fin yellow. Submargins of caudal fin may be black or may be identical to body color. HABITAT: FL, northern and southern Gulf of Mexico, Bermuda, Bahamas, and Caribbean Sea to southern Brazil. Occur over coral reef tops, reef slopes, and patch reefs from near shore to about 300 ft. BIOLOGY: Form aggregations while feeding on zooplankton. NOTE: Previously known as *Chromis multilineata*.

Bermuda Chromis - *Chromis bermudae* Nichols, 1920

FEATURES: Head, body, and dorsal- and anal-fin bases dark cobalt blue. Pectoral, pelvic, outer dorsal, and anal fins; caudal peduncle; and caudal fin yellow. Front of head and snout blackish with paler blue lines around eyes to snout tip and on upper jaw. Black spot on pectoral-fin base. HABITAT: Bermuda. Occur over level bottoms usually with coralline algae from about 165 to 200 ft.

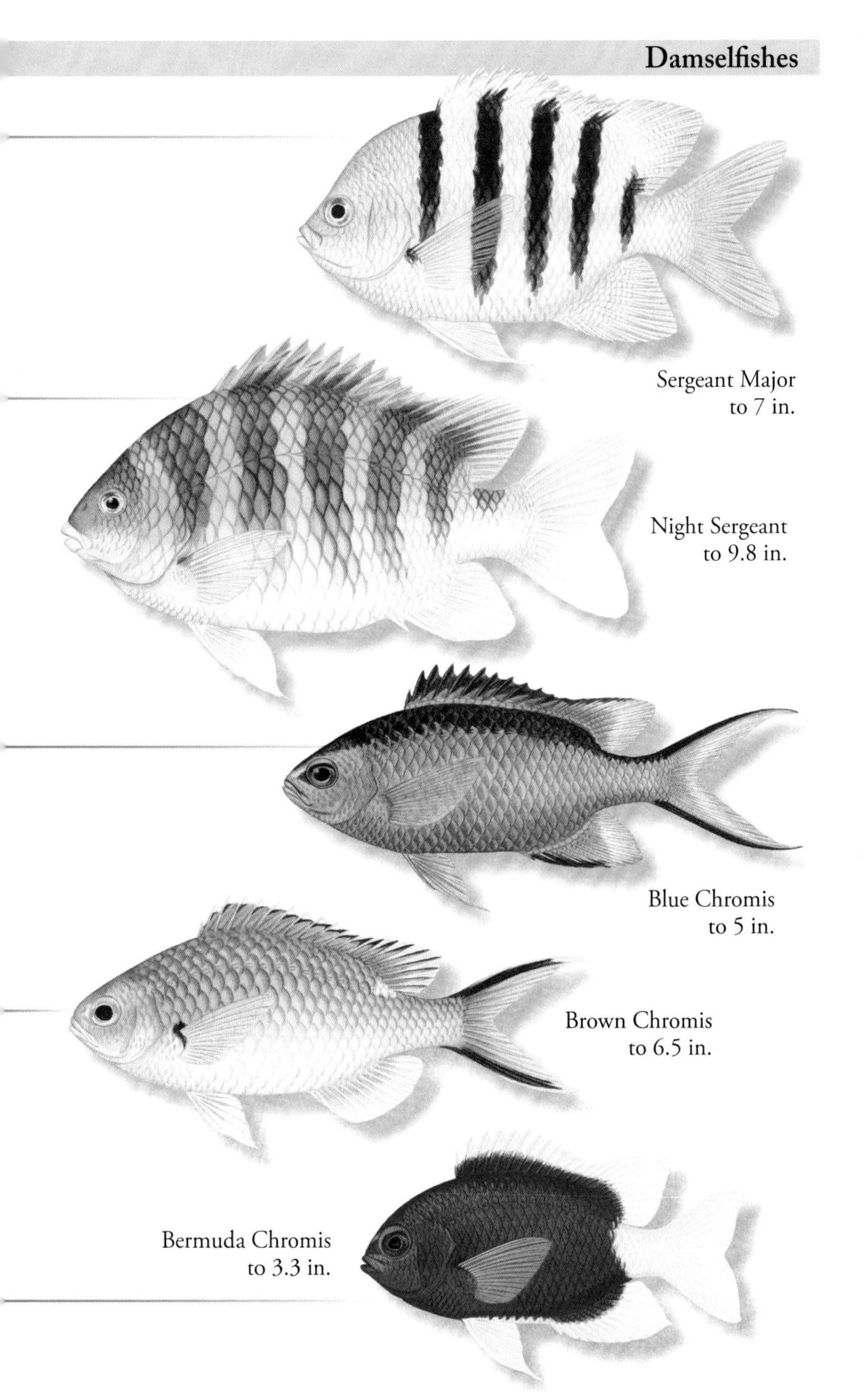
Sergeant Major
to 7 in.
Night Sergeant
to 9.8 in.
Blue Chromis
to 5 in.
Brown Chromis
to 6.5 in.
Bermuda Chromis
to 3.3 in.

Yellowtail Reeffish - *Chromis enchrysurus* Jordan & Gilbert, 1882

FEATURES: Dark blue to gray or brown dorsally from snout to posterior base of soft dorsal fin. Ventral area bluish white to grayish white. Outer portions of soft dorsal and anal fins yellow. Caudal fin yellow. Bright blue band runs from tip of snout to first dorsal-fin base. HABITAT: NC to FL, Gulf of Mexico, Bermuda, Bahamas, Caribbean Sea to Guyana. Also off southern Brazil. Occur over steep slopes of coral and patch reefs from about 65 to 330 ft. BIOLOGY: Feed on algae and zooplankton. Form small groups.

Sunshinefish - *Chromis insolata* (Cuvier, 1830)

FEATURES: Brownish to olivaceous on upper two-thirds of body. Ventral area whitish to brownish. Blue line from tip of snout and above eyes may be present. Small dark spot at upper pectoral-fin base usually present. Outer portions of soft dorsal and caudal fins yellowish to pale. Juveniles yellow dorsally, bluish purple along midbody, whitish below. HABITAT: NC to FL, Gulf of Mexico, Bermuda, Bahamas, and Caribbean Sea. Found over outer and seaward reefs. BIOLOGY: Feed on small invertebrates. Form clusters close to bottom.

Purple Reeffish - *Chromis scotti* Emery, 1968

FEATURES: Adults bluish green to grayish blue with few to many bright blue spots on scales. Blue spots may blend to form V-shape on snout. Eyes always with blue to bright blue crescent. Smaller specimens dusky blue dorsally, pale below, with bright blue spots. Pectoral fins always transparent. HABITAT: NC to FL, Gulf of Mexico, Bahamas, and Caribbean Sea to Brazil. Found over steep, outer reef slopes and patch reefs to about 164 ft.

Whitetail Reeffish - *Chromis vanbebberae* McFarlan, Baldwin, Robertson, Rocha & Tornabene, 2020

FEATURES: Dark purplish blue dorsally from snout to posterior end of spiny dorsal fin. Ventral area pearly white to grayish white. Cobalt blue lines through and around eyes. Soft dorsal, caudal, anal, pelvic, and pectoral fins pearly white. Juveniles dark brown dorsally from snout to posterior end of spiny dorsal fin; pearly white below and on all remaining fins; cobalt blue band through eyes. HABITAT: Bermuda, Bahamas, and Caribbean Sea to Brazil. Absent from Cuba. Occur over outer reef edges from about 160 to 580 ft.

Yellowtail Damselfish - *Microspathodon chrysurus* (Cuvier, 1830)

FEATURES: Body and all fins, except caudal fin, bluish brown to yellowish brown with dark-edged scales. Caudal fin always yellow or pale yellow. Small, bright blue spots scattered on head, upper body, and dorsal fin. Juveniles dark blue to brownish blue with bright blue spots on head, body, dorsal and anal fins; caudal fin transparent, becoming yellow with age. HABITAT: FL, Gulf of Mexico, Bermuda, Bahamas, Caribbean Sea to Venezuela. Occur over shallow coral reefs. BIOLOGY: Territorial. Feed on algae, coral polyps, and a wide variety of invertebrates.

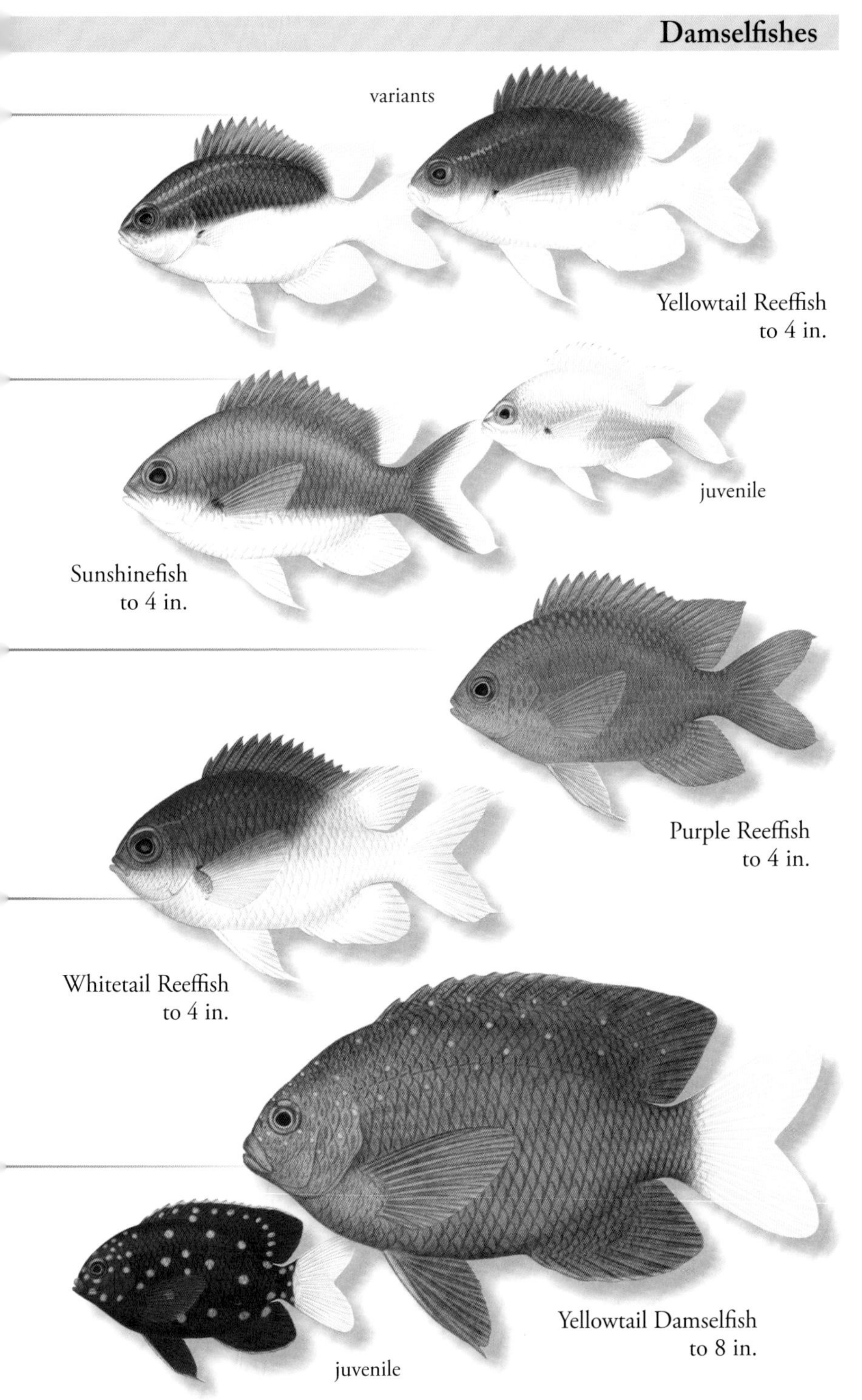
variants
Yellowtail Reeffish
to 4 in.
juvenile
Sunshinefish
to 4 in.
Purple Reeffish
to 4 in.
Whitetail Reeffish
to 4 in.
Yellowtail Damselfish
to 8 in.
juvenile

Regal Demoiselle - *Neopomacentrus cyanomos* (Bleeker, 1856)

FEATURES: Shades of brown or gray to purplish brown. Top of head and nape may be greenish. Centers of scales may have a blue spot. All with a large black spot with some iridescence at upper opercular margin. Black spot at upper pectoral-fin base. Rear lower soft dorsal fin and inner portion of caudal fin white to yellow. Rear soft dorsal fin rays and outer caudal fin rays long, blackish, and trailing. HABITAT: Introduced from the Indo-West Pacific. Established populations in scattered areas of the Gulf of Mexico and southeastern Caribbean Sea. Occur near bottom over outer and inshore sheltered reefs to about 175 ft. Also in currents above dropoffs.

Dusky Damselfish - *Stegastes adustus* (Troschel, 1865)

FEATURES: Color varies. Body and fins dark bluish brown or purplish brown to reddish brown. Scale edges dark. May have faint blue spots on head, chest, and abdomen. Small blue spot at pectoral-fin base. Narrow blue line along dorsal and anal fin margins. Soft dorsal and anal fins comparatively short with tips reaching to or slightly beyond caudal-fin base. Juveniles bright orange dorsally; purplish blue on sides and below; bright blue spots on scales; a large, black ocellated spot on base of dorsal fin; and a small black spot on peduncle. HABITAT: S FL, Gulf of Mexico, Bahamas, Caribbean Sea to Venezuela. Possibly Bermuda. Occur around shallow and turbulent coral and rocky shores.

Longfin Damselfish - *Stegastes diencaeus* (Jordan & Rutter, 1897)

FEATURES: Color varies. Body and fins dark bluish brown or gray brown to blackish. May have greenish cast on head. Scale edges dark. Dark spot at pectoral-fin base. Narrow blue line along dorsal and anal fin margins. Dorsal and anal fins comparatively long, with tips extending well beyond caudal-fin base. Juveniles shades of yellow with blue lines on head that merge with blue spots along scale rows on upper body; blue lines on snout form a V-shape; black ocellated spot present on dorsal fin. HABITAT: S FL, southern Gulf of Mexico, Bahamas, and Caribbean Sea. Found around coral and rocky reefs of sheltered lagoons and inshore areas from near shore to about 150 ft. BIOLOGY: Territorial and pugnacious. Feed on algae and seaweeds.

Beaugregory - *Stegastes leucostictus* (Müller & Troschel, 1848)

FEATURES: Color varies. Dark olive brown to dusky gray. May be pale below. Caudal fin yellowish to pale. Usually with rows of blue spots on head and upper sides. Narrow blue line along dorsal and anal fin margins. Juveniles with blue spots over dark dusky blue to gray on upper head and body; yellow below; a black ocellated spot present on dorsal fin. Spot on caudal peduncle absent in all. Body comparatively shallow. HABITAT: FL, Gulf of Mexico, Bermuda, Bahamas, and Caribbean Sea. Found around shallow, quiet seagrass beds, coral and rocky reefs, and sandy areas. Also around mangroves and sponge beds. BIOLOGY: Territorial. Feed on algae and invertebrates.

Regal Demoiselle
to 5 in.
Dusky Damselfish
to 6 in.
juvenile
Longfin Damselfish
to 5 in.
juvenile
Beaugregory
to 4 in.
juvenile

Freshwater Gregory - *Stegastes otophorus* (Poey, 1868)

FEATURES: Adults shades of brown to bluish gray with yellow to yellow orange on posterior soft dorsal, caudal, anal, and pelvic fins. Pectoral fins yellowish. Juveniles dark bluish to blackish with a series of bright blue spots along head, along scale rows, and on spiny dorsal fin; yellow on posterior soft dorsal and anal fins, and all or part of caudal fin. HABITAT: Greater Antilles (except Puerto Rico) and southern Caribbean Sea from Honduras to eastern Colombia. Occur in river mouths and low-salinity waters from shore to about 10 ft. Also enter fresh water.

Bicolor Damselfish - *Stegastes partitus* (Poey, 1868)

FEATURES: Several color phases exist with variation in each: dark brown or blackish to grayish on anterior half to three-quarters of body with pearly white to pale gray on posterior portion; dark brown to blackish on anterior portion of body, yellow below or behind, with posterior portion pearly; overall pearly to grayish with or without black on caudal fin. Caudal fin may be pale to grayish. Pectoral fins yellowish. Juveniles similar to adults. HABITAT: NC to FL, Gulf of Mexico, Bermuda, Bahamas, and Caribbean Sea. Occur around shallow coral reefs and patch reefs in deeper water. BIOLOGY: Aggressive and territorial. Feed on algae and invertebrates.

Yellow-tipped Damselfish - *Stegastes pictus* (Castelnau, 1855)

FEATURES: Shades of dark brown to almost black with yellow pectoral fins, upper peduncle, and caudal fin lobe. Juveniles with yellow on most of caudal fin and onto rear soft dorsal fin, may have a black spot on soft dorsal-fin base. Dorsal and anal fins long, pointed and extend past caudal-fin base. HABITAT: Southeastern Caribbean islands and off Brazil. Reef associated from about 10 to 280 ft.

Threespot Damselfish - *Stegastes planifrons* (Cuvier, 1830)

FEATURES: Purplish brown to brownish gray dorsally; yellowish below. Always with large black spot at upper pectoral-fin base and on upper caudal peduncle. Upper rim of eye always yellowish. Snout may have a blue wash. Juveniles yellow with a small black spot on upper pectoral-fin base, a large black spot on rear upper side, and a black spot on caudal peduncle. Body comparatively deep. HABITAT: NC (possible) to FL, Gulf of Mexico, Bermuda, Bahamas, and Caribbean Sea. Occur over inshore and offshore shallow coral reefs. Found in caves at night.

Cocoa Damselfish - *Stegastes xanthurus* (Poey, 1860)

FEATURES: Variable. Dark bluish brown to greenish brown dorsally; yellow below. May have blue spots on head and upper body. Pectoral fins yellow. Rear portion of dorsal fin yellowish. Small spot at upper pectoral-fin base. May have a small dark spot on upper caudal peduncle. Juveniles with blue spots over dark dusky blue on upper head and body; yellow below; black spot present on rear dorsal fin; may also have a black spot on upper peduncle. Body comparatively deep. HABITAT: NC to FL, Gulf of Mexico, Bermuda, Bahamas, Caribbean Sea to Brazil. Coral reef associated to about 100 ft. NOTE: Previously known as *Stegastes variabilis.*

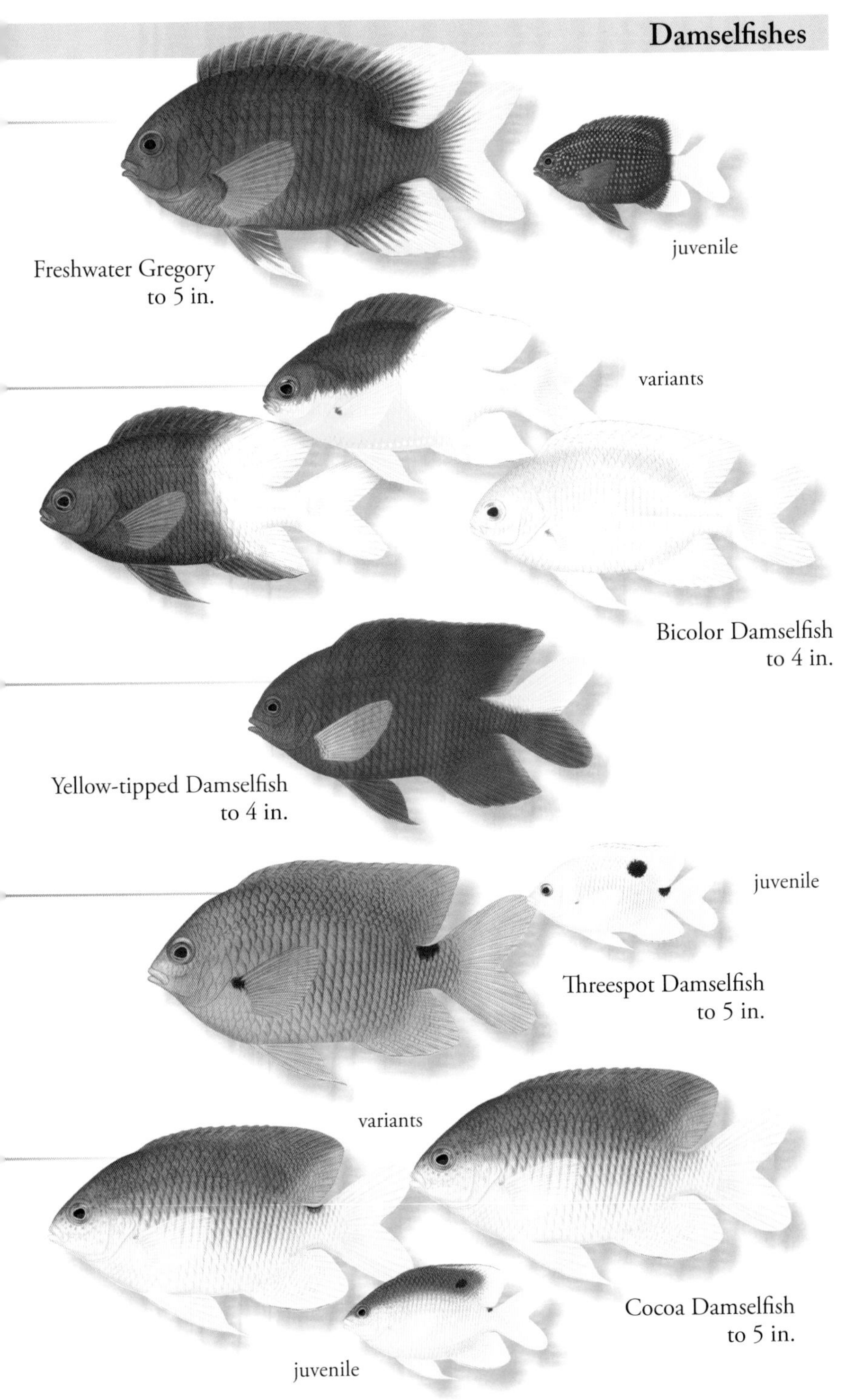
juvenile
Freshwater Gregory
to 5 in.
variants
Bicolor Damselfish
to 4 in.
Yellow-tipped Damselfish
to 4 in.
juvenile
Threespot Damselfish
to 5 in.
variants
Cocoa Damselfish
to 5 in.
juvenile

Labridae - Wrasses and Parrotfishes

Spotfin Hogfish - *Bodianus pulchellus* (Poey, 1860)

FEATURES: Red on anterior two-thirds of body with white area on lower part of head. White area on head may blend with a faint to distinct white streak on lower body. Posterior portion of dorsal fin, upper caudal peduncle, and caudal fin yellow. Two dark red to blackish streaks behind eyes. First two dorsal-fin spines and tips of upper pectoral-fin rays black. Intermediate specimens blackish anteriorly, yellow posteriorly, with white streak extending from lower portion of head. Juveniles yellow with black on anterior dorsal fin. HABITAT: SC to FL, Gulf of Mexico, Bahamas, and Caribbean Sea to southern Brazil. Occur over rocky and coral bottoms from about 30 to 400 ft.

Spanish Hogfish - *Bodianus rufus* (Linnaeus, 1758)

FEATURES: Color and pattern vary with age. Purplish blue to reddish area on upper anterior portion of body; margins of area may be distinct or diffuse and may extend partially or entirely into dorsal fin. Lower body and fins yellow with areas of blue. Blue areas may be small or may extend over ventral area and fins. Adults become more mottled and blue with age. Some specimens almost entirely bluish. HABITAT: NC to FL, Gulf of Mexico, Bermuda, Bahamas, and Caribbean Sea to southern Brazil. Found over coral reefs and offshore banks to about 130 ft.

Creole Wrasse - *Clepticus parrae* (Bloch & Schneider, 1801)

FEATURES: Color and pattern highly variable. Initial phase purple dorsally, lavender on sides, paler below; head dark above eyes; may have irregular pale areas along back. Terminal phase purple to lavender anteriorly, yellowish posteriorly; body variably mottled with paler and darker blotches; head dark above eyes. Juveniles purple to lavender with regularly spaced pale areas along back. HABITAT: FL, Gulf of Mexico, Bermuda, Bahamas, and Caribbean Sea. Over patch reefs and outer reef edges. Also around oil pylons.

Bluelip Parrotfish - *Cryptotomus roseus* Cope, 1871

FEATURES: Initial phase may be overall pale, iridescent salmon red, or dull brown dorsally, pearly white ventrally; may have broken whitish stripes and pale flecks and spots on upper body. Terminal phase dull to iridescent greenish dorsally, with broken salmon stripe from opercle to caudal fin; blue and pinkish bands run from mouth to eyes; dark spot at upper pectoral-fin base. HABITAT: FL, southern Gulf of Mexico, Bermuda, Bahamas, and Caribbean Sea to southern Brazil. Found over weedy or sandy bottoms to about 190 ft.

Red Hogfish - *Decodon puellaris* (Poey, 1860)

FEATURES: Color, pattern vary. Reddish dorsally; whitish below. May display bars or may be bicolored. Lips yellow. Yellow lines radiate from eyes. Yellow spots follow scales on sides. Fins pale reddish with yellow spots and lines. Upper and lower caudal-fin rays elongate. HABITAT: S FL, Gulf of Mexico, Bermuda, Bahamas, Caribbean Sea to southern Brazil. Over rocky bottoms and deep reefs from about 60 to 900 ft. NOTE: *Decodon* taxonomy is currently unresolved and may undergo revision.

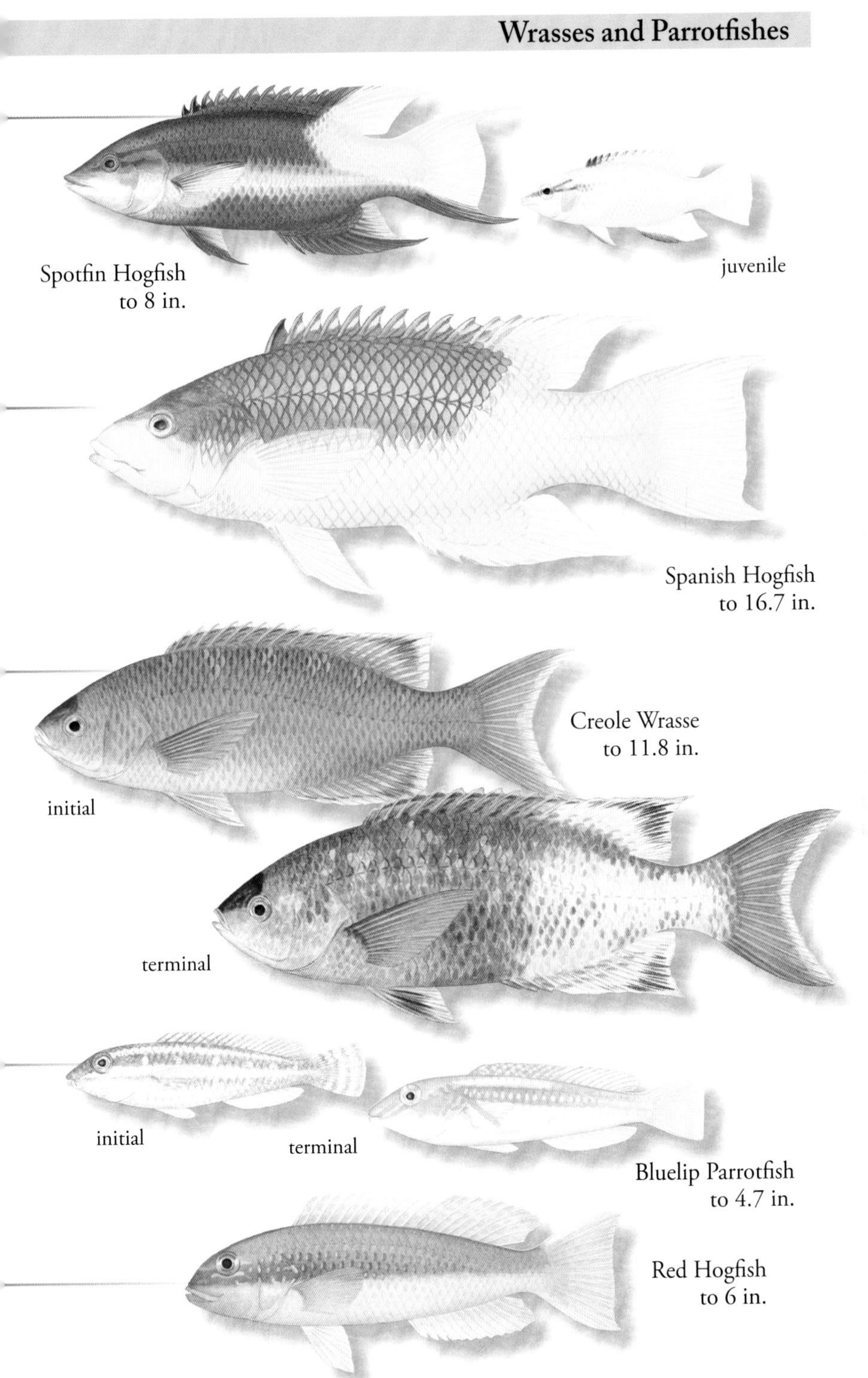

Spotfin Hogfish
to 8 in.

Spanish Hogfish
to 16.7 in.

Creole Wrasse
to 11.8 in.

Bluelip Parrotfish
to 4.7 in.

Red Hogfish
to 6 in.

Dwarf Wrasse - *Doratonotus megalepis* Günther, 1862

FEATURES: Color and pattern highly variable. Shades of green to reddish brown with irregular reddish bands, lines, and spots on head and body. Oblique white band below eyes. Head profile may be white. Fins mottled. May also be uniformly green with white along fin margins. Snout pointed. Anterior dorsal-fin spines tall. Caudal fin rounded. Males usually larger than females. HABITAT: FL, southern Gulf of Mexico, Bahamas, and Caribbean Sea to Brazil. Occur over shallow seagrass beds from shore to about 50 ft.

Greenband Wrasse - *Halichoeres bathyphilus* (Beebe & Tee-Van, 1932)

FEATURES: Initial phase reddish dorsally, fading to pinkish below; olive green band from snout to eye; distinct broken yellow stripe on sides and small dark spot on caudal peduncle; may have small dark spot above pectoral fin. Terminal phase pinkish green dorsally, becoming pink at midline; olive green band from snout to eye and from eye to opercular margin; distinct broken yellow stripe on sides; dark oblong spot above pectoral fin. HABITAT: NC to FL, scattered in Gulf of Mexico, and in southern Caribbean Sea. Reef associated from about 165 to 390 ft.

Slippery Dick - *Halichoeres bivittatus* (Bloch, 1791)

FEATURES: All phases variable. Initial phase with black stripe from snout to caudal peduncle that becomes single row of spots; bicolored spot on opercular margin; small black spot on dorsal fin. Terminal phase greenish with pinkish wavy stripes on head and body; purplish stripe on sides follows scale rows; bicolored spot on opercular margin; fins with pinkish stripes and bands. Intermediate phases share attributes of mature phases. HABITAT: NC to FL, Gulf of Mexico, Bermuda, Bahamas, and Caribbean Sea to northern Brazil. Over shallow reefs, rocky bottoms, and seagrass beds.

Mardi Gras Wrasse - *Halichoeres burekae* Weaver & Rocha, 2007

FEATURES: Initial phase salmon pink dorsally with white stripe on lower sides; snout yellow; large black blotch on caudal peduncle. Terminal phase with purplish head and anterior upper sides; bright blue lines and spots on head, body, and fins; bright blue ocellated spot above pectoral fin; large bright yellow area on sides; dorsal fin with blackish band anteriorly. Snout comparatively short. HABITAT: In the Gulf of Mexico known from Stetson Bank, Flower Garden Banks, and Veracruz. Reef associated to about 160 ft. BIOLOGY: Form small mixed schools. IUCN: Endangered.

Painted Wrasse - *Halichoeres caudalis* (Poey, 1860)

FEATURES: Initial phase pale salmon pink dorsally, with diffuse pinkish and yellowish stripes on lower sides; head with pale bluish lines; pale blue spots follow scale rows on sides; small black spot on lower dorsal fin. Terminal phase greenish dorsally, fading to pale reddish on sides; head with bright blue lines and a dark spot behind eyes; blue spots follow scale rows on sides; may have small black spot on lower dorsal fin. HABITAT: NC to FL, Gulf of Mexico, and Caribbean Sea to eastern Venezuela. Reef associated from about 85 to 240 ft.

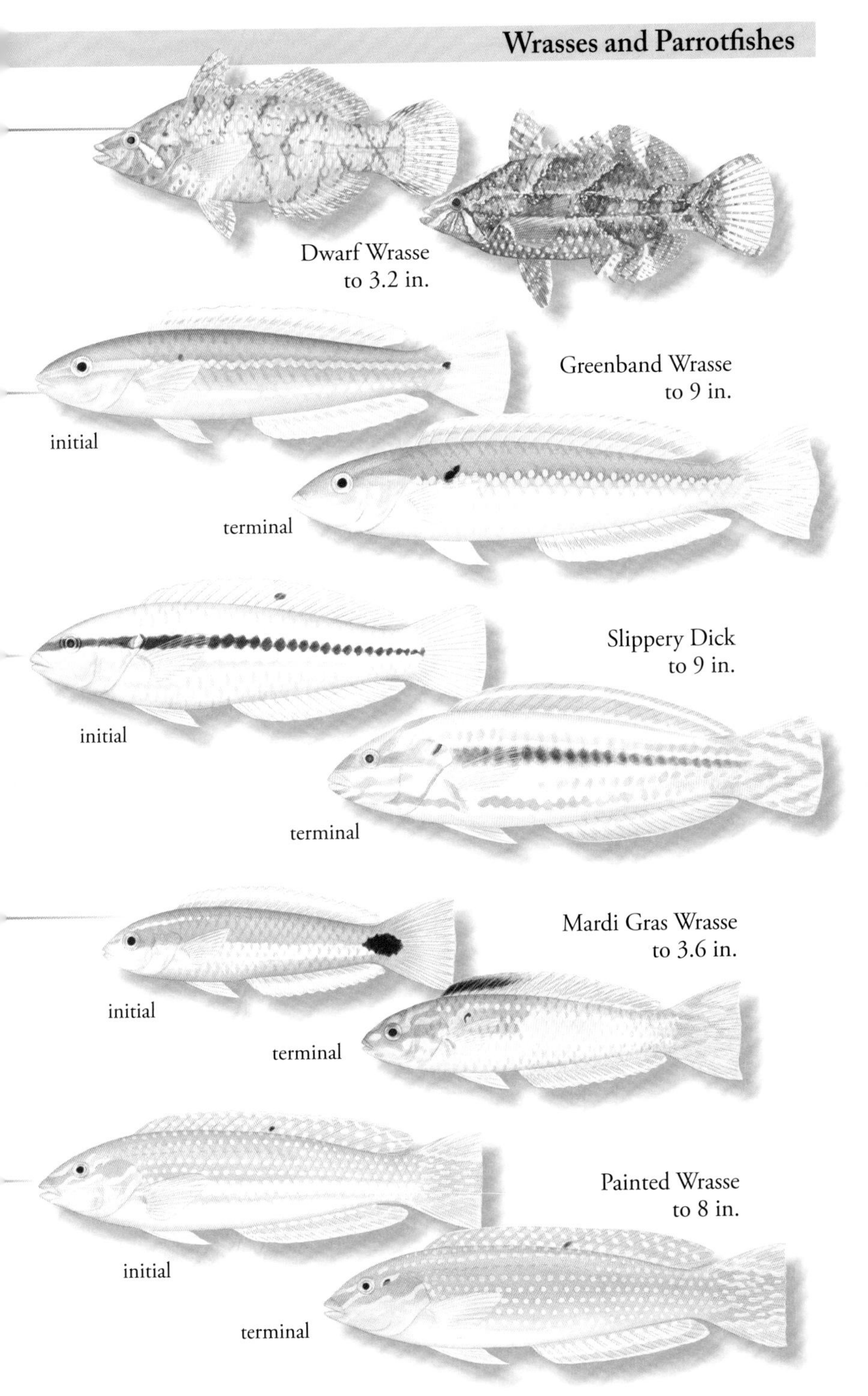
Dwarf Wrasse
to 3.2 in.
Greenband Wrasse
to 9 in.
initial
terminal
Slippery Dick
to 9 in.
initial
terminal
Mardi Gras Wrasse
to 3.6 in.
initial
terminal
Painted Wrasse
to 8 in.
initial
terminal

Yellowcheek Wrasse - *Halichoeres cyanocephalus* (Bloch, 1791)

FEATURES: Initial phase yellow on upper portion of head, along back, and onto dorsal fin; a broad blue stripe runs from opercle to caudal fin; white below. Terminal phase with blue along head profile, yellow from upper portion of head and along back; a broad dark blue to blackish stripe runs from opercle to caudal fin; white to pale turquoise below; dusky yellow band behind eye surrounded by blue, wavy lines; lower portion of dorsal fin blue to blackish. Juveniles yellow to orange on upper portion of head, upper back, and onto dorsal fin; shades of blue to purple on sides, paler below; a small black ocellated spot on lower portion of dorsal fin and on peduncle. HABITAT: FL, Bahamas, and Caribbean Sea. Found over rocky reefs and rubble from about 90 to 300 ft.

Yellowhead Wrasse - *Halichoeres garnoti* (Valenciennes, 1839)

FEATURES: Initial phase with a brown swath on upper portions of head and forebody, yellow on sides, whitish below; chin white. Terminal phase variable: yellow, tannish, or greenish on head and forebody; posterior portion of body green to bluish; always with a dark midbody bar that joins a dark stripe along posterior portion of back that can be dark blue, pink, or red; chin white. Initial and terminal phases with narrow black lines and spots radiating behind eyes. Juveniles yellow to orange with a bright blue stripe along midline. HABITAT: S FL, Gulf of Mexico, Bermuda, Bahamas, and Caribbean Sea. Occur over rocky and coral reefs and rubble and nearby sandy bottoms to about 390 ft.

Clown Wrasse - *Halichoeres maculipinna* (Müller & Troschel, 1848)

FEATURES: Initial phase greenish, with pinkish to orangish spots and wavy lines on sides. Terminal phase greenish to greenish blue on anterior upper sides, becoming greenish yellow posteriorly with orangish spots and wavy lines; large, black blotch on lower sides behind anal-fin tip. Initial and terminal phases with small blue spot at upper pectoral-fin base. Juveniles greenish yellow dorsally with broad blackish stripe on sides. All whitish below with pinkish bands; orange to reddish stripes on head and a triangular mark behind eyes; black blotch on anterior dorsal fin. HABITAT: NC to FL, Gulf of Mexico, Bermuda, Bahamas, and Caribbean Sea. Over shallow coral and rocky reefs and adjacent sandy bottoms to about 160 ft.

Rainbow Wrasse - *Halichoeres pictus* (Poey, 1860)

FEATURES: Initial phase tannish yellow dorsally, with brownish stripe from snout to opercle; stripe becomes a double row of spots on sides; whitish ventrally. Terminal phase lavender green or bluish green to yellowish green dorsally; nape rosy; whitish below; upper and lower portions separated by blue lines and spots; always with a large, black ocellated blotch on caudal peduncle. Juveniles similar to initial phase. HABITAT: FL, Bermuda, Bahamas, southern Gulf of Mexico, and Caribbean Sea to Guyana. Reef associated from about 15 to 180 ft. Usually occur well off the bottom.

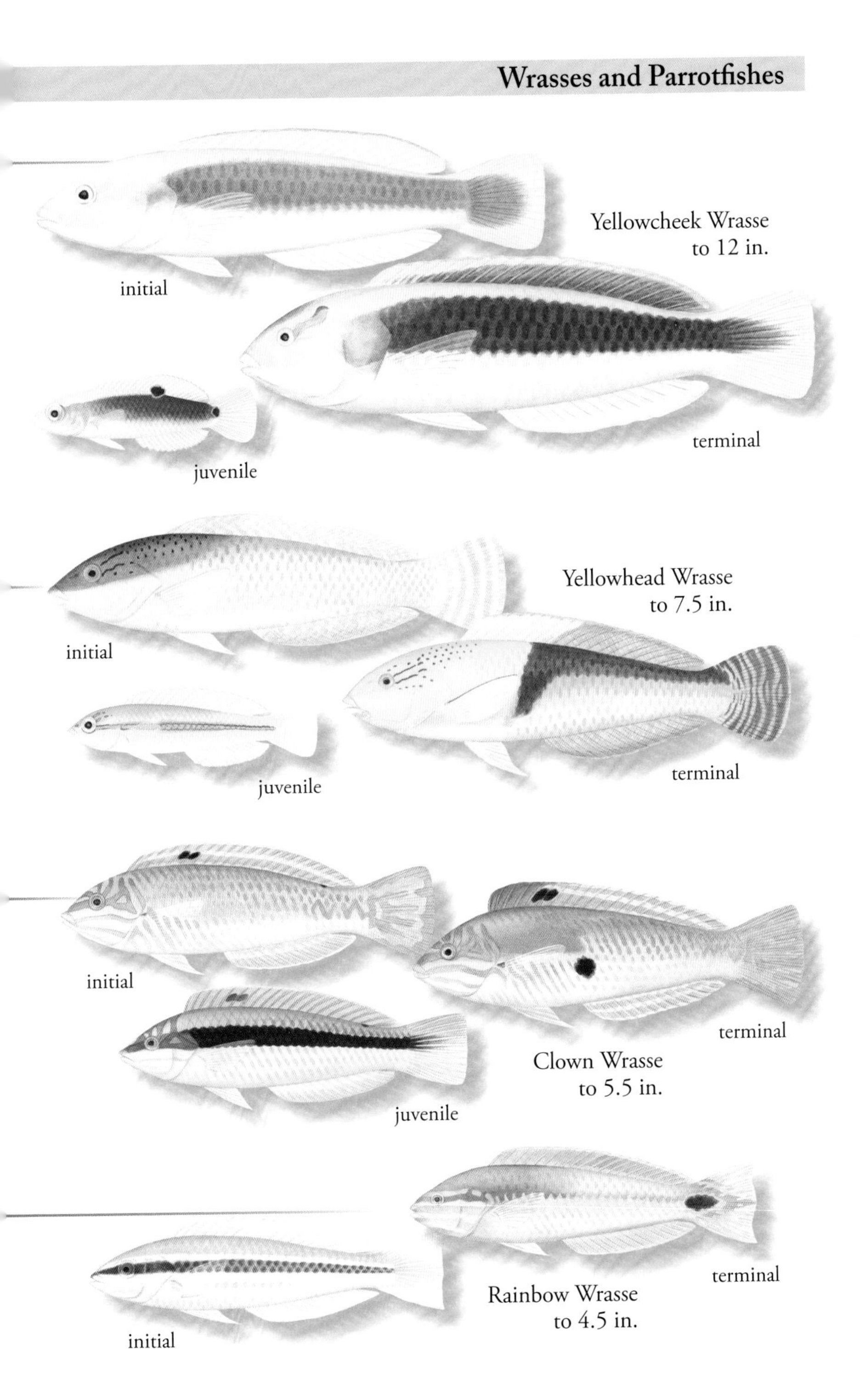
Yellowcheek Wrasse
to 12 in.
initial
terminal
juvenile
Yellowhead Wrasse
to 7.5 in.
initial
terminal
juvenile
initial
terminal
Clown Wrasse
to 5.5 in.
juvenile
terminal
Rainbow Wrasse
to 4.5 in.
initial

Blackear Wrasse - *Halichoeres poeyi* (Steindachner, 1867)

FEATURES: Initial phase yellowish to greenish yellow; may be almost uniformly colored, with a pale stripe on lower sides, or with several pinkish to blackish bands on upper sides; black spot on rear dorsal-fin base, and sometimes a black spot on rear portion of dorsal fin. Terminal phase greenish yellow, with pinkish and bluish wavy lines and spots on head and body; small greenish black spot behind eyes; small black spot on rear dorsal-fin base; caudal fin with pink V-shaped mark. Juveniles similar to initial phase but more intensely colored. HABITAT: GA to FL, southern Gulf of Mexico, Bahamas, Caribbean Sea to southern Brazil. Occur over shallow seagrass beds, coral and rocky reefs, and rubble to about 230 ft

Puddingwife - *Halichoeres radiatus* (Linnaeus, 1758)

FEATURES: Color and pattern highly variable. Initial phase greenish orange with whitish saddles on upper sides; interspaces shades of orange to black. Terminal phase purplish on head, olive to yellowish green on body with pale bluish saddle or bar near midflank. Both phases with bright blue lines and spots on head, blue spots along scale rows and fins, a small dark spot at upper pectoral-fin base, and yellowish caudal-fin margin. Juveniles orangish with bluish white bars dorsally, a bluish white stripe on lower side, a large black ocellus on dorsal-fin base, and a small black spot on caudal-fin base. Initial and terminal robust and deep-bodied. HABITAT: NC to FL, Gulf of Mexico, Bermuda, Bahamas, and Caribbean Sea. Also off eastern Brazil. Reef associated from near shore to about 164 ft.

Social Wrasse - *Halichoeres socialis* Randall & Lobel, 2003

FEATURES: Initial phase greenish gray dorsally, white below; lower sides with two brownish stripes; snout yellowish; large black spot on upper caudal-fin base. Terminal phase grayish green to grayish blue on upper portion of head and back, becoming orangish posteriorly; lower sides yellow; an orange to brownish blotch is present over pectoral fins; a bright blue stripe runs from under eye to caudal fin where it becomes broken spots; may have a dark orange to black spot at upper caudal-fin base; dorsal, caudal, and anal fins orangish with bright blue spots along bases. HABITAT: Belize. Occur over inshore barrier reefs, rubble, and seagrass beds. IUCN: Endangered.

Hogfish - *Lachnolaimus maximus* (Walbaum, 1792)

FEATURES: Color highly variable. Initial phase shades of orange to pearly white on lower portions of head and body; upper portion of head and nape reddish with paler lines and spots; may also appear nearly uniformly colored. Terminal phase pearly white on lower head and body with dark brownish to blackish on upper head and nape; snout elongate and upturned. Juveniles variably mottled and barred in pearly white and reddish orange. All with dark spot at rear portion of second dorsal-fin base. Anterior dorsal-fin spines free, elongate. All may pale or darken. HABITAT: NC to FL, Gulf of Mexico, Bermuda, Bahamas, Caribbean Sea to northern Brazil. Over soft bottoms and around shallow reefs to about 130 ft. Juveniles commonly over seagrass and inshore reefs. BIOLOGY: Feed on variety of invertebrates. IUCN: Vulnerable.

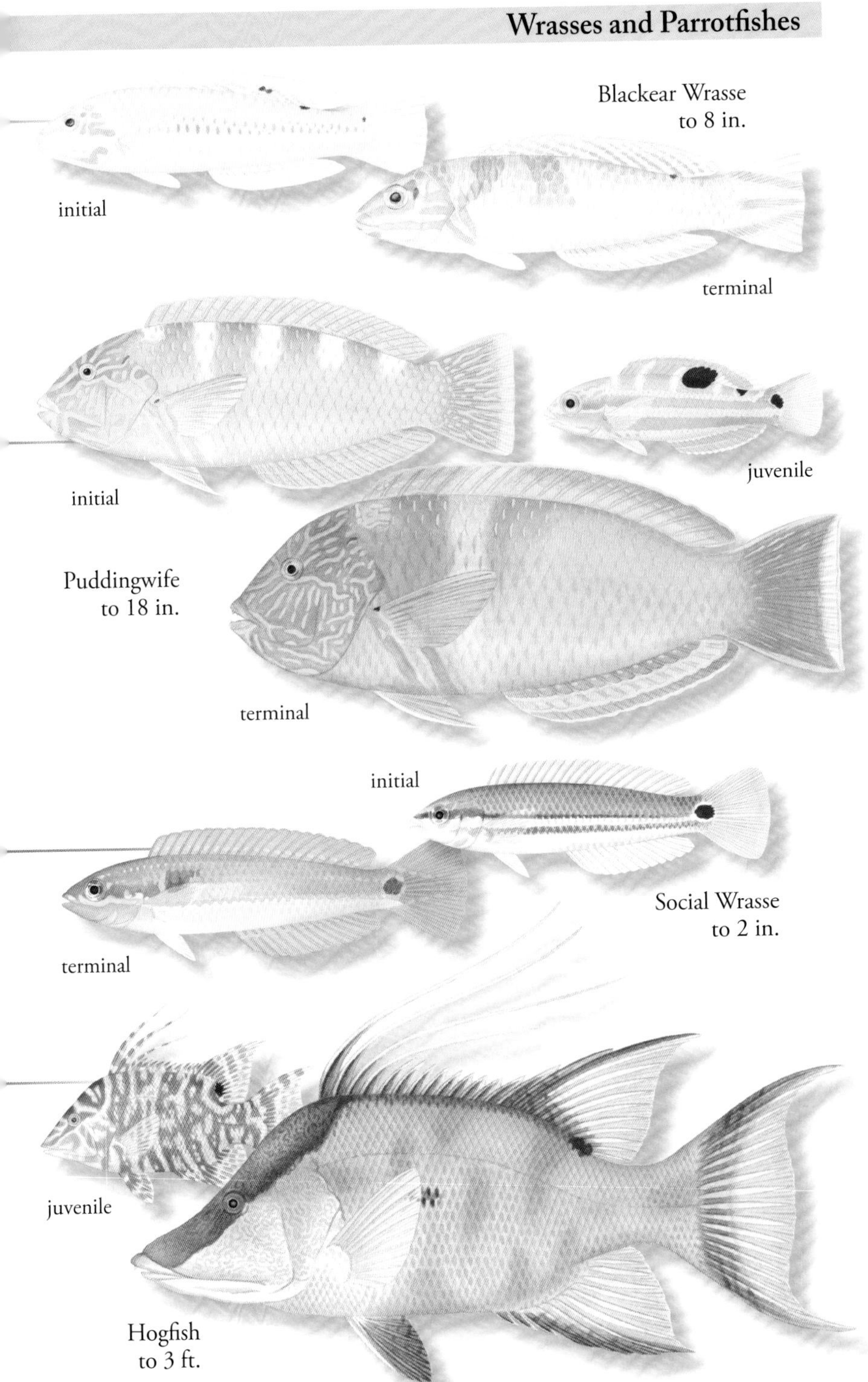
Blackear Wrasse
to 8 in.
initial
terminal
initial
juvenile
Puddingwife
to 18 in.
terminal
initial
terminal
Social Wrasse
to 2 in.
juvenile
Hogfish
to 3 ft.

Emerald Parrotfish - *Nicholsina usta* (Valenciennes, 1840)

FEATURES: Color and pattern of both phases highly variable. Initial phase shades of green, red, tan, or brown with vague to distinct pale and dark bars, or with broad stripes on sides; dark spot on first dorsal-fin membrane. Terminal phase shades of red, green, or yellow, or brown dorsally, variably striped below; irises red; bright blue lines on scales; fins banded in red or green. All with one or two vague to bright bluish white bands from lower jaw to eyes, and two bluish white spots behind eyes. Teeth separate with flattened tips and arranged in rows. HABITAT: NJ to FL, Gulf of Mexico, and Caribbean Sea to southeastern Brazil. Found over algae-covered reefs, rubble bottoms, and seagrass beds from near shore to about 240 ft. Primarily in shallow water.

Midnight Parrotfish - *Scarus coelestinus* Valenciennes, 1840

FEATURES: All phases and ages mottled in shades of indigo blue. Pale blue blotches on head. Scale centers variably pale blue. All fins dark blue. Snout gently sloping in both phases. Upper and lower caudal-fin tips somewhat to greatly elongate. Teeth fused. HABITAT: S FL, Gulf of Mexico, Bermuda, Bahamas, and Caribbean Sea. Reef associated to about 65 ft. Most common along inshore reef flats. BIOLOGY: Use teeth to scrape algae from rocks and coral polyps from coral heads.

Blue Parrotfish - *Scarus coeruleus* (Edwards, 1771)

FEATURES: Head, body, and fins entirely bluish in initial and terminal phases. Lips may be blackish. Initial phase with evenly rounded snout. Terminal phase with distinct hump on snout. Juveniles pearly blue with yellow over upper portion of head and anterior dorsal area. Teeth fused. HABITAT: MD (rare) to FL, southern Gulf of Mexico, Bermuda, Bahamas, Caribbean Sea. Over shallow coral reefs to about 65 ft.; also over shallow sandy and rubble flats. Juveniles found over seagrass beds and around mangroves. BIOLOGY: Use teeth to scrape algae from hard substrates and coral polyps from coral heads.

Rainbow Parrotfish - *Scarus guacamaia* Cuvier, 1829

FEATURES: Juveniles and initial phase rusty orange with green lines and spots radiating from eyes and green scale centers on body. Lower portion of head and chest rusty orange. Upper and lower profiles similar in shape. Terminal phase rusty orange anteriorly, with purplish wash. Posterior portion of body green. Demarcation between anterior and posterior portions variable. Head profile concave above eyes. Body robust, comparatively deep. Teeth fused. HABITAT: S FL, Bermuda, Bahamas, and Caribbean Sea. Associated with shallow coral reefs. Juveniles occur around mangroves. BIOLOGY: Retreat into a "home" cave at night. Use fused teeth to scrape algae and coral polyps from reef surfaces. IUCN: Near Threatened.

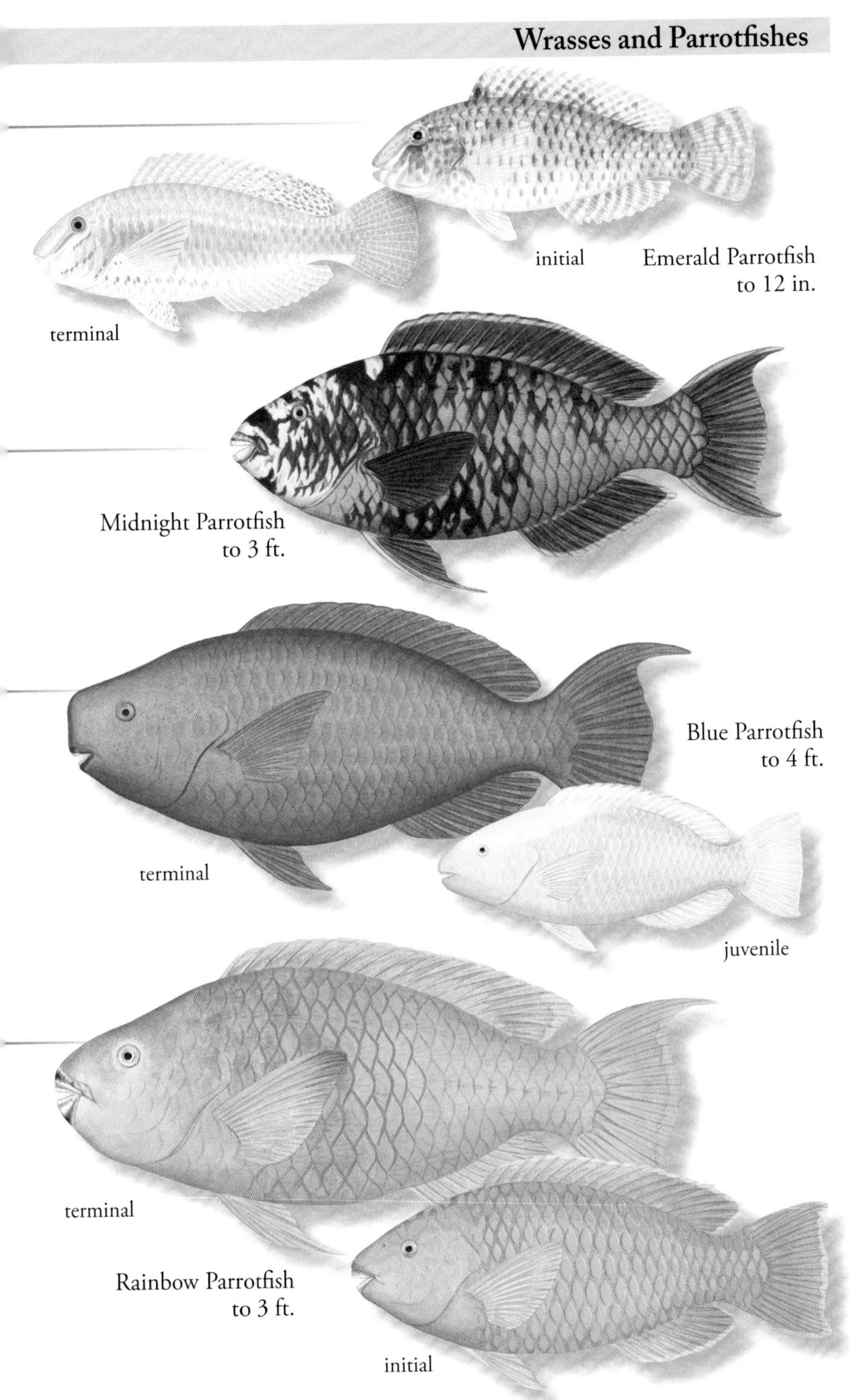
initial
Emerald Parrotfish
to 12 in.
terminal
Midnight Parrotfish
to 3 ft.
Blue Parrotfish
to 4 ft.
terminal
juvenile
terminal
Rainbow Parrotfish
to 3 ft.
initial

Striped Parrotfish - *Scarus iseri* (Bloch, 1789)

FEATURES: Initial phase with three broad, dark brown stripes separated by white stripes running from head to caudal fin; abdomen pale with white streaks; snout usually yellow; dorsal, caudal, and anal fins yellowish. Terminal phase somewhat variable; always with two wavy, blue green bands above and below eyes; pinkish on sides of head; upper opercular margin dark; may have a broad yellow, red, pink, or grayish stripe behind opercle; posterior portions of body with blue green scale centers; caudal fin with blue upper and lower margins, and pinkish to orange inner stripes. HABITAT: S FL, scattered in Gulf of Mexico, Bermuda, Bahamas, and Caribbean Sea. Strays to NC. Coral reef associated. BIOLOGY: Scrape algae from coral surfaces with fused teeth. Feed in groups. May school with Princess Parrotfish.

Princess Parrotfish - *Scarus taeniopterus* Lesson, 1829

FEATURES: Initial phase with three brownish stripes separated by yellow to white running from the head to caudal fin; abdomen pale with white streaks; snout brownish; dorsal, caudal, and anal fins yellowish; upper and lower margins of caudal fin usually brownish. Terminal phase with two blue, wavy stripes above and below eyes; sides of head orange; a broad, yellow patch behind pectoral fins; dorsal and anal fins with broad, orange inner band; caudal fin with orange upper and lower inner margins. Both phases become mottled while on the bottom or at night. HABITAT: FL, scattered in Gulf of Mexico, Bermuda, Bahamas, and Caribbean Sea. Occur over shallow coral reefs and seagrass beds. BIOLOGY: Form feeding aggregations. Take bites of algae from reef surfaces. Form a mucous cocoon at night.

Queen Parrotfish - *Scarus vetula* Bloch & Schneider, 1801

FEATURES: Initial phase reddish to purplish brown on upper body. Broad whitish stripe on lower sides. Ventral area whitish with reddish brown scale centers. Head whitish with pale green hues. Terminal phase green to blue green with dusky orange to rosy scale margins on body. Green to blue green bands above and below jaws. Pectoral fins with orange to rosy inner band. Caudal fin with orange on inner upper and lower lobes. HABITAT: S FL, southern Gulf of Mexico, Bermuda, Bahamas, and Caribbean Sea. Associated with shallow coral and rocky reefs. BIOLOGY: Use teeth to scrape algae from reef surfaces. Create mucous cocoon at night.

Greenblotch Parrotfish - *Sparisoma atomarium* (Poey, 1861)

FEATURES: Initial phase entirely reddish to yellowish red or with broken whitish stripes ventrally; eyes and pectoral and anal fins yellowish. Terminal phase greenish yellow to greenish dorsally; sides may be reddish to greenish; abdomen with bright blue along scales; narrow yellow band from corner of mouth to eye; greenish blotch above pectoral fin; small dark blotch between first and second dorsal-fin spines. HABITAT: NC to FL, Gulf of Mexico, Bermuda, Bahamas, and Caribbean Sea. Around soft coral, and over seagrass beds and around mangroves from about 55 to 250 ft.

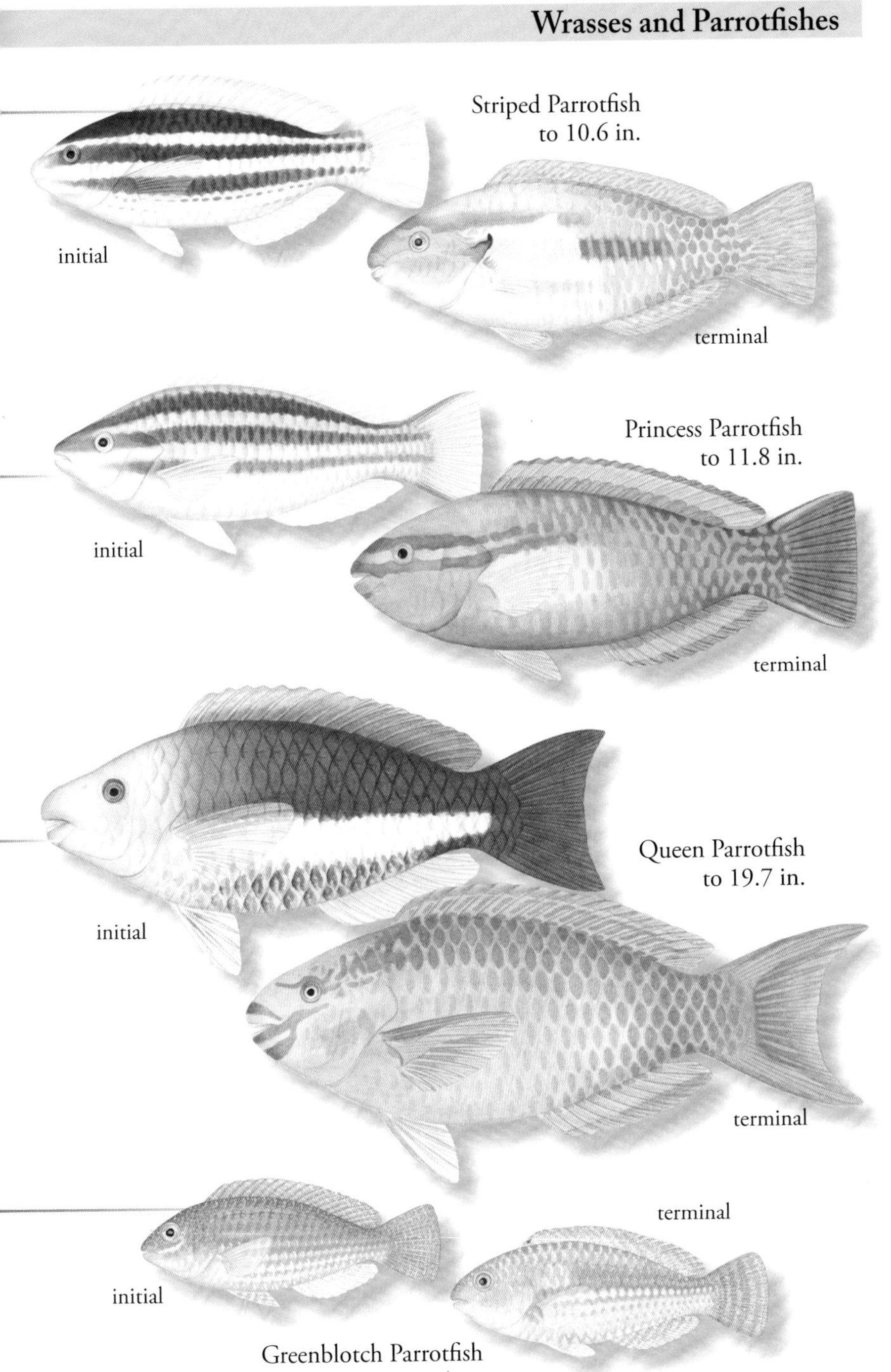
Striped Parrotfish
to 10.6 in.
initial
terminal
Princess Parrotfish
to 11.8 in.
initial
terminal
Queen Parrotfish
to 19.7 in.
initial
terminal
terminal
initial
Greenblotch Parrotfish
to 4 in.

Redband Parrotfish - *Sparisoma aurofrenatum* (Valenciennes, 1840)

FEATURES: Initial phase variable. May be dark olivaceous overall, greenish dorsally, and reddish below, or greenish with two broken whitish stripes on sides; may be slightly to highly mottled; always with distinct white blotch on upper caudal peduncle. Terminal phase greenish with pinkish area on lower sides; upper portion of head and forebody may be darker green; always with reddish band from corner of mouth to opercle and small black spots above larger yellow blotch on upper side above pectoral fins; anal fin reddish; caudal-fin tips blackish. HABITAT: NC to FL, Gulf of Mexico, Bermuda, Bahamas, and Caribbean Sea. Occur around coral reefs and over seagrass beds from near shore to about 350 ft. Usually in shallow water.

Redtail Parrotfish - *Sparisoma chrysopterum* (Bloch & Schneider, 1801)

FEATURES: Initial phase faintly or intensely mottled dusky reddish to dusky olivaceous; always with distinct blackish blotch on upper pectoral-fin base; colors may quickly fade or intensify. Terminal phase greenish to brownish gray, often with blue area on lower side; diffuse blue line below eyes; chin blue; dark spot at upper pectoral-fin base; pectoral fins with yellowish rays; inner portion of caudal fin yellowish to white, bordered in reddish; may become mottled while on the bottom or at night. HABITAT: S FL, southern Gulf of Mexico, Bermuda, Bahamas, and Caribbean Sea. Occur over shallow coral reefs and seagrass beds. BIOLOGY: Feed on benthic algae and seagrasses by taking single large bites. Swim by flapping pectoral fins.

Grey Parrotfish - *Sparisoma griseorubrum* Cervigón, 1982

FEATURES: Initial phase blotched in shades of brick red, grayish pink, or gray; may appear mottled; fins shades of orange red to orange pink; black blotch at pectoral-fin base; yellow to white blotch on upper peduncle. Terminal phase greenish with a reddish cast on sides; a white band extends from chin to eyes; irises red; scales with pale green margins; pectoral fins yellow at base with a black blotch above; whitish on upper peduncle. HABITAT: Southeastern Caribbean Sea off the coast of Venezuela. Occur over coral reefs and seagrass beds to about 100 ft.

Bucktooth Parrotfish - *Sparisoma radians* (Valenciennes, 1840)

FEATURES: Initial phase olivaceous to yellowish brown; dorsal area densely mottled and speckled; may show irregular bars and diffuse stripes; ventral area pale with dense speckles that follow scale rows; opercular margin and pectoral-fin base with patch of blue green; may also have blue green band from mouth to behind eyes. Terminal phase highly variable: uniform to mottled olivaceous to greenish brown; dorsal area densely speckled; always with a blue green and pinkish band from mouth to behind eyes; pectoral-fin base always with blackish blotch; caudal-fin margin always blackish. All with distinct tentacle on each nostril. HABITAT: S FL, Gulf of Mexico, Bermuda, Bahamas, and Caribbean Sea to southern Brazil. Found over seagrass beds, also around mangroves, rarely around reefs. BIOLOGY: Swim rapidly away when disturbed; assume a mottled color pattern on the bottom.

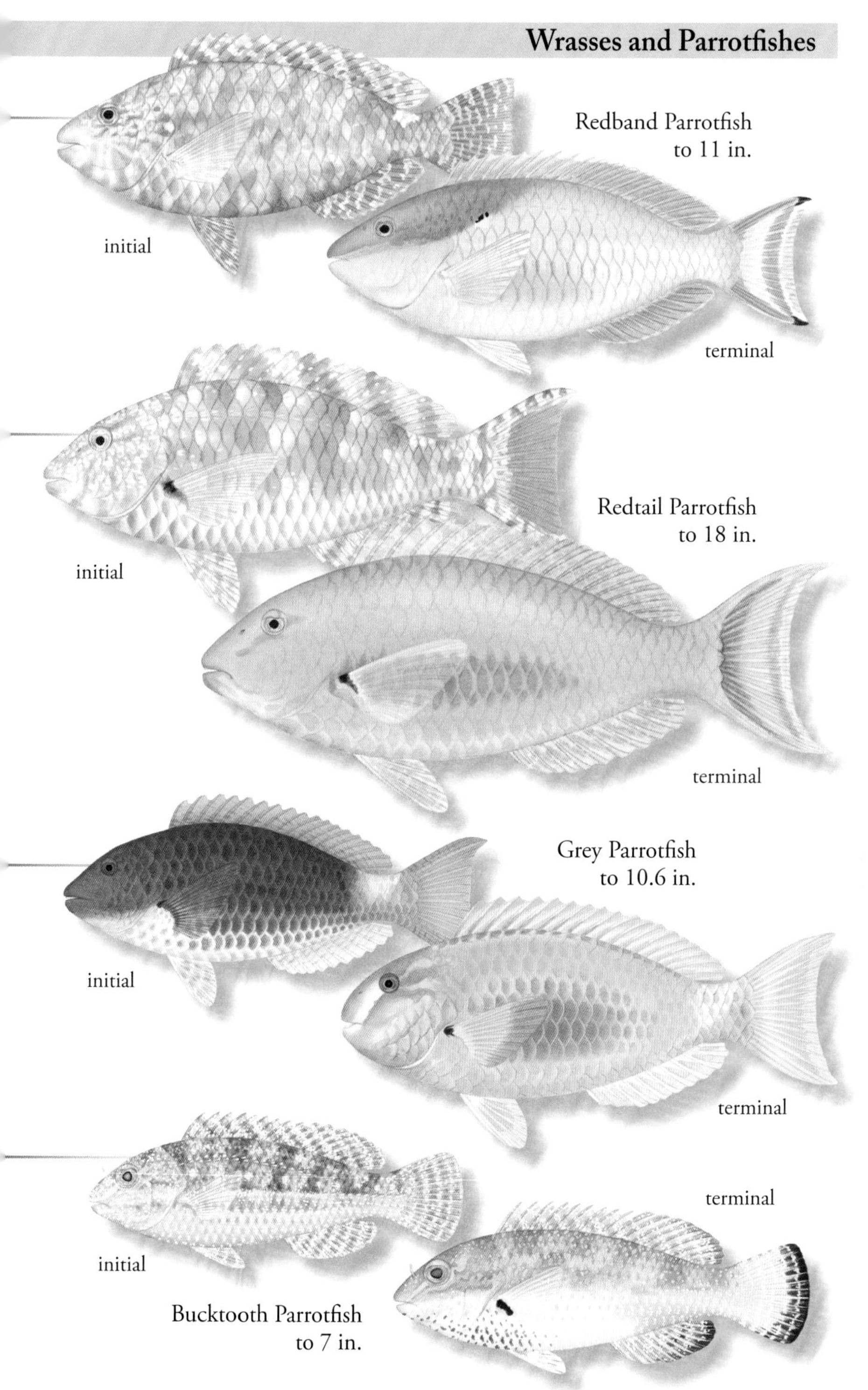
Redband Parrotfish
to 11 in.
initial
terminal
Redtail Parrotfish
to 18 in.
initial
terminal
Grey Parrotfish
to 10.6 in.
initial
terminal
terminal
initial
Bucktooth Parrotfish
to 7 in.

Yellowtail Parrotfish - *Sparisoma rubripinne* (Valenciennes, 1840)

FEATURES: Initial phase pale brownish gray with dark-rimmed scales and irregular bars on body; dark bars on chin; caudal peduncle and caudal fin yellow; pelvic and anal fins pinkish. Terminal phase overall greenish blue on head and body. Dark blotch at upper pectoral-fin base. Inner portion of caudal fin yellow. Assume a mottled pattern while resting on the bottom or at night. HABITAT: MA to FL, Gulf of Mexico, Bermuda, Bahamas, Caribbean Sea. Also tropical West Africa. Occur over shallow seagrass beds and coral reefs. BIOLOGY: Feed on benthic vegetation. When pursued may swim into surf or lie on bottom.

Stoplight Parrotfish - *Sparisoma viride* (Bonnaterre, 1788)

FEATURES: Initial phase mottled blackish brown and white on head and upper body; abdomen and fins reddish; inner portion of caudal fin whitish. Terminal phase blue to greenish; upper portion of head and scale margins dusky pink to purplish; dusky pinkish to purplish stripe runs from mouth to opercle; always with bright yellow spot at upper opercular margin, yellow blotch at caudal-fin base, and yellow crescent on caudal-fin rays. Juveniles (not shown) reddish brown with evenly spaced white spots on sides and white blotch at caudal-fin base. HABITAT: S FL, Gulf of Mexico, Bermuda, Bahamas, Caribbean Sea. Occur over shallow coral reefs and seagrass beds. BIOLOGY: Feed on benthic vegetation.

Bluehead - *Thalassoma bifasciatum* (Bloch, 1791)

FEATURES: Initial phase with purplish band through eyes and purplish blotch on opercle; diffuse, bicolored bars on sides; black blotch on anterior dorsal-fin membranes; dark blotch above pectoral-fin base. Terminal phase with bluish head and greenish body; always with a black and white V-shaped wedge on forebody. Juveniles yellow dorsally, white ventrally, with or without a dark lateral stripe; markings on head and dorsal fin similar to initial phase. HABITAT: NC to FL, Gulf of Mexico, Bermuda, Bahamas, and Caribbean Sea. Found over shallow coral reefs and seagrass beds. BIOLOGY: Feed on zooplankton and bottom-dwelling crustaceans. Spawns throughout the year.

Rosy Razorfish - *Xyrichtys martinicensis* Valenciennes, 1840

FEATURES: Initial phase white to yellow on head, pink to salmon on sides; dark blotch on opercle; scales on chest white with dark centers; large white blotch and vertical red lines on abdomen; may display faint bars on sides. Terminal phase pearly yellow anteriorly, pinkish posteriorly; head with pale blue bars; scales with pale blue centers; base of pectoral fins with dark, multicolored band; body comparatively deep. HABITAT: S FL, southern Gulf of Mexico, Bermuda, Bahamas, and Caribbean Sea to southern Brazil. Demersal over sandy and rubble bottoms adjacent to reefs and over shallow seagrass beds. BIOLOGY: Feed on small, sand-dwelling invertebrates. Burrow into bottom sediment when threatened.

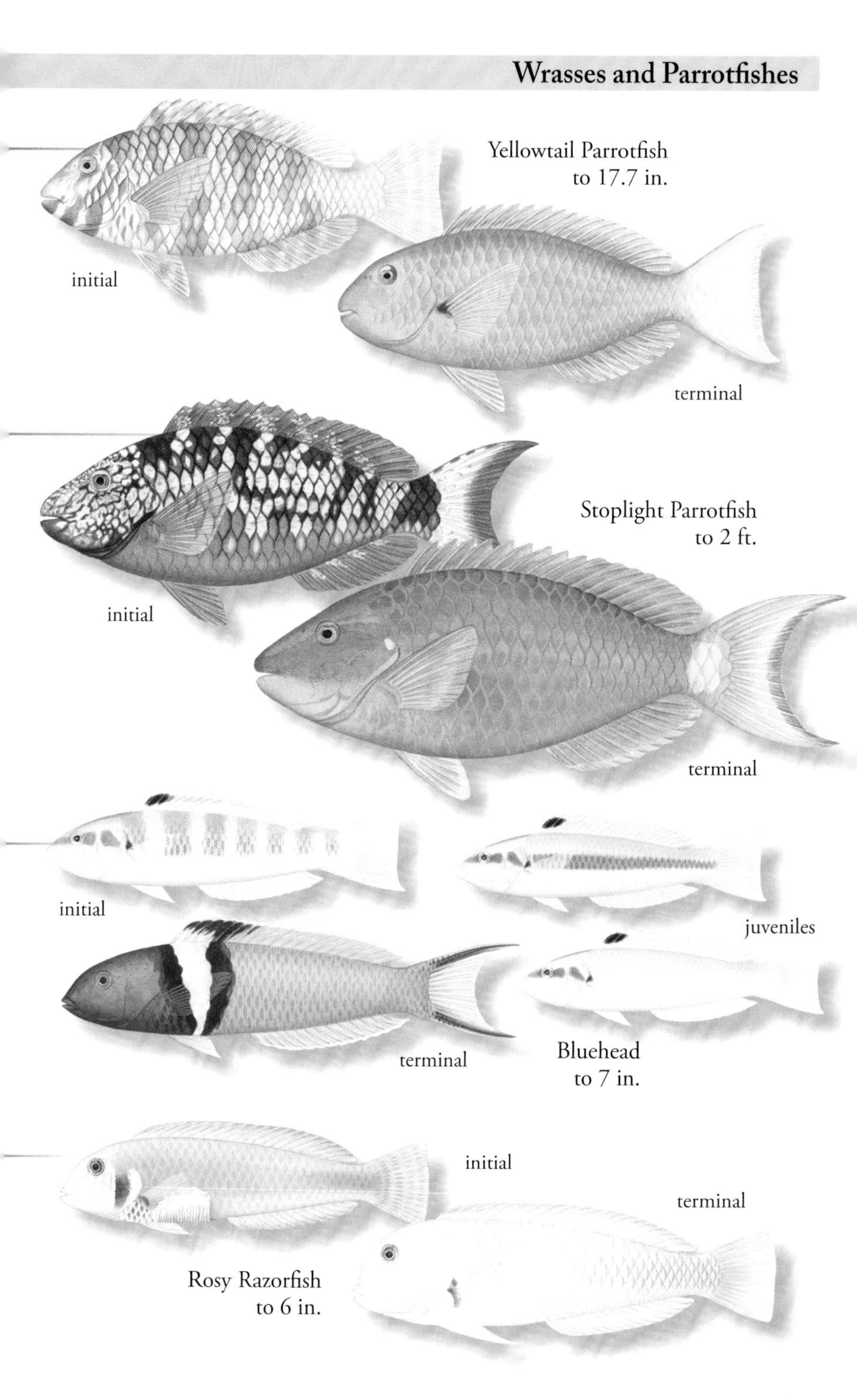
Yellowtail Parrotfish
to 17.7 in.
initial
terminal
Stoplight Parrotfish
to 2 ft.
initial
terminal
initial
juveniles
terminal
Bluehead
to 7 in.
initial
terminal
Rosy Razorfish
to 6 in.

Labridae - Wrasses and Parrotfishes, *cont.*

Pearly Razorfish - *Xyrichtys novacula* (Linnaeus, 1758)

FEATURES: Initial phase pearly pinkish or orangish to whitish with bright blue lines on head; distinct pearly patch on abdomen; first dorsal-fin spine taller than second. Terminal phase pearly greenish yellow with bright blue lines on head; diffuse dusky purplish to reddish blotch on sides. Both with blue lines on fins and scales rows, very steep snout profile, and laterally compressed body. HABITAT: NC (rare) to FL, Gulf of Mexico, Bermuda, Bahamas, Caribbean Sea to Brazil. Also eastern Atlantic. Over sandy bottoms adjacent to reefs and near seagrass beds from shore to about 300 ft.

Green Razorfish - *Xyrichtys splendens* Castelnau, 1855

FEATURES: Initial phase greenish to pearly or reddish; sides of head with pale green and orange bars; first two dorsal-fin spines tall. Terminal phase greenish to orange; sides of head with greenish and orangish bars; one to several blackish ocellated spots on sides; caudal fin with pinkish outer margin. Both with reddish irises and bluish spot between first and second dorsal-fin spines. Both may display pale bars on sides. HABITAT: S FL, southern Gulf of Mexico, Bermuda, Bahamas, and Caribbean Sea to southern Brazil. Over shallow rubble, sandy bottoms adjacent to reefs and over seagrass beds.

ALSO IN THE AREA: *Sparisoma axillare*, see p. 547.

Percophidae - Duckbills

Duckbill Flathead - *Bembrops anatirostris* Ginsburg, 1955

FEATURES: Tannish dorsally; yellowish tan to whitish ventrally. Scales darkly outlined. Irregular, brownish blotches on sides. Spiny dorsal and anal fins with dark margins. Head flattened. Upper jaw with tentacle at rear corner. Eyes comparatively small. Second dorsal-fin spine elongate. HABITAT: Gulf of Mexico and Caribbean Sea. Demersal from about 270 to 1,700 ft.

Goby Flathead - *Bembrops gobioides* (Goode, 1880)

FEATURES: Tannish dorsally; yellowish tan to whitish ventrally. Scales darkly outlined. Spiny dorsal fin with dark anterior margin. Soft dorsal fin dark along anterior margin and base. Dark spot on upper caudal-fin base, and dark margin on caudal fin. Head and anterior body flattened. Upper jaw with short tentacle at rear corner. Eyes comparatively large. HABITAT: NY to FL, Gulf of Mexico, Bahamas, and Antilles. On outer continental shelves and slopes from about 270 to 1,800 ft.

Scaly-chinned Duckbill - *Chrionema squamentum* (Ginsburg, 1955)

FEATURES: Tannish to creamy white with a series of alternating large and small brown oval patches overlaid with yellow on heads and sides. Spiny dorsal fin yellow at base, soft dorsal fin yellow at base and on margin, caudal fin with yellow upper and lower margins, anal fin yellow along margin. Irises with a yellow inner ring. First dorsal-fin spine elongate. HABITAT: NC to FL, Bahamas, and Caribbean Sea. Demersal over soft and rubble bottoms from about 380 to 1,000 ft.

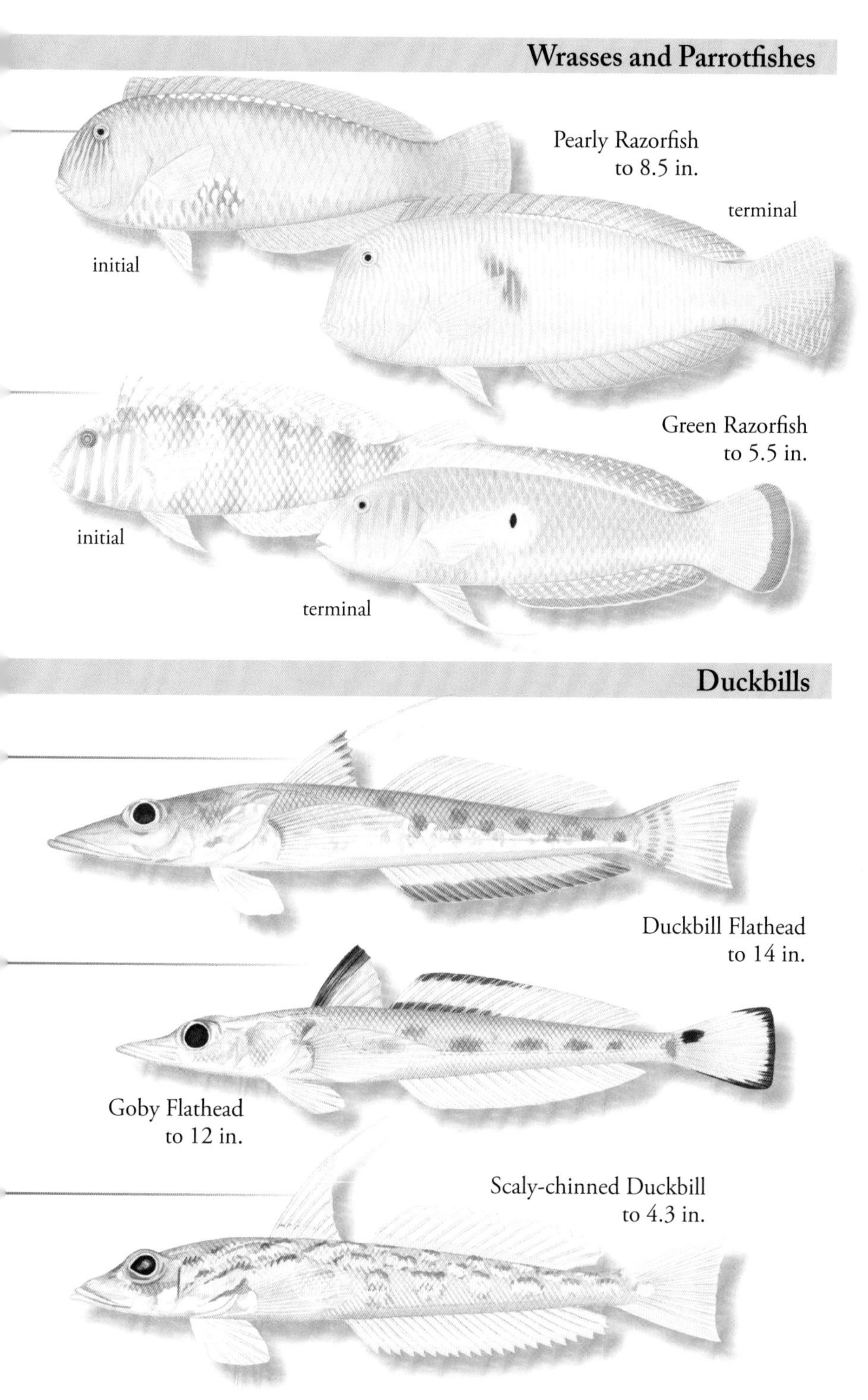
Wrasses and Parrotfishes
Pearly Razorfish
to 8.5 in.
terminal
initial
Green Razorfish
to 5.5 in.
initial
terminal
Duckbills
Duckbill Flathead
to 14 in.
Goby Flathead
to 12 in.
Scaly-chinned Duckbill
to 4.3 in.

Uranoscopidae - Stargazers

Southern Stargazer - *Astroscopus ygraecum* (Cuvier, 1829)

FEATURES: Brownish dorsally and on sides with a dense covering of small dark-ringed spots. Whitish below. First dorsal fin blackish. Soft dorsal fin with one to three dark bands. Pectoral fins dark, pale at margin. Mouth broad, nearly vertical. Head flattened dorsally. Eyes small, set on top of head. Small venomous spine above each pectoral fin. Body robust. HABITAT: NC to FL, southern Caribbean Sea to southern Brazil. Found coastally over soft bottoms from near shore to about 230 ft. BIOLOGY: Lie buried in bottom sediment with only top of head exposed. Use electric organ to stun prey. Feed on fishes. NOTE: Previously known as *Astroscopus y--graecum.*

Lancer Stargazer - *Kathetostoma albigutta* Bean, 1892

FEATURES: Brownish dorsally to about midline. Whitish below. Small, white spots on upper head. Spots become much larger posteriorly, forming blotches. Soft dorsal fin with several blackish blotches. Caudal fin with blackish blotches. Pectoral fins blackish, white along lower margin. Mouth broad, nearly vertical. Head flattened dorsally. Eyes small, set on top of head. Lower preopercular corner with three spines. Long, pointed, venomous spine above each pectoral fin. Spiny dorsal fin absent. Body robust. HABITAT: NC to FL and Gulf of Mexico to Yucatán. Demersal over soft bottoms in offshore waters from about 90 to 1,400 ft. BIOLOGY: Adults found in deeper waters; juveniles occur in shallow waters.

Spiny Stargazer - *Kathetostoma cubana* Barbour, 1914

FEATURES: Color varies. Shades of brown, black, or greenish brown to gray dorsally; white to yellowish below. Top and sides of head densely spotted and mottled. Spots become larger posteriorly. Dorsal fin with a dark blotch. Caudal fin with a broad blackish inner bar. Pectoral fins brownish with white margins. Mouth broad, nearly vertical. Head flattened dorsally. Eyes small, set on top of head. Lower preopercular corner with three spines. Long, pointed, venomous spine above each pectoral fin. Spiny dorsal fin absent. Body robust. HABITAT: GA to FL, Bahamas, and Caribbean Sea. Demersal over soft bottoms of in offshore waters on continental shelves and slopes from about 330 to 2,300 ft.

Tripterygiidae - Triplefin Blennies

Lofty Triplefin - *Enneanectes altivelis* Rosenblatt, 1960

FEATURES: Translucent with five oblique, reddish to brownish bars on sides. Bar on peduncle nearly vertical and not usually darker than others. Reddish brown bar below eyes, and diffuse reddish bands radiate from eyes. Anal fin with reddish bands. Patch of scales present on operculum. Pectoral-fin base and abdomen scaled. First dorsal fin considerably tall, banner-like, with first spine reaching past second spine of second dorsal fin when depressed. Pectoral fins usually with 14 rays. HABITAT: S FL, Bahamas, Greater and Lesser Antilles to Nicaragua. Also Campeche Bay. Demersal over clear and shallow coral reefs and rocky shores to about 40 ft.

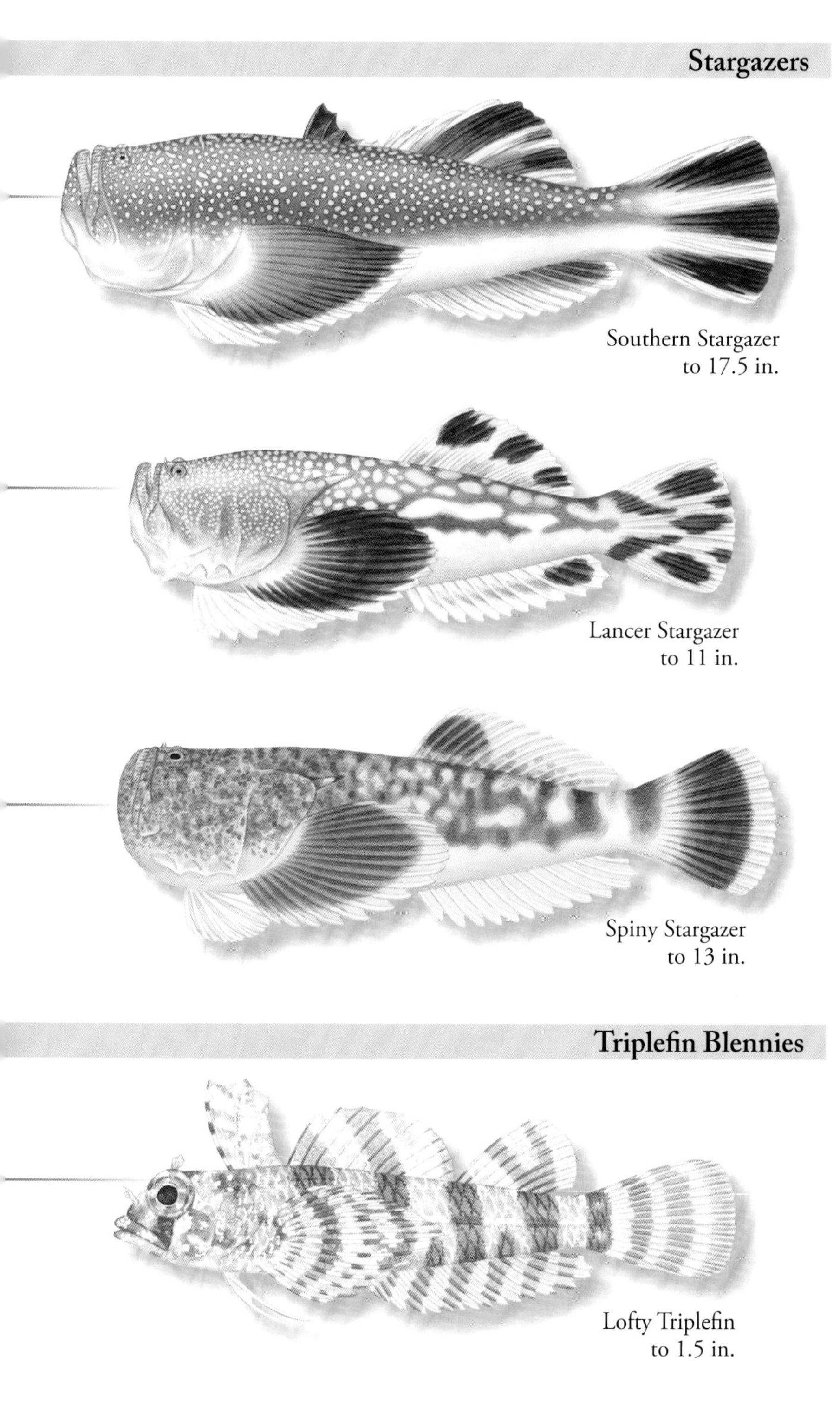

Southern Stargazer
to 17.5 in.

Lancer Stargazer
to 11 in.

Spiny Stargazer
to 13 in.

Triplefin Blennies

Lofty Triplefin
to 1.5 in.

Blackedge Triplefin - *Enneanectes atrorus* Rosenblatt, 1960

FEATURES: Translucent with five oblique brownish to blackish bars on sides. Bar on peduncle wedge-shaped and darkest. Head mottled and spotted in shades of red, orange, and white. Anal fin with red or black bands. Snout comparatively long. Operculum may be scaled. Pectoral-fin base and abdomen scaleless. First dorsal fin short in females, tall in males. Pectoral fins usually with 15 rays. HABITAT: Bahamas and Caribbean Sea. Demersal over rocky reefs and patch coral reefs from near shore to about 110 ft.

Roughhead Triplefin - *Enneanectes boehlkei* Rosenblatt, 1960

FEATURES: Translucent with five oblique, reddish to brownish bars on sides. Anterior bars usually indistinct; bar on peduncle usually darker than others and often blackish. A whitish bar precedes bar on peduncle. Diffuse reddish bands radiate from eyes. Anal fin banded. Pectoral-fin base and abdomen scaleless. First dorsal-fin spine shorter than anterior portion of second dorsal fin. Pectoral fins usually with 15 rays. HABITAT: S FL, southern Gulf of Mexico, Bahamas, and Caribbean Sea. Demersal over patch coral, rocky reefs, and rocky bottoms from near shore to about 50 ft.

Two-bar Triplefin - *Enneanectes deloachorum* Victor, 2013

FEATURES: Translucent with five brownish gray bars on sides. Bar under soft dorsal fin darker than anterior bars; bar on peduncle blackish. Underside of head white to yellow; sides of head spotted. Three dark spots on second dorsal-fin base. Caudal fin red basally, reddish to reddish black submarginally, with a white central bar. Pectoral-fin base and abdomen scaleless. First dorsal fin short. Pectoral fins usually with 15 rays. HABITAT: S FL, Dry Tortugas, Bahamas, Greater and Lesser Antilles to Venezuela. Also Veracruz, Mexico. Reef associated. Observed perching on sponges and corals.

Redeye Triplefin - *Enneanectes jordani* (Evermann & Marsh, 1899)

FEATURES: Translucent with four oblique, reddish to brownish bars on body. Broad, nearly rectangular black bar on peduncle. Caudal fin red at base with a reddish inner band. Dark bar below eyes. Anal fin mostly evenly reddish or with red bands. First dorsal-fin spine not reaching past second spine of second dorsal fin when depressed. Pectoral-fin base and abdomen scaled. Pectoral fins usually with 15 rays. HABITAT: S FL, Bahamas, and Caribbean Sea. Demersal over shallow, sheltered reefs, often on undersides of ledges. NOTE: Previously known as *Enneanectes pectoralis*.

Matador Triplefin - *Enneanectes matador* Victor, 2013

FEATURES: Translucent with four oblique reddish black bars on body. Wedge-shaped black bar on peduncle. Caudal fin red, or with a white inner band. Head shades of red with scattered white spots. Anal fin banded. Males with black on first and second dorsal fins. Snout comparatively short and steep. Pectoral-fin base and abdomen scaleless. First dorsal fin tall in large males. Pectoral fins usually with 15 rays. HABITAT: Bahamas to western and eastern Caribbean Sea. Demersal over shallow-water coral formations.

ALSO IN THE AREA: *Enneanectes flavus*, *E. quadra*, and *E. wilki*, see p. 547.

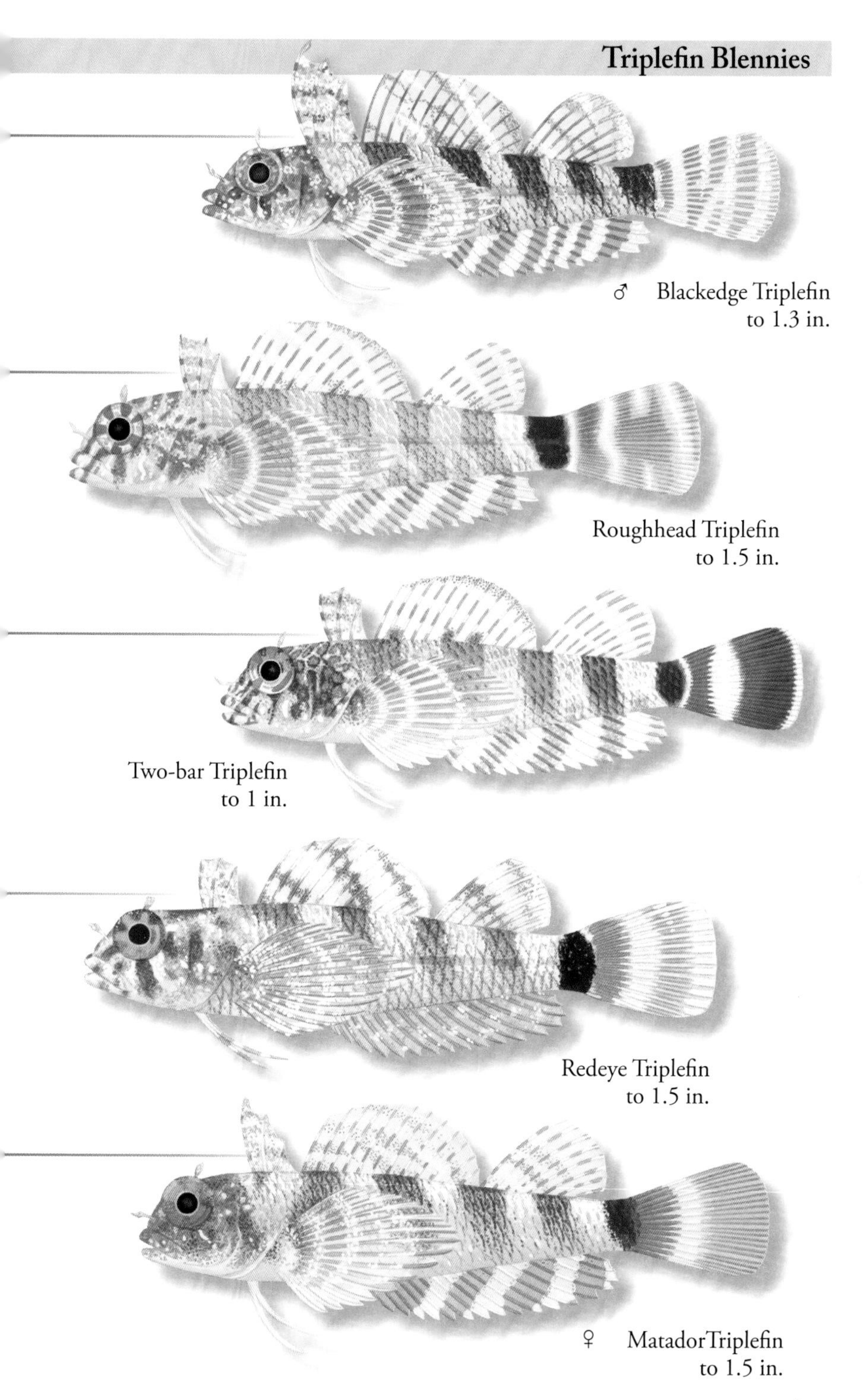

♂ Blackedge Triplefin
to 1.3 in.

Roughhead Triplefin
to 1.5 in.

Two-bar Triplefin
to 1 in.

Redeye Triplefin
to 1.5 in.

♀ MatadorTriplefin
to 1.5 in.

Dactyloscopidae - Sand Stargazers

Bigeye Stargazer - *Dactyloscopus crossotus* Starks, 1913

FEATURES: Pale tannish. Eyes usually blackish. Body may be unmarked or may have 8–12 indistinct brownish bars along dorsal profile. Cheek and opercle may be iridescent in some specimens. Eyes large, unstalked. First several dorsal-fin spines free. Body comparatively shallow. HABITAT: S FL, Bahamas, and Greater and Lesser Antilles, to southern Brazil. Demersal in shallow water along sandy beaches to about 30 ft. BIOLOGY: Burrow into sandy bottoms to lie in wait for prey.

Shortchin Stargazer - *Dactyloscopus poeyi* Gill, 1861

FEATURES: Whitish with a network of irregular brownish rings on top of head, and scattered blackish and brown speckles on upper sides. Speckles may be dense dorsally and extend onto lower sides. Eyes stalked. Lips fringed, upper lip with 13–17 fringes. Upper opercular margin with 17–20 skin flaps. First two to five dorsal fin spines are free and separate from rest of fin. Lateral line dips downward under last dorsal-fin ray. HABITAT: S FL, Bahamas, and Caribbean Sea. Demersal on shallow-water sandy bottoms, often in high surge areas to about 30 ft.

Sand Stargazer - *Dactyloscopus tridigitatus* Gill, 1859

FEATURES: Pale whitish with irregular brownish and tannish mottling on head. Upper body with brownish flecks. Some specimens display 11 to 14 short brownish bars along dorsal profile. Eyes very small, on tall stalks. First three to four dorsal-fin spines separate. Arched portion of lateral line comparatively short. HABITAT: FL, Bahamas, Bermuda, and Caribbean Sea to Brazil. Demersal over sandy bottoms from shore to about 95 ft., often in high surge areas. BIOLOGY: Burrow into soft bottoms to await prey.

Arrow Stargazer - *Gillellus greyae* Kanazawa, 1952

FEATURES: Tannish dorsally; translucent below. Upper portion of body mottled in white and tan with small dark flecks and five narrow brownish bars. A sixth bar is present on caudal peduncle. Dark band across eyes. First dorsal fin reddish brown. Dorsal fin with a three-spined finlet and a notch at midlength. HABITAT: S FL, southern Gulf of Mexico, Bermuda, Bahamas, and Caribbean Sea to southern Brazil. Demersal over shallow sandy areas around coral reefs, seagrass beds, boulders, and pilings from shore to about 90 ft. Avoids open beach bottoms. BIOLOGY: Burrow into soft bottoms.

Masked Stargazer - *Gillellus healae* Dawson, 1982

FEATURES: Pale with a brownish bar blow and between eye. Seven brownish bars on upper sides that are generally broader at midline and paler at centers. May have a dark blotch between bars. Eyes elevated but not stalked and with a circle of dermal flaps. Dorsal fin with a three-spined finlet, sometimes with one or two free spines behind. HABITAT: SC to FL Keys, and in Gulf of Mexico from Dry Tortugas to LA, also off northern Yucatán and around Aruba. Demersal over soft bottoms from about 70 to 240 ft. BIOLOGY: Burrow into soft bottoms to await prey.

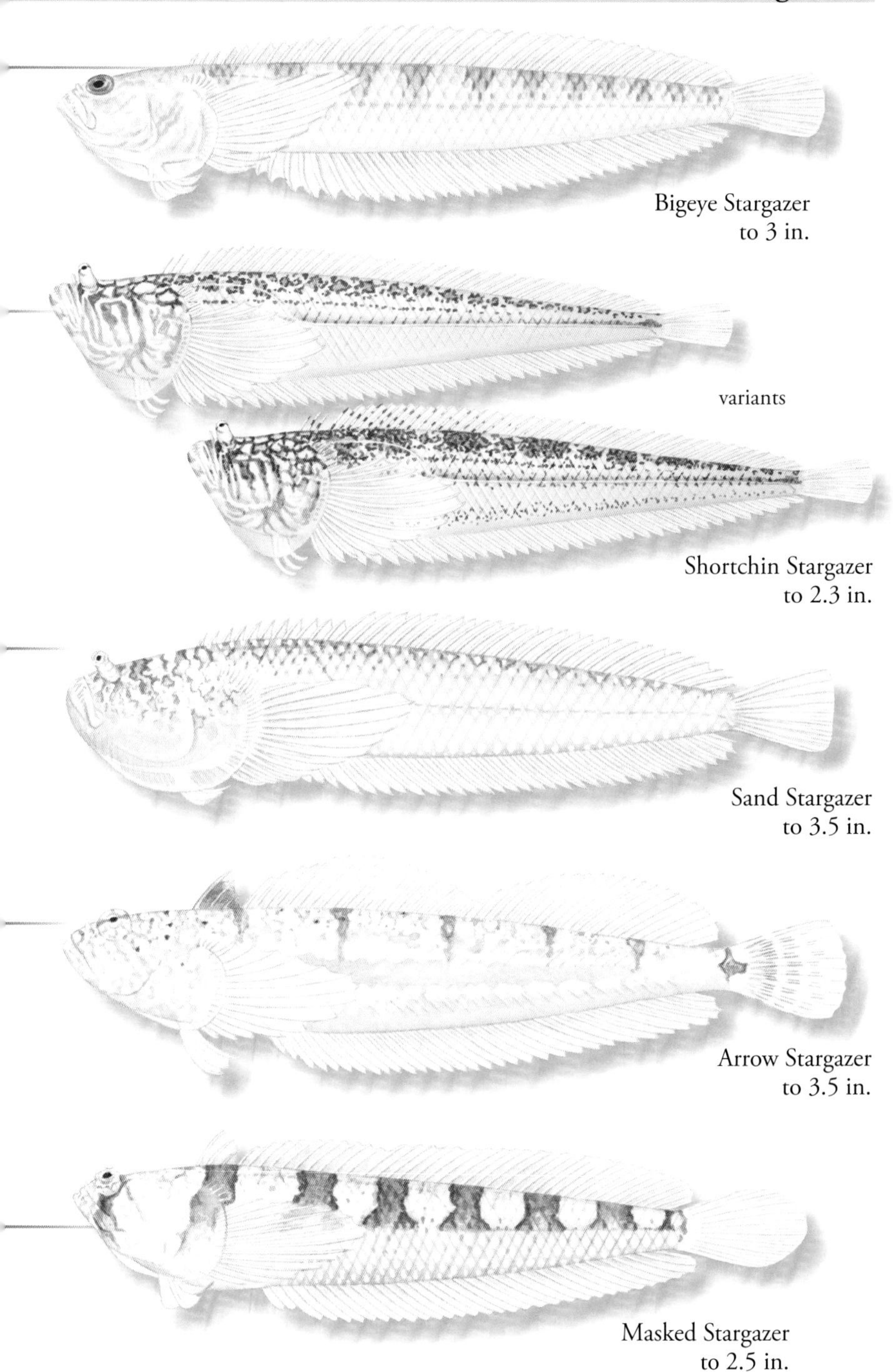

Bigeye Stargazer
to 3 in.

Shortchin Stargazer
to 2.3 in.

Sand Stargazer
to 3.5 in.

Arrow Stargazer
to 3.5 in.

Masked Stargazer
to 2.5 in.

Flagfin Stargazer - *Gillellus inescatus* Williams, 2002

FEATURES: White with tiny dark dots on snout, cheeks, and along back. Brownish bands radiate from eyes and on cheeks. Upper sides with eight narrow reddish brown bars that are paler at centers. Eyes elevated with a ring of small flaps. Dorsal fin with a single, separate, and elongate anterior spine with a fleshy tab at tip. Sometimes with one or two separate spines behind. HABITAT: Known only from off Navassa Island. Demersal over soft and rubble bottoms to about 100 ft.

Jackson's Stargazer - *Gillellus jacksoni* Dawson, 1982

FEATURES: Tannish dorsally with about seven irregular hourglass-shaped bars on upper sides. Lower sides creamy yellow without marks. Head variably blotched and mottled in brown, orange, and white. Eyes elevated with a ring of small flaps. Dorsal fin with a separate and dark three-spined finlet and a notch at midlength. HABITAT: Eastern and southeastern Caribbean Sea. Demersal over sandy bottoms from near shore to about 55 ft. BIOLOGY: Burrow into soft bottoms.

Warteye Stargazer - *Gillellus uranidea* Böhlke, 1968

FEATURES: Pale with four to five narrow brown to red saddles on upper sides, or variably plain with traces of brown dorsally and on sides. Head with dark spots and blotches. Eyes elevated but not stalked and with a circle of dermal flaps, and a posterior tentacle. Fimbriae absent from upper lip. Dorsal fin with a three-spined finlet and a notch at midlength. HABITAT: SE FL, Bahamas, and Caribbean Sea. Demersal over shallow-water sandy bottoms around rocks and patch reefs to about 40 ft.

Saddle Stargazer - *Platygillellus rubrocinctus* (Longley, 1934)

FEATURES: Pale with four broad brown to reddish brown bars from first dorsal fin to caudal peduncle. Bars generally reach lower profile. Brownish mottling on head. Pectoral fin may have brownish flecks. Eyes elevated with an incomplete circle of dermal flaps. Dorsal fin with a three-spined finlet and a notch at midlength. HABITAT: S FL, Bahamas, and Caribbean Sea. Demersal over sandy and sand-rubble bottoms around reefs from near shore to about 100 ft.

Sailfinned Stargazer - *Platygillellus smithi* Dawson, 1982

FEATURES: Whitish with about four brown bars from first dorsal fin to caudal peduncle. Brown bands radiate. Head with brown mottling that merges with first brown bar. Eyes elevated but not stalked. Dorsal fin with a tall, three-spined finlet and a notch at midlength. HABITAT: Bahamas to Venezuela. Demersal over sandy and sand-rubble bottoms and around reefs from near shore to about 120 ft.

NOTE: There are five other poorly known Stargazers in the area.

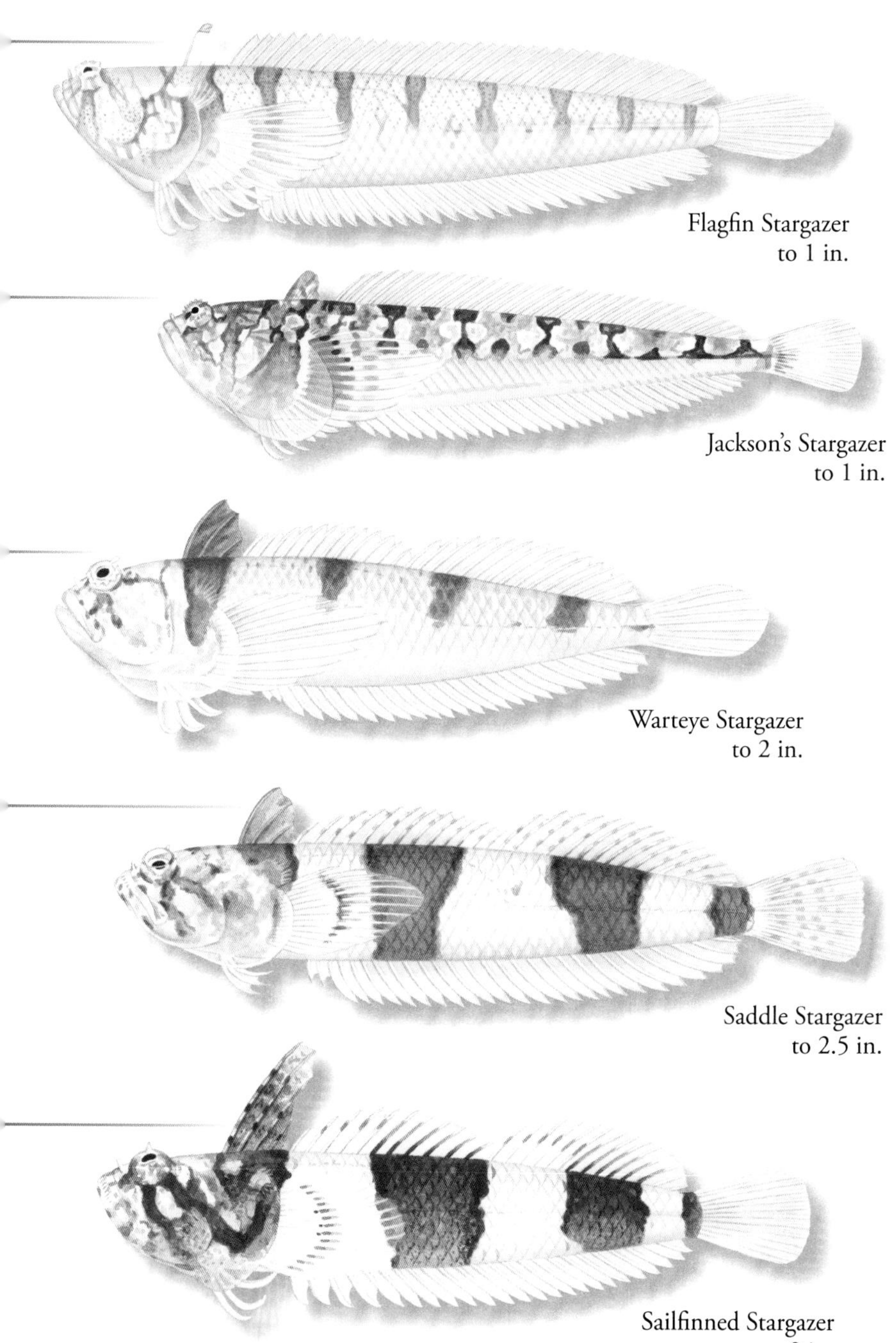

Flagfin Stargazer
to 1 in.

Jackson's Stargazer
to 1 in.

Warteye Stargazer
to 2 in.

Saddle Stargazer
to 2.5 in.

Sailfinned Stargazer
to 2 in.

Blenniidae - Combtooth Blennies

Pearl Blenny - *Entomacrodus nigricans* Gill, 1859

FEATURES: Color and pattern vary. Tannish with dark brownish to reddish brown bars on sides. Bars often subdivided. Bluish to white spots pepper pale areas and edges of bars. Snout and upper lip with narrow dark lines. Diffuse dark spot on head behind eyes. Head blunt. Snout steeply sloping. Tuft of cirri over nostrils and eyes. Single cirrus on nape. Body elongate. Dorsal fin distinctly notched. HABITAT: S FL, S Gulf of Mexico, Bermuda, Bahamas, and Caribbean Sea. Occur in tide pools and over shallow-water rocky and hard bottoms in high surge areas to about 30 ft.

Barred Blenny - *Hypleurochilus bermudensis* Beebe & Tee-Van, 1933

FEATURES: Color and pattern vary. Head grayish orange to red or brown with darker and paler smudges. Body pale grayish to pale tannish with wide, brownish bars on sides. Bars merge along midline and may be covered with spots. Pale areas with dark smudges. Cirri on eyes may have one to six branches. HABITAT: S FL, Gulf of Mexico, Bermuda, and Bahamas. Occur over rocky bottoms, jetties, and coral reefs to about 90 ft.

Featherduster Blenny - *Hypleurochilus multifilis* (Girard, 1858)

FEATURES: Color variable. Grayish to tan with diffuse large and small brownish spots that merge to form irregular bars on body. Narrow bars radiate from eyes. Anal fin with a blackish inner margin. Females generally with a pale or dark spot on membrane between first and second dorsal-fin spines and a dark spot at caudal-fin base. Cirri on eyes tall and bushy to very tall and multi-branched with a thick central stalk. HABITAT: Gulf of Mexico from Marco Island, FL, to Tabasco, Mexico. Occur over shallow-water oyster beds, rocks, pilings, jetties, and coral reefs to about 60 ft.

Oyster Blenny - *Hypleurochilus pseudoaequipinnis* Bath, 1994

FEATURES: Females tannish to whitish with about six pairs of diffuse, dark brown blotches that form saddles on upper sides, and diffuse brownish spots on pale interspaces. Males greenish gray with small orange spots on head and larger orange spots on body. Both with dark bars on chin and spots on head; dark area below a pale stripe present behind eye; dorsal fin with a dark spot on membrane between first and second spines. Cirri on eyes with a tall central tentacle and small branches at base. HABITAT: SE FL, Yucatán, and Caribbean Sea to Brazil. Demersal around mangrove roots, rocky outcroppings, and pilings to about 80 ft.

Orangespotted Blenny - *Hypleurochilus springeri* Randall, 1966

FEATURES: Pale bluish to creamy, with orange spots on head and forebody. Spots cluster to form bars on sides and become darker and larger posteriorly. Dorsal, caudal, and anal fins evenly spotted. Distinct spot between first and second dorsal-fin spines absent. Cirri above eyes short, with few short branches. HABITAT: S FL, Bahamas, and Caribbean Sea. Found over shallow rocky and reef areas along clear and quiet shorelines to about 15 ft. BIOLOGY: Feed on worms, crustaceans, and algae.

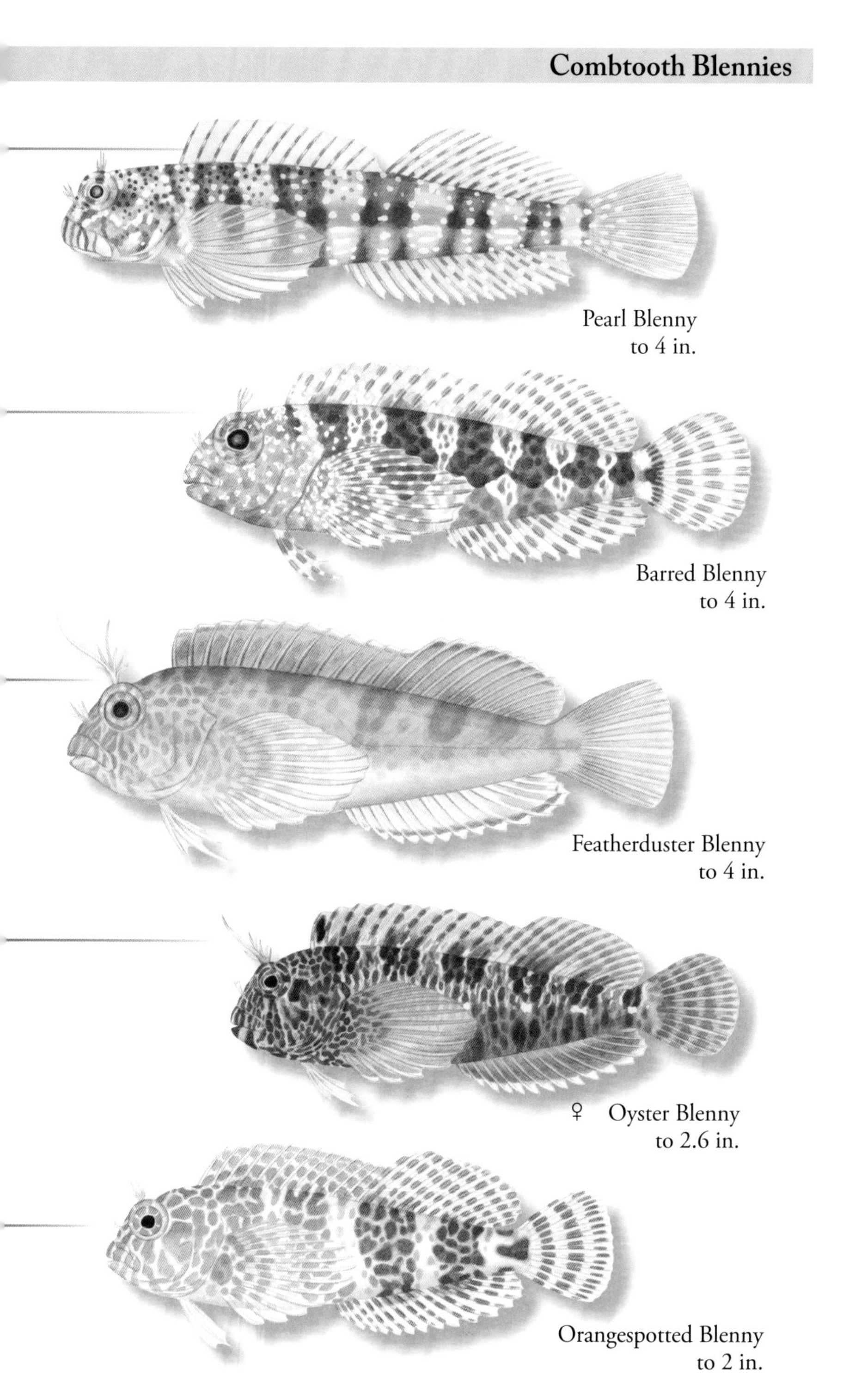

Pearl Blenny
to 4 in.

Barred Blenny
to 4 in.

Featherduster Blenny
to 4 in.

♀ Oyster Blenny
to 2.6 in.

Orangespotted Blenny
to 2 in.

Longhorn Blenny - *Hypsoblennius exstochilus* Böhlke, 1959

FEATURES: Mottled and blotched in white and brown anteriorly, becoming yellowish posteriorly. Wide blackish bar through eyes. Bluish spot on second dorsal-fin spine. Fins yellowish. Very tall and thick cirrus with smaller branches along rear edge over eyes. HABITAT: Bahamas to the southern Caribbean Sea. Demersal over shallow-water, exposed rocky bottoms with wave action, and eroded bottoms with small corals and sea fans.

Feather Blenny - *Hypsoblennius hentz* (Lesueur, 1825)

FEATURES: Grayish or tannish to pale greenish with dark brown spots and smudges that cluster to form oblique bars on head and body. Bluish spot usually present on second dorsal-fin spine. Dark crescent shape usually behind eyes. Pelvic fins dark. Anal fin dark in males, banded in females. Cirrus above eyes in males tall and fleshy with slender branches, short and tufted in females. HABITAT: Nova Scotia to FL, and Gulf of Mexico to eastern Yucatán. Occur over shallow-water soft and muddy bottoms, oyster reefs, and seagrass beds to about 30 ft.

Tessellated Blenny - *Hypsoblennius invemar* Smith-Vaniz & Acero P., 1980

FEATURES: Head and body covered with reddish to orange spots and polygons. Spots surrounded by blackish rings dorsally. Interspaces bluish to greenish blue. May have pale yellowish green saddles. Black blotch behind eyes. Dorsal and anal fins spotted. Fleshy cirri above each eye with up to four branches. HABITAT: S FL, Gulf of Mexico, southern Caribbean Sea to Brazil. Occur in large, empty barnacle shells that are attached to pilings, buoys, and rocks in shallow, clear waters to about 80 ft.

Highfin Blenny - *Lupinoblennius nicholsi* (Tavolga, 1954)

FEATURES: Color and pattern variable. Females and juveniles tannish with numerous brownish blotches on sides and onto dorsal fin. Mature males dark brown with distinct to very faint pale blotches along back. All with two dark brown bars below eyes; may also have brownish spots on gill membranes. A long, single cirrus is present over eyes in males, absent in females. Head compressed and very steep, particularly in males. HABITAT: St. Augustine to St. Petersburg, FL, and from Tuxpan to Cancún, Mexico. Possibly to Quintana Roo. Demersal over shallow-water rocks, oyster beds, and pilings.

Mangrove Blenny - *Lupinoblennius vinctus* (Poey, 1867)

FEATURES: Shades of brown to dark purplish brown with irregular pale yellowish blotches and bands on sides. Bands may merge onto fins. Cheeks very dark with small yellow to white blotches. Dorsal fin with white to yellow on first membrane. Males with a dark blotch on first to third dorsal-fin spines. A single, unbranched cirrus present over eyes. Cirri absent to nostrils and nape. Snout comparatively short and steep. Gill opening joined at throat. HABITAT: S FL, southern Gulf of Mexico Greater Antilles, and Caribbean Sea. Found in shallow waters around mangroves and in estuaries to about 15 ft. Also enter fresh water.

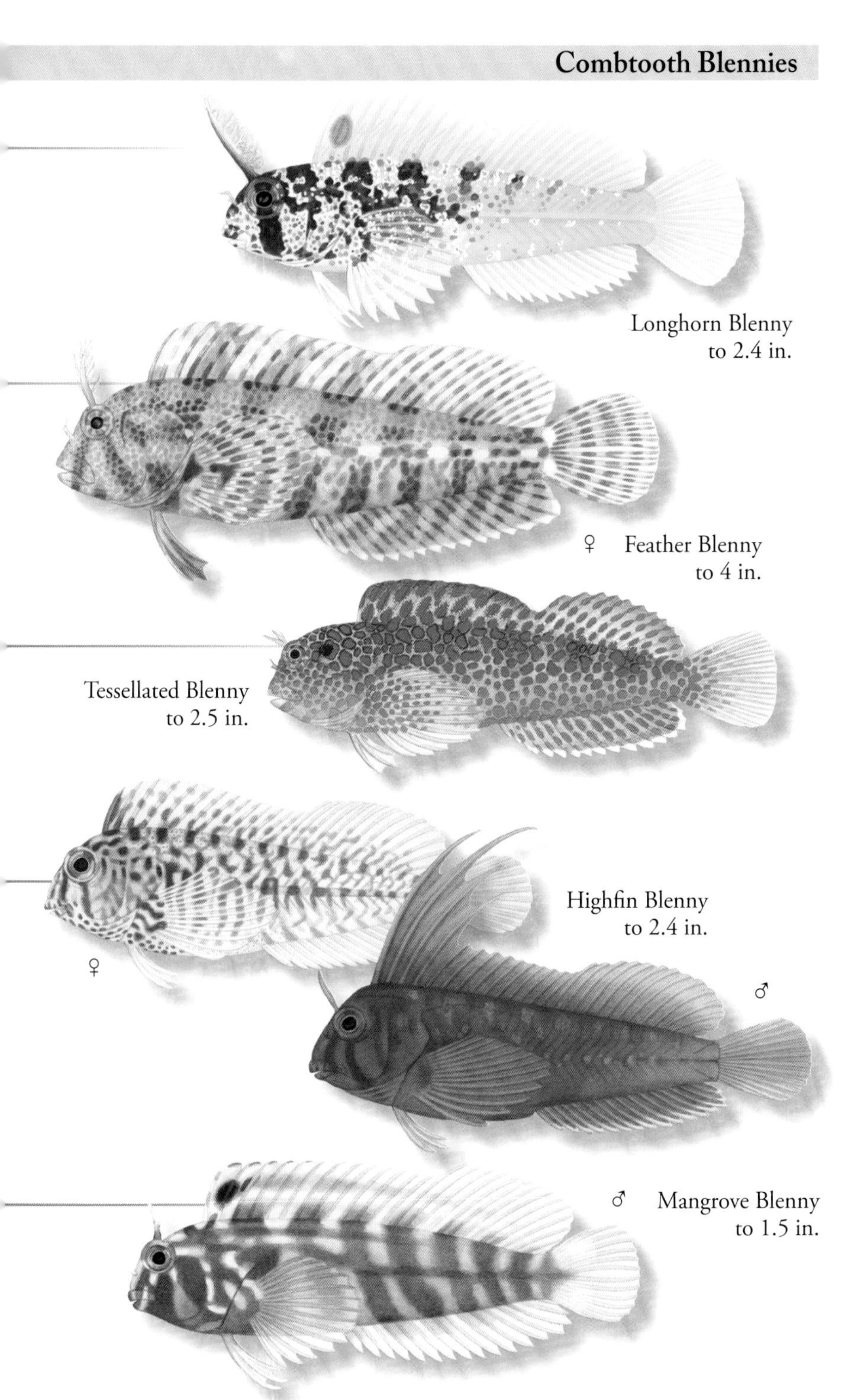
Longhorn Blenny
to 2.4 in.
♀ Feather Blenny
to 4 in.
Tessellated Blenny
to 2.5 in.
Highfin Blenny
to 2.4 in.
♀
♂
♂ Mangrove Blenny
to 1.5 in.

Muzzled Blenny - *Omobranchus punctatus* (Valenciennes, 1836)

FEATURES: Color varies. Shades of gray or greenish gray to tan with 11 dark saddles that appear as rows of short stripes on upper sides. Saddles become faint posteriorly. Several grayish stripes behind pectoral fins. Underside of head with broad, wavy dark bands that may be distinct to obscure. Reddish brown patch on nape. Dorsal and anal fins may be dark posteriorly. Head rounded in profile. Snout overhangs lower jaw. Body elongate. Scales absent. HABITAT: Introduced from Indo-Pacific. Scattered from Panama to Brazil. Demersal in shallow coastal waters over a variety of soft and hard bottoms and in estuaries to about 10 ft.

Redlip Blenny - *Ophioblennius macclurei* (Silvester, 1915)

FEATURES: Color and pattern highly variable. Overall reddish brown, overall grayish, or reddish brown to gray anteriorly and grayish to pearly white posteriorly. May appear blotched and barred. Lower lip reddish to pinkish or orange. Lower pectoral-fin rays and dorsal- and anal-fin margins may be reddish to pinkish. Simple cirrus over eyes. Nostrils with branched cirri. A pair of small and simple cirri on each side of nape. Eyes large. Snout profile almost vertical. HABITAT: NC (rare) to FL Keys, northwestern and southern Gulf of Mexico, Bermuda, Bahamas, and Caribbean Sea. Demersal over shallow coral reefs and rocky shores in clear water to about 240 ft. BIOLOGY: Feed almost exclusively on filamentous algae. Territorial. NOTE: Previously *O. atlanticus*.

Seaweed Blenny - *Parablennius marmoreus* (Poey, 1876)

FEATURES: Color and pattern highly variable. Tannish with series of brownish spots; brownish with series of pale spots; heavily spotted along sides or overall brownish to yellowish. Always with bluish lines on snout and lower cheeks. Lines may be straight, wavy, or broken. Pale to bright blue spot may be present on anterior dorsal fin. Cirri above eyes long with several branches at base. Cirri on nape absent. Anterior nostril with a fleshy flap. Notch between spiny and soft dorsal fins low. HABITAT: NY to FL, Gulf of Mexico, Bermuda, Bahamas, and Caribbean Sea to southern Brazil. Occur over coral and rocky reefs, limestone boulders, and eroded basins to about 120 ft. Sometimes around mangrove roots. Also in empty barnacle shells. BIOLOGY: Omnivorous, but feed mostly on algae. Males are territorial.

Molly Miller - *Scartella cristata* (Linnaeus, 1758)

FEATURES: Color and pattern highly variable. Tannish to grayish or reddish or orange with a series of diffuse, darker smudges along back and sides interlaced with pale wavy lines. Blotches along back merge onto dorsal fins. May appear overall brownish. Red spots usually present on head. Tuft of cirri present over eyes. Row of cirri present along nape profile. Cirri over eyes and nape short to long and usually banded in red and white. Specimens from Eleuthera reported to have few to no cirri on nape. HABITAT: FL, Gulf of Mexico, Bermuda, Bahamas, and Caribbean Sea to southern Brazil. Also eastern Atlantic and Mediterranean Sea. Demersal along shallow rocky shores, in tide pools, crevices, and empty barnacles, and on jetties to about 70 ft. Also on *Sargassum* rafts.

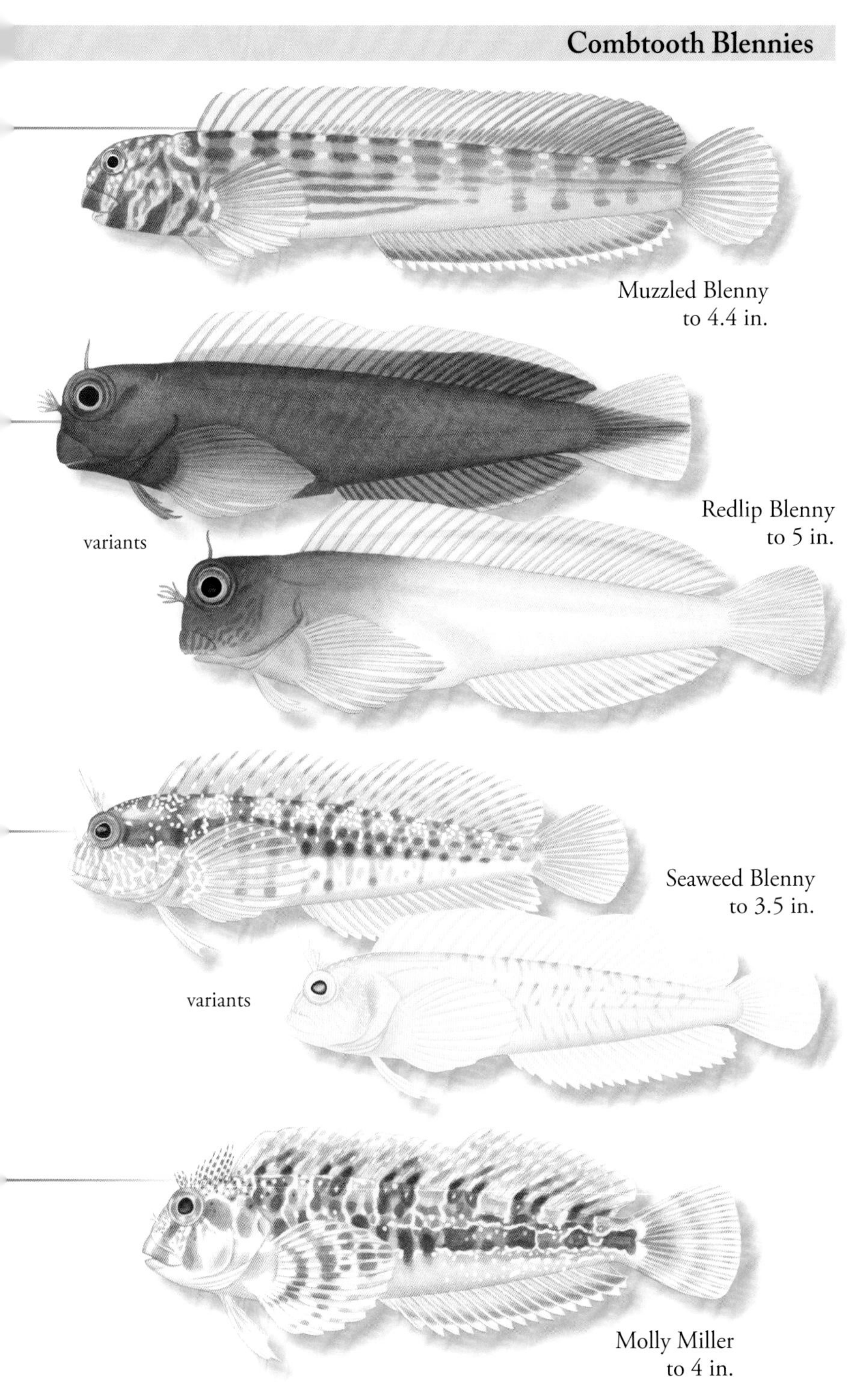
Muzzled Blenny
to 4.4 in.
Redlip Blenny
to 5 in.
variants
Seaweed Blenny
to 3.5 in.
variants
Molly Miller
to 4 in.

Whitecheek Blenny - *Brockius albigenys* (Beebe & Tee-Van, 1928)

FEATURES: Tannish brown with about eight rows of irregular, darker blotches that loosely form bars. Upper row merges onto dorsal fins. A whitish bar radiates below eyes—may appear as an inverted V-shape. Whitish bar at pectoral-fin base. Fins shades of yellow, with or without brownish bands. Cirri on nostrils, branched cirrus present over eyes, and two or more branched cirri on nape. Body comparatively deep and robust. HABITAT: Bahamas and Caribbean Sea. Demersal over exposed coral reefs to about 50 ft. NOTE: Previously known as *Labrisomus albigenys.*

Spotcheek Blenny - *Brockius nigricinctus* (Howell Rivero, 1936)

FEATURES: Females whitish to buff with blackish to brownish bars on sides. Males yellowish to buff with orange bars on sides. Bars extend into dorsal fins. Dark bands radiate from eyes. All with a dark ocellated spot on opercle. Snout pointed. First dorsal-fin spine shorter than remaining spines. Tuft of cirri on nostrils and over eyes. Transverse row of cirri on nape. HABITAT: S FL, Bahamas, and Caribbean Sea. Demersal in clear, shallow waters over mixed bottoms, reefs, eroded limestone, and in tide pools. Occur from near shore to about 40 ft. NOTE: Previously known as *Labrisomus nigricinctus.*

Puffcheek Blenny - *Gobioclinus bucciferus* (Poey, 1868)

FEATURES: Females shades of brown, tan, or green; fins variably spotted and banded. Males shades of red on head; fins spotted and banded in shades of red. All with four to five dark bars on sides that extend onto dorsal fins and are bordered by pale blotches; two small dark spots behind, and a dark bar below, eyes. Opercle lacks ocellus. Tuft of cirri on nostrils and over eyes. Transverse row of cirri on nape. First dorsal-fin spine taller than fifth. HABITAT: S FL, Bermuda, Bahamas, and Caribbean Sea. Demersal in tide pools and over shallow mixed sand and rocky bottoms, seagrass beds, patch reefs, and other coral formations. NOTE: Previously known as *Labrisomus bucciferus.*

Quillfin Blenny - *Gobioclinus filamentosus* (Springer, 1960)

FEATURES: Shades of brown to gray with four irregular dark bars on sides that extend onto fins. Opercle with a large black ocellus. Fins banded and blotched. Tuft of cirri on nostrils, a fan-like cirrus over eyes. Tufts of cirri on nape. Anterior dorsal fin very tall and incised. Pectoral and pelvic-fin rays long and deeply incised. HABITAT: Bahamas and Caribbean Sea. Demersal over reefs with abundant algae from about 40 to 400 ft. NOTE: Previously known as *Labrisomus filamentosus.*

Palehead Blenny - *Gobioclinus gobio* (Valenciennes, 1836)

FEATURES: Pale with four or five blackish brown to brownish bars on sides that are darker above midline, widest at midline, and bordered by pale areas. Sides of head spotted and mottled. Females with more distinct spots on dorsal fins. Opercle lacks ocellus. Tuft of cirri on nostrils and over eyes. Transverse row of cirri on nape. First dorsal-fin spine slightly shorter than fourth and fifth spines. HABITAT: SE FL, Bahamas, and Caribbean Sea to Brazil. Demersal from over rocky and sandy bottoms, coral formations, and seagrass beds to about 50 ft. NOTE: Previously known as *Labrisomus gobio.*

Whitecheek Blenny
to 2.5 in.

Spotcheek Blenny
to 3 in.

Puffcheek Blenny
to 3.5 in.

Quillfin Blenny
to 4.7 in.

Palehead Blenny
to 2.5 in.

Mimic Blenny - *Gobioclinus guppyi* (Norman, 1922)

FEATURES: Females greenish gray to buff with darker bars on sides. Males grayish green with grayish reddish bars; may be very dark and quickly change color. Bars extend into lower dorsal fins. Opercle with a pale to dark ocellus. Head and fins variably spotted and banded. Tuft of cirri on nostrils and over eyes. Transverse row of cirri on nape. HABITAT: S FL, Campeche Bay, Bahamas, Caribbean Sea. Possibly to Brazil. Demersal in tide pools and over mixed sand and rocky bottoms, coral and patch reefs, and limestone slopes to about 25 ft. NOTE: Previously known as *Labrisomus guppyi.*

Longfin Blenny - *Gobioclinus haitiensis* (Beebe & Tee-Van, 1928)

FEATURES: Tannish with six to seven brownish, broken bars on sides. Anterior bars extend onto dorsal fin. Opercle lacks an ocellus. Fins banded or spotted. Dark bar below eyes. Tuft of cirri on nostrils and over eyes. Transverse row of cirri on nape. First dorsal-fin spine taller than third and fourth spines. Pelvic fins long, with third ray half, or less than half, the length of the longest ray. HABITAT: S and W FL, off LA, N Yucatán, Bahamas, and Caribbean Sea. Demersal over patch reefs, coral formations, rocky and rubble bottoms with algae to about 50 ft. NOTE: Previously as *Labrisomus haitiensis.*

Downy Blenny - *Gobioclinus kalisherae* (Jordan, 1904)

FEATURES: Head reddish to pinkish. Body with six to seven wide brown or reddish brown to greenish brown bars that extend onto dorsal fin. Bars separated by diffuse, pale bars. Fins densely spotted. Opercle lacks an ocellus; small specimens with a dark blotch on rear opercle. Tuft of cirri on nostrils and over eyes. Transverse row of cirri on nape. First dorsal-fin spine taller than third and fourth spines. HABITAT: S FL, southern Gulf of Mexico, Bahamas, and Caribbean Sea to Brazil. Demersal over shallow rocky and rubble bottoms, reefs, seagrass beds, and in lagoons. NOTE: Previously as *L. kalisherae.*

Mock Blenny - *Labrisomus cricota* Sazima, Gasparini & Moura, 2002

FEATURES: Males reddish on head and pelvic fins; olive gray on body and fins. Females grayish brown on head and pelvic fins; violet gray to grayish brown on body. All with four to five dark bars on body that extend onto fins. Bars may have secondary bars above and below. Opercle with a large blackish blotch surrounded by an indistinct pale ring. Tuft of cirri on nostrils and over eyes. Transverse rows of cirri on nape. HABITAT: Reported from E FL, Belize, Panama, off Venezuela, and Lesser Virgin Islands. Demersal over shallow reefs, sand around rocks, and rocky bottoms with algal growth.

Hairy Blenny - *Labrisomus nuchipinnis* (Quoy & Gaimard, 1824)

FEATURES: Females tannish, densely spotted, and with irregular brownish bars that extend onto dorsal fins. Males diffusely mottled and blotched in shades of green; reddish on lower head and on abdomen; outer margins of fins reddish. All with ocellated spot on opercle. Breeding males with bright reddish head and alternating blackish and grayish side bars. Tuft of cirri on nostrils and over eyes. Transverse row of cirri on nape. HABITAT: FL, Gulf of Mexico, Bermuda, Bahamas, and Caribbean Sea to Brazil. Demersal in shallow coastal waters over sand and seagrass beds, rocky areas, and coral rubble.

♀
Mimic Blenny
to 4.5 in.
♂
Longfin Blenny
to 3 in.
Downy Blenny
to 3 in.
♂
Mock Blenny
to 6.3 in.
♀
Hairy Blenny
to 9 in.
♂

Goldline Blenny - *Malacoctenus aurolineatus* Smith, 1957

FEATURES: Tannish to whitish with dark brownish to blackish bars on sides. Anterior bars wide, defined, may merge. Orange to golden wavy stripes and spots overlay bars. Lower head banded. Anal fin golden. Males more distinctly marked. Females may have spotting on dorsal fin. Tuft of cirri over eyes and row of cirri on nape. HABITAT: S FL, Yucatán Peninsula, Bahamas, and Caribbean Sea. Demersal in shallow coastal waters over eroded limestone, sea grass beds, boulders, coral reefs, and in tide pools. Occur from near shore to about 15 ft. Often associated with sea urchins.

Diamond Blenny - *Malacoctenus boehlkei* Springer, 1959

FEATURES: Females with blackish bands through eyes, greenish honeycomb pattern on head, grayish honeycomb pattern on body, and blackish blotches along back. Males with reddish bands through eyes, orange honeycomb pattern on head, orange to reddish honeycomb pattern on body, and brownish blotches or saddles along back. All with a blue ocellated spot on anterior dorsal fin, and a row of diamond shapes along ventral profile. Snout pointed. HABITAT: S FL, Bahamas, Yucatán Peninsula, and Caribbean Sea. Demersal over coral reefs, reef dropoffs, and patch reefs to about 100 ft.

Delaland Blenny - *Malacoctenus delalandii* (Valenciennes, 1836)

FEATURES: Color and pattern highly variable. Shades of pale gray to tan with a series of five irregular dark saddles over series of dark blotches along midline. Pale areas between dark areas merge onto dorsal fins. Lower sides unmarked or with loose rows of pale spots. Fins unmarked or spotted. Males with an open ocellus on lower operculum. Two to five cirri over eyes; 12–18 cirri on sides of nape. HABITAT: Caribbean Sea to Brazil. Absent from Cuba and Jamaica. Demersal over shallow, sheltered, turbid, rocky reefs and rubble, algae, and seagrass beds to about 10 ft.

Imitator Blenny - *Malacoctenus erdmani* Smith, 1957

FEATURES: Color varies. Shades of tan, gray, or brown with darker saddles that are bordered by darker margins. Saddles do not extend onto fins. Saddle under soft dorsal-fin base with a very dark margin and a blue inner spot. Lower sides with about 10 oblique dark bars. Body scales with whitish centers. Head with dark bands through and below eyes. Anterior dorsal fin distinctly banded. Usually with two cirri on nostrils and over eyes. Branched cirri on nape. HABITAT: Bahamas and Caribbean Sea. Demersal over shallow-water rocks, depressions, coral debris, and algae.

Dusky Blenny - *Malacoctenus gilli* (Steindachner, 1867)

FEATURES: Color and pattern highly variable. Shades of tan to gray with wavy broken bars on body. May be very dark. A black-ringed brown mark behind eyes often followed by another dark mark. Distinct to indistinct semi-ocellus on lower first dorsal-fin membrane. A large ocellus with a blue upper blotch at dorsal-fin notch that extends onto back. Body scales with white centers. HABITAT: Bermuda, Bahamas, southern Gulf of Mexico, and Caribbean Sea. Demersal over shallow-water sand, rocks. seagrass beds, patch reefs, and in tide pools. Associates with sea anemones.

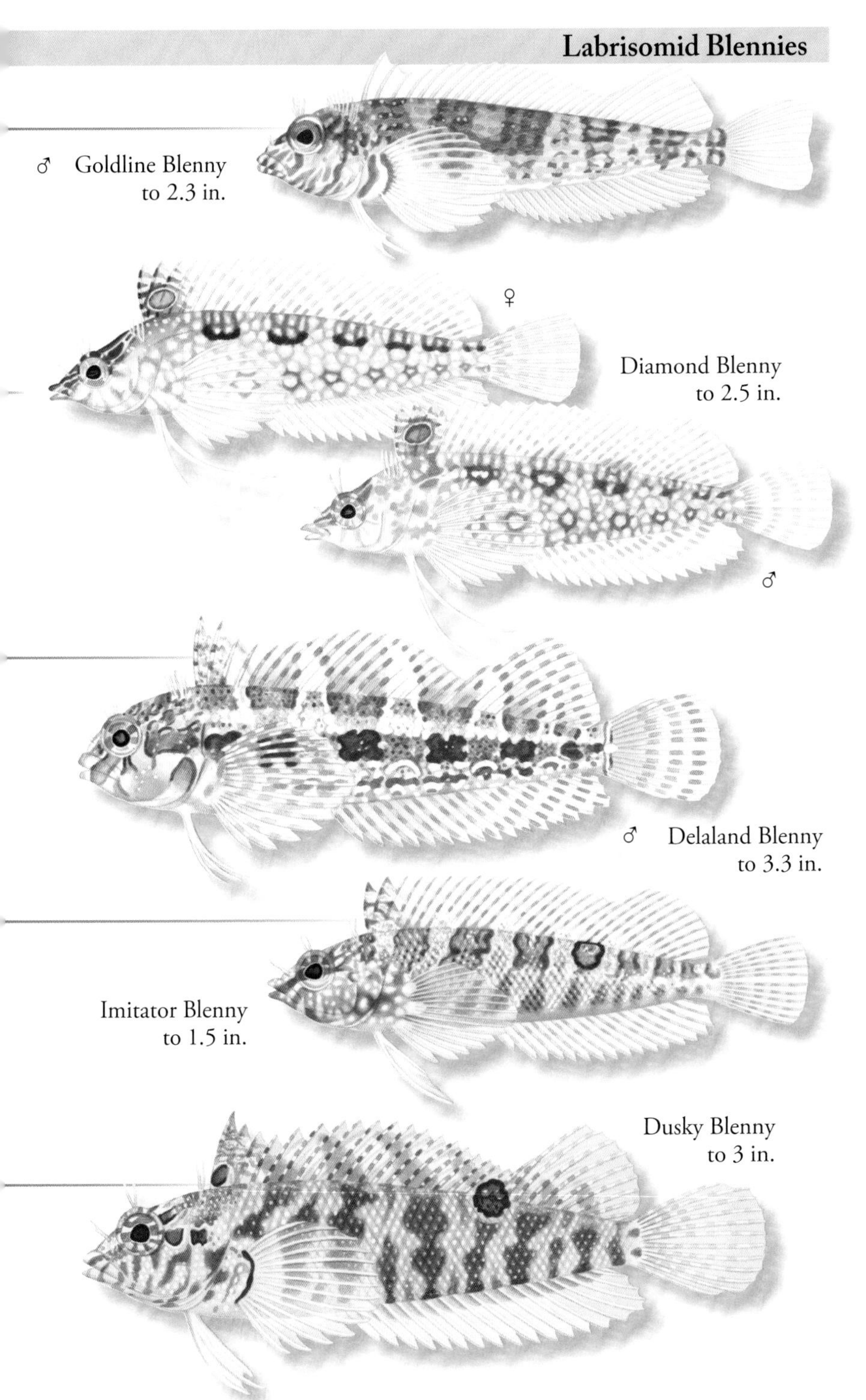

Goldline Blenny
to 2.3 in.

Diamond Blenny
to 2.5 in.

Delaland Blenny
to 3.3 in.

Imitator Blenny
to 1.5 in.

Dusky Blenny
to 3 in.

Rosy Blenny - *Malacoctenus macropus* (Poey, 1868)

FEATURES: Color, pattern highly variable. Females pale tannish to brown; densely spotted and diffusely blotched; pinkish ventrally. Males grayish to reddish or brownish, with diffuse to well-defined dark bars on upper sides; bars may blend together; spotted ventrally. All with spots on head and dorsal fins. Single cirrus on nostril and over eyes. Single, erect cirrus on both sides of nape. HABITAT: S FL, southern Gulf of Mexico, Bermuda, Bahamas, and Caribbean Sea. Demersal in shallow coastal waters over patch reefs, mixed sand, rubble, seagrass beds, and with sponges to about 30 ft.

Saddled Blenny - *Malacoctenus triangulatus* Springer, 1959

FEATURES: Color, pattern variable. Three to five saddles on upper sides. Saddles may be pale, dark, uniformly colored, reticulating, or nearly absent. May have dark blotches on lower sides between saddles. Lower portion of head banded or blotched. May have dark blotch on anterior dorsal fin that merges with first bar. Females with spotted fins. Tuft of cirri over eyes and row of cirri on nape. HABITAT: S FL, southern Gulf of Mexico, Bahamas, and Caribbean Sea. Demersal over shallow rocky and reef areas, in crevices, and among weed and rubble from shore to about 50 ft.

Barfin Blenny - *Malacoctenus versicolor* (Poey, 1876)

FEATURES: Color variable. Females yellowish to pale grayish with five olivaceous to brownish bars on sides; small, scattered, dark spots on areas between bars; dorsal, caudal, and anal fins may have scattered spots. Males shades of brown with brownish bars; dark spots between bars and on fins absent. Males darker than females, bars more defined. Bars extend to dorsal-fin margins in all. All with tuft of cirri above eyes and a transverse row of cirri on both sides of nape. HABITAT: S FL, southern Gulf of Mexico, Bahamas, and Caribbean Sea. Demersal over coral reefs and rocky, sandy, and rubble areas from shore to about 130 ft.

Threadfin Blenny - *Nemaclinus atelestos* Böhlke & Springer, 1975

FEATURES: Reddish orange to orange brown. Males with dark spot on anterior dorsal fin and dark anterior anal fin. Females without dark markings. White bands below eyes. Nostrils with short tube. Long, flattened cirrus above eyes—shorter in females. Simple cirrus on both sides of nape. Middle pectoral-fin rays elongate. Pelvic fins very long. HABITAT: Flower Garden Banks, north of Dry Tortugas, Bermuda, Bahamas, Antilles, and off Costa Rica. Demersal over rocks and coral from about 110 to 840 ft.

Goatee Blenny - *Paraclinus barbatus* Springer, 1955

FEATURES: Color and pattern highly variable. Shades of reddish brown with a large, solid blotch behind pectoral-fin base and a diffuse area behind. Head with whitish net-like lines. Three large, pale to dark blue ocellated spots at soft dorsal-fin base. Caudal and anal fins banded in brown and black. Tip of lower jaw with a long, fleshy barbel. Nostrils with a long cirrus; paddle-like cirri over eyes and on nape. First three dorsal-fin spines tall and incised. HABITAT: Bahamas, Greater Quintana Roo, Belize, Antilles, and off Costa Rica and Venezuela. Demersal over reefs from about 40 to 155 ft.

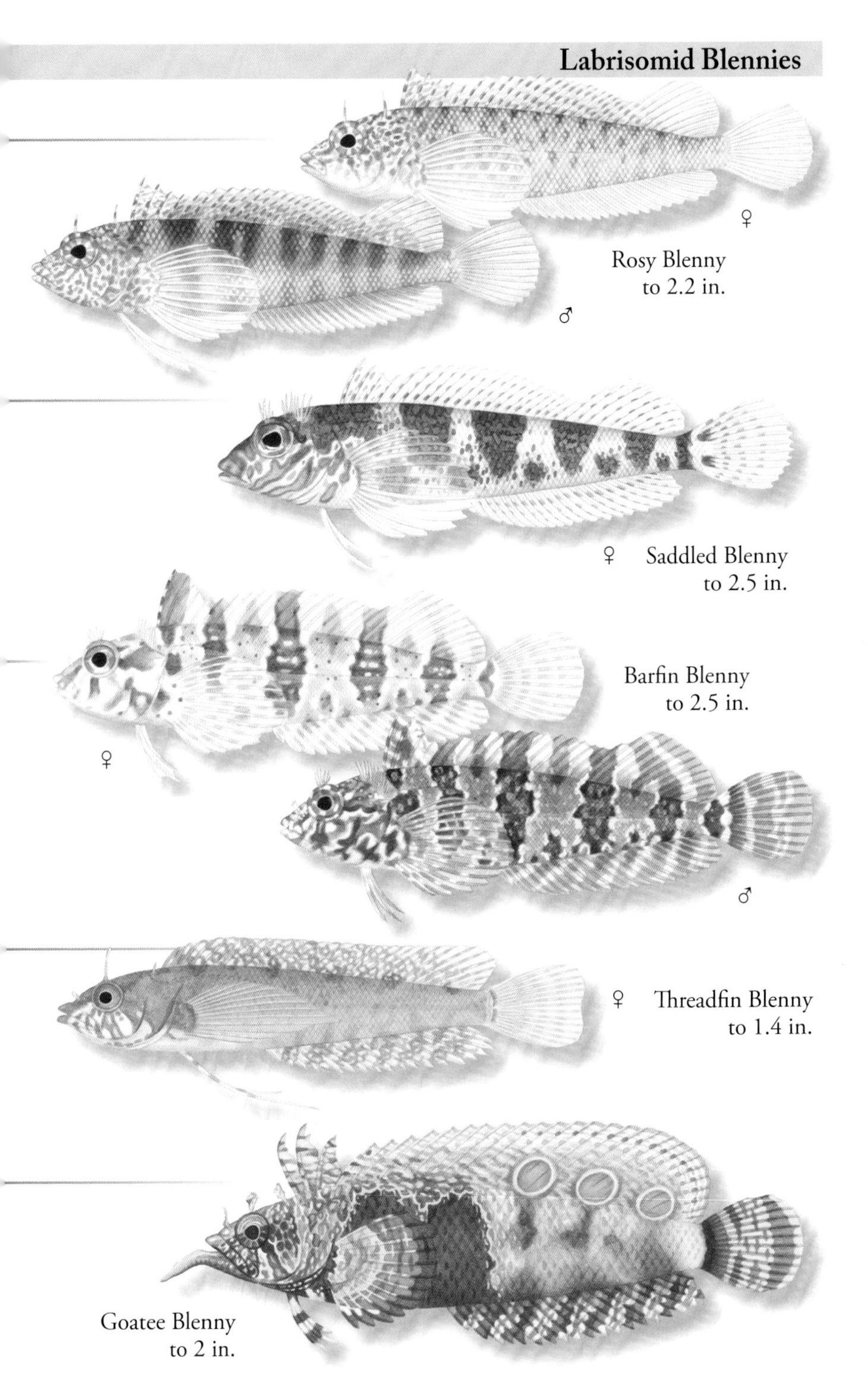
♀
Rosy Blenny
to 2.2 in.
♂
♀
Saddled Blenny
to 2.5 in.
Barfin Blenny
to 2.5 in.
♀
♂
♀
Threadfin Blenny
to 1.4 in.
Goatee Blenny
to 2 in.

Coral Blenny - *Paraclinus cingulatus* (Evermann & Marsh, 1899)

FEATURES: Mottled whitish to tan or brown with five broad brown bars on sides. Anterior bars may be well-defined, posterior bars more diffuse. All extend onto fins. Anterior head reddish. Dark reddish bar below eyes. Lappet-like, pronged cirrus over eyes and on both sides of nape. Pectoral-fin base scaleless. HABITAT: S FL, Dry Tortugas, Bahamas, and Caribbean Sea. Demersal over reefs and coral rubble and in tide pools from shore to about 30 ft.

Banded Blenny - *Paraclinus fasciatus* (Steindachner, 1876)

FEATURES: Color and pattern highly variable. Pale to dark or densely mottled in shades of buff, gray, green, orange, ochre to dark brown. Faint, diffuse, darker bars or rows of blotches on sides. Dark bar at caudal-fin base. Fins spotted and banded. Zero to four (usually one) pale to dark ocelli on rear portion of dorsal fin. One to four cirri above eyes. Lappet-like, pronged cirri on nape—dark in males. HABITAT: FL, southern Gulf of Mexico, Bahamas, and Caribbean Sea. Demersal over shallow sandy, rocky, rubble, coral, and seagrass bottoms to about 40 ft.

Horned Blenny - *Paraclinus grandicomis* (Rosén, 1911)

FEATURES: Shades of brown or orange brown to reddish brown with an irregular pale streak below eyes. Anterior dorsal fin with a dark blotch. Dorsal and anal fins irregularly banded with transparent margins. Ocelli on dorsal fin absent. Caudal fin transparent. Cirri over eyes very tall and fringed posteriorly, reach first dorsal-fin spine when depressed. Simple cirri on nape. Opercle with a simple spine. First through third dorsal-fin membranes incised. HABITAT: S FL; Greater and Lesser Antilles. Demersal over shallow reefs and seagrass beds, and inside sponges and shells to about 10 ft.

Bald Blenny - *Paraclinus infrons* Böhlke, 1960

FEATURES: Color and pattern vary. Mottled and blotched in shades of red or orange to violet. Top of head whitish. Broad whitish area below eyes, and a whitish band from eyes to pectoral-fin base. Dorsal and anal fins banded. Caudal fin translucent with faint bands. Two small ocelli at rear dorsal-fin base. Lappet-like cirrus over eyes. Cirri absent on nape. First two dorsal-fin spines form a tall finlet. Opercle with a spiny projection. HABITAT: FL Keys, Bahamas, central Caribbean Sea, and Belize. Demersal over coral reefs and dropoffs from about 40 to 160 ft.

Marbled Blenny - *Starksia weigti* Baldwin & Castillo, 2011

FEATURES: Color and pattern highly variable. Tannish, brownish, reddish, or olivaceous. May be uniformly to irregularly marbled. Irregular whitish band behind eyes. May have about six bars or rows of pale blotches on sides. Bars and blotches may be well defined or obscure. Variable number of ocelli may be present on rear dorsal and anal fins. One to four joined cirri above eyes. Cirri on nape flat, erect, and fringed. First two dorsal-fin spines form a very tall finlet. Opercle with a simple spine. HABITAT: FL, Bahamas, southern Gulf of Mexico, Nicaragua, and off Venezuela. Demersal over shallow seagrass beds and coral reefs, and inside sponges from about 15 to 30 ft.

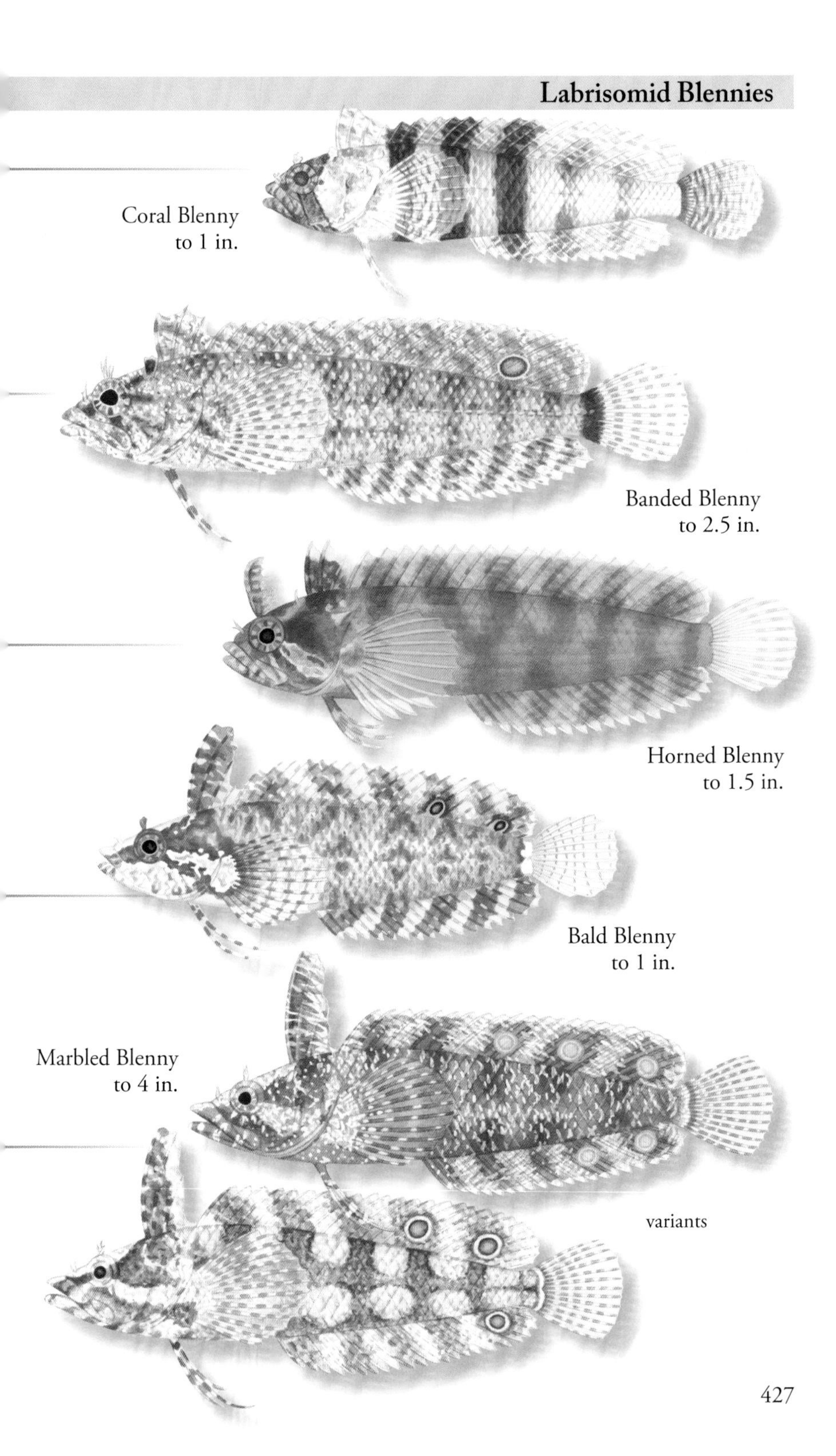
Coral Blenny
to 1 in.
Banded Blenny
to 2.5 in.
Horned Blenny
to 1.5 in.
Bald Blenny
to 1 in.
Marbled Blenny
to 4 in.
variants

Surf Blenny - *Paraclinus naeorhegmis* Böhlke, 1960

FEATURES: Color and pattern variable. Mottled and blotched in shades of violet brown to brown with three to four irregular darker bars on body. Top and underside of head whitish and mottled. Irregular whitish blotches from eyes to behind pectoral-fin base. Dorsal and anal fins banded. Single ocellus on rear soft dorsal fin. Cirri over nostrils and a lobed, lappet-like cirrus over eyes. Nape with two to six clustered cirri. Opercle with spiny projection. First two dorsal-fin spines form a tall finlet. HABITAT: Bahamas, Belize, and southern Caribbean Sea. Demersal over shallow water eroded limestone and rocky bottoms, with algae and sea urchins to about 30 ft.

Blackfin Blenny - *Paraclinus nigripinnis* (Steindachner, 1867)

FEATURES: Shades of greenish brown or reddish brown to blackish. May be pale to very dark. Most with six or seven bars on sides that extend onto dorsal fin. Bars may be obscure. All with a dark bar at caudal-fin base. Usually with one ocellus on rear dorsal fin. Several cirri over each eye. Nape with pale, lappet-like cirri; cirri may have several prongs. HABITAT: S FL, Dry Tortugas, Bahamas, and Antilles. Possibly Bermuda. Occur in a variety of shallow-water coastal habitats to about 30 ft.

Smootheye Blenny - *Starksia atlantica* Longley, 1934

FEATURES: Pale grayish with dark brownish red blotches (female) or irregular orange blotches (male) loosely arranged in two to three rows; top row as saddles. Cheeks with an irregular blotch that lacks a dark posterior margin. Pectoral-fin base with wavy semi-circular marks. Rear dorsal-fin base with a blackish spot. Sometimes with a black spot at rear anal-fin base. Cirri on eyes absent. Simple cirrus on nostrils and nape. HABITAT: Bahamas and Greater and Lesser Antilles. Demersal over rocks and coral reefs from near surface to about 80 ft.

Culebra Blenny - *Starksia culebrae* (Evermann & Marsh, 1899)

FEATURES: Pale tan with dark bands on jaws and radiating from eyes. Body with alternating brown and tan blotches that loosely form bars. Caudal peduncle with two half-crescent marks. Small red spots around eyes, dark-ringed red spots at pectoral-fin base, and pairs of red spots along dorsal-fin base. Fins may be banded with red spots. Simple cirri on nostrils, over eyes, and on nape. HABITAT: Greater and Lesser Antilles. Absent from Cuba. Demersal over lagoonal reefs from near surface to about 30 ft.

Elongate Blenny - *Starksia elongata* Gilbert, 1971

FEATURES: Shades of tan with seven brownish bars on sides (female), or grayish with seven blackish bars on body that extend onto dorsal-fin base (male). Dark bars narrower than pale interspaces and bordered with whitish speckles. Caudal peduncle with two half-crescent marks. Cheeks with one or two small white spots surrounded by a large dark blotch. Simple cirri on nostrils and nape. Cirri absent over eyes. Body comparatively long and slender. HABITAT: Bahamas and in small scattered areas in Caribbean Sea. Demersal over coral and in channels of shallow, windward reefs to about 20 ft.

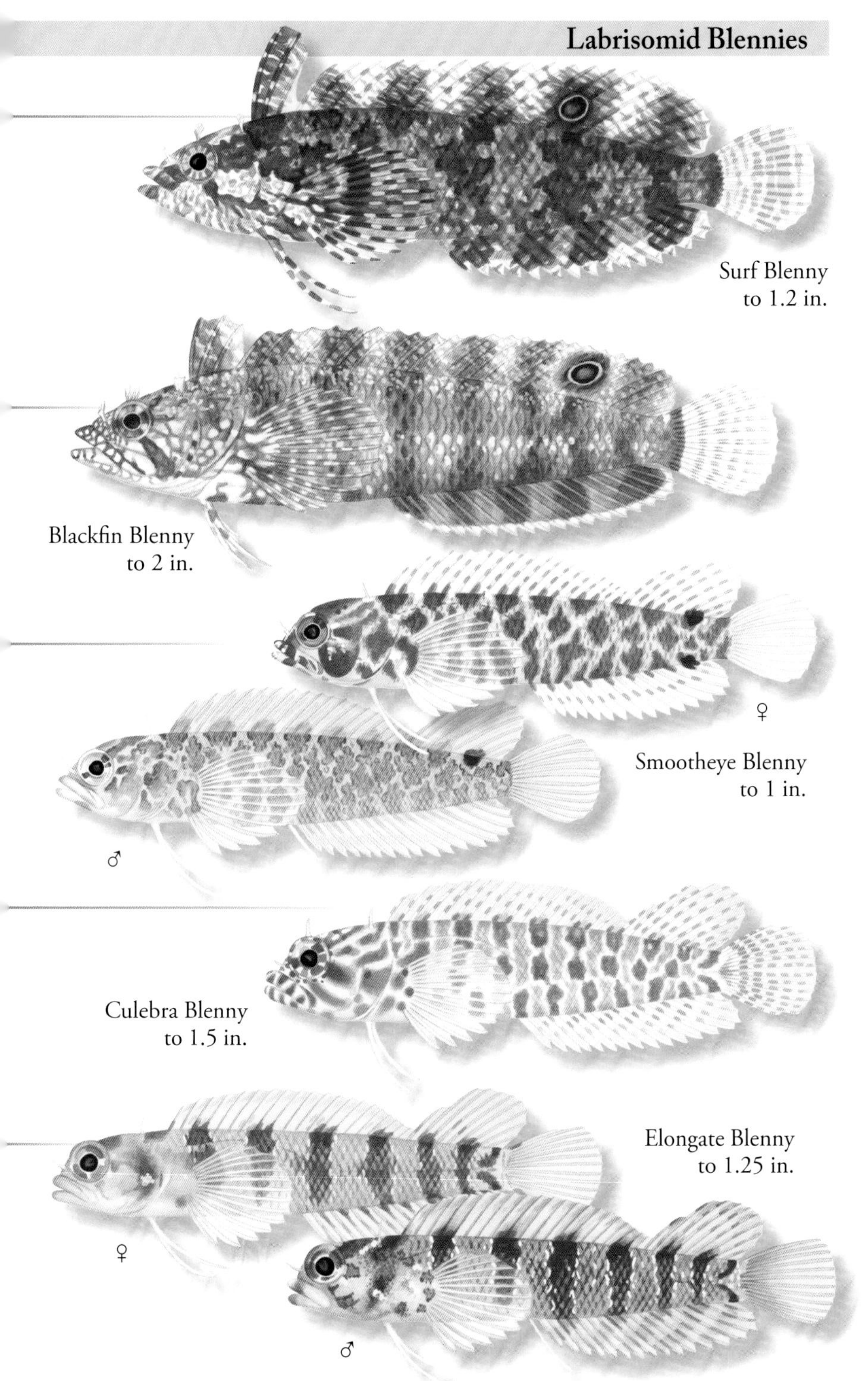

Surf Blenny
to 1.2 in.

Blackfin Blenny
to 2 in.

Smootheye Blenny
to 1 in.

Culebra Blenny
to 1.5 in.

Elongate Blenny
to 1.25 in.

Blackbar Blenny - *Starksia fasciata* (Longley, 1934)

FEATURES: Shades of tan to dusky orange with seven broad black bars on body. May have a dark saddle on nape. Head with areas of red, orange, or yellow. Dorsal fin with bright red spots at base and above dark bars. Dorsal- and anal-fin margins may be dark. Females with variable rows of red spots below and behind eyes and along cheek margins. Males with a small to large, dark semi-circle on lower cheeks. Single, simple cirrus on nostrils, over eyes, and on nape. HABITAT: Bahamas and Cuba. Demersal over shallow weedy and rocky bottoms to about 25 ft.

Greenfield's Blenny - *Starksia greenfieldi* Baldwin & Castillo, 2011

FEATURES: Shades of tan to orange with two to three rows of round brownish to blackish spots on upper body; uppermost row merges onto dorsal fins. Females with vague bands on jaws and preopercular margin, few tiny whitish spots on head, and a dark area on lower opercle. Males with large, close-set yellowish spots on head and a large black blotch on anterior dorsal fin. All may have spots on soft dorsal, caudal, and posterior anal fins. Single, simple cirrus on nostrils, over eyes, and on nape. HABITAT: Lesser Antilles. Demersal over shallow-water reef crests, coral rubble, and lagoonal reefs to about 30 ft.

Ringed Blenny - *Starksia hassi* Klausewitz, 1958

FEATURES: Shades of dark pink to orange or purplish with about nine narrow, pale bars on sides that can be distinct to very faint. Reddish spots on upper jaw, a white blotch behind eyes, small white spots on head, and a white semi-circle at pectoral-fin base ringed with reddish spots. Dorsal and caudal fins spotted; anal fin banded. Nostrils with a long cirrus. Short, simple cirrus over eyes and on nape. HABITAT: Bahamas and Caribbean Sea. Demersal over corals and inside tube sponges to about 65 ft.

Lang's Blenny - *Starksia langi* Baldwin & Castillo, 2011

FEATURES: Shades of pale brown with two rows of 9–10 dark brown to blackish oval blotches bordered by whitish spots on upper sides. Orange to reddish orange spots along dorsal-fin base. Females with scattered blackish spots on lower head and on pectoral-fin base. Males with a large blackish crescent-shaped mark on lower preopercle. Single, simple cirrus on nostrils, over eyes, and on nape. HABITAT: Southern Quintana Roo to Panama. Demersal over shallow-water coral heads surrounded by sand, and on reef crests to about 20 ft.

Blackcheek Blenny - *Starksia lepicoelia* Böhlke & Springer, 1961

FEATURES: Shades of reddish orange to brownish orange with faint to vaguely distinct darker blotches on sides. Reddish orange spots along dorsal- and anal-fin margins. Females with bands on jaws and large spots on sides of head. Males with a large blackish blotch on lower preopercle. All with an oblong blotch followed by two spots on pectoral-fin base. Single, simple cirrus on nostrils, over eyes, and on nape. HABITAT: FL Keys, Bahamas, and Greater Antilles. Demersal over coral reefs, usually on dropoffs, from near surface to about 90 ft.

♀
♂
Blackbar Blenny
to 1 in.
♀
♂
Greenfield's Blenny
to 1 in.
Ringed Blenny
to 1.5 in.
♀
Lang's Blenny
to 1 in.
♂
♀
Blackcheek Blenny
to 1 in.
♂

Dwarf Blenny - *Starksia nanodes* Böhlke & Springer, 1961

FEATURES: Color, pattern highly variable. Shades of brown, orange, or red to purplish. May be sparely to densely spotted or with about seven pale bars or wedge-shaped saddles. Females with a distinct dark band radiating below eyes and several white bands at pectoral-fin base. Males with a large dark blotch on preopercle and a white band at pectoral-fin base. All with a blackish, bisected, horizontal heart shape. Simple, single cirrus on nostrils, over eyes, and on nape. HABITAT: Bahamas and Caribbean Sea. Absent from Cuba. Demersal over coral reefs and dropoffs from about 35 to 100 ft.

Occidental Blenny - *Starksia occidentalis* Greenfield, 1979

FEATURES: Shades of tan with three rows of oblong brownish blotches with pale margins that may form lose bars. A whitish Y-shaped mark below eyes over a row of orange red spots or lines. Several orange red spots on head, opercle, and pectoral-fin base. Reddish spots along dorsal-fin base. Simple, single cirrus on nostrils, over eyes, and on nape. HABITAT: Quintana Roo to Panama. Demersal over shallow-water coral, coral rock, and rubble on inner and outer reefs.

Checkered Blenny - *Starksia ocellata* (Steindachner, 1876)

FEATURES: Almost uniformly brownish to grayish, or with three rows of evenly spaced blotches on sides; uppermost row merges onto dorsal fins. All with dark-ringed, golden to orange spots on sides of head and at pectoral-fin base. May have ringed spots along dorsal-fin base. Simple cirrus on nostrils, over eyes, and on nape. HABITAT: NC to FL, Gulf of Mexico, northern Cuba, and Bahamas. Demersal over rocky and coral bottoms and inside sponges to about 80 ft.

Robertson's Blenny - *Starksia robertsoni* Baldwin, Victor & Castillo, 2011

FEATURES: Shades of bright red to dark red with three rows of indistinct darker blotches on sides. Irises with a red pinwheel pattern. Lower head with numerous white bands, spots, and wavy lines. White wavy lines on pectoral-fin base. Rows of reddish spots along dorsal and anal fin bases. Dorsal, caudal, and anal fins spotted. Single, simple cirrus on nostrils, over eyes, and on nape. HABITAT: Nicaragua to Panama, including San Andreas and Providencia islands. Demersal over coral reefs and hard substrates, usually on dropoffs to about 90 ft.

Sangrey's Blenny - *Starksia sangreyae* Castillo & Baldwin, 2011

FEATURES: Female pale brown; head pale with a brownish horseshoe-shaped mark behind eyes. Males dark brown; head shades of brick red to red with vague darker areas and a few pale spots. All with about 10 narrow pale bars on sides. Second-to-last bar forms an hourglass-shape with a black spot at top and a small dark spot at bottom; bottom spot may be indistinct. Single, simple cirrus on nostrils and on both sides of nape. Cirri absent over eyes. HABITAT: Quintana Roo to Honduras, and Jamaica. Demersal over outer reefs and reef flats from near surface to about 70 ft.

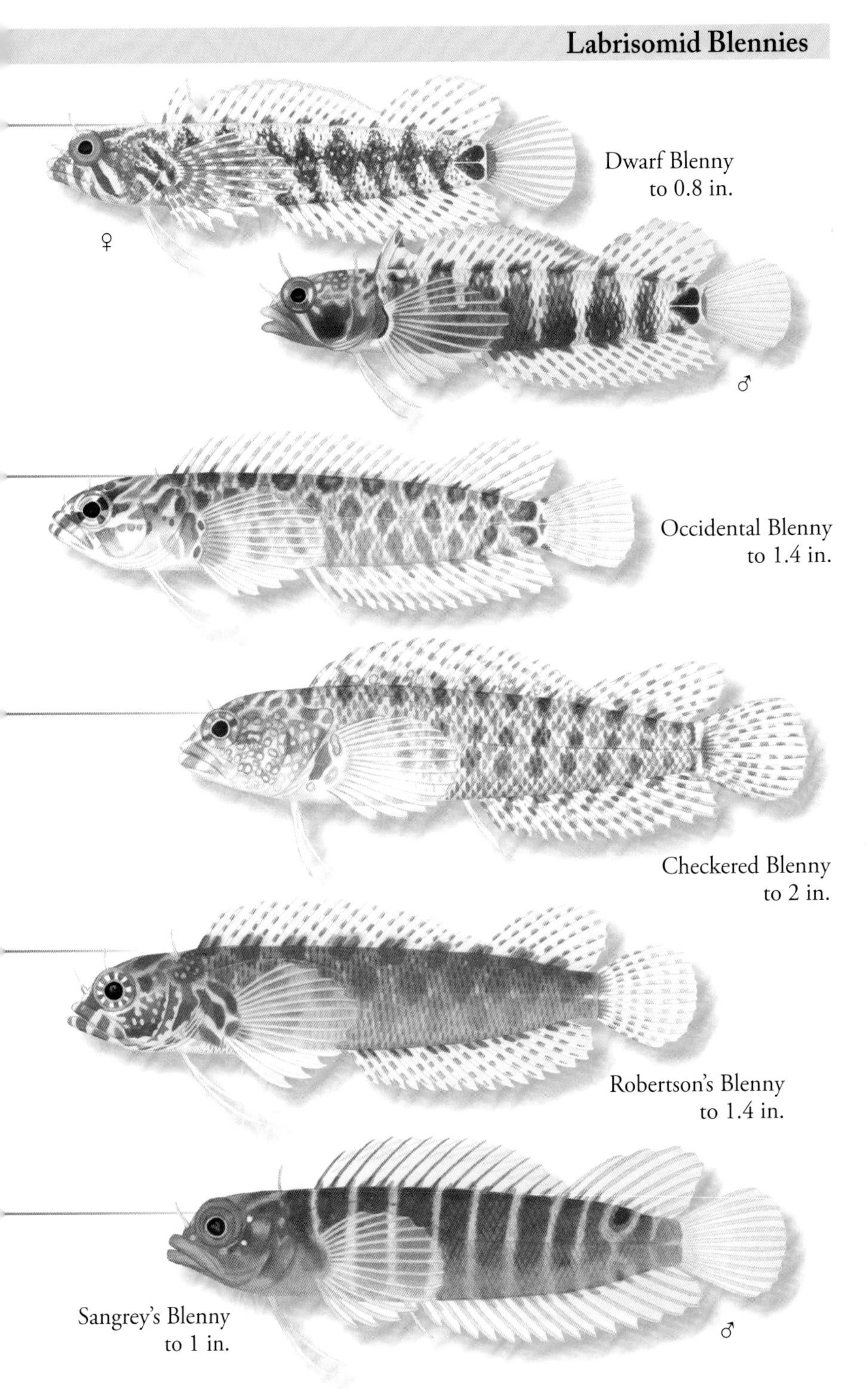

Dwarf Blenny
to 0.8 in.

Occidental Blenny
to 1.4 in.

Checkered Blenny
to 2 in.

Robertson's Blenny
to 1.4 in.

Sangrey's Blenny
to 1 in.

Chessboard Blenny - *Starksia sluiteri* (Metzelaar, 1919)

FEATURES: Shades of white, gray, or tan with two rows of black spots along back; some with a third row of black spots on lower sides. Irregular patches of pale to dark orange or brown between black spots. Top of head with a red to black heart-shaped mark. Males with a dark blotch on anterior dorsal fin. Spots or bars absent from pectoral-fin base. Dorsal, caudal, and anal fins variably spotted. Simple, single cirrus on nostrils, over eyes, and on nape. HABITAT: Islands of the southern Caribbean Sea. Possibly St. Croix. Demersal in reef crevices and over forereefs to about 150 ft.

Brokenbar Blenny - *Starksia smithvanizi* Williams & Mounts, 2003

FEATURES: Grayish tan to reddish with some darker red on top of head. Eight to nine blackish bars from nape to caudal peduncle extend from dorsal-fin base to lower body. Bars are narrow at top, wide at middle, and bordered by a few bluish white spots. Third to last bar as a spot at lower dorsal-fin base. Jaws banded. Males with bluish white spots on head and a dark blotch on anterior dorsal fin. Simple, single cirrus on nostrils, over eyes, and on nape. HABITAT: Greater and Lesser Antilles. Absent from Cuba. Demersal over rocks, boulders, and seagrass beds to about 20 ft.

Splendid Shy Blenny - *Starksia splendens* Victor, 2018

FEATURES: Females shades of red dorsally and on sides with about nine darker blotches on sides and five white blotches along dorsal profile; abdomen white. Males grayish tan with about nine darker blotches on sides and about five white blotches along dorsal profile; abdomen white. All with a short white blotch behind eyes and three blackish dots on lower jaw. Cirri on nostrils and over eyes very long in males, shorter in females. Cirri on nape short in all. HABITAT: Grand Cayman Island; possibly more widely distributed. Demersal in small crevices of shallow-water hard coral and rocky reefs from about 15 to 50 ft.

Springer's Blenny - *Starksia springeri* Castillo & Baldwin, 2011

FEATURES: Pale gray to whitish with a network of irregular triangular blotches on sides interspersed with pale spots. May be nearly uniformly reddish to reddish brown. A large dark blotch on cheeks with an anterior pale notch. A large black spot at rear dorsal-fin base and a smaller black spot at rear anal-fin base. Dark bars below eyes. Simple, single cirrus on nostril and nape. Cirri absent over eyes. HABITAT: Curaçao, Bonaire, Isla de Alves, and Isla Los Roques. Demersal over reef flats and outer reefs to about 80 ft.

Weigt's Blenny - *Starksia weigti* Baldwin & Castillo, 2011

FEATURES: Color and pattern slightly vary. Shades of red with clusters of whitish between nearly indistinct to distinct darker bars on body. Jaws with white to pale spots. Short white bar behind eyes. Dorsal, caudal, and anal fins spotted in red and white with larger spots at base. Males with a large blackish blotch at lower preopercle. Single, simple cirrus on nostrils, over eyes, and on nape. HABITAT: Bay of Campeche to Honduras. Demersal over dropoffs of rocky and coral reefs from near shore to about 90 ft.

Chessboard Blenny
to 1 in.

Brokenbar Blenny
to 1 in.

Splendid Shy Blenny
to 0.9 in.

Springer's Blenny
to 1 in.

Weigt's Blenny
to 1.5 in.

Williams' Blenny - *Starksia williamsi* Baldwin & Castillo, 2011

FEATURES: Females bright red with tiny white spots on body; jaws banded; white bar behind eyes; white semi-circle at pectoral-fin base. Males shades of red to orange, sometimes with faint irregular bars on body; jaws uniformly colored; preopercle with a large dark blotch at lower corner; whitish band below eyes; white semi-circle at pectoral-fin base. Fins spotted. Single, simple cirrus on nostrils, over eyes, and on nape. HABITAT: St. Croix to Tobago, including Isla Aves. Perhaps to Bonaire. Demersal over coral and rocky bottoms, usually on shallow-water dropoffs to about 90 ft.

Naked Blenny - *Stathmonotus gymnodermis* Springer, 1955

FEATURES: Color and pattern vary. Shades of brown to green. May be uniformly colored or with sparse to dense spots and mottling. Usually with a white bar bordered by dark lines below eyes; bar may be faint. Dorsal, caudal, and anal fins may be banded and/or with dark margins. Small, simple cirrus on nostrils, nape, and lower preoperculum. Flap-like cirrus over eyes. Pectoral fins well developed. Dorsal fin entirely spiny. Body scaleless. HABITAT: Bahamas and Caribbean Sea. Demersal over shallow-water reefs and hard bottoms along exposed shorelines to about 20 ft.

Blackbelly Blenny - *Stathmonotus hemphillii* Bean, 1885

FEATURES: Males may be pale, reddish orange, blackish, or greenish. Some with black-margined white bands on head and/or a white ocellated spot behind eyes. Fins may have irregular white bands and a reddish margin. Females whitish with a network of dark lines below and behind eyes. Cirri on head absent. Pectoral fins greatly reduced. Dorsal fin entirely spiny. Body scaleless. HABITAT: S FL, Bahamas, Greater and Lesser Antilles, and Yucatán to Nicaragua. Demersal over rocks and rubble near coral reefs and with seagrass beds from shore to about 100 ft.

Eelgrass Blenny - *Stathmonotus stahli* (Evermann & Marsh, 1899)

FEATURES: Shades of brown to grayish green; sometimes green. May have rows of whitish or reddish to blackish spots or mottling on sides. Sides of head and chest with wavy bands. Single, flap-like cirrus over eyes. Simple cirrus on nape. Pectoral fins large. Dorsal fin entirely spiny. Body scaled. Caudal fin with 12 rays. HABITAT: Puerto Rico and Lesser Antilles to Venezuela. Demersal over shallow-water hard bottoms, coral, and sponges to about 30 ft. OTHER NAME: Southern Eelgrass Blenny.

Northern Eelgrass Blenny - *Stathmonotus tekla* Nichols, 1910

FEATURES: Shades of green to brown. May be nearly uniformly colored or densely mottled. White band behind eyes, and faint to distinct blotches on lower head and chest. Single, flap-like cirrus above eyes. Simple cirrus on nape. Pectoral fins large. Dorsal fin entirely spiny. Body scaled. Caudal fin with 11 rays. HABITAT: S FL, Bahamas, Cuba, Haiti, Jamaica, Cayman Islands, and Yucatán to Colombia. Demersal over shallow-water hard bottoms, rubble, coral, and sponges to about 30 ft.

NOTE: There are nine other rare Labrisomid Blennies in the area.

♀
Williams' Blenny
to 1.4 in.
variants
Naked Blenny
to 1.6 in.
♀
♂
Blackbelly Blenny
to 2 in.
Eelgrass Blenny
to 1.6 in.
Northern Eelgrass Blenny
to 1.6 in.

Chaenopsidae - Tube Blennies

Roughhead Blenny - *Acanthemblemaria aspera* (Longley, 1927)

FEATURES: Females yellow to pale pinkish with pale spots anteriorly. Dark spots or broken lines on lower portion of head. Males brownish or reddish to blackish with scattered pale spots anteriorly; some are pale overall; dark ocellated spot on anterior dorsal fin. Both with spines on top of head forming a V-shape from nape to eyes. Complex, highly branched cirrus over nostrils and eyes. HABITAT: S FL, Yucatán, Bay of Campeche, Bahamas, and Caribbean Sea. Demersal over shallow-water live and dead coral and rubble. Inhabit abandoned holes to about 90 ft.

Speckled Blenny - *Acanthemblemaria betinensis* Smith-Vaniz & Palacio, 1974

FEATURES: Shades of reddish brown to brown with rows of white to creamy white spots on sides; males darker than females. Spots smaller and more numerous in males. May have a faint to distinct white stripe below and behind eyes. Dark blotch on anterior dorsal-fin membranes. Short row of spines behind eyes, two rows of spines between eyes, and spines around eyes. Branched cirri on nostrils and over eyes. HABITAT: Costa Rica to Colombia. Demersal over shallow-water rocks and dead coral; often in turbid and brackish water. Inhabit large and empty worm holes.

Papillose Blenny - *Acanthemblemaria chaplini* Böhlke, 1957

FEATURES: Yellowish to orangish anteriorly. Tannish to yellowish posteriorly, with a series of diffuse pale and dark spots on sides. First few dorsal-fin membranes with white below, blackish above. A few minute spines present below eyes. Few to numerous simple cirri on head. Cirri on nostrils and over eyes branched, but on a single plane. HABITAT: S FL, Bahamas, Cuba, Puerto Rico, and Panama. Demersal over shallow-water limestone slopes with coral heads. Inhabit abandoned holes.

False Papillose Blenny - *Acanthemblemaria greenfieldi* Smith-Vaniz & Palacio, 1974

FEATURES: Shades of red or greenish to blackish anteriorly with irregular, dense, pale and dark spots on head. Translucent posteriorly with dark spots along dorsal and ventral profiles and along midline. Anterior dorsal fin with an orange margin and black spots on membranes that are bordered with orange. Short, knobby spines on a pair of ridges on snout. Short, moderately branched cirri over eyes. HABITAT: Western and central Caribbean Sea. Demersal over shallow-water, silt-free patch reefs and forereef areas. Inhabit abandoned holes to about 50 ft.

Thorny-bush Blenny - *Acanthemblemaria harpeza* Williams, 2002

FEATURES: Brown to reddish brown with blue spots on head. Posterior body translucent greenish with darker areas along dorsal and ventral profiles and along spinal column. Irises bright orange. Lower jaw barred. Underside of head blackish. Top of head with a triangular patch of spines. Snout with spines. First dorsal-fin membrane with a flap. Cirri on nostrils densely branched; cirri over eyes with up to 30 branches. HABITAT: Navassa Island and St. Kitt's. Possibly more widely distributed. Demersal over shallow-water coral rock bottoms to about 15 ft.

Roughhead Blenny
to 1.4 in.

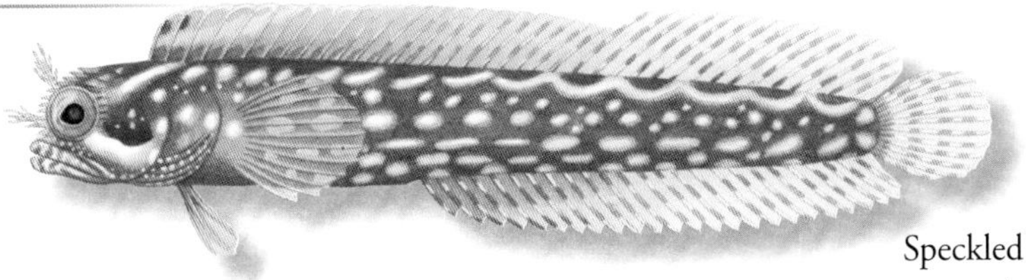

♀

♂

Speckled Blenny
to 1.4 in.

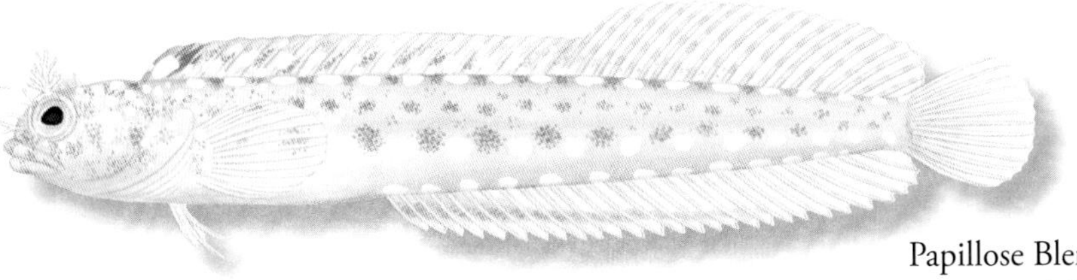

Papillose Blenny
to 1.8 in.

variants

False Papillose Blenny
to 1.5 in.

Thorny-bush Blenny
to .75 in.

Secretary Blenny - *Acanthemblemaria maria* Böhlke, 1961

FEATURES: Dark reddish brown to black with large white blotches along dorsal and ventral profiles that may merge at midline. Jaws with tiny white spots. Irregular white from eyes to pectoral-fin base. Anterior dorsal fin finely spotted. Snout and top of head with short, pointed, forward projecting spines; dorsal patch orange. Cirri on nostrils and eyes finely branched. HABITAT: Bahamas and Caribbean Sea. Demersal over shallow-water limestone and corals in high surge areas. Inhabit abandoned holes.

Medusa Blenny - *Acanthemblemaria medusa* Smith-Vaniz & Palacio, 1974

FEATURES: Color, pattern highly variable. Shades of brown, green, orange to nearly white with rows of darker spots along sides. A wavy white band extends from eyes to opercle. Head may be heavily spotted. Large black blotch between second and sixth dorsal-fin spines. Translucent posteriorly. Snout and top of head with many slender, single papillae. Nostrils with a branched cirrus. Cirri over eyes complexly branched. First dorsal-fin membrane with a flap. HABITAT: Lesser Antilles to Venezuela. Demersal over shallow-water rocks and coral reefs to about 15 ft. Inhabit abandoned holes.

Dwarf Spinyhead Blenny - *Acanthemblemaria paula* Johnson & Brothers, 1989

FEATURES: Grayish green to brownish anteriorly with dense white spots. Underside of head very dark. Translucent posteriorly with oblique dark blotches along dorsal and ventral margins. Dark blotches along spinal column. Anterior dorsal fin mottled and spotted. Forward-facing spines on snout, behind eyes, and on top of head. Nostrils with two to three branched cirri. Cirri over eyes with up to 30 branches. First dorsal-fin membrane with a flap. HABITAT: New Providence, Bahamas, Belize, and western Honduras. Possibly more widely distributed. Demersal over shallow-water dead coral and spur-and-groove coral areas to about 5 ft. Inhabit abandoned holes.

Spotjaw Blenny - *Acanthemblemaria rivasi* Stephens, 1970

FEATURES: Shades of green, brown, or red to orange with rows of white spots along dorsal profile and abdomen. Blackish spots under and behind jaws. Irises bright red with a white to bright blue inner ring. Black spot on first dorsal-fin membrane. Simple cirri on nostrils and over eyes. HABITAT: Southwestern Caribbean Sea. Demersal over coral heads in heavy surge areas to about 100 ft. Inhabit abandoned barnacles and worm holes.

Spinyhead Blenny - *Acanthemblemaria spinosa* Metzelaar, 1919

FEATURES: Snout whitish; eyes yellowish green, usually with a dark blotch behind eyes and white bar below eyes. Head variable spotted. Body translucent with brownish and white spots. Numerous small, close-set spines on snout, around eyes, and on top of head. Nostrils with simple cirri. Cirri over eyes simple to weakly branched. HABITAT: Bahamas and Caribbean Sea. Demersal over small patch reefs, on dead and live coral from near surface to about 100 ft. Inhabit abandoned holes.

Secretary Blenny
to 2 in.

Medusa Blenny
to 1.5 in.

Dwarf
Spinyhead Blenny
to 1 in.

Spotjaw Blenny
to 1.4 in.

Spinyhead Blenny
to 1.1 in.

Yellowface Pikeblenny - *Chaenopsis limbaughi* Robins & Randall, 1965

FEATURES: Shades of tan with pale speckles on head and body. Dark areas may form wide, diffuse bars. Midlateral stripe always present; may be faint and pale to dark. Males more variable in color; always with a black ocellated spot below orange blotch in first dorsal-fin membrane. Females with low, spotted, anterior dorsal fin; males with tall anterior dorsal fin. Snout U-shaped from above. HABITAT: Bahamas and Caribbean Sea. Also reported from S FL. Reef associated over sand and rubble bottoms to about 70 ft. Burrow in sand and inhabit holes and abandoned worm tubes.

Bluethroat Pikeblenny - *Chaenopsis ocellata* Poey, 1865

FEATURES: Shades of tan with pale speckles on head and body. Dark areas may form diffuse bars. Midlateral stripe absent. Males variable in color; always with a curved black mark partially enclosing an orange spot on first dorsal-fin membrane, and blue gill membranes. Females with low dorsal fin; males with tall dorsal fin. Snout bluntly V-shaped from above. HABITAT: S FL, Bahamas, NW Cuba, N Yucatán, Puerto Rico, Virgin Islands, and off Venezuela. Demersal over sandy areas with grasses and algae to about 40 ft. Inhabit holes and abandoned worm tubes.

Resh Pikeblenny - *Chaenopsis resh* Robins & Randall, 1965

FEATURES: Shades of brown dorsally with dense yellowish spots; sides with a brownish stripe; lower sides with large whitish spots; black mark behind eyes. Females with low banded and spotted dorsal fin. Males with relatively high and spotted dorsal fin and a large black ocellated spot on second dorsal-fin membrane. Snout bluntly V-shaped from above. HABITAT: Southern Caribbean Sea from Colombia to Carúpano, Venezuela. Demersal over shallow-water sandy bottoms around reefs to about 30 ft. Usually inhabit holes and vertical burrows.

Twinhorn Blenny - *Coralliozetus cardonae* Evermann & Marsh, 1899

FEATURES: Females green on top of head and green or blue on cheeks and abdomen; lower head with white and black flecks. Males shades of black to dark brown anteriorly with an oblique bluish white band below eyes and tiny bluish white dots on head and forebody. All translucent posteriorly with dark bands on spinal column and rows of dark blotches along dorsal and ventral margins. Simple cirrus on nostrils. One large and one tiny cirrus over eyes. HABITAT: Bahamas and Caribbean Sea. Demersal over rocks and reefs in shallow surge areas to about 15 ft. Inhabit holes and empty barnacles.

Moth Blenny - *Ekemblemaria nigra* (Meek & Hildebrand, 1928)

FEATURES: Shades of brown to greenish with vague to distinct blotches on head and sides. Cheeks with a large dark blotch bordered anteriorly by a narrow, white C-shaped mark. Head with small white spots. Body with distinct to faint whitish wavy lines. Anterior dorsal-fin margin orange. Rear dorsal, caudal, and anal fins translucent. Slender cirrus on nostrils. Thick, branched cirrus over eyes. HABITAT: Costa Rica to Colombia. Demersal over shallow-water reefs and boulders with algae to about 15 ft. Inhabit abandoned tube worm holes.

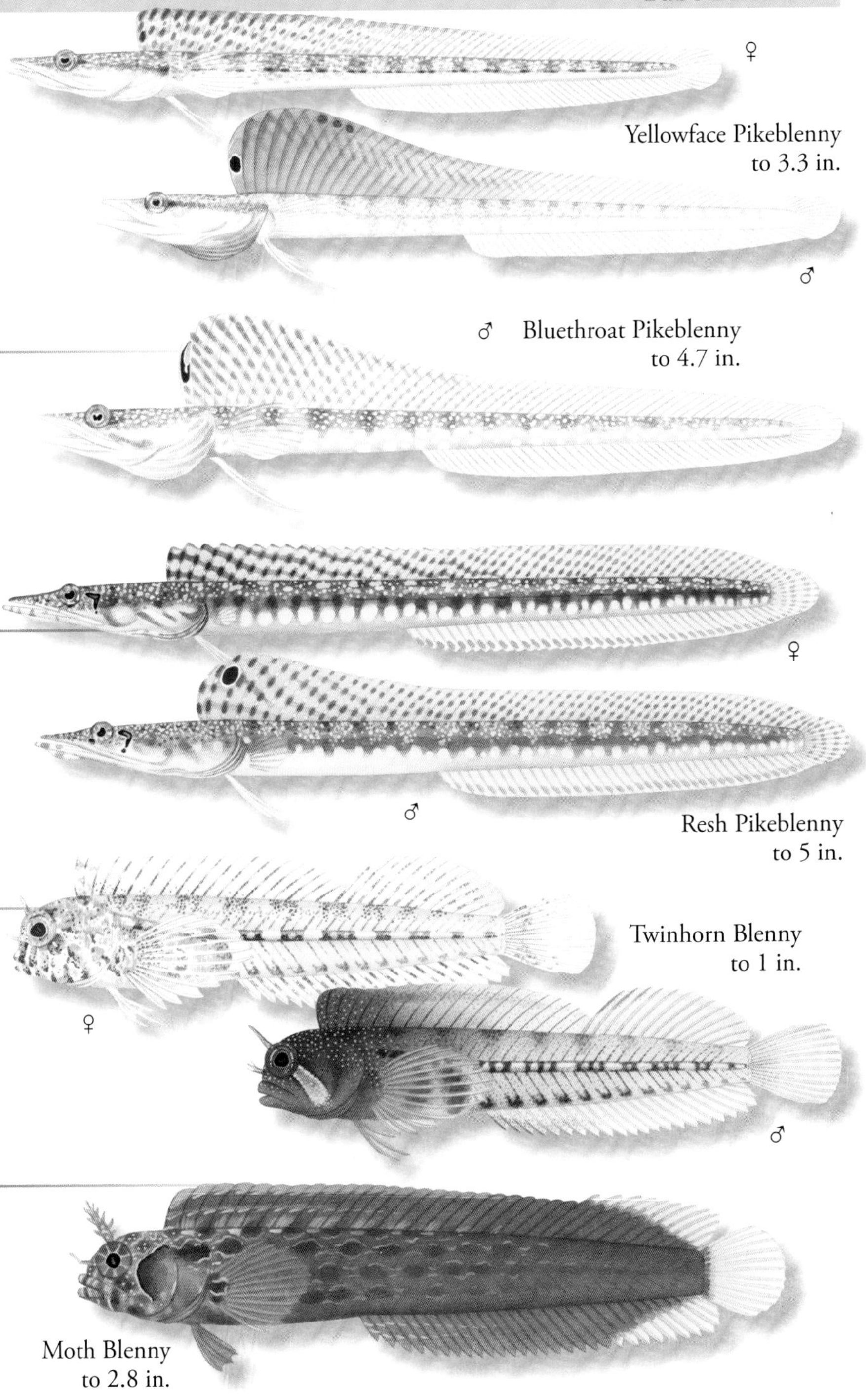

♀

Yellowface Pikeblenny
to 3.3 in.

♂

♂ Bluethroat Pikeblenny
to 4.7 in.

♀

♂

Resh Pikeblenny
to 5 in.

Twinhorn Blenny
to 1 in.

♀

♂

Moth Blenny
to 2.8 in.

Banner Blenny - *Emblemaria atlantica* Jordan & Evermann, 1898

FEATURES: Color and pattern vary. Females brownish with six pale bars or series of irregular blotches on sides that are outlined with dark brown. Males more uniformly colored, may have faint to distinct pale blotches on sides. Both with sloping snout. Simple, banded cirrus on nostrils and over eyes; cirri over eyes long. Pectoral fins with 14 rays. Males with sail-like dorsal fin. HABITAT: GA to FL. Eastern and northern Gulf of Mexico and Bermuda. Demersal over rocky reefs and rubble from about 100 to 360 ft.; much shallower around Bermuda.

Caribbean Blenny - *Emblemaria caldwelli* Stephens, 1970

FEATURES: Females yellowish orange on head with red to reddish orange spots that form bands radiating from eyes; irises red; cheeks orange; abdomen red; rear body translucent; anterior dorsal fin banded in orange and black. Males (not shown) rusty on head with black flecks; irises yellow; body creamy, peppered with black; anterior dorsal fin black with white stripes. Long cirrus on nostrils; very long cirrus over eyes. Anterior dorsal fin bluntly pointed. HABITAT: Bahamas and eastern and western Caribbean Sea. Demersal over sand and pebble bottoms around reef dropoffs from about 30 to 150 ft.

Spiny Sailfin Blenny - *Emblemaria diphyodontis* Stephens & Cervigón, 1970

FEATURES: Females pale to yellowish with about nine brownish bars on back and blotches along sides; pale blotches on cheeks; anterior dorsal fin pointed and banded. Males shades of grayish green to brown with darker spots and vermiculations that form about nine bars on back and blotches along sides; pale blotches on cheeks; anterior dorsal fin spotted and deeply incised. Simple, banded cirrus over eyes, longer in males. HABITAT: St. Vincent Island to Cubagua Island, and Margarita Island off Venezuela. Demersal over shallow-water shelly and rubble bottoms. Inhabit empty shells.

Filament Blenny - *Emblemaria hyltoni* Johnson & Greenfield, 1976

FEATURES: Females translucent to yellowish with or without reddish bands below eyes, reddish orange blotches on cheeks, and large red blotches on abdomen; fins spotted. Males shades of reddish brown with greenish brown anteriorly, becoming somewhat translucent posteriorly; irises bright red; vague reddish bands below eyes. All with long, simple, banded cirrus on nostrils and over eyes. First two dorsal-fin spines very long and filamentous. HABITAT: Roatan Island, Honduras. Demersal over sandy and shelly bottoms from about 50 to 65 ft. Inhabit holes.

Sailfin Blenny - *Emblemaria pandionis* Evermann & Marsh, 1900

FEATURES: Females and immature males pale brownish to grayish with dark spots and clusters of pale spots. Males dark brown to bluish black; dorsal fin banded; anterior dorsal fin ray may have blue to yellowish streaks. Both with steep snout, a long cirrus over each eye, and 13 pectoral-fin rays. Cirri over eyes usually branched at tip. Males with sail-like dorsal fin. HABITAT: S and W FL, Flower Garden Banks, southern Gulf of Mexico, Bahamas, and Caribbean Sea. Demersal in clear waters over rocky shores, coral rubble, and sandy bottoms to about 120 ft. Inhabit holes and abandoned worm tubes.

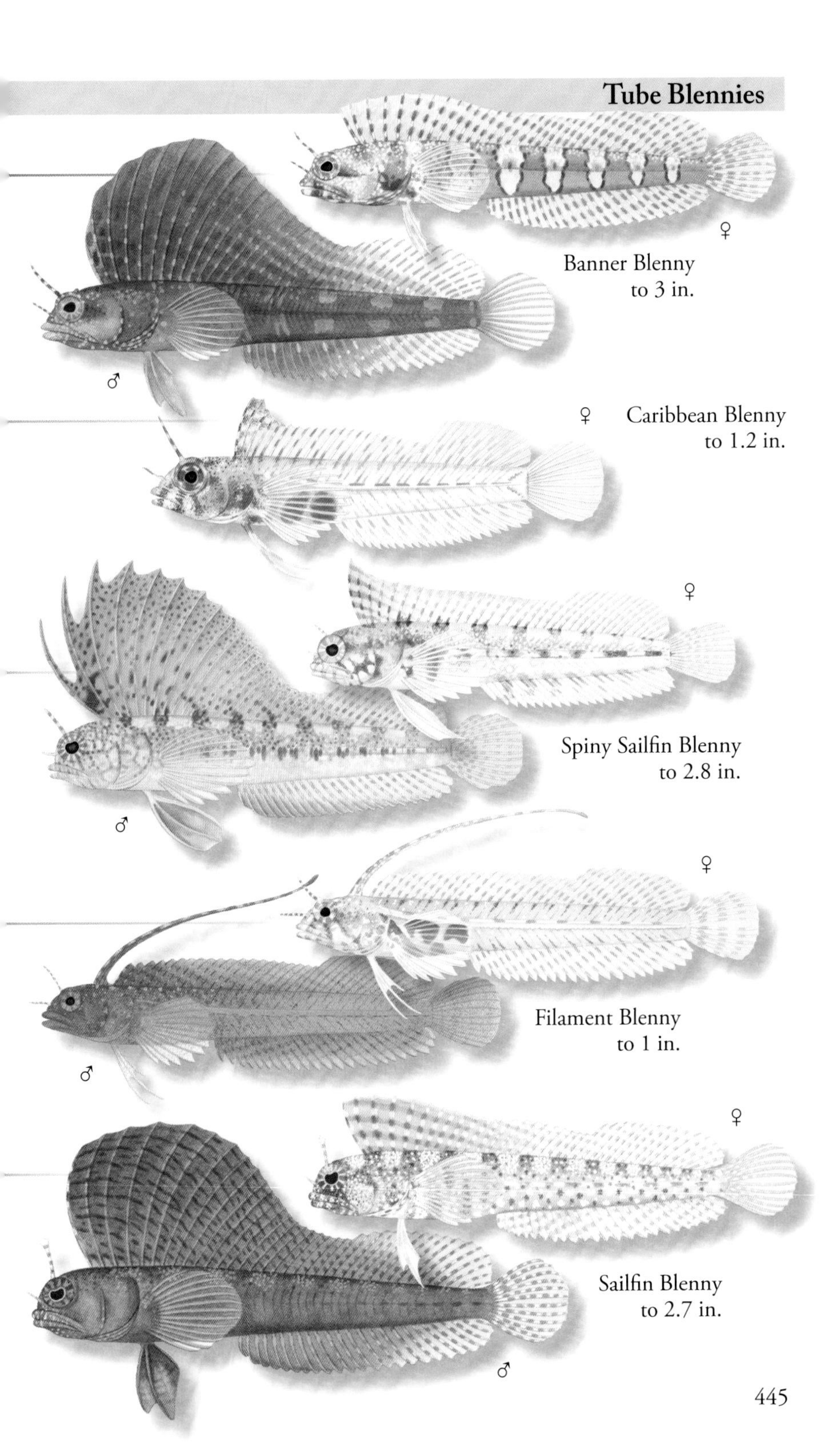
♀
Banner Blenny
to 3 in.
♂
♀
Caribbean Blenny
to 1.2 in.
♀
Spiny Sailfin Blenny
to 2.8 in.
♂
♀
Filament Blenny
to 1 in.
♂
♀
Sailfin Blenny
to 2.7 in.
♂

Ribbon Blenny - *Emblemaria vitta* Williams, 2002

FEATURES: Color varies. Females translucent greenish gray with dense white and blackish spotting on head; clusters of whitish spots along dorsal and ventral profiles. Males range from solid brown or speckled brown to bluish anteriorly; blue specimens with dark spots; brown speckled specimens with clusters of white spots; lower anterior dorsal fin with a bright yellow patch; anterior anal fin blackish to brownish. Female anterior dorsal fin moderately tall. Male anterior dorsal fin sail-like. Slender cirrus over nostrils. Long, simple cirrus over eyes. HABITAT: Bahamas, Haiti to US Virgin Islands, and Belize to Honduras. Demersal on sandy and rubble bottoms at bases of cliffs and spur-and-groove formations from about 30 to 105 ft.

Blackhead Blenny - *Emblemariopsis bahamensis* Stephens, 1961

FEATURES: Females translucent with a white, narrow mark behind eyes; rows of brownish blotches along dorsal and ventral profiles. Males dark brown to blackish on head and forebody; translucent posteriorly; anterior dorsal fin brownish with white spots that may form bands. All with reddish patch on cheeks, pale spots on head, and alternating reddish to brown and white bands on spinal column. Slender cirrus on nostrils. Cirri absent over eyes. Anterior dorsal fin low, and not lobed. HABITAT: Bahamas, Cuba to Barbuda, Cayman Islands, and Nicaragua to Costa Rica. Demersal over shallow-water sheltered patch reefs to about 40 ft.

Bottome's Blenny - *Emblemariopsis bottomei* Stephens, 1961

FEATURES: Females translucent pale gray to pale pink with a series of blackish spots on cheeks and dark spots at pectoral-fin base; upper and lower jaws banded; rows of dark flecks along dorsal and ventral profiles. Males blackish anteriorly, translucent posteriorly; head with or without small pale blue spots. All with short red and white bands on spinal column and a series of dark flecks along ventral margin. Female anterior dorsal fin with a small white finlet. Male anterior dorsal fin with a nearly straight, pale blue margin. Nostrils with a fine, thin cirrus. Cirri absent over eyes. HABITAT: Lesser Antilles to Aruba, and Venezuela. Demersal over corals, sponges, dead coral, and encrusting substrate from about 15 to 60 ft.

Carib Blenny - *Emblemariopsis carib* Victor, 2010

FEATURES: Females and immature males translucent with a series of orange to brown spots on lower head; top of head with minute, dark specks and an orange spot behind eyes. Males reddish black anteriorly, becoming translucent posteriorly. All with long red and short white streaks on spinal column. Male anterior dorsal fin very tall, pointed with orange along margin bordered by thin white lines; female anterior dorsal fin relatively tall and blunt. HABITAT: Haiti, Dominican Republic, Puerto Rico, and Lesser Antilles. Demersal over live and dead corals to about 30 ft. Inhabit empty holes.

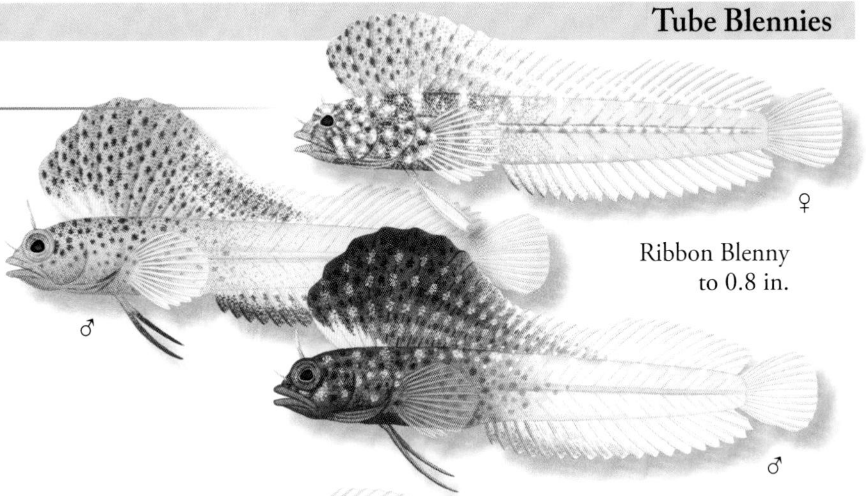

♀

Ribbon Blenny
to 0.8 in.

♂

♂

Blackhead Blenny
to 1 in.

♀

♂

♀

♀

Bottome's Blenny
to 1 in.

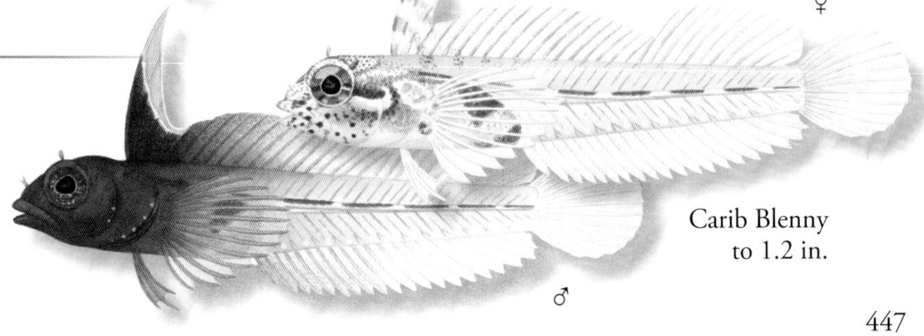

♂

♀

Carib Blenny
to 1.2 in.

♂

Glass Blenny - *Emblemariopsis diaphana* Longley, 1927

FEATURES: Translucent. Females sparsely pigmented on head and anterior dorsal fin. Males similar to females but more densely pigmented on head and anterior dorsal fin; head becoming blackish while inhabiting burrows. Organs, spinal column, and ribs pigmented. First dorsal-fin spine taller than second and third. Together, first three spines form a spike. Cirri above eyes always absent. HABITAT: FL Keys, Dry Tortugas, Cayman Islands, Barbados, and Merida, Mexico, to Belize. Demersal over large and live corals in shallow water to about 70 ft.

Fine-cirrus Blenny - *Emblemariopsis leptocirris* Stephens, 1970

FEATURES: Females translucent with small reddish brown spots on lower portion of head and tiny brown specks on top of head that form a U-shape; abdomen with a peach to reddish blotch and brownish spots; spinal column red or pink to peach. Males blackish anteriorly, semi-translucent posteriorly; spinal column mostly reddish. Female anterior dorsal fin pointed; male anterior dorsal fin rounded. All with a very small cirrus over eyes. HABITAT: Haiti to Barbuda, Cayman Islands, and Quintana Roo to Panama. Demersal over reefs in lagoon, on dropoffs, and offshore to about 60 ft.

Redspine Blenny - *Emblemariopsis occidentalis* Stephens, 1970

FEATURES: Females (not shown) translucent with small reddish to blackish spots on lower head, a spot behind upper eyes, and a bar below eyes; abdomen with a wedge-shaped mark. Males blackish anteriorly, semi-translucent posteriorly. Female anterior dorsal fin tall and pointed. Male anterior dorsal fin pointed, with a narrow red to orange outer band bordered by white below. Spinal column with long peach to red marks. Tiny cirrus over on nostrils and over eyes. HABITAT: Bahamas. Demersal over limestone and patch reefs from about 15 to 80 ft.

Seafan Blenny - *Emblemariopsis pricei* Greenfield, 1975

FEATURES: Females and pale males with a brownish band below eyes and small, bright white spots on sides of head; top of head with a bullseye mark; upper and lower sides of body with alternating brown and white marks; spinal column with alternating red and white separated by brown. Males blackish brown anteriorly, becoming semi-translucent posteriorly; often with all bluish white dots on cheeks. All with a long, slender cirrus on nostrils; cirri absent over eyes. Anterior dorsal fin low. HABITAT: Quintana Roo to Honduras. Demersal on corals to about 100 ft. Perch over and lives in holes.

Randall's Blenny - *Emblemariopsis randalli* Cervigón, 1965

FEATURES: Females translucent with a pink wash on cheeks, a series of dark spots along back, and yellowish spots on top of reddish to brownish spinal column. Males blackish anteriorly, semi-translucent gray posteriorly. Female anterior dorsal fin pale brownish yellow with an inner white band and a pinkish margin. Male anterior fin dark brown with a broad, red margin and a narrow inner band. Both with rounded anterior dorsal fin. HABITAT: Venezuela. Demersal over shallow-water rocky and coral bottoms from about 15 to 50 ft.

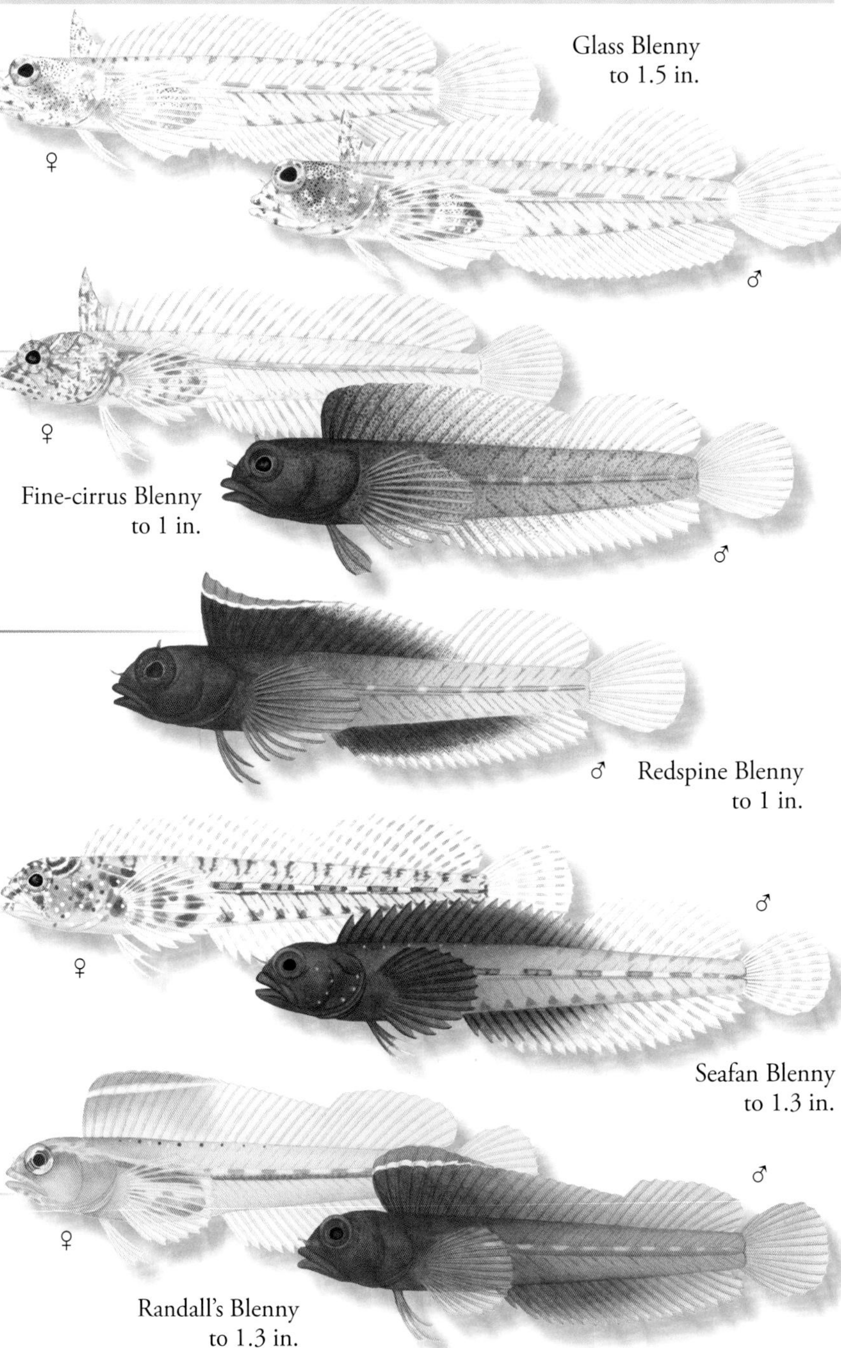
Glass Blenny
to 1.5 in.
♀
♂
♀
Fine-cirrus Blenny
to 1 in.
♂
♂ Redspine Blenny
to 1 in.
♂
♀
Seafan Blenny
to 1.3 in.
♂
♀
Randall's Blenny
to 1.3 in.

Ruetzler's Blenny - *Emblemariopsis ruetzleri* Tyler & Tyler, 1997

FEATURES: Females translucent; head grayish with large, irregular white spots or blotches and tiny greenish and brownish flecks. Young males with irregular brown spots on head and anterior dorsal fin. Adult males dark brown on head and anterior dorsal fin. All with a dark wedge-shape mark on abdomen, and brown and white marks along spinal column. Anterior dorsal fin tall and bluntly pointed. Tiny cirrus on nostrils and over eyes. HABITAT: Puerto Rico to St. Croix, Cayman Islands, and Quintana Roo to Panama. Demersal over shallow-water dead and living coral to about 30 ft.

Tayrona Blenny - *Emblemariopsis tayrona* (Acero, 1987)

FEATURES: Females translucent; red and brown or greenish bands on head and anterior dorsal fin; abdomen striped; red spots along dorsal and ventral profiles. Males blackish on head with white spots or lines on cheeks; anterior dorsal fin with a broad red outer band bordered by white lines and small dark spots along base. All with red and white marks along spinal column. Anterior dorsal-fin spines tall, pointed. Nostrils with a fringed cirrus. Single, short cirrus over eyes. HABITAT: Colombia to Trinidad and Tobago. Demersal over coral and rocks to about 50 ft. NOTE: Previously *E. ramirezi.*

Wrasse Blenny - *Hemiemblemaria simula* Longley & Hildebrand, 1940

FEATURES: Three color phases: yellow dorsally, fading to white below; yellow to greenish yellow dorsally with dark midlateral stripe; pale with midlateral stripe broken into segments. All with purplish to blackish stripe through eyes and a black spot on anterior dorsal fin. Juveniles translucent with dark midlateral stripe. All stages with protruding lower jaw. HABITAT: S FL, Bahamas, Cuba, Yucatán, and western Caribbean Sea. Demersal over shallow-water reefs to about 60 ft. Inhabit empty holes.

Arrow Blenny - *Lucayablennius zingaro* (Böhlke, 1957)

FEATURES: Color varies. Translucent with a reddish to purplish stripe from snout, through eyes, and onto lower sides. Narrow stripe on top of head. Stripes bordered by white to yellow. Three black ocellated spots on rear dorsal-fin base, and two black ocellated spots on rear anal-fin base. Snout long and pointed. Jaws long. HABITAT: Bahamas and Caribbean Sea. Occur over coral reefs and dropoffs, in caves and crevices and inside tube sponges from about 20 to 360 ft.

Chameleon Blenny - *Protemblemaria punctata* Cervigón, 1966

FEATURES: Females and juveniles shades of brown on head with a white bar below and behind eyes; about 10 brownish saddles along back, and numerous reddish spots posteriorly. Non-breeding males greenish on head; breeding males reddish brown on head; vague to distinct white bar behind eyes; dark and pale saddles along back; reddish to brownish spots posteriorly. All with red banded irises and two crests on top of head. Female anterior dorsal fin elevated; male anterior dorsal fin low. HABITAT: Eastern Venezuela. Demersal over shallow-water reefs to about 70 ft. Live in empty holes.

NOTE: There are 11 other rare Tube Blennies in the area.

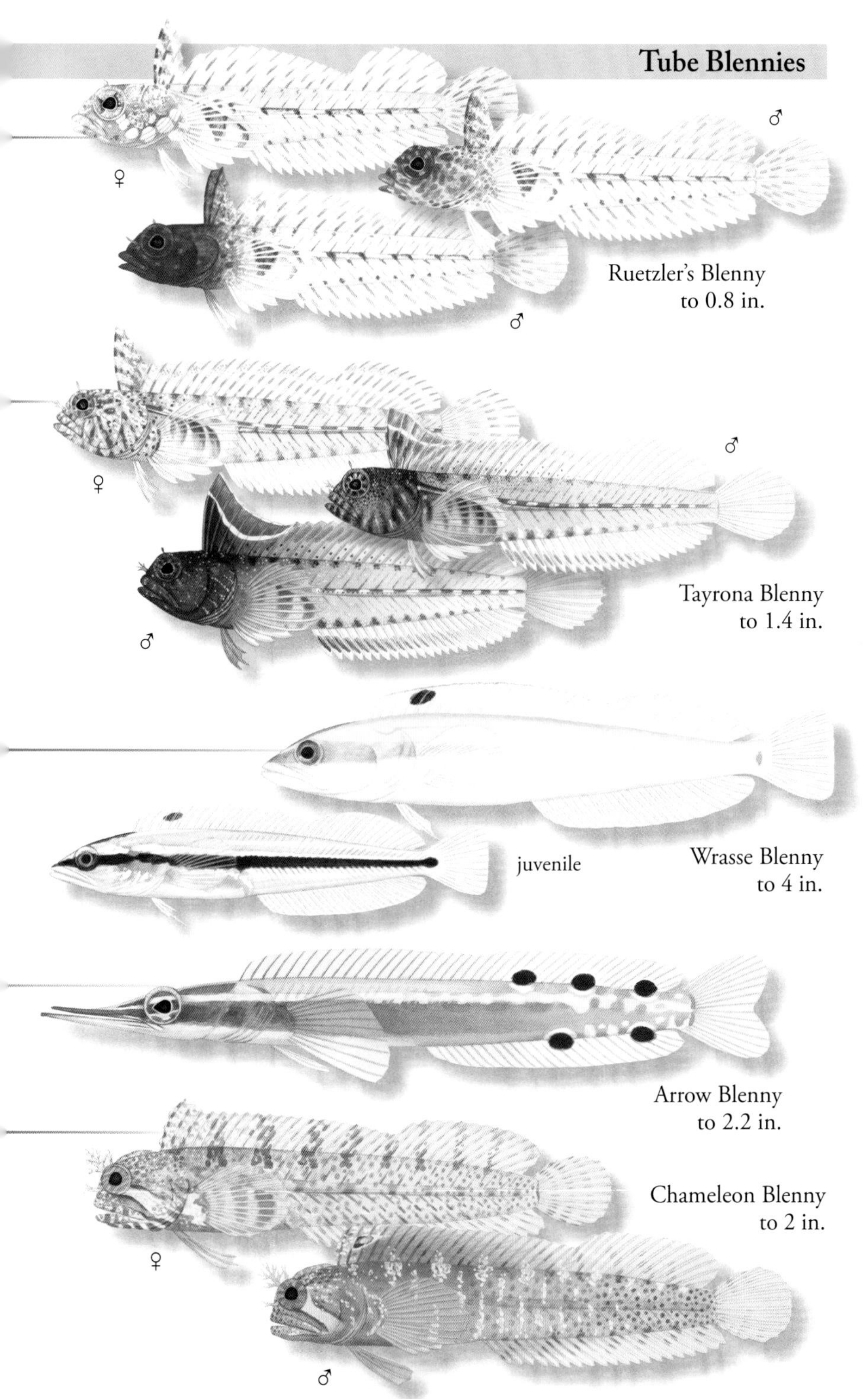
♂
♀
Ruetzler's Blenny
to 0.8 in.
♂
♂
♀
Tayrona Blenny
to 1.4 in.
♂
juvenile
Wrasse Blenny
to 4 in.
Arrow Blenny
to 2.2 in.
Chameleon Blenny
to 2 in.
♀
♂

Gobiesocidae - Clingfishes

Flarenostril Clingfish - *Acyrtops amplicirrus* Briggs, 1955

FEATURES: Shades of green with scattered bluish white spots or reticulations on head and body. May also have scattered brownish dots and spots. Dark bands radiate from eyes. Nostrils with a large flap. Eyes small and set far apart. Dorsal fin with five to six rays. Sucking disk relatively small with short papillae anteriorly and posteriorly, and a small patch at center. HABITAT: Puerto Rico, Lesser Antilles, and Yucatán Peninsula. Demersal over shallow-water seagrass beds, rocks, and corals to about 10 ft. NOTE: Similar to, and possibly confused with, *Acyrtops beryllinus*.

Emerald Clingfish - *Acyrtops beryllinus* (Hildebrand & Ginsburg, 1927)

FEATURES: Color and pattern highly variable. Usually emerald green; may also be brownish to yellowish. May have diffuse, pale whitish spots and lines dorsally or may show a reticulating pattern. A few golden to brassy lines may radiate from eyes. Nostrils with a short flap. Dorsal fin with five and seven rays. Sucking disk relatively small with short papillae anteriorly and posteriorly, and a small patch at center. HABITAT: S FL, Bahamas, and Caribbean Sea to Brazil. Associated with turtle-grass beds to about 10 ft. NOTE: Similar to, and possibly confused with, *Acyrtops amplicirrus*.

Papillate Clingfish - *Acyrtus artius* Briggs, 1955

FEATURES: Ground color tan to purple or brick red, becoming semi-transparent posteriorly. Head covered in dense, tiny white dots and blackish to purplish clusters. Greenish to brownish bands and lines of dark spots posteriorly. Eyes large with a red iris and thin, white inner ring. Sucking disk large with seven or eight rows of papillae anteriorly, and a W-shaped patch posteriorly. HABITAT: Bahamas and Caribbean Sea. Demersal over shallow-water coral spur-and-groove areas, coral walls, and outer ridges.

Orange-spotted Clingfish - *Acyrtus lanthanum* Conway, Baldwin & White, 2014

FEATURES: Color and pattern vary. Translucent whitish to pale bluish with irregular brownish blotches on head, several brownish bands on midbody, and a brownish band at caudal-fin base. Clusters of white and orange spots pepper dorsal surface. Eyes large with radiating dark bands. Sucking disk with a large C-shaped patch of small papillae anteriorly, two small patches in middle, and C-shaped patches posteriorly. HABITAT: Bahamas, Puerto Rico, and Caribbean Sea. Demersal over coral rubble in shallow coastal lagoons to about 20 ft.

Red Clingfish - *Acyrtus rubiginosus* (Poey, 1868)

FEATURES: Color and pattern vary. May be uniformly tannish or whitish to pinkish with or without brick red or purplish red to brownish bands that may be intersected with pale spots. Small spots and blotches may pepper pale bands. Bands may radiate from eyes. Irises often red with a thin, white inner ring. Sucking disk with a large and wide C-shaped patch of papillae anteriorly, and a short semi-circular patch posteriorly. HABITAT: Bahamas and Caribbean Sea. Demersal over boulders and rocks in high surge areas and in tide pools to about 15 ft. Often associated with sea urchins.

Flarenostril Clingfish
to 0.75 in.
Emerald Clingfish
to 1 in.
variants
Papillate Clingfish
to 1.3 in.
Orange-spotted Clingfish
to 1.3 in.
Red Clingfish
to 1.4 in.

Padded Clingfish - *Arcos nudus* (Linnaeus, 1758)

FEATURES: Whitish to tannish with clusters of small, irregular brownish to grayish spots that form larger blotches and bands. Interspaces with paler spots. Sucking disk very large with broad bands of papillae anteriorly and posteriorly, with two close-set patches at rear center. Dorsal fin with seven or eight rays. HABITAT: Bahamas and Caribbean Sea. Demersal over shallow-water rocky shores in high surge areas to about 30 ft.

Lappetlip Clingfish - *Gobiesox barbatulus* Starks, 1913

FEATURES: Brownish with darker spots and lines that radiate from eyes, and broader, irregular darker lines that run along back and sides. Dorsal, caudal, and anal fins nearly black. Short, fleshy lappets present around snout and upper lip. Dorsal fin with 10–12 rays. Sucking disk with a large C-shaped band of papillae anteriorly, two short patches of kidney-shaped papillae at sides, and a broad semi-circle of papillae posteriorly. HABITAT: Belize and Venezuela to southern Brazil. Demersal over rocks in tide pools and around mangrove roots to about 5 ft.

Riverine Clingfish - *Gobiesox cephalus* Lacepède, 1800

FEATURES: Color and pattern highly variable. Shades of gray, tan, or brown with vague or heavy spotting and mottling. Can be very dark or pale. Usually with dark stripes on rear body that may be obscure or very dark. All with a dark blotch at dorsal-fin insertion. Snout comparatively narrow and depressed. Dorsal fin with 10–12 rays. Sucking disk with a narrow semi-circle of papillae anteriorly and posteriorly, two long patches at sides, and a pair of small patches at center. HABITAT: Antilles and southern Caribbean Sea. Demersal in shallow coastal freshwater streams and brackish waters to about 10 ft.

Bahama Clingfish - *Gobiesox lucayanus* Briggs, 1963

FEATURES: Pale gray to pinkish with rows of large brownish to blackish spots that radiate from eyes, form bands on back, and form lines along sides. Spots on snout and under eyes vague. Spots in bands elongate. Dorsal and caudal fins with dark inner margins. Dorsal fin with 8–10 rays (usually 9). Sucking disk large, with a broad band of papillae anteriorly, two widely spaced patches at center, and five or six rows posteriorly. HABITAT: Bahamas and Greater Antilles. Demersal over shallow-water rocky and limestone ledges to about 10 ft.

Dark-finned Clingfish - *Gobiesox nigripinnis* (Peters, 1859)

FEATURES: Pale tannish to grayish with diffuse to very dark brownish to blackish spots on dorsal surface that become stripes on rear lower sides. Dark bands radiate from eyes. Dorsal, caudal, and anal fins blackish with narrow white margins. Dorsal fin with 13 rays. Sucking disk large, with a broad semi-circular band of papillae anteriorly, two long patches at sides, and a semi-circular patch posteriorly. HABITAT: Hispaniola, Lesser Antilles, Venezuela, and Panama. Demersal over shallow-water boulders with algae. Also reef associated to about 70 ft.

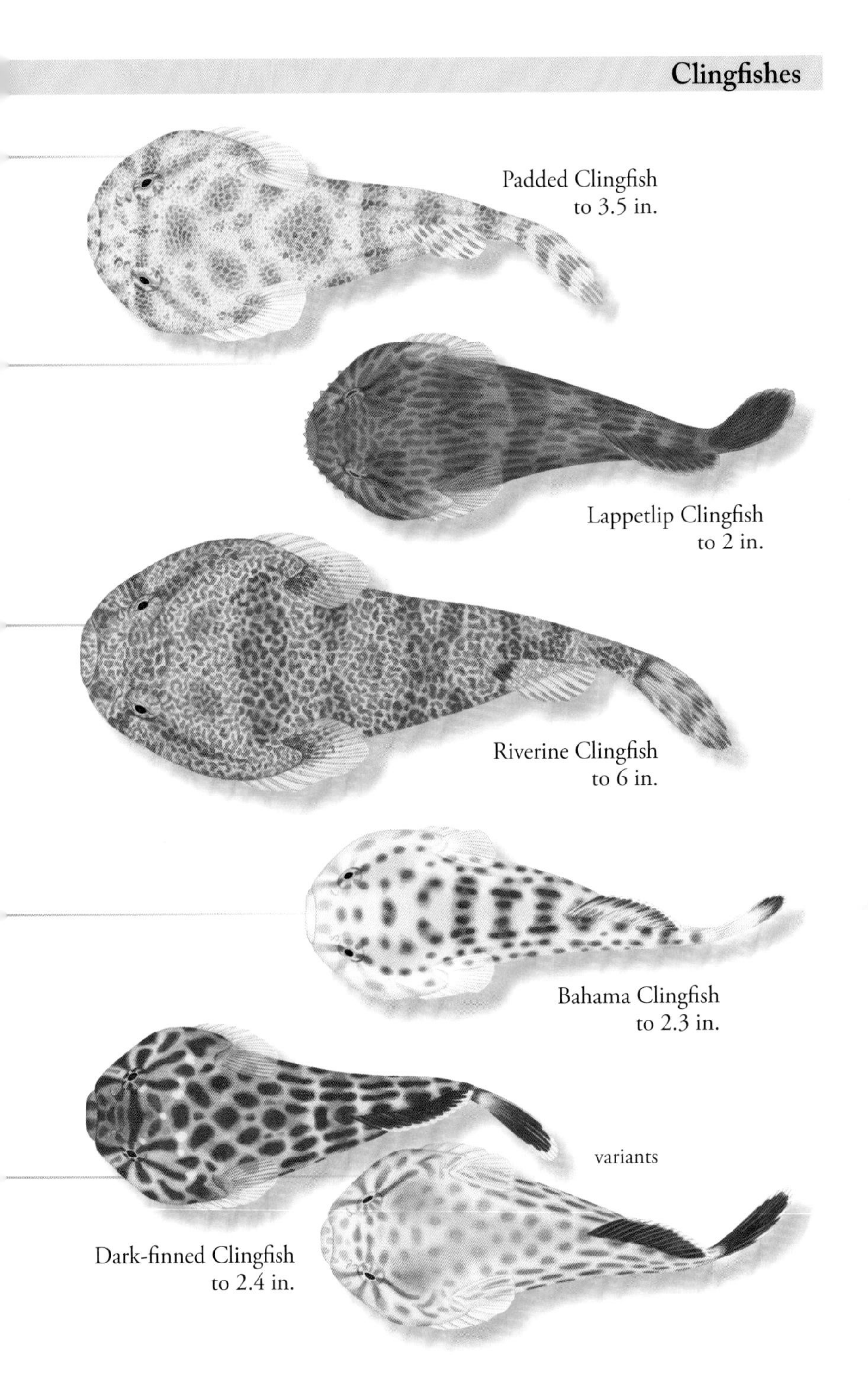
Padded Clingfish
to 3.5 in.
Lappetlip Clingfish
to 2 in.
Riverine Clingfish
to 6 in.
Bahama Clingfish
to 2.3 in.
variants
Dark-finned Clingfish
to 2.4 in.

Stippled Clingfish - *Gobiesox punctulatus* (Poey, 1876)

FEATURES: Color and pattern highly variable. Shades of tan, gray, or olive. Dorsal surface stippled with small black spots. May have diffuse or distinct bands radiating from eyes. May have a few diffuse to dark bands across back. Upper lip broad. Dorsal fin with 11–12 (usually 11) rays. Sucking disk with a broad row of papillae anteriorly and posteriorly, and with two widely spaced patches at center. HABITAT: Bahamas and Caribbean Sea. (Records from northern Gulf of Mexico are erroneous.) Demersal over limestone rocks and ledges in clear water with high surge and algae to about 10 ft.

Skilletfish - *Gobiesox strumosus* Cope, 1870

FEATURES: Color and pattern highly variable. Shades of brown, olive brown, or blackish. May have a mottled, lined, banded, or a net-like pattern or may be uniformly colored with pale spots and margins. Lines may radiate from eyes. Upper lip broad. Short, fleshy lappets present around snout and upper lip. Dorsal fin usually with 12 rays. Sucking disk with a broad row of papillae anteriorly and posteriorly, and with two small, widely spaced patches at center. HABITAT: NJ to FL, Gulf of Mexico, and N Caribbean Sea. Occur in rocky tide pools and over grassy areas to about 60 ft.

Broadhead Clingfish - *Tomicodon briggsi* Williams & Tyler, 2003

FEATURES: Females whitish to tannish with hourglass-shaped clusters of brownish vermiculations on back; clusters of vermiculations on sides. Males mottled and speckled in shades of tan; about six dark-margined, hourglass-shaped marks on back; lower sides with dark V-shaped marks. All with dark bands radiating from eyes. Dorsal fin with seven rays. Sucking disk with broad rows of papillae anteriorly and posteriorly. Anus closer to anal fin than sucking disk. HABITAT: Lesser Antilles and off Belize. Demersal over shallow-water mixed sandy and coral bottoms to about 40 ft.

Barred Clingfish - *Tomicodon fasciatus* (Peters, 1859)

FEATURES: Dorsal surface with six to eight wide, dark, reddish brown bands separated by narrow ochre to pale bands. Pale bands may have dark reddish centers. Dark bands radiate from eyes. Dark blotch at pectoral-fin base and anterior dorsal-fin base. Dorsal fin with seven to eight rays. Sucking disk with broad patches of papillae anteriorly and posteriorly. Papillae absent at center. HABITAT: Bahamas, Puerto Rico, Lesser Antilles, and Nicaragua to Venezuela. Demersal over shallow-water rocky bottoms coral reefs to about 20 ft., around mangroves, and in tide pools.

Reitz's Clingfish - *Tomicodon reitzae* Briggs, 2001

FEATURES: Color and pattern vary. Shades of tan to dark brown dorsally with six distinct to vague hourglass-shaped blotches with pale spots inside and around edges. Dark lines radiate from eyes. Sides with 8–13 dark bars with narrow, pale interspaces. Caudal fin with a dark blotch. Dorsal fin with a dark blotch and seven to eight rays. Sucking disk with a broad patch of papillae anteriorly and posteriorly; anterior edge with fleshy tabs. Anus closer to sucking disk than anal fin. HABITAT: Bahamas and Caribbean Sea. Demersal over shallow water eroded in areas of surge to about 70 ft.

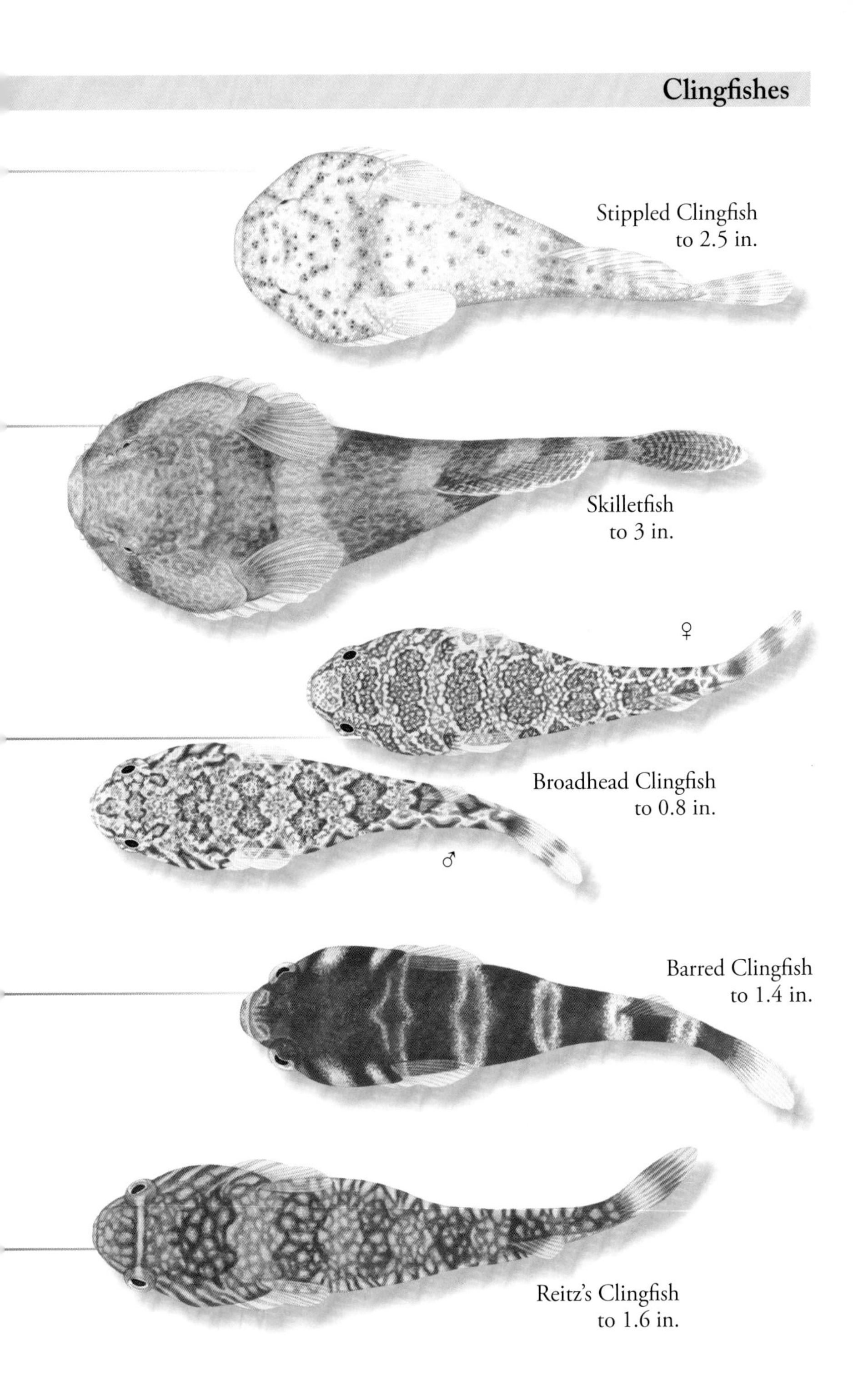
Stippled Clingfish
to 2.5 in.
Skilletfish
to 3 in.
♀
Broadhead Clingfish
to 0.8 in.
♂
Barred Clingfish
to 1.4 in.
Reitz's Clingfish
to 1.6 in.

Gobiesocidae - Clingfishes, *cont.*

Antillean Clingfish - *Tomicodon rhabdotus* Smith-Vaniz, 1969

FEATURES: Shades of tan with and brownish red bands—some with paler centers—from snout to caudal fin. Dark bands separated by narrower, speckled, tannish bands. Dark bands radiate below and behind eyes. Nostrils with a large flap. Dorsal fin with eight to nine rays. Sucking disk with six to seven rows of broad papillae anteriorly and posteriorly; papillae absent from center. HABITAT: Lesser Antilles and coast of Venezuela. Demersal over shallow-water rocky surge areas to about 3 ft.

Surge Clingfish - *Tomicodon rupestris* (Poey, 1860)

FEATURES: Color, pattern vary. Whitish to tannish with 5–6 dark hourglass-shaped marks dorsally. Sides with dark bars and/or V-shaped marks. A heart- to diamond-shaped mark at caudal-fin base. Dark bands below and behind eyes. Dorsal fin with 8–9 rays. Sucking disk with broad patches of papillae anteriorly and posteriorly. HABITAT: Bahamas and Caribbean Sea. Demersal over rocky bottoms in shallow-water surge areas.

NOTE: There are 10 other small and poorly recorded Clingfishes in the area.

Callionymidae - Dragonets

Lancer Dragonet - *Callionymus bairdi* Jordan, 1888

FEATURES: Variably mottled and spotted in shades of tannish to grayish. Lower head faintly to vividly barred. Spiny dorsal fin of females pale anteriorly, dark posteriorly. Spiny dorsal fin of males is tall, expanded, and variably spotted and banded in yellow to gray. Preopercle with one forward-facing lower spine and three to nine upward-facing prongs. Soft dorsal fin with nine rays. Caudal keels absent. HABITAT: Cape Hatteras, NC, to FL, Bermuda, Bahamas, and Caribbean Sea. Demersal over sandy bottoms around coral reefs to about 30 ft. NOTE: Previously known as *Paradiplogrammus bairdi*.

Spotted Dragonet - *Chalinops pauciradiatus* (Gill, 1865)

FEATURES: Females tannish to whitish and densely mottled and speckled. Males similarly patterned but usually darker; may have orange patches and blackish spots on head. First dorsal-fin spines long and filamentous in males. Preopercle with three upward-facing prongs. Soft dorsal fin with six rays. All with a spotted lateral keel from abdomen to caudal fin. HABITAT: NC to LA, Flower Garden Banks, Bermuda, Bahamas, and Caribbean Sea. Demersal over shallow-water seagrass beds to about 30 ft. NOTE: Previously known as *Diplogrammus pauciradiatus*.

Spotfin Dragonet - *Synchiropus agassizii* (Goode & Bean, 1888)

FEATURES: Reddish with greenish to yellowish mottling dorsally. White below. Spiny dorsal fin with black ocellated spot. Soft dorsal fin and caudal fins with yellow bands. Anal fin with blackish submarginal band. Soft dorsal-fin margin straight in females, taller and slightly convex in males. Males with elongate middle caudal-fin rays. HABITAT: Nova Scotia to FL, Gulf of Mexico, Bermuda, and Caribbean Sea to Brazil. Demersal over sand and mud from about 300 to 2,100 ft.

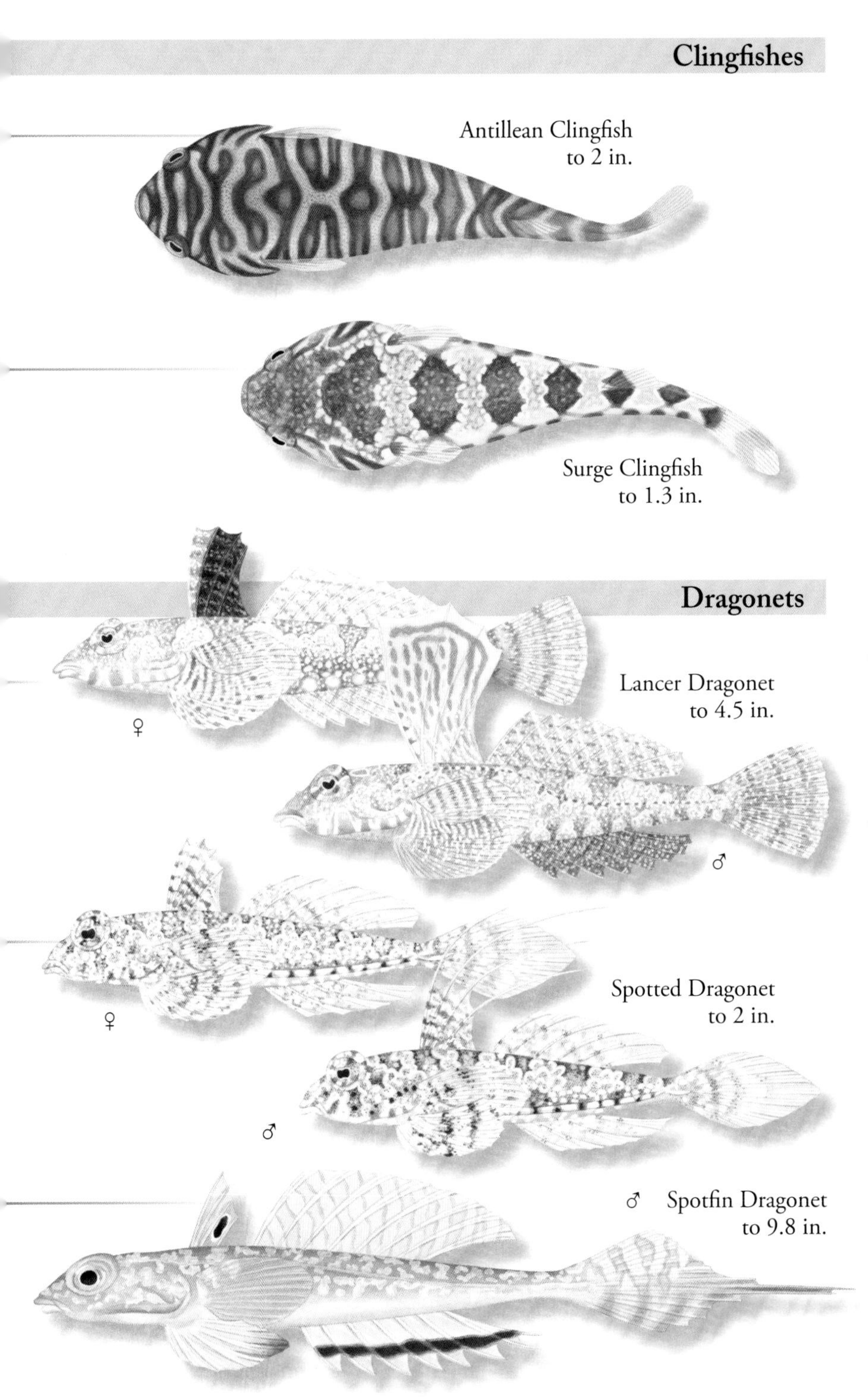
Clingfishes
Antillean Clingfish
to 2 in.
Surge Clingfish
to 1.3 in.
Dragonets
Lancer Dragonet
to 4.5 in.
♀
♂
Spotted Dragonet
to 2 in.
♀
♂
♂ Spotfin Dragonet
to 9.8 in.

Eleotridae - Sleepers

Fat Sleeper - *Dormitator maculatus* (Bloch, 1792)

FEATURES: Color variable. Tan or brown to bluish with dark reddish to blackish scales. Dark bands radiate from eyes. Dorsal and anal fins banded. Always with a bright to faint, bluish ocellated blotch above pectoral-fin base. Juveniles with obscure bars and blotches along midline. Mouth oblique and relatively small. Snout profile convex. Scales large. Body relatively deep. HABITAT: NC, FL, Gulf of Mexico, Bahamas, and Caribbean Sea to southern Brazil. Occur in a wide variety of habitats in shallow coastal marine, brackish, and fresh waters to about 20 ft.

Largescaled Spinycheek Sleeper - *Eleotris amblyopsis* (Cope, 1871)

FEATURES: Head and body shades of tan to brown and densely speckled and blotched. Dorsal portion may be paler than sides and ventral area. Several dark bands radiate from eyes. Fins banded. Mouth large, oblique. Head broad, flattened dorsally. Preopercle with a sharp, concealed spine. Spiny dorsal fin with six spines. Caudal fin rounded. HABITAT: NC to FL, Gulf of Mexico, and Caribbean Sea to Brazil. Mostly absent from Lesser Antilles. Occur around rocks and vegetation in shallow coastal marine, brackish, and fresh waters to about 15 ft. Also in ponds and ditches.

Emerald Sleeper - *Erotelis smaragdus* (Valenciennes, 1837)

FEATURES: Dark olivaceous to brown above. Tan below. Dark spot at upper pectoral-fin base present. Jaws large and oblique with thin lips. Head laterally expanded with dorsal profile almost straight. Spiny dorsal fin with six weak spines. Caudal fin pointed to bluntly pointed, with anterior rays originating anterior to peduncle. HABITAT: FL, Gulf of Mexico, Bahamas, and Caribbean Sea to Brazil. Demersal over sandy and silty bottoms in marine and brackish coastal waters and around mangroves to about 15 ft.

Bigmouth Sleeper - *Gobiomorus dormitor* Lacepède, 1800

FEATURES: Olivaceous to yellowish or olive brown dorsally. Paler below. Obscure bands radiate from eyes. An obscure stripe runs from pectoral-fin base to caudal fin. Dorsal, caudal, and pectoral fins banded. Mouth very large and moderately oblique, with bands of sharp teeth. Head is large and flattened dorsally. Spiny dorsal fin with six weak spines. Preopercular margin is fleshy. HABITAT: FL, Gulf of Mexico, Bermuda, and Caribbean Sea to Brazil. Demersal over soft bottoms in coastal marine, brackish, and fresh waters over soft bottoms to about 20 ft. Adults may be found far inland.

Guavina - *Guavina guavina* (Valenciennes, 1837)

FEATURES: Shades of dark grayish brown to dark bluish gray with vague, darker saddles dorsally, and paler vermiculations on sides. Three vague bands radiate behind eyes. Dorsal fins banded. All fins with yellow margins. Mouth large and oblique, with bands of conical teeth. Head long, flattened, and wide. Spiny dorsal fin with seven spines. HABITAT: E FL, southern Gulf of Mexico to Panama, Cuba, Puerto Rico and adjacent islands, Martinique, and Venezuela to Brazil. Demersal over shallow muddy bottoms in brackish estuaries, hypersaline waters, and in fresh water to about 30 ft. BIOLOGY: Migrate to and from fresh and salt water. Tolerant of low oxygen levels.

Fat Sleeper
to 11.8 in.
Largescaled Spinycheek Sleeper
to 12 in.
Emerald Sleeper
to 9.6 in.
Bigmouth Sleeper
to 2.4 ft.
Guavina
to 12 in.

Sabre Goby - *Antilligobius nikkiae* Van Tassell & Colin, 2012

FEATURES: Pale grayish dorsally with a blue stripe along profile. Rear jaws reddish. Broad yellow stripe from lower eyes to caudal fin. Inner dorsal fins yellow. Caudal and anal fins with narrow yellow stripes. Spiny dorsal fin very tall, pointed. Caudal fin pointed at tip. HABITAT: Bahamas, Cuba, Belize, Honduras, Puerto Rico, and off Venezuela. Demersal over reefs, steep walls, ledges, and slopes from about 240 to 670 ft.

Whiskered Goby - *Barbulifer antennatus* Böhlke & Robins, 1968

FEATURES: Whitish to creamy with dense brownish spots on head, scattered to dense brown spots on body, and a large brownish blotch on caudal-fin base. May be almost uniformly dark. Numerous long, narrow barbels around upper and lower jaws and in front of eyes. Gill opening an S-shaped slit. HABITAT: Bahamas, Quintana Roo to Belize, Jamaica, and Puerto Rico to Venezuela. Demersal over shallow-water rocky bottoms.

Bearded Goby - *Barbulifer ceuthoecus* (Jordan & Gilbert, 1884)

FEATURES: Color and pattern somewhat variable. Head and body with pale patches and dark speckles and spots. Dark band below eye. Dark bar at pectoral-fin base. A few short barbels on chin, snout, and below eyes. Gill opening an S-shaped slit. HABITAT: S FL, Bahamas, and Caribbean Sea to Brazil. Demersal over nearshore, shallow-water, silty rocks and rubble, and among algae of seagrass beds to about 15 ft.

Antilles Frillfin - *Bathygobius antilliensis* Tornabene et al., 2010

FEATURES: Shades of grayish brown to brown with vague darker saddles and alternating rows of dark, oblong blotches on sides. Rows of dark spots behind mouth. Three dark blotches on abdomen. Pectoral fins with 18–21 rays; upper 3–5 filamentous, free at tips. Midline scales number 38–42. HABITAT: Bermuda, Bahamas, Cuba to Venezuela, and Quintana Roo to Honduras. Demersal in tide pools, on reef crests, and exposed areas.

Notchtongue Goby - *Bathygobius curacao* (Metzelaar, 1919)

FEATURES: Tan to brownish with pale scale centers. Irregular bands radiate from eyes. Sides with somewhat regularly spaced, dark blotches. Second dorsal and anal fins with tannish inner margin. Tongue deeply notched at tip. Pectoral fins with 15–18 rays; upper four or five filamentous, free at tips. Midline scales number 31–34. HABITAT: S FL, southern Gulf of Mexico, Bermuda, Bahamas, and Caribbean Sea. Demersal in tide pools, around mangroves, and over sheltered seagrass beds to about 15 ft.

Twin-spotted Frillfin - *Bathygobius geminatus* Tornabene et al., 2010

FEATURES: Greenish gray to brownish gray with pale scale centers. Vague dark saddles on back and irregular blotches along back. Pairs of dark oblong blotches from gills to peduncle. A dark stripe may be present along midline. Caudal-fin base with two or three dark spots. Pectoral fins with 17–18 rays; upper three to five filamentous, free at tips. Midline scales number 36–38. HABITAT: E FL and Puerto Rico. Demersal over coastal shallow-water, sandy, and rocky bottoms to about 10 ft.

Sabre Goby
to 1.7 in.
Whiskered Goby
to 1 in.
Bearded Goby
to 1.4 in.
Antilles Frillfin
to 3.5 in.
Notchtongue Goby
to 3 in.
Twin-spotted Fillfin
to 1.8 in.

Checkerboard Frillfin - *Bathygobius lacertus* (Poey, 1860)

FEATURES: Shades of tan with irregular dark brown saddles and two rows of brown blotches on lower sides that form a distinct to indistinct checkerboard pattern. Narrow to wide bar on caudal-fin base. Pectoral fins with 18–20 rays; upper three to five filamentous threads. Midline scales number 38–42. HABITAT: FL Keys, Bermuda, Bahamas, eastern Caribbean Sea, Cayman Is., Puerto Rico, Virgin Is., Curaçao to Isla Orchila and Tobago. Demersal in tide pools, on reef crests, along rocky beaches, and around mangroves.

Island Frillfin - *Bathygobius mystacium* Ginsburg, 1947

FEATURES: Tannish to grayish with pale to white scale centers. Irregular bands radiate from eyes. Upper sides with obscure saddles; lower sides with series of dark to blackish blotches. Tongue rounded at tip. Pectoral fins with 19–20 rays; upper 4–5 filamentous, free at tips. Midline scales number 33–36. HABITAT: S FL, southern Gulf of Mexico, Bahamas, and Caribbean Sea. Demersal over unsheltered rocky and sandy bottoms.

Frillfin Goby - *Bathygobius soporator* (Valenciennes, 1837)

FEATURES: Color, pattern highly variable. Shades of brown to gray; may be very dark. Head blotched. Five dark saddles along back; saddle below first dorsal fin broadest. Dark blotches on sides. Tongue slightly notched at tip. Pectoral fins with 18–21 rays; upper 4–5 filamentous, free at tips. Midline scales number 37–41. HABITAT: NC to FL, Gulf of Mexico, Bermuda, Bahamas, and Caribbean Sea to Brazil. Demersal in shallow-water habitats from muddy river bottoms to estuaries and rocky tide pools.

White-eye Goby - *Bollmannia boqueronensis* Evermann & Marsh, 1899

FEATURES: Pale with orange brown blotches along midline. Blotch at caudal-fin base darker than other blotches. Eyes with a bright white, reflective iris. Dorsal and upper caudal fins with orange brown stripes. Spiny dorsal fin with a black ocellated spot. Soft dorsal fin with one spine and 12 rays. HABITAT: S FL, Jamaica, Puerto Rico, and Yucatán to Guyana. Demersal over soft bottoms from about 88 to 180 ft.

Translucent Goby - *Chriolepis fisheri* Herre, 1942

FEATURES: Color and pattern vary. Mottled and blotched in translucent shades of whitish or yellowish to reddish brown. All with orange red irises, vague to distinct saddles along back, and a dark bar at caudal-fin base. Dorsal fins with pale margins. Pelvic fins separate. HABITAT: Bahamas, Grand Cayman, Hispaniola to Venezuela, Belize, Honduras, and western Brazil. Demersal over reefs from near surface to about 270 ft.

Barfin Goby - *Coryphopterus alloides* Böhlke & Robins, 1960

FEATURES: Translucent. Pale anteriorly; yellowish posteriorly. Dark bar below first dorsal fin. Dark smudges on head and body. Cheeks reddish. Dark smudge at caudal-fin base. Females with dark smudge between second and third dorsal-fin spines. In males, dark smudge on dorsal fin merges with a dark band. HABITAT: S FL, Bahamas, and scattered in Caribbean Sea. Over coral heads in clear water from about 30 to 90 ft.

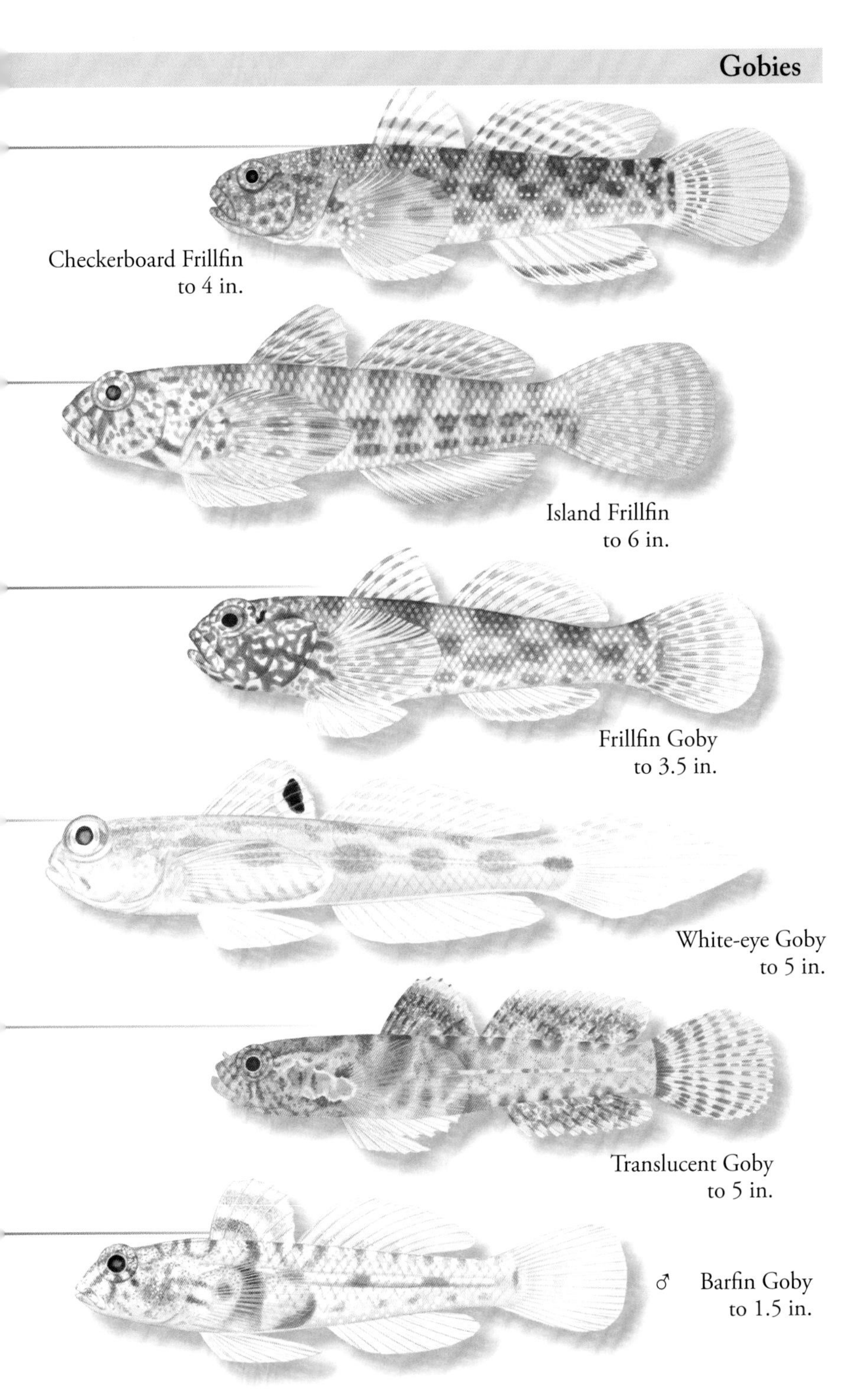
Checkerboard Frillfin
to 4 in.
Island Frillfin
to 6 in.
Frillfin Goby
to 3.5 in.
White-eye Goby
to 5 in.
Translucent Goby
to 5 in.
♂ Barfin Goby
to 1.5 in.

Colon Goby - *Coryphopterus dicrus* Böhlke & Robins, 1960

FEATURES: Translucent with brownish and whitish stripes below and behind eyes. Sides with rows of fairly regularly spaced brown and whitish spots. Usually two distinct brown spots at pectoral-fin base. Brownish bar or hourglass-shaped mark at caudal-fin base. HABITAT: S FL, southern Gulf of Mexico, Bahamas, and Caribbean Sea to Brazil. Demersal over rocky bottoms between coral colonies to about 180 ft.

Pallid Goby - *Coryphopterus eidolon* Böhlke & Robins, 1960

FEATURES: Translucent bluish gray. Orange to yellow stripe behind eye bordered with dark pigment. Orange to yellow blotches below eyes and in series along back and on sides. Small black bar at caudal-fin base. Spinal column with dark smudges. Pelvic-fin spines joined. HABITAT: NC to FL, southern Gulf of Mexico, and Caribbean Sea to Brazil. Demersal over fine sand around coral and limestone bottoms to about 180 ft.

Bridled Goby - *Coryphopterus glaucofraenum* Gill, 1863

FEATURES: Color corresponds to habitat. Translucent whitish to tannish and sparsely to densely marked. All with spotted eyes, variable orangish spots or stripes on head, white "bridle" from upper jaw to preopercle, a two-pronged spot or bar above opercle, few to many white spots and dark spots that form Xs on sides, and a pair of elongate spots or a barbell-shaped mark at caudal-fin base. HABITAT: NC to FL, Flower Garden Banks, S Gulf of Mexico, Bermuda, Bahamas, and Caribbean Sea to Brazil. Demersal in shallow-water sandy bays and around mangroves and reefs to about 200 ft.

Glass Goby - *Coryphopterus hyalinus* Böhlke & Robins, 1962

FEATURES: Translucent. Upper snout usually blue; lower snout and upper lip dark brownish. Patch of melanophores form a variable dark wedge behind and below eyes. Large orange area from head to midbody. Orange spots along spinal column. Anus near rear of large black blotch. HABITAT: S FL, Bermuda, Bahamas, and Caribbean Sea. In aggregations over reef dropoffs and spur-and-groove areas to about 170 ft.

Kuna Goby - *Coryphopterus kuna* Victor, 2007

FEATURES: Translucent grayish white. Two rows of small spots on upper eyes. White and brownish to reddish brown flecks and spots on head and body. Dark bar or cluster of spots below eyes. Dark spots cluster below first dorsal fin; may form a broken arc. Fins may be spotted. HABITAT: SE FL, Quintana Roo to Panama, Cayman Is., Bonaire, Dominica, and Guadeloupe. Demersal over sandy bottoms in relatively deep water.

Peppermint Goby - *Coryphopterus lipernes* Böhlke & Robins, 1962

FEATURES: Translucent yellowish to gold. Bright blue markings on snout and over eyes, and pale blue lines behind eyes that may extend to forebody. Dark ring around anus. Several pale blotches along spinal column. Second dorsal-fin spine elongate. HABITAT: FL Keys, Bahamas, and Caribbean Sea. Demersal over live coral heads, reef dropoffs, and spur-and-groove areas from about 20 to 200 ft.

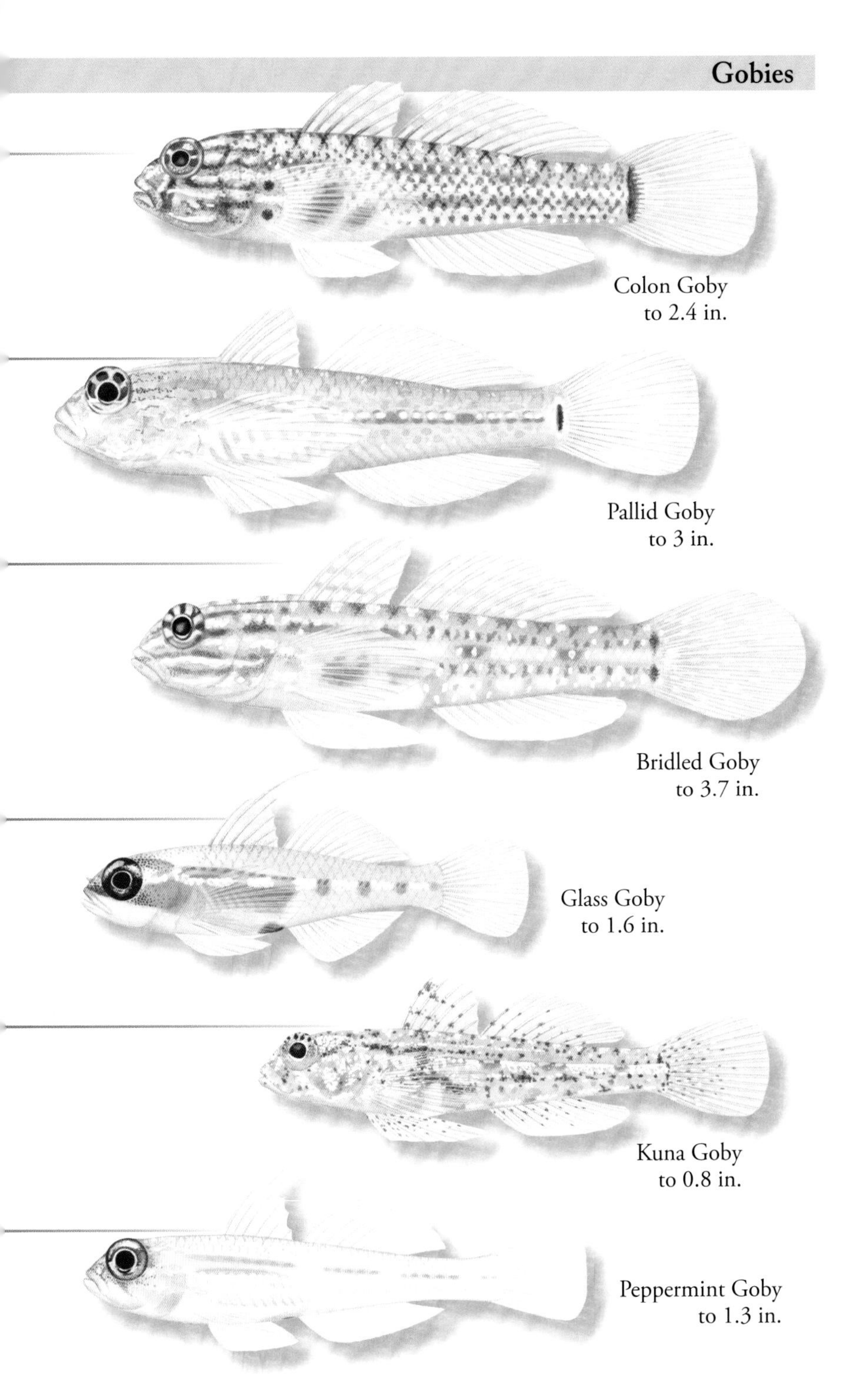

Colon Goby
to 2.4 in.

Pallid Goby
to 3 in.

Bridled Goby
to 3.7 in.

Glass Goby
to 1.6 in.

Kuna Goby
to 0.8 in.

Peppermint Goby
to 1.3 in.

Masked Goby - *Coryphopterus personatus* (Jordan & Thompson, 1905)

FEATURES: Translucent. Upper snout usually blue. A patch of dark melanophores form a variable dark wedge behind and below eyes. Pale blue spot at pectoral-fin base. Large orange area from head to midbody. Orange spots along spinal column. Large black blotch around anus. HABITAT: S FL, Bermuda, Bahamas, and Caribbean Sea. Near bottom over corals and in caves 6 to 100 ft. BIOLOGY: Hovers in schools.

Spotted Goby - *Coryphopterus punctipectophorus* Springer, 1960

FEATURES: Whitish to bluish white with a series of orange brown blotches on sides. Several dark-margined lines extend from snout through eyes to below first dorsal fin. Lower pectoral-fin base with a distinct dark spot. Dorsal fins banded. Caudal fin spotted. HABITAT: NC to FL, and Gulf of Mexico. Demersal over sandy and silty bottoms near rocky reef bases from about 10 to 120 ft.

Bartail Goby - *Coryphopterus thrix* Böhlke & Robins, 1960

FEATURES: Translucent whitish. Eyes spotted. Body with fairly evenly spaced brownish and white spots. Intensity of spots may vary. Usually with a large dark spot on upper pectoral-fin base. Spot blackish in males. Dark bar at caudal-fin base. Second dorsal-fin spine elongate. HABITAT: NC to FL, Flower Garden Banks, Bahamas, and Caribbean Sea to Brazil. Demersal over sandy bottoms around coral heads from about 5 to 300 ft.

Patch-reef Goby - *Coryphopterus tortugae* (Jordan, 1904)

FEATURES: Color corresponds to habitat. Translucent to tannish and sparsely to densely marked. All with spotted eyes, a white "bridle" from jaw to preopercle, a dark triangular spot or single-pronged bar over opercle, few to many white and orangish spots on sides, dark spots that may form Xs, and a variable dark bar at caudal-fin base. Usually with no spot on lower pectoral-fin base. HABITAT: FL Keys, Bermuda, Bahamas, and Caribbean Sea. Demersal over clear- and shallow-water sandy bottoms around patch reefs to 100 ft.

Venezuelan Goby - *Coryphopterus venezuelae* Cervigón, 1966

FEATURES: Color corresponds to habitat. Translucent whitish to tannish and sparsely to densely marked. All with spotted eyes, variable orangish spots or stripes on head, white "bridle" from upper jaw to preopercle, pale to dark patch above opercle, few to many white and orangish spots on sides that may form Xs, and a dark bar or pair of spots on caudal-fin base. Most with a distinct spot on lower pectoral-fin base. HABITAT: SE FL, Bermuda, Bahamas, and Caribbean Sea. Demersal over sandy and muddy bottoms.

Toadfish Goby - *Cryptopsilotris batrachodes* (Böhlke, 1963)

FEATURES: Head with a brown band from between eyes to pectoral-fin base and a white patch on top. Pectoral fins with a dark semi-circle. Body translucent with irregular dark spots and saddles. Dorsal and caudal fins translucent with broad, dark bands. HABITAT: Bahamas and Caribbean Sea. Demersal over shallow-water sandy bottoms with small rubble to 20 ft. NOTE: Previously known as *Ctenogobius batrachodes*.

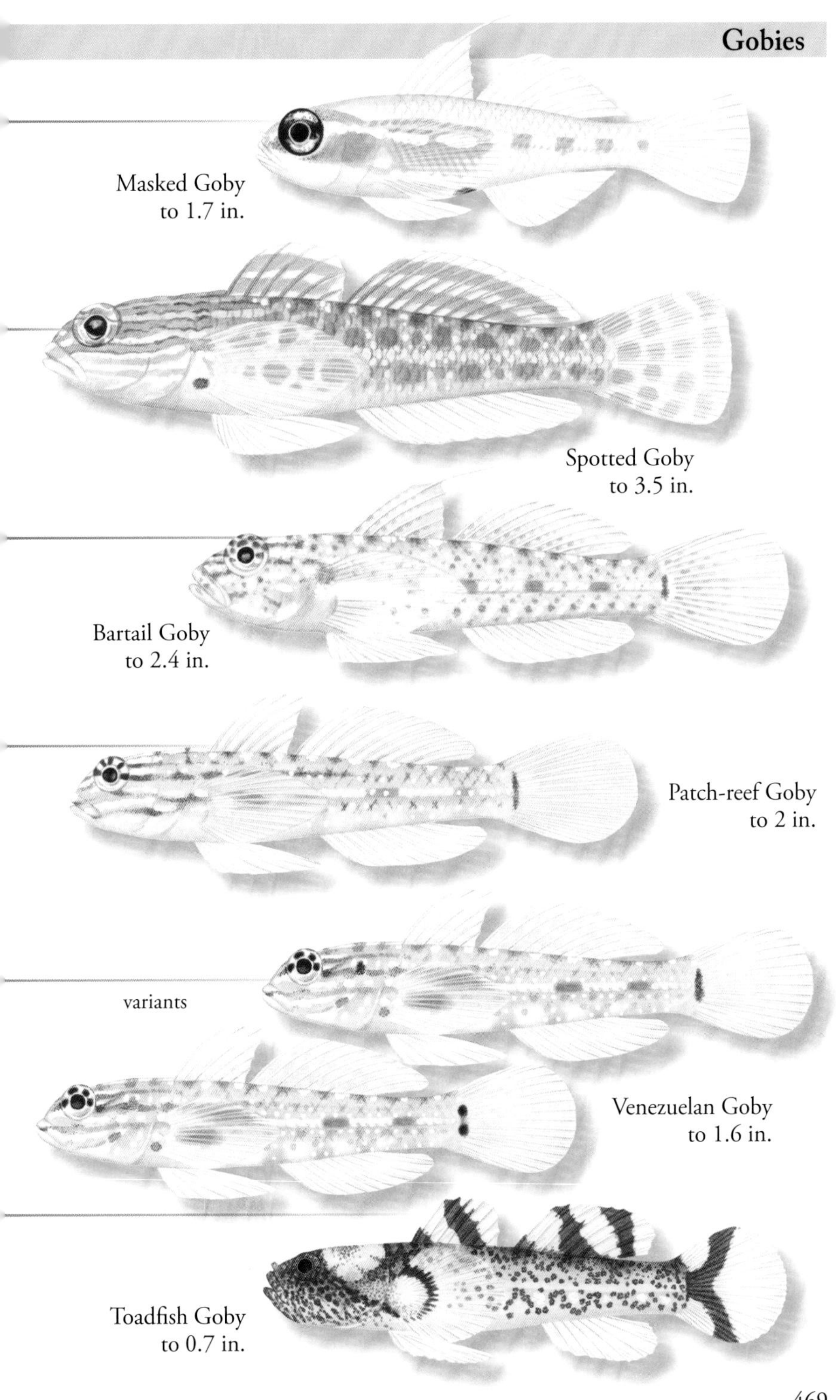
Masked Goby
to 1.7 in.
Spotted Goby
to 3.5 in.
Bartail Goby
to 2.4 in.
Patch-reef Goby
to 2 in.
variants
Venezuelan Goby
to 1.6 in.
Toadfish Goby
to 0.7 in.

Darter Goby - *Ctenogobius boleosoma* (Jordan & Gilbert, 1882)

FEATURES: Color highly variable. Dark bars on snout. Dark blotch above pectoral-fin base. Four to five dark blotches along midline that merge with V-shaped saddles. Dark spot at caudal-fin base—blackish in males. Dorsal and caudal fins banded. Upper caudal fin orange in males. HABITAT: DE to FL, Gulf of Mexico, Bermuda, Bahamas, and Caribbean Sea to southern Brazil. Demersal over muddy bottoms of brackish bays and estuaries and in fresh water. Also in lower estuaries and around barrier islands.

Blotchcheek Goby - *Ctenogobius fasciatus* Gill, 1858

FEATURES: Tannish dorsally; whitish below. Brownish red streaks from mouth through eyes. Lower preopercular corner brown to black. Brownish mark at pectoral-fin base. Upper sides with wavy, broken brownish red streaks and spots. Lower sides with a series of black blotches. Dorsal fins and upper caudal fin banded. Lower caudal-fin lobe blackish. Third dorsal-fin spine elongate in males. HABITAT: E FL; eastern and southern Caribbean Sea. Demersal in low-salinity estuaries and lower rivers to about 30 ft.

Slashcheek Goby - *Ctenogobius pseudofasciatus* (Gilbert & Randall, 1971)

FEATURES: Tannish. Angular dark mark from corner of mouth to preopercular corner. Dark bands on head. Body with irregular cross-hatching dorsally and four dark blotches along midline. Distinct dark spot on caudal peduncle. Dorsal and caudal fins banded. HABITAT: E FL and Belize to Trinidad. Demersal in shallow brackish and marine waters to about 50 ft. Also in rivers.

Dash Goby - *Ctenogobius saepepallens* (Gilbert & Randall, 1968)

FEATURES: Translucent whitish gray to bluish white with iridescent reflections. May have diffuse to distinct brownish to reddish brown spots dorsally. All with a variable dark bar below eye, a dark triangle-shaped mark on lower portion of opercle, and five elongate "dashes" along midline. Dashes may be large to very faint. Fins variably streaked. Third dorsal-fin spine elongate. HABITAT: FL, Bahamas, and Caribbean Sea. Demersal over sandy bottoms to about 150 ft. BIOLOGY: Share burrows with shrimps.

Freshwater Goby - *Ctenogobius shufeldti* (Jordan & Eigenmann, 1887)

FEATURES: Tannish yellow to light brown. Dark band from corner of mouth to upper portion of opercle. Head and upper sides with dark flecks. Four squarish blotches along midline and a dark spot at caudal-fin base. Dorsal and caudal fins banded. HABITAT: NC to Veracruz, Mexico. Demersal in low-salinity bays, estuaries, and in fresh water.

Emerald Goby - *Ctenogobius smaragdus* (Valenciennes, 1837)

FEATURES: Tannish with irregular dark smudges dorsally and along midline. Dark blotch above pectoral-fin base. Dark-ringed greenish spots scattered on head and sides. Dorsal, caudal, and pectoral fins banded. Males with elongate third dorsal-fin spine and very long caudal fin. HABITAT: SC to SW FL, Cuba, Puerto Rico to Guadeloupe, and Cumana, Venezuela, to southern Brazil. Demersal over shallow-water muddy bottoms.

♀ Darter Goby
to 3 in.

Blotchcheek Goby
to 3.6 in.

Slashcheek Goby
to 2.2 in.

Dash Goby
to 2 in.

Freshwater Goby
to 4.3 in.

♀ Emerald Goby
to 4 in.

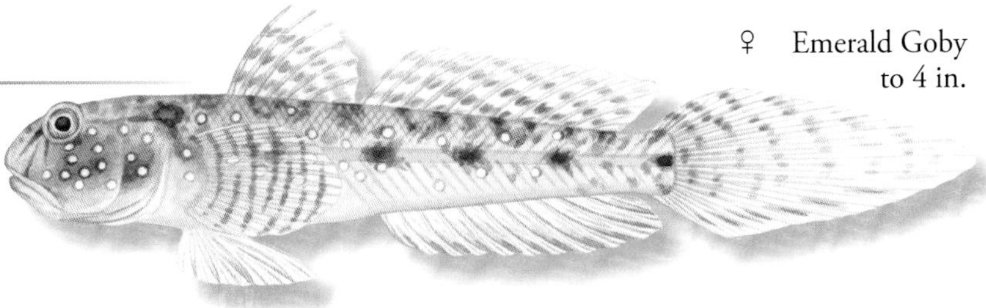

Marked Goby - *Ctenogobius stigmaticus* (Poey, 1860)

FEATURES: Tannish. Cheeks with three to four elongate, dark spots. A large dark blotch present on shoulder above pectoral-fin base. Reddish smudges on upper head, along back, and along midline. Pale bars on sides. First dorsal fin with at least one elongate spine. HABITAT: SC to S FL, Yucatán, Cuba to Guadeloupe, and Panama to S Brazil. Demersal over shallow-water, coastal, muddy, and sandy bottoms to 30 ft.

Spottail Goby - *Ctenogobius stigmaturus* (Goode & Bean, 1882)

FEATURES: Pale tannish to grayish with irregular, darker smudges and lines on head and body. Dark bar below eyes. Preopercular margin dark. Dark spot at upper pectoral-fin base. Dark blotches along midline resemble commas. Dark spot at caudal-fin base. Fins banded. HABITAT: SE FL, Bahamas, N Cuba, Yucatán, Belize, Puerto Rico to Anguilla, Curaçao, and Bermuda. Demersal in a variety of shallow-water habitats.

Blacknose Goby - *Elacatinus atronasus* (Böhlke & Robins, 1968)

FEATURES: Snout black with a bright yellow oval mark on top. A narrow yellow stripe runs from top of eye to caudal-fin base, often with a break at rear. Lower sides with a wide blue black stripe that ends in a large black spot on caudal-fin base. Upper lip separated from snout. Pectoral fins with16 rays. HABITAT: Bahamas. Demersal over vertical coral ledges with holes and around undercut coral ledges from about 80 to 115 ft.

Cayman Cleaner Goby - *Elacatinus cayman* Victor, 2014

FEATURES: Snout black with a bright yellow V-shaped mark between eyes. Yellow stripe runs from top of eye to mid-caudal fin and is boarded above and below by narrow bluish stripes. Sides with a moderately broad black stripe. Top lip separated from snout. Pectoral fins with 16–17 rays. HABITAT: Cayman Islands. Demersal over coral heads, rocky rubble, and large sponges from near surface to about 165 ft.

Cayman Sponge Goby - *Elacatinus centralis* Victor, 2014

FEATURES: Snout grayish to black—lacks marks on top. Narrow, pale yellow stripe runs from top of eye to caudal-fin bas; stripe becomes whitish and very narrow posteriorly. Sides with a broad blackish gray stripe. Top lip separated from snout. Pectoral fins with 17–19 rays. HABITAT: Cayman Islands. Demersal in tubular sponges around reefs from near surface to about 165 ft.

Shortstripe Goby - *Elacatinus chancei* (Beebe & Hollister, 1933)

FEATURES: Snout and underside of head yellow to grayish yellow. Cheeks pinkish to red. Irises yellow with a black inner stripe. Short yellow stripe boarded above and below by black stripe runs behind eye and fades over pectoral fins. Sides with a very broad blackish stripe that terminates on caudal fin. Top lip separated from snout. Pectoral fins with 19–20 rays. HABITAT: Southern Bahamas and Antilles to Venezuela. Demersal inside or just outside tube sponges around reefs from about 15 to 390 ft.

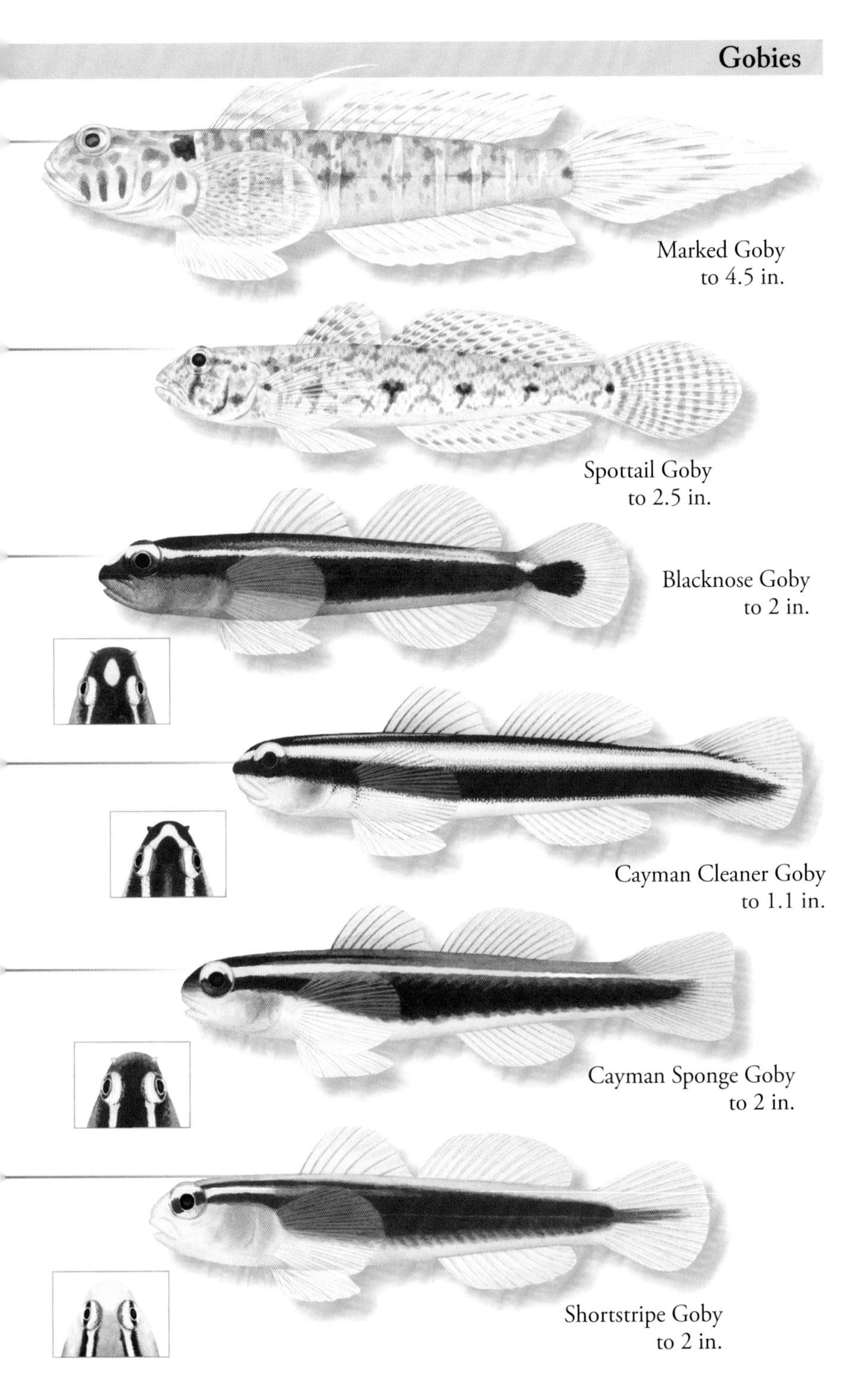

Marked Goby
to 4.5 in.

Spottail Goby
to 2.5 in.

Blacknose Goby
to 2 in.

Cayman Cleaner Goby
to 1.1 in.

Cayman Sponge Goby
to 2 in.

Shortstripe Goby
to 2 in.

Belize Sponge Goby - *Elacatinus colini* Randall & Lobel, 2009

FEATURES: Snout yellow brown to yellowish green with a bright yellow bar on top. A narrow stripe extends from top of eye to upper caudal fin; stripe yellow just behind eyes becoming bright white. Broad black stripe on lower sides that extends onto caudal fin. Upper lip separated from snout. Pectoral fins with 17–18 rays. HABITAT: Belize to Honduras. Demersal over sponges in lagoons and on reefs from about 5 to 55 ft.

Sharknose Goby - *Elacatinus evelynae* (Böhlke & Robins, 1968)

FEATURES: Snout black with a white or yellow V-shaped mark that extends through upper margins of eyes and onto a broad pale or dark blue stripe on upper sides. A broad black stripe extends from lower snout onto lower caudal fin. Upper lip connected to snout. Pectoral fins with15–18 rays; usually 16. HABITAT: Bahamas and Caribbean Sea. Demersal over coral heads, rocky rubble, and large sponges to about 240 ft.

Cleaner Goby - *Elacatinus genie* (Böhlke & Robins, 1968)

FEATURES: Snout black with a bright yellow V-mark that extends through upper eyes and onto a broad white stripe on upper sides. A moderately broad black stripe extends from lower snout and along lower sides and onto caudal fin. Upper lip separated from snout by a deep groove. Pectoral fins with 16–18 rays. HABITAT: Bahamas and Cuba. Demersal over coral, rocks, sponges, and under ledges from near surface to about 100 ft.

Yellowline Goby - *Elacatinus horsti* (Metzelaar, 1922)

FEATURES: Snout yellowish brown to brown without marks on top. Dark brown dorsally; slightly paler below. A distinct yellow to whitish stripe extends from upper portion of eye to caudal-fin base. HABITAT: Bahamas and Caribbean Sea. Yellow form occurs in the northeastern Bahamas, Cayman Islands, Curaçao, Bonaire, and Panama; the white form occurs off Hispaniola to Swan Island and Navassa Island. Demersal in tube sponges and massive barrel sponges from near surface to about 130 ft.

Barsnout Goby - *Elacatinus illecebrosus* (Böhlke & Robins, 1968)

FEATURES: Snout black with a bright yellow or white bar on top. Bright yellow or white stripe from upper portion of eye and onto a grayish or bluish stripe on upper side. A moderately broad to broad black stripe extends from lower snout and along lower sides and onto caudal fin. Upper lip connected to snout. Pectoral fins with 16–18 rays. HABITAT: Quintana Roo to Belize, and Nicaragua to Colombia. Demersal from near surface to about 100 ft.

Caribbean Neon Goby - *Elacatinus lobeli* Randall & Colin, 2009

FEATURES: Snout black with a bright blue bar in front of eyes. Bright blue stripe extends from top of eye onto a bluish stripe on upper side. A black stripe extends from lower snout to lower caudal fin; stripe narrow anteriorly, very broad posteriorly. Upper lip connected to snout. Pectoral fins with 15–17 rays. HABITAT: Belize to Honduras. Demersal over corals on rocky bottoms from near surface to about 90 ft.

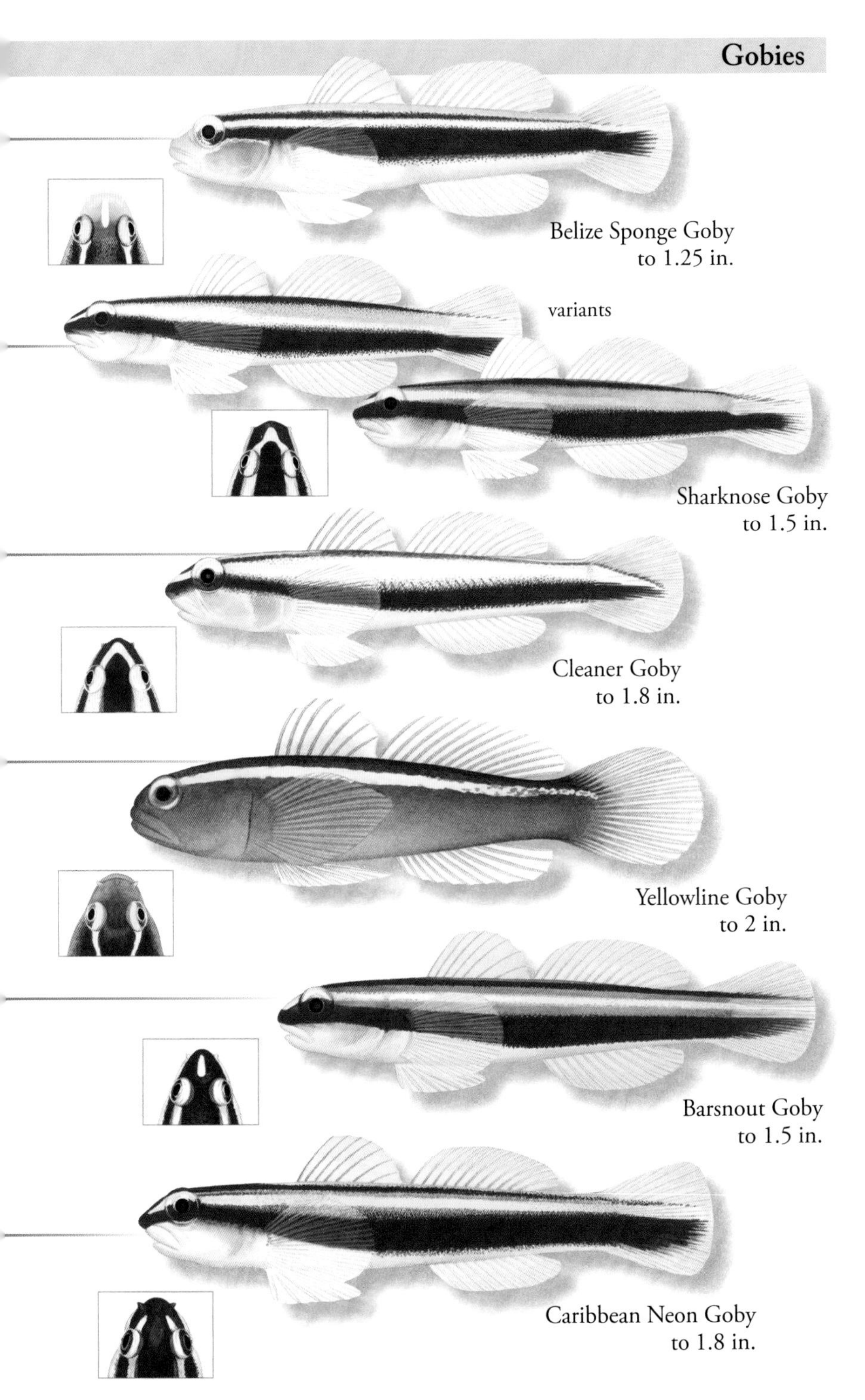

Belize Sponge Goby
to 1.25 in.

Sharknose Goby
to 1.5 in.

Cleaner Goby
to 1.8 in.

Yellowline Goby
to 2 in.

Barsnout Goby
to 1.5 in.

Caribbean Neon Goby
to 1.8 in.

Lori's Goby - *Elacatinus lori* Colin, 2002

FEATURES: Snout blackish with a narrow bright blue to whitish bar on top. A narrow bright blue to whitish stripe runs from top of eye to upper caudal-fin base. A broad black stripe runs from snout to caudal-fin base. Ventral area pale to purplish. Upper lip connected to snout. Pectoral fins with 18–19 rays. HABITAT: Belize to Honduras. Demersal inside sponges around reefs from near surface to about 100 ft.

Spotlight Goby - *Elacatinus louisae* (Böhlke & Robins, 1968)

FEATURES: Snout black with a small, round, white spot on top. A narrow yellow to bluish white stripe runs from top of eye onto a grayish stripe on upper side. A moderately broad blackish stripe runs from snout to caudal-fin base where it becomes a large spot. Upper lip separated from snout. Pectoral fins with 17–19 rays. HABITAT: Bahamas, Cuba, Haiti, Dominican Republic, Jamaica, Cayman Islands, Belize, and Honduras. Demersal in large to massive tube sponges around reefs from about 50 to 360 ft.

Neon Goby - *Elacatinus oceanops* Jordan, 1904

FEATURES: Snout blackish with a bright blue wedge in front of eyes. A broad bright blue stripe runs from top of eye to upper caudal fin. A broad black stripe runs from snout to lower caudal fin. Upper lip connected to snout. Pectoral fins with 16–18 rays. HABITAT: S FL, Flower Garden Banks, Campeche Bay, and Cuba. Demersal over corals, rocks, and sometimes sponges to about 130 ft.

Broadstripe Goby - *Elacatinus prochilos* (Böhlke & Robins, 1968)

FEATURES: Snout blackish with a white V-shape mark that extends through upper eyes and onto upper caudal fin. A broad black stripe runs along lower sides onto caudal fin; stripe very broad behind pectoral fins. Upper lip separated from snout. Pectoral fins with 17–19 rays. HABITAT: Veracruz, Mexico, Quintana Roo to Honduras, and Jamaica to Trinidad. Demersal over corals and sponges from near surface to about 180 ft.

Yellownose Goby - *Elacatinus randalli* (Böhlke & Robins, 1968)

FEATURES: Snout tannish yellow to brownish yellow with a bright yellow bar on top. Lower head yellowish. A bright yellow stripe runs from top of eye to upper caudal fin. A broad blackish stripe runs from lower eye to caudal fin. A narrow blue stripe runs from below eye to pectoral fin. Upper lip not connected to snout. Pectoral fins with 17–19 rays. HABITAT: Puerto Rico, Dominica to Tobago, coast of Venezuela, and recorded from S Belize. Demersal over coral heads from near surface to about 170 ft.

Serranilla Goby - *Elacatinus serranilla* Randall & Lobel, 2009

FEATURES: Snout black with a bright white bar on top. Narrow white stripe runs from top of eye to upper caudal fin. A narrow black stripe runs from below eye to lower caudal fin. Upper lip separated from snout. Pectoral fins with 17–18 rays. HABITAT: Southern Dominican Republic and Haiti, and Jamaica to off Honduras. Demersal in sponges from about 50 to 80 ft.

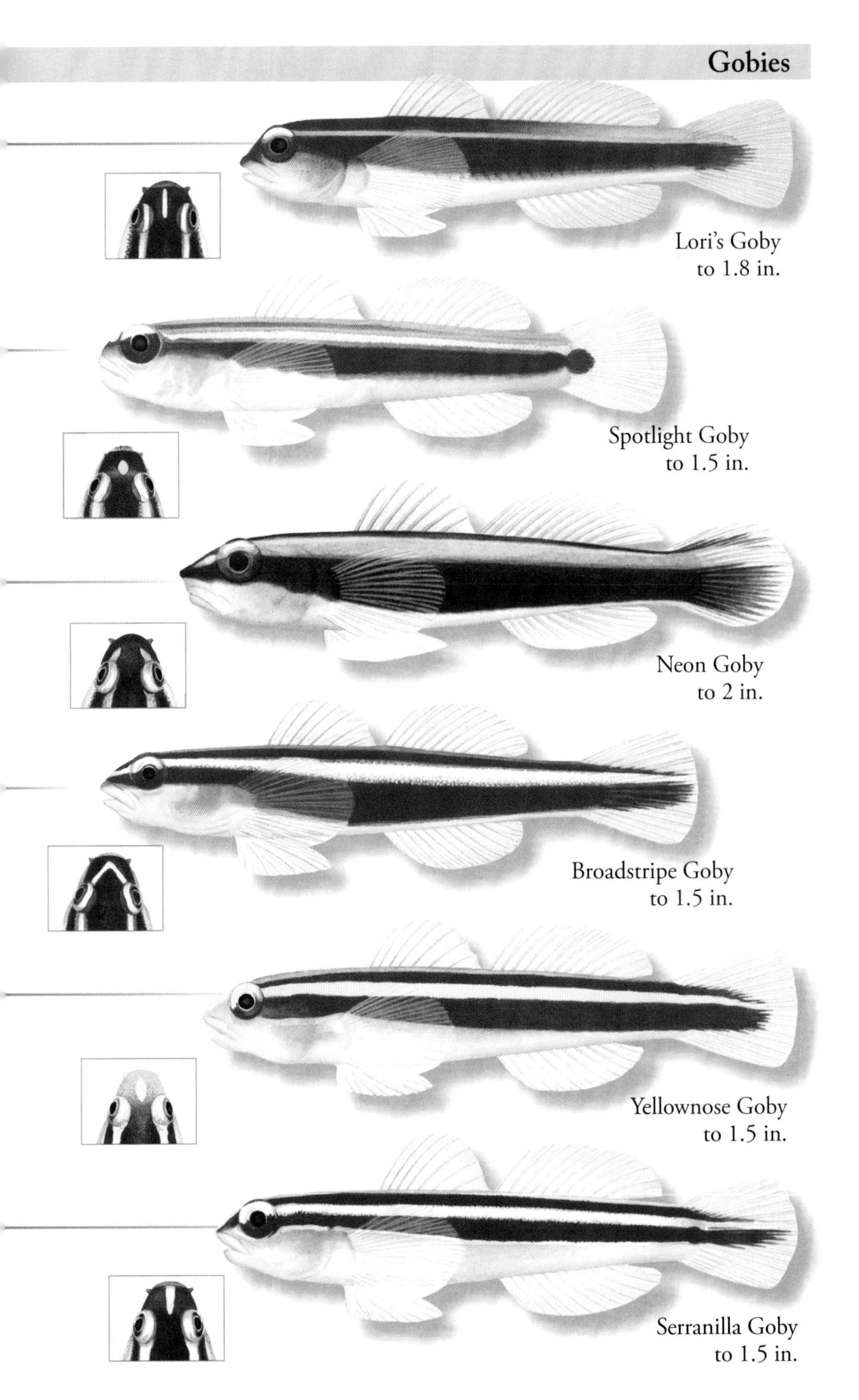
Lori's Goby
to 1.8 in.
Spotlight Goby
to 1.5 in.
Neon Goby
to 2 in.
Broadstripe Goby
to 1.5 in.
Yellownose Goby
to 1.5 in.
Serranilla Goby
to 1.5 in.

Yellowprow Goby - *Elacatinus xanthiprora* (Böhlke & Robins, 1968)

FEATURES: Snout blackish to brownish with a yellow bar on top. A narrow yellow stripe runs from top of eye to upper caudal fin. A broad black stripe runs from eye to caudal fin. Upper lip separated from snout. Pectoral fins with 19–20 rays. HABITAT: NC to Crystal River, FL, Flower Gardens Banks, and Campeche, Mexico, to the Yucatán Peninsula. Demersal inside and on tube sponges around reefs from about 60 to 80 ft.

Bicolored Sponge Goby - *Evermannichthys bicolor* Thacker, 2001

FEATURES: Uniformly black to purplish black dorsally; whitish below midline. Pink wash on cheeks. Fins translucent. Mouth large and oblique. First dorsal fin with five to seven spines. Three to four rough scales along ventral tail base. HABITAT: Western Haiti, Navassa Island, and Jamaica. Demersal inside barrel sponges from about 60 to 135 ft.

Roughtail Goby - *Evermannichthys metzelaari* Hubbs, 1923

FEATURES: Translucent with irregular dark reddish brown bars and smudges along sides. First dorsal fin with four or five spines. Rough scales along ventral tail base. HABITAT: Bahamas, Belize, and Caribbean Sea. Absent from Cuba and Cayman Islands. Demersal inside loggerhead sponges from near surface to about 135 ft.

Pugnose Goby - *Evermannichthys silus* Böhlke & Robins, 1969

FEATURES: Reddish orange and peppered with minute darker flecks. Irises may be blackish. Dorsal and anal fin margins with blackish margins. Snout short and blunt. First dorsal fin with six or seven spines. Two to four rough scales along ventral tail base. HABITAT: Samana Cay, Bahamas, and Venezuelan coast from Curaçao, Bonaire, and Los Roques. Demersal inside sponges from near surface to about 100 ft.

Sponge Goby - *Evermannichthys spongicola* (Radcliff, 1917)

FEATURES: Translucent with irregular blackish gray bands, bars, and smudges on head and body. Markings become more broken posteriorly. First dorsal fin with six or seven spines. Scales absent along ventral tail base. HABITAT: NC to St. Joseph Bay, FL. Also Campeche, Mexico. Demersal inside and over loggerhead and other similar sponges from about 40 to 190 ft.

Lyre Goby - *Evorthodus lyricus* (Girard, 1858)

FEATURES: Tannish to grayish. Iridescent area at pectoral-fin base. Five or six dark smudges along dorsal profile that merge with smudges along midline. Smudge under first dorsal fin darkest, merges into lower dorsal-fin base. Two distinct, squarish to teardrop-shaped blotches at caudal-fin base. Snout short, rounded. Males with elongate dorsal-fin spines, long caudal fin, and pinkish stripe on upper and lower caudal-fin lobes. HABITAT: VA to FL, Gulf of Mexico, and Caribbean Sea to southern Brazil. Absent from Bahamas and Cayman Islands. Demersal over shallow-water muddy bottoms in bays and estuaries and in tidal fresh water.

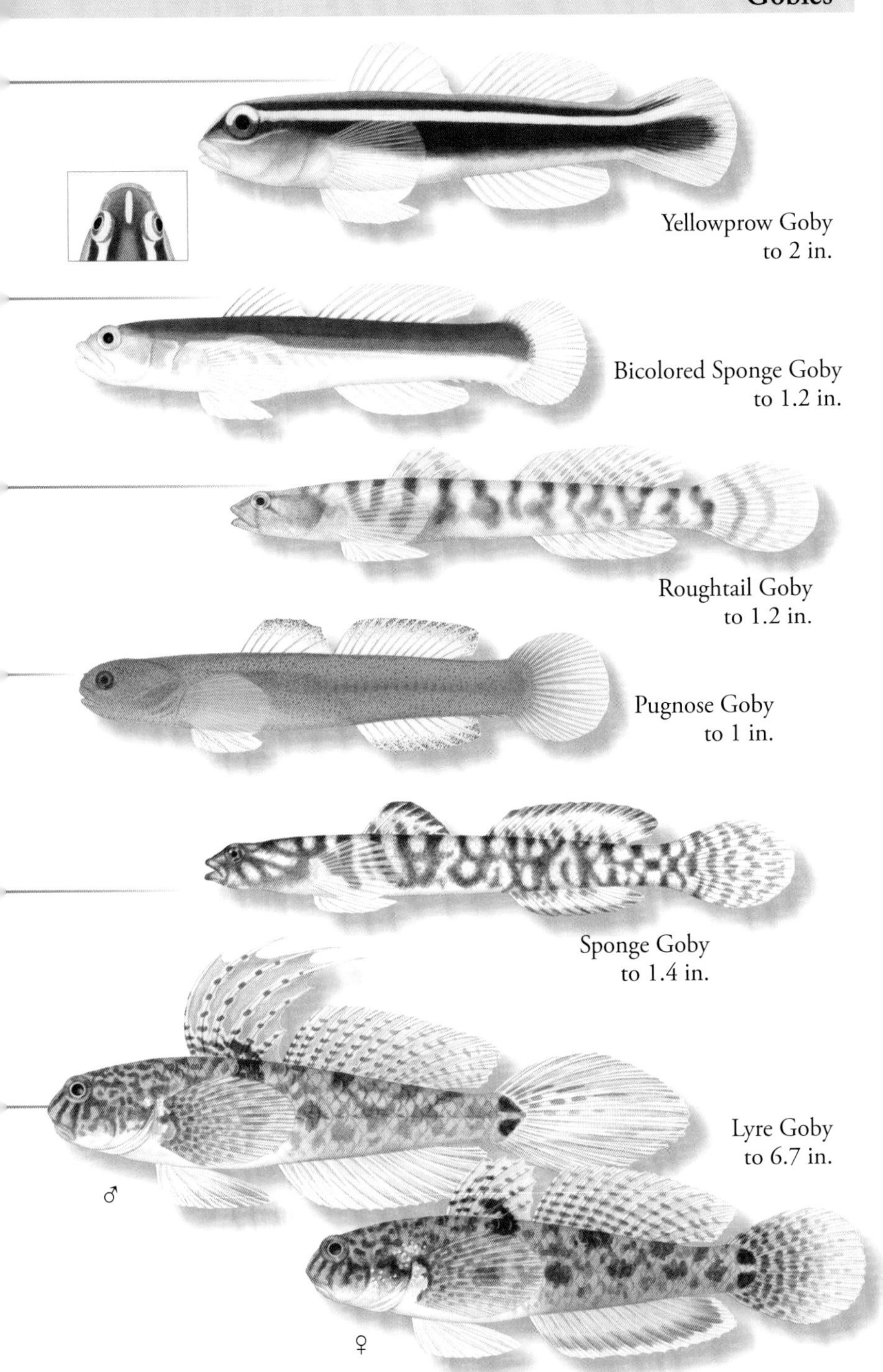

Yellowprow Goby
to 2 in.

Bicolored Sponge Goby
to 1.2 in.

Roughtail Goby
to 1.2 in.

Pugnose Goby
to 1 in.

Sponge Goby
to 1.4 in.

Lyre Goby
to 6.7 in.

Nineline Goby - *Ginsburgellus novemlineatus* (Fowler, 1950)

FEATURES: Color, pattern vary slightly. Head shades of gray to brown. Body blackish brown to blackish blue with nine bright blue stripes from top of head to caudal peduncle. Cheeks reddish. Irises radiating yellow to blackish. HABITAT: Bahamas and Antilles to coast of Venezuela. Also Belize and Panama. Demersal under sea urchins in clear-water upper tidal zones.

Goldspot Goby - *Gnatholepis thompsoni* Jordan, 1904

FEATURES: Shades of pale gray to tan or pinkish with rows of brownish spots on sides. Spots may overlie dark smudges. Narrow dark bar between and below eyes. Typically with a bright golden spot almost encircled by a blackish brown to pale brown ring over pectoral-fin base. Snout short, steeply sloping. HABITAT: FL, Flower Garden Banks, southern Gulf of Mexico, Bermuda, Bahamas, and Caribbean Sea. Also in eastern Atlantic. Demersal over sand and rubble bottoms from near surface to about 290 ft.

Violet Goby - *Gobioides broussonnetii* Lacepède, 1800

FEATURES: Bluish gray to purplish brown dorsally. Variously creamy to whitish on sides and below. Dark chevron-like marks along midline. Mouth large, gaping. Eyes small. Dorsal and anal fins continuous with pointed caudal fin. Body elongate. HABITAT: SC to FL, Gulf of Mexico, Greater Antilles, and Panama to S Brazil. Demersal over shallow-water muddy bottoms in bays, estuaries, and river mouths to fresh water.

Highfin Goby - *Gobionellus oceanicus* (Pallas, 1770)

FEATURES: Pale brown to gray dorsally; white on abdomen. Dark bar below eyes. Large, blackish ocellated spot under first dorsal fin. Smaller, paler ocellated spots anterior and posterior to large spot. Faint smudges along sides. Dark spot at caudal-fin base. First dorsal fin spines elongate. HABITAT: VA to FL, Gulf of Mexico, and Caribbean Sea to Brazil. Demersal over muddy bottoms from near shore to about 150 ft.

Rockcut Goby - *Gobiosoma grosvenori* (Robins, 1964)

FEATURES: Tannish to whitish with nine irregular speckled and blotched bars from nape to caudal peduncle. Bars may be very obscure. Dark dashes bisect bars along midline. Variable dark bars radiate from eyes. Head moderately depressed. Second dorsal fin with one spine and nine rays. Scales cover posterior body. HABITAT: S FL, Veracruz to Yucatán, Bahamas, Jamaica, Puerto Rico, Virgin Islands, and Golfo de Cariaco and Golfo de Paria, Venezuela. Demersal over nearshore soft and mixed bottoms.

Hildebrand's Goby - *Gobiosoma hildebrandi* (Ginsburg, 1939)

FEATURES: Yellowish tan with about six oblique and broken bars on sides that are bisected with darker dashes along midline. Brownish speckles and blotches on interspaces. Brownish bands radiate from eyes. First dorsal-fin spine elongate in males. HABITAT: Bocas del Toro, Panama, and the Caribbean and Pacific Panama Canal zone. Demersal over shallow-water brackish and freshwater muddy bottoms.

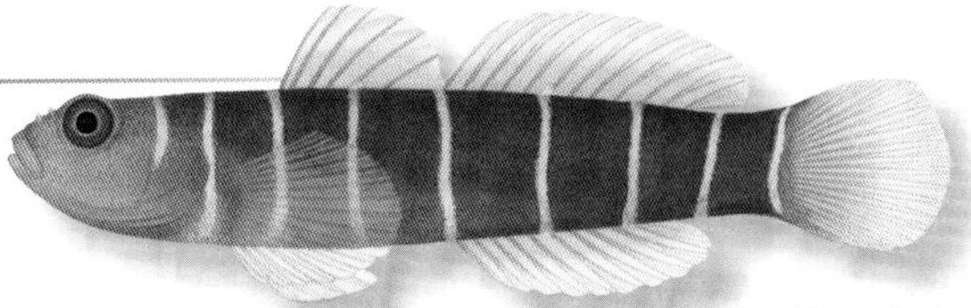

Nineline Goby
to 1 in.

Goldspot Goby
to 3 in.

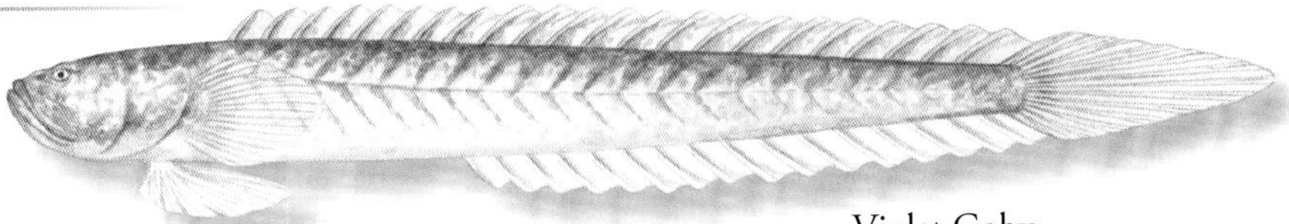

Violet Goby
to 2 ft.

Highfin Goby
to 8 in.

Rockcut Goby
to 1.5 in.

♀ Hildebrand's Goby
to 1.8 in.

Code Goby - *Gobiosoma robustum* Ginsburg, 1933

FEATURES: Color, pattern highly variable. Shades of pale tan to pale gray with brown, gray, or blackish vermiculated bars on sides that are intersected along midline with darker dots and dashes. Head variably mottled, spotted, and barred. Scales absent. HABITAT: Jacksonville, FL, to FL Keys, and Gulf of Mexico to Yucatán Peninsula. Demersal over seagrass beds, oyster beds, and algal mats in shallow coastal waters.

Maracaibo Goby - *Gobiosoma schultzi* (Ginsburg, 1944)

FEATURES: Greenish brown to grayish brown with irregular paler blotches, spots, and broken bars, or almost uniformly colored. Anal fin may be reddish at base. Dark bar at caudal-fin base. A small patch of scales present behind pectoral fins, and 7–14 rows of scales on tail base. HABITAT: Lake Maracaibo, Venezuela. Demersal over sand, gravel, rubble, and rocks in shallow-water fresh and brackish coastal zones.

Vermiculated Goby - *Gobiosoma spes* (Ginsburg, 1939)

FEATURES: Females mottled in shades of brown with about five irregular blackish bars on sides; dark bands radiate from eyes; blackish bar at caudal-fin base. Males dark brown to almost black, with irregular paler blotches and vermiculations; may have faint bars; first dorsal-fin spine elongate. A small patch of scales present behind pectoral fins, and 7–16 rows of scales on tail base. HABITAT: Lake Maracaibo, Venezuela. Demersal over shallow-water muddy bottoms in coastal bays, estuaries, and nearby fresh waters.

Isthmian Goby - *Gobiosoma spilotum* (Ginsburg, 1939)

FEATURES: Females yellowish anteriorly, becoming dark posteriorly; blackish spots, blotches and speckles cover body; dark bands radiate from eyes; fins irregularly spotted and banded. Males yellowish anteriorly, grayish to blackish posteriorly; blackish spots and bands on head and along bases of dorsal fins. All with blackish dashes along midline. Scales on posterior body form a wedge-shape behind pectoral fins. HABITAT: Entrance to the Panama Canal. Demersal over shallow-water sandy, muddy, and rocky bottoms.

Yucatan Goby - *Gobiosoma yucatanum* Dawson, 1971

FEATURES: Tannish to pale grayish tan with irregular dark blotches, tiny dark specks, and four irregular blackish to brownish saddles. One or two dark bands below eyes. Blackish bar at caudal-fin base. A patch of scales behind pectoral-fin base, and 3–11 rows of scales on tail base. HABITAT: Chetumal, Quintana Roo, to Brus Laguna, Honduras. Demersal in shallow-water estuaries, lagoons, rivers, and around mangroves.

Paleback Goby - *Gobulus myersi* Ginsburg, 1939

FEATURES: Pale tannish to creamy white dorsally; brownish below. Upper sides with scattered, dark flecks and two poorly defined saddles. May have obscure, dark blotch on upper pectoral-fin base. Dark semi-circle on caudal-fin base. Body scaleless. HABITAT: NW FL, Grand Bahama Island, Lesser Antilles to Venezuela, Belize, and Panama to Colombia. Demersal around coral reefs over sandy bottoms from about 15 to 155 ft.

Code Goby
to 1 in.
Maracaibo Goby
to 1 in.
Vermiculated Goby
to 1.6 in.
♀
♂
Isthmian Goby
to 1 in.
♀
♂
Yucatan Goby
to 1 in.
Paleback Goby
to 7 in.

Crested Goby - *Lophogobius cyprinoides* (Pallas, 1770)

FEATURES: Females tannish with brownish bands radiating from eyes and onto back and sides; sides with brownish saddles and blotches; first dorsal fin with a blackish smudge and an orange blotch. Males dark grayish green; become blackish while breeding; dorsal fin nearly uniformly dark. All with prominent crest on head. HABITAT: S FL, western and southern Gulf of Mexico, Bermuda, Bahamas, and Caribbean Sea. Demersal in shallow, protected, estuarine, and coastal waters, including bays and mangrove areas.

Mahogany Goby - *Lythrypnus crocodilus* (Beebe & Tee-Van, 1928)

FEATURES: Whitish with blackish spots on head and blackish bars on body that lack pale centers. Narrow lines on interspaces between spots and bars. Large blackish blotch at pectoral-fin base. Dorsal, caudal, and anal fins reddish with pale margins. HABITAT: Southern Bahamas, and Caribbean Sea. Demersal over coral in spur-and-groove areas and shallow-water bottoms with surge to about 75 ft.

Dwarf Goby - *Lythrypnus elasson* Böhlke & Robins, 1960

FEATURES: Rusty red with dense dark speckling and purple to orange spots and blotches on lower head. May also have scattered white spots. Larger specimens may be uniformly dark brown to blackish. First two dorsal-fin spines elongate. HABITAT: NC, NW FL, Bahamas, and Caribbean Sea. Demersal over isolated coral heads, under ledges, and on reef dropoffs from about 30 to 500 ft.

Diphasic Goby - *Lythrypnus heterochroma* Ginsburg, 1939

FEATURES: Whitish irregular orange spots and bars on head, orange saddles along back, broad orange bars on abdomen, and orange bars on lower posterior body. Orange areas speckled and bordered with black. White bar on abdomen. Scales with black margins. HABITAT: Bahamas, Cayman Is., Jamaica, Haiti, Belize, and off Colombia. Demersal over spur-and-groove coral bottoms and isolated patch reefs to about 50 ft.

Pygmy Goby - *Lythrypnus minimus* Garzón-Ferreira & Acero P., 1988

FEATURES: Semi-translucent orange to reddish orange with about 10 white, narrow, broken saddles or bars on sides. Lower head with orange to reddish orange blotches. Orange areas overlaid with darker dots. Dorsal and caudal fins banded. Males with a large black blotch at pectoral-fin base; reduced to absent in juveniles and females. HABITAT: Bahamas, Saba Bank to Dominica, Belize, Isla de Providencia, Colombia, and Curaçao to Bonaire. Demersal over coral reefs from near surface to about 360 ft.

Island Goby - *Lythrypnus nesiotes* Böhlke & Robins, 1960

FEATURES: Whitish to tannish with red or black spots on head, and red or black bars on body. Spots and bars peppered with black flecks. Narrow lines between spots and bars. Large black blotch at pectoral-fin base. Dorsal, caudal, and anal fins often reddish. HABITAT: GA to FL, Flower Garden Banks, southern Gulf of Mexico, Bahamas, and Caribbean Sea. Demersal over rocky bottoms and coral reefs to about 30 ft.

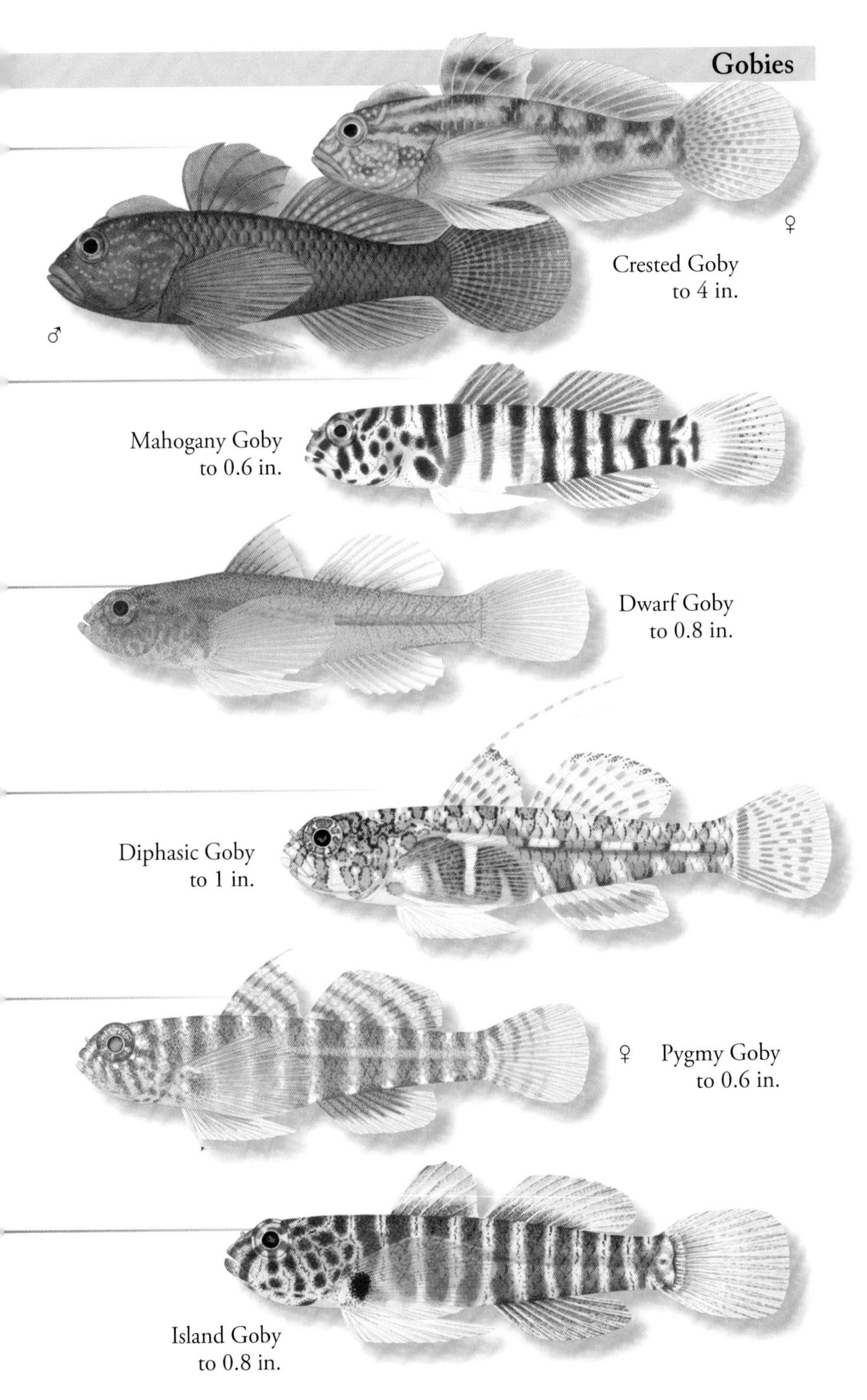
♀
Crested Goby
to 4 in.
♂
Mahogany Goby
to 0.6 in.
Dwarf Goby
to 0.8 in.
Diphasic Goby
to 1 in.
♀
Pygmy Goby
to 0.6 in.
Island Goby
to 0.8 in.

Okapi Goby - *Lythrypnus okapia* Böhlke & Robins, 1964

FEATURES: Translucent white with orange to brown spots on head and five to eight oblique and forked orange to brown bars on body. Bars may appear as broken spots. Spots and bars with small brown to black flecks along margins. Spiny dorsal fin red at margin. Dark-margined scales on posterior body. HABITAT: Bahamas, Cayman Islands, Haiti, US Virgin Islands, Turks and Caicos, and off Colombia. Demersal over patch reefs surrounded by sand from near surface to about 95 ft.

Convict Goby - *Lythrypnus phorellus* Böhlke & Robins, 1960

FEATURES: Translucent with three or four rows of brownish orange spots below eyes and irregular brownish orange bars on body that have pale centers and narrow interspaces. Spots and bars with blackish margins. Two indistinct brownish spots at pectoral-fin base. HABITAT: NC to FL Keys, Flower Garden Banks, N Cuba, Panama, Curaçao, Bonaire, and Santa Margarita Island. Demersal under coral ledges and on rubble to about 140 ft.

Bluegold Goby - *Lythrypnus spilus* Böhlke & Robins, 1960

FEATURES: Alternating orangish to gold and blue gray bars on head and body. Blue gray bars bisected by a dark line that may appear reflective. A large dark blotch covers most of pectoral-fin base. Dorsal and caudal fins spotted. First two dorsal-fin spines may be very long. HABITAT: NC to FL Keys, Flower Garden Banks, Bermuda, Bahamas, and Caribbean Sea. Demersal over coral reefs from about 3 to 320 ft.

Seminole Goby - *Microgobius carri* Fowler, 1945

FEATURES: Pinkish white with iridescent highlights. Bright blue stripe behind eyes. Dusky green stripe along dorsal profile. Broad yellow orange stripe on midline bordered above and below by blue stripes. Anterior dorsal-fin spines elongate. HABITAT: NC to NW FL, Yucatán, Bahamas, Antilles to Curaçao, Belize, and Panama to Colombia. Demersal over sandy bottoms from about 20 to 70 ft. BIOLOGY: Build burrows.

Banner Goby - *Microgobius microlepis* Longley & Hildebrand, 1940

FEATURES: Males grayish green dorsally; cheeks yellowish orange with iridescent blue streaks; dorsal fins rosy below, with a yellow streak above; some with a dusky to brownish stripe on upper sides. Females similarly colored on cheeks and dorsal fins, but with a white to yellowish oblong triangular area above anal fin. All with pinkish nuchal crest and pointed caudal fin. HABITAT: S FL, Bahamas, NE Cuba, Quintana Roo to Belize. Over shallow, sandy bottoms. BIOLOGY: Build burrows.

Signal Goby - *Microgobius signatus* Poey, 1876

FEATURES: Tannish dorsally; brownish along midline. Rows of bright blue spots on cheeks. A vertical white stripe bordered by black to brown above pectoral-fin base. Dorsal and anal fins pink with inner yellow and blue stripes. All with low nuchal crest. HABITAT: NW Cuba, Puerto Rico to Virgin Islands, E Panama to E Venezuela. Also off Honduras and Nicaragua. Demersal over shallow-water muddy and sandy bottoms.

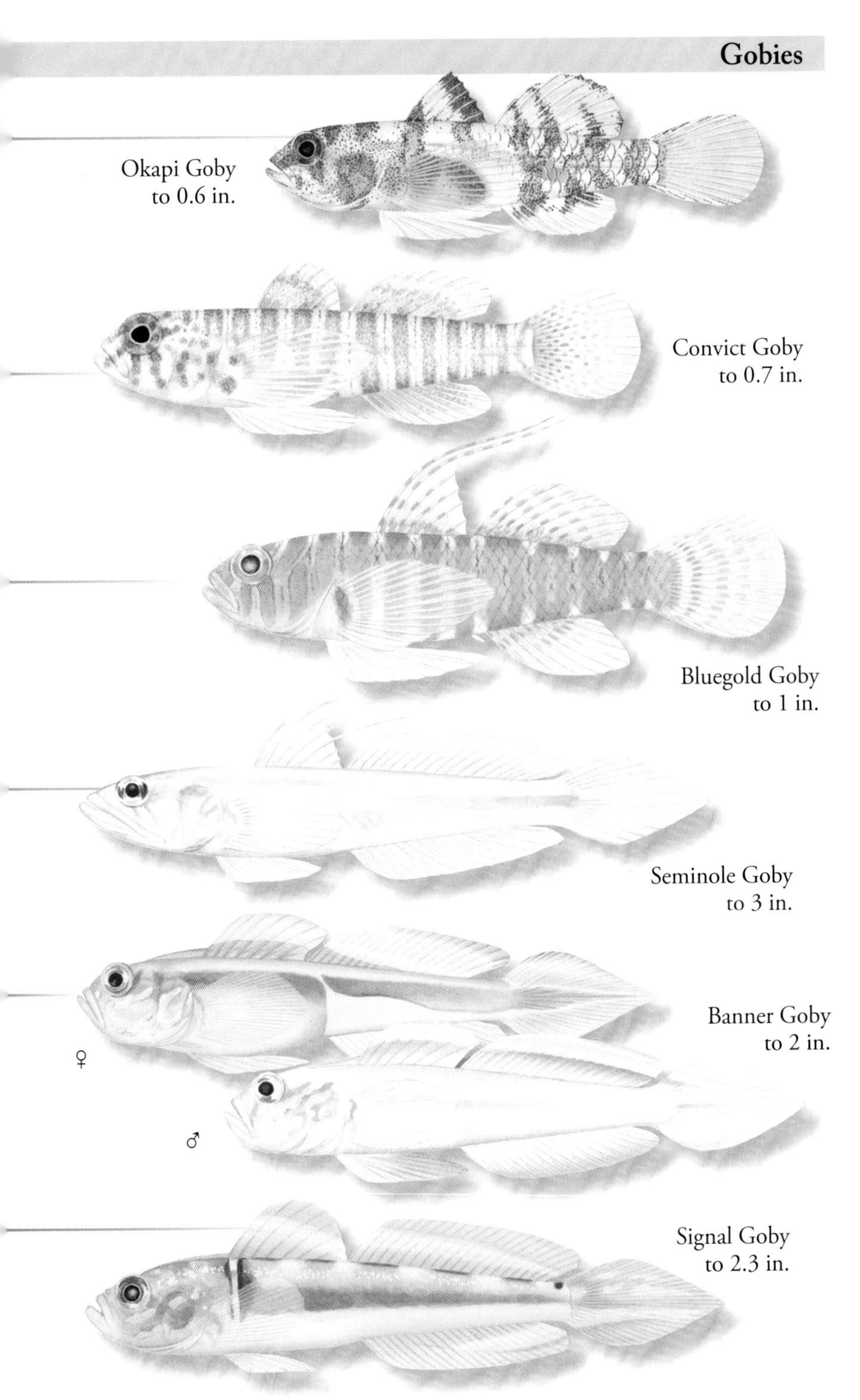
Okapi Goby
to 0.6 in.
Convict Goby
to 0.7 in.
Bluegold Goby
to 1 in.
Seminole Goby
to 3 in.
Banner Goby
to 2 in.
♀
♂
Signal Goby
to 2.3 in.

Orangespotted Goby - *Nes longus* (Nichols, 1914)

FEATURES: Color, pattern correspond to habitat. Pale with pairs of irregular brown blotches or bars along sides. May have few to many yellowish blotches between dark blotches. Head spotted and banded. Fins spotted to banded. First dorsal-fin spine elongate. Body elongate. HABITAT: S FL, Bermuda, Bahamas, and Caribbean Sea. Demersal over sandy bottoms from shore to about 30 ft. Usually in shrimp burrows.

Spotfin Goby - *Oxyurichthys stigmalophius* (Mead & Böhlke, 1958)

FEATURES: Pearly white to pearly gray with oblong orange to brown blotches dorsally. Four to five large, darker orange or brown blotches along midline. Head with spots and bands. Dorsal and upper caudal fins spotted. Spiny dorsal fin with large black blotch at rear. HABITAT: S FL, Bahamas, N Cuba, Puerto Rico to Curaçao, Veracruz to Honduras, Cayman Islands, and off Nicaragua and Colombia. Demersal over sandy and muddy bottoms from about 6 to 330 ft. May share shrimp burrows.

Dawson's Goby - *Priolepis dawsoni* Greenfield, 1989

FEATURES: Shades of white to cream with seven orange to brownish bars from anterior dorsal fin to peduncle. First two bars wider than pale bars. Bands above and below eyes, and across opercle. Dorsal fins spotted. Caudal fin spotted dorsally. Mouth large and oblique. Second dorsal-fin spine elongate. HABITAT: Dominica, Lesser Antilles, and Venezuela to Brazil. Demersal over sandy, rocky, and algal bottoms from near surface to about 215 ft. SIMILAR SPECIES: *Priolepis robinsi.*

Rusty Goby - *Priolepis hipoliti* (Metzelaar, 1922)

FEATURES: Dark reddish orange with narrow, pale bluish bars on head and body. Body densely peppered with minute, dark melanophores. Dorsal and caudal fins with spots forming bands. Mouth large and oblique. Second dorsal-fin spine elongate. HABITAT: S FL, FL Keys, Flower Garden Banks, Bermuda, Bahamas, and Caribbean Sea to Brazil. Demersal over reefs and under ledges about near surface to about 380 ft.

Scaleless Goby - *Psilotris alepis* Ginsburg, 1953

FEATURES: Pale gray with clusters of brownish black spots that form about eight saddles and blotches on sides. Spots form broad bands on head. White blotches between dark areas. Dorsal and caudal fins weakly banded. Pelvic fins separate. Scales and lateral line absent. HABITAT: Bahamas, Cuba, Cayman Islands, Jamaica, Yucatán to Belize, Panama, off Venezuela, and off St. Croix. Demersal over shallow-water corals and rocks.

Highspine Goby - *Psilotris celsa* Böhlke, 1963

FEATURES: Pale grayish to tannish with blackish to brownish bars on sides and pale spots on interspaces. Head reddish to brownish with large pale spots. Black flecks pepper head and body. Dark bar at caudal-fin base. First dorsal-fin spine elongate. Pelvic fins separate. Scales absent. HABITAT: Bermuda, Bahamas, Puerto Rico to Virgin Islands, scattered Belize to Colombia. Demersal over coral and rocky bottoms to about 100 ft.

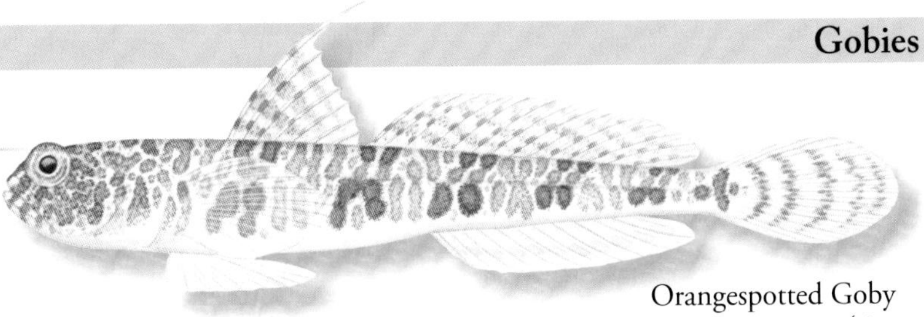

Orangespotted Goby
to 4 in.

Spotfin Goby
to 6.5 in.

Dawson's Goby
to 1 in.

Rusty Goby
to 1.6 in.

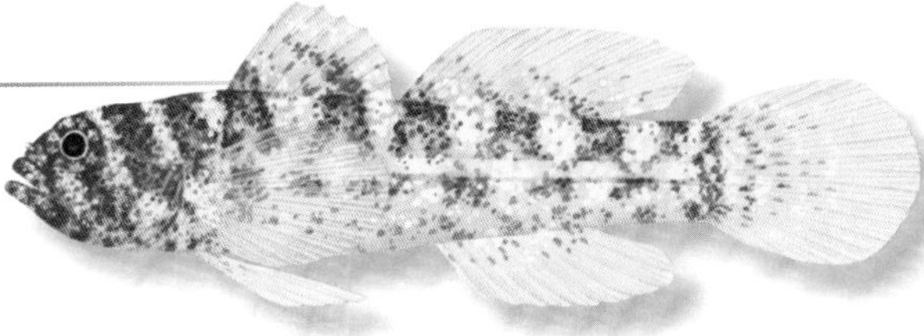

Scaleless Goby
to 1 in.

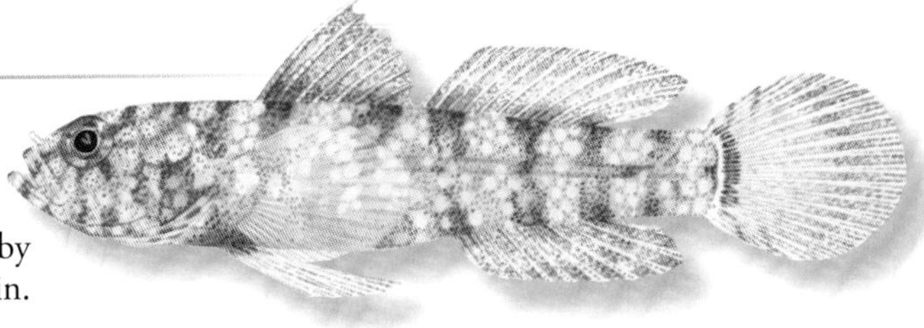

Highspine Goby
to 1.6 in.

Kaufman's Goby - *Psilotris kaufmani* Greenfield, Findley & Johnson, 1993

FEATURES: Pale bluish with golden brown spots, blotches, bars, and saddles that are sparsely to densely covered with blackish flecks. Whitish lines on interspaces. Dorsal and upper caudal-fin lobe spotted. Pectoral fins bluish. Dark bar on caudal-fin base. First two dorsal-fin spines elongate. Pelvic fins separate. Scales absent. HABITAT: Bahamas, Jamaica, Honduras, Isla de Providencia, and Bocas del Toro. Demersal over coral and soft bottoms from about 35 to 90 ft.

Clementine Splitfin Goby - *Psilotris vantasselli* Tornabene & Baldwin, 2019

FEATURES: Pearly with irregular yellow bars on head and sides that extend onto dorsal fins. Bars may be bordered by tiny black flecks. Caudal fin with two broad yellow bands. Eyes large with yellow to green irises. Body long and sender. Pelvic fins separate. Two scales present on caudal-fin base. HABITAT: Bahamas, Aruba, Curaçao, and Bonaire. Demersal over coral and rocks from about 50 to 520 ft.

Tusked Goby - *Risor ruber* (Rosén, 1911)

FEATURES: Translucent grayish to brownish with darker speckles that cluster to form saddles and blotches. Abdomen may have wide dark and pale bars. May have white spots on head. Mouth very small, located under blunt snout, and with protruding teeth. HABITAT: GA to FL, Flower Garden Banks, Yucatán Peninsula, Bahamas, and Caribbean Sea to Brazil. Demersal inside large sponges from near surface to about 400 ft.

Orangesided Goby - *Tigrigobius dilepis* (Robins & Böhlke, 1964)

FEATURES: Translucent yellowish to buff dorsally and posteriorly. Top of head white with red bands. Lower head with red spots. Abdomen white with two large black-bordered red blotches. Spinal column with alternating red, black, and white bars. HABITAT: Bahamas, Veracruz to Yucatán, and Caribbean Sea. Absent from Cuba. Demersal over coral heads, sponges, and rocky bottoms from about 16 to 100 ft.

Frecklefin Goby - *Tigrigobius gemmatus* (Ginsburg, 1939)

FEATURES: Blackish brown with eight to nine diffuse dark bars on sides. Head mottled with or without whitish spots on interspaces. May have clusters of white spots along back. Fins with dense speckling that may form bands on caudal fin, dark bases on dorsal fins, and dark a margin on anal fin. First dorsal-fin spine elongate in males. HABITAT: Bahamas and Caribbean Sea. Demersal in shallow-water holes over rocks and reefs.

Cayman Greenbanded Goby - *Tigrigobius harveyi* Victor, 2014

FEATURES: Head whitish to buff with a bright red band from snout through eyes to nape. Body shades of bright green to grayish green with 20–28 (usually 25) very narrow, pale green bars. Fins pale yellow to dusky yellow. First dorsal-fin spine elongate. HABITAT: Cayman Islands. Demersal over shallow-water coral reefs. Often under sea urchins.

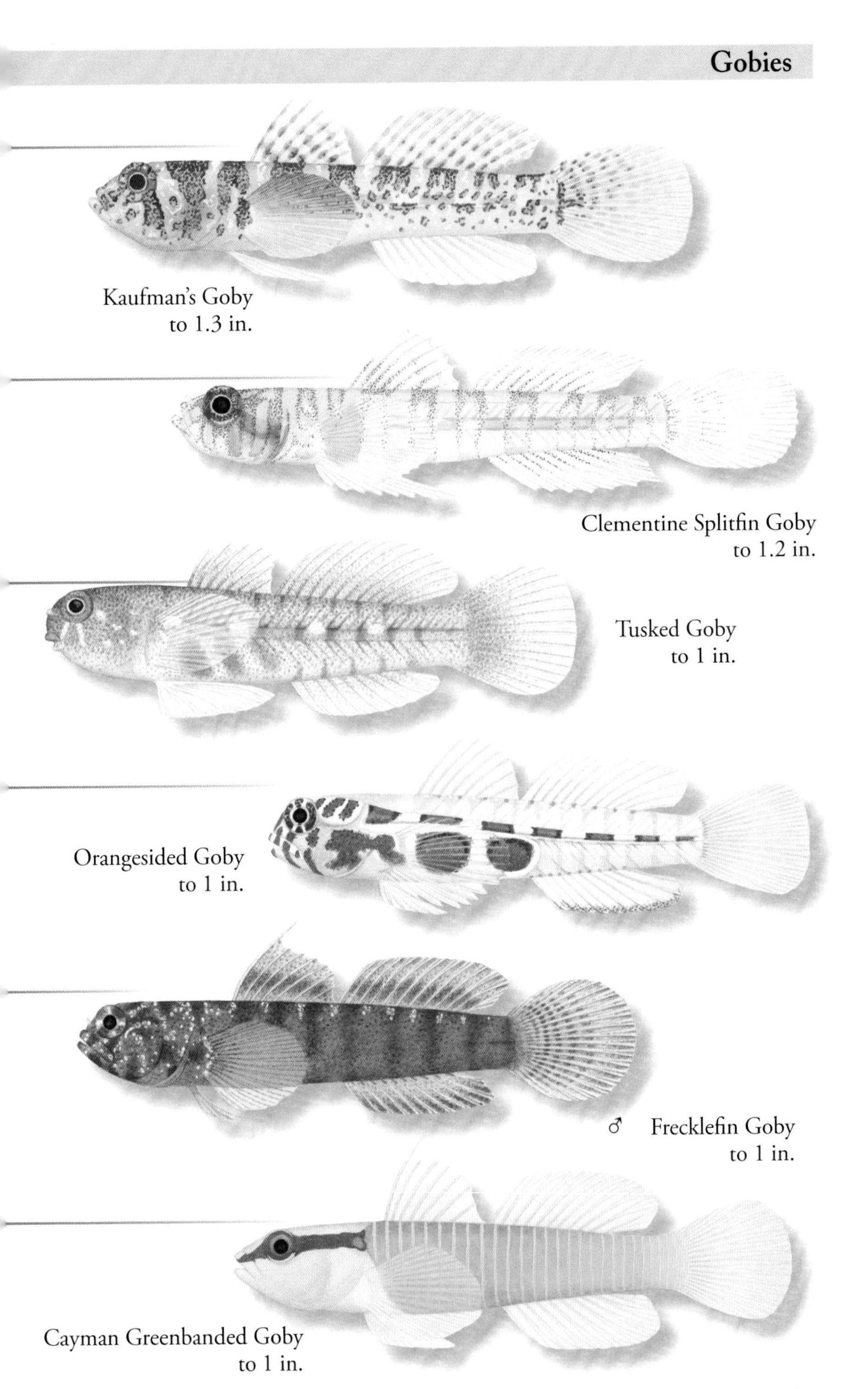
Kaufman's Goby
to 1.3 in.
Clementine Splitfin Goby
to 1.2 in.
Tusked Goby
to 1 in.
Orangesided Goby
to 1 in.
♂ Frecklefin Goby
to 1 in.
Cayman Greenbanded Goby
to 1 in.

Tiger Goby - *Tigrigobius macrodon* (Beebe & Tee-Van, 1928)

FEATURES: Translucent with narrow, dark, blackish brown stripes encircling head and body. Cheeks reddish. Spinal column with alternating dark and pale smudges. Fins translucent. First dorsal-fin spine elongate in males. Body scaleless except on caudal peduncle. HABITAT: FL, Bermuda, Bahamas, Cuba to Virgin Islands. Demersal over rock faces, coral rubble, pilings, and large sponges. Also in tide pools. NOTE: Previously known as *Elacatinus macrodon*.

Greenbanded Goby - *Tigrigobius multifasciatus* (Steindachner, 1876)

FEATURES: Head shades of white or tan to yellow, with a red stripe from snout through eyes to nape. Band usually followed by an iridescent orange to red blotch and black area. Body greenish or grayish to bluish green with 15–21 very narrow, bright green bars. Fins dusky ochre. First dorsal-fin spine often elongate. HABITAT: Bahamas, Cayman Islands, and Cuba to Curaçao. Demersal over shallow-water coral reefs and under sea urchins.

Semiscaled Goby - *Tigrigobius pallens* (Ginsburg, 1939)

FEATURES: Shades of translucent straw with about 8–10 dark wavy bars from nape to peduncle. Bars darker and wider anteriorly. Blackish and white bands radiate around eyes. Cheeks rosy. Dorsal fins unmarked to faintly spotted. First dorsal-fin spine elongate in males. HABITAT: Bahamas and Caribbean Sea. Absent from Cuba. Demersal over shallow-water rocky and coral reefs, in crevices, and on coral walls in high-energy zones.

Ribbon Goby - *Tigrigobius redimiculus* (Taylor & Akins, 2007)

FEATURES: Translucent pinkish on head with rows of red spots below and wavy red bars behind eyes. Translucent tannish posteriorly with greenish stripes encircling body from nape to peduncle. Abdomen and spinal column with alternating black and white bars. First dorsal-fin spine elongate in males. HABITAT: Gulf of Mexico from Isla de Lobos to Isla Verde, Mexico. Demersal over very shallow-water brain coral and algae encrusted coral rubble near reefs.

Redcheek Goby - *Tigrigobius rubrigenis* (Victor, 2010)

FEATURES: Head whitish with a red band from snout through eyes to nape, and a red stripe from lower jaw to pectoral-fin base. Body blackish green with 22–26 very narrow, pale green bars. Fins pale yellow to dusky yellow. First dorsal-fin spine elongate. HABITAT: Belize and Honduras islands in the Gulf of Honduras. Demersal over shallow-water coral reefs under sea urchins.

Leopard Goby - *Tigrigobius saucrus* (Robins, 1960)

FEATURES: Pale, translucent whitish. Cheeks rosy. Head and body with brownish orange or dark brown to reddish brown spots overlaid by darker flecks. Spots form rows posteriorly. Abdomen with large, darker, squarish blotches. Spinal column alternating black and white. HABITAT: FL Keys, N Bahamas, and Caribbean Sea. Demersal over coral heads from near surface to about 100 ft. NOTE: Previously *Elacatinus saucrus*.

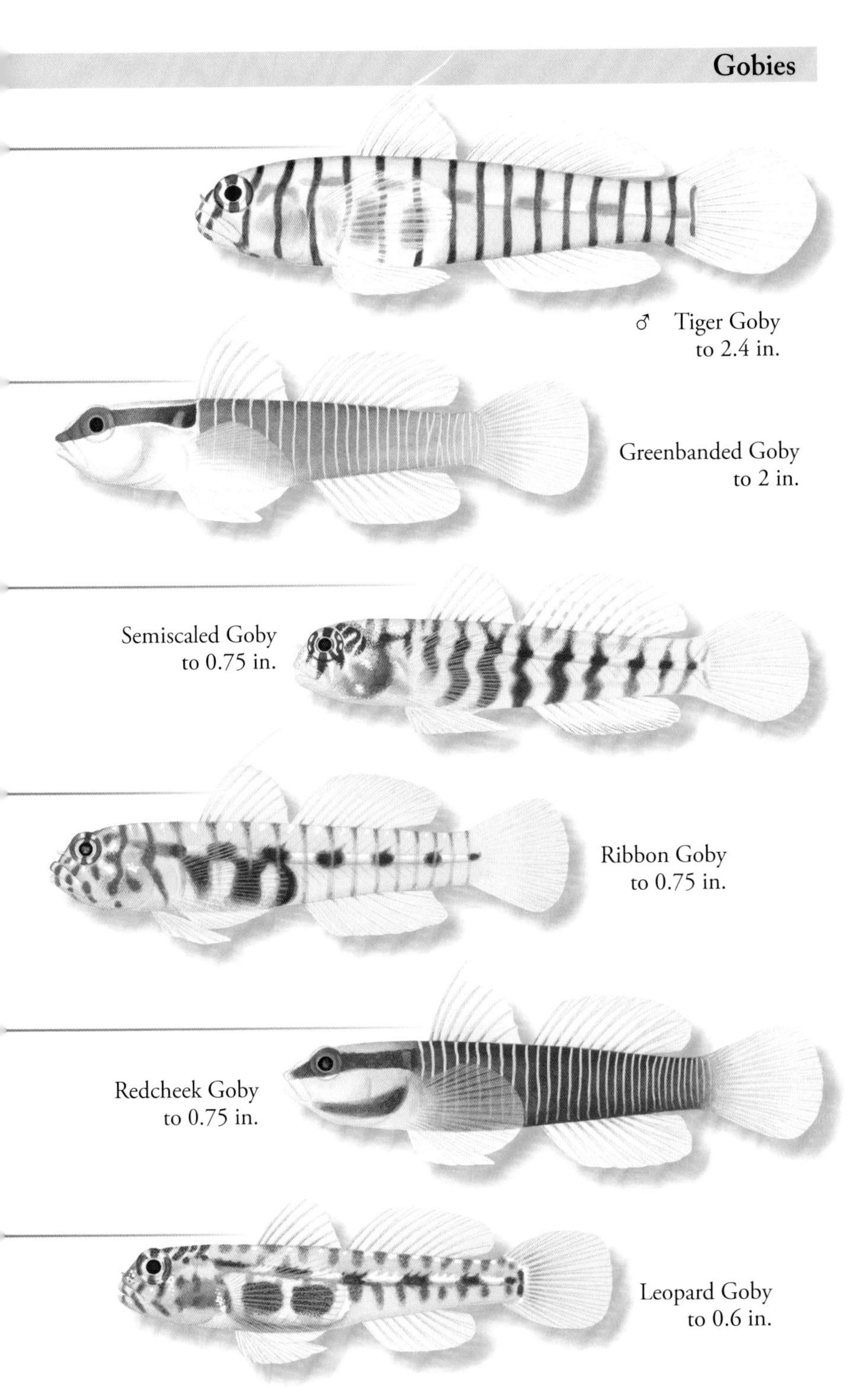

♂ Tiger Goby
to 2.4 in.
Greenbanded Goby
to 2 in.
Semiscaled Goby
to 0.75 in.
Ribbon Goby
to 0.75 in.
Redcheek Goby
to 0.75 in.
Leopard Goby
to 0.6 in.

Gobiidae - Gobies, *cont.*

Zebrette Goby - *Tigrigobius zebrellus* (Robins, 1958)

FEATURES: Color and pattern vary. Head yellowish; body translucent white to blue. Cheeks rosy. Black bars encircle head and body. Bars may be as wider or wider than interspaces. Fins translucent. First dorsal-fin spine may be elongate. HABITAT: Southern Caribbean Sea from Gulf of Venezuela to Tobago. Demersal over a variety of soft and hard bottoms from near surface to about 65 ft.

Lemon Goby - *Vomerogobius flavus* Gilbert, 1971

FEATURES: Snout and cheeks rosy with purplish blue spots and lines. Irises bright yellow. Body translucent yellowish orange. Spinal column and inner caudal fin lemon yellow. Dorsal fin spines filamentous. Body elongate. HABITAT: Bahamas and islands of the Gulf of Honduras. Occur near bottom in schools above steep reef dropoffs and in recessed in steep walls from about 45 to 140 ft.

NOTE: There are 28 other rare or poorly recorded Gobies in the area, see p. 547.

Microdesmidae - Wormfishes

Pugjaw Wormfish - *Cerdale floridana* Longley, 1934

FEATURES: Pale tannish to whitish; somewhat translucent. Minute to tiny, dark flecks pepper upper portions of head and body. Irises white. Snout short. Mouth oblique with protruding lower jaw. Body elongate and somewhat laterally compressed. Dorsal and anal fins connected to caudal fin. HABITAT: SE FL to FL Keys, Bermuda, Bahamas, and Caribbean Sea. Absent from Cayman Islands, Jamaica, and most of the Venezuelan coast. Demersal in crevices and burrows around coral reefs to about 100 ft.

Bahia Wormfish - *Microdesmus bahianus* Dawson, 1973

FEATURES: Whitish to tannish with two brownish to blackish stripes that run along dorsal profile and on sides from snout and lower jaw to caudal-fin base. Irises bright white to bright yellow. Snout short. Mouth oblique with protruding lower jaw. Dorsal-fin origin midway between snout and anal-fin origin. Dorsal and anal fins connected to caudal fin. Body elongate, round in cross-section, and worm-like. HABITAT: Cape Canaveral, FL, Belize to Panama, Martinique, and SW Brazil. Demersal over shallow-water soft bottoms, in burrows, and in tide pools.

Pink Wormfish - *Microdesmus longipinnis* (Weymouth, 1910)

FEATURES: Tannish with an orange to pinkish cast and iridescent sheen. Lower jaw may be barred. Snout short. Mouth oblique; lower jaw protrudes. Dorsal-fin origin just behind pectoral fins. Dorsal and anal fins connected to caudal fin. Body elongate, laterally compressed, and eel-like. HABITAT: NC to FL, in Gulf of Mexico from FL to Yucatán, in Caribbean Sea from Yucatán to Panama, Bermuda, Puerto Rico, and Tobago. Demersal over shallow-water sandy and muddy bottoms, in burrows, and in tide pools.

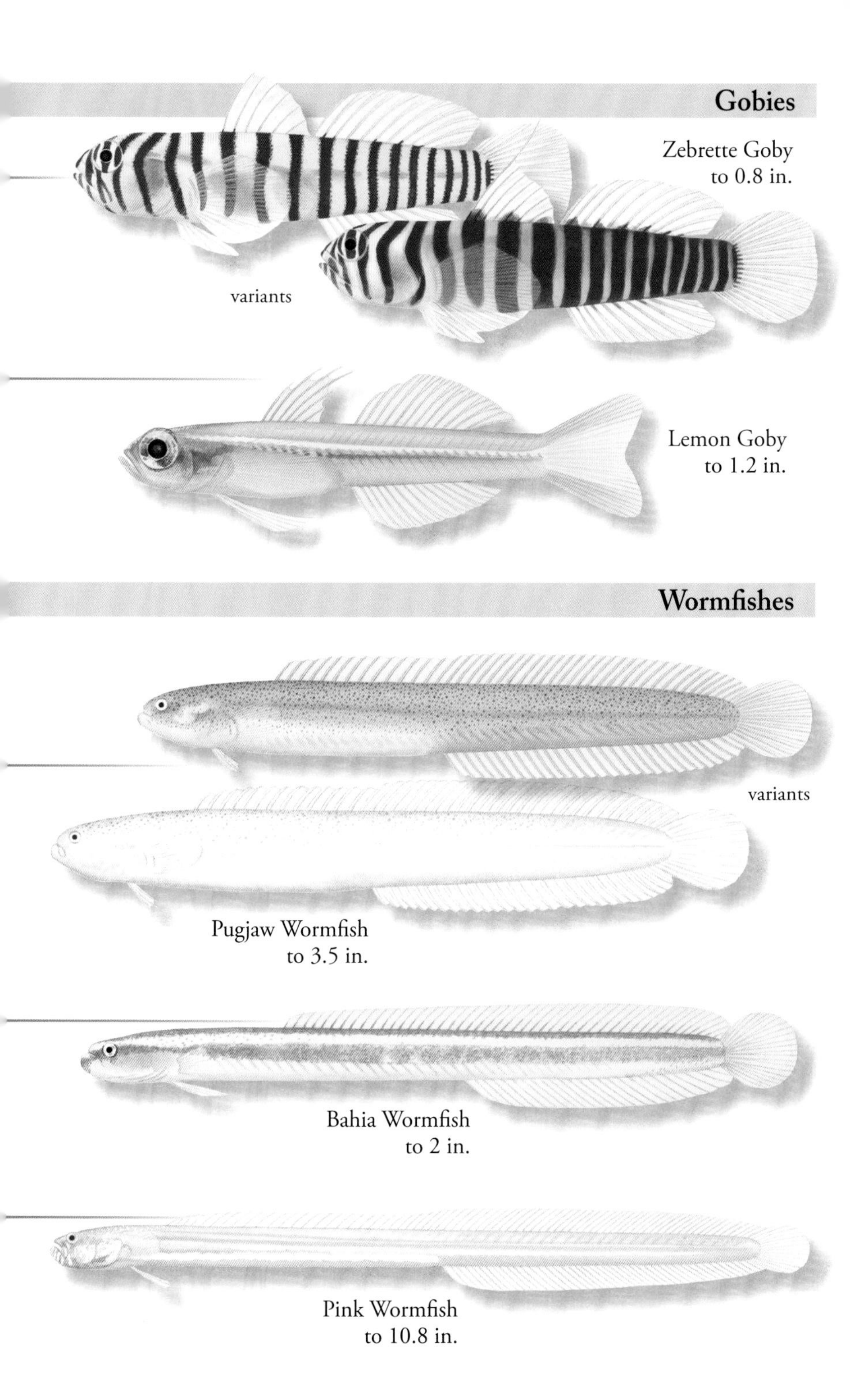
Gobies
Zebrette Goby
to 0.8 in.
variants
Lemon Goby
to 1.2 in.
Wormfishes
variants
Pugjaw Wormfish
to 3.5 in.
Bahia Wormfish
to 2 in.
Pink Wormfish
to 10.8 in.

Pterelectridae - Dartfishes

Hovering Dartfish - *Ptereleotris helenae* (Randall, 1968)

FEATURES: Pale bluish gray with lavender highlights. Blue line from snout dorsal-fin origin and behind eyes to opercle. Vertical fins yellowish with blue outer margins and red inner margins. Caudal fin oblong. HABITAT: FL Keys, Bahamas, Yucatán Peninsula, and Caribbean Sea. Occur near bottoms and in U-shaped burrows over coral rubble and sandy and silty bottoms around reefs.

ALSO IN THE AREA: *Ptereleotris calliura*, *Ptereleotris randalli*, see p. 547.

Ephippidae - Spadefishes

Atlantic Spadefish - *Chaetodipterus faber* (Broussonet, 1782)

FEATURES: Silvery with broad, brownish gray, or brownish to blackish bars from eyes to peduncle. Bars may be faint to absent, especially in large adults. Juveniles dark brown or blackish with some white. Preopercular margin finely serrated, lacks long spine at corner. Opercle with blunt spine at posterior margin. Second dorsal and anal fins with elongate anterior rays in adults. Body disk-shaped in profile. HABITAT: MA to FL, Gulf of Mexico, Bermuda, Bahamas, and Caribbean Sea to Brazil. Occur coastally over wrecks, reefs, pilings. Also around buoys and mangroves, in harbors, and under bridges.

Acanthuridae - Surgeonfishes

Doctorfish - *Acanthurus chirurgus* (Bloch, 1787)

FEATURES: Shades of brown to gray. Narrow blue lines radiate from eyes. Opercular margin blackish. Usually with 8–12 dark bars on sides. Vertical fins with narrow blue margins. Caudal-fin base may be pale. Caudal-fin margin somewhat to moderately concave. HABITAT: MA (rare) to FL, Gulf of Mexico, Bermuda, Bahamas, and Caribbean Sea to Brazil. Over coral reefs and rocky bottoms with sandy areas.

Blue Tang - *Acanthurus coeruleus* Bloch & Schneider, 1801

FEATURES: Pale to dark blue or purplish. May appear uniformly colored or with narrow, wavy lines on sides. May also display pale bars on sides. Can pale or darken. Yellow spine on caudal peduncle. Juveniles yellow with blue on eyes and margins of dorsal and anal fins. Body very deep, laterally compressed. HABITAT: NY (rare) to FL, Gulf of Mexico, Bermuda, Bahamas, and Caribbean Sea to Brazil. Found over shallow coral reefs and rocky bottoms. BIOLOGY: Occur singly or in large schools.

Northern Ocean Surgeon - *Acanthurus tractus* Poey, 1860

FEATURES: Yellowish, bluish gray, or grayish brown to brown. Distinct bars absent from sides. Narrow blue lines radiate from eyes. Opercular margin blackish. Vertical fins with narrow blue margins. Caudal-fin base often whitish. Pelvic-fin membranes blackish. Caudal-fin margin concave. HABITAT: MA to FL, Gulf of Mexico, Bermuda, Bahamas, and Caribbean Sea. Occur over coral reefs, rocky bottoms, and seagrass beds. NOTE: Previously known as *Acanthurus bahianus*.

Dartfishes
Hovering Dartfish
to 4.3 in.
Spadefishes
Atlantic
Spadefish
to 3 ft.
juvenile
Surgeonfishes
Doctorfish
to 13.8 in.
Blue Tang
to 14 in.
juvenile
Northern Ocean
Surgeon
to 14 in.

Sphyraenidae - Barracudas

Great Barracuda - *Sphyraena barracuda* (Edwards, 1771)

FEATURES: Bluish gray to gray dorsally. Upper sides with oblique bars. Silvery below. Lower sides with few to many variably sized, gray to black blotches. Fins may be tipped white. Head large. Lower jaw protrudes. Mouth large with large canine-like teeth. Caudal fin slightly forked. HABITAT: Circumglobal. In western Atlantic from MA to FL, Gulf of Mexico, Bermuda, Bahamas, Caribbean Sea to Brazil. Found from inshore to offshore waters over structures and reefs. BIOLOGY: Adults usually occur singly. Juveniles schooling. Voracious predator.

Northern Sennet - *Sphyraena borealis* DeKay, 1842

FEATURES: Olivaceous dorsally; silvery on sides and below. Fins often yellowish. Head large. Mouth large and toothy. Lower jaw protrudes. Pectoral fins do not reach pelvic-fin origin. Pelvic-fin base under or slightly posterior to dorsal-fin origin. Caudal fin forked. HABITAT: Nova Scotia to FL, Gulf of Mexico, Bermuda, northern Bahamas, and Caribbean Sea to Brazil. Found over a variety of bottoms in coastal waters to about 215 ft. BIOLOGY: Schooling. Feed on small fishes, squids, and shrimps.

Guaguanche - *Sphyraena guachancho* Cuvier, 1829

FEATURES: Olivaceous dorsally; silvery on sides and below. A yellow stripe is present on sides. Head large. Mouth large and toothy. Lower jaw protrudes. Margins of caudal, anal, and pelvic fins blackish. Pectoral fins reach past pelvic-fin origin. Pelvic-fin base under or slightly anterior to first dorsal-fin origin. Caudal fin forked. HABITAT: MA (rare) to FL, Gulf of Mexico, Bahamas, Caribbean Sea to Brazil. Occur in turbid coastal waters over muddy bottoms. Also in bays and estuaries.

Gempylidae - Snake Mackerels

Snake Mackerel - *Gempylus serpens* Cuvier, 1829

FEATURES: Steely gray to grayish brown with silvery luster. Fins with dark margins. Jaws large, toothy. Lower jaw with pointed tip. First dorsal fin long-based, with 26–32 spines. Second dorsal fin followed by five or six finlets. Anal fin followed by six or seven finlets. Body very elongate and laterally compressed. HABITAT: Circumglobal in tropical to temperate seas from surface to about 650 ft. In western Atlantic from NY to FL, Gulf of Mexico, Bermuda, Bahamas, and Caribbean Sea to Brazil. Offshore, oceanic from surface to about 3,300 ft.

Escolar - *Lepidocybium flavobrunneum* (Smith, 1843)

FEATURES: Almost uniformly dark brown, becoming blackish with age. First dorsal fin low, with eight or nine spines. Second dorsal fin followed by four to six finlets. Anal fin followed by four or five finlets. Lateral line sinuous. Caudal peduncle with one strong keel and two smaller keels. Body deep, robust. HABITAT: Circumglobal in tropical to warm temperate seas. In western Atlantic from Georges Bank to FL, Gulf of Mexico, Bermuda, Bahamas, and Caribbean Sea to Brazil. Over continental slopes from about 650 ft to 4,300 ft.

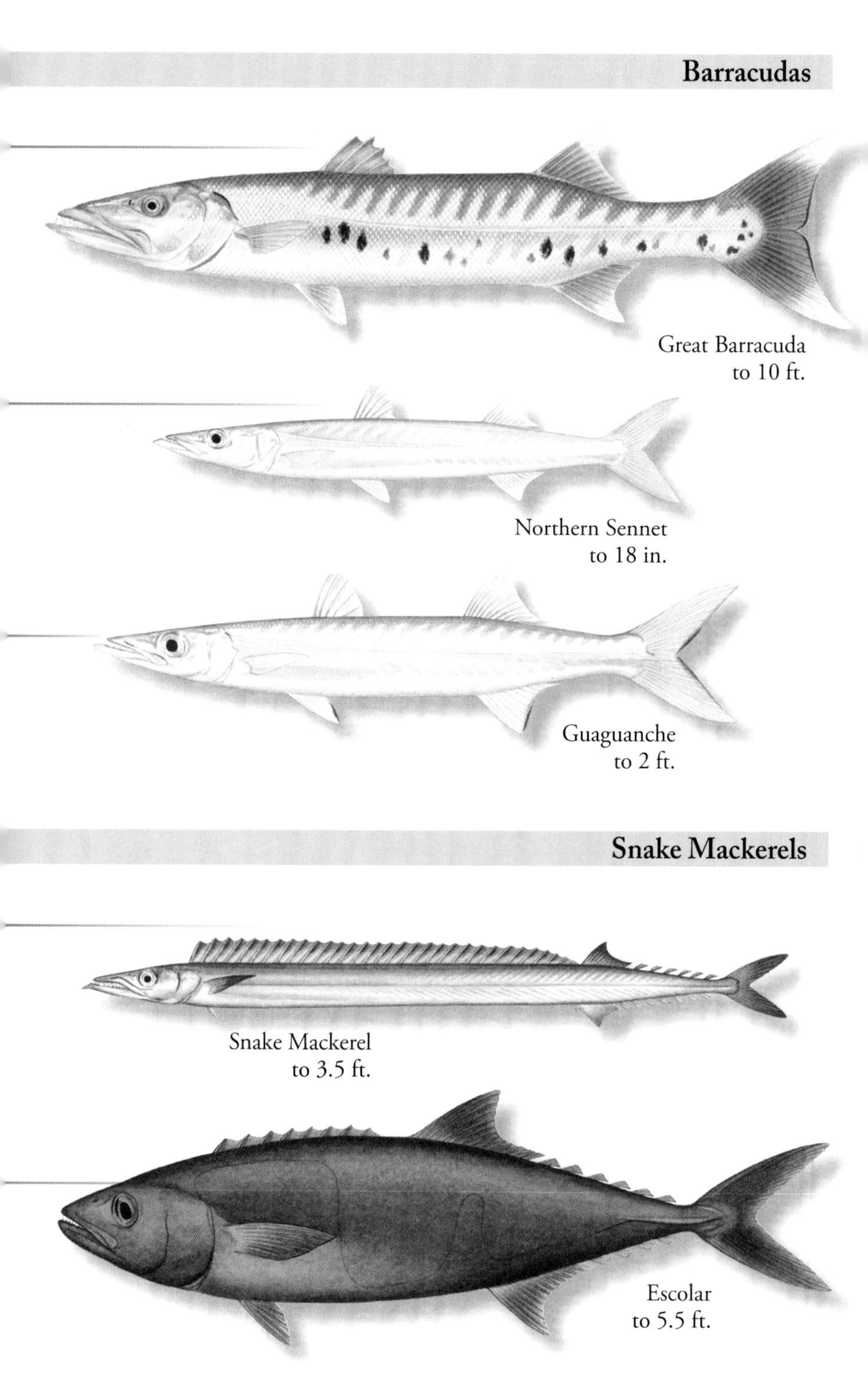
Barracudas
Great Barracuda
to 10 ft.
Northern Sennet
to 18 in.
Guaguanche
to 2 ft.
Snake Mackerels
Snake Mackerel
to 3.5 ft.
Escolar
to 5.5 ft.

Gempylidae - Snake Mackerels, *cont.*

Black Snake Mackerel - *Nealotus tripes* Johnson, 1865

FEATURES: Dark brownish gray to blackish brown. Dorsal and anal fins pale. Jaws large. First dorsal fin with 20–21 spines. Second dorsal and anal fins followed by two finlets. Body elongate, compressed. HABITAT: Circumglobal in tropical to warm temperate seas. Oceanic from surface to about 5,400 ft. In western Atlantic from Newfoundland to FL, Gulf of Mexico, Bermuda, Bahamas, and Caribbean Sea to Uruguay.

American Sackfish - *Neoepinnula americana* (Grey, 1953)

FEATURES: Dark brownish gray dorsally; silvery on sides and below. Fins dark. Two flat spines at opercular margin. First dorsal fin originates over opercle. Lateral line is double; both branches originate above upper gill opening. Body moderately deep, compressed. HABITAT: Bermuda, FL, Gulf of Mexico, and Caribbean Sea to northern Brazil. Reported to New England. Occur near bottom from about 545 to 1,500 ft.

Black Gemfish - *Nesiarchus nasutus* Johnson, 1862

FEATURES: Dark brown with violet sheen. Fins blackish. Jaws large with conical, fleshy tips. Dorsal fin with 19–21 spines. Second dorsal and anal fins followed by two finlets. Body long and laterally compressed. HABITAT: Nearly circumglobal in tropical to warm temperate seas. In western Atlantic from NC to FL, Gulf of Mexico, Bermuda, Bahamas, and Caribbean Sea to northern Brazil. Near bottom from about 655 to 4,000 ft.

Roudi Escolar - *Promethichthys prometheus* (Cuvier, 1832)

FEATURES: Silvery gray to coppery brown. Fins blackish in large specimens, yellowish with dark tips in small specimens. First dorsal fin with 17–19 spines. Second dorsal fin and anal fins followed by two finlets. Lateral line single, arched anteriorly. HABITAT: Circumglobal in tropical to warm temperate seas. Eastern United States. Gulf of Mexico to Brazil. Near bottom over continental slopes to about 2,500 ft.

Oilfish - *Ruvettus pretiosus* Cocco, 1833

FEATURES: Brown to dark brown. Tips of pectoral and pelvic fins blackish. First dorsal fin low, with 13–15 spines. Second dorsal fin followed by two finlets. Lateral line straight posteriorly. Abdomen keeled with bony scales. HABITAT: Worldwide in tropical to warm temperate seas. Newfoundland to FL, Gulf of Mexico, Caribbean Sea, and Bermuda. Near bottom over continental slopes and rises to about 2,300 ft.

Trichiuridae - Cutlassfishes

Atlantic Cutlassfish - *Trichiurus lepturus* Linnaeus, 1758

FEATURES: Iridescent silvery blue to steely blue. Dorsal fin pale dusky yellow. Mouth large; lower jaw protrudes. Dorsal fin originates anterior to opercular opening and tapers posteriorly. Body tapers to a pointed tail. Caudal fin absent. HABITAT: Circumglobal in tropical to warm temperate seas. ME to FL, Gulf of Mexico, Bermuda, Bahamas, and Caribbean Sea to Argentina. Near bottom on continental shelves to about 330 ft.

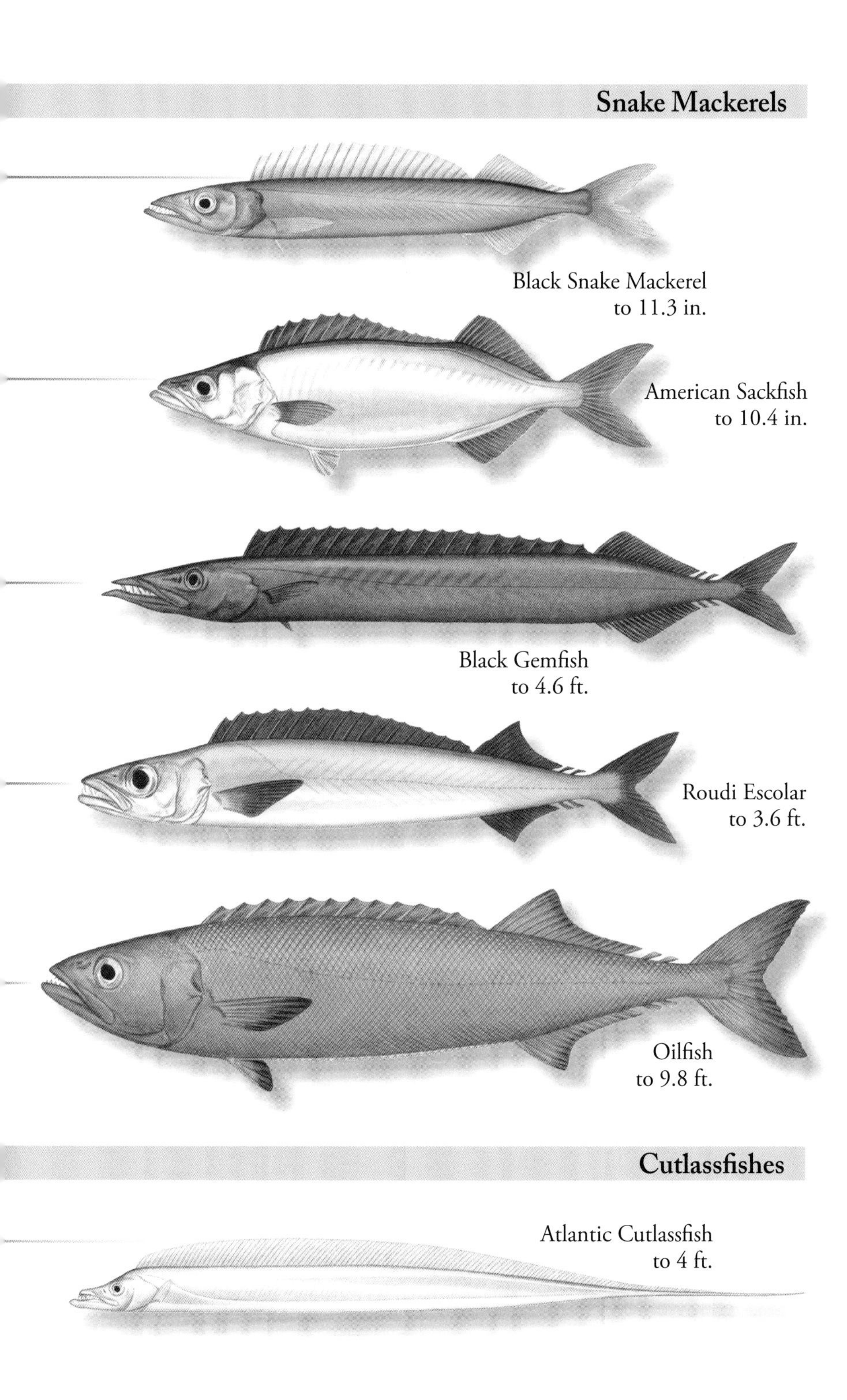
Snake Mackerels
Black Snake Mackerel
to 11.3 in.
American Sackfish
to 10.4 in.
Black Gemfish
to 4.6 ft.
Roudi Escolar
to 3.6 ft.
Oilfish
to 9.8 ft.
Cutlassfishes
Atlantic Cutlassfish
to 4 ft.

Scombridae - Mackerels and Tunas

Wahoo - *Acanthocybium solandri* (Cuvier, 1832)

FEATURES: Iridescent bluish green dorsally; silvery below. Numerous irregular coppery bars on sides. Snout long, pointed. First dorsal fin with 23–27 spines; originates over pectoral-fin base. Second dorsal fin followed by nine finlets. Mouth large; body elongate, slightly compressed. HABITAT: Circumglobal in tropical to subtropical seas. In western Atlantic from NJ to FL, Gulf of Mexico, Bermuda, Bahamas, and Caribbean Sea to southern Brazil. Offshore in open ocean from surface to midwater depths. BIOLOGY: Feed on squids and a variety of fishes. Sought commercially and for sport.

Bullet Tuna - *Auxis rochei* (Risso, 1810)

FEATURES: Dark blue dorsally; deep purple to blackish over head. Upper scaleless area with about 15 irregular, moderately broad, oblique bars; some may merge. Dorsal fins widely separated. Pectoral fins short, not extending past corselet. Corselet comparatively wide under second dorsal fin. Body comparatively shallow. HABITAT: Circumglobal in tropical to warm temperate seas. MA to FL, Gulf of Mexico, Bahamas, and Caribbean Sea to Uruguay. Pelagic and oceanic over continental shelves and slopes. BIOLOGY: Form large schools. Sought commercially. OTHER NAME: Bullet Mackerel.

Frigate Tuna - *Auxis thazard* (Lacepède, 1800)

FEATURES: Dark blue dorsally; deep purple to blackish over head. Upper scaleless area with 15 or more irregular, oblique, wavy bars. Bars may be whole, broken, or converging. Dorsal fins widely separated. Pectoral fins short, reaching slightly beyond corselet. Corselet comparatively narrow under second dorsal fin. Body comparatively deep. HABITAT: Circumglobal in tropical to subtropical seas. In western Atlantic from MA to FL, Gulf of Mexico, Bahamas, and Caribbean Sea to Uruguay. Pelagic and oceanic over continental shelves and slopes. BIOLOGY: Schooling. OTHER NAME: Frigate Mackerel.

Little Tunny - *Euthynnus alletteratus* (Rafinesque, 1810)

FEATURES: Greenish to dark blue dorsally. Silvery below. Upper scaleless area with highly variable cluster of narrow, oblique, wavy dark lines. Several faint to dark spots on abdomen below pectoral fins. May display dark bars when distressed. Dorsal fins connected at base. HABITAT: Eastern and western Atlantic and Mediterranean Sea. In western Atlantic from MA to FL, E Gulf of Mexico, Bermuda, Bahamas, and Caribbean Sea to E Brazil. Pelagic and at surface mainly over continental shelves. Also found coastally and inshore. BIOLOGY: Schooling.

Skipjack Tuna - *Katsuwonus pelamis* (Linnaeus, 1758)

FEATURES: Bluish black dorsally; silvery below. Upper scaleless area usually with several oblique, cobalt blue bands. Lower sides with four to six grayish stripes that may appear as discontinuous lines of dark blotches. Dorsal fins separated by a small interspace. HABITAT: Circumglobal in tropical to warm temperate seas. In western Atlantic from Nova Scotia to FL, Gulf of Mexico, Bermuda, Bahamas, and Caribbean Sea to Uruguay. Pelagic and oceanic from continental shelves to mid-ocean, generally above the thermocline. BIOLOGY: Form large schools. Often mix with Blackfin Tuna.

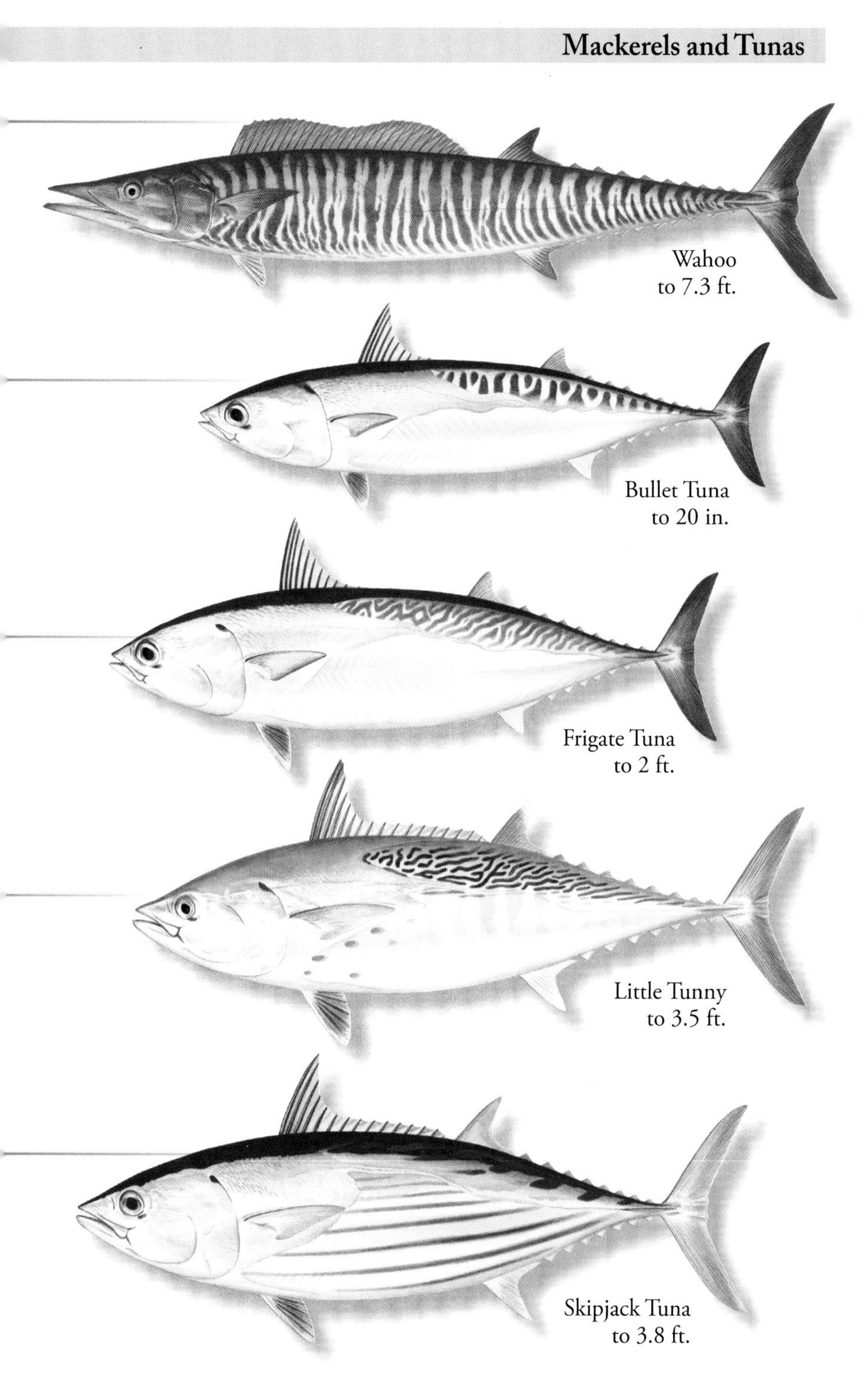
Wahoo
to 7.3 ft.
Bullet Tuna
to 20 in.
Frigate Tuna
to 2 ft.
Little Tunny
to 3.5 ft.
Skipjack Tuna
to 3.8 ft.

Atlantic Bonito - *Sarda sarda* (Bloch, 1793)

FEATURES: Bluish to greenish blue dorsally, fading to silvery below. Upper sides with numerous oblique, dark stripes. May also have broad, diffuse bars underlying dark stripes. First dorsal fin with straight to slightly concave margin; separated from second dorsal fin by small interspace. Body entirely covered in minute scales. Scales on corselet well developed. HABITAT: Eastern and western Atlantic and Mediterranean Sea. In western Atlantic from Nova Scotia to FL, N Gulf of Mexico, Campeche Bay, Colombia, Venezuela, and Argentina. Pelagic and at surface of coastal and inshore waters. BIOLOGY: School near the surface. Migratory. Sought commercially and as sportfish.

Atlantic Chub Mackerel - *Scomber colias* Gmelin, 1789

FEATURES: Turquoise blue dorsally; silvery below. Back and upper sides with many oblique, faint to blackish wavy lines. Rows of dusky spots along midline. Ventral area with numerous grayish spots. Eyes large and with adipose lids. Dorsal fins separate. Two small keels at caudal-fin base. HABITAT: Eastern and western Atlantic. In western Atlantic from Gulf of St. Lawrence to FL, Gulf of Mexico, Venezuela to southern Brazil and Uruguay. Pelagic in coastal waters. BIOLOGY: Schooling, migratory.

Serra Spanish Mackerel - *Scomberomorus brasiliensis* Collette, Russo & Zavala-Camin, 1978

FEATURES: Iridescent greenish to blue green dorsally. Silvery below. Sides with several rows of fairly closely spaced rusty orange to yellowish oval spots. Anterior portion of first dorsal fin blackish, white at base. Pectoral fin scaled at base. Second dorsal fin with 15–19 rays. Lateral line gently sloping. HABITAT: Central and South American coasts from Belize to southern Brazil. Occur from tidal estuaries and along rocky and sandy coasts. BIOLOGY: Form schools. Sought commercially and for sport.

King Mackerel - *Scomberomorus cavalla* (Cuvier, 1829)

FEATURES: Iridescent bluish with greenish reflections dorsally. Silvery below. Juveniles with irregular rows of gray to yellowish spots on sides. First dorsal fin uniformly colored. Dorsal fins scarcely separated. Pectoral fins scaled at base. Lateral line dips downward under second dorsal fin. HABITAT: MA (rare) to FL, Gulf of Mexico, Bahamas, and Caribbean Sea to southern Brazil. Pelagic over continental shelves and slopes. Also over outer reefs, wrecks, and hard structures. Prefer clear water. BIOLOGY: Occur singly or in small groups. Migratory. Sought commercially and as sportfish.

Atlantic Spanish Mackerel - *Scomberomorus maculatus* (Mitchill, 1815)

FEATURES: Iridescent greenish to blue green dorsally. Silvery below. Sides with several rows of fairly evenly spaced rusty orange to yellowish oval spots. Anterior portion of first dorsal fin blackish, white at base. Pectoral fin scaled at base. Second dorsal fin with 17–20 rays. Lateral line gently sloping. HABITAT: MA to FL and Gulf of Mexico to Yucatán. Pelagic over continental shelves and slopes. Also at surface and near shore. Enter estuaries. BIOLOGY: Form large and small schools. Migratory. Sought commercially and as sportfish.

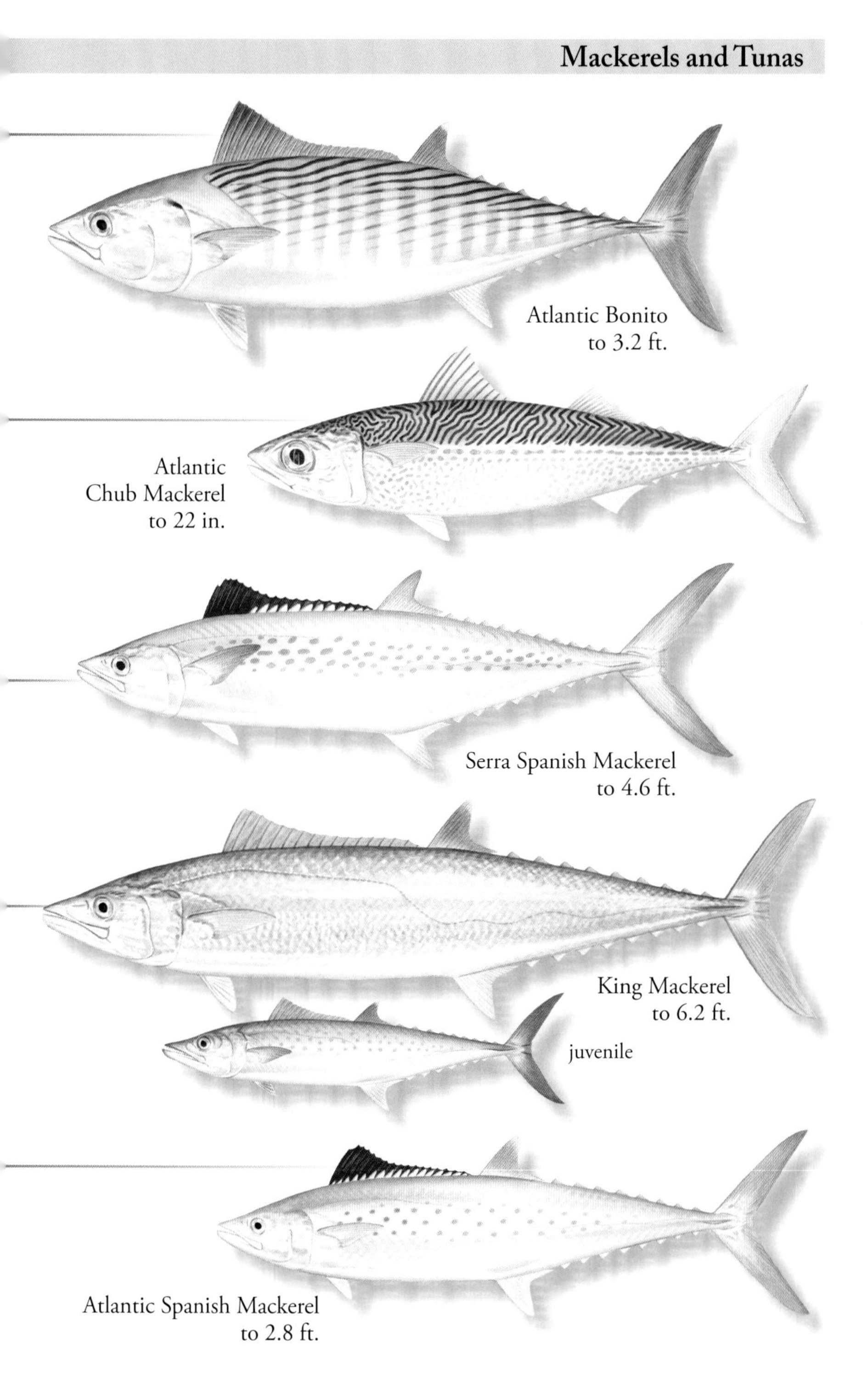
Atlantic Bonito
to 3.2 ft.
Atlantic
Chub Mackerel
to 22 in.
Serra Spanish Mackerel
to 4.6 ft.
King Mackerel
to 6.2 ft.
juvenile
Atlantic Spanish Mackerel
to 2.8 ft.

Cero - *Scomberomorus regalis* (Bloch, 1793)

FEATURES: Iridescent greenish to blue green dorsally. Silvery below. Sides with rusty orange to yellowish streak along midline; similarly colored oblong spots present above and below streak. Anterior portion of dorsal fin blackish, white at base. Inner portion of second dorsal fin dark. Pectoral fins scaled. Lateral line gradually sloping. HABITAT: MA to FL, Bahamas, Antilles to Venezuela, and SE Brazil. Occur in clear water around reefs. Also reported from over seagrass beds. BIOLOGY: Usually solitary or in small groups. Sought as sportfish.

Albacore - *Thunnus alalunga* (Bonnaterre, 1788)

FEATURES: Metallic blackish blue dorsally; silvery on sides and below. Caudal-fin with a narrow white margin. Pectoral fins extremely long, reaching well beyond second dorsal-fin origin, usually to or beyond second dorsal finlet. HABITAT: Circumglobal in tropical to temperate seas. In western Atlantic from Nova Scotia to FL, Gulf of Mexico, Bermuda, Bahamas, and Caribbean Sea to Uruguay. Oceanic, usually below thermocline. BIOLOGY: Highly migratory. Sought commercially. IUCN: Near Threatened.

Yellowfin Tuna - *Thunnus albacares* (Bonnaterre, 1788)

FEATURES: Metallic blackish blue dorsally, yellow on upper sides, silvery below. Lower sides often with about 20 rows of pale spots and nearly vertical lines. Second dorsal and anal fins yellow. Finlets yellow with black margin. Juveniles spotted. Pectoral fins moderately long. Second dorsal and anal fins tall, become very tall and long with age. HABITAT: Circumglobal in tropical to subtropical seas. In western Atlantic from Nova Scotia to FL, Gulf of Mexico, Bermuda, Bahamas, and Caribbean Sea to Brazil. Oceanic, above and below thermocline. BIOLOGY: School near surface, sometimes with other tunas of similar size. Sought commercially. IUCN: Near Threatened.

Blackfin Tuna - *Thunnus atlanticus* (Lesson, 1831)

FEATURES: Metallic blackish blue dorsally. May have yellow luster on upper sides. Lower sides silvery; may have almost vertical pale streaks. Second dorsal and anal fins dusky with silvery luster. Finlets dusky with a trace of yellow. Pectoral fins moderately long. HABITAT: In western Atlantic from MA to FL, Gulf of Mexico, Bahamas, and Caribbean Sea to Brazil. Oceanic, in warm water; not usually below 68°F. BIOLOGY: Migrate northward in summer. School with Skipjack Tuna of the same size.

Bigeye Tuna - *Thunnus obesus* (Lowe, 1839)

FEATURES: Metallic blackish blue dorsally. Silvery below; live specimens lack pale spots or streaks. May have golden streak on sides. Second dorsal and anal fins silvery yellow and moderately tall. Finlets yellow with black margin. Juveniles barred. Pectoral fins very long in small specimens, reach just beyond second dorsal-fin origin in adults. HABITAT: Circumglobal in tropical to subtropical seas. In western Atlantic from Newfoundland to FL, Gulf of Mexico, Bermuda, Bahamas, and Caribbean Sea to Uruguay. Pelagic and oceanic from surface to about 820 ft. BIOLOGY: Juveniles and small adults school at surface; may school with Yellowfin and Skipjack tunas. IUCN: Vulnerable.

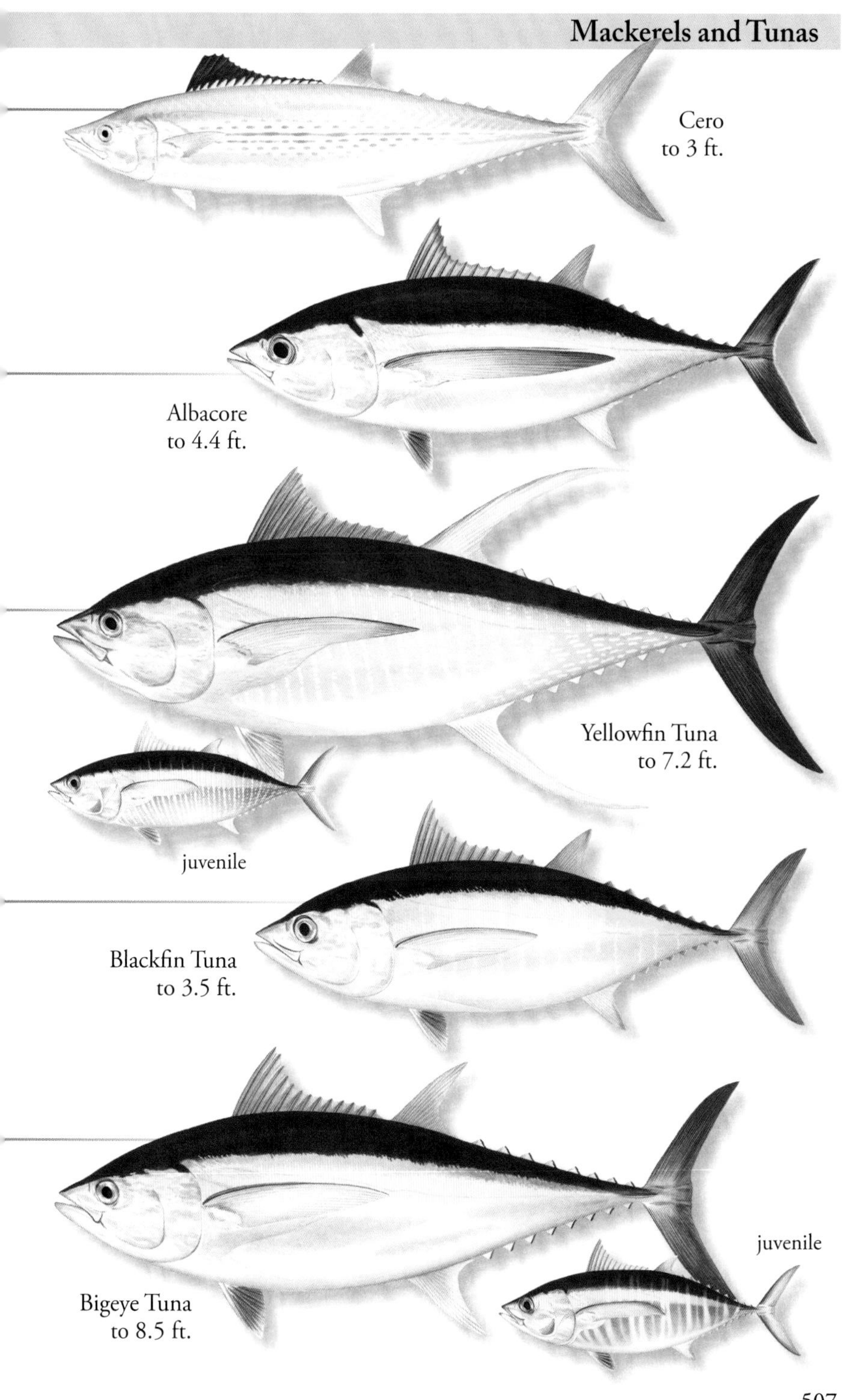
Cero
to 3 ft.
Albacore
to 4.4 ft.
Yellowfin Tuna
to 7.2 ft.
juvenile
Blackfin Tuna
to 3.5 ft.
Bigeye Tuna
to 8.5 ft.
juvenile

Scombridae - Mackerels and Tunas, *cont.*

Atlantic Bluefin Tuna - *Thunnus thynnus* (Linnaeus, 1758)

FEATURES: Metallic dark blue to black dorsally. Silvery to iridescent coppery below. Oblique pale lines and spots on lower sides. Second dorsal fin brownish with golden luster. Anal fin with dark margin. Finlets dusky yellow with dark margin. Lateral keel blackish in adults. Juveniles barred. Pectoral fins short. Body very robust. HABITAT: Atlantic Ocean. In western Atlantic from Newfoundland to FL, Gulf of Mexico, Bermuda, Bahamas, and Caribbean Sea to northeastern Brazil. Pelagic and oceanic. Seasonally close to shore. BIOLOGY: Very fast; highly migratory. Juveniles in warm water. Adults enter cold water. Populations declining. IUCN: Endangered.

Xiphiidae - Swordfish

Swordfish - *Xiphias gladius* Linnaeus, 1758

FEATURES: Back and upper sides metallic bluish black to brownish black. Coppery on sides fading to silvery below. Bars absent, but may change color. Bill flattened in cross-section. Eyes large. First dorsal fin short-based in adults, long-based in juveniles. Second dorsal and anal fins small. Pelvic fins absent. Single, broad lateral keel present. Body robust. HABITAT: Worldwide in tropical to cold seas. In western Atlantic from Greenland to FL, and Caribbean Sea to Argentina. Oceanic, generally above the thermocline. BIOLOGY: Migrate to temperate or cold waters in summer and to warm waters in winter. Opportunistic feeders. Highly important game and foodfish.

Istiophoridae - Billfishes

Sailfish - *Istiophorus platypterus* (Shaw, 1792)

FEATURES: Dark blue dorsally, coppery on sides, silvery below. May display iridescent blue to greenish bars and spots on upper sides. Dorsal fin bluish black to dark brownish blue with dark spots; tall and sail-like. Capable of rapid color change and may become overall dark brownish to bluish brown, particularly while hunting. Pelvic fins very long, almost reaching anus. HABITAT: Circumglobal in tropical to warm temperate seas. In western Atlantic from NY to FL, Gulf of Mexico, Bermuda, Bahamas, and Caribbean Sea to northern Argentina. Occur from near shore to open ocean; usually above the thermocline. BIOLOGY: Very swift and migratory. May hunt in coordinated groups. Sought commercially and for sport. NOTE: Previously known as *Istiophorus albicans.*

White Marlin - *Kajikia albida* (Poey, 1860)

FEATURES: Dark blue to brown dorsally, coppery on sides, silvery white below. May display iridescent bars on sides. Dorsal fin dark blue with blackish spots. Dorsal and first anal fins rounded at tips. Pectoral fins usually rounded at tips. Anus located comparatively close to first anal-fin origin. Lateral scales pointed anteriorly, stiff and one- to two-pronged posteriorly. HABITAT: Atlantic Ocean. Nova Scotia to FL, Gulf of Mexico, Bermuda, Bahamas, and Caribbean Sea to S Brazil. Oceanic. Usually found above the thermocline in water deeper than 164 ft. BIOLOGY: Highly migratory. Sought commercially and for sport. IUCN: Vulnerable.

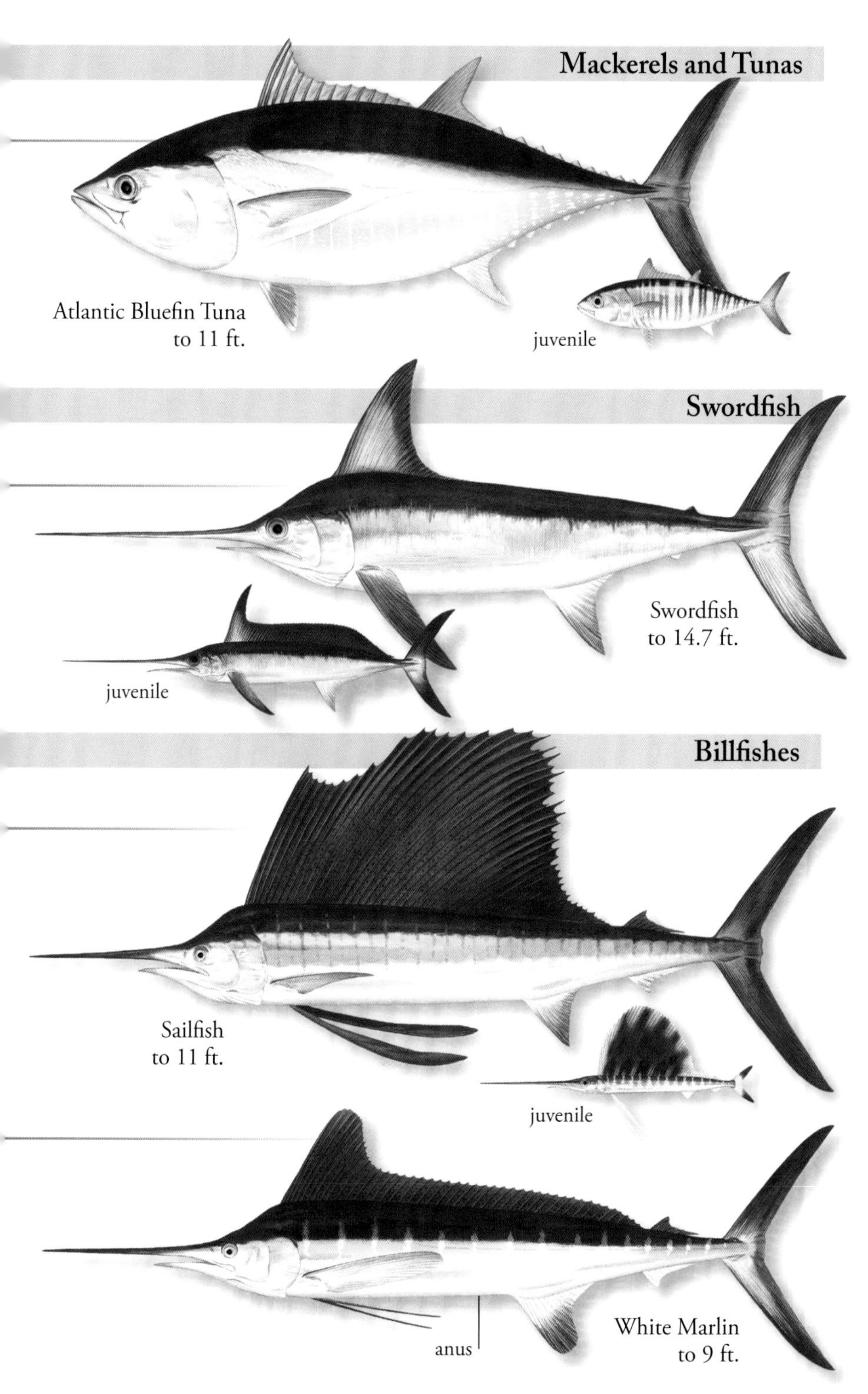
Mackerels and Tunas
Atlantic Bluefin Tuna
to 11 ft.
juvenile
Swordfish
Swordfish
to 14.7 ft.
juvenile
Billfishes
Sailfish
to 11 ft.
juvenile
anus
White Marlin
to 9 ft.

Istiophoridae - Billfishes, *cont.*

Blue Marlin - *Makaira nigricans* Lacepède, 1802

FEATURES: Dark blackish blue to blackish brown dorsally. Silvery white below. Bright blue bars or rows of vertical spots on sides. First dorsal fin dark blue to black. Nape comparatively tall. First dorsal-fin lobe tall and pointed. Juveniles lack pointed bill; dorsal fin sail-like. Pelvic fins shorter than pectoral fins. Lateral line reticulating. HABITAT: Circumglobal in tropical to temperate seas. In western Atlantic from ME to FL, Gulf of Mexico, Bermuda, Bahamas, and Caribbean Sea to S Brazil. Oceanic in clear water. Usually in water of 71–88°F. BIOLOGY: Highly migratory. Enter colder water to feed. Sought commercially and for sport. IUCN: Vulnerable.

Roundscale Spearfish - *Tetrapturus georgii* Lowe, 1841

FEATURES: Dark blue dorsally. Coppery on sides. Silvery white below. Usually displays iridescent inverted V-shaped bars on sides. Dorsal fin dark blue; spots absent. Dorsal and first anal fins usually blunt at tips. Pectoral fins usually pointed at tips. Anus located comparatively far from first anal-fin origin. Lateral scales rounded anteriorly, soft, and two- to three-pronged posteriorly. HABITAT: Atlantic Ocean and Mediterranean Sea. In western Atlantic from MA to FL, Gulf of Mexico, Bermuda, Bahamas, and Caribbean Sea to S Brazil. Oceanic; pelagic. BIOLOGY: May be confused with White Marlin.

Longbill Spearfish - *Tetrapturus pfluegeri* Robins & de Sylva, 1963

FEATURES: Dark blue dorsally. Coppery to iridescent grayish on sides. Silvery white below. Sides lack bars or spots. First dorsal fin dark blue; spots absent. Anterior dorsal-fin lobe taller than depth of body. Bill comparatively short. Pelvic fins longer than pecotral fins. Body comparatively shallow. HABITAT: Atlantic Ocean. In western Atlantic from MA to FL, Gulf of Mexico, Bermuda, Bahamas, and Caribbean Sea to S Brazil. Occur in tropical to subtropical waters, usually above the thermocline. Oceanic and pelagic.

Nomeidae - Driftfishes

Bigeye Cigarfish - *Cubiceps pauciradiatus* Günther, 1872

FEATURES: Blackish to blackish brown. Snout blunt. Mouth small. Eyes very large, surrounded by adipose tissue. Bony keel present on breast. Pectoral fins long, wing-like. Lateral line high on the body. HABITAT: Circumglobal in warm seas. Nova Scotia to FL, Gulf of Mexico, Bermuda, Bahamas, and Caribbean Sea to Argentina. Oceanic. Occur from near surface at night to about 3,280 ft during the day.

Man-of-war Fish - *Nomeus gronovii* (Gmelin, 1789)

FEATURES: Juveniles iridescent blue dorsally; silvery below; sides with blackish blue bars and blotches; pelvic bluish black with small to large white areas. Adults (not shown) dark bluish or brown. Snout blunt. Mouth small. Eyes moderate, surrounded by adipose tissue. Pectoral fins large, fan-like. HABITAT: Circumglobal in tropical to warm temperate seas. Newfoundland to FL, Gulf of Mexico, Bermuda, Bahamas, Caribbean Sea to Brazil. Oceanic. Juveniles associated with Man-of-war Jelly.

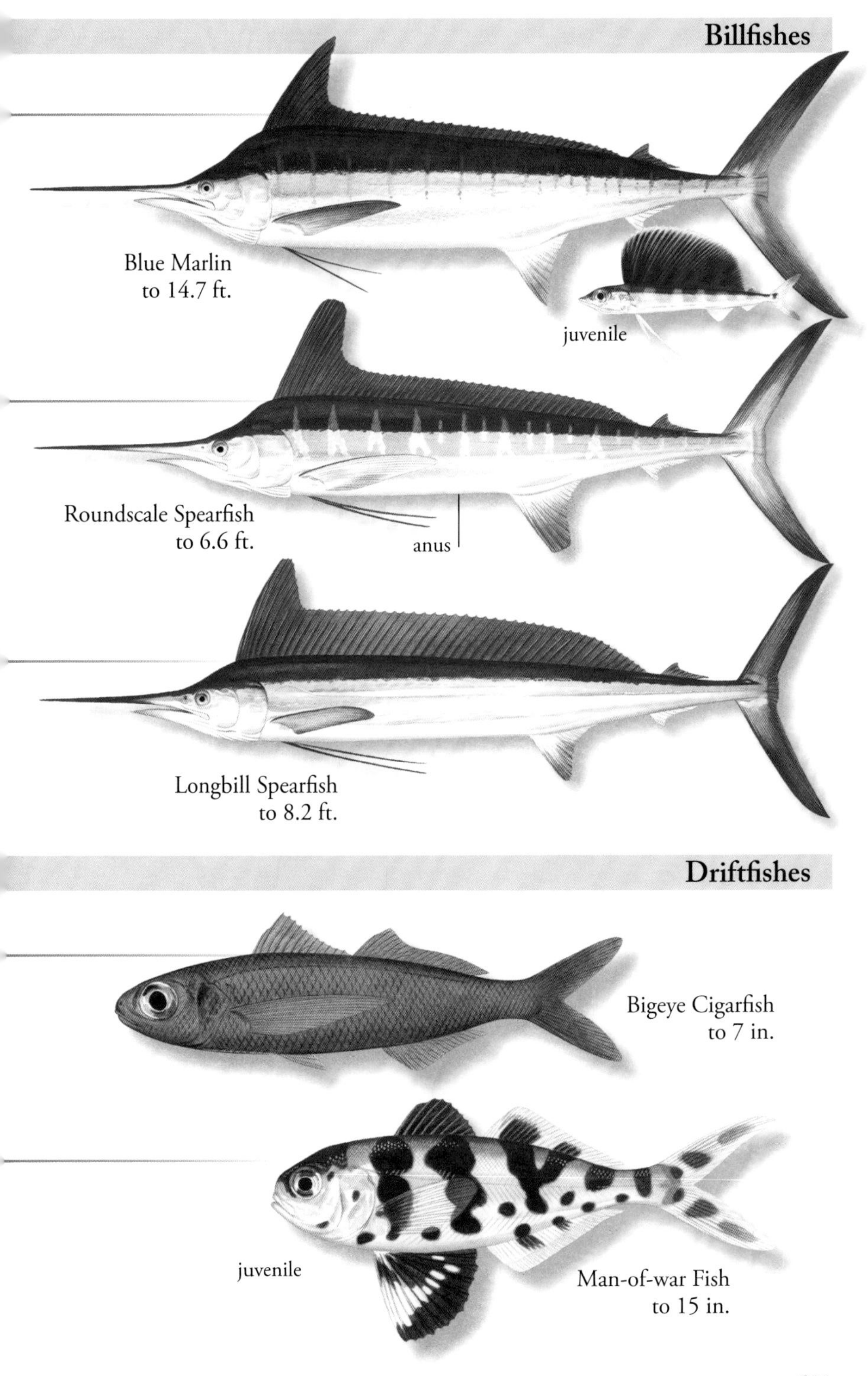
Blue Marlin
to 14.7 ft.
juvenile
Roundscale Spearfish
to 6.6 ft.
anus
Longbill Spearfish
to 8.2 ft.
Driftfishes
Bigeye Cigarfish
to 7 in.
juvenile
Man-of-war Fish
to 15 in.

Nomeidae - Driftfishes, *cont.*

Freckled Driftfish - *Psenes cyanophrys* Valenciennes, 1833

FEATURES: Juveniles silvery with distinct dark spots forming narrow stripes on sides. Observed specimens blue or brown dorsally with dark fins. Adults described as yellowish with lines along scale rows. Snout blunt. Eyes moderately large, surrounded by adipose tissue. Lateral line high on body. HABITAT: Circumglobal in tropical to subtropical seas. MA to FL, Gulf of Mexico, Bermuda, Bahamas, and Caribbean Sea to Brazil. Adults occur to 1,800 ft. Juveniles associated with sea jellies, flotsam, and *Sargassum* weed.

ALSO IN THE AREA: *Cubiceps gracilis, Psenes arafurensis*, see p. 547.

Ariommatidae - Ariommatids

Silver-rag - *Ariomma bondi* Fowler, 1930

FEATURES: Bluish dorsally; silvery on sides and below. Others described as bluish brown dorsally. Lining of body cavity pale. Snout blunt. Mouth small. Eyes large and surrounded by adipose tissue. Lateral line high on body. Caudal peduncle with two weak keels. Body elongate. HABITAT: In tropical and warm waters of Atlantic. Nova Scotia to FL, Gulf of Mexico, Bermuda, Bahamas, Greater Antilles, and Colombia to Uruguay. Demersal over outer continental shelves from about 80 to 2,100 ft.

Spotted Driftfish - *Ariomma regulus* (Poey, 1868)

FEATURES: Silvery to pale brown. Numerous dark spots on upper sides. Opercle, first dorsal fin, and pelvic fins blackish. Snout blunt. Mouth small. Eyes moderately large, surrounded by adipose tissue. Lateral line high on the body. Caudal peduncle with two weak keels. Body deep, oval in profile. HABITAT: NJ to FL, Gulf of Mexico, E and S Caribbean Sea. Occur near bottom from about 600 to 1,600 ft.

Stromateidae - Butterfishes

Gulf Butterfish - *Peprilus burti* Fowler, 1944

FEATURES: Iridescent blue dorsally; silvery below. Fins dusky. Snout blunt. Mouth small. Dorsal and anal fins long-based; anterior lobes moderately tall. Pelvic fins absent. Series of pores along back. Lateral line high on the body. HABITAT: Gulf of Mexico from S FL to Campeche, Mexico. Pelagic from near surface to about 900 ft. BIOLOGY: Form large, loose schools. Juveniles associated with floating seaweeds and sea jellies.

Harvestfish - *Peprilus paru* (Linnaeus, 1758)

FEATURES: Iridescent blue to green dorsally; silvery below. Fins dusky. Tips of dorsal, caudal, and anal fins may be blackish. Mouth small. Snout short, blunt. Dorsal and anal fins long-based; anterior lobes elongate to very elongate. Pelvic fins absent. Pores under dorsal fin absent. Lateral line high on the body. Body very deep, laterally compressed. HABITAT: VA to FL, Gulf of Mexico, Bermuda, and Caribbean Sea. Pelagic. Along coast in bays and inshore waters to about 400 ft.

Driftfishes

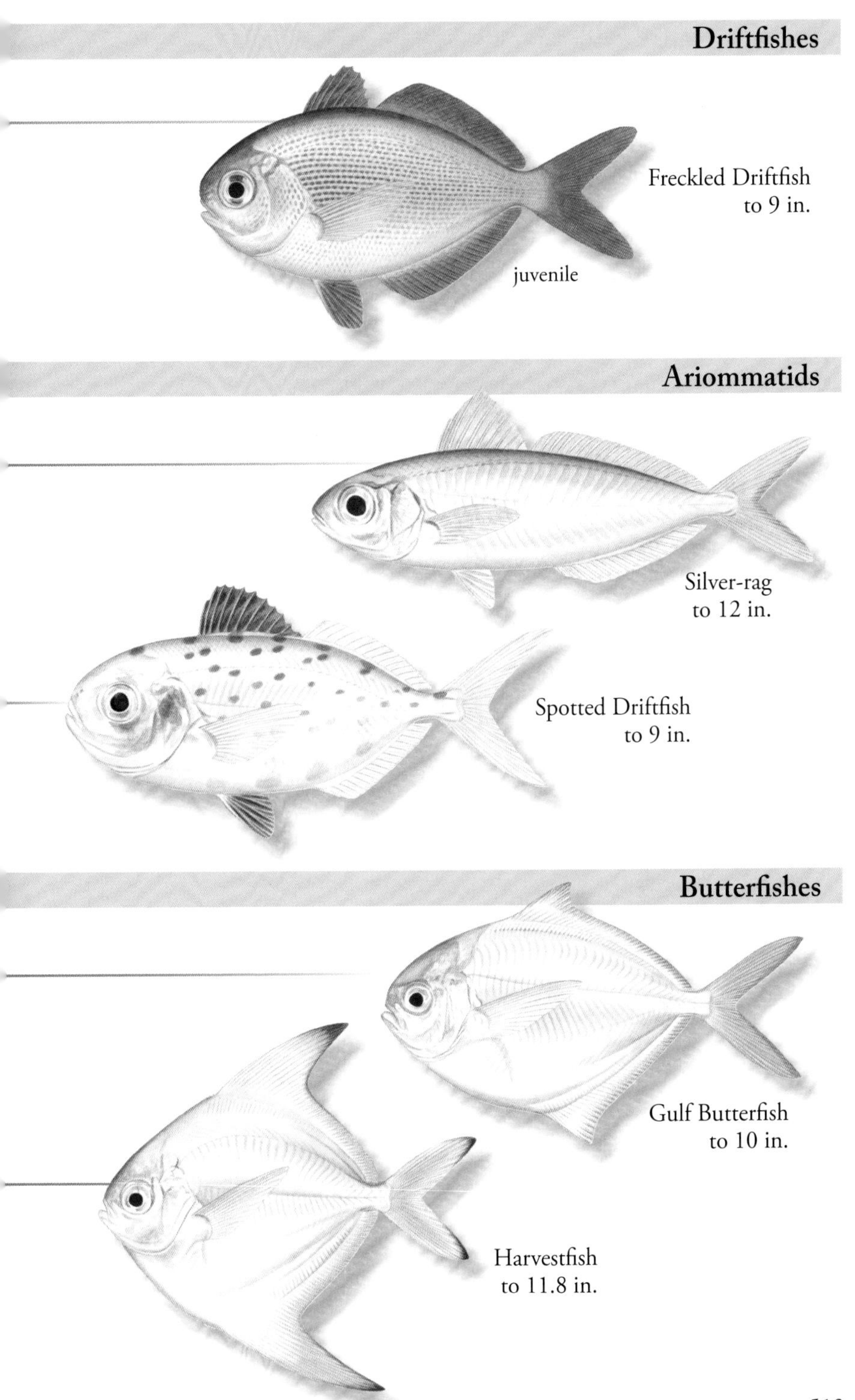

Paralichthyidae - Sand Flounders

Sand Whiff - *Citharichthys arenaceus* Evermann & Marsh, 1900

FEATURES: Eyed side shades of tan to brown and densely covered with darker and paler speckles. Upper jaw extends to rear edge of lower eye. Lateral line nearly straight, with a very slight arch over pectoral fins. Upper eye slightly anterior to lower eye. Space between eyes with a bony ridge. HABITAT: E FL and Caribbean Sea to southern Brazil. Absent from Cayman Islands and Jamaica. Demersal over shallow-water sandy and muddy bottoms from estuaries, bays, lagoons, and mangrove tidal creeks to river tributaries.

Anglefin Whiff - *Citharichthys gymnorhinus* Gutherz & Blackman, 1970

FEATURES: Eyed side tannish, with several somewhat symmetrically arranged, dark-ringed blotches. Caudal fin with two dark blotches near base. Eyes large. Dorsal and anal fins expanded, angular. Males with several spiny projections on head and two large dark blotches on dorsal and anal fins. Females with reduced spines on head and dark flecks on dorsal and anal fins. HABITAT: NC to FL, Bahamas, Gulf of Mexico, and Caribbean Sea to Guyana. Demersal over sandy and muddy bottoms from about 65 to 660 ft. More common in shallower range of depth.

Spotted Whiff - *Citharichthys macrops* Dresel, 1885

FEATURES: Eyed side tan to brown with prominent dark spots and blotches on body and fins. Spots may coalesce into larger blotches. Caudal fin with two dark spots near base. Blind side is white. Upper jaw extends to about middle of lower eye. Leaf-like cirri on margin of lower opercle on blind side. Lateral line nearly straight. HABITAT: NC to FL; Gulf of Mexico to Brazil. Also northern Bahamas, Cuba, and Jamaica. Demersal over hard sand bottoms of continental shelves from shore to about 330 ft.

Bay Whiff - *Citharichthys spilopterus* Günther, 1862

FEATURES: Eyed side pale to dark brown, with or without numerous spots and blotches on body and fins. Small dark spot on caudal peduncle. May display obscure, dark chevron-shaped mark on posterior body. Blind side white. Upper jaw extends nearly to posterior portion of lower pupil. Eyes separated by a low and narrow ridge. Lateral line with a slight arch over pectoral fin HABITAT: NJ to FL, Gulf of Mexico, and Caribbean Sea to southern Brazil. Demersal over soft bottoms in estuaries, lagoons, and brackish and hypersaline waters to from shore to about 250 ft.

Mexican Flounder - *Cyclopsetta chittendeni* Bean, 1895

FEATURES: Eyed side brown with a blackish blotch under pectoral fin. Dorsal and anal fins with two large, dark blotches that are partially enclosed by a pale C-shaped mark; also with several small, dark, partially ocellated spots. Caudal fin with three dark, partially ocellated blotches at margin. Blind side white. Upper jaw extends past rear margin of lower eye. Eyes separated by a moderately narrow ridge. HABITAT: NW FL to Yucatán and southern Caribbean Sea to Brazil. Also Jamaica. Found over soft bottoms on inner continental shelves from about 20 to 490 ft.

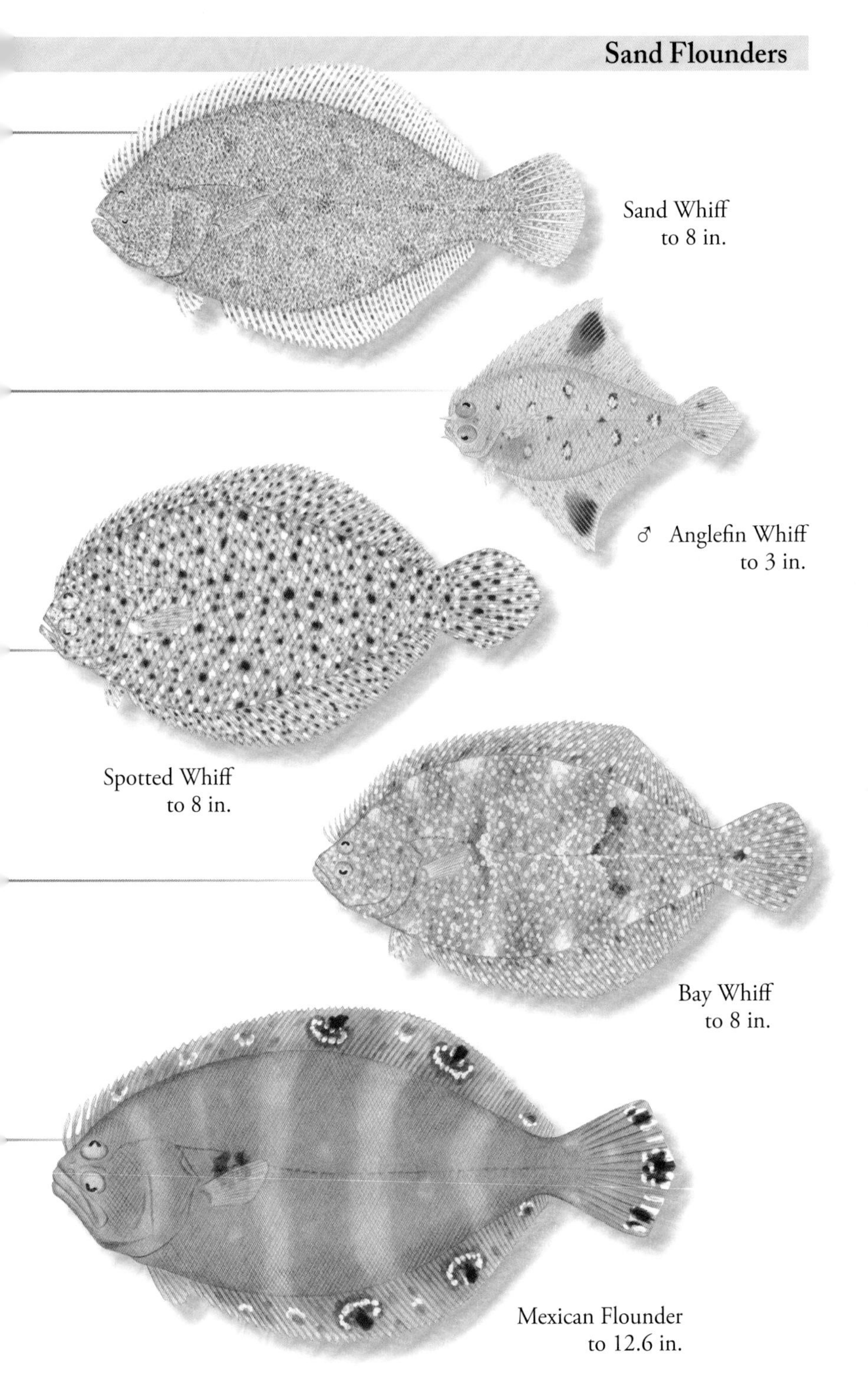
Sand Whiff
to 8 in.
♂ Anglefin Whiff
to 3 in.
Spotted Whiff
to 8 in.
Bay Whiff
to 8 in.
Mexican Flounder
to 12.6 in.

Spotfin Flounder - *Cyclopsetta fimbriata* (Goode & Bean, 1885)

FEATURES: Eyed side brown with numerous small, pale, and dark speckles that form loose clusters. Pectoral fin with a broad, dark margin. Dorsal and anal fins with two large, partially ocellated dark blotches. Center of caudal fin with a large, partially ocellated spot; sometimes with dark blotches at margin. Blind side white. Mouth large. Upper jaw extends to rear margin of lower eye. Eyes separated by a narrow ridge. HABITAT: NC to FL, Gulf of Mexico, and Caribbean Sea to Brazil. Absent from Cuba, Haiti, and Dominican Republic. Demersal over soft bottoms from about 60 to 750 ft.

Fringed Flounder - *Etropus crossotus* Jordan & Gilbert, 1882

FEATURES: Eyed side shades of brown with dense specks and minute flecks. Two or three dusky blotches sometimes present on lateral line. Caudal fin may have a dark margin. Mouth very small, extending to anterior portion of lower eye. Eyes separated by a slightly elevated ridge. Upper margin of head slightly concave. Body comparatively deep. HABITAT: VA to FL, Gulf of Mexico, and Caribbean Sea to Brazil. Demersal over soft mud and sandy bottoms from shore to about 360 ft. BIOLOGY: Enter bays during summer months.

Shrimp Flounder - *Gastropsetta frontalis* Bean, 1895

FEATURES: Eyed side tan to brown with two ocellated spots on upper side, one ocellated spot on lower side. Several dark lines on and between eyes. Body variably flecked and marbled. Anterior dorsal-fin spines tall, with third spine tallest. Lateral line strongly arched over pectoral fin. HABITAT: NC to FL, Gulf of Mexico, and scattered in Caribbean Sea. Demersal over soft bottoms from about 115 to 700 ft. BIOLOGY: Feed on crustaceans and fishes.

Gulf Flounder - *Paralichthys albigutta* Jordan & Gilbert, 1882

FEATURES: Eyed side shades of brown to gray with three ocellated spots in a triangular pattern on midbody: one on upper body, one on lower body, and one on posterior lateral line. Spots may be faint in large specimens. Remainder of dorsal surface with numerous small spots and flecks. Jaws large with upper extending to beyond rear margin of lower eye. Upper eye located directly above lower eye. Lateral line strongly arched over pectoral fin. HABITAT: NC to FL, Gulf of Mexico to Yucatán, and northern Bahamas. Demersal over soft bottoms from about 115 to 700 ft. BIOLOGY: Adults migrate offshore in fall and winter to spawn.

Tropical Flounder - *Paralichthys tropicus* Ginsburg, 1933

FEATURES: Eyed side shades of tan to brown and covered with numerous pale and dark spots and regularly spaced moderate and small asymmetric ocellated spots and eye spots. Fins with rows of spots and asymmetric eye spots. Eyes moderately large and separated by a flat space. Mouth large. Upper jaw extends to rear margin of lower eye. Lateral line strongly arched over pectoral fin. HABITAT: Northern coast of South America from eastern Colombia to Suriname. Also off Costa Rica. Demersal over sandy and muddy bottoms from near shore to about 600 ft.

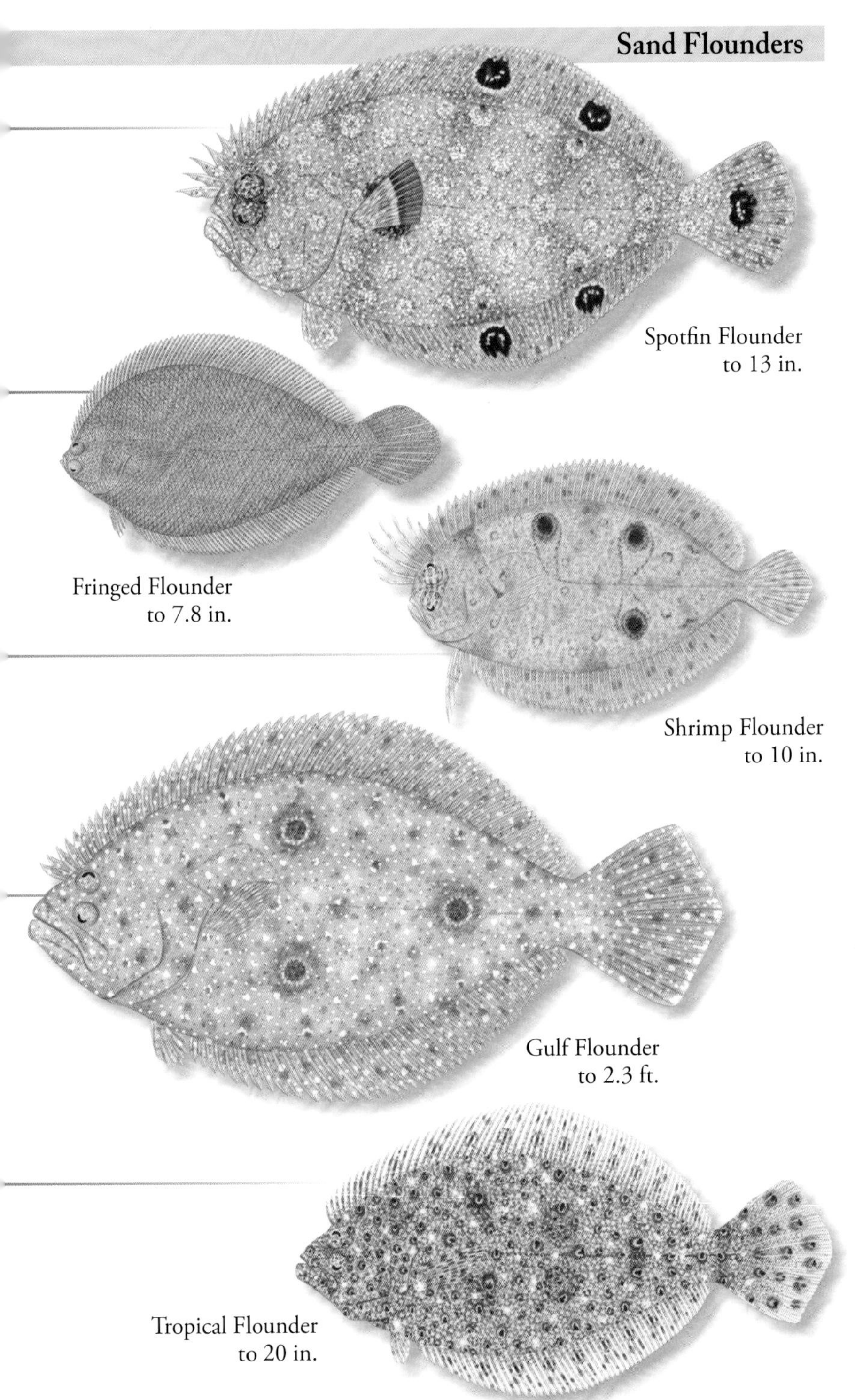

Spotfin Flounder
to 13 in.

Fringed Flounder
to 7.8 in.

Shrimp Flounder
to 10 in.

Gulf Flounder
to 2.3 ft.

Tropical Flounder
to 20 in.

Paralichthyidae - Sand Flounders, *cont.*

Shoal Flounder - *Syacium gunteri* Ginsburg, 1933

FEATURES: Eyed side tannish, with or without ocellated spots or blotches. Usually with a large diffuse blotch on peduncle. Bases of dorsal and anal fins with a series of dark spots. Lower eye distinctly to upper eye. Eyes widely separated in both sexes, more so in males. Males with elongate upper pectoral-fin rays. HABITAT: FL, Gulf of Mexico, and Caribbean Sea. Absent from Cuba and Cayman Islands. Demersal over sandy and muddy bottoms from about 30 to 570 ft.

Channel Flounder - *Syacium micrurum* Ranzani, 1842

FEATURES: Eyed side tan to brown with small dark and pale spots that form loose ocellated clusters. Dark clusters along lateral line and at center of caudal fin. Bases of dorsal and anal fins with a series of dark spots. Upper jaw extends to middle of lower eye. Males with moderately separated eyes and elongate pectoral-fin rays. Lateral line nearly straight. HABITAT: SC to FL, Gulf of Mexico, Bahamas, and Caribbean Sea to Brazil. Demersal over sandy and muddy bottoms from about 30 to 1,350 ft.

Dusky Flounder - *Syacium papillosum* (Linnaeus, 1758)

FEATURES: Eyed side shades of gray to brown with small, dark, and pale spots and faint to distinct ocellated clusters. Bases of dorsal and anal fins with a series of dark spots. May also appear fairly uniformly colored. Several dark lines on snout and between eyes. Males with widely separated eyes, iridescent bluish white lines on head, elongate pectoral-fin rays. Lateral line nearly straight. HABITAT: NC to FL, Gulf of Mexico, Caribbean Sea to Brazil. Demersal over sandy and muddy bottoms from about 10 to 490 ft.

NOTE: There are 11 other rare or poorly recorded Sand Flounders in the area.

Bothidae - Lefteye Flounders

Peacock Flounder - *Bothus lunatus* (Linnaeus, 1758)

FEATURES: Eyed side shades of tan, greenish, yellow, or gray to whitish with numerous pale to dark blue circles, semi-circles, and spots. May be heavily mottled or almost uniformly pale or dark. Usually with two or three dark clusters on lateral line. Eyes widely separated—more so in mature males—with a circle of tentacles. Males with elongate pectoral-fin rays. All with a notch above snout. HABITAT: E FL, Gulf of Mexico, Bermuda, Bahamas, Caribbean Sea to Brazil. Demersal over sand, coral, rocky, and seagrass bottoms, and around mangroves from shore to about 400 ft.

Mottled Flounder - *Bothus maculiferus* (Linnaeus, 1758)

FEATURES: Eyed side shades of tan, greenish, yellow, or gray to whitish with numerous, pale to dark blue circles, semi-circles, and spots. May be heavily mottled or almost uniformly pale or dark. Usually with a black heart-shaped mark on lateral line. Eyes widely separated—more so in mature males—with fleshy tentacles. Males with elongate upper pectoral-fin rays. All with a notch above snout. HABITAT: FL Keys, Bermuda, Bahamas, and Caribbean Sea to Brazil. Demersal over softs and hard bottoms near reefs and over seagrass beds from shore to about 150 ft.

Sand Flounders

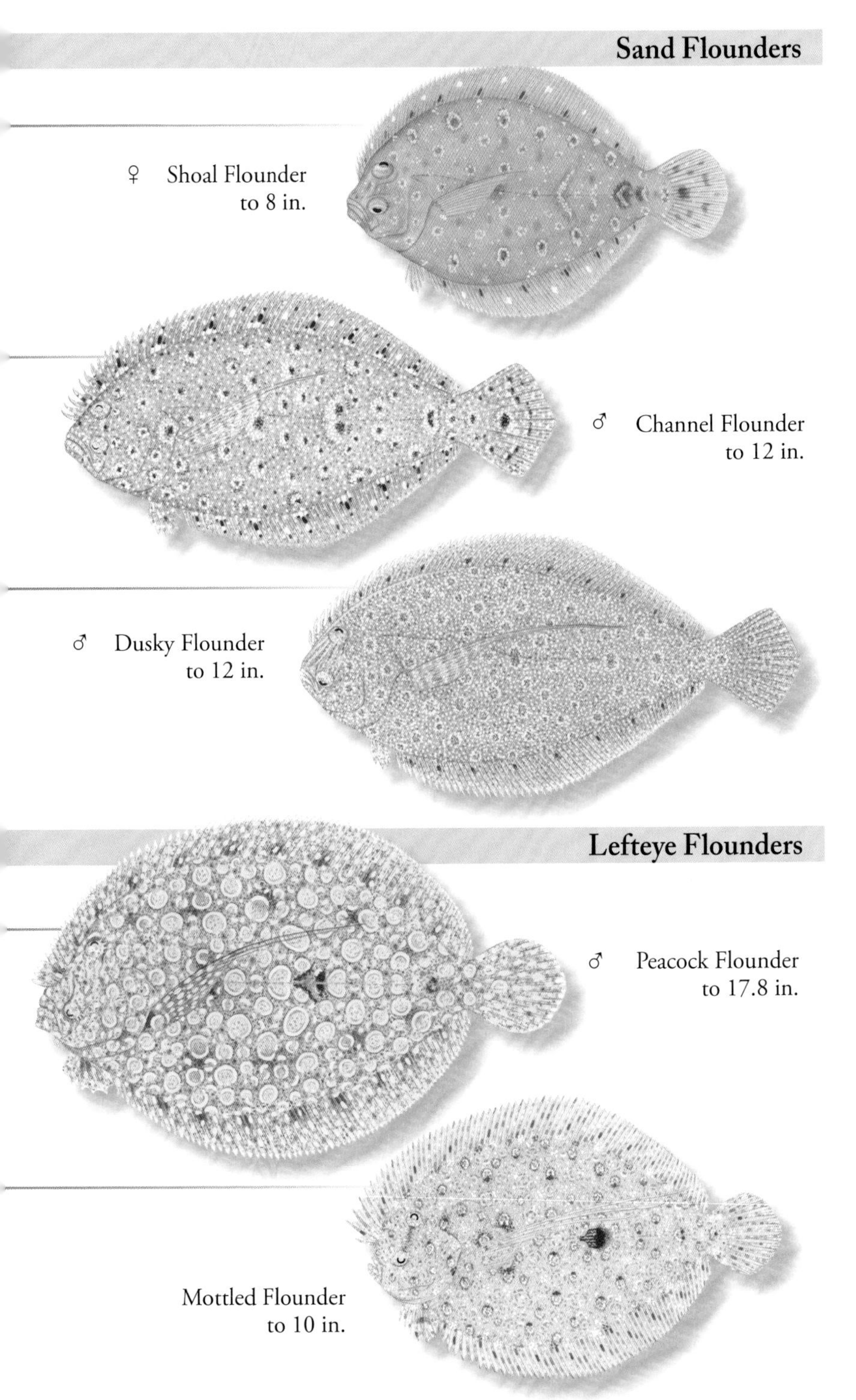

Lefteye Flounders

Bothidae - Lefteye Flounders, *cont.*

Eyed Flounder - *Bothus ocellatus* (Agassiz, 1831)

FEATURES: Eyed side shades of pale tan or brown to gray with white rings, rosettes, and ocellated spots. Three dark clusters on lateral line, middle cluster wedge-shaped. Caudal fin with a dark cluster on upper and lower lobes. Eyes large, widely separated—more so in males—and with a small tentacle. Males with a spine at tip of snout and elongate upper pectoral-fin rays. HABITAT: NY to FL, Gulf of Mexico, Bermuda, Bahamas, Caribbean Sea to Brazil. Demersal over sandy and shelly bottoms from about 15 to 390 ft.

Twospot Flounder - *Bothus robinsi* Topp & Hoff, 1972

FEATURES: Eyed side shades of tan, pink, or gray with numerous darker spots and pale rings. Several dark blotches along lateral line. Caudal fin with two dark spots on middle rays. Eyes large, widely separated—more so in males. Males also with a sharp spine on snout and elongate upper pectoral-fin rays. HABITAT: NC to FL, Gulf of Mexico, Bermuda, Bahamas, and Caribbean Sea to Brazil. Demersal over soft bottoms in bays, lagoons, and along the coast from about 30 to 295 ft.

Spiny Flounder - *Engyophrys senta* Ginsburg, 1933

FEATURES: Eyed side shades of tan to grayish with dark speckles, white rosettes, and irregular ocellated spots. Three dark clusters along lateral line; middle cluster heart-shaped. Small, dark clusters along dorsal and ventral profiles, and along dorsal and anal fin bases. Spiny ridge between eyes. Single long to short tentacle on eyes; may be absent in males. HABITAT: NC to FL, Gulf of Mexico, northern Bahamas, southern Caribbean Sea to Brazil. Also Bahamas. Over soft bottoms from about 100 to 600 ft.

NOTE: There are 10 other poorly recorded or deep-water Lefteye Flounders in the area.

Achiridae - American Soles

Plainfin Sole - *Achirus declivis* Chabanaud, 1940

FEATURES: Eyed side shades of brown to blackish brown with about 8–10 very narrow blackish stripes on sides. Caudal fin unmarked. Blind side whitish. Upper head margin with short tentacles. Pectoral fins small. HABITAT: Belize to southern Brazil. Also Jamaica and Puerto Rico to St. Barthélemy. Demersal over soft bottoms from estuaries to nearshore waters.

Lined Sole - *Achirus lineatus* (Linnaeus, 1758)

FEATURES: Eyed side shades of brown to olive brown with small and large dark spots on body and fins. About eight narrow, faint to dark bars may be present. Blind side white anteriorly, brownish posteriorly. Fimbriae present around mouth on eyed side and on blind side of head. Tufts of dark cirri on both sides of body. Pectoral fins very small. HABITAT: SC to FL, Gulf of Mexico, Caribbean Sea to Argentina. Demersal over sandy mud bottoms in estuaries and in hypersaline lagoons. Also in closed lagoons and over seagrass beds.

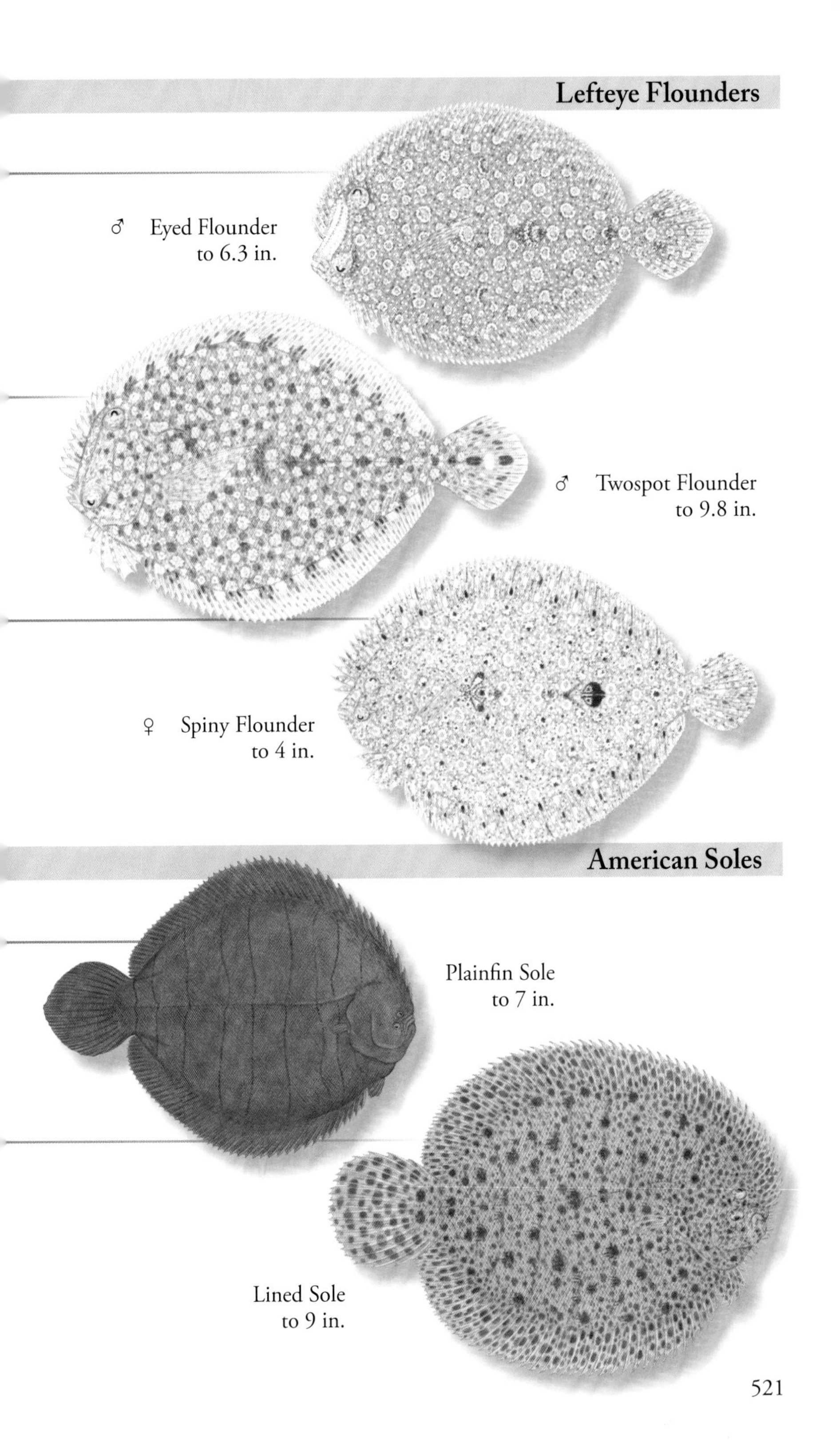
Lefteye Flounders
♂ Eyed Flounder
to 6.3 in.
♂ Twospot Flounder
to 9.8 in.
♀ Spiny Flounder
to 4 in.
American Soles
Plainfin Sole
to 7 in.
Lined Sole
to 9 in.

Nude Sole - *Gymnachirus nudus* Kaup, 1858

FEATURES: Eyed side shades of tan to brown with 13–21 wavy blackish bars that are usually narrower than interspaces. Bars may be distinct to very faint. Juveniles very dark brown to almost black, dorsally and ventrally; some may be gray. Fimbriae around head profile and around eyes. Scales absent. Short cirri on eyed side present or absent. Pectoral fins present or absent. HABITAT: Bay of Campeche, Cayman Islands, Jamaica to St. Barthélemy, and southern Caribbean Sea to southern Brazil. Demersal over soft bottoms from shore to about 330 ft.

Scrawled Sole - *Trinectes inscriptus* (Gosse, 1851)

FEATURES: Eyed side tannish to grayish with a dark, reticulating pattern from head to caudal fin. Several small, dark, scattered blotches may be present. Fimbriae present around mouth on eyed side and are broadly distributed on blind side of head. Pectoral fins rudimentary. HABITAT: S FL, Bahamas, and Caribbean Sea to Guyana. Demersal over soft bottoms in clear, coastal waters from estuaries and bays to mangrove-lined lagoons.

Slipper Sole - *Trinectes paulistanus* (Miranda Ribeiro, 1915)

FEATURES: Eyed side shades of pale brown to nearly uniform and very dark brown with about eight blackish, very narrow dotted bars and diffuse, scattered dark blotches. Blotches may form a loose net-like pattern. Fins spotted. Fimbriae present around head profile. Clusters of cirri present on dark blotches. Pectoral fin on eyed side usually a single ray. HABITAT: Southern Caribbean Sea to southern Brazil. Demersal over soft bottoms in shallow-water estuaries, bays, and hypersaline lagoons.

ALSO IN THE AREA: *Gymnachirus texae*, *Trinectes maculatus*, see p. 547.

Cynoglossidae - Tonguefishes

Coral Reef Tonguefish - *Symphurus arawak* Robins & Randall, 1965

FEATURES: Eyed side grayish to yellowish with two to seven (usually four to five) oblique, vague, broken brownish bars. May also have variable dark blotches. Narrow, dark bar above upper eye. Rear margin of tail dark. Blind side flecked. Dorsal fin with 70–76 rays. Caudal fin bluntly rounded. HABITAT: FL Keys, Bahamas, and Caribbean Sea. Demersal over sandy bottoms around coral reefs from near shore to about 130 ft. OTHER NAME: Caribbean Tonguefish.

Caribbean Tonguefish - *Symphurus caribbeanus* Munroe, 1991

FEATURES: Eyed side grayish to yellowish with 9–15 irregular, broken, dark grayish bars. Bases of dorsal and anal fins with a row of large, blackish blotches. Pinkish on cheeks. Dorsal fin with 89–96 rays. Caudal fin bluntly pointed. HABITAT: NW Cuba, Haiti to Saint Maarten, and Honduras to Colombia. Demersal over shallow-water sandy and muddy bottoms.

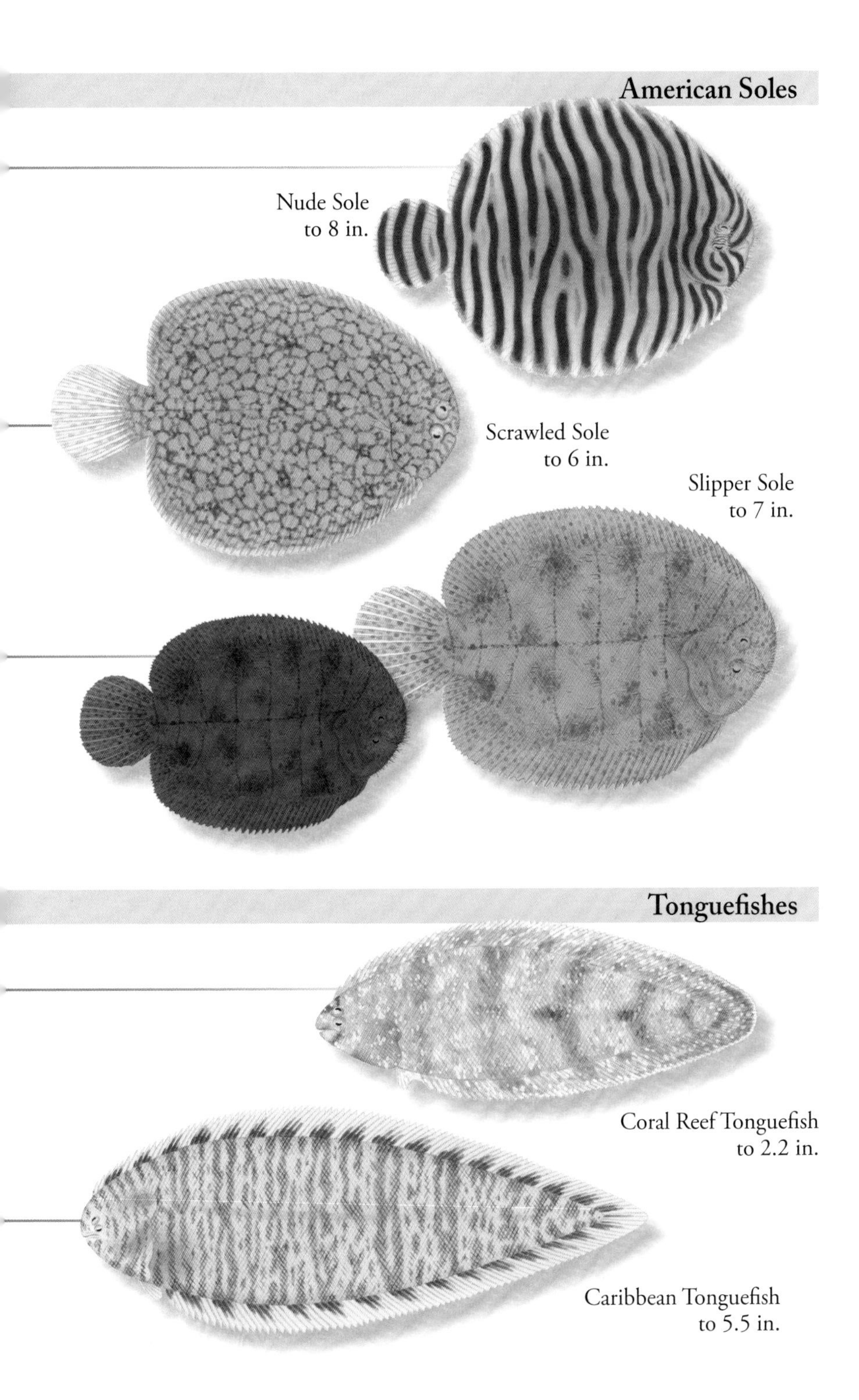
American Soles
Nude Sole
to 8 in.
Scrawled Sole
to 6 in.
Slipper Sole
to 7 in.
Tonguefishes
Coral Reef Tonguefish
to 2.2 in.
Caribbean Tonguefish
to 5.5 in.

Offshore Tonguefish - *Symphurus civitatum* Ginsburg, 1951

FEATURES: Eyed side pale to dark brown. May have 6–14 faint to distinct bars. Opercle often with a dark patch. Posterior third of dorsal and anal fins dark. Caudal fin dark. Blind side whitish. Dorsal fin with 86–93 rays. HABITAT: NC to FL Keys, Apalachicola Bay, FL, to Yucatán. Over sandy and silty bottoms from about 3 to 240 ft.

Spottedfin Tonguefish - *Symphurus diomedeanus* (Goode & Bean, 1885)

FEATURES: Eyed side uniformly dark brown to heavily mottled and blotched. Specimens from light-colored bottoms usually pale brown or yellowish on eyed side. Diffuse spots may be present along anterior dorsal and anal fin bases. Posterior dorsal and anal fins always with one to five distinct round spots. Dorsal fin with 86–96 rays. HABITAT: NC to FL, Gulf of Mexico, and Caribbean Sea to Brazil. Also Jamaica and Puerto Rico. Demersal over shelly mud and sand bottoms from about 20 to 1,000 ft.

Ocellated Tonguefish - *Symphurus ommaspilus* Böhlke, 1961

FEATURES: Eyed side creamy to tan with dense speckles and blotches. May display several irregular dark bars and brownish blotches along dorsal and anal fin bases. All with a single black ocellated spot on rear dorsal and anal fins. Cheeks pinkish. Dorsal fin with 75–79 rays. HABITAT: Bahamas and scattered in Caribbean Sea. Demersal over sandy bottoms around coral reefs in shallow, clear water.

Blackcheek Tonguefish - *Symphurus plagiusa* (Linnaeus, 1766)

FEATURES: Eyed side shades of tan to brown and faintly to densely mottled, blotched, and barred. Usually with a dark blotch at upper opercle. Fins blotched or nearly unpigmented. Opercular cavity and throat region dark. Dorsal fin with 81–91 rays. HABITAT: NY to FL, Gulf of Mexico, Cuba, and Bahamas. Demersal over sandy and muddy bottom from shallow-water estuaries and tidal creeks to about 98 ft.

Tessellated Tonguefish - *Symphurus tessellatus* (Quoy & Gaimard, 1824)

FEATURES: Eyed side shades of tan to brown with five to nine dark bars; may have faint dark bars on interspaces. Dorsal and anal fins blotched at midlength, blackish on tail. Opercular cavity and throat region dark. Dorsal fin with 91–102 rays. HABITAT: Caribbean Sea to Uruguay. Demersal over soft silt and muddy sand bottoms from shore to about 280 ft.

Spottail Tonguefish - *Symphurus urospilus* Ginsburg, 1951

FEATURES: Eyed side shades of brown to gray with 4–11 (usually 6–11) dark bars. Dorsal and anal fins faintly to densely spotted. Caudal fin with a distinct black ocellated spot. Dorsal fin with 82–90 rays. HABITAT: NC to FL, Gulf of Mexico, and southern Cuba. Demersal over vegetated bottoms from about 15 to 1,060 ft. Usually above 130 ft.

NOTE: There are five other rare or deep-water Tonguefishes in the area.

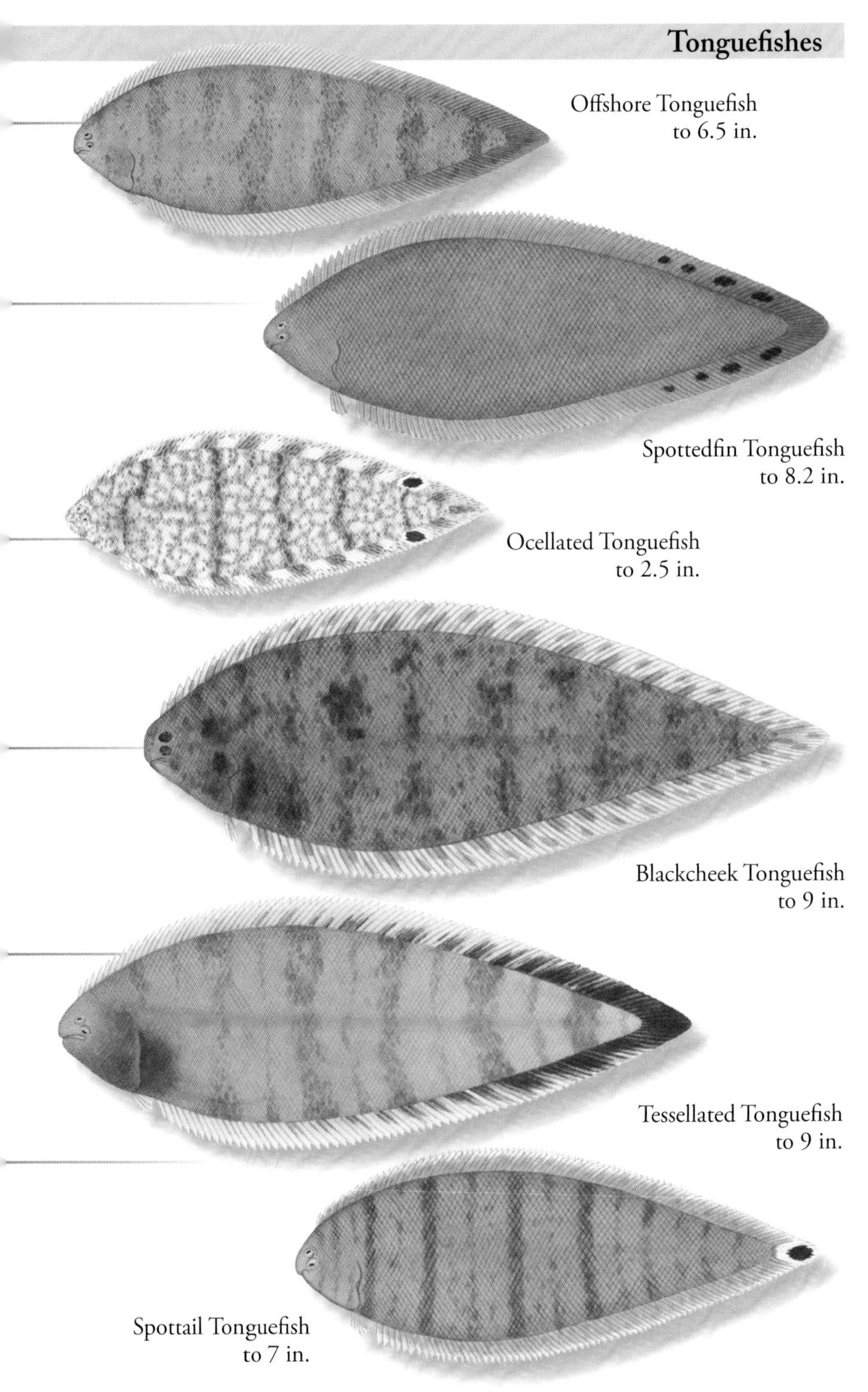
Offshore Tonguefish
to 6.5 in.
Spottedfin Tonguefish
to 8.2 in.
Ocellated Tonguefish
to 2.5 in.
Blackcheek Tonguefish
to 9 in.
Tessellated Tonguefish
to 9 in.
Spottail Tonguefish
to 7 in.

Balistidae - Triggerfishes

Gray Triggerfish - *Balistes capriscus* Gmelin, 1789

FEATURES: Color and pattern vary. Grayish, greenish gray, or ochre to brownish. May be nearly uniformly colored or with obscure to very dark, irregular bars on upper sides. Whitish below. Fins spotted and banded. May have bright blue spots or lines on sides and fins. Caudal-fin margin convex in juveniles; upper and lower rays long and trailing in adults. HABITAT: Nova Scotia to FL, Gulf of Mexico, Bermuda, Bahamas, and Caribbean Sea to Argentina. Also in eastern Atlantic. Found over coral reefs and rocky, sandy, and grassy bottoms to about 360 ft. Juveniles associated with *Sargassum*.

Queen Triggerfish - *Balistes vetula* Linnaeus, 1758

FEATURES: Greenish to bluish gray, or yellowish brown dorsally. Lower portion of head and abdomen yellowish. Two curved, blue bands on cheeks. Dark lines radiate from eyes. Upper body with dark, diagonal lines. Dorsal, caudal, and anal fins with blue inner and outer margins. May pale, darken, or change color. Anterior dorsal-fin ray, upper and lower caudal-fin rays long and trailing in large specimens. HABITAT: MA to FL, Gulf of Mexico, Bermuda, Bahamas, and Caribbean Sea to Brazil. Also in eastern Atlantic. Occur over rocky and coral reefs and adjacent sandy and grassy bottoms to about 900 ft.

Rough Triggerfish - *Canthidermis maculata* (Bloch, 1786)

FEATURES: Blackish or grayish to pale or dark blue with irregular, pearly white to blue spots on body. Spots may be distinct to faint. Spots larger and fewer in juveniles. A deep groove is present in front of eyes. Dorsal fin with 23–25 rays. Body comparatively elongate. HABITAT: Circumglobal in tropical to warm temperate seas. In western Atlantic from NY to FL, Gulf of Mexico, Bermuda, Bahamas, and Caribbean Sea to Argentina. Pelagic. Occur from near shore to well offshore. Usually associated with flotsam and *Sargassum* seaweed.

Ocean Triggerfish - *Canthidermis sufflamen* (Mitchill, 1815)

FEATURES: Brownish to grayish dorsally and on sides. Paler below. May be very pale overall. Usually with a black to brownish black blotch at pectoral-fin base. Margins of dorsal, caudal, and anal fins may be dark. A deep groove is present in front of eyes. Dorsal fin with 26–27 rays. Body comparatively deep. HABITAT: MA to FL, Gulf of Mexico, Bermuda, Bahamas, and Caribbean Sea to Brazil. Pelagic. Occur around dropoffs of offshore reefs in clear water from near surface to about 980 ft.

Black Durgon - *Melichthys niger* (Bloch, 1786)

FEATURES: Dark blackish brown to almost uniformly blackish. Cheeks may have yellowish to orange tinge, and blue lines may radiate from eyes. Scale edges often edged in blue, forming a distinct cross-hatched and zigzag pattern. All with a bright, pale blue band at dorsal- and anal-fin bases. Scales on posterior body are keeled. HABITAT: Circumtropical. In western Atlantic from S FL, N and S Gulf of Mexico, Bermuda, Bahamas, and Caribbean Sea to Brazil. Found in clear water over rocky and coral reefs from surface to about 1,300 ft.

Gray Triggerfish
to 2 ft.
Queen Triggerfish
to 20 in.
Rough Triggerfish
to 20 in.
Ocean Triggerfish
to 2 ft.
Black Durgon
to 20 in.

Balistidae - Triggerfishes, *cont.*

Sargassum Triggerfish - *Xanthichthys ringens* (Linnaeus, 1758)

FEATURES: Color varies and may pale or darken. Shades of blue to gray with blackish to brownish spots that follow scale rows on sides. Dorsal- and anal-fin base black to reddish brown. Caudal-fin margins black to reddish brown. All with three narrow, blue to blackish, grooved lines on lower cheeks. HABITAT: NC to FL, northern and southern Gulf of Mexico, Bermuda, Bahamas, and Caribbean Sea to Brazil. Adults occur over deep reefs. Juveniles with floating *Sargassum* seaweed.

Monacanthidae - Filefishes

Dotterel Filefish - *Aluterus heudelotii* Hollard, 1855

FEATURES: Shades of olivaceous or tan to brownish. May have a pale lateral band and pale blotches anteriorly and ventrally. Bright blue spots and wavy lines present on head and body. Can pale, darken, or change color. Juveniles nearly uniform dark brown to black. Second dorsal fin with 36–41 rays. Body comparatively deep. HABITAT: MA to FL, Gulf of Mexico, Bermuda, southern Caribbean Sea to Brazil. Occur over seagrass beds and sandy and muddy bottoms to about 100 ft. BIOLOGY: Feed vertically on a variety of plants and invertebrates.

Unicorn Filefish - *Aluterus monoceros* (Linnaeus, 1758)

FEATURES: Metallic pale grayish to brownish. May be uniformly colored or may display a reticulating pattern of small, pale spots and lines. Second dorsal and anal fins yellowish. Juveniles variably blotched and spotted in shades of brown to black. Second dorsal fin with 46–50 rays. HABITAT: Circumtropical. In western Atlantic from MA to FL, Gulf of Mexico, Bermuda, Bahamas, and Caribbean Sea to Brazil. Occur over soft and rocky bottoms from surface to about 490 ft.

Orange Filefish - *Aluterus schoepfii* (Walbaum, 1792)

FEATURES: Metallic pale gray to silvery. All with few to numerous brown or orange to yellowish spots that may cluster into blotches or bands. May also display brownish bands or stripes. Second dorsal fin with 32–40 rays. Body oval to rhomboid in profile. HABITAT: Nova Scotia to FL, Gulf of Mexico, Bermuda, Bahamas, and Caribbean Sea to Brazil. Occur over seagrass beds and sandy and muddy bottoms from surface to about 165 ft. Also around reefs. BIOLOGY: Feed vertically on a variety of plants.

Scrawled Filefish - *Aluterus scriptus* (Osbeck, 1765)

FEATURES: Color and pattern highly variable. Shades of greenish, yellowish or bluish to brownish. Ground color may be uniform or blotched and banded. Black spots and bright blue spots and lines on head and body. May pale, darken, or change color. Juveniles ochre to blackish with a pale, net-like pattern. Second dorsal fin with 43–49 rays. Body comparatively shallow. HABITAT: Circumglobal in warm seas. In western Atlantic from MA to FL, Gulf of Mexico, Bermuda, Bahamas, and Caribbean Sea to Brazil. Occur over coral and rocky reefs to about 400 ft. Juveniles around *Sargassum*.

Triggerfishes

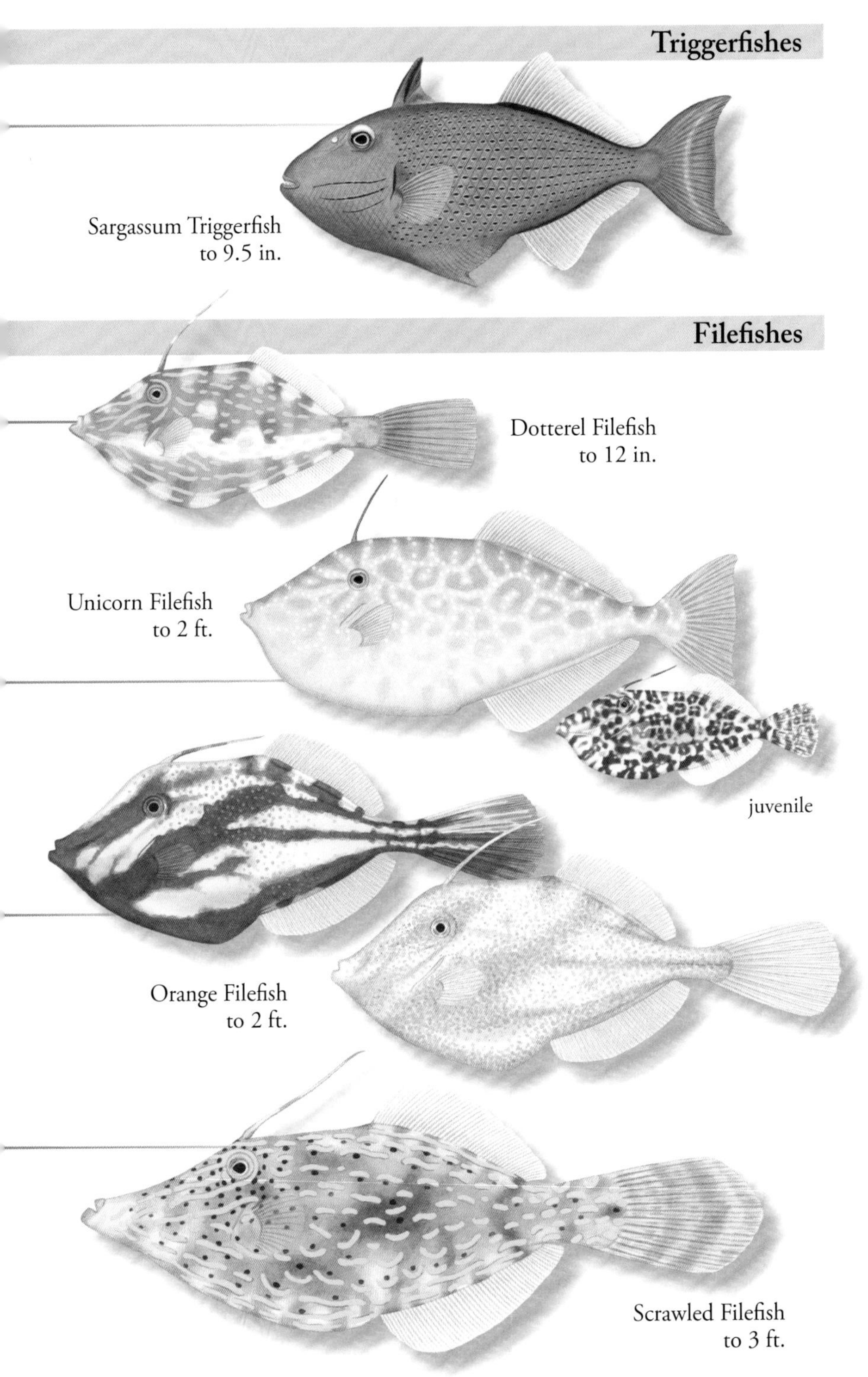

Whitespotted Filefish - *Cantherhines macrocerus* (Hollard, 1853)

FEATURES: Color and pattern vary. May be grayish anteriorly and orangish posteriorly or almost entirely grayish or brownish. Most with a yellowish to white saddle on mid-back. Large diffuse to distinct pale spots may cover body. Caudal fin always dark; may have a pale inner bar. Capable of quick color change. Adults with two or three pairs of enlarged spines on caudal peduncle. HABITAT: FL, Gulf of Mexico, Bermuda, Bahamas, Caribbean Sea to Brazil. Occur in clear water over coral and rocky reefs.

Orangespotted Filefish - *Cantherhines pullus* (Ranzani, 1842)

FEATURES: Color and pattern vary. Shades of brown to gray. Most with two or more faint to distinct pale stripes and orange spots on sides. Blue to orange lines usually present on snout. One or two distinct white spots on caudal peduncle. Second dorsal fin with 33–36 rays. HABITAT: MA to FL, Gulf of Mexico, Bermuda, Bahamas, and Caribbean Sea to southern Brazil. Also eastern Atlantic. Adults over shallow-water coral and rocky reefs. Juveniles pelagic.

Fringed Filefish - *Monacanthus ciliatus* (Mitchill, 1818)

FEATURES: Highly variable. Pale to dark green or tannish to brown. Body variably marked with pale and dark spots, lines, and blotches. May be uniformly dark or pale. Snout somewhat upturned. Pelvic flap very large, expandable. Second dorsal fin with 29–37 rays. Body comparatively deep. HABITAT: Newfoundland to FL, Gulf of Mexico, Bahamas, Caribbean Sea to Argentina. Also Bermuda and eastern Atlantic.

Slender Filefish - *Monacanthus tuckeri* Bean, 1906

FEATURES: Highly variable. Tannish, grayish, brownish, or yellowish to greenish. Body variably marked with spots, blotches, bands, and pale reticulating pattern. May be bicolored. Snout long and upturned. Second dorsal fin with 32–37 rays. Body comparatively elongate. HABITAT: NC to FL, Gulf of Mexico, Bermuda, Bahamas, and Caribbean Sea. Occur over shallow-water sandy, rocky, and grassy bottoms.

Planehead Filefish - *Stephanolepis hispidus* (Linnaeus, 1766)

FEATURES: Variable. Shades of white, tan, ochre, or brown to olivaceous. Irregularly spotted, blotched, or banded. Snout comparatively short. Second dorsal fin with 29–35 (usually 31–34) rays; second ray filamentous in males. HABITAT: Nova Scotia to FL, Gulf of Mexico, Bermuda, Bahamas, and Caribbean Sea to Brazil. Adults over sandy, muddy, and grassy bottoms. Juveniles with floating *Sargassum* seaweed.

Pygmy Filefish - *Stephanolepis setifer* (Bennett, 1831)

FEATURES: Variable. Tan to brown or olivaceous with irregular oblique bars, dark oblong blotches, and pale spots. Caudal fin with two dark bands. Second dorsal fin with 27–29 (rarely 30) rays; second ray filamentous in males. HABITAT: NC to FL, Gulf of Mexico, Bermuda, Bahamas, and Caribbean Sea to Brazil. Adults over sandy, muddy, and grassy bottoms. Juveniles with floating *Sargassum* seaweed.

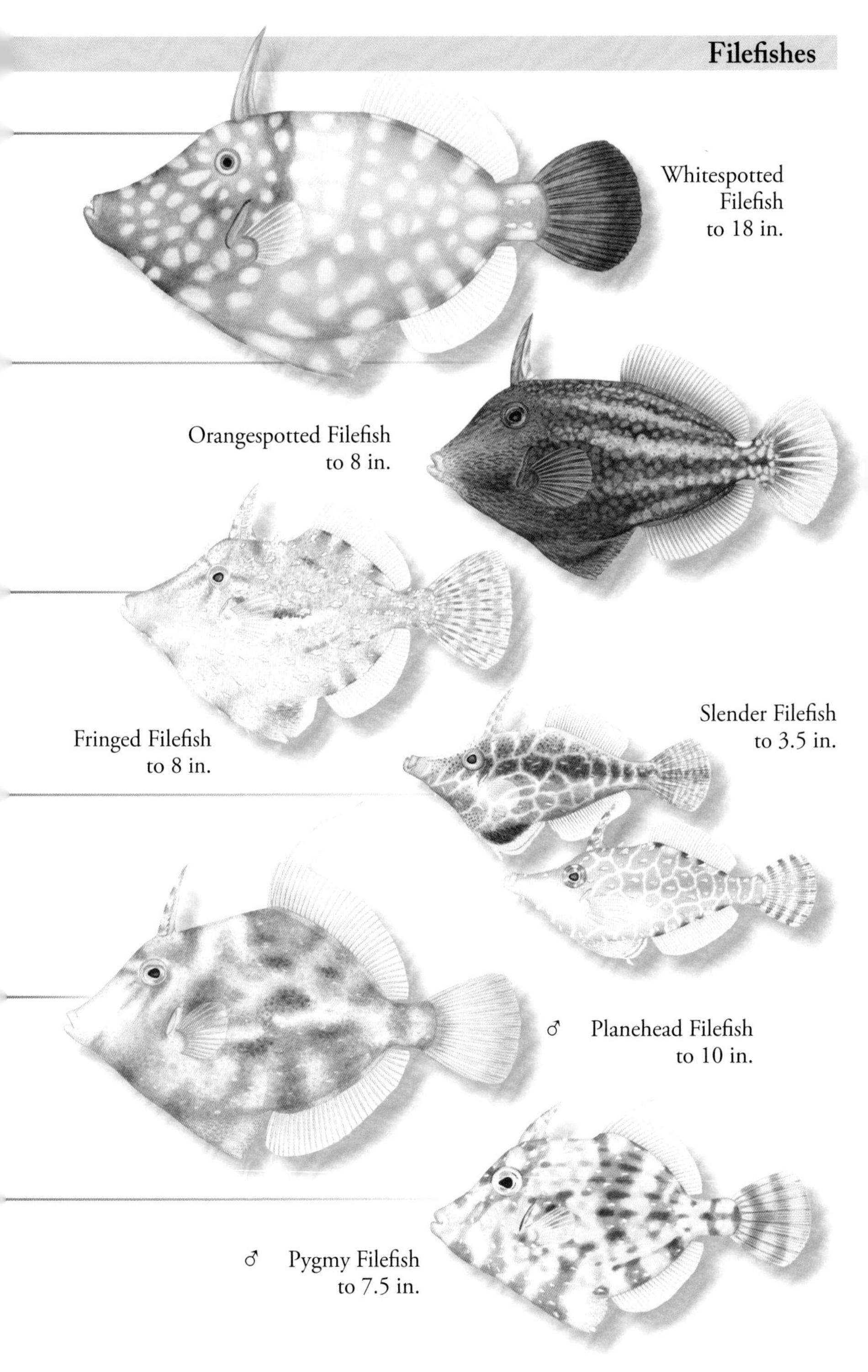
Whitespotted
Filefish
to 18 in.
Orangespotted Filefish
to 8 in.
Fringed Filefish
to 8 in.
Slender Filefish
to 3.5 in.
♂ Planehead Filefish
to 10 in.
♂ Pygmy Filefish
to 7.5 in.

Ostraciidae - Boxfishes

Honeycomb Cowfish - *Acanthostracion polygonium* Poey, 1876

FEATURES: Color and pattern highly variable. Yellowish, greenish, or brownish to bluish with dark brownish to blackish hexagonal marks on sides. Head with a net-like pattern of dark, wavy lines. May quickly change color. Scales in front of eyes and anterior to anal fin expanded into spine-like projections. Pectoral fins usually with 12 rays (rarely 11). HABITAT: NJ to FL, Gulf of Mexico, Bermuda, Bahamas, and Caribbean Sea to Brazil. Occur over coral reefs and in estuaries to about 260 ft.

Scrawled Cowfish - *Acanthostracion quadricornis* (Linnaeus, 1758)

FEATURES: Color and pattern variable. Yellowish, greenish, brown, or gray with bright blue to black spots and wavy lines on body. Lines may be broad to narrow and few to many. Usually with an unbroken line or a series of spots along dorsal and ventral profiles. Spots and lines extend onto caudal fin. Scales in front of eyes and anterior to anal fin expand into spine-like projections. Pectoral fin usually with 11 rays (rarely 10 or 12). HABITAT: MA to FL, Gulf of Mexico, Bermuda, Bahamas, and Caribbean Sea to Brazil. Rarely stray to South Africa. Found primarily over shallow seagrass beds. BIOLOGY: Feed on soft corals, sponges, tunicates, and shrimps.

Spotted Trunkfish - *Lactophrys bicaudalis* (Linnaeus, 1758)

FEATURES: Color is variable. Tannish or whitish to pale gray. Evenly spaced, relatively large, brown to blackish spots cover head and body. Spots extend onto caudal fin and may merge to form a dense network of wavy lines. May display diffuse dark areas. Scales anterior to anal fin expanded into spine-like projections. Pectoral fins with 12 rays. HABITAT: GA to FL, Flower Garden Banks, southern Gulf of Mexico, Bahamas, and Caribbean Sea to Brazil. Occur over seagrass beds and coral reefs in clear water to about 310 ft. NOTE: Previously *Rhinesomus bicaudalis*.

Trunkfish - *Lactophrys trigonus* (Linnaeus, 1758)

FEATURES: Color and pattern highly variable. Tannish, olivaceous, or brownish to grayish. May display numerous pale spots, dark blotches, or bands, or a network of dark wavy lines. Often with a blackish chain-like patch behind pectoral fins. Scales anterior to anal fin expanded into spine-like projections. Pectoral fins usually with 12 rays (rarely 11 or 13). HABITAT: MA to FL, Gulf of Mexico, Bermuda, Bahamas, and Caribbean Sea to Brazil. Occur over seagrass beds, rubble areas, and offshore reefs. BIOLOGY: Feed on worms, bivalves, crabs, and tunicates. OTHER NAME: Buffalo Trunkfish.

Smooth Trunkfish - *Lactophrys triqueter* (Linnaeus, 1758)

FEATURES: Variable. Blackish to brown or golden with numerous white spots on head and body. Spots may merge to form large white areas. Usually with a large blackish blotch behind pectoral fins. May have a large tannish blotch on sides. Pectoral, dorsal, and anal fins yellowish. Body lacks prolonged spines. Pectoral fins with 12 rays. HABITAT: MA to FL, Gulf of Mexico, Bermuda, Bahamas, and Caribbean Sea to Brazil. Occur over coral reefs to about 260 ft. NOTE: Previously *Rhinesomus triqueter*.

Honeycomb Cowfish
to 16 in.

juvenile

Scrawled Cowfish
to 18 in.

Spotted Trunkfish
to 18 in.

juvenile

Trunkfish
to 18 in.

Smooth Trunkfish
to 12 in.

juvenile

Tetraodontidae - Puffers

Two-stripe Sharpnose Puffer - *Canthigaster figueiredoi* Moura & Castro, 2002

FEATURES: Uniform pale brown to greenish tan dorsally. White below. Blue lines on snout and around eyes. A black stripe runs from behind eyes to upper caudal-fin margin. A black, broken, zigzag stripe runs from below gill opening to lower caudal-fin margin. Stripes bordered by blue. Dorsal and caudal fins tannish. HABITAT: Guyana to Santa Catarina, Brazil, including offshore islands. May stray to Venezuela and Trinidad. Occur over coral and rocky reefs from near shore to about 180 ft. BIOLOGY: Often in pairs, hovering above reefs. Feed on plants, sponges, and invertebrates.

Goldface Toby - *Canthigaster jamestyleri* Moura & Castro, 2002

FEATURES: Tan to orange brown dorsally with dark spots and scrawls. White below. Blue lines on snout and blue spots on sides and abdomen. A blackish stripe runs from behind eyes to upper caudal-fin margin. A black to brownish stripe runs from below gill opening to lower caudal-fin margin. Dark blotch at dorsal-fin base. Caudal fin yellowish with blue spots. HABITAT: NC to FL; Gulf of Mexico to southern Caribbean Sea. Occur over reefs and hard bottoms from about 45 to 715 ft.

Sharpnose Puffer - *Canthigaster rostrata* (Bloch, 1786)

FEATURES: Dark tan to brown dorsally. Sides yellowish to white. Bright blue lines radiate from eyes. Blue lines and spots on snout, sides, back, and abdomen. Blackish to dark brown on upper and lower peduncle and onto caudal-fin margins. Caudal fin yellowish with blue lines and dark upper and lower margins. HABITAT: NC to FL, Gulf of Mexico, Bermuda, Bahamas, and Caribbean Sea. Occur over coral reefs, seagrass beds, and in mangrove creeks from near shore to about 510 ft.

Banded Puffer - *Colomesus psittacus* (Bloch & Schneider, 1801)

FEATURES: Shades of tan to greenish or yellowish with five to six wide brownish to blackish saddles along back. May be heavily blotched on sides. White below. Caudal fin usually dark. Body robust and covered with tiny prickles from eyes to dorsal fin. HABITAT: Colombia to eastern Brazil. Also Cuba. Demersal over shallow-water soft bottoms in estuaries and around mangroves. Also in freshwater streams and rivers.

Smooth Puffer - *Lagocephalus laevigatus* (Linnaeus, 1766)

FEATURES: Greenish gray, gray, or brownish gray dorsally. Silvery on sides; may have a golden luster. White below. Spots absent on sides. Small specimens with three or four dark saddles on upper sides. Pectoral fin uniformly colored. Snout blunt, rounded. Pectoral fins usually with 17–18 rays. Caudal-fin lobes about equal in length. HABITAT: MA to FL, Gulf of Mexico, Bermuda, Bahamas, and Caribbean Sea to Argentina. Also in eastern Atlantic. Pelagic and demersal over sandy and muddy bottoms from near shore to about 600 ft.

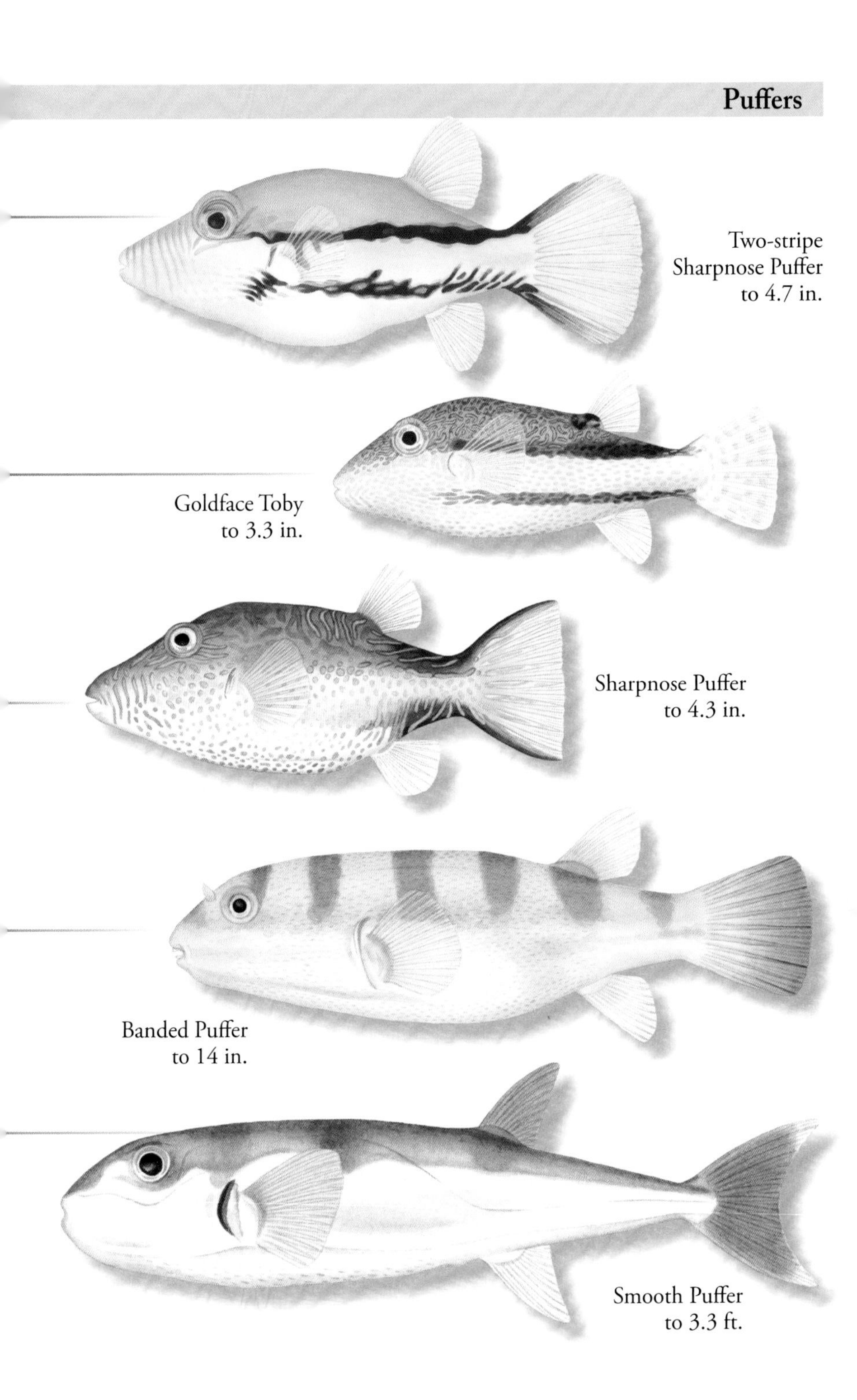
Two-stripe
Sharpnose Puffer
to 4.7 in.
Goldface Toby
to 3.3 in.
Sharpnose Puffer
to 4.3 in.
Banded Puffer
to 14 in.
Smooth Puffer
to 3.3 ft.

Oceanic Puffer - *Lagocephalus lagocephalus* (Linnaeus, 1758)

FEATURES: Dark bluish black dorsally. Silvery on sides; white below. Dark and pale areas distinctly delineated. Upper portion of pectoral fins bluish black. Several rows of faint to dark spots under pectoral-fin base. Juveniles with large spots on back and sides. Snout blunt, rounded. Pectoral fins with 13–16 rays. Lower caudal-fin lobe longer than upper lobe. HABITAT: Circumglobal in tropical to temperate seas. In western Atlantic from Newfoundland to FL, Gulf of Mexico, Bermuda, Bahamas, and Caribbean Sea to southern Brazil. Oceanic and pelagic.

Marbled Puffer - *Sphoeroides dorsalis* Longley, 1934

FEATURES: Variable. Mottled to uniformly brown to gray with small, pale speckles dorsally. One to five dark blotches on lower sides. Caudal fin dark at base and along margin. Some with a network of pale blue lines. Males with yellow scrawls on lower head and sides. All with pair of tiny lappets on back. HABITAT: NC to FL, Gulf of Mexico, Bahamas, Caribbean Sea to Suriname. Demersal over soft bottoms from about 65 to 330 ft.

Green Puffer - *Sphoeroides greeleyi* Gilbert, 1900

FEATURES: Color and pattern vary. Whitish to creamy with clusters of brownish, blackish, or orange brown spots dorsally. Clusters may merge to form large blotches. May display irregular dark bands while resting. Irises bright orange. Caudal fin with pale bars. Usually with small skin flaps along lateral girdle. HABITAT: Caribbean Sea to southern Brazil. Absent from Cayman Islands and Cuba. Demersal over shallow-water soft bottoms in estuaries and tidal creeks, along beaches, and around mangroves.

Southern Puffer - *Sphoeroides nephelus* (Goode & Bean, 1882)

FEATURES: Color and pattern vary. Brown dorsally, with darker spots and blotches, and pale bluish white spots loosely arranged into rings and lines. Dark bar between eyes. Ventral area white. Some males with small, bright red spots. Caudal fin uniformly colored to vaguely barred. Prickles on abdomen end anterior to anus. HABITAT: SC to FL, Gulf of Mexico, Bahamas, and northern and western Caribbean Sea. Demersal over soft bottoms in shallow-water bays, estuaries, and hypersaline lagoons.

Blunthead Puffer - *Sphoeroides pachygaster* (Müller & Troschel, 1848)

FEATURES: Shades of brown or gray dorsally and on sides. Usually with obscure, dark blotches on sides. White ventrally. Caudal fin dark with pale tips in large specimens. Snout very blunt, rounded in profile. Dorsal and anal fins short-based. Prickles and skin flaps entirely absent. HABITAT: Circumglobal except eastern Pacific. In western Atlantic from NJ to FL, Gulf of Mexico, southern Caribbean Sea to Argentina. Demersal over sandy, muddy, and rocky bottoms from surface to about 1,640 ft; usually deeper than 330 ft. Juveniles pelagic.

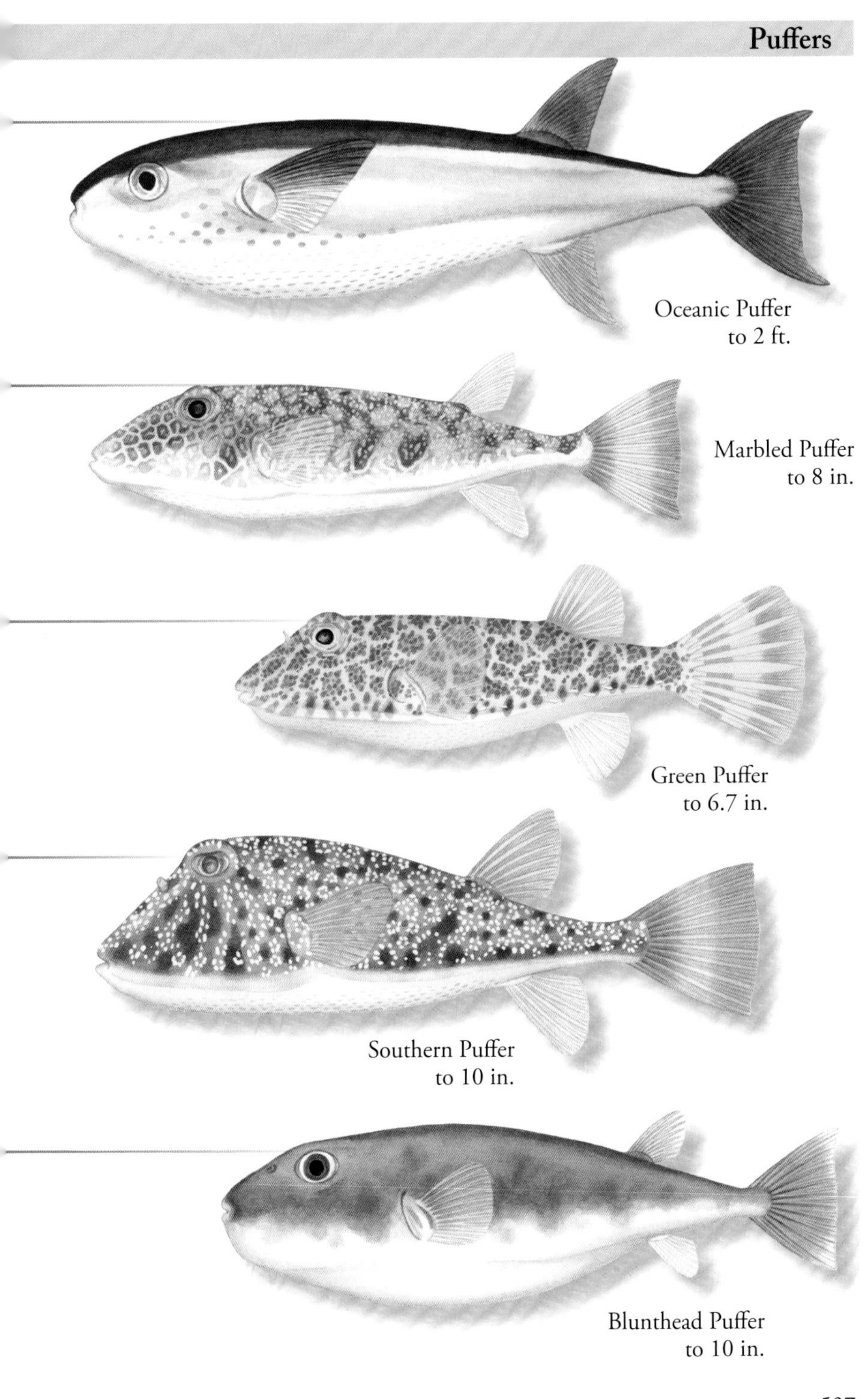
Oceanic Puffer
to 2 ft.
Marbled Puffer
to 8 in.
Green Puffer
to 6.7 in.
Southern Puffer
to 10 in.
Blunthead Puffer
to 10 in.

Least Puffer - *Sphoeroides parvus* Shipp & Yerger, 1969

FEATURES: Pale brown to gray with darker spots, blotches, and mottling dorsally and on sides. Tiny, pale blue green spots pepper upper body. Blotch behind pectoral fins, when present, not darker than other blotches. White ventrally. Snout blunt. Prickles cover body from snout to under dorsal fin. Dorsal and anal fins short-based. Lappets and skin flaps absent. HABITAT: In Gulf of Mexico from Apalachicola Bay, FL, to Bay of Campeche. Demersal over sandy and muddy bottoms along the coast and in bays and estuaries from about 15 to 160 ft.

Bandtail Puffer - *Sphoeroides spengleri* (Bloch, 1785)

FEATURES: Highly variable. Shades of brown dorsally with pale and dark spots, mottling, or a dark lateral stripe. All with a row of large, dark spots from chin to caudal peduncle that do not usually merge with color above. Usually white ventrally. Caudal fin distinctly banded. Snout moderately pointed. Upper body with many small pale lappets. Prickles are present or absent dorsally and on cheeks. Abdomen with prickles almost to anus. HABITAT: MA to FL, Gulf of Mexico, Bermuda, Bahamas, and Caribbean Sea to Brazil. Demersal over reefs and seagrass beds to about 240 ft.

Checkered Puffer - *Sphoeroides testudineus* (Linnaeus, 1758)

FEATURES: Pattern varies. Tannish dorsally with a network of brownish blotches and bands. Small to large darker spots overlay blotches and bands. Bands loosely arranged into a bull's-eye pattern on midback. Lower sides spotted. White to creamy below. Caudal fin may be barred. Snout moderately blunt. Prickles cover most of body and are usually embedded. HABITAT: RI to FL, southern Gulf of Mexico, Bahamas, and Caribbean Sea to Brazil. Demersal over shallow-water seagrass beds, around mangroves, and in bays and estuaries to about 65 ft.

Bearded Puffer - *Sphoeroides tyleri* Shipp, 1972

FEATURES: Tannish to pale greenish gray dorsally with irregular brownish to grayish interconnected spots and marbling. May have larger, scattered, darker blotches. White below. All with a diffuse, dark blotch behind mouth. Dark spots present along lower sides. Caudal fin with a broad, dark margin. Several fleshy tabs along rear lower sides. HABITAT: Colombia to southern Brazil. Demersal along sandy beaches, over shelly bottoms, and in areas with sponges from about 30 to 260 ft.

Speckled Puffer - *Sphoeroides yergeri* Shipp, 1972

FEATURES: Shades of greenish gray to tan dorsally with numerous, irregular reddish speckles. Speckles often for rosettes. White below. Caudal fin dark at base with a wide, moderately dark margin. Skin flaps scattered on back and sides. HABITAT: Belize to eastern Venezuela. Demersal over soft bottoms in clear water from near shore to about 200 ft.

ALSO IN THE AREA: *Sphoeroides georgemilleri*, see p. 547.

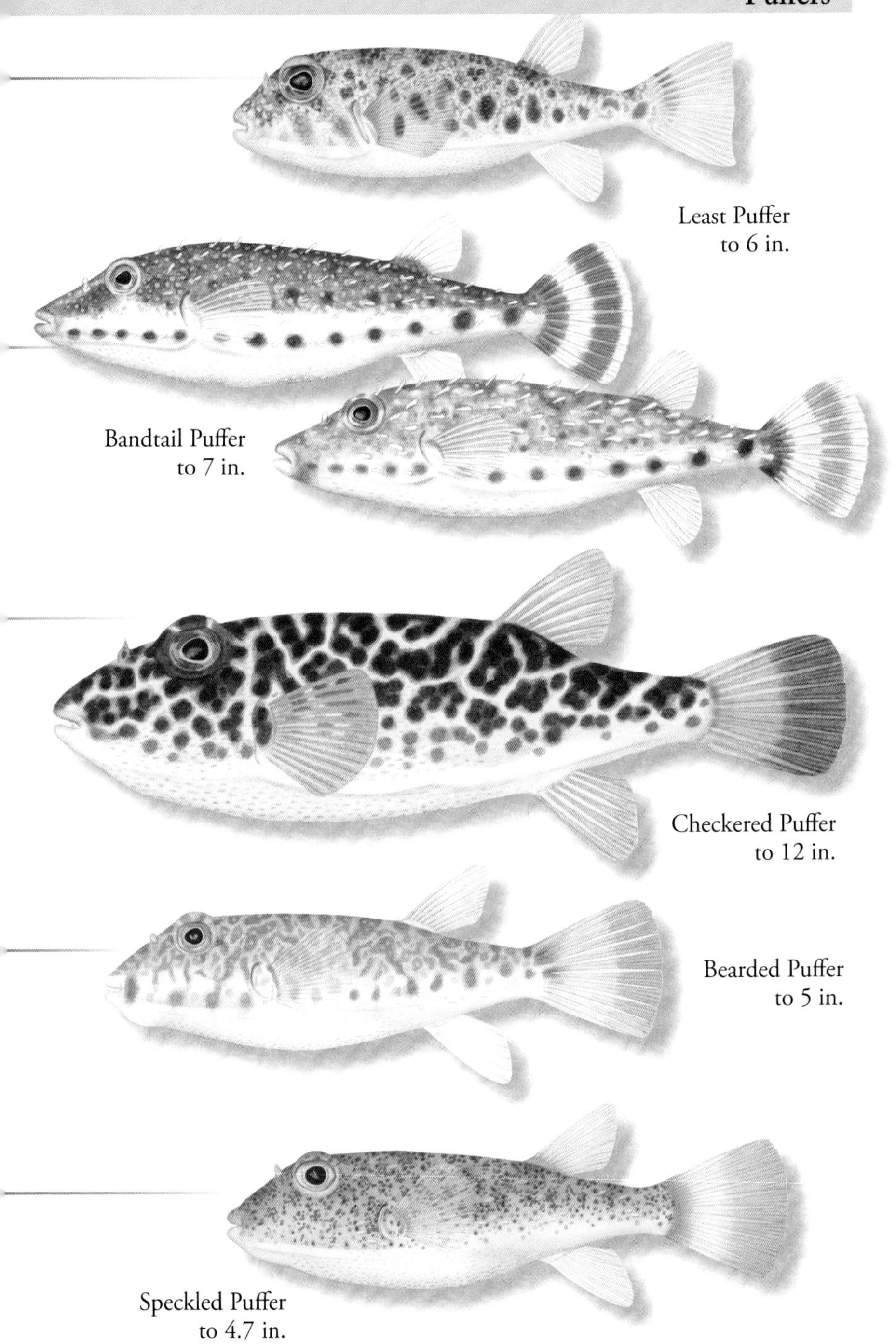
Least Puffer
to 6 in.
Bandtail Puffer
to 7 in.
Checkered Puffer
to 12 in.
Bearded Puffer
to 5 in.
Speckled Puffer
to 4.7 in.

Diodontidae - Porcupinefishes

Bridled Burrfish - *Chilomycterus antennatus* (Cuvier, 1816)

FEATURES: Color and pattern vary. Tannish with small, close-set, black to brown spots on head and body. Large, dark brownish to greenish blotch above pectoral-fin base and at dorsal-fin base. May also have irregular, large blotches on head and sides. Caudal fin may be spotted. A short to tall fleshy tentacle present above eyes. Spines immovable. HABITAT: S FL, northern Gulf of Mexico Bahamas, and Caribbean Sea. Found over seagrass beds and adjacent coral reefs from near surface to about 170 ft.

Web Burrfish - *Chilomycterus antillarum* Jordan & Rutter, 1897

FEATURES: Variable. Whitish to pale tan with a brownish to reddish net-like pattern on head and body. Pale interspaces may be large to very small or appear as pale spots. All with a large, dark blotch above and behind pectoral fins and at dorsal-fin base. May also have a dark blotch above anal-fin base. Short to tall fleshy tentacle present above eyes. Spines immovable, those on abdomen with fleshy filaments. HABITAT: FL, Bermuda, Bahamas, and Caribbean Sea to Brazil. Also reported from Flower Garden Banks. Near reefs and over seagrass beds.

Spotted Burrfish - *Chilomycterus reticulatus* (Linnaeus, 1758)

FEATURES: Tannish dorsally and on sides. May have darker areas on upper body. Relatively small, close-set, blackish spots cover head, upper body, and fins. Anal fin may lack spots. Ventral area whitish. Juveniles blue dorsally, white below, with comparatively large blackish spots dorsally and on sides. Eyes lack a tentacle. Spines immovable. Peduncle with one or two spines dorsally. HABITAT: Circumtropical. In western Atlantic from NC to FL, northern Gulf of Mexico, Campeche Bay, and scattered in Caribbean Sea to Brazil. Demersal over coral and rocky reefs to about 460 ft.

Striped Burrfish - *Chilomycterus schoepfii* (Walbaum, 1792)

FEATURES: Whitish to tannish dorsally and on sides with numerous brown, close-set, and wavy lines. Lines are narrow to broad but are generally evenly spaced. Five to seven blackish blotches present on back and sides. Ventral area white to yellow, sometimes blackish. Four to six fleshy tentacles present on chin. Small tentacle above eyes present or absent. Spines immovable. HABITAT: Nova Scotia to FL, Gulf of Mexico, Bahamas, and NW Caribbean Sea. Demersal over seagrass beds in protected bays, estuaries, and coastal lagoons; move to reefs in winter.

Pelagic Porcupinefish - *Diodon eydouxii* Brisout de Barneville, 1846

FEATURES: Blue to purplish blue dorsally and on sides. May appear blotched. White below. Moderately large, blackish blue spots on head and body. Spots may be present on fin bases. Spines on head and body comparatively long. One spine present on upper caudal peduncle. Dorsal and anal fins with bluntly pointed tips. Body relatively slender. HABITAT: Circumtropical. In western Atlantic from Bermuda, western Gulf of Mexico, Nicaragua, Lesser Antilles, and off central Brazil. Pelagic in upper water column.

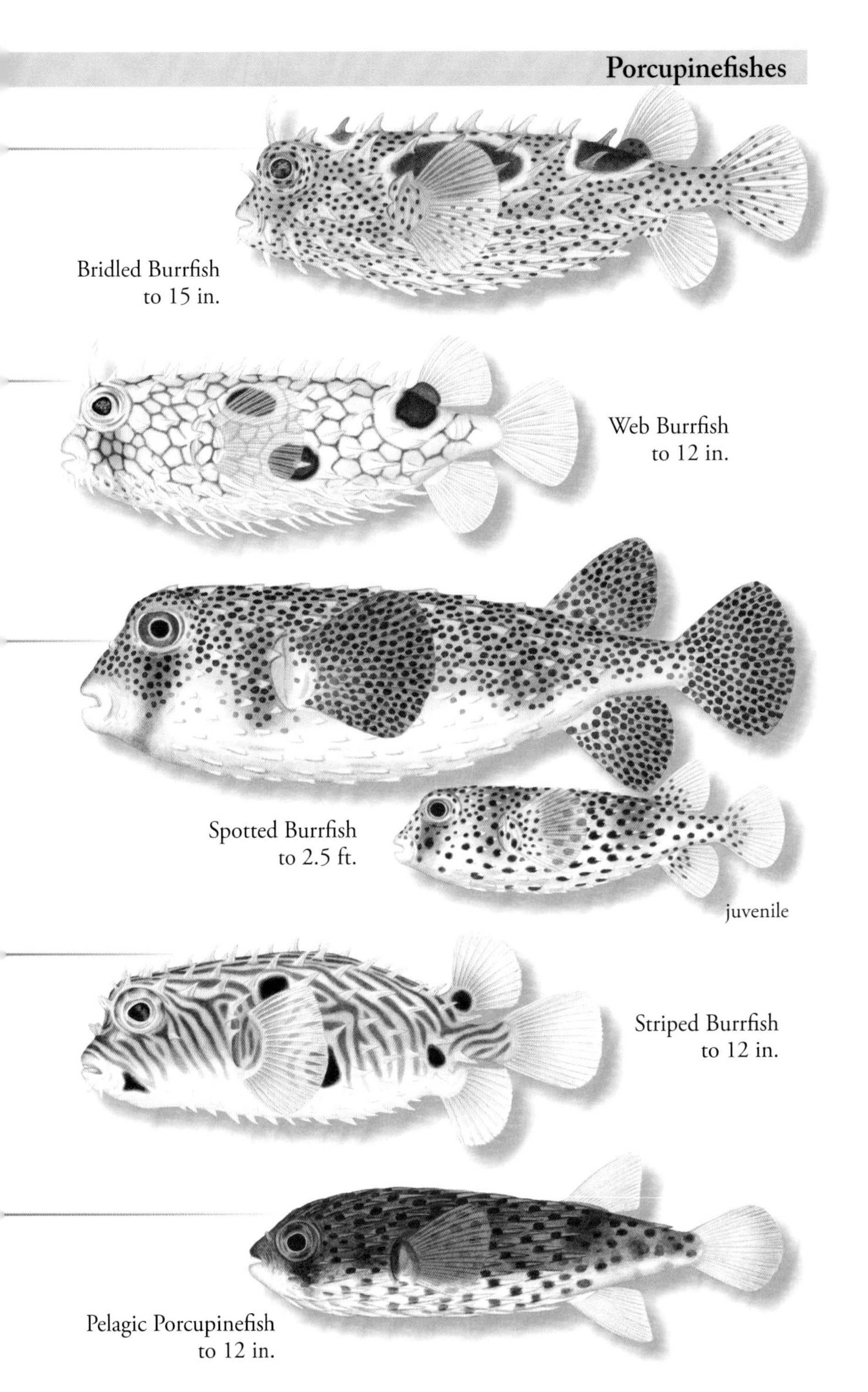
Bridled Burrfish
to 15 in.
Web Burrfish
to 12 in.
Spotted Burrfish
to 2.5 ft.
juvenile
Striped Burrfish
to 12 in.
Pelagic Porcupinefish
to 12 in.

Diodontidae - Porcupinefishes, *cont.*

Balloonfish - *Diodon holocanthus* Linnaeus, 1758

FEATURES: Tannish dorsally; pale below with moderately sized, dark spots on upper body. A dark bar extends between and below eyes. Several dark blotches on body. Fins usually unspotted; if spots are present, only on bases. Spines long; those on forehead are longer than those on body. Caudal peduncle lacks spines. HABITAT: Circumtropical. In western Atlantic from MA to FL, Gulf of Mexico, Bermuda, Bahamas, and Caribbean Sea to Brazil. Demersal over soft bottoms and reefs from near shore to about 340 ft.

Porcupinefish - *Diodon hystrix* Linnaeus, 1758

FEATURES: Tannish to grayish dorsally with numerous small, dark spots on head, upper body, and fins. Whitish below. May have a darkly pigmented ring around abdomen. Spines long; those on forehead shorter than those on body. One or two spines present on upper peduncle. HABITAT: Circumtropical. MA to FL, Gulf of Mexico, Bermuda, Bahamas, and Caribbean Sea to Brazil. Adults occur over coral and rocky reefs. Juveniles occur in estuaries.

Molidae - Molas

Sharptail Mola - *Masturus lanceolatus* (Liénard, 1840)

FEATURES: Shades of brown to gray dorsally and posteriorly—may be nearly uniformly colored or with white spots and blotches. Whitish ventrally. Mouth small. Gill opening round. Dorsal and anal fins tall, nearly vertical. Posterior margin of body with a blunt to pointed "tail." Body comparatively oblong in profile. HABITAT: Circumglobal in tropical to temperate seas. In western Atlantic from MA to FL, Gulf of Mexico, Bermuda, and Bahamas to Brazil. Oceanic. Occur from surface to about 2,200 ft.

Ocean Sunfish - *Mola mola* (Linnaeus, 1758)

FEATURES: Shades of brown to gray dorsally and posteriorly—may be nearly uniformly colored, or with white spots and blotches. Whitish to dusky ventrally. Mouth small. Gill opening round. Dorsal and anal fins tall, nearly vertical. Posterior margin of body rounded and somewhat to distinctly scalloped. Body round to oval in profile. May have a hump over head. HABITAT: Circumglobal in tropical to warm temperate seas. Newfoundland to FL, Gulf of Mexico, eastern Caribbean Sea to Argentina. From inshore water to open ocean. Occur from surface to about 1,300 ft.

Slender Mola - *Ranzania laevis* (Pennant, 1776)

FEATURES: Dark blackish blue dorsally; iridescent silvery to coppery on sides and below. Head and body with wavy, silver bars and spots. Bars on head outlined with dark lines and spots. Mouth small, vertically oriented. Gill openings round. Dorsal and anal fins tall, nearly vertical. Posterior margin of body almost straight. Body at least twice as long as it is deep. Small specimens very elongate. HABITAT: Circumglobal in tropical to warm temperate seas. In western Atlantic from DE to FL, Gulf of Mexico, Bermuda, Bahamas, and Caribbean Sea to Brazil. Oceanic. Occur from surface to about 460 ft.

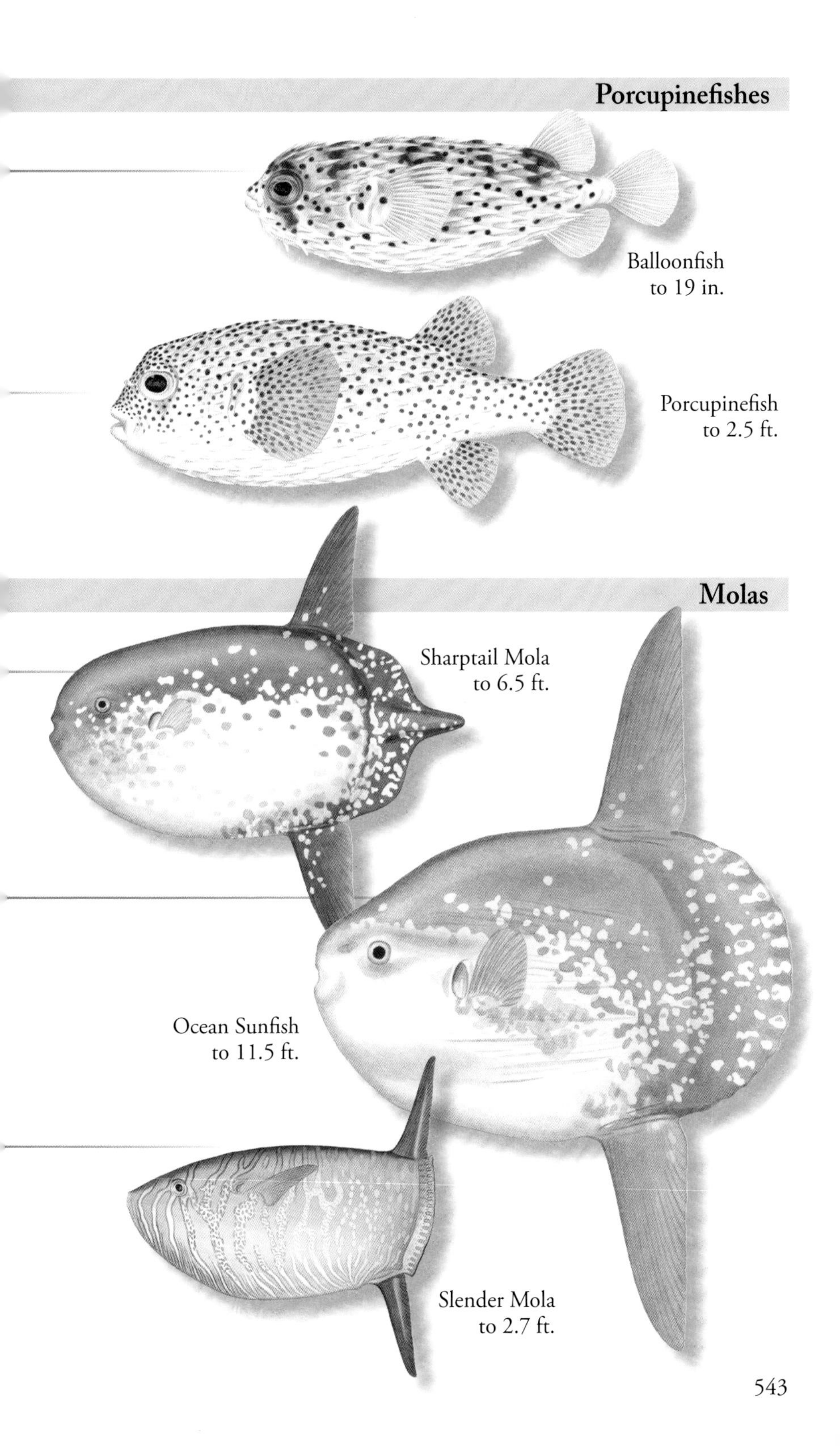
Porcupinefishes
Balloonfish
to 19 in.
Porcupinefish
to 2.5 ft.
Molas
Sharptail Mola
to 6.5 ft.
Ocean Sunfish
to 11.5 ft.
Slender Mola
to 2.7 ft.

Glossary

Below are common technical terms used in this book to describe fishes and their environment. Many of the terms referring to color patterns, anatomical features, and proportions are illustrated in the front of this book on pages 16–20.

Adipose eyelid: A fatty, transparent tissue that covers the eyes in some fishes.
Adipose fin: A fleshy, rayless fin on the dorsal midline between the dorsal and caudal fin in some fishes. In some species it is fused to the tail and separated from it by a slight notch.
Anadromous: A fish is anadromous when it lives primarily in the ocean and migrates into fresh water to spawn.
Anal fin: The fin on the ventral surface of fishes located between the anus and caudal fin.
Anterior: At or toward the front, or head, of a fish.
Band: A stripe or elongated area of a different color from the surrounding area.
Barbel: A fleshy, tentacle-like appendage usually on the chin, on the snout, or near the mouth of some fishes that is typically sensitive to chemicals or to touch.
Base: The point or area where a fin emerges from the body.
Benthic: Bottom-dwelling; occurring on or attached to the sea floor.
Brackish: Describes water that is a mixture of fresh and salt water and is less salty than ocean water and more salty than fresh water.
Buckler: A bony structure, often with a projecting spine, located on the sides, on the dorsal surface, or along the abdomen of some fishes.
Canine: A prominent, conical, and pointed tooth, usually larger than other teeth.
Catadromous: A fish is catadromous when it lives primarily in fresh water and migrates to the ocean to spawn.
Caudal fin: The fin at the rear end of the body, also called the tail fin.
Caudal peduncle: The portion of the fish body between the end of the anal fin and the base of the caudal fin.
Cephalic disk: A modified dorsal fin on the head of remoras that forms an oval-shaped adhesive disk.
Cephalic fin: A portion of modified pectoral fin on the head of manta rays that is elongated, flexible, and used to direct water into the mouth.
Chest: The ventral portion of the body in front of the pelvic fins; sometimes referred to as the breast.
Chevron: A diffuse, V-shaped mark that may be dark or pale.
Circumglobal: Occurring around the world in all oceans.
Cirrus (pl. cirri): A fleshy filament or tab on the head, body, or fins; sometimes occurring as fringe-like, as branched, or as a series.
Clasper: An elongate modification of the pelvic fin of chimaeras, sharks, skates, and rays that is used for internal fertilization.
Commensal: Describes a relationship between two species in which one species benefits and the other species is neither helped nor harmed.
Compressed: Describes a fish body that is flattened from side to side.
Concave: Arched or rounded inward.
Confluent: Refers to fins that are joined or meet without a notch.
Continental shelf: The submerged portion of a continent that is relatively flat, gently sloping, and extends from shore to a depth of about 660 ft (200 m).

Glossary

Continental slope: The submerged portion of a continent that is steeply sloping and extends from a depth of about 660 ft. (200 m) to 13,200 ft. (4,000 m).
Continuous: Unbroken; usually refers to fins that are joined without a gap or notch.
Convex: Arched or rounded outward.
Corselet: Region behind the head and surrounding the pectoral fins that is covered in specially modified scales.
Demersal: Living unattached at or near the bottom.
Denticle: A small tooth-like or thorn-like projection.
Depressed: Describes a fish body that is flattened from top to bottom.
Detritus: Disintegrating organic material—usually refers to small particulate plant or animal debris.
Dorsal fin: A fin located on the back of a fish; often notched or separated into two or more fins.
Dorsally: Refers to the back or upper portion of a fish.
Esca: Fleshy growth on an anglerfish's head that acts as a lure.
Estuary: An area partially enclosed by land where fresh water combines with salt water.
Filamentous: Refers to a long, slender, thread-like structure.
Filter-feed: Describes a method of feeding by removing plankton from the water using gill rakers.
Finlet: A short, isolated, separate fin segment between the dorsal and caudal fins and between the anal and caudal fins.
Flotsam: Floating wreckage or debris—usually artifical and synthetic.
Hermaphroditic: A fish that possesses both male and female reproductive organs either at the same time or sequentially.
Hypersaline: An aquatic environment that is saltier than typical seawater.
Initial (phase): The color phase before sex change in a species that is hermaphroditic, with female and male reproductive organs at different periods in its life. *See also* **Terminal.**
Inshore: Refers to waters over the continental shelf.
Intertidal: The shore zone between high and low tide lines.
Keel: A raised ridge on a scale or on either side of the caudal peduncle.
Laminae: Thin, flat structures of the cephalic disk on remoras.
Lateral line: A sensory organ along the side of the body that perceives vibrations and is formed by a series of tubes, canals, or pores and motion-sensing cells. The lateral line is complete when it extends from the head to the caudal fin; it is incomplete when it does not reach the caudal fin.
Laterally: Refers to the side of a fish.
Lure: A fleshy appendage at the tip of a modified dorsal-fin spine that is used to attract prey.
Margin: The outermost edge of a fin.
Maxilla: The outermost or hindmost main bone of the two bones that form the upper jaw.
Midline: The horizontal median line or plane of a body.
Nape: The part of the back immediately behind the head, or the area on the dorsal surface between the head and the dorsal-fin origin.
Ocellated: Having a pattern of a single ocellus or multiple ocelli.

Ocellus (pl. ocelli): A spot of color inside a ring of another color.
Offshore: Refers to waters beyond the edge of the continental shelf.
Omnivore: An animal that feeds on both animals and plants.
Opercle: The portion of the side of the head, typically composed of large, flattened bones, that covers the gills.
Origin: Refers to the front of a fin, where the first spine or ray emerges from the body.
Papillae: Small, fleshy projections.
Pectoral fins: The paired fins behind the gill opening on each side of the body.
Pelagic: Living in open waters or open ocean, away from the bottom.
Pelvic fins: The paired fins on the lower part of the body and in front of the anus. The position of the pelvic fins varies.
Planktivore: An animal that feeds exclusively on planktonic (small drifting) matter.
Posterior: At or toward the rear, or tail, of a fish.
Preopercle: The portion of the side of the head composed of a bone lying anterior to the opercle. The preopercle is often defined by a groove in the skin.
Profile: Side view; describes the outline of a fish body.
Protrusible: Refers to a mouth that is able to extend and lengthen forward to catch prey.
Ray: Supporting elements of the fins. Soft rays are segmented, flexible, and branched, while spiny rays are unsegmented, hard, and unbranched.
Reticulating: A pattern or network of irregular, repeating, and interconnecting dark or pale lines.
Rhomboid: A shape similar to a square but with oblique angles.
Rostrum: The area of the snout that is sometimes flattened or elongate.
Saddle: A patch of color extending across the back and onto the upper sides that may form a Y-, V-, or U-shape.
Saline: Containing salt.
Scute: A bony plate or modified scale that is usually keeled.
Serrated: Having saw-like notches along the edge.
Simple: Not divided or branched.
Snout: The region of the head in front of the eyes and above the mouth.
Spine: A sharp, unbranched, unsegmented, rigid fin ray; also a sharp bony projection on the head or body.
Spiracle: The circular opening between the eye and first gill opening in some sharks, rays, and sturgeons.
Subtropical: Describes the regions between the tropical and temperate latitudes.
Temperate: Describes the regions between the subtropical and polar latitudes.
Terminal (phase): The color phase after sex change in a species that is hermaphroditic, with female and male reproductive organs at different periods in its life. *See also* **Initial.**
Thorn: A sharp, bony projection.
Tropical: The world's warmest region, between the equator and the tropics.
Tubercle: A hardened, conical projection on the surface of the body.
Turbid: Refers to water that is murky due to suspended silt or sediment.
Ventrally: Refers to the abdominal or lower portion of a fish.
Vermiculation: Cluster of wavy lines or blotches.

Select Bibliography and Additional Resources

NOTE: A printable appendix of species also found in the area can be viewed at www.press.jhu.edu. In the search box type the title of this book.

Böhlke, J. E., and C. C. G. Chaplin. 1968. *Fishes of the Bahamas and Adjacent Tropical Waters.* Wynnewood: Livingston Publishing Company.

Campagno, L., and S. Fowler. 2005. *Sharks of the World.* Princeton: Princeton University Press.

Carpenter, K. E., ed. 2002. *FAO Species Identification Guide for Fishery Purposes: The Living Marine Resources of the Western Central Atlantic.* Vol. 1–3. Rome: Food and Agriculture Organization of the United Nations.

Collette, B., and J. Graves. 2019. *Tunas and Billfishes of the World.* Baltimore: Johns Hopkins University Press.

Greenfield, D. W., and E. Thomerson. 1997. *Fishes of the Continental Waters of Belize.* Gainesville: University Press of Florida.

Hoese, H. D., and R. Moore. 1977. *Fishes of the Gulf of Mexico: Texas, Louisiana, and Adjacent Waters.* 2nd ed. College Station: Texas A&M University Press.

McEachran, J. D., and J. D. Fechhelm. 2005. *Fishes of the Gulf of Mexico.* Vols. 1 and 2. Austin: University of Texas Press.

Murdy, E. O., and J. A. Musick. 2013. *Field Guide to Fishes of the Chesapeake Bay.* Baltimore: Johns Hopkins University Press.

Nelson, J. S. 2006. *Fishes of the World.* 4th ed. Hoboken: John Wiley & Sons.

Page, L. M., et al. 2013. *Common and Scientific Names of Fishes from the United States, Canada, and Mexico.* 7th ed. Bethesda: American Fisheries Society, Special Publication 34.

Randall, J. E. 1996. *Caribbean Reef Fishes.* 3rd ed. Neptune: T. F. H. Publications, Inc.

Smith-Vaniz, W. F., et al. 1999. *Fishes of Bermuda: History, Zoogeography, Annotated Checklist, and Identification Key.* Bethesda: American Society of Ichthyologists and Herpetologists, Special Publication Number 4.

The websites listed below have searchable databases that are useful sources of taxonomic and ichthyological information:

California Academy of Sciences, Eschmeyer's Catalog of Fishes: https://researcharchive.calacademy.org/research/ichthyology/catalog/fishcatmain.asp
Encyclopedia of Life: http://www.eol.org/
Fishbase: www.fishbase.org/
Florida Museum of Natural History, Ichthyology: www.flmnh.ufl.edu/fish/
Gobies: http://gobiidae.com/
Integrated Taxonomic Information System: www.itis.gov/
IUCN Red List of Threatened Species: www.iucnredlist.org/
Smithsonian National Museum of Natural History, Research and Collections, Vertebrate Zoology: http://vertebrates.si.edu/fishes/
Smithsonian Tropical Research Institute: http://www.stri.org/sftep/

Index

Index

Index

Index

Index

Index

Index

Index

Index

Index

Index

Index

About the Authors

Val Kells is an award-winning freelance Marine Science Illustrator. She received her B.S. in Environmental Studies and completed her formal training in Scientific Communication and Illustration at the University of California, Santa Cruz, in 1985 and 1986. She has worked with designers, educators, and curators to develop a wide variety of illustrations for interpretive and educational displays in public aquariums, museums, and nature centers. Her artwork has been displayed in over 30 aquariums, including the Long Beach Aquarium of the Pacific, Monterey Bay Aquarium, the North Carolina Aquariums, and the Texas State Aquarium, among others. Val also collaborates with editors, researchers, and scientists, and her work has been published in many books and periodicals. She recently illustrated and co-authored *A Field Guide to Coastal Fishes: From Maine to Texas* and *A Field Guide to Coastal Fishes: From Alaska to California* and illustrated *Tunas and Billfishes of the World*, all published by Johns Hopkins University Press. Val lives in on the Outer Banks of North Carolina. She is an avid fisherman and naturalist and spends her off time exploring, fishing, seining, and documenting the aquatic habitats along the coasts.

Luiz A. Rocha is the Follett Chair of Ichthyology at the California Academy of Sciences. He received his B.S. and M.S. at the Universidade Federal da Paraíba, Brazil, in 1996 and 1999, and his Ph.D. at the University of Florida in 2003. His major research interests include evolution, conservation, taxonomy, and community ecology of marine fishes. He frequently tries to combine these fields, linking ecology to evolution and using molecular tools to answer biogeographic and taxonomic questions. The overall objective of this interdisciplinary research is to explain what generates and maintains the extremely high biodiversity in coastal marine regions, both shallow and deep water. He has spent more than 6,000 hours underwater collecting and observing fishes, and he has published more than 150 peer-reviewed articles. In addition, his work has been featured in many popular media outlets, including the *New York Times*, *Scientific American*, and *National Geographic* magazine, and has supported conservation efforts across the globe. Luiz lives with his family in Marin County, California, and spends his weekends hiking and mountain biking with his children and their very active dog.

About the Authors

Carole C. Baldwin is Curator of Fishes and Chair of the Department of Vertebrate Zoology at the Smithsonian Institution's National Museum of Natural History. She received a B.S. from James Madison University, M.S. from the College of Charleston, and Ph.D. from the College of William and Mary. Her current research is focused on diversity and eco-evolution of Caribbean reef fishes through integrative genetic and morphological investigation. This work has recently involved submersible diving to 1,000 feet in depth in the Caribbean as part of the Deep Reef Observation Project (DROP), an initiative that Carole established in 2011 to explore and monitor long-term changes in poorly studied tropical deep-reef ecosystems. DROP sub diving has resulted in the discovery of over seven new genera and 60 new species of fishes and invertebrates and the identification of a previously unrecognized deep-reef zone below the mesophotic, which she and colleagues named the rariphotic. Carole co-authored *One Fish, Two Fish, Crawfish, Bluefish: The Smithsonian Sustainable Seafood Cookbook*, was a lead curator of the Smithsonian's Sant Ocean Hall, and was featured in the IMAX film *Galapagos 3D*.

About the Art. The illustrations were created with watercolor and gouache on 300 lb. hotpress Arches® watercolor paper. Brushes included sizes 2 to 14, round, oval, and flat. Techniques included the use of liquid rubber mask, blotters, and gesso.

Notes

Notes

Notes

Also from Johns Hopkins University Press

A Field Guide to Coastal Fishes

FROM MAINE TO TEXAS

Val Kells and Kent Carpenter

A Field Guide to Coastal Fishes is a comprehensive, current, and accurate identification guide to the more than 1,000 nearshore and offshore fishes that live in brackish and marine waters from Maine to Texas.

Reliable and up to date, this is the most complete book ever published on East and Gulf Coast fishes—perfect for boat, home, or classroom. Its beautiful design and accessible format make it an ideal guide for fishermen, divers, students, scientists, naturalists, and fish enthusiasts alike.

Paperback; 448 pages; 1079 color illustrations; 1 map

ISBN: 978-0-8018-9838-9

From the Arctic waters of Alaska to the southern tip of California, *A Field Guide to Coastal Fishes: From Alaska to California* is this region's most current and thorough fish identification guide. This fully illustrated guide captures the stunning diversity of fishes along the western coastlines of the United States and Canada, from shallow, brackish waters to depths of about 200 meters.

Whether you are an angler, scuba diver, naturalist, student, or teacher, you will find this to be your go-to guide and a welcome addition to your boat, in your backpack, or on your bookshelf.

Paperback; 360 pages; 950+ color illustrations; 1 map

ISBN: 978-1-4214-1832-2

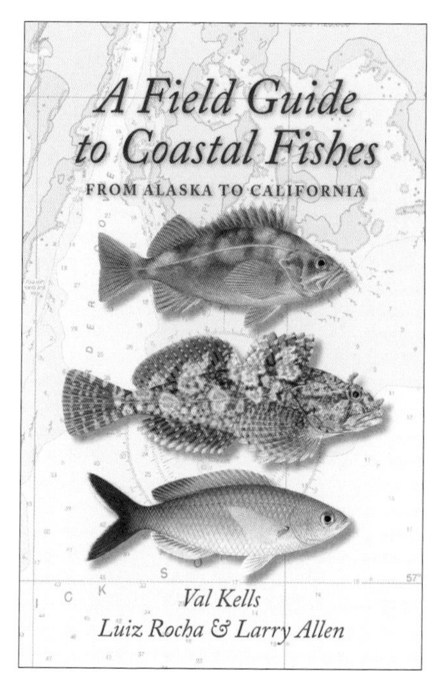

Notes